Carbon-Rich Compounds

Edited by
Michael M. Haley
and Rik R. Tykwinski

Related Titles

Diederich, F., Stang, P. J., Tykwinski, R. R. (eds.)

Acetylene Chemistry

Chemistry, Biology, and Material Science

528 pages
2005
Hardcover
ISBN 3-527-30781-8

Hirsch, A., Brettreich, M.

Fullerenes

Chemistry and Reactions

435 pages
2005
Hardcover
ISBN 3-527-30820-2

Hopf, H.

Classics in Hydrocarbon Chemistry

Syntheses, Concepts, Perspectives

560 pages
2000
Softcover
ISBN 3-527-29606-9

Carbon-Rich Compounds

From Molecules to Materials

Edited by Michael M. Haley and Rik R. Tykwinski

WILEY-VCH Verlag GmbH & Co. KGaA

The Editors

Prof. Michael M. Haley
Department of Chemistry
University of Oregon
Eugene, Oregon 97403-1253
USA

Prof. Rik R. Tykwinski
Department of Chemistry
University of Alberta
Edmonton, Alberta T6G 2G2
Canada

■ All books published by Wiley-VCH are carefully produced. Nevertheless, authors, editors, and publisher do not warrant the information contained in these books, including this book, to be free of errors. Readers are advised to keep in mind that statements, data, illustrations, procedural details or other items may inadvertently be inaccurate.

Library of Congress Card No.: applied for
British Library Cataloguing-in-Publication Data
A catalogue record for this book is available from the British Library.

Bibliographic information published by Die Deutsche Bibliothek
Die Deutsche Bibliothek lists this publication in the Deutsche Nationalbibliografie; detailed bibliographic data is available in the Internet at ⟨http://dnb.ddb.de⟩.

© 2006 WILEY-VCH Verlag GmbH & Co. KGaA, Weinheim

All rights reserved (including those of translation into other languages). No part of this book may be reproduced in any form – by photoprinting, microfilm, or any other means – nor transmitted or translated into a machine language without written permission from the publishers. Registered names, trademarks, etc. used in this book, even when not specifically marked as such, are not to be considered unprotected by law.

Printed in the Federal Republic of Germany.
Printed on acid-free paper.

Typesetting Asco Typesetters, Hong Kong
Printing betz-druck GmbH, Darmstadt
Binding J. Schäffer GmbH i.G., Grünstadt

ISBN-13: 978-3-527-31224-5
ISBN-10: 3-527-31224-2

Foreword

Ever since *Curl*, *Kroto*, and *Smalley* were awarded the Nobel Prize for their discovery of the fullerenes, I open the oral exam for the bachelor in organic chemistry with the question: "How many allotropes of carbon are there?" The upright student will give "Three" as an answer, but the good one will say "At least three". In fact, the number of conceivable scaffolds that can be constructed with this basic building block is endless: just take a piece of paper and your carbon-filled pencil! On the other hand, that this question is new in organic chemistry, can be maintained only by those who know little about the history of our discipline.

Adolf von Baeyer, one of the giants of organic chemistry, was interested in acetylenes, ever since his first Ph.D. student in Berlin (a young man called *Max Berend*) studied the reaction of iodine with silver acetylide. Later, when *Baeyer* had moved to Munich to become the successor of *Justus von Liebig*, he took up his old acetylene studies again and asked the question whether the property of carbon atoms to form long chains could also be realized for "pure carbon", i.e., when it does not carry hydrogen substituents. In this case ring-like carbon arrangements should also be possible, and this particular form of carbon should be colorless, volatile, and extremely explosive. *Baeyer* was successful in preparing carbon-rich systems such as "tetraacetylene dicarboxylic acid", although he failed in finding "explosive diamonds". However, as a stable side-product of his considerations the Baeyer strain theory survived until the present.

Clearly, the notion of carbon-rich compounds resists definition: From what degree on is an organic compound carbon rich? Just before it becomes a specimen of hydrogen polluted carbon? As so often in organic chemistry – the longest C–C bond, the lowest chemical shift for an aromatic compounds – concepts are best illustrated by practical examples. These, of course, can always be surpassed by other examples. The different chapters of this book demonstrate that under the roof of the ill-defined concept of carbon richness there is an amazing variety of structural diversity encompassing all types of linear, flat, or three-dimensional carbon scaffolds. Whether these will ever be of any practical value – i.e., as novel materials, electronic devices, information storage compounds (just add your own favorite buzz word) – is not really of importance presently (and may be not at all).

The systems discussed by the 13 experts of the book are prime examples of outstanding, imaginative, and far-reaching fundamental research. As such, these

Carbon-Rich Compounds. Edited by Michael M. Haley and Rik R. Tykwinski
Copyright © 2006 WILEY-VCH Verlag GmbH & Co. KGaA, Weinheim
ISBN: 3-527-31224-2

works not only contribute to our constantly growing arsenal of preparative methods – always a worthwhile venture for a compound maker – but also shed light on one of the most basic concepts of organic chemistry: unsaturation. It is the behavior of π-electrons that determines organic structure and reactivity to a large extent, and to learn more about it, carbon-rich compounds can serve as excellent model molecules. After all, where can one expect to learn more about organic chemistry than by studying its defining element?

Braunschweig, Germany *Henning Hopf*
October 2005

Contents

Foreword   V

Preface   XVII

List of Contributors   XIX

1	**Pioneers of Carbon-rich Compounds** *1*	
	Charles A. Johnson II and Michael M. Haley	
1.1	Introduction *1*	
1.2	19th Century Achievements *1*	
1.3	1900–1928: Dawn of the Twentieth Century *4*	
1.4	1929–1949: Rise of the Polycyclic Aromatic Hydrocarbon *7*	
1.5	1950–1969 *12*	
1.5.1	The Annulenes, Dehydrobenzoannulenes, and Phenylacetylene Scaffolding *13*	
1.5.2	Fused Polycyclic and *peri*-Condensed Benzenoid Systems *16*	
1.5.3	The Helicenes, Radialenes, Fulvalenes, and Circulenes *19*	
1.6	1970–Present: The Way Ahead *22*	
	References *23*	
2	**Electronic Conduction in Photoactive Metallo-wires** *26*	
	Anthony Harriman and Raymond Ziessel	
2.1	Introduction *26*	
2.2	Attenuation along Molecular Bridges *32*	
2.3	Information Transfer *38*	
2.3.1	Intramolecular Triplet Energy Transfer *38*	
2.3.2	Short Covalent Bridges *42*	
2.3.3	Supramolecular Systems *52*	
2.3.4	Prolonging the Excited State Lifetime *55*	
2.3.5	Long-range Triplet Energy Transfer *64*	
2.4	Molecular-scale Switches *70*	
2.5	Perspectives *75*	
2.6	Experimental: Selected Procedures *77*	

Carbon-Rich Compounds. Edited by Michael M. Haley and Rik R. Tykwinski
Copyright © 2006 WILEY-VCH Verlag GmbH & Co. KGaA, Weinheim
ISBN: 3-527-31224-2

2.6.1	General Procedure for the Preparation of the Mononuclear [RuL$_n$]	77
2.6.2	General Procedure for the Preparation of the Hetero-Dinuclear Complexes **81**$_1$ and **81**$_5$ 77	
2.6.3	Synthesis at the Complex 78	
	Acknowledgements 81	
	Abbreviations 81	
	References 82	

3	**All-benzenoid Polycyclic Aromatic Hydrocarbons: Synthesis, Self-assembly and Applications in Organic Electronics** 90	
	Jishan Wu and Klaus Müllen	
3.1	A Brief Introduction to Polycyclic Aromatic Hydrocarbons 90	
3.2	All-benzenoid PAHs – Synthesis, Structural Characterizations and Electronic Properties 96	
3.2.1	Hexa-*peri*-hexabenzocoronene – An Old Story with New Discoveries	96
3.2.2	All-benzenoid Graphitic PAHs Larger than HBCs 106	
3.2.3	PAHs with Varying Peripheries 114	
3.2.4	"Superbenzene" Chemistry and Others 117	
3.3	Self-assembly and Application of Columnar Liquid Crystals based on PBAHs 120	
3.3.1	Columnar Superstructures in the Bulk State 120	
3.3.2	Alignment on Substrates and Device Applications of Columnar Liquid Crystals 123	
3.3.3	Controlled Self-assembly in Solution 128	
3.3.4	Two-dimensional Crystals at the Solid/Liquid Interface 129	
3.4	Conclusion 131	
3.5	Experimental: Selected Procedures 132	
3.5.1	Synthesis of hexa-*peri*-hexabenzocoronene 10 by Cu(II)-mediated oxidative cyclodehydrogenation – a general procedure to prepare unsubstituted graphitic molecules 132	
3.5.2	Synthesis of hexakis(4-n-dodecylphenyl)-*peri*-hexabenzocoronene (HBC-PhC12) – a general synthetic method towards six-fold alkyl- and alkylphenyl-substituted HBCs 132	
3.5.3	Functionalization of insoluble HBC building blocks 30–32 by Sonogashira coupling reactions. Synthesis of hexakis(1-dodecynylphenyl)-*peri*-hexabenzocoronene (34a) as a representative example 133	
3.5.4	Synthesis of C96-C12 precursor 1,3,5-tris[3′,4′-di(4″-dodecylphenyl)-2′,5′-diphenylphenyl]benzene (44a) by Diels-Alder cycloaddition reaction – a representative procedure for the synthesis of branched oligophenylenes 133	
3.5.5	Hydrogenation of hexakis-dodecyl-*peri*-hexabenzo-coronenes 74 134	
3.5.6	Synthesis of HBC dimer 38 by Yamamoto coupling reaction 134	
	Acknowledgements 134	

Abbreviations   135
References   135

4   [N]Phenylenes: a Novel Class of Cyclohexatrienoid Hydrocarbons   140
Ognjen Š. Miljanić and K. Peter C. Vollhardt
4.1   Introduction   140
4.2   Preparation of Phenylenes   145
4.2.1   Early Synthetic Strategies   145
4.2.2   Syntheses of New Phenylenes   149
4.2.2.1   Angular and Helical Phenylenes   149
4.2.2.2   Zigzag Phenylenes   151
4.2.2.3   Phenylenes with Mixed Topology: the "Bent" Isomers   153
4.2.2.4   Branched Phenylenes   155
4.2.2.5   Circular Phenylenes   160
4.3   Comparative Reactivity of the Phenylenes   162
4.3.1   Hydrogenation   162
4.3.2   Oxacyclopropanation and Cyclopropanation   164
4.3.3   [4+2]Cycloadditions   167
4.3.4   Flash Vacuum Pyrolysis   171
4.3.5   Interaction with Organometallic Fragments   172
4.4   Physical Properties of the Phenylenes   175
4.4.1   Structural Properties   175
4.4.2   Magnetic Properties   182
4.4.3   Energetic Properties   184
4.5   Conclusion and Outlook   187
4.6   Experimental: Selected Procedures   188
4.6.1   Preparation of Bis(trimethylsilyl) Zigzag [4]Phenylene: A One-pot Intra- and Intermolecular [CpCo(CO)$_2$]-Mediated Cyclization   188
4.6.2   Preparation of syn-Doublebent [5]Phenylene: A [CpCo(eth)$_2$]-Mediated Cyclization   188
4.6.3   Preparation of Dehydrobenz[12]annulene 74a: A Stephens-Castro Coupling   188
4.6.4   Hydrogenation of Bis(trimethylsilyl) Bent [4]Phenylene   189
4.6.5   Cyclopropanation of Triangular [4]Phenylene   189
4.6.6   Reaction of Angular [3]Phenylene with TCNE   189
4.6.7   Reaction of Angular [3]Phenylene with [CpCo(eth)$_2$]   189
    Acknowledgements   190
    Abbreviations   190
    References   191

5   A Solid State Strategy for the Preparation of Carbon-rich Polymers   198
Frank W. Fowler and Joseph W. Lauher
5.1   Introduction   198
5.2   Topochemical Diacetylene Polymerizations   199
5.2.1   Diacetylene Polymerizations   202

5.2.1.1	Dipyridyldiacetylenes *206*
5.2.1.2	Monosubstituted Diacetylenes *208*
5.2.1.3	Summary of Diacetylene Polymerization *214*
5.3	Topochemical Triacetylene Polymerization *215*
5.4	Topochemical Diene and Triene Polymerization *220*
5.5	Summary *224*
5.6	Experimental: Selected Procedures *224*
5.6.1	3-[(Pyridin-4-ylmethyl)-amino]-cyclohex-2-enone (20a) *224*
5.6.2	Dodeca-4,6,8-triynedioic acid (17b) *225*
5.6.3	Poly(20a–17b) *225*
	Acknowledgments *226*
	Abbreviations *226*
	References *226*

6	**Chiral Carbon-rich Macrocycles and Cyclophanes** *229*
	Katie Campbell and Rik R. Tykwinski
6.1	Introduction *229*
6.2	Strategies for Achieving Chiral Macrocycles or Cyclophanes *229*
6.2.1	Helical Chirality Using Achiral Building Blocks *231*
6.2.2	Chiral Macrocycles Using Chiral Carbon-rich Building Blocks *255*
6.2.3	Metal-containing Chiral Macrocycles *268*
6.3	Conclusions *286*
6.4	Experimental: Selected Procedures *287*
6.4.1	Cyclophane 12 *287*
6.4.2	Cyclophane 43 *288*
6.4.3	Cyclophane 86 *288*
6.4.4	(R)-(+)-124 *289*
6.4.5	Complex 262 *289*
6.4.6	Complex 264 *289*
	Acknowledgments *290*
	Abbreviations *290*
	References *291*

7	**Carbon-rich Cycles with Two and More 1,3-Butadiyne Units – Syntheses, Structures and Reactivities** *295*
	Rolf Gleiter and Daniel B. Werz
7.1	Introduction *295*
7.2	Syntheses *296*
7.3	1,3-Butadiyne Moieties as Spacers and Framework *301*
7.4	Reactivity *304*
7.4.1	Base-induced Prototropic Rearrangements *304*
7.4.2	Elimination Reactions *308*
7.4.3	Reductions of Cyclic 1,3-Butadiynes *309*
7.4.4	Acid-induced Transannular Reactions *311*

7.5	Spectroscopic and Structural Investigations	316
7.5.1	Structural Investigations	316
7.5.2	NMR Spectroscopic Investigations	324
7.5.3	Photoelectron and Electron Absorption Spectra of Cyclic 1,3-Diynes	326
7.6	Concluding Remarks and Outlook	327
7.7	Experimental: Selected Procedures	328
7.7.1	General Procedure for the Preparation of N,N'-Dialkyl-1,8-diazacyclotetradeca-3,5,10,12-tetraynes (27a–e)	328
7.7.2	General Procedure for the Cadiot–Chodkiewicz Coupling During the Syntheses of Azacyclotetraynes 10(n)	328
7.7.3	General Procedure for the Preparation of the Tetrachalcogenacyclotetraynes 29(n) and 33(n) and Hexachalcogenacyclohexaynes 34(n) and 35(n)	329
7.7.4	General Procedures for the Preparation of Hexadiyne-bridged Macrobicycles	329
7.7.4.1	Alternative A	329
7.7.4.2	Alternative B	329
	Acknowledgments	330
	Abbreviations	330
	References	330
8	**Carbon-rich Compounds: Computational Considerations**	**334**
	Peter R. Schreiner	
8.1	Introduction	334
8.2	Basic Aspects	338
8.2.1	Molecular Mechanics (MM)	339
8.2.2	Wavefunction Theory	339
8.2.3	Semiempirical Methods	340
8.2.4	Hartree-Fock Theory	342
8.2.5	Symmetry	344
8.2.6	Basis Sets	344
8.2.7	Electron Correlation Methods	345
8.2.8	Density Functional Theory	348
8.2.9	Validation	348
8.3	Applications	350
8.3.1	Alkynes, Cumulenes, and the $sp-sp^2$ Transition	350
8.3.2	Cyclization Reactions of Polyunsaturated Systems	358
8.3.3	Expanded (Super)molecules	363
8.3.4	Diamondoids and Related sp^3-Structures	367
8.3.4.1	Diamondoidyl Cations	368
8.3.4.2	Diamondoidyl Radicals	368
8.3.4.3	Diamondoid Radical Cations	370
8.3.4.4	Diamondoid Supermolecules	370

8.4	Conclusions and Outlook	373
	Acknowledgments 373	
	Abbreviations 373	
	References 376	
9	**Fullerene Reactivity – Fullerene Cations and Open-Cage Fullerenes**	**383**
	Toshikazu Kitagawa, Yasujiro Murata, and Koichi Komatsu	
9.1	Introduction 383	
9.2	Functionalized Fullerene Cations 383	
9.2.1	Derivatized C_{60} and C_{70} Cations 384	
9.2.1.1	Generation of Derivatized C_{60} and C_{70} Cations as Long-lived Species 384	
9.2.1.2	Reactions via RC_{60}^+ and RC_{70}^+ Intermediates 388	
9.2.1.3	Thermodynamic Stabilities of RC_{60}^+ and RC_{70}^+ 392	
9.2.2	Aza[60]fullerene Cation $C_{59}N^+$ 393	
9.2.2.1	Generation of $C_{59}N^+$ 393	
9.2.2.2	Reactions via $C_{59}N^+$ 395	
9.2.3	Electrophilic Functionalization of Carbon Nanotubes 398	
9.3	Open-Cage Fullerenes to Endohedral Fullerenes 398	
9.3.1	Synthesis of Open-Cage Fullerene Derivatives 401	
9.3.1.1	An Open-Cage Fullerene with a 14-Membered-Ring Orifice 401	
9.3.1.2	Open-Cage Fullerenes with a 12-Membered-Ring Orifice 401	
9.3.1.3	Enlargement of the Orifice by Insertion of a Sulfur Atom 402	
9.3.1.4	Enlargement of the Orifice by Reaction with Aromatic Hydrazine or Diamine 404	
9.3.2	Insertion of a Small Guest into Open-Cage Fullerene Derivatives 408	
9.3.2.1	Insertion of a Helium Atom and a Hydrogen Molecule 408	
9.3.2.2	Insertion of a Water Molecule 411	
9.3.3	Synthesis of an Endohedral Fullerene by Closure of the Orifice 412	
9.4	Summary and Conclusion 414	
9.5	Experimental: Selected Procedures 415	
9.5.1	Synthesis of 1-Chloro-4-dichloromethyl-1,4-dihydro[60]fullerene (12b) 415	
9.5.2	Synthesis of 1-Dichloromethyl-4-hydroxy-1,4-dihydro[60]fullerene (2b) 415	
9.5.3	Generation of Dichloromethylated [60]Fullerene Cation (3b) 415	
9.5.4	Synthesis of an Open-Cage Fullerene with an Eight-membered-ring Orifice (49) 416	
9.5.5	Synthesis of an Open-Cage Fullerene with a 12-Membered-ring Orifice (54) 416	
9.5.6	Synthesis of an Open-Cage Fullerene with a 13-Membered-ring Orifice (56) 416	
	Abbreviations 416	
	References 417	

10	**Polyynes** *421*	
	Vivian W. W. Yam and C. H. Tao	
10.	Introduction *421*	
10.1.1	Organic Polyynes *422*	
10.1.2	Organometallic Polyynyl Complexes *424*	
10.2	Synthetic Routes Towards Transition Metal Alkynyl Complexes *425*	
10.2.1	Copper-catalyzed Dehydrohalogenation *426*	
10.2.2	Metal Halides with Alkynylating Agents *426*	
10.2.3	Metal-free Methods with Terminal Alkynes *427*	
10.3	Rhenium(I) Alkynyls *427*	
10.4	Platinum(II) Alkynyls *432*	
10.4.1	Platinum(II) Phosphine Systems *432*	
10.4.2	Platinum(II) Polypyridine Systems *440*	
10.5	Copper(I) and Silver(I) Alkynyls *448*	
10.6	Gold(I) Alkynyls *453*	
10.7	Mixed-metal Alkynyls *457*	
10.8	Conclusions *464*	
10.9	Experimental: Selected Procedures *466*	
10.9.1	Synthesis of $[(^tBu_2bpy)(CO)_3ReC\equiv C-C\equiv C-C\equiv CPh]$ *466*	
10.9.2	Synthesis of $[1,3,5-\{PhC\equiv CPt(PEt_3)_2C\equiv CC_6H_4C\equiv C\}_3C_6H_3]$ *466*	
10.9.3	Synthesis of $[Pt(trpy)C\equiv C-Ph]PF_6$ *466*	
10.9.4	Synthesis of $[Cu_3(\mu\text{-dppm})_3(\mu_3\text{-}\eta^1\text{-}C\equiv C-{}^tBu)_2]PF_6$ *467*	
10.9.5	Synthesis of $[\{\eta^2\text{-}Re(CO)_3(bpy)(C\equiv CPh)\}_2Ag]PF_6$ *467*	
10.9.6	Synthesis of $[Ag_6(\mu\text{-dppm})_4\{\mu_3\text{-}C\equiv C-C\equiv C-Re(Me_2bpy)(CO)_3\}_4](PF_6)_2$ *467*	
10.9.7	Synthesis of $[1,3-\{Cl(PEt_3)_2PdC\equiv C\}_2-5-\{(bpy)(CO)_3ReC\equiv C\}C_6H_3]$ *467*	
	Acknowledgments *468*	
	Abbreviations *468*	
	References *469*	
11	**Defined-length Carbon-rich Conjugated Oligomers** *476*	
	Herbert Meier	
11.1	Introduction *476*	
11.2	Oligo(1,4-phenylenevinylene)s OPV *479*	
11.2.1	Synthesis *480*	
11.2.2	Absorption and Fluorescence *488*	
11.2.3	Applications in Materials Science *492*	
11.3	Oligo(1,4-phenyleneethynylene)s OPE *499*	
11.3.1	Synthesis *499*	
11.3.2	Absorption and Fluorescence *504*	
11.3.3	Applications in Materials Science *508*	
11.4	OPVs and OPEs with Terminal Donor–Acceptor Substitution *509*	
11.4.1	Synthesis *510*	
11.4.2	Absorption and Fluorescence *511*	

11.5	Summary and Outlook	516
11.6	Experimental: Selected Procedures	518
11.6.1	General Procedure for the Extension of an OPV Aldehyde by Two Styryl Units (Scheme 11.7)	518
11.6.2	all-(E)-2,2′,5,5′-Tetrapropoxy-4,4′-bis(2,5-dipropoxy-4-{2,5-dipropoxy-4-[2,5-dipropoxy-4-(2,5-dipropoxy-4-{2,5-dipropoxy-4-[2,5-dipropoxy-4-(2,5-dipropoxystyryl)styryl]styryl}-styryl)styryl]styryl}styryl)stilbene (2j)	518
11.6.3	General Procedure for the Cation–Anion Coupling Reaction (Scheme 11.10)	519
11.6.4	Preparation of 1,4-Bis(2,5-dipropoxyphenylethynyl)-2,5-dipropoxybenzene (1b)	519
11.6.5	General Procedure for the Preparation of the Series 52c ($n = 1–4$) by Wittig–Horner Reactions	520
	Abbreviations	520
	References	521
12	**Synthesis and Chemistry of Polycyclic Aromatic Hydrocarbons with Curved Surfaces: Buckybowls**	**529**
	Andrzej Sygula and Peter W. Rabideau	
12.1	Introduction	529
12.2	Synthesis of Buckybowls	530
12.2.1	Corannulene: A Synthetic Travel from "Wet" Chemistry ...	530
12.2.2	... to Flash Vacuum Pyrolysis ...	531
12.2.3	... and Back	533
12.2.4	Practical Synthesis of Corannulene by Intramolecular Carbenoid Coupling of Dibromomethyl Groups	535
12.2.5	Beyond Corannulene	538
12.2.5.1	On the Surface of C_{60}	538
12.2.5.2	Molecular Architectures with Corannulene Scaffolds	544
12.2.5.3	Large Systems with Corannulene Fragments	547
12.3	Bowl-to Bowl Inversion in Buckybowls	549
12.4	Organometallic Derivatives of Buckybowls	553
12.4.1	Alkali Metal Reduction	553
12.4.2	Transition Metal Complexes	555
12.5	Conclusions	560
12.6	Experimental: Selected Procedures	560
12.6.1	Preparation of Octabromide 11 from 10	560
12.6.2	Preparation of Corannulene from 11 by Low-valent Vanadium Coupling	560
12.6.3	Preparation of Tetrabromocorannulene from 11 by Carbenoid Coupling	561
12.6.4	Preparation of 43 by Double Diels–Alder Reaction	561
12.6.5	Preparation of 44 by Aluminum Cyclohexoxide Reduction of 43	561
12.6.6	Preparation of 45	561

12.6.7	Preparation of Dodecabromide 46	*562*
12.6.8	Preparation of Dicorannulenobarrelene 47	*562*
	Acknowledgments *562*	
	Abbreviations *562*	
	References *563*	

13 Reduction of Carbon-rich Compounds *566*
Tamar Sternfeld and Mordecai Rabinovitz

13.1	Introduction *566*	
13.2	Electron Transfer Methods and Detection of the Products	*567*
13.2.1	Reduction via Alkali Metals *567*	
13.2.2	The Application of Electron Shuttles *567*	
13.2.3	Potassium Graphite Intercalates as Reducing Agents	*568*
13.2.4	Electrochemical Studies *570*	
13.2.5	Ion Solvation Equilibria in π-Conjugated Reduced Systems	*571*
13.2.6	Magnetic Resonance Spectroscopy – Tool of Choice for the Study of Reduced Carbon-rich Compounds *571*	
13.2.6.1	NMR *571*	
13.2.6.2	EPR *572*	
13.3	Monocyclic π-Conjugated Anions *573*	
13.3.1	Annulenes: General *573*	
13.3.2	"Classical Annulenes" *574*	
13.3.2.1	Cyclooctatetraene–[8]Annulene *574*	
13.3.2.2	[12]Annulene *575*	
13.3.2.3	[16]Annulene *577*	
13.3.2.4	[18]Annulene *578*	
13.3.3	Dehydroannulenes *579*	
13.3.4	Cyclophanes *581*	
13.4	Reduction of Nonplanar Polycyclic Aromatic Hydrocarbons	*586*
13.4.1	Nonplanar PAHs *586*	
13.4.2	PAHs with Five-membered Rings *587*	
13.4.2.1	Corannulene *587*	
13.4.2.2	Reduction of Corannulene with Lithium *587*	
13.4.2.3	Reduction of Corannulene with Different Alkali Metals	*589*
13.4.2.4	Substituted Corannulenes *590*	
13.4.2.5	Extended Corannulenes *592*	
13.4.3	Strain in PAHs: The Story of Curved Pyrene *596*	
13.4.3.1	First Reduction Process – Dimerization *598*	
13.4.3.2	The Second Reduction Stage *598*	
13.5	Reduced Fullerenes *599*	
13.5.1	General *599*	
13.5.2	Reduction Stages of Fullerenes *600*	
13.5.2.1	Anions of C_{60} *602*	
13.5.2.2	Anions of C_{70} *603*	
13.5.2.3	Reduced Higher Fullerenes *604*	

13.5.3 The Aromatic Character of Reduced Fullerenes 605
13.5.3.1 Endohedral Helium in Fullerenes – an Internal Probe 606
13.5.3.2 ^{13}C NMR of Reduced Fullerenes 607
13.5.3.3 Methylene Bridged Fullerenes – an External Probe 608
13.5.4 Fullerene Derivatives 610
13.5.4.1 Metallofullerene Complexes 611
13.5.5 Nanotubes 611
13.6 Conclusions 612
13.7 Experimental: Selected Procedures 613
13.7.1 Reduction with Li and K 613
13.7.2 Quench Reaction with Water 614
13.7.3 Reduction with C_8K 614
Acknowledgments 614
Abbreviations 614
References 615

Index *625*

Preface

It was not long ago that the mention of carbon-rich materials conjured up images of only diamond and graphite, while carbon-rich molecules referred primarily to polycyclic aromatic hydrocarbons. It is amazing how times have changed. A plethora of new synthetic break-throughs, coupled with an unprecedented range of futuristic applications, has radically transformed this realm of science. While the discovery of the fullerenes and carbon nanotubes did much to bring the science of carbon-rich molecules and materials into the limelight over the past couple of decades, seminal work in this area has been ongoing for well over 125 years.

This monograph was designed to offer an overview of the spectacular modern face of the chemistry associated with carbon-rich compounds. Recent advances in methodologies for the synthesis of carbon-rich molecules are at the forefront of several chapters, including many examples of molecules of previously unimaginable proportions, such as superbenzenes, polyyne molecular wires, and conjugated oligomers that are nanometers in diameter/length. As a practical aspect, each individual chapter contains a handful of important experimental protocols selected by the authors to provide a starting point for the reader. Synthetic advancements have naturally led to new carbon-rich topologies. The relationship between structure and property is thus a recurring theme throughout this book, and the expert authors offer fundamental insight and perspective into the unique, and sometimes unpredictable, properties of these materials from both a theoretical and experimental perspective. Finally, with new properties come new applications. Never before has the future of carbon-rich materials been so bright, literally, in some cases. Many carbon-rich molecules are being incorporated into increasingly sophisticated luminescent devices (OLEDs), field effect transistors (FETs), and solar cells, while others are being explored as active components in molecular electronics. Alternatively, the study of carbon-rich molecules and oligomers can often lead to a better understanding of larger systems that are themselves difficult if impossible to study easily, such as nanotubes, polymers like poly(phenylene ethynylene), or photosynthetic processes. Such investigations shed light on both the potential and limitations of new materials.

This book was the collective work of a number of people. Most importantly, we appreciate the contributions of the authors, whose time, effort and expertise have made this monograph both an interesting read and a stimulating scientific

Carbon-Rich Compounds. Edited by Michael M. Haley and Rik R. Tykwinski
Copyright © 2006 WILEY-VCH Verlag GmbH & Co. KGaA, Weinheim
ISBN: 3-527-31224-2

resource. We express our gratitude to Ms. Annie Tykwinski for designing the cover art, and we would like to thank Drs. Elke Maase and Manfred Köhl at Wiley-VCH for an enjoyable collaboration in the preparation of this book.

Michael M. Haley
Eugene, Oregon
October, 2005

Rik R. Tykwinski
Edmonton, Alberta

List of Contributors

Katie Campbell
Department of Chemistry
University of Alberta
Edmonton
Alberta, T6G 2G2
Canada

Frank W. Fowler
Department of Chemistry
State University of New York
Stony Brook, NY 11794
USA

Rolf Gleiter
Institut für Organische Chemie
Universität Heidelberg
Im Neuenheimer Feld 270
69120 Heidelberg
Germany

Michael M. Haley
Department of Chemistry
University of Oregon
Eugene, OR 97403-1253
USA

Anthony Harriman
University of Newcastle
Molecular Photonics Laboratory
School of Natural Sciences
Bedson Building, Newcastle upon Tyne
NE1 7RU
United Kingdom

Charles A. Johnson II
Department of Chemistry
University of Oregon
Eugene, OR 97403-1253
USA

Toshikazu Kitagawa
Institute for Chemical Research (ICR)
Kyoto University
Gokasho
Uji
Kyoto 611-0011
Japan

Koichi Komatsu
Institute for Chemical Research (ICR)
Kyoto University
Gokasho
Uji
Kyoto 611-0011
Japan

Joseph W. Lauher
Department of Chemistry
State University of New York
Stony Brook, NY 11794
USA

Herbert Meier
Institut für Organische Chemie
Johannes Gutenberg-Universität
Düsbergweg 10-14
55099 Mainz
Germany

Ognjen Š. Miljanić
Center for New Directions in
Organic Synthesis
Department of Chemistry
University of California at Berkeley
and the Chemical Sciences Division
Lawrence Berkeley National Laboratory
Berkeley, California 94720-1460
USA

List of Contributors

Klaus Müllen
Max-Planck-Institut für Polymerforschung
Ackermannweg 10
55128 Mainz
Germany

Yasujiro Murata
Institute for Chemical Research (ICR)
Kyoto University
Gokasho
Uji
Kyoto 611-0011
Japan

Peter W. Rabideau
Department of Chemistry
Mississippi State University
Mississippi State, MS 39762-9573
USA

Mordecai Rabinovitz
Department of Organic Chemistry
The Hebrew University of Jerusalem
Givat Ram
Jerusalem, 91904
Israel

Peter R. Schreiner
Institute of Organic Chemistry
Justus-Liebig University
Heinrich-Buff-Ring 58
35392 Giessen
Germany
and
Department of Chemistry
The University of Georgia
1001 Cedar Street
Athens
GA 30602-2525
USA

Tamar Sternfeld
Department of Organic Chemistry
The Hebrew University of Jerusalem
Givat Ram
Jerusalem, 91904
Israel

Andrzej Sygula
Department of Chemistry
Mississippi State University
Mississippi State, MS 39762-9573
USA

Chi-Hang Tao
Department of Chemistry
The University of Hong Kong
Pokfulam Road
Hong Kong
P. R. China

Rik R. Tykwinski
Department of Chemistry
University of Alberta
Edmonton
Alberta, T6G 2G2
Canada

K. Peter C. Vollhardt
Center for New Directions in
Organic Synthesis
Department of Chemistry
University of California at Berkeley
and the Chemical Sciences Division
Lawrence Berkeley National Laboratory
Berkeley, California 94720-1460
USA

Daniel B. Werz
Laboratory for Organic Chemistry
Swiss Federal Institute of Technology Zürich
ETH-Hönggerberg
Wolfgang-Pauli-Strasse 10
8093 Zürich
Switzerland

Jishan Wu
Max-Planck-Institute for Polymer Research
Ackermannweg 10
55128 Mainz
Germany

Vivian W. W. Yam
Department of Chemistry
The University of Hong Kong
Pokfulam Road
Hong Kong
P. R. China

Raymond Ziessel
Laboratoire de Chimie Moléculaire
Ecole Européenne de Chimie Polymères
et Matériaux (ECPM)
25, rue Becquerel
67087 Strasbourg Cedex 2
France

1
Pioneers of Carbon-rich Compounds

Charles A. Johnson II and Michael M. Haley

1.1
Introduction

This chapter provides a historical perspective on the discovery of and classical approaches to carbon-rich molecules prior to 1970. Carbon-rich is defined here as a ratio of at least one C atom per H atom. Due to the tremendous amount of research in this field, this chapter focuses on the pioneering synthetic efforts and key successes thereafter and will not attempt to serve as an inclusive document on the subject. Several excellent references have been compiled which can provide further information [1]. Additionally, the overwhelming plethora of substituted variants, while important in relation to derivatization and molecular diversity, will not be addressed unless unique in application or properties. Examples include the numerous alkyl-substituted or partially reduced derivatives of polycyclic aromatic hydrocarbons described in detail by Clar [1a]. Finally, fused benzenoid structures are described with classical numerical nomenclature instead of the alphabetical IUPAC methodology used by Harvey [1c].

1.2
19th Century Achievements

Initial research on carbon-rich systems began with the isolation of polycycles from coal-tar and other petroleum products in the mid 19th century. Separation, purification, and characterization of several different macrocycle topologies by German chemists, such as Graebe and Glaser, provided key insight into existing hydrocarbon diversity as well as potential future topologies. Such efforts normally included fractional distillation of the crude fuel, treatment with concentrated acid and base, final recrystallization or distillation, and derivatization to assist in identification. Naphthalene **1**, phenanthrene **2**, anthracene **3**, biphenyl **4**, fluorene **5**, picene **6**, pyrene **7**, and fluoranthene **8** were all isolated prior to 1900 in this fashion (Fig. 1.1) [1a].

In addition to petroleum component analyses, the late 19th century spawned the

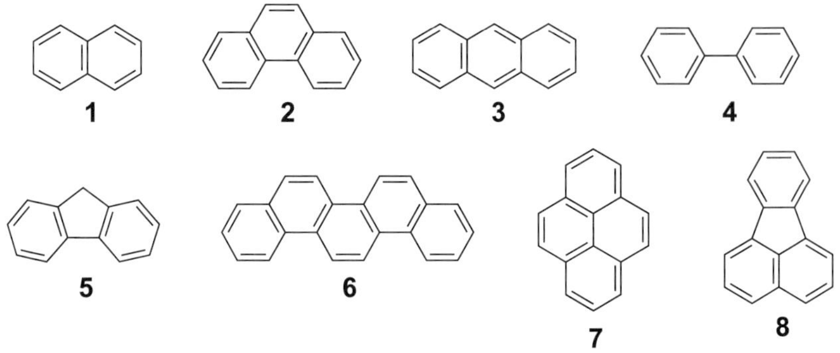

Figure 1.1. Structures **1–8**.

first synthetic efforts on two, three, and four linearly-annelated polycyclic systems, most of which had been previously isolated from fossil fuels, as well as *peri*-condensed benzenoid systems such as terphenyl and tetraphenyl. The majority of syntheses used pyrolysis or pyro-condensation and consisted of passing precursor molecules through red-hot tubes or distillation over broken glass at red heat [1a]. Examples include Graebe's formation of fluorene from diphenylmethane in 1874 (Scheme 1.1a) [2] and Radziszewski's production of naphthalene from (α,β-dibromobutyl)benzene **9** in 1876 (Scheme 1.1b) [3]. Pyrolysis at high temperature though, routinely resulted in poor product yield due to formation of multiple byproducts and different isomers. The reinvestigation of naphthalene pyrolysis by Lang et al. in 1957 showed the lack of specificity as three dinaphthyls, perylene, two benzofluoranthenes, ternaphthyl, and terrylene were all produced [4].

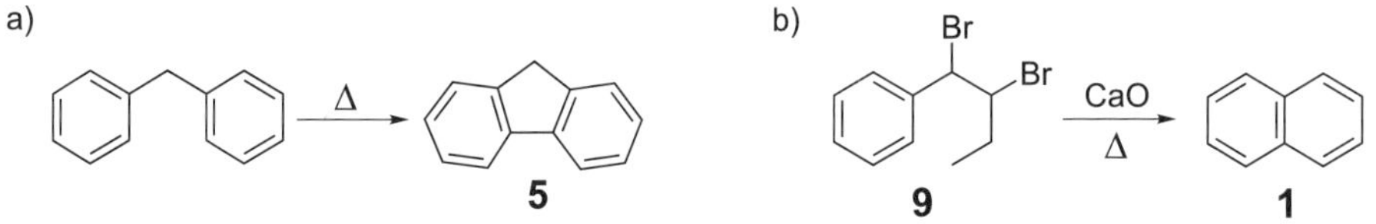

Scheme 1.1

The most important early cyclization technique developed was von Baeyer's Zn-dust distillation in 1866 [5]. The procedure effectively reduced carbon-rich quinone or oxygen-containing precursors, as shown by the formation of benzene from phenol [1a]. Also of significance was the multistep production of phenanthrene by Pschorr in 1896 (Scheme 1.2) [6]. Condensation of *o*-nitrobenzaldehyde and phenylacetic acid produced the uniquely substituted intermediate **10**. Sequential nitro group reduction and diazotization followed by Cu-induced ring closure and thermal decarboxylation afforded the polycycle. Pschorr's methodology, important because of its low temperature conditions, was later used to produce several other condensed polyarenes in the coming decades, such as chrysene and picene

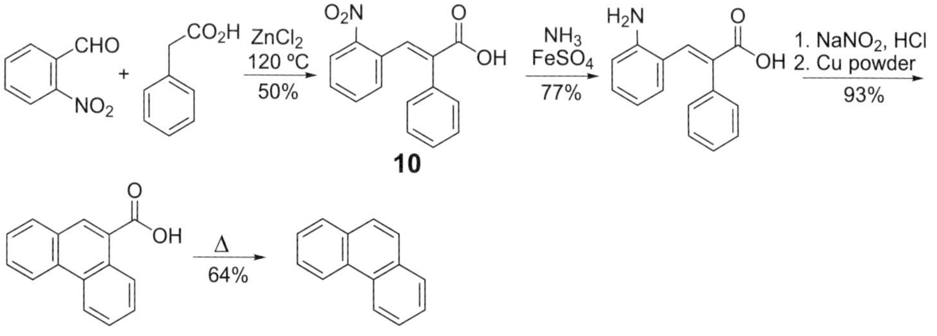

Scheme 1.2

[1a]. Examples of other pre-1900 ring closures include Klinger's reduction of a substituted phenanthrene to 1.2,7.8-dibenzochrysene [7] and Riese's production of p-terphenyl from 1,4-dibromobenzene, bromobenzene, and Na in ether [8].

The first synthesis of a tri-annelated system occurred 30 years after its initial discovery from coal-tar in 1832 by Dumas and Laurent when Limpricht hydrolyzed benzyl chloride under heat to afford anthracene [1a]. The most significant improvement on anthracene production, however, came in 1868 when Liebermann and Graebe obtained it via Zn-dust distillation of the natural product alizarin [1a]. An interesting dimer, known as dianthracene or p-anthracene **11**, was obtained by several groups in the 1890s by irradiation of anthracene solutions (Scheme 1.3) [9]. Irradiation of several anthracene derivatives afforded similarly novel structures [9].

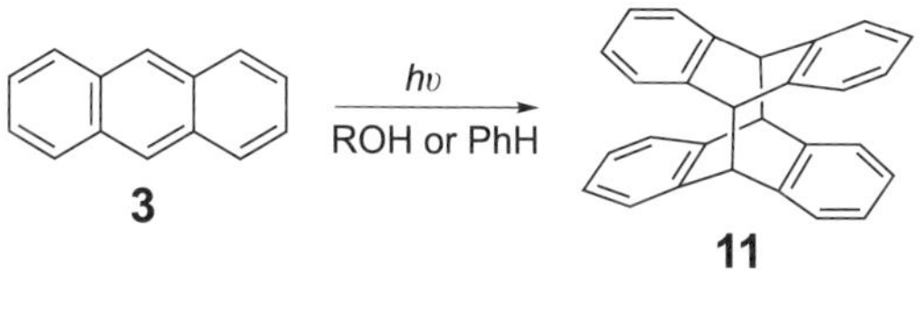

Scheme 1.3

The production of tetracene **12** prior to the turn of the century marked the largest linearly-annellated polyarene at the time [10]. The synthetic pathway began with condensation of phthalic anhydride and succinic acid with NaOAc (Scheme 1.4). Rearrangement of **13** with NaOEt and subsequent Zn-dust distillation afforded a combination of **12** and dihydro-derivative **14**.

Assembly of carbon-rich molecules by means of acetylene chemistry is also of note prior to 1900. Although isolated from petroleum products in 1825 by Faraday [1a], the first total synthesis of benzene by Berthelot in 1867 via cyclotrimerization of acetylene [11] set a precedent for future thermal trimerization reactions, a key procedure to produce uniquely substituted arene rings [1e]. The first ethynyl ho-

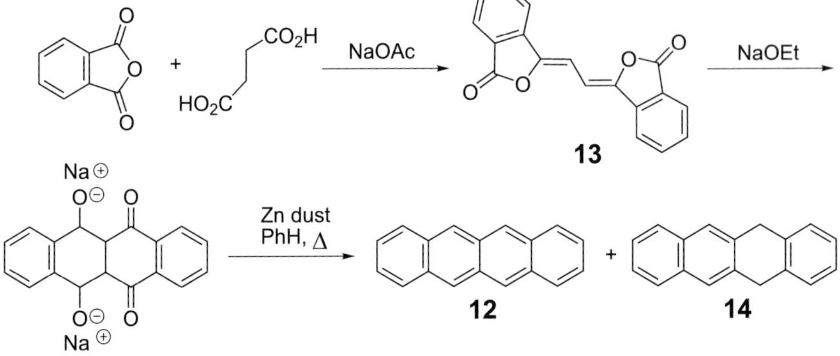

Scheme 1.4

mocoupling occurred in 1869 when Glaser formed 1,4-diphenylbutadiyne **15** from phenylacetylene and CuCl in ammonia and alcohol (Scheme 1.5) [12]. In the process he isolated the intermediate copper complex **16**. The first cross-coupling of asymmetrical acetylides occurred in 1882 when Baeyer obtained an unsymmetrical butadiyne as well as the two symmetrical diethynyl dimers from o-nitrophenylacetylene and phenylacetylene [13]. Although Straus and Moreau improved alkyne coupling conditions in the early 20th century, Glaser's oxidative conditions would serve as a standard technique until the 1950s when Eglinton and Galbraith, Chodkiewicz and Cadiot, Hay, and Brockman developed alternative methods [14].

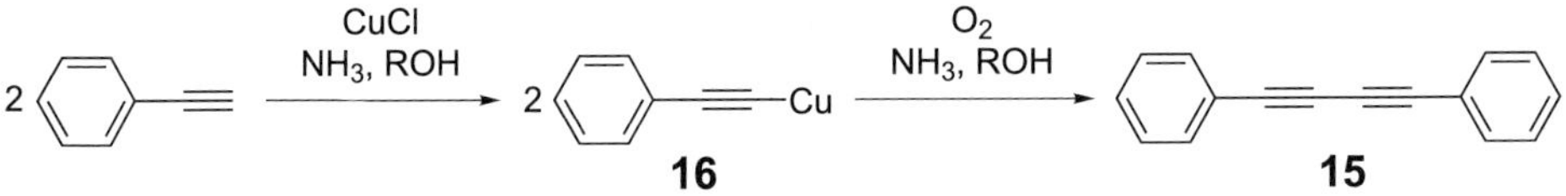

Scheme 1.5

1.3
1900–1928: Dawn of the Twentieth Century

The turn of the century and the following three decades focused mainly on the development of new, efficient routes to previously isolated structures. The replacement of pyrolysis with tailored pathways afforded increased yields and convincing proof of structure through derivatization of intermediates and products. Two examples include Mannich's production of triphenylene from an acid-catalyzed, threefold condensation of cyclohexanone followed by reduction over freshly reduced, hot Cu, [15] as well as Ullmann and Meyer's condensation of iodobenzene with Cu to afford biphenyl [16]. Of note was a significant simplification in pyrolysis for low molecular weight hydrocarbons, which consisted of inserting an electrically heated glowing wire into boiling hydrocarbon vapor [17]. Löb successfully produced biphenyl from benzene in the early 1900s via this new method.

Tetraphenylallene **17**, one of the first carbon-rich allenes, was produced in the early 1900s by Vorländer and Siebert via halogenation/dehydrohalogenation of phenyl-substituted olefin **18** (Scheme 1.6) [18]. This classical method set a standard for the preparation of arylallenes as the resultant product could not isomerize due to lack of an abstractable hydrogen [1d].

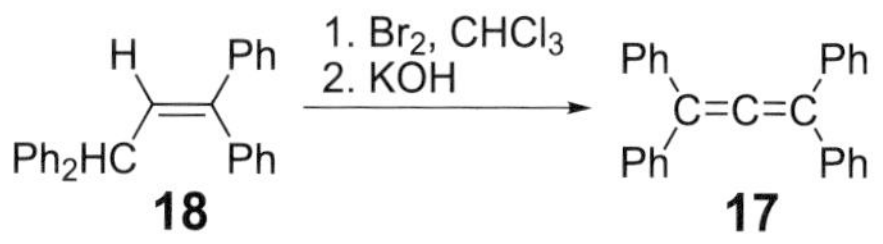

$$\underset{\textbf{18}}{\overset{Ph_2HC}{\underset{H}{>}}=\overset{Ph}{\underset{Ph}{<}}} \quad\xrightarrow[\text{2. KOH}]{\text{1. Br}_2\text{, CHCl}_3}\quad \underset{\textbf{17}}{\overset{Ph}{\underset{Ph}{>}}C=C=C\overset{Ph}{\underset{Ph}{<}}}$$

Scheme 1.6

Two important polycycles produced prior to 1920 were perylene **19** (Fig. 1.2a) and pyrene **7**, both of note as parent structures for numerous fused benzenoid analogs produced in the following decades. In 1910, Scholl, Seer, and Weitzenböck first produced perylene from an AlCl$_3$-mediated cyclization of naphthalene and/or 1,1′-binaphthyl under heat in low yield [19]. A subsequent success with improved yield occurred via treatment of 1,8-diiodonaphthalene with Cu powder. Weitzenböck was responsible for the first synthesis of pyrene in 1913 [20]. The five-step synthesis began with bromination of o,o′-ditolyl followed by conversion into dicarboxylic acid **20** via a dinitrile intermediate (Scheme 1.7). Cyclization and Zn-dust distillation afforded the tetra-fused structure in a well-designed synthesis, confirming its structure through intermediate analysis.

In relation to size, the most notable molecules of the early 20th century include decacyclene **21** and tetraphenyltetracene **22**, also known as rubrene (Schemes 1.8 and 1.9, respectively). Decacyclene was produced in 1903 by Dziewoński and Rehländer by heating acenaphthene **23** with sulfur (Scheme 1.8) [21]. Ten years later a new route with lead dioxide provided **21** as well as the byproduct fluorocyclene **24** [22]. Rubrene **22**, produced by Moureu, Dufraisse, and Dean in 1926, is a red hydrocarbon obtained from dimerization of phenylethynyldiphenylmethyl chloride **25** under heat (Scheme 1.9) [23]. The structure is remarkable in its acid-catalyzed

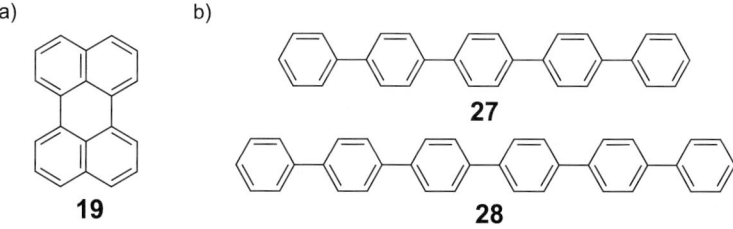

Figure 1.2. (a) Structure **19**, (b) structures **27** and **28**.

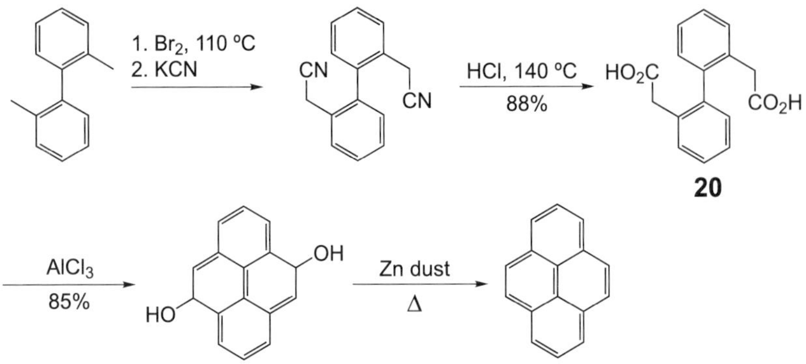

Scheme 1.7

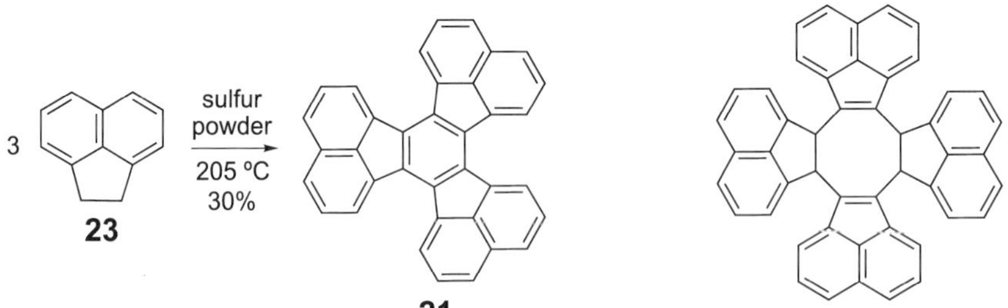

Scheme 1.8

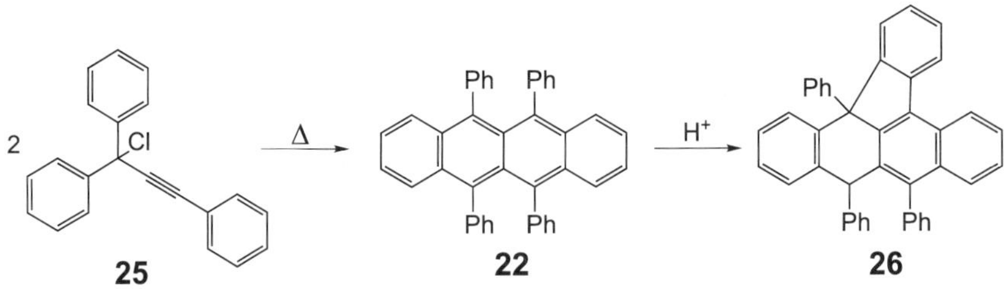

Scheme 1.9

isomerization to the colorless pseudorubrene **26** [24] and a photo-induced peroxide derivative that reformed rubrene upon heating [23].

Evolution to the two next largest *peri*-condensed benzenoid polyphenyls occurred in 1924 with the production of quinquephenyl **27** by Gerngross and coworkers (Fig. 1.2b) [25]. Two distinct pathways afforded the pentamer: (i) reduction of a benzenediazonium sulfate in acid with Cu powder, and (ii) Ag-mediated cross-coupling of 4-iodobiphenyl and 4-iodoterphenyl. In the same year, Pummerer and Bittner

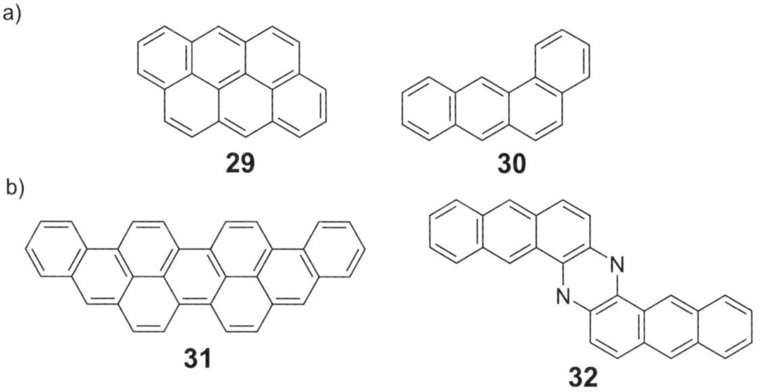

Figure 1.3. (a) Structures **29** and **30**, (b) structures **31** and **32**.

produced sexiphenyl **28** (Fig. 1.2b) via a similar Ag-mediated homocoupling of 4-iodoterphenyl [26]. Although improvements on the above two polyarenes continued in the following decades, sexiphenyl marked the largest linear, *peri*-condensed hydrocarbon prior to 1970.

1.4
1929–1949: Rise of the Polycyclic Aromatic Hydrocarbon

The 1930s marked a significant increase in new and larger polycyclic aromatic hydrocarbons with the first appearance of hexa-annelated systems and initial work on benzenoid isomers and derivatives of anthanthrene **29**, tetraphene **30**, pyrene, and perylene (Fig. 1.3a). In addition to continued research from distinguished chemists such as Scholl, the pioneering efforts of Erich Clar and James W. Cook were a key factor for this development. Müllen and Wu discuss recent advances in polycyclic aromatic hydrocarbons in Chapter 3.

Indisputably, the most significant contributions to polycyclic aromatic hydrocarbons originated with Clar at the beginning of this decade. His efforts were published in the journal series *Polynuclear Aromatic Hydrocarbons* and/or *Aromatic Hydrocarbons* [27] as well as in an excellent two volume series, *Polycyclic Hydrocarbons*, [1a] which provided a comprehensive look at hydrocarbon diversity. Not only did he and his coworkers produce an exceeding number of new carbon-rich structures, but his adaptation and development of new synthetic methodologies allowed confirmation of structure as well as improved yields to provide sufficient quantities for physical property studies. Clar and coworkers continued their work well into the latter parts of the 20th century. His early work focused on benzo-, naphtho-, and anthraceno-derivatives of anthracene and phenanthrene [1a].

An example of the numerous, new synthetic methods developed by Clar was the Zn-dust melt [1a]. Published in 1939, the method provided a more practical one for reduction of polycycle quinones and oxygen-containing precursors than von

Baeyer's traditional solid-state, Zn-dust distillation. The key advantage was lower decomposition rates due to incorporation of solvent. The new technique was applied to several previously known systems, such as violanthrene **31** (Fig. 1.3b) from violanthrone and tetracene from tetracenequinone [1a]. It was also shown to work on five-membered systems, such as dihydropentalene derivatives, and heterocycles, such as anthrazine **32** (Fig. 1.3b) [1a].

One of Clar's most notable early successes of the decade was pentacene **33**. Clar and John isolated **33** via dehydrogenation of a dihydropentacene isomer, several of which had been laboriously produced since 1911 when Phillipi first claimed one as **33** [28]. The combination of two groups' efforts provided a far simpler route to pentacene two decades later. The highly efficient route began with condensation of o-phthalaldehyde and cyclohexane-1,4-dione to afford quinone **34** (Scheme 1.10) [29]. Reduction with Al powder afforded **33** in two steps [30]. Clar and coworkers also synthesized several pentacene derivatives in the 1940s: 1.2-benzopentacene from pseduocumene and three additional dibenzopentacene derivatives from naphthalene and/or phenanthrene starting materials [1a]. The syntheses focused on condensation cyclization reactions of keto-acids to form the penultimate quinones and ultimate polycycles.

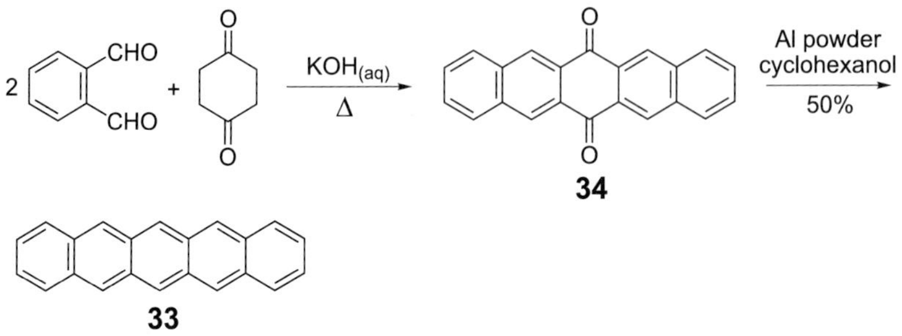

Scheme 1.10

Cook, another key personality of the decade, began delineation of his work on carbon-rich systems in two published journal series, *Polycyclic Aromatic Hydrocarbons* [31] and *Production of Cancer by Pure Hydrocarbons,* [32] in the early 1930s. The motivation for the research was the structural basis for cancer-causing compounds and his methodology consisted of administering vapor and *in vivo* doses to laboratory rats. A large part of his synthetic work focused on derivatization of parent hydrocarbons to determine the extent of activity. A typical example was increasing the alkyl chain from methyl to isopropyl on benzanthracene [33] and monitoring tumor formation in rats. His elucidation of carcinogenic activity for his own structures as well as other compounds was a major step in the acknowledgement of health issues associated with highly fused polycycles.

Cook's early research efforts focused on isolating coal-tar components [34], such as perylene and benzoperylene, as well as developing synthetic routes to previously isolated compounds. In contrast to the numerous Zn-dust distillation cyclizations,

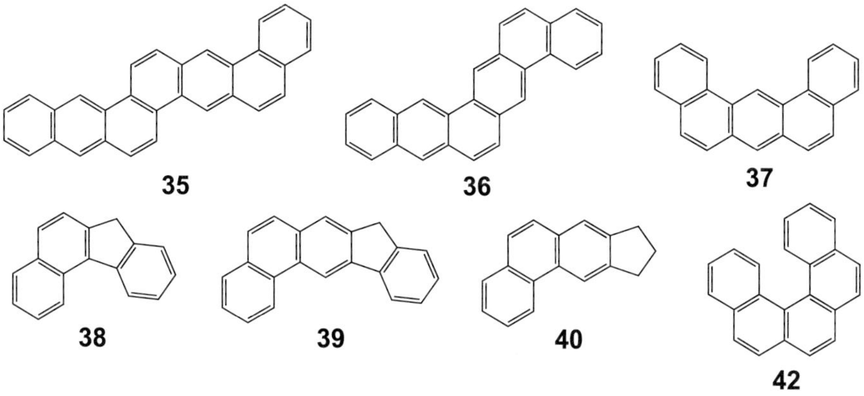

Figure 1.4. Structures **35–40** and **42**.

Cook utilized an atypical selenium dehydrogenation of a partially saturated keto-derivative to afford pyrene in 1934 [35]. Cook and coworkers also successfully synthesized several new polycycles and numerous derivatives through the process of their research. A few examples of novel systems produced include new isomers of benzopyrene, anthraceneotetraphene **35**, benzopentaphene **36**, dibenzanthracene **37**, as well as several condensed fluorene structures such as **38** and **39** (Fig. 1.4) [36]. Cook's formation of cyclopentenophenanthrene **40**, an important phenanthrene derivative used to produce several active natural products, was an early highlight of his career [37].

Through his synthetic research, Cook correctly elucidated the structure of several previously reported intermediates and hydrocarbons. His work on perylene and the first helicenes is of note. Although described in earlier 20th century literature by Weitzenböck and coworkers, as well as Mayer and Oppenheimer [38], confirmed isolation of the first two helicenes, tetra- **41** and pentahelicene **42**, did not occur until the early 1930s when Cook reinvestigated the work [39]. Early efforts with Pschorr methodology were shown to provide the desired product but also resulted in a mixture of polycyclic isomers when analyzed by increasingly detailed structural elucidation techniques. Cook determined that a Pschorr-like intermediate **43** of [4]helicene provided two distinct tetrabenzenoid fused carboxylic acids **44**, **45**, which lead to [4]helicene but also tetraphene (Scheme 1.11). He effectively separated **44** and **45** via their K and Na salts and characterized **30** and **41** via their picrate and quinone derivatives. For [5]helicene, Cook overcame the previous difficulties by separating the concurrently formed isomer, 1.2,5.6-dibenzanthracene, through the picrate. Although less significant in the [4]- and [5]helicenes, future helicenes would present a new facet in hydrocarbon chemistry due to their chirality resulting from steric overcrowding.

Optimization of an unsuccessful route initiated by Bentley et al. in 1907 ultimately led to the synthesis of the first hexa-annelated structure, hexacene, in 1939 by both Clar and Marschalk via two different routes [1a, 40]. In search of a hepta-

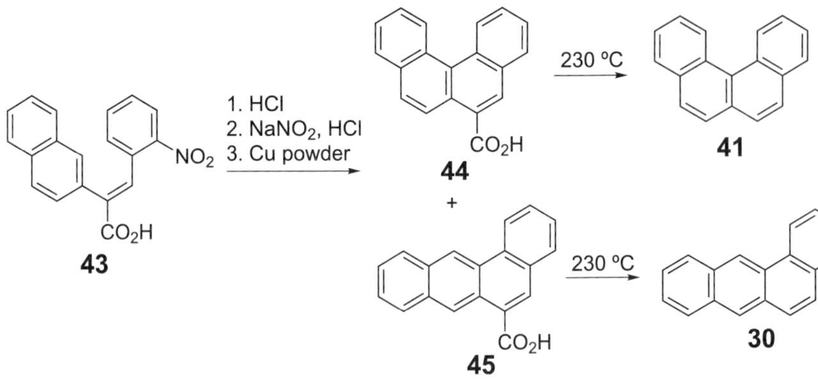

Scheme 1.11

annelated compound, Clar's initial efforts resulted in several quinone and hydroheptacene derivatives and ultimately to heptacene in 1942 in very low yield [1a]. An improved pathway confirming Clar's earlier success did not appear until a decade later when Bailey and Liao dehydrogenated a dihydroheptacene isomer with Pd/C at high temperature to afford the highly reactive heptacene in 76% yield [41]. Other early noteworthy efforts on linearly annelated structures include Marschalk's production of several hydroxyquinone and quinone derivatives of the unattainable octa-, nona-, and undecacenes in the early 1950s [42].

Scholl was the first to produce two fundamental polyarenes in the early 1930s that served as the basis for future benzenoid derivatives. Oxidation of dibenzoperopyrenequinone **46** with HNO_3 afforded intermediate **47** and subsequent decarboxylation with soda-lime provided coronene **48** (Scheme 1.12a) [43]. Two years later, effective reduction and dehydrogenation of his previously isolated bisanthenequinone **49** with HI and red phosphorus followed by Cu-mediated sublimation afforded bisanthene **50** (Scheme 1.12b) [44]. Clar effectively developed more efficient routes for both coronene (Scheme 1.12c) and bisanthene in the next decade [1a].

Clar further modified his Zn-dust reduction technique for quinone reduction in the late 1940s with the addition of pyridine and AcOH [1a]. The method proved much milder than previous Zn-reductions and allowed isolation of large polyarenes with decreased decomposition due to reduced reaction duration. For example, dibenzoperylenequinone was effectively reduced to dibenzoperylene [1a].

Lothrop's production of biphenylene **51** in 1941 from distillation of o,o'-dibromobiphenyl **52** with Cu_2O marked the first conclusive evidence for a stable, fused carbon-rich structure containing both four- and six-membered rings (Scheme 1.13) [45]. The history of biphenylene dates back to its first attempted manufacture in 1893 by Hosaeus, who isolated biphenyl instead [45]. All subsequent attempts prior to Lothrop utilizing Wurtz chemistry resulted in isomeric cycles or irreproducible results. Of the several routes developed after Lothrop, Wittig and Sondheimer completed the two most intriguing (Scheme 1.13). In 1961, Wittig successfully py-

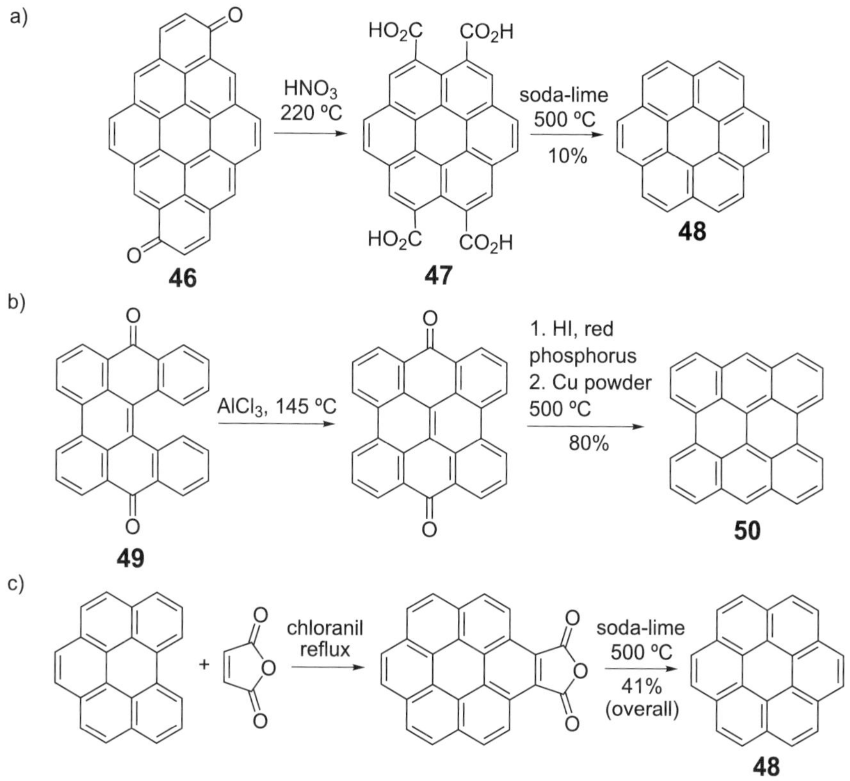

Scheme 1.12

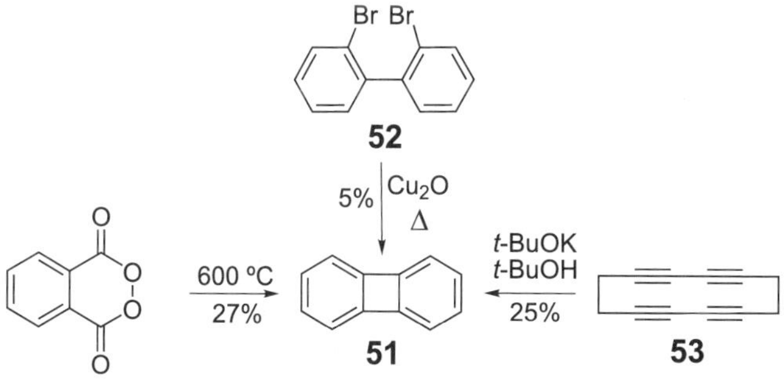

Scheme 1.13

rolyzed phthalyl peroxide as well as subliming a mercury intermediate obtained from dilithiobiphenyl with Ag powder [46]. In 1962, Sondheimer treated cyclic tetraalkyne **53** with *tert*-BuOK [47]. The chemistry of biphenylene and larger derivatives is discussed by Miljanic and Vollhardt in Chapter 4.

One of the most remarkable molecules to originate in the 1940s was trypticene **54**, a 3-fold symmetrical structure named for the triptych of antiquity, a book with three leaves hinged on a common axis. Although Clar began initial work on the seven-step synthesis in the 1930s [1a], the hydrocarbon was finally isolated by Bartlett and coworkers in 1942 and originated from their efforts to elucidate the special properties of bridgehead substituents on bicyclic systems [48]. Diels-Alder reaction of anthracene and *p*-benzoquinone followed by reduction afforded dihydroxytriptycene **55**, which was finally reduced to triptycene after significant optimization (Scheme 1.14). Conversion of the hydroxy groups to halogen substitutents through the diamine intermediate **56** followed by Busch's heterogeneous reduction afforded **54** in low but unspecified yield.

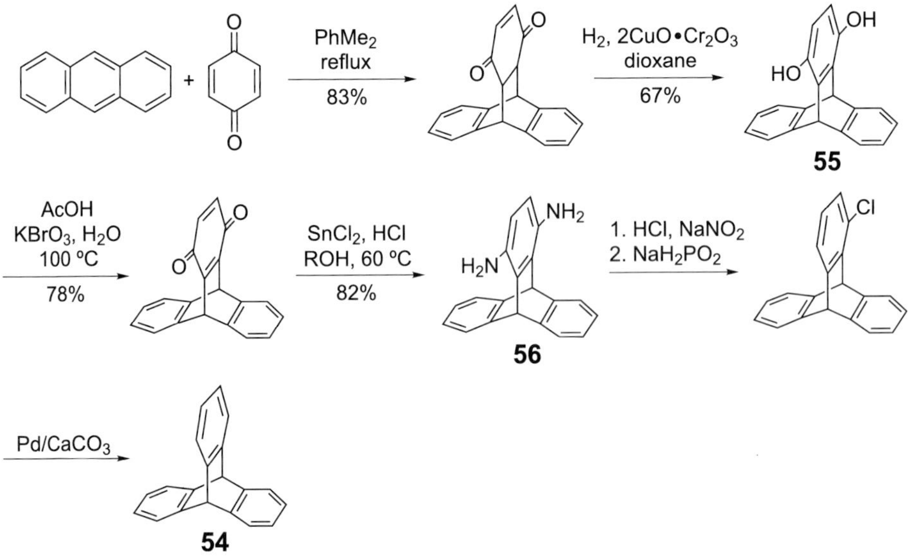

Scheme 1.14

1.5
1950–1969

After a decrease in new system development in the 1940s, the following two decades showed a sharp increase in polycyclic aromatic hydrocarbon production and diversity. The new benzenoid systems, mainly pyrene and perylene derivatives, represent the largest structures isolated prior to 1970. The introduction of the annulenes, radialenes, and fulvalenes are also of note during this time period. In addition to continued efforts by Clar, key personalities for these decades include Franz Sondheimer, Geoffrey Eglinton, Shōichi Misumi, Masazumi Nakagawa, and Georg Wittig.

1.5.1
The Annulenes, Dehydrobenzoannulenes, and Phenylacetylene Scaffolding

In the mid-1950s, Sondheimer and coworkers initiated their groundbreaking work on the annulenes and dehydroannulenes. The work was described in a publication series, *Unsaturated Macrocyclic Compounds*, [49] as well as several reviews [50]. The ensuing research probed the concept of aromaticity and set precedent for future alkyne chemistry. In addition to use of previously developed photolytic ring-opening routes by Schröder and Oth, Masamune, and van Tamelen, Sondheimer's group developed an oxidative coupling of terminal diacetylenes leading to macrocyclic polyacetylenes [50c]. An example of the previous method was photolysis of cyclooctatetraene dimer **57** to afford [16]annulene **58** (Scheme 1.15a). The synthesis of the first higher annulene, [18]annulene **59**, is an example of the latter.

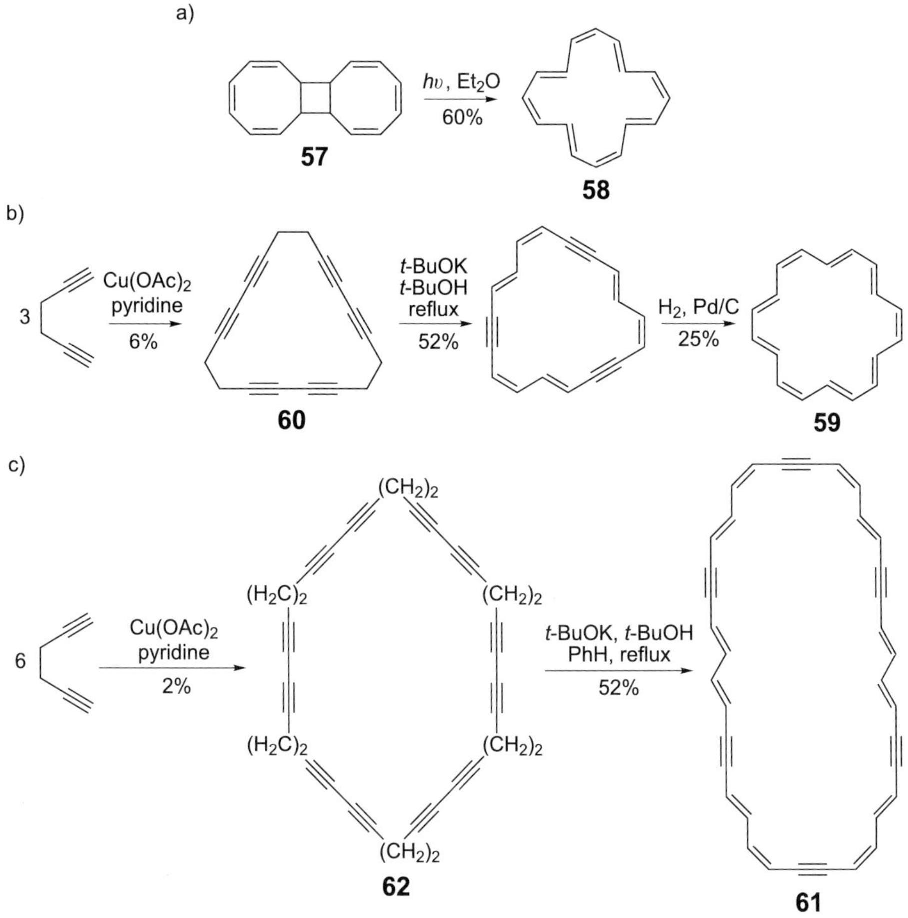

Scheme 1.15

It occurred from an initial three-fold oxidative coupling of 1,5-hexadiyne to form macrocycle **60** followed by subsequent rearrangement and catalytic hydrogenation (Scheme 1.15b). With their new method, the group produced annulenes and dehydroannulenes with ring sizes ranging from 12 to 30 atoms. The configurational isomerism for Sondheimer's structures was probed through synthetic methods as well as interconversion of conformational isomerism via σ-bond rotation. One of the more spectacular systems developed by Sondheimer and coworkers was the large ring polyacetylene **61**, obtained from treatment of hexamer **62** with base (Scheme 1.15c) [51].

Concurrent with Sondheimer's annulenes, work on macrocyclic structures with ethynyl spacers between phenyl rings, known as dehydrobenzoannulenes, emerged in the chemical record. In the hope of isolating a dehydrobenzo[18]annulene, Eglinton and coworkers instead produced strained [12]annulene dimer **63** via high dilution oxidative conditions in 1960 (Fig. 1.5a) [52]. Isolation of a second dehydrobenzo[12]annulene **64** was reported in 1966 via successful trimerization of the cuprous salt of o-iodophenylacetylene in pyridine in 26% yield [53]. Staab et al. simultaneously reported production of **64** via an alternative, multistep pathway [54]. Several other novel phenylacetylene macrocycles appeared in the 1960s, a representative example being Akiyama and Nakagawa's successful dimerization of 1,8-diethynylanthracene with $Cu(OAc)_2$ and pyridine to afford **65** (Fig. 1.5a) [55]. Werz and Gleiter discuss current advances in diyne macrocycle chemistry in Chap-

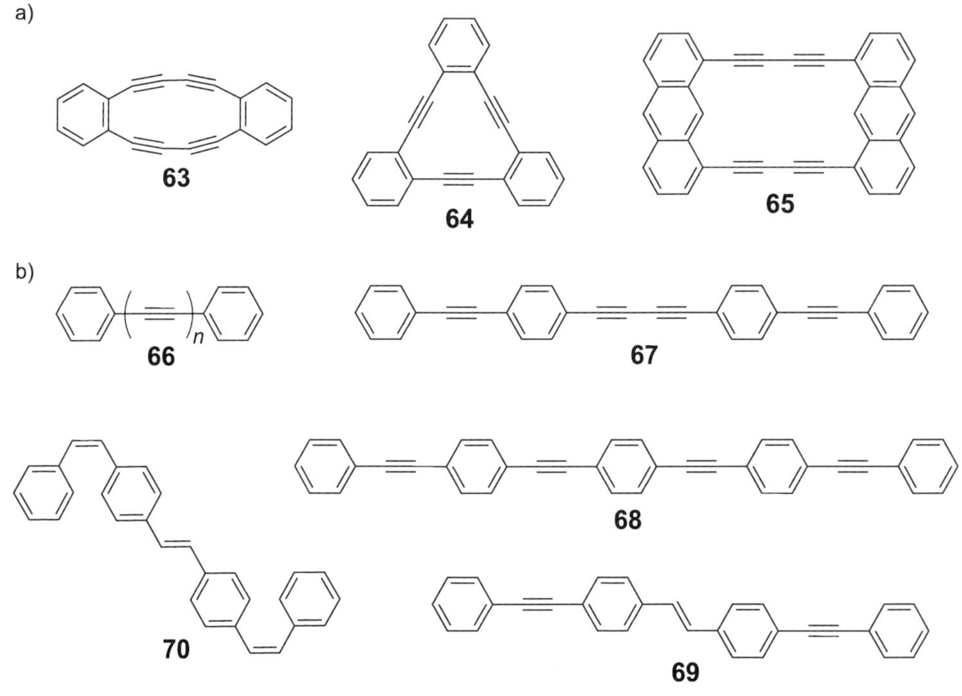

Figure 1.5. (a) Structures **63–65**, (b) structures **66–70**.

1.5 1950–1969 | 15

ter 7, whereas Campbell and Tykwinski outline chiral carbon-rich macrocycles in Chapter 6.

Expansion in the diphenylpolyacetylene series past Graebe's original dimer began in the early 1950s with the near simultaneous efforts of two groups. In 1951, both Schlubach and coworkers [56] and Nakagawa [57] reported isolation of diphenyltriyne **66** ($n = 3$) and diphenyltetrayne **66** ($n = 4$) (Fig. 1.5b). Schlubach used a combination of Grignard homocoupling and polyacetylenediol condensations to produce the products. In the same year, Schlubach and coworkers also reported synthesis of the pentamer **66** ($n = 5$) [58]. Three years later, Whiting and coworkers produced hexayne ($n = 6$) and octayne ($n = 8$) derivatives of **66** via either oxidative homocoupling of triyne precursors with CuCl or treatment of ene-yne-glycol intermediates with $SOCl_2$, respectively [59]. Modern polyyne chemistry is summarized by Yam and Tao in Chapter 10.

Even though the octayne derivative marked the longest linear diphenylpolyacetylene, multiple phenylacetylene scaffold topologies were developed in the following decade. Representative examples include carbon rods **67** and **68** produced by Misumi in 1961 (Fig. 1.5b) [60]. Misumi and coworkers also produced several mixed hybrid scaffolds combining ethenyl and ethynyl spacers, such as **69** (Fig. 1.5b) as well as conformational isomeric ethenyl rods, such as **70**, in 1962 [61]. Meier further discusses defined-length carbon-rich oligomers in Chapter 11, and the use of carbon-rich molecules as molecular wires is presented by Harriman and Ziessel in Chapter 2.

In 1969, Wegner reported the formation of polydiacetylene **71** via topochemical polymerization of substituted diacetylenes in the solid state (Scheme 1.16) [62]. The ene-yne polymeric structure was the result of multiple 1,4-additions between adjacent molecules. Wegner originally utilized different monomeric functional groups and determined that those capable of H-bonding, such as urethanes and diols, provided the necessary intermolecular alignment to facilitate the conjugate addition. Fowler and Lauher further elaborate on solid-state formation and chemistry of polydiacetylene and polytriacetylene in Chapter 5.

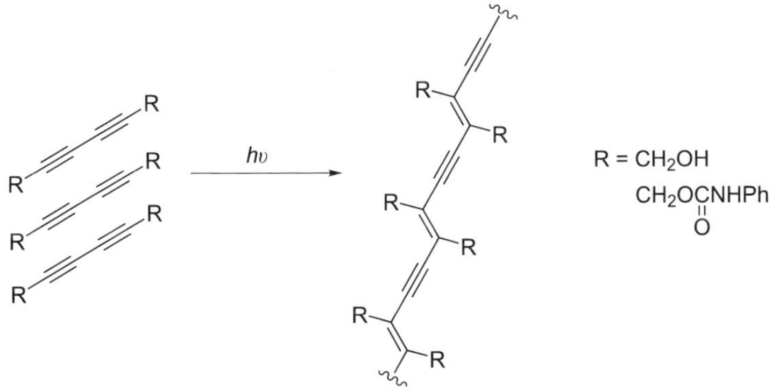

$R = CH_2OH$
$CH_2OCNHPh$
      $\|$
      O

71

Scheme 1.16

1.5.2
Fused Polycyclic and *peri*-Condensed Benzenoid Systems

In addition to his numerous other achievements, application of benzyne chemistry to carbon-rich polyarenes in the 1950s was a signature mark of Georg Wittig. Fourteen years after the first lengthy production of triptycene, Wittig and Ludwig effectively prepared triptycene in 28% yield in one step from anthracene, o-bromofluorobenzene, and Mg [63]. Another excellent example was the synthesis of 1.2,6.7-dibenzopyrene from 2,2′-dilithiobiphenyl and $TiCl_4$ via a suspected tetrabenzyne intermediate in 1957 [64]. Although produced in 1943 via Grignard reaction [65], Wittig and coworkers reacted dilithiobiphenyl **72** with $CoCl_2$ to afford tetraphenylene **73** in 3% yield (Scheme 1.17) [64]. The reaction simultaneously produced hexaphenylene **74** (1%), octaphenylene **75** (17.5%), biphenyl (16%), o-quaterphenyl (<1%), and o-sexiphenyl (1%). Substitution of $NiCl_2$ for the metal complex provided only **73** and biphenyl in 46% and 12% yield, respectively. Other notable successes include phenanthryne **76** (Scheme 1.18a), obtained from BuLi and 9-fluorophenanthrene, and its subsequent reaction with anthracene and tetraphenylfuran to form interesting adducts [66]. In 1959, Carey and Millar trimerized a phenanthryne intermediate obtained from 9,10-dichlorophenanthrene to afford hexabenzotriphenylene **77** via a similar route to Wittig's triphenylene route (Scheme 18b) [67].

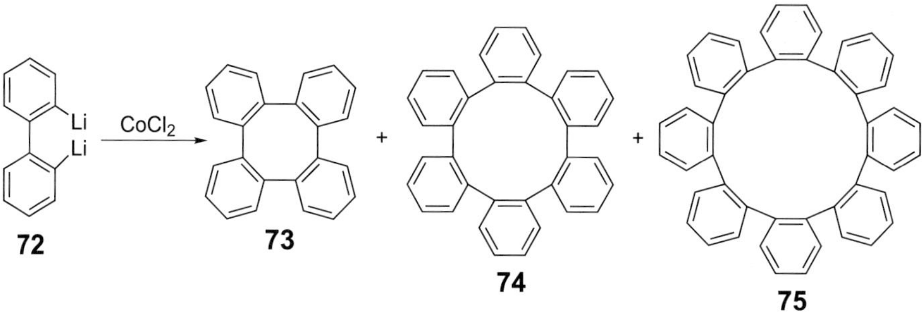

Scheme 1.17

The late 1950s saw several expanded biphenylene topologies. Some of the more notable include Cava's 1,2-benzobiphenylene **78** (Fig. 1.6a) obtained through a highly reactive benzocyclobutadiene [68] as well as his production of 1.2,7.8-dibenzobiphenylene **79**, a structure similar to [4]helicene [69]. One of the more spectacular biphenylene derivatives was Nenitzescu and coworkers' bulky 1,2,5,8-tetraphenyl-2.3,6.7-dibenzobiphenylene **80** produced by Diels-Alder chemistry in 1962 (Scheme 1.19) [70].

With the exception of fluoranthene, which was synthetically isolated in the 1930s by von Braun and Anton [71], and decacyclene, a key signature of 1950s chemistry was the rapid expansion of conjugated, fused systems with five-membered rings.

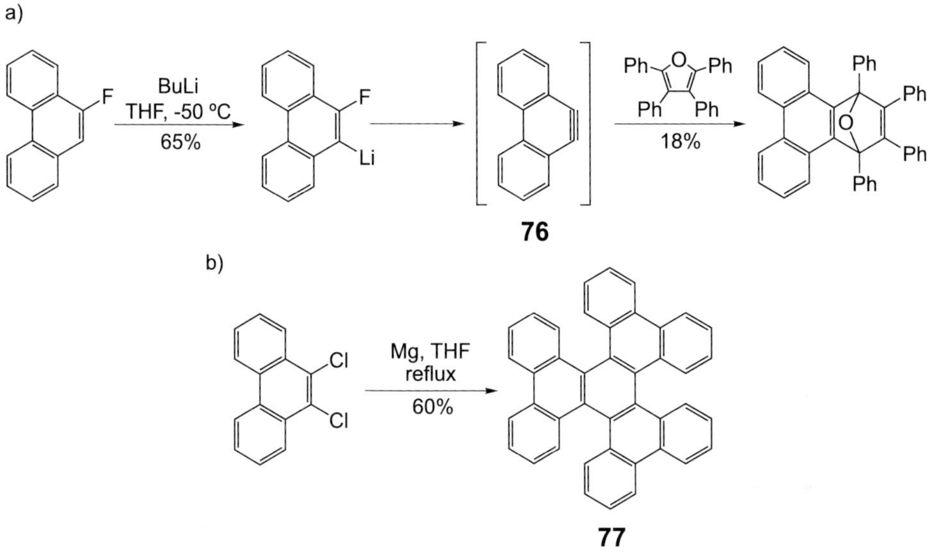

Scheme 1.18

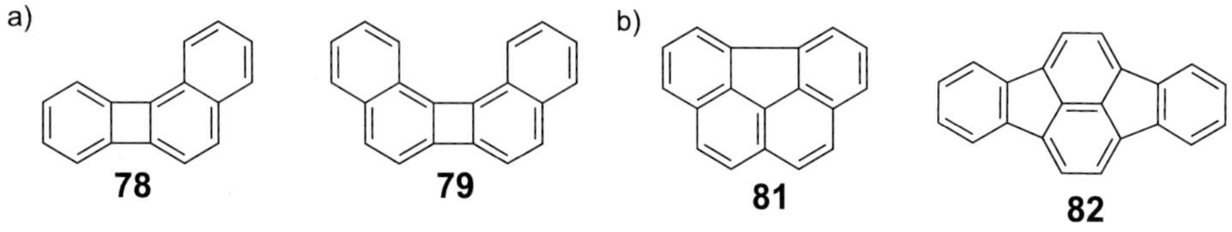

Figure 1.6. (a) Structures **78** and **79**, (b) structures **81** and **82**.

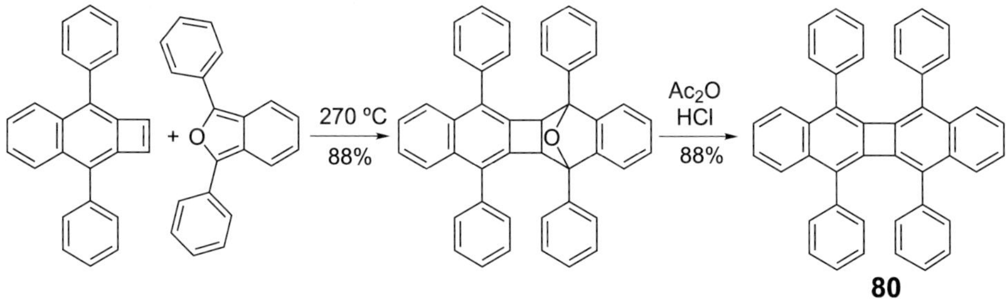

Scheme 1.19

The noteworthy investigators of these systems include Clar, Campbell, Lang, and Aitken. The majority of systems developed were benzo-derviatives of fluoranthene, benzofluoranthene **81** (Fig. 1.6b), and phenylenefluoranthene **82** [1a, 72]. Some of the more novel systems included an additional fused seven-membered ring. In 1956, Boekelheide and coworkers produced acepleiadylene **83**, unique in structure due to fusion of three ring sizes [73]. Starting from previously isolated diketone **84**,

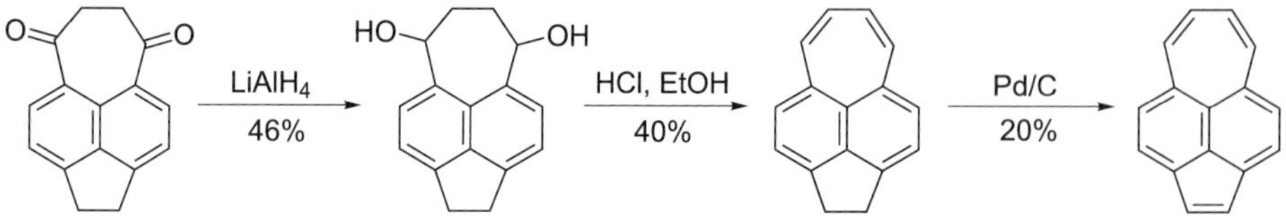

Scheme 1.20

reduction and acid-catalyzed dehydration afforded acepleiadiene **85** (Scheme 1.20). Subsequent dehydrogenation with Pd/C produced the unsaturated **83** in 20% yield. Reid and coworkers produced a constitutional isomer of **83** one year earlier [74].

In addition to development of several novel benzofluoranthene systems, including **81**, 2.13,11.12-dibenzofluoranthene, and multiple naphthofluoranthene isomers, Campbell's key contribution was that of obtaining new routes to older fused systems, such as 10.11-benzofluoranthene and 2.3,6.7-dibenzofluoranthene [75]. The largest and most spectacular example of the fused systems, though, occurred in 1959 when Zander produced 1.2,3.4,5.6,7.8-tetra-(peri-naphthylene)-anthracene **86** via a two-fold Diels-Alder cyclization (Scheme 1.21) [76].

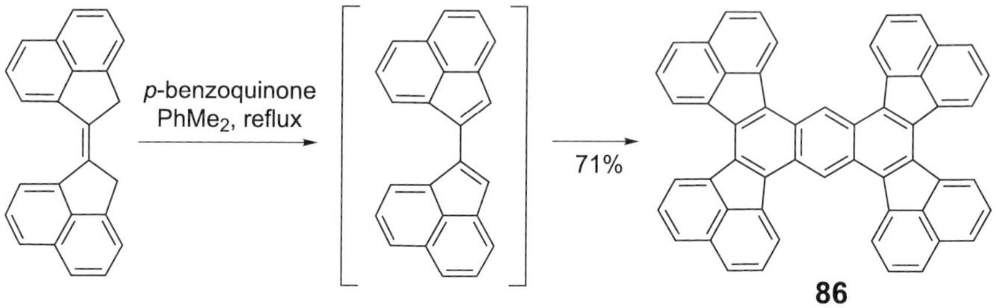

Scheme 1.21

Although the majority of their work focused on saturated systems, Blood and Linstead produced dibenzopentalene **87**, one of the first stable, fully conjugated analogs containing two fused five-membered rings (Fig. 1.7a). Published in their series *Fused Carbon Rings* in 1952, the authors successfully converted a dihydropentalene intermediate into a dibromide and subsequently reduced the ultimate precursor with silver acetate [77]. The next stable carbon-rich pentalene derivative did not occur until 1962 when Le Goff produced hexaphenylpentalene **88** in excellent yield (77%) from NBS oxidation of a hexaphenyldihydropentalene precursor (Fig. 1.7a) [78]. Le Goff's structure remained the only nonbenzenoid stabilized pentalene until the early 1970s.

The 1950s and 1960s also saw work on a number of linearly-annelated benzenoid derivatives of the highly reactive hexa- and heptacene as well as the first octacene derivative. One notable structural motif present in all of the analogs is a pyr-

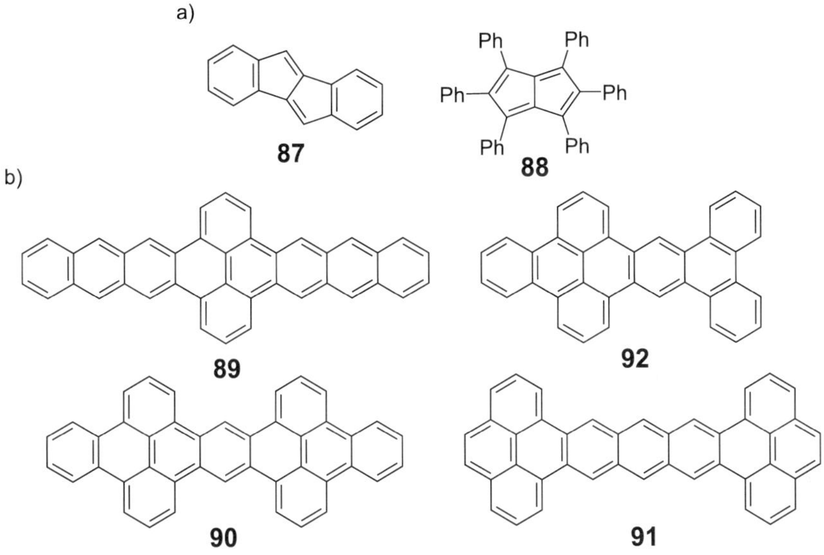

Figure 1.7. (a) Structures **87** and **88**, (b) structures **89–92**.

ene ring, which likely added stability over purely linear structures. Key proprietors for these molecules were Clar, Boggaino, Zander, Guye-Vuilleme, MacPherson, and McCallum. Dibenzoctacene **89** (Fig. 1.7b) was produced in 1963 by Clar and coworkers via two-fold condensation of derivatized naphthalene compounds to a central substituted-pyrene [1a]. Di- and tetrabenzoheptacene, **90** and **91**, both produced by Clar and coworkers in the 1960s, represent the two heptacene derivatives [1a]. Clar also produced the bulkiest pentacene derivative, tetrabenzopentacene **92**, in 1956 in a similar fashion to his other pentacene derivatives in the preceding decade [1a].

The largest fused benzenoid systems developed prior to 1970 were based on perylene, peropyrene **93**, terrylene **94**, coronene, and bisanthene (Fig. 1.8). Also responsible for terrylene, Clar and coworkers produced three additional benzenoid derivatives as well as circumanthrene **95**, isolated from two-fold condensation of tetracene derivatives in 1956 [1a]. Halleux and coworkers obtained tetrabenzobisanthene **96** from Zn-dust dimerization of dibenzoperinaphthone [79]. Synthesized by Clar in the same year [1a], hexabenzocoronene **97** was produced by Halleux and coworkers in 1958 from a NaCl–ZnCl melt of hexaphenylbenzene [79].

1.5.3
The Helicenes, Radialenes, Fulvalenes, and Circulenes

A resurgence in helicene chemistry occurred in 1956 when Newman and Lednicer reported production of phenanthro[3,4-c]phenanthrene or hexahelicene **98**, which displayed the stereotypical chirality resultant from steric overcrowding [80]. The

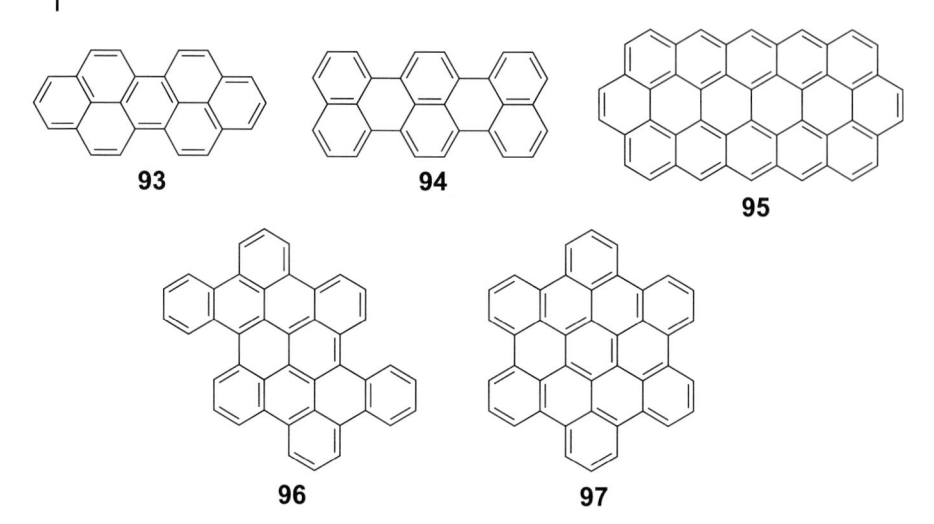

Figure 1.8. Structures **93–97**.

lengthy ten-step synthesis quickly introduced four of the six final benzene rings but required considerable optimization for the final steps resulting in an overall yield of less than 2%.

A key basis for modern helicene cyclization chemistry occurred in 1964 when Wood and Mallory effectively prepared phenanthrene via photochemical induced cyclization of stilbene [81]. The new method effectively provided efficient pathways to substituted phenanthrene derivatives, easier access to existing and new helicene structures, as well as sufficient quantities for property studies. In the mid 1960s, Martin and coworkers successfully isolated the previously made hexahelicene as well as new hepta- **99**, octa- **100**, and nonahelicenes **101** via highly efficient syntheses (Scheme 1.22) [82]. These preparations culminated in photo-induced cyclization of bisarylethylenes with an I_2 oxidant and Hg-lamp radiation source. The key improvements over previous efforts included production of a phenanthrene derivative via photo-induced cyclization of intermediate **102** and subsequent conversion to Wittig ylide **103**, a common intermediate to the four helicenes.

The 1960s also saw the appearance of the alicyclic radialenes, unique in that all ring carbons are sp^2 hybridized and double bonds are semicyclic. The production of tetramethylenecyclobutane or [4]radialene **104** from pyrolysis of a tetraaminoxide-substituted cyclobutane **105** in 1962 by Griffin and Peterson marked the first appearance of the non-alkyl substituted **104** (Scheme 1.23a) [83]. The first parent [3]radialene or trimethylenecyclopropane **106** was produced by Dorko in 1965 [84] and Griffin and coworkers in 1966 [85]. Dorko reduced Feist's acid **107** to arrive at the dibromide **108** and then applied a base-induced β-elimination to obtain **106** (Scheme 1.23b). Griffin and coworkers obtained **106** from a triiodide as well as the trisquaternary hydroxide in lower yields.

Pentafulvalene and heptafulvalene **109**, the first symmetrical fulvalenes, were iso-

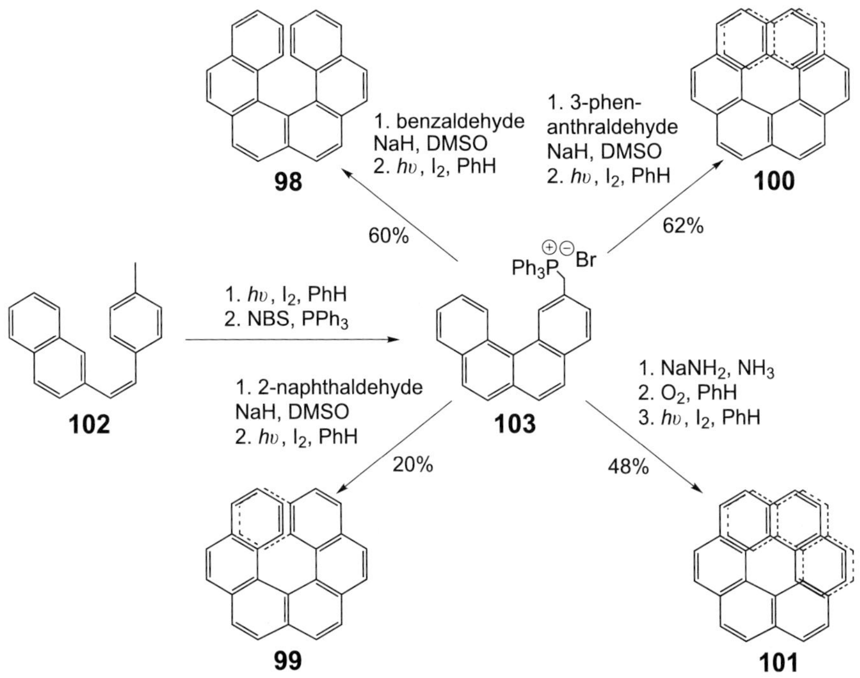

Scheme 1.22

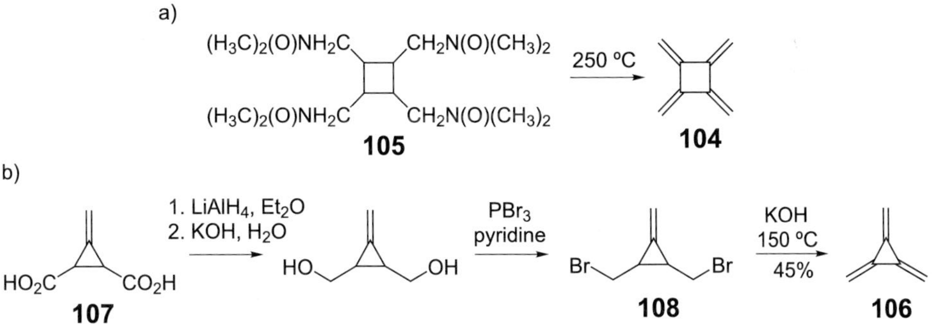

Scheme 1.23

lated in the late 1950s by Doering [86]. A more complicated example, 1,2,3,1′,2′,3′-hexaphenylpentafulvalene **110**, was produced by Pauson and Williams in 1961 [87]. Ketone **111** was reduced and then dehydrogenated to triphenylcyclopentadiene **112** with HCl (Scheme 1.24a). Treatment of **112** with BuLi and then oxidation with iodine afforded the substituted fulvalene. Jones and Ennis developed a significant improvement for **109** in the late 1960s. Tosyl hydrazone salt **113** was obtained from 7,7-dichlorocycloheptatriene and subsequently photolyzed with cyclohexene

1 Pioneers of Carbon-rich Compounds

a)

Scheme 1.24

to afford the fulvalene (Scheme 1.24b) [88]. Initial work on asymmetrical fulvalenes, such as heptapentafulvalene, began in the 1960s by Prinzbach and coworkers [89].

Barth and Lawton's production of [5]circulene or corannulene **114** (Fig. 1.9) in 1966 [90] was a significant achievement due to the fullerene-based, nonplanar structure of **114**. The 17-step synthesis began with acenaphthene and relied on alkylation, condensation, and Pd-catalyzed aromatization reactions to afford the macrocycle. The following decade would see continued efforts to isolate [7]circulene **115** (Fig. 1.9) but another decade would pass before **115** was successfully produced [1d]. Sygula and Rabideau discuss more recent work on buckybowls and fullerene fragments in Chapter 12, whereas fullerene reactivity is outlined by Kitagawa, Murata, and Komatsu in Chapter 9.

1.6
1970–Present: The Way Ahead

Chemistry research on carbon-rich molecules following that outlined here has produced and continues to produce remarkable structures in size, topology, and spatial orientation. The accomplishments of early chemists though are truly remarkable in the light of current synthetic and analytical techniques that modern

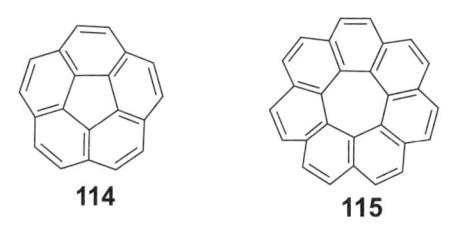

Figure 1.9. Structures **114** and **115**.

chemists take for granted. It is unfortunate that the majority of the pioneers' efforts resulted in low yields and thus insufficient quantities for system property studies – a common theme in early synthetic chemistry. Regardless, without the groundbreaking work of early chemists such as Glaser, Scholl, and Clar, current carbon-rich molecule methodology and progress would not have achieved its present status or its future potential.

References

1 (a) E. Clar, *Polycyclic Hydrocarbons*, Vol. *I–II*, Academic Press, London, **1964**; (b) A. T. Balaban, M. Banciu, V. Ciorba, *Annulenes, Benzo-, Hetero-, Homo-, Derivatives, and their Valence Isomers*, Volumes *I–III*, CRC, Boca-Raton, **1987**; (c) R. G. Harvey, *Polycyclic Aromatic Hydrocarbons*, Wiley-VCH, New York, **1997**; (d) H. Hopf, *Classics in Hydrocarbon Chemistry*, Wiley-VCH, Weinheim, **2000**; (e) D. Astruc (Ed.), *Modern Arene Chemistry*, Wiley-VCH, Weinheim, **2002**.

2 C. Graebe, *Ber. Dtsch. Chem. Ges.* **1874**, *7*, 1623–1628, 1629–1630.

3 B. Radziszewski, *Ber. Dtsch. Chem. Ges.* **1876**, *9*, 260–262.

4 K. F. Lang, H. Buffleb, J. Kalowy, *Chem. Ber.* **1957**, *90*, 2888–2893.

5 A. v. Baeyer, *Ann. Chem. Pharm.* **1866**, *140*, 295–313.

6 R. Pschorr, *Ber. Dtsch. Chem. Ges.* **1896**, *29*, 496–501.

7 H. Klinger, C. Lonnes, *Ber. Dtsch. Chem. Ges.* **1896**, *29*, 2152–2156, 2157–2158, 2158–2160.

8 F. Riese, *Ann. Chem. Pharm.* **1872**, *164*, 161–176.

9 (a) C. E. Linebarger, *Am. Chem. J.* **1892**, *14*, 597–602; (b) W. R. Orndorff, F. K. Cameron, *Am. Chem. J.* **1895**, *17*, 659–681; (c) C. E. Linebarger, *J. Am. Chem. Soc.* **1895**, *17*, 354–358.

10 (a) S. Gabriel, A. Michael, *Ber. Dtsch. Chem. Ges.* **1877**, *10*, 2199–2210; (b) F. Nathanson, *Ber. Dtsch. Chem. Ges.* **1893**, *26*, 2576–2582; (c) S. Gabriel, E. Leupold, *Ber. Dtsch. Chem. Ges.* **1898**, *31*, 1159–1174.

11 Cited in Ref 1e: M. Berthelot, *Ann. Chem. Pharm.* **1867**, *141*, 173–184.

12 C. Glaser, *Ber. Dtsch. Chem. Ges.* **1869**, *2*, 422–424.

13 A. Baeyer, L. Landsberg, *Ber. Dtsch. Chem. Ges.* **1882**, *15*, 50–56, 57–61.

14 (a) T. F. Rutledge, *Acetylenic Compounds*, Reinhold Book Corp., New York, **1968**; (b) P. Siemsen, R. C. Livingston, F. Diederich, *Angew. Chem.* **2000**, *112*, 2740–2767; *Angew. Chem. Int. Ed. Engl.* **2000**, *39*, 2632–2657.

15 C. Mannich, *Ber. Dtsch. Chem. Ges.* **1907**, *40*, 159–165.

16 F. Ullmann, G. M. Meyer, O. Loewenthal, E. Gilli, *Liebigs Ann.* **1904**, *332*, 38–81.

17 Cited in Ref 1a: (a) W. Löb, *Z. Elektrochem.* **1901**, *7*, 903; (b) W. Löb, *Z. Elektrochem.* **1902**, *8*, 777.

18 D. Vorländer, C. Siebert, *Ber. Dtsch. Chem. Ges.* **1906**, *39*, 1024–1035.

19 R. Scholl, C. Seer, R. Weitzenböck, *Ber. Dtsch. Chem. Ges.* **1910**, *43*, 2202–2209.

20 R. Weitzenböck, *Monatsh. Chem.* **1913**, *34*, 193–223.

21 (a) K. Dziewoński, *Ber. Dtsch. Chem. Ges.* **1903**, *36*, 962–971; (b) P. Rehländer, *Ber. Dtsch. Chem. Ges.* **1903**, *36*, 1583–1587.

22 K. Dziewoński, S. Suknarowski, *Ber. Dtsch. Chem. Ges.* **1918**, *51*, 457–465.

23 C. Moureu, C. Dufraisse, P. M. Dean, *C. R. Acad. Sci., Paris* **1926**, *182*, 1440–1443.

24 (a) C. Moureu, C. Dufraisse, G. Berchet, *C. R. Acad. Sci., Paris* **1927**, *185*, 1085–1087; (b) C. Dufraisse, *Bull. Soc. Chim. Fr.* **1936**, *3*, 1847–1857, 1857–1872; (c) C. Dufraisse, R.

Hoclois, *Bull. Soc. Chim. Fr.* **1936**, *3*, 1873–1880, 1880–1983, 1894–1905; (d) C. Dufraisse, L. Veluz, *Bull. Soc. Chim. Fr.* **1936**, *3*, 1905–1913.

25 (a) O. Gerngross, M. Dunkel, *Ber. Dtsch. Chem. Ges.* **1924**, *57*, 739–746; (b) O. Gerngross, C. Schachnow, R. Jonas, *Ber. Dtsch. Chem. Ges.* **1924**, *57*, 747–750.

26 R. Pummerer, K. Bittner, *Ber. Dtsch. Chem. Ges.* **1924**, *57*, 84–88.

27 E. Clar, D. G. Stewart, *J. Am. Chem. Soc.* **1954**, *76*, 3504–3507 and references contained therein.

28 E. Clar, F. John, *Ber. Dtsch. Chem. Ges.* **1930**, *63*, 2967–2977.

29 W. Ried, F. Anthöfer, *Angew. Chem.* **1953**, *65*, 601.

30 V. Brückner, A. Karczag, K. Körmendy, M. Meszag, J. Tomasz, *Tetrahedron Lett.* **1960**, 5–6.

31 J. W. Cook, R. S. Ludwiczak, R. Schoental, *J. Chem. Soc.* **1950**, 1112–1121 and references contained therein.

32 G. M. Badger, J. W. Cook, C. L. Hewett, E. L. Kennaway, N. M. Kennaway, R. H. Martin, *Proc. R. Soc. London, Ser. B* **1942**, *131*, 170–182 and references contained therein.

33 (a) G. M. Badger, J. W. Cook, F. Goulden, *J. Chem. Soc.* **1940**, 16–18; (b) J. W. Cook, A. M. Robinson, *J. Chem. Soc.* **1940**, 303–304.

34 J. W. Cook, C. Hewett, I. Hieger, *Nature* **1932**, *130*, 926.

35 J. W. Cook, C. L. Hewett, *J. Chem. Soc.* **1934**, 365–377.

36 (a) J. W. Cook, *J. Chem. Soc.* **1931**, 489–499, 499–507; **1932**, 1472–1484; (b) J. W. Cook, C. L. Hewett, I. Hieger, *J. Chem. Soc.* **1933**, 395–396, 396–398; (c) J. W. Cook, C. L. Hewett, *J. Chem. Soc.* **1933**, 398–405; (d) J. W. Cook, A. Dansi, C. L. Hewett, J. Iball, W. V. Mayneord, E. Roe, *J. Chem. Soc.* **1935**, 1319–1325.

37 J. W. Cook, C. L. Hewett, *J. Chem. Soc.* **1933**, 1098–1111.

38 (a) R. Weitzenböck, H. Lieb, *Monatsh. Chem.* **1912**, *33*, 549–565; (b) R. Weitzenböck, A. Klinger, *Monatsh. Chem.* **1918**, *39*, 315–324.

39 J. W. Cook, *J. Chem. Soc.* **1931**, 2524–2528, 2529–2532; **1933**, 1592–1597.

40 (a) W. H. Bentley, A. Friedl, F. Thomas, C. Weizmann, *J. Chem. Soc.* **1907**, 411–435; (b) W. H. Bentley, A. Friedl, C. Weizmann, *J. Chem. Soc.* **1907**, 1588–1593; (c) C. Marschalk, *Bull. Soc. Chim. Fr.* **1939**, *6*, 1112–1121, 1122–1125.

41 W. J. Bailey, C.-W. Liao, *J. Am. Chem. Soc.* **1955**, *77*, 992–993.

42 (a) C. Marschalk, *Bull. Soc. Chim. Fr.* **1938**, *5*, 306–308; **1941**, *8*, 354–369; **1950**, *16*, 311–317.

43 R. Scholl, K. Meyer, *Ber. Dtsch. Chem. Ges.* **1932**, *65*, 902–915.

44 (a) R. Scholl, J. Mansfeld, *Ber. Dtsch. Chem. Ges.* **1910**, *43*, 1734–1746; (b) R. Scholl, K. Meyer, *Ber. Dtsch. Chem. Ges.* **1934**, *67*, 1236–1238.

45 W. C. Lothrop, *J. Am. Chem. Soc.* **1941**, *63*, 1187–1191.

46 (a) G. Wittig, H. F. Ebel, *Liebigs Ann.* **1961**, *650*, 20–34; (b) G. Wittig, W. Herwig, *Chem. Ber.* **1954**, *87*, 1511–1512.

47 R. Wolovsky, F. Sondheimer, *J. Am. Chem. Soc.* **1962**, *84*, 2844–2845.

48 P. D. Bartlett, M. J. Ryan, S. G. Cohen, *J. Am. Chem. Soc.* **1942**, *64*, 2649–2653.

49 I. T. Storie, F. Sondheimer, *Tetrahedron Lett.* **1978**, 4567–4568 and references contained therein.

50 (a) F. Sondheimer, *Proc. R. Soc. London, Ser. A* **1967**, *297*, 173–204; (b) F. Sondheimer, *Pure Appl. Chem.* **1971**, *23*, 331–353; (c) F. Sondheimer, *Acc. Chem. Res.* **1972**, *5*, 81–91.

51 (a) F. Sondheimer, Y. Amiel, R. Wolovsky, *J. Am. Chem. Soc.* **1959**, *81*, 4600–4606; (b) F. Sondheimer, R. Wolovsky, *J. Am. Chem. Soc.* **1962**, *84*, 260–269.

52 O. M. Behr, G. Eglinton, A. R. Galbraith, R. A. Raphael, *J. Chem. Soc.* **1960**, 3614–3625.

53 I. D. Campbell, G. Eglinton, W. Henderson, R. A. Raphael, *Chem. Commun.* **1966**, 87–89.

54 H. A. Staab, F. Graf, *Tetrahedron Lett.* **1966**, 751–757.

55 S. Akiyama, M. Nakagawa, *Chem. Ind.* **1960**, 346–347.

56 H. H. Schlubach, V. Franzen, *Liebigs Ann.* **1951**, *573*, 115–120.
57 M. Nakagawa, *Chem. Abstr.* **1951**, *45*, 7081, 8486.
58 H. H. Schlubach, V. Franzen, *Liebigs Ann.* **1951**, *572*, 105–110.
59 J. B. Armitage, N. Entwistle, E. R. H. Jones, M. C. Whiting, *J. Chem. Soc.* **1954**, 147–154.
60 S. Misumi, *Bull. Chem. Soc. Jpn.* **1961**, *34*, 1827–1832.
61 S. Misumi, M. Kuwana, M. Nakagawa, *Bull. Chem. Soc. Jpn.* **1962**, *35*, 135–142.
62 G. Wegner, *Z. Naturforsch. B: Chem. Sci.* **1969**, *24*, 824–832; *Makromol. Chem.* **1970**, *134*, 219–229.
63 G. Wittig, R. Ludwig, *Angew. Chem.* **1956**, *68*, 40.
64 G. Wittig, G. Lehmann, *Chem. Ber.* **1957**, *90*, 875–892.
65 W. S. Rapson, R. G. Shuttleworth, J. N. van Niekerk, *J. Chem. Soc.* **1943**, 326–327.
66 G. Wittig, W. Uhlenbrock, P. Weinhold, *Chem. Ber.* **1962**, *95*, 1692–1702.
67 J. G. Carey, I. T. Millar, *J. Chem. Soc.* **1959**, 3144–3146.
68 M. P. Cava, D. R. Napier, *J. Am. Chem. Soc.* **1957**, *79*, 1701–1705; **1958**, *80*, 2255–2257.
69 M. P. Cava, J. F. Stucker, *J. Am. Chem. Soc.* **1955**, *77*, 6022–6026.
70 C. D. Nenitzescu, M. Avram, I. G. Dinulescu, G. Mateescu, *Liebigs Ann.* **1962**, *653*, 79–85.
71 J. v. Braun, E. Anton, *Ber. Dtsch. Chem. Ges.* **1929**, *62*, 145–151.
72 (a) N. Campbell, D. H. Reid, *J. Chem. Soc.* **1952**, 3281–3284; (b) H. W. D. Stubbs, S. H. Tucker, *J. Chem. Soc.* **1951**, 2936–2939.
73 (a) V. Boekelheide, W. E. Langeland, C.-T. Liu, *J. Am. Chem. Soc.* **1951**, *73*, 2432–2435; (b) V. Boekelheide, G. K. Vick, *J. Am Chem. Soc.* **1956**, *78*, 653–658.
74 D. H. Reid, W. H. Stafford, J. P. Ward, *J. Chem. Soc.* **1955**, 1193–1201.
75 (a) N. Campbell, H. Wang, *J. Chem. Soc.* **1949**, 1513–1515; (b) N. Campbell, N. M. Khanna, A. Marks, *J. Chem. Soc.* **1951**, 2511–2512; (c) N. Campbell, A. Marks, D. H. Reid, *J. Chem. Soc.* **1950**, 3466–3470; (d) N. Campbell, A. Marks, G. V. McHattie, *J. Chem. Soc.* **1955**, 1190–1192.
76 M. Zander, *Chem. Ber.* **1959**, *92*, 2740–2743.
77 C. T. Blood, R. P. Linstead, *J. Chem. Soc.* **1952**, 2263–2268.
78 E. Le Goff, *J. Am. Chem. Soc.* **1962**, *84*, 3975–3976.
79 A. Halleux, R. H. Martin, G. S. D. King, *Helv. Chim. Acta* **1958**, *41*, 1177–1183.
80 M. S. Newman, D. Lednicer, *J. Am. Chem. Soc.* **1956**, *78*, 4765–4770.
81 C. S. Wood, F. B. Mallory, *J. Org. Chem.* **1964**, *29*, 3373–3377.
82 (a) M. Flammang-Barbieux, J. Nasielski, R. H. Martin, *Tetrahedron Lett.* **1967**, 743–744; (b) R. H. Martin, M. Flammang-Barbieux, J. P. Cosyn, M. Gelbcke, *Tetrahedron Lett.* **1968**, 3507–3510.
83 G. W. Griffin, L. I. Peterson, *J. Am. Chem. Soc.* **1962**, *84*, 3398–3400.
84 E. A. Dorko, *J. Am. Chem. Soc.* **1965**, *87*, 5518–5520.
85 P. A. Waitkus, L. I. Peterson, G. W. Griffin, *J. Am. Chem. Soc.* **1966**, *88*, 181–183.
86 Cited in Ref 1d: (a) W. v. E. Doering, *The Kekule-Symposium*, London, **1958**, Butterworths and Co., London **1959**, 45; (b) W. v. E. Doering, *U.S. Dep. Comm. Off. Tech. Serv. Rep.* **1958**, 147, 858; (c) J. R. Mayer, Ph. D. Thesis, Yale University, **1955**.
87 P. L. Pauson, B. J. Williams, *J. Chem. Soc.* **1961**, 4153–4157.
88 W. M. Jones, C. L. Ennis, *J. Am. Chem. Soc.* **1967**, *89*, 3069–3071; **1969**, *91*, 6391–6397.
89 (a) H. Prinzbach, W. Rosswog, *Angew. Chem.* **1961**, *73*, 543; *Tetrahedron Lett.* **1963**, 1217–1221; (b) M. Neuenschwander, W. K. Schenk, *Chimia* **1972**, *26*, 194–197.
90 W. E. Barth, R. G. Lawton, *J. Am. Chem. Soc.* **1966**, *88*, 380–381; **1971**, *93*, 1730–1745.

2
Electronic Conduction in Photoactive Metallo-wires

Anthony Harriman and Raymond Ziessel

2.1
Introduction

Research in the emerging field of molecular electronics is intended [1–4] to identify molecules able to duplicate the functions of active and passive components (e.g., wires, switches, relays, sensors, resistors, capacitors, LEDs, etc.) of electronic circuitry and integrated circuits. In many cases, the devices are simply molecular-based analogs of conventional silicon-based circuitry. Such molecular-scale systems have the great advantage, however, of less manufacturing complexity, lower power consumption and easier scaling. Because of their small size and diverse properties, molecular electronic components offer genuine benefits over their larger silicon-based competitors, although it should be stressed that there are no commercial applications as yet. Since molecular electronics borrows heavily from more conventional, and better developed, technologies progress in the field has been extremely rapid over the past five years. It has now reached the point at which several of the major manufacturers of semiconductor components are launching their own research programs into molecular electronics. A number of important goals on the approach to true miniaturization have been met already.

The simplest, and perhaps the most elegant, demonstration of molecular electronics is to attach a single molecule to two metallic electrodes and monitor current flow along the molecular axis [5–7]. Indeed, charge transport through organic molecules has been studied for about three decades, but mostly in solution phase [8] or via quantum chemical computational studies [9–11]. Conventional theories and analytical methods do not hold for single molecules sandwiched between two electrodes. The situation is further exacerbated by the realization that certain critical aspects of basic solid-state physics do not easily translate to the molecular world. Thus, when two disparate materials are brought together their Fermi levels align such that a conducting wire of length L has a resistance R. Increasing the length of the wire to $2L$ has the effect of doubling the resistance to $2R$. This is not the case with molecular-scale wires because electrons must move between well-defined orbitals. Simply linking together a series of molecules via covalent bonding does not facilitate proper blending of the orbitals and it is likely

Carbon-Rich Compounds. Edited by Michael M. Haley and Rik R. Tykwinski
Copyright © 2006 WILEY-VCH Verlag GmbH & Co. KGaA, Weinheim
ISBN: 3-527-31224-2

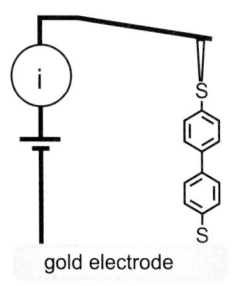

Figure 2.1. Pictorial representation of the measurement of electrical conductivity at the single molecule level.

that insulating units will be introduced into the system. Careful attention has to be paid to the precise spatial arrangement of individual orbitals [12] and to proper matching of their electronic levels [13]. There is an additional problem in that molecular properties tend to depend on the nature of the environment and systems that operate nicely in polar solvents may fail to work in the solid state [14, 15]. In particular, attaching molecules to active surfaces can cause profound changes in their electronic properties [16]. To realize molecular assemblies with predetermined functionality, it is necessary to consider simultaneously (i) the design of the molecule, (ii) the molecule/electrode interface, (iii) how to address individual molecules, (iv) how to align several hundred such molecules in 2- or 3-dimensional space, and (v) the overall compatibility of the system. Organic molecules tend to be less robust than inorganic architectures! Despite the severity of these and related problems, the future development of organic-based molecular electronics seems inevitable [1–4].

It has been demonstrated that a single transition metal complex can span the nanometer-scale gap between remote electrodes and promote current flow [17]. Certain organic molecules are known to operate in a similar fashion (Fig. 2.1) [18–20]. Related work has shown that individual molecules can operate as the active elements in electronic switches [21]. Analytical techniques have been developed to monitor such processes [22] and to image single molecules [23–25]. Bundling together several thousand such molecules can lead to the construction of random-access memory devices capable of write, read and erase cycles [26]. Several research groups have described nano-wires formed by controlled polymerization [27–29] or functionalization of polymeric micelles [30]. Protocols have been devised that permit the exact positioning of such wires on macroscopic surfaces with unparalleled precision [31]. The small carbon nano-tubes produced in recent years have attracted considerable attention and new techniques have been developed that remove trace metallic impurities [32]. These materials are prime candidates to replace silicon in the near future. The conductivity of a single nano-particle has been demonstrated by the familiar approach of attaching it to microscopic electrodes [33]. Carbon nano-tubes have also been used to house single-electron transistors [34] whilst a wide diversity of functional molecules has been directly attached to single-wall carbon nano-tubes [35]. Single carbon nano-wires have been protected with an insulating coat [36]. A molecular-scale transistor has been built from monolayers of conjugated organic molecules bound to a doped silicon wafer [37]. Two such molecules operate cooperatively to form a logic gate.

The field has developed to yield miniaturized light-emitting diodes [38, 39] and chemical sensors [40, 41] able to monitor changes in pH and ion concentrations. In fact, the design of molecular electronic devices that function under illumination – molecular photonics – has become an extremely popular spin-off from the general field of molecular electronics. Light can be used to initiate long-range electron- or energy-transfer processes [42–44] or to induce substantial conformational changes [45]. A number of photonic wires have been reported in which selective illumination into one chromophore causes rapid information transfer along the molecular axis. Such systems require cooperativity between a chromophore, donor and acceptor terminals and a connector (or wire) [46]. With long connectors, like DNA, it becomes possible to transfer photonic energy over unusually large distances [47] and artificial wires, such as oligomeric porphyrins [48, 49], are becoming available. Optical communications and biophotonics stand out as being the two major applications for this new technology, although countless luminescent sensors have been developed over the past few decades. Many of the earlier photonic-based molecular systems were intended to duplicate important features of the natural photosynthetic reaction center [50–53], but more recent attention has focused primarily on the design of smart materials for particular applications. Additional opportunities arise from the use of two-color [54] or multiphoton effects [55]. Over the years, this field has progressively shifted from well-defined supermolecules to loosely associated supramolecular systems in which critical points such as the exact nature of the photoactive module, the stability, and the conformational motion under illumination have not yet been addressed.

The development of intelligent molecular photonic devices is made possible from our detailed understanding of the mechanism of light-induced energy or electron transfer and from the comprehensive unraveling of molecular photophysics that predates the present subject. Indeed, we now have a thorough grasp of the many factors that combine to limit the rates of energy and electron transfer and suitable control elements can be built into new prototypic molecular-scale devices. This point can be emphasized by reference to two molecular arrays. Firstly, Lindsey et al. [56, 57] have developed several molecular photonic wires around porphyrin-based modules **1** and **2** (Fig. 2.2). A pyrromethene dye is used as the primary light harvester. Upon excitation, this dye injects photons into a linear array of metallo-porphyrins where the photon is transferred rapidly along the array until it is trapped by a low-energy acceptor. Careful balancing of the rates of forward and reverse energy transfer [58] allow energy migration to proceed with almost 100% efficiency. By adding alternative branches of metallo-porphyrins, it has been possible to build a simple logic gate that operates by way of competitive electron transfer between 2_{neu} and 2_{oxy}. The gate can be switched by an electrochemical mechanism [59].

Secondly, Gust and Moore and their coworkers [60] have successfully designed a

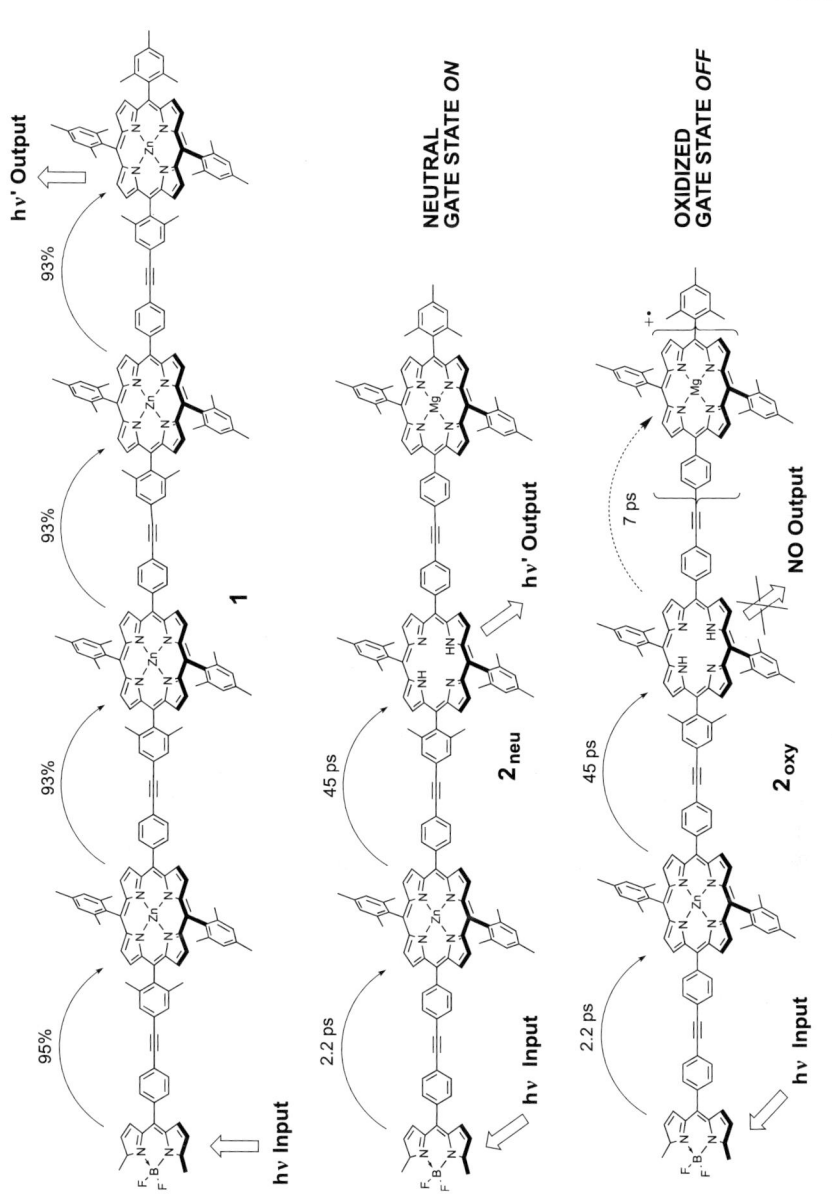

Figure 2.2. Examples of porphyrin-based molecular wires studied by Lindsey et al.

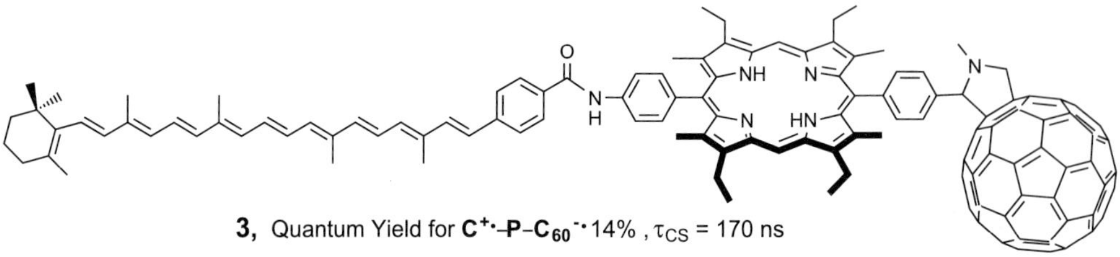

3, Quantum Yield for $C^{+\cdot}$-P-$C_{60}^{-\cdot}$ 14%, τ_{CS} = 170 ns

Figure 2.3. Example of a porphyrin–carotene–fullerene triad studied by Gust, Moore et al.

porphyrin-based triad that mimics most of the essential features of the natural photosynthetic reaction center (Fig. 2.3). The triad **3** comprises a porphyrin-based donor attached to a fullerene-based acceptor with a carotene as secondary donor. Light-induced electron transfer occurs even at 8 K to give a long-lived charge-separated (CS) state. Subsequent charge recombination occurs to form the triplet excited state localized on the carotenoid. This type of cascade effect has been used by various researchers to built molecular systems where the CS state survives for microseconds [61] milliseconds [62] or even seconds [63]. There seems no other way to ensure that the CS state can be stabilized against rapid charge recombination [64]. It is interesting to note that similar photosystems have been extended to function as a molecular AND gate and as an XOR gate [65]. Although far from practical, these latter systems operate merely by changing the frequency of the excitation source.

At the same time, advances in synthetic chemistry mean that it is possible to build complicated molecular architectures from rather sophisticated modules in such a way that the shape, conformation, stereochemistry and electronic organization remain well defined. Such systems can be broadly classified as purely organic (i.e., no metal fragment), hybrid (i.e., containing a single metal), or as coordination complexes containing at least two metals. The formation of these molecular devices can be purely covalent or induced by self-organization. In the latter case, assemblage is driven by the directionality and cooperativity of the noncovalent bonding vectors and by the number of attachment points. In covalent systems, the efficiency of information transfer along the molecular axis is controlled by the extent of electronic coupling between the various cofactors. The connectors, that both facilitate electronic communication and provide the necessary scaffold by which to hold the active subunits in place, can be provided by saturated (e.g., linear or cyclic templates), unsaturated (e.g., alkenes, alkynes, or aromatics) or hybrid groups. Fine-tuning of the electronic properties of the connector becomes possible by careful optimization of the geometry or by attaching secondary fragments [66].

In the following discussion we have chosen to describe the development of optoelectronic systems based on transition metal complexes [67–70]. Whilst such systems differ from organic devices, most notably because triplet excited states dominate the photophysics, it has to be realized that this is a rather artificial distinction. The compounds to be described herein possess chelating units such as

2,2′-bipyridine, 1,10-phenanthroline, benzimidazole, 2,2′:6′,2″-terpyridine, ortho-metallated phenyl-bispyridine, or ortho-metallated 6-phenyl-bipyridine fragments. The synthetic strategies and procedures employed for the engineering of such systems can be classified generally as:

1. Linear procedures involving the stepwise construction of the ligands by appropriate methods such as Kröhnke synthesis or metal-catalyzed Suzuki, Stille or Sonogashira cross-coupling reactions, followed by successive steps involving attachment of appropriate metal precursors (e.g., Ru, Os, Re, Ir ...), thereby affording heteronuclear complexes.
2. Convergent procedures where the initial building blocks are engineered in such a way that appropriate functions are present to enable the subsequent formation of the target multi-nuclear metal complexes.

These systems are designed either to convey information (e.g., photon, electron, shape, proton) transfer between distinct sites under illumination or to recognize certain events (e.g., cation binding, changes in temperature or pressure). In all such systems, light-induced electron transfer or electron exchange play important roles. A prime motivation for the development of such molecular entities is to transfer information rapidly when stimulated by a photon pulse. In turn, this demands the identification of connectors able to promote long-range transfer at high speed. Because of the steep attenuation of molecular conductivity with increasing length [71], the design of suitable connectors becomes critical. In the following discussion we focus on the role of the connector.

Early work indicated that certain connectors could induce modest electronic coupling over relatively large distances. Thus, Lehn et al. [72] used a polyene to span a ruthenium(II) tris(2,2′-bipyridine) binuclear complex **4** such that the metal centers were separated by ca. 24 Å (Fig. 2.4). Selective oxidation of one metal center gave rise to the corresponding mixed-valence Ru^{II}/Ru^{III} complex that displayed a strong intervalence charge-transfer absorption band in the near IR region. Spectral analysis in terms of Hush theory showed that the coupling element was an impressive

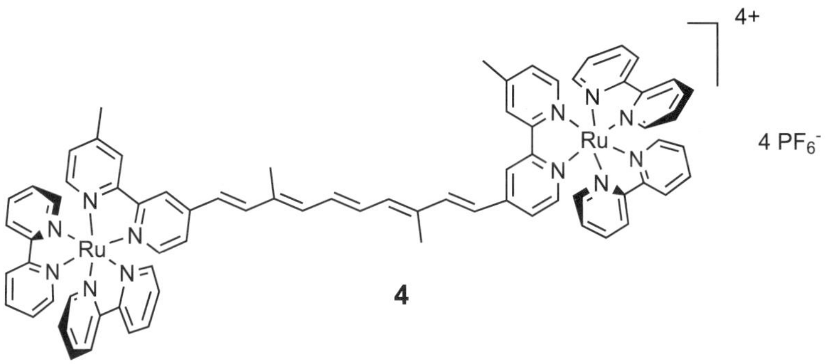

Figure 2.4. The elongated binuclear complex studied by Lehn et al.

0.26 eV. Of major concern for the future development of molecular electronic devices is the observation that the size of the coupling element decreases smoothly with increasing length of the polyene [73], as expected for a metallic wire. In this system, the polyene connector acts as both spacer and molecular-scale wire. It prevents folding motions in which the terminals might approach each other and thereby cause a short circuit [74]. The polyenic wire is fairly rigid but susceptible to oxidative attack; one of its important roles in Nature is to operate as an antioxidant. Even so, the realization that simple molecular components can promote long-range electronic coupling is of extreme importance.

2.2
Attenuation along Molecular Bridges

One of the easiest ways to engineer wire-like molecular systems is to build up a two center supermolecule in which one terminal plays the role of the donor (D) and the other terminal functions as the acceptor (A). One terminal is selected to be the chromophore. The connector has to be rigid if mechanistic information is to be gathered from subsequent kinetic studies. Historically, the first systems were constructed from purely organic molecules; notably, porphyrins [75], **5** or **6**, or 1,4-dimethoxynaphthalene [76], **7**, **8** or **9**, which acted as electron donor centers, and with simple quinones (**5** and **6**) or 1,1-dicyanovinyl units (**7**, **8** or **9**) as electron acceptors (Fig. 2.5). In these prefabricated molecular dyads, the light-active donor groups are separated from the electron acceptor by an increasing number of rigid modules; in some cases the orientation of the acceptor has been tuned during the synthetic protocol (e.g., compare **7** and **8**). Under photo-excitation, the donor provides an electron to the quinone or 1,1-dicyanovinyl acceptor since there is a large thermodynamic driving force for this process from the excited state, but not for the corresponding ground state. The rate of electron transfer, measured by time-resolved fluorescence or transient absorption spectroscopy, shows a significant dependence on the number of interspersed saturated modules and decreases by three to six orders of magnitude, respectively, in the **5** and **6** series of compounds. Moreover, the rate of electron transfer decreases by an order of magnitude on shifting from a linear structure in **7** to a bent geometry in **8** [77]. By increasing the number of saturated functions from 8e(sp) to 12e(sp), on moving from **7** to **9**, the rate of electron transfer decreases by two orders of magnitude [78]. These results have been discussed within the framework of a hyperconjugation process between donor and acceptor centers favoring super-exchange interactions. The variety of rate constants demonstrates firmly the importance of the bridge in promoting long-range electron transfer. This work also illustrates the difficulty in building systems capable of very long distance electron transfer, since the attenuation in the degree of electronic coupling falls rapidly with increasing length of the bridge. Note the subtlety of how slight changes in orientation affect the rate of electron transfer. There have been numerous such systems reported in the literature with over 200 porphyrin-quinone dyads described in the past 20 years.

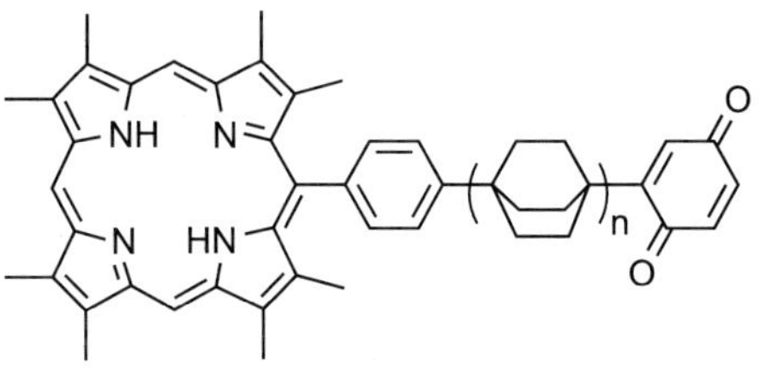

5₀, n = 0 ; $R_{Por/Q}$ = 6.5 Å ; k_{ET} = 5.8 10⁹ s⁻¹; ΔG = -0.55 eV
5₁, n = 1 ; $R_{Por/Q}$ = 14.8 Å ; k_{ET} = 1.5 10⁷ s⁻¹; ΔG = -0.55 eV
5₂, n = 2 ; $R_{Por/Q}$ = 18.8 Å ; k_{ET} < 4 10⁶ s⁻¹; ΔG = -0.55 eV

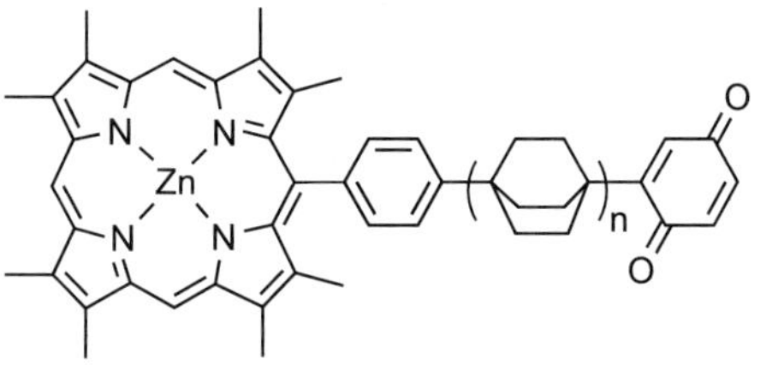

6₀, n = 0 ; $R_{Por/Q}$ = 6.5 Å ; k_{ET} > 10¹² s⁻¹; ΔG = -0.97 eV
6₁, n = 1 ; $R_{Por/Q}$ = 14.8 Å ; k_{ET} = 1.5 10¹⁰ s⁻¹; ΔG = -0.97 eV
6₂, n = 2 ; $R_{Por/Q}$ = 18.8 Å ; k_{ET} < 9 10⁶ s⁻¹; ΔG = -0.97 eV

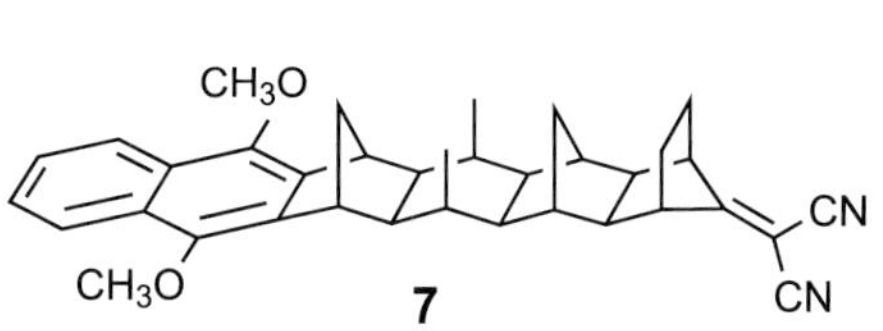

7
$R_{D/A}$ = 11.4 Å ; k_{ET} = 3.0 10⁹ s⁻¹; ΔG = -0.41 eV

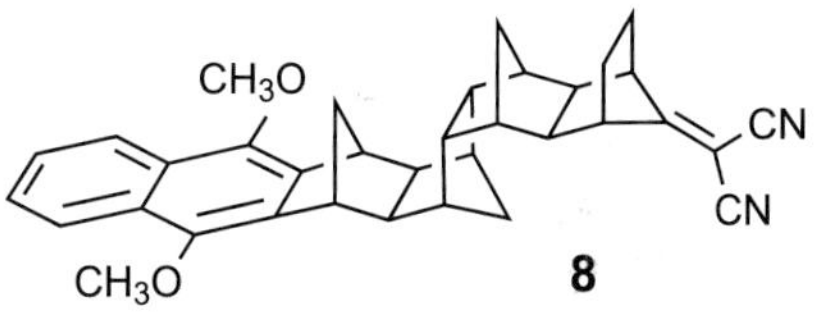

8
$R_{D/A}$ = 10.3 Å ; k_{ET} = 2.2 10⁸ s⁻¹; ΔG = -0.36 eV

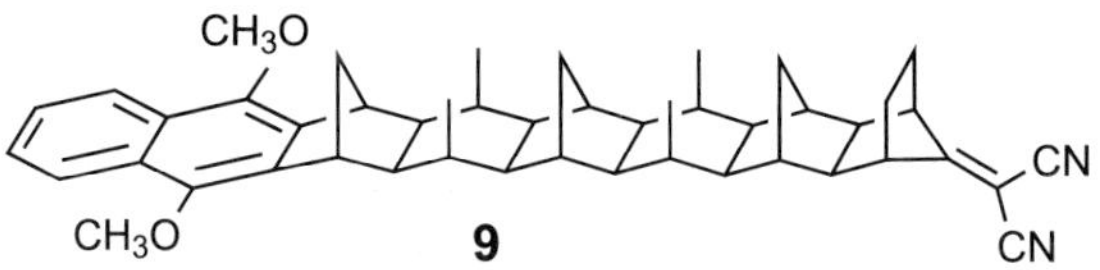

9
$R_{D/A}$ = 14.9 Å ; k_{ET} = 1.6 10⁷ s⁻¹; ΔG = -0.55 eV

Figure 2.5. Examples of some of the early molecular dyads used to examine light-induced electron transfer.

The use of *p*-phenylenevinylene threads in well-defined oligomers (**10₁** to **10₅**) allows the formation of a photo-induced CS state in processes that are only weakly dependent on the separation distance (Fig. 2.6) [79]. An interesting switch in mechanism was discovered when the length of the bridge was increased from a single phenylene unit in **10₁** to five phenylene rings in **10₅** [80]. For short linkages, the distance dependence followed the super-exchange phenomenon, with the at-

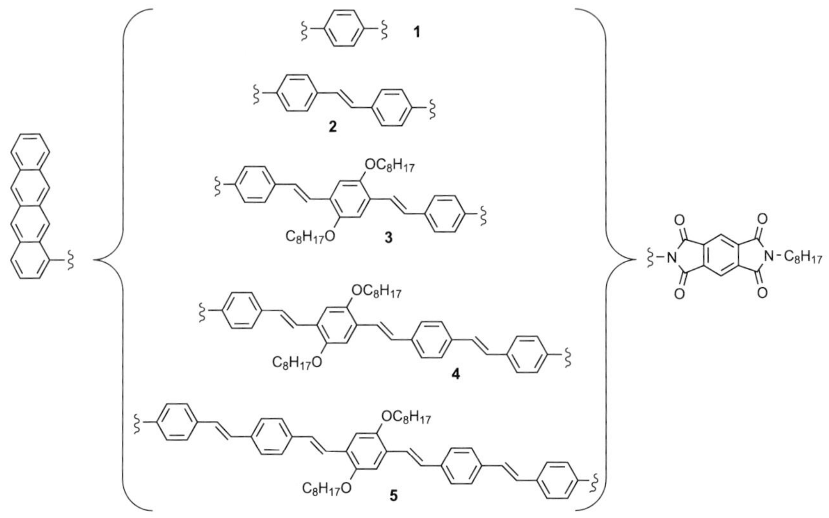

10₁, R_{DA} = 11.1 Å ; k_{CS} = 1.3 10¹¹ s⁻¹ ; ΔG_{CS} = -0.84 eV
10₂, R_{DA} = 17.7 Å ; k_{CS} = 2.3 10¹⁰ s⁻¹ ; ΔG_{CS} = -0.77 eV
10₃, R_{DA} = 24.3 Å ; k_{CS} = 3.9 10¹¹ s⁻¹ ; ΔG_{CS} = -0.74 eV
10₄, R_{DA} = 30.9 Å ; k_{CS} = 2.6 10¹¹ s⁻¹ ; ΔG_{CS} = -0.72 eV
10₅, R_{DA} = 38.0 Å ; k_{CS} = 2.2 10¹¹ s⁻¹ ; ΔG_{CS} = -0.70 eV

Figure 2.6. Molecular dyads found to exhibit a switch in electron-transfer mechanism.

tenuation factor that measures the resistivity of the bridge being $\beta = 0.57\,\text{Å}^{-1}$ for **10₁** and **10₂**. For longer bridges, an abrupt change in mechanism occurred and the rate of electron transfer was found to be less affected by incremental increases in separation distance [80, 81] with $\beta = 0.04\,\text{Å}^{-1}$ for **10₃** and **10₅**. This effect was explained in terms of the decrease in energy of the bridging LUMO with increased π-electron conjugation. Beyond a critical length, the LUMO of the bridge becomes accessible to the electron donor, and the mechanism switches from super-exchange to a type of intramolecular charge-transfer process. The importance of this work is that it opens the way to engineer molecular dyads capable of very long distance charge transfer. The study also shows that the bridge does not have to take a passive role in the transfer process [81].

Closely related work by Guldi et al. [82] has shown that oligo(p-phenylenevinylenes) can promote electron transfer over distances in excess of 40 Å; see for example **11** (Fig. 2.7). Here, the donor is a π-extended tetrathiafulvalene and C_{60} acts as the electron acceptor. Light-induced electron transfer is believed to take place via long-range super-exchange interactions with an attenuation coefficient β of $0.01\,\text{Å}^{-1}$. This is an exceptionally small attenuation coefficient and provides for an electronic coupling matrix element of 5.5 cm⁻¹ over a distance of 40 Å. There is no indication that the bridge enters directly into the electron-transfer chemistry.

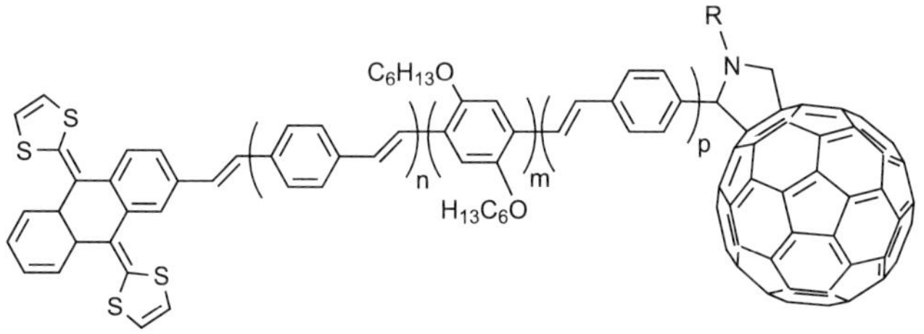

11_1, n = 1, m = 0, p = 0, R = (CH$_2$CH$_2$O)$_3$CH$_3$
11_2, n = 0, m = 1, p = 1, R = C$_8$H$_{17}$
11_3, n = 1, m = 1, p = 1, R = CH$_3$
11_4, n = 2, m = 1, p = 2, R = (CH$_2$CH$_2$O)$_3$CH$_3$

TTF$^{+\cdot}$–Wire–C$_{60}$$^{\cdot-}$: ΔG_{CS} = -0.7 eV,
τ_{CS} = 430 ns for 11_1
β = 0.01Å^{-1} for k_{CS} in compds 11_1 to 11_4

Figure 2.7. Structure of the molecular dyad found to display an exceptionally small attenuation factor for through-bond electron transfer.

A popular way in which to construct wire-like molecular systems involves building up dinuclear complexes in which one metal center plays the role of donor and the other acts as the acceptor under visible light illumination [67–70]. The construction of such architectures is usually based on the prior synthesis of a polytopic ligand of well-defined geometry followed by capping of the terminal chelation sites with suitable metallo-fragments. From a conceptual point of view, the bridging ligand will control the separation distance, the orientation of the chromophores and the extent of electronic interaction between the photo- and/or redox-active terminals. The bridging ligand can comprise various spacers, these being saturated or conjugated, some of which possess low-lying π^* orbitals (i.e., good electron acceptors) or act as good π donor or poor π^* acceptor moieties. Some of these bridges can themselves be photo-active and behave as a switch by which to control energy or electron flow. In some cases, fine tuning of the energetics of the bridge is feasible by the attachment of adventitious cations [66]. In Fig. 2.8–2.10 are shown selected Ru/Os complexes where the bridging ligand is flexible (e.g., complexes **12** and **13**) [83, 84], semi-rigid (e.g., **14**) [85], hybrid (e.g., **15** to **17**) [86–88], or rigid (e.g., **18** to **20**) [89–91]. The flexible hydrocarbon chain (Fig. 2.8) does not provide for good electronic interactions between the donor (Ru^{II}) and acceptor (Os^{II}) centers. In such systems, electronic energy transfer takes place by way of diffusional encounter between the terminals and involves Förster-type coulombic interactions. The rate of triplet energy transfer is only weakly dependent on the length of the bridge [83]. It should be noted that there is reasonable spectral overlap between emission from the Ru-based donor and absorption by the Os-based acceptor that favors the Förster mechanism. This realization has resulted in the design of several hundred such molecular dyads. In principle, the kinetic information pertaining to the rate of intramolecular triplet energy transfer can be used to measure average diffusion coefficients and diffusion lengths. Such work is rendered difficult, however, by the poor emission yields of donor and/or acceptor units.

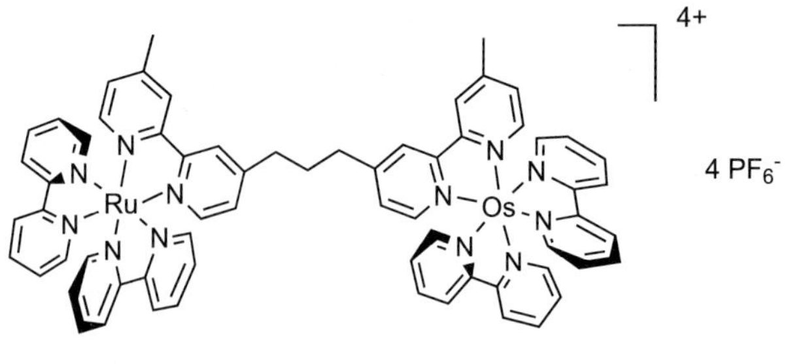

12, R_{MM} = 12.8 Å, k_{ET} = 4.5 10^8 s^{-1}, ΔG = -0.32 eV

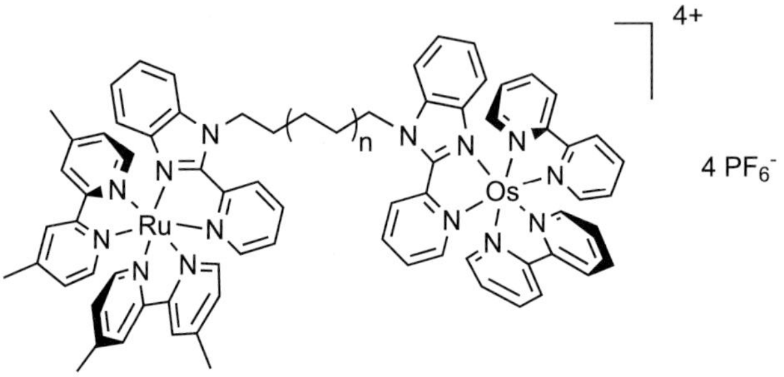

13$_0$, n = 0, R_{MM} = 9.2 Å, k_{ET} = 4.7 10^8 s^{-1}, ΔG = -0.24 eV
13$_1$, n = 1, R_{MM} = 13.0 Å, k_{ET} = 1.7 10^8 s^{-1}, ΔG = -0.24 eV

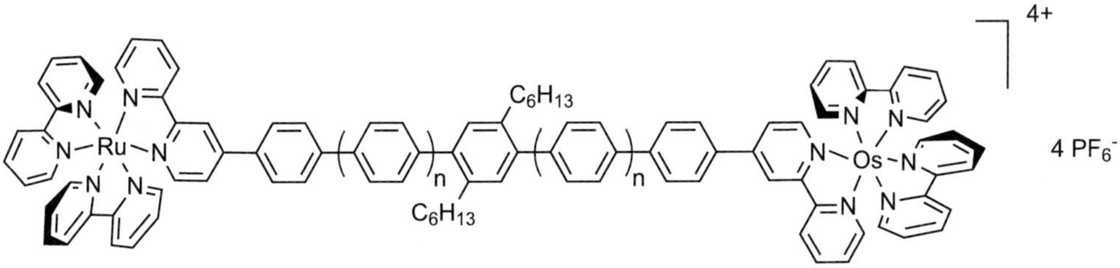

14$_0$, n = 0, R_{MM} = 24.0 Å, k_{ET} = 6.7 10^8 s^{-1}, ΔG = -0.37 eV
14$_1$, n = 1, R_{MM} = 32.5 Å, k_{ET} = 0.1 10^8 s^{-1}, ΔG = -0.37 eV
14$_2$, n = 2, R_{MM} = 42.0 Å, k_{ET} = 0.1 10^8 s^{-1}, ΔG = -0.37 eV

Figure 2.8. Examples of molecular dyads used to study intramolecular triplet energy transfer.

By increasing the number of spanning phenylene subunits in complexes **14$_0$**, **14$_1$** and **14$_2$**, excitation energy transfer has been observed over distances up to 42 Å [85]. On the basis of an exponential decrease in the rate of triplet energy transfer with increasing separation distance, it was concluded that transfer occurs by way of through-bond electron exchange. An attenuation factor of $\beta = 0.32$ Å^{-1} was deduced at room temperature and also at 77 K, showing that this particular energy-transfer step is not affected much by temperature or by the state of the solvent (i.e., fluid or frozen). Unlike electron transfer, electronic energy transfer is characterized by a rather small reorganization energy and weak dependence on

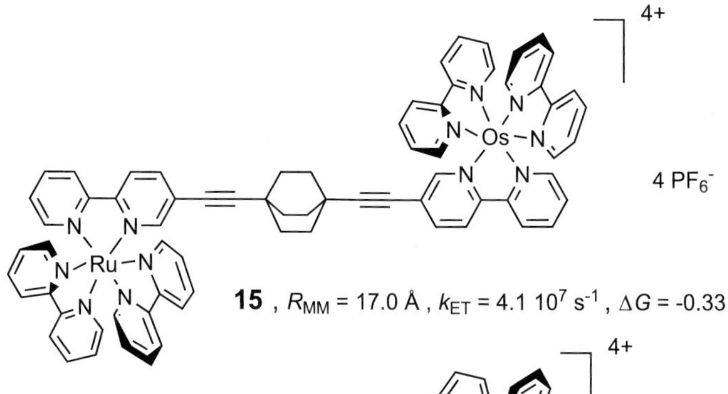

15, R_{MM} = 17.0 Å, k_{ET} = 4.1 10^7 s^{-1}, ΔG = -0.33 eV

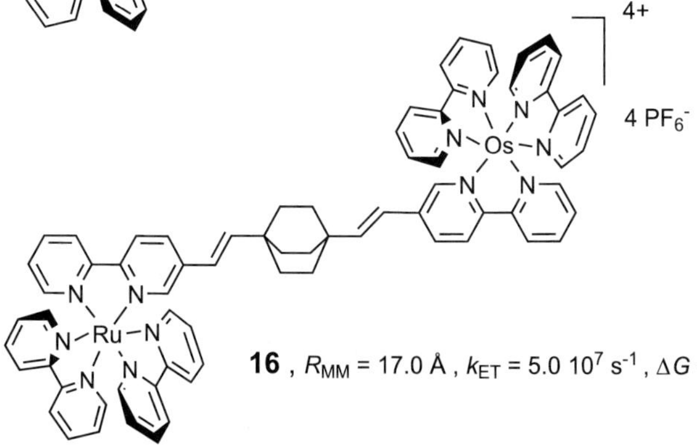

16, R_{MM} = 17.0 Å, k_{ET} = 5.0 10^7 s^{-1}, ΔG = -0.36 eV

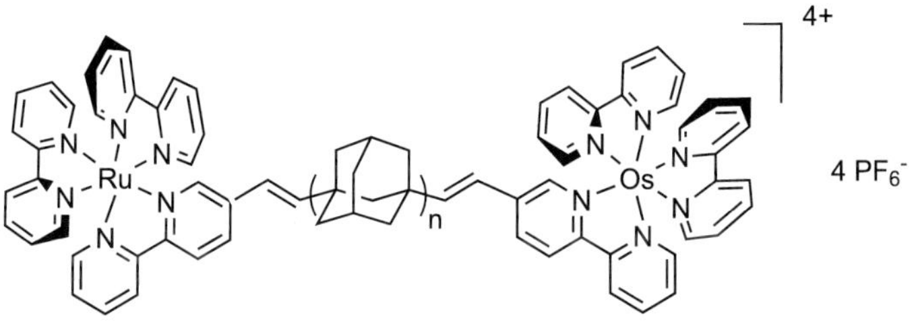

17$_1$, n = 1, R_{MM} = 18.1 Å, k_{ET} = 5.8 10^8 s^{-1}, ΔG = -0.36 eV
17$_2$, n = 2, R_{MM} = 22.3 Å, k_{ET} = 4.0 10^7 s^{-1}, ΔG = -0.36 eV

Figure 2.9. Additional examples of molecular dyads used for the investigation of intramolecular triplet energy transfer.

solvent polarity. Indeed, cases are known where the rate of energy transfer is essentially the same in both fluid solution at room temperature and in the solid state at low temperature [92, 93].

Triplet energy transfer has been examined in a variety of hybrid molecular dyads where unsaturated groups are attached to a central bicyclooctane framework [86]. Some of these systems are illustrated in Fig. 2.9. In each case, there is a reasonably large thermodynamic driving force for triplet energy transfer from the Ru-based donor to the Os-based acceptor such that electron exchange should occur at a rate near the apex of a Marcus-type rate vs. energy gap profile. However, the rates are relatively slow and essentially insensitive to the nature of the unsaturated group (triple bond vs. double bond, respectively, in complexes **15** and **16**). In **15**, where the internuclear separation distance is about 17 Å, the measured rate of intramolecular triplet energy transfer is comparable to the rate calculated for through-space transfer [86]. The driving force for energy transfer was estimated to be ca. 0.33 eV whilst the total reorganization energy accompanying energy transfer was estimated at 0.2 eV. This places electron exchange within the Marcus inverted region. Assuming energy transfer is exclusively via Dexter-type electron exchange, the electronic coupling matrix element for **15** would have a value of only 0.4 cm^{-1}. The corresponding calculation for **16** predicts a coupling element of 0.6 cm^{-1} [87].

Increasing the number of adamantane spacer units from one in complex **17$_1$** to two in complex **17$_2$** decreases the rate of triplet energy transfer by one order of magnitude whilst increasing the average separation distance by ca. 3 Å [88]. There is no direct spectroscopic evidence for light-induced isomerization in the ethene-based spacers but, in the absence of detailed information about how changes in temperature affect the rate of energy transfer, the mechanism remains unknown. It is possible that both through-space (i.e., Förster-type coulombic energy transfer) and through-bond (i.e., Dexter-type electron exchange) interactions contribute to the overall process. In this case, the ability of the central bicyclooctane unit to function as a regulator, or insulator, cannot be judged. It is interesting to note that selective oxidation of the Os^{II} unit switches on the possibility for intramolecular electron transfer between the terminals [88]. Such a situation provides an opportunity to compare the relative efficacy of similar bridges to conduct electrons or photons.

In marked contrast to the above, rigidification of the bridge using tetrapyridinophenazine or adamantane linkers (Fig. 2.10) ensures better control of the donor/acceptor separation distance and of the geometry of the spacer unit. In particular, for complex **18** an efficient energy-transfer process is observed [89]. Here, the bridge remains coplanar with the coordinated groups of the multi-topic ligand. It is likely that illumination results in the formation of the lowest-energy metal-to-ligand, charge transfer state localized on the multitopic ligand, rather than on the capping bipyridine ligands [89, 90]. Triplet energy transfer, therefore, involves only short-range interactions and is expected to be rapid. Incorporation of a saturated ring in the bridge, as in complex **19**, again leads to a serious attenuation in the rate of energy transfer (Fig. 2.10) [90, 91]. Further increases in the separation distance, as in complex **20**, decrease the rate approximately 6-fold.

2.3
Information Transfer

2.3.1
Intramolecular Triplet Energy Transfer

The photophysical properties of the luminescent complexes formed between 2,2'-bipyridine (bipy) and certain transition metals, most notably Re^{I}, Ru^{II} or Os^{II} but

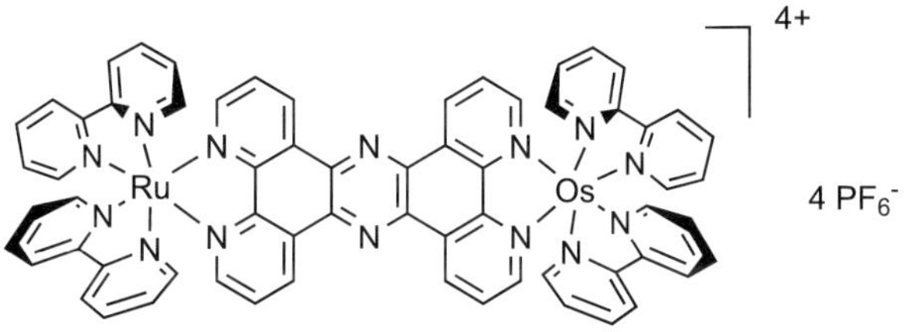

Figure 2.10. Mixed-metal dyads used to study triplet energy transfer.

18, R_{MM} = 12.8 Å, k_{ET} > 1.0 10^9 s^{-1}, ΔG = -0.37 eV

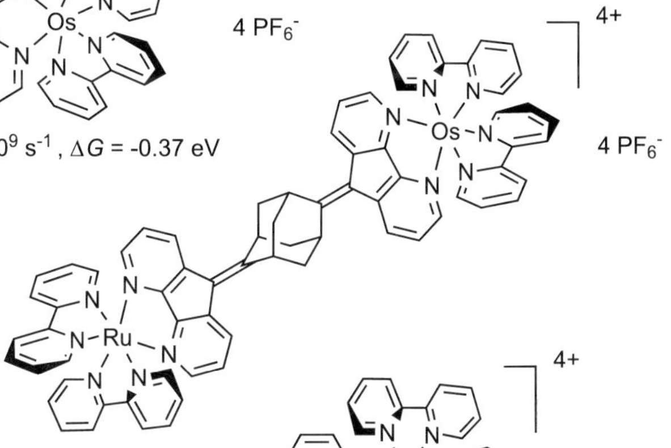

19, R_{MM} = 16.0 Å, k_{ET} = 2.6 10^8 s^{-1}, ΔG = -0.40 eV

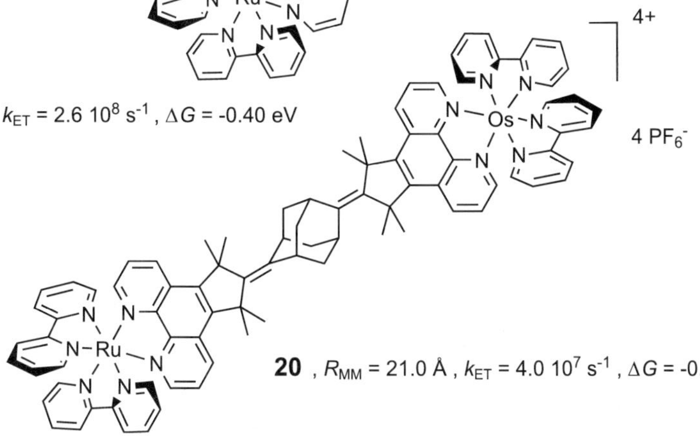

20, R_{MM} = 21.0 Å, k_{ET} = 4.0 10^7 s^{-1}, ΔG = -0.38 eV

also Pt^{II}, have been reported in great detail for both mono- and multinuclear species. The absorption spectra show prominent bands ascribed to metal-to-ligand, charge-transfer (MLCT) transitions in the near UV or visible regions. It is realized that intersystem crossing to the triplet manifold is extremely rapid [94] and probably quantitative [95], at least in the mononuclear complexes. Triplet lifetimes recorded for simple analogs tend to be rather long, typically some hundreds of nanoseconds, and temperature dependent [96, 97]. The synthesis of functionalized complexes of this type necessarily generates a mixture of enantiomers that can be resolved in some cases [98]; this particular problem is overcome by using the corresponding 2,2′:6′,2″-terpyridine ligands [99]. A further difficulty with the mixed-metal complexes, especially Ru^{II}/Os^{II} binuclear complexes, is that the MLCT transitions overlap to such an extent that it is not possible to selectively excite only the donor chromophore. Despite this important realization, the study of intramolecular triplet energy transfer from the Ru^{II}-based donor to the Os^{II}-based acceptor has become standard policy for evaluating the resistivity of putative molecular-scale bridges.

As mentioned earlier, triplet energy transfer can occur by way of Förster-type through-space interactions [100] or via Dexter-type electron exchange [101]. In the former case, the rate constant for Förster-type triplet energy transfer (k_F) can be expressed as follows:

$$k_F = \frac{8.8 \times 10^{-25} K^2 \Phi_L J_L}{n^4 \tau_L R_{CC}^6} \tag{1}$$

Here, J_L is a spectral overlap integral that accounts for the overlap between emission from the donor and absorption by the acceptor whilst K is an orientation factor that describes the mutual spatial arrangement of donor, acceptor and bridge. The terms Φ_L and τ_L, respectively, are the emission quantum yield and lifetime measured for the donor in the absence of acceptor, n is the refractive index of the surrounding medium and R_{CC} is the separation distance. Usually, R_{CC} is taken as the metal–metal separation. According to Eq. (1), the rate of through-space energy transfer is expected to be only weakly dependent on temperature and solvent polarity but sensitive to changes in geometry, especially the distance between donor and acceptor (Fig. 2.11). It is important to note that all the parameters needed to calculate k_F with good precision can be obtained from independent measurements. The only real uncertainty relates to estimating R for flexibly linked dyads.

In the context of designing molecular-scale wires, the Dexter mechanism is more relevant since it provides direct information on the propensity of the connector to conduct electrons. This mechanism involves the simultaneous transfer of an electron, by way of bridging LUMOs, and a positive hole, via the bridging HOMOs, from donor to acceptor [102]. It is a relatively short-range effect (Fig. 2.11) and, to be efficient, requires orbital overlap between donor and acceptor. The rate constant

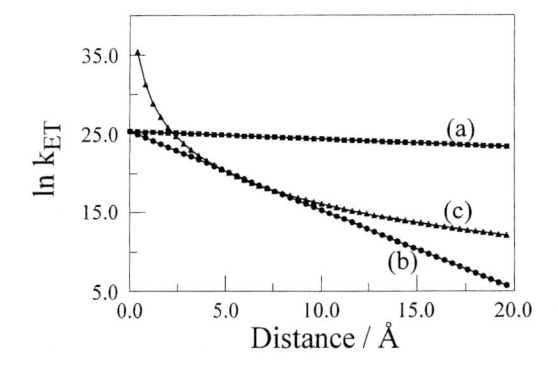

Figure 2.11. Effect of separation distance on the rate constant for intramolecular triplet energy transfer as calculated for (a) Dexter electron exchange with $\beta = 0.1\,\text{Å}^{-1}$, (b) Dexter electron exchange with $\beta = 1\,\text{Å}^{-1}$, and (c) Forster dipole–dipole energy transfer.

2.3 Information Transfer

for Dexter-type electron exchange (k_D) can be expressed in terms of the Fermi Golden rule [103]:

$$k_D = \frac{4\pi^2 V_{DA}^2 J_D}{h} \quad (2)$$

Here, V_{DA} refers to the electronic coupling matrix element and J_D is the Franck–Condon weighted density of states. This latter term can be related to the change in free energy of activation ($\Delta G^{\#}$) and the reorganization energy (λ) accompanying electron exchange. In order to account for energy transfer taking place in the Marcus inverted region [104], it is necessary to include the effects of a medium-frequency vibrational mode ($\hbar\omega_M$) in calculating the activation energy [105]. Again, it is important to note that the J_D factor can be calculated precisely for Ru^{II}/Os^{II} dyads from emission spectra recorded for the corresponding donor and acceptor units [106]. Such procedures allow determination of the magnitude of the electronic coupling matrix element (V_{DA}) for electron exchange and the Huang–Rhys factor S. The activation energy can be measured simply by following the rate of energy transfer as a function of temperature. Provided the energy gap (ΔE) between triplet states localized on donor and acceptor units can be derived from emission spectra, the activation energy allows determination of λ (where k_B is the Boltzmann constant).

$$J_D = \frac{1}{\sqrt{4\pi\lambda k_B T}} \sum_{n=0}^{\infty} \exp(-S) \left(\frac{S^n}{n!}\right) \exp\left(-\frac{\Delta G^{\#}}{k_B T}\right) \quad (3)$$

$$\Delta G^{\#} = \frac{(\Delta E + \lambda + n\hbar\omega_M)^2}{4\lambda} \quad (4)$$

Knowledge of V_{DA} is of great importance for the rational design of new bridges and for understanding the role of the bridge in electron-exchange processes [107–110]. According to super-exchange theory, the coupling element will decrease exponentially with increasing distance (R) between donor and acceptor [111].

$$V_{DA} = V_{DA}^0 \exp(-\beta R) \quad (5)$$

Here, β is an attenuation coefficient that describes the resistivity of the bridge and the term V_{DA}^0 refers to the coupling element at orbital contact. The coupling element can also be related to the energy gap (E_{DB}) between orbitals on the donor and on the bridge (Fig. 2.12) [107–110].

$$V_{DA} = \frac{\alpha_{DB}\alpha_{BA}}{E_{DB}} \quad (6)$$

In this latter expression, α_{DB} and α_{BA} refer, respectively, to atomic orbital coefficients describing coupling between the donor and the first member of the bridge

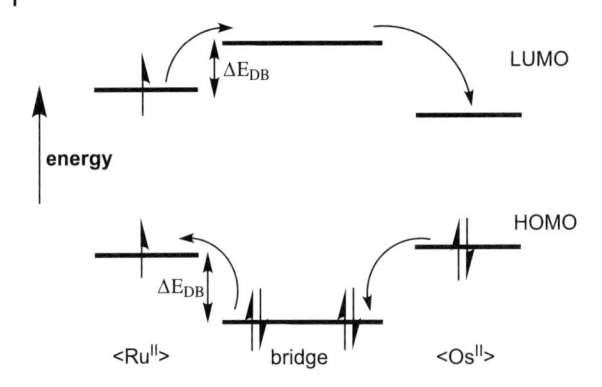

Figure 2.12. Pictorial representation of the super-exchange model whereby electron transfer from donor to acceptor occurs via the LUMO on the bridge and hole transfer takes place via the bridge HOMO.

and between the last member of the bridge and the acceptor. The important point of this expression is that the rate of Dexter-type electron exchange will depend on properties of the bridge. It is this realization that opens up the possibility to design effective molecular-scale bridges.

2.3.2
Short Covalent Bridges

In order to gauge the ability of acetylene-based bridges to mediate electron exchange, the Ru^{II}/Os^{II} mixed-metal complexes shown in Fig. 2.13 were synthesized [112]. The choice of acetylenic bridges was made following the extensive discussion in the literature regarding conductive polymers, but at the time of starting this project virtually nothing was known about how such connectors functioned in simple molecular dyads. Early work established that the acetylene group could be polymerized to yield highly colored films that adhered strongly to surfaces [113]. It was shown that the polymerization process was extremely sensitive to the structure of the initial complex but unfortunately the resultant films were nonluminescent. This work clearly demonstrated that electron delocalization takes place along the acetylenic wire, especially in the terpyridine-based complexes. The importance of cumulene-type structures, as a means to extend electron delocalization, was stressed.

A set of bipy-based Ru^{II}/Os^{II} binuclear complexes, comprising compounds 21–23 (Fig. 2.13), was prepared such that the energy gap between triplet states localized on donor and acceptor units remained at ca. 0.34 eV [112]. These complexes differed in the site of attachment of the acetylene bridge and used both ethynylene and butadiynylene connectors. Of course, with such systems it is difficult to be definitive about the exact nature of the bridge. It was found, however, that intramolecular triplet energy transfer was extremely fast in each case. There was little

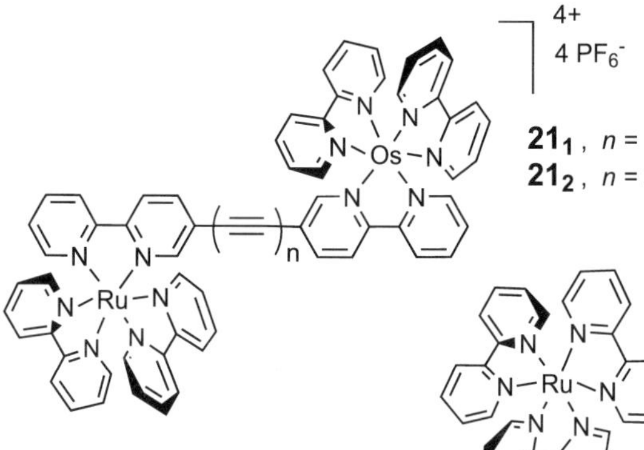

21₁, $n = 1$, $R_{MM} = 11.8$ Å, $k_{ET} = 2.5 \cdot 10^{11}$ s^{-1}, $\Delta G = -0.29$ eV
21₂, $n = 2$, $R_{MM} = 14.8$ Å, $k_{ET} = 1.3 \cdot 10^{11}$ s^{-1}, $\Delta G = -0.29$ eV

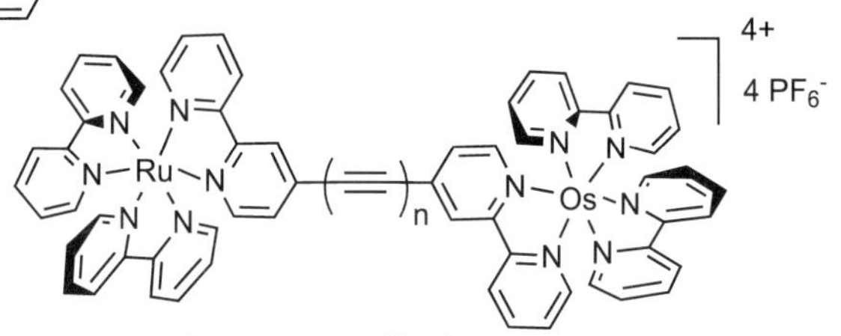

22₁, $n = 1$, $R_{MM} = 13.8$ Å, $k_{ET} = 6.0 \cdot 10^{10}$ s^{-1}, $\Delta G = -0.24$ eV
22₂, $n = 2$, $R_{MM} = 16.4$ Å, $k_{ET} = 2.7 \cdot 10^{10}$ s^{-1}, $\Delta G = -0.24$ eV

23₁, $n = 1$, $R_{MM} = 6.8$ Å, $k_{ET} = 1.7 \cdot 10^{11}$ s^{-1}, $\Delta G = -0.55$ eV
23₂, $n = 2$, $R_{MM} = 8.6$ Å, $k_{ET} = 0.5 \cdot 10^{11}$ s^{-1}, $\Delta G = -0.55$ eV

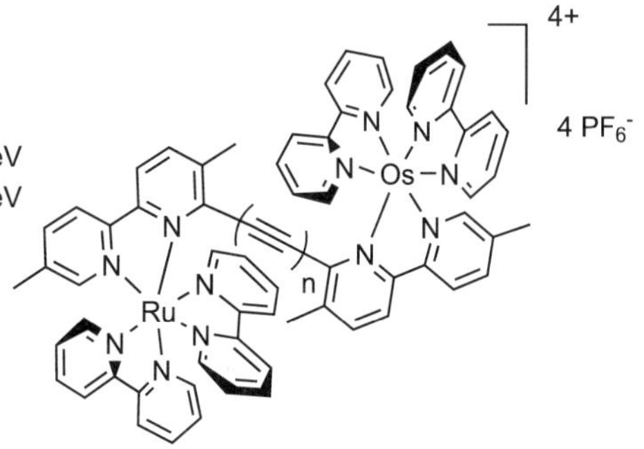

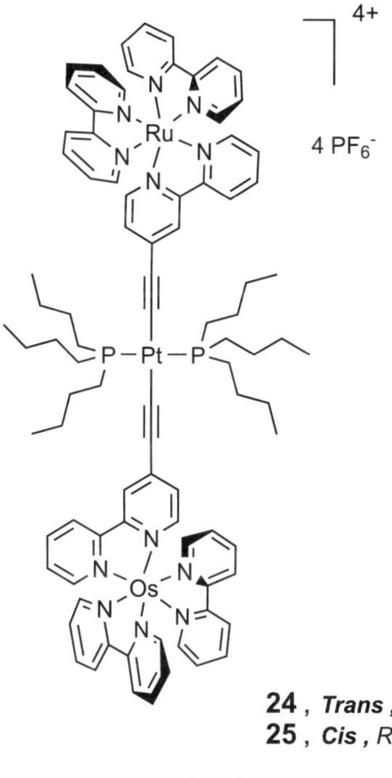

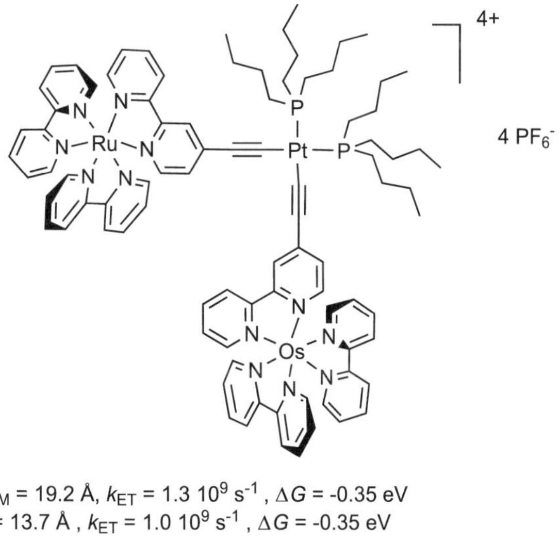

24, *Trans*, $R_{MM} = 19.2$ Å, $k_{ET} = 1.3 \cdot 10^{9}$ s^{-1}, $\Delta G = -0.35$ eV
25, *Cis*, $R_{MM} = 13.7$ Å, $k_{ET} = 1.0 \cdot 10^{9}$ s^{-1}, $\Delta G = -0.35$ eV

Figure 2.13. Selected Ru^{II}/Os^{II} molecular dyads studied by Ziessel, Harriman et al.

sensitivity towards the site of attachment; even the ortho,ortho-bridged systems undergo quantitative energy transfer. These latter complexes possess distorted geometries such that the mononuclear complexes exhibit relatively short triplet lifetimes. Increasing the length of the connector caused a slight decrease in the rate of energy transfer but there was no spectroscopic evidence to suggest direct involvement of a triplet state localized on the bridge. Calculations indicated that the measured rate of energy transfer was far too fast to be consistent with the Förster mechanism and, as a consequence, the results were interpreted in terms of the Dexter mechanism. It was concluded that electronic coupling between donor and acceptor was very strong. Even so, transient absorption spectroscopy indicated that intramolecular energy transfer could not compete with rapid intersystem crossing within the MLCT state of the donor. That is to say, despite the strong coupling, energy transfer takes place from a thermally relaxed triplet state.

In order to reduce the extent of electronic coupling along the molecular axis, a Pt^{II} bis-acetylide complex was used as bridge (see compound **24**) [114]. Intramolecular triplet energy transfer was quantitative in this linear binuclear complex, where the metal centers are separated by ca. 19.2 Å. It is interesting to compare the energy-transfer process in **24** with that elaborated for **22$_2$**. Thus, in **22$_2$** detailed analysis indicates that the coupling element V_{DA} has a value of ca. 12 cm^{-1}. This is indicative of strong coupling along the molecular axis and it should be noted that the MLCT triplet states localized on both donor and acceptor are formed by charge injection into the substituted polytopic ligand. As such, the electron is poised to move along the bridging LUMO. In contrast, the presence of the central Pt^{II} residue causes the corresponding MLCT triplet states to involve charge donation to the unsubstituted bipy ligands (Fig. 2.14). This has a profound effect on the separation distance. The Pt^{II} complex also acts as a high-energy barrier since its HOMO and LUMO are well removed from those of either the acetylene or the metallo-terminals. As a result, V_{DA} falls to ca. 0.2 cm^{-1}. For **24**, it appears that ca. 15% of the overall energy transfer occurs by way of the Förster-type mechanism

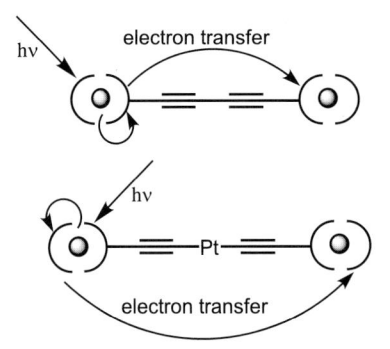

Figure 2.14. Schematic representation of energy transfer in the poly(acetylene)- and Pt^{II} bis(acetylide)-based dyads. In the former case, the MLCT triplet state is formed by initial charge injection into the bridging ligand but in the latter case the lowest-energy MLCT triplet involves charge injection into the capping ligand.

whereas Dexter-type electron exchange is exclusively responsible for rapid electron exchange in 23_2 [114].

The Pt^{II}-bridged *cis*-isomer **25** also displays efficient intramolecular triplet energy transfer [114]. Here, the Ru^{II}/Os^{II} separation is reduced to 13.7 Å but there are shorter separations between the capping bipy ligand on the donor and the accepting ligands on the Os^{II}-based complex. This situation favors Förster-type interactions. Indeed, energy transfer in **25** is dominated by the Förster mechanism and there is little indication for competing Dexter-type electron exchange. The energy levels for the *cis*-Pt^{II} complex are pushed even further from blending with those of the reactants and this has the effect of reducing V_{DA} to an unworkable level. This system was the first artificial prototype to exhibit switching between dipole–dipole and electron-exchange mechanisms [114].

The use of Pt^{II} bis-acetylide as an unusual spacer unit was continued by virtue of the synthesis of bipy-based complexes linked to a zinc(II) porphyrin or to pyrene (Fig. 2.15) [115, 116]. The porphyrin-based system, complex **26**, exhibits intramolecular triplet energy transfer from the Ru^{II} complex to the appended metalloporphyrin. This process involves electron exchange and is characterized by a modest energy gap of 0.38 eV. The rate constant for triplet–triplet energy transfer is 2×10^8 s^{-1} at room temperature. Selective excitation into the lowest-energy singlet state localized on the porphyrin, this being a π,π^* excited state, results in fast singlet–triplet energy transfer to the Ru^{II} complex, followed by triplet triplet energy transfer (Fig. 2.16). The rate constant for this spin-forbidden process is 7×10^8 s^{-1} at room temperature. This dyad is one of several known examples of Ru^{II} poly(pyridine) complexes linked to a porphyrin [93, 117–121]. In all cases, it has been demonstrated that highly efficient intramolecular energy-transfer steps lead to population of the π,π^* excited triplet state localized on the porphyrin. As such, the Ru^{II} complex serves to promote intersystem crossing within the porphyrin. The presence of the central Pt^{II} bis-acetylide has little, if any, effect on the photophysical properties of the porphyrin as demonstrated by the results collected for complex **27** [115].

In contrast to all other reported Ru^{II} poly(pyridine)-porphyrin dyads, complex **26** shows efficient singlet–singlet energy transfer from the second excited singlet state localized on the porphyrin to the MLCT state of the Ru^{II} complex (Fig. 2.16) [115]. The lifetime of the S_2 state is only ca. 3 ps but there is good spectral overlap between emission from S_2 and MLCT absorption in the 420–500 nm region. The close positioning of the reactants favors Förster-type energy transfer. The measured rate constant is 3×10^{11} s^{-1}, such that the efficiency of intramolecular energy transfer is ca. 50%. This study has been followed by a virtual avalance of reports describing light-induced electron or energy transfer from the S_2 state of a metalloporphyrin [122–127]. Indeed, Mataga and coworkers [128] have used the S_2 state to study electron transfer in the Marcus inverted region and thereby provide the first clear demonstration of the full bell-shaped rate vs. energy gap profile for charge separation. Other studies have shown that certain carotenoids transfer excitation energy from their S_2 state to closely-bound porphyrins [129, 130]. An important point to emerge from this work is that it is possible to compete with very fast non-

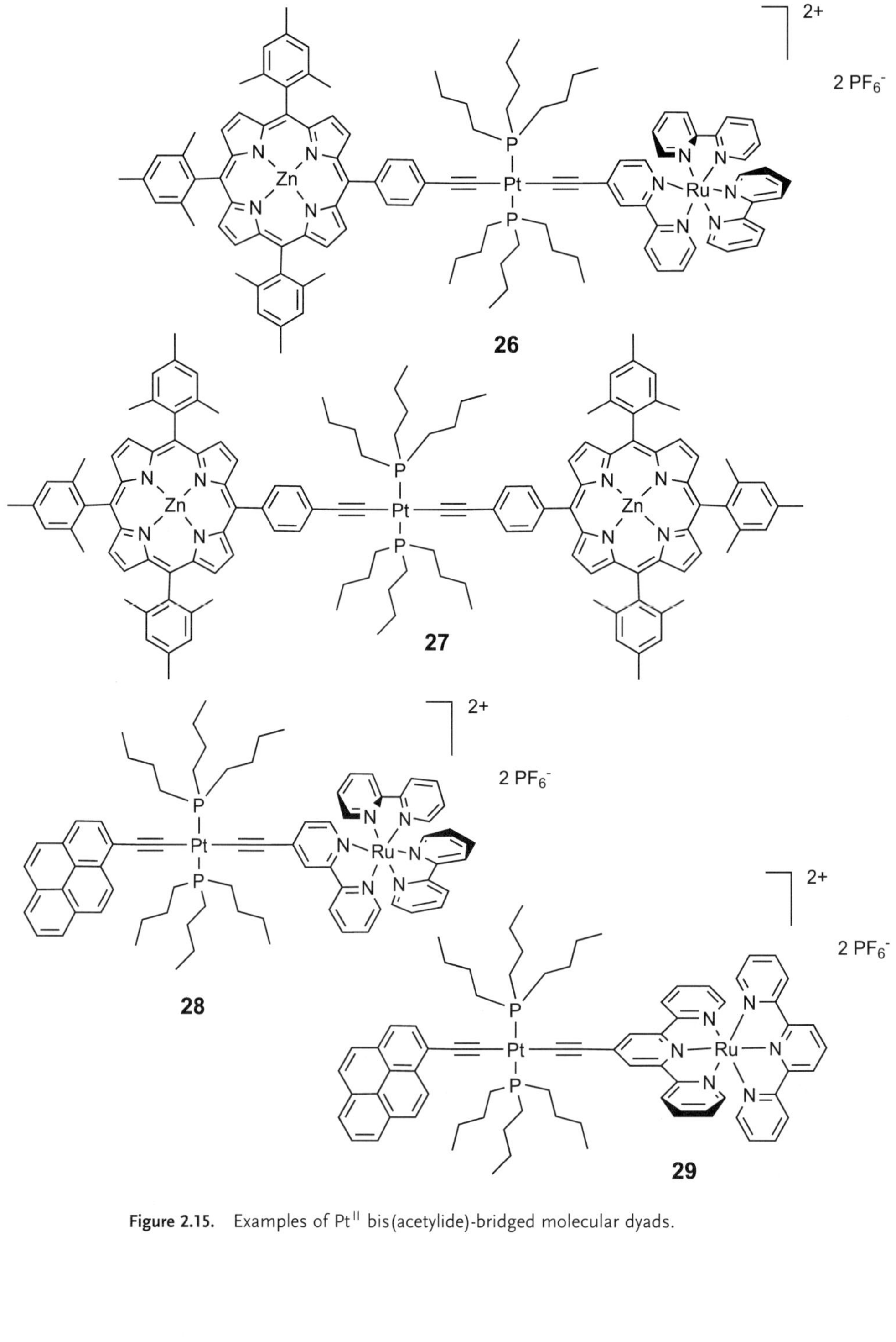

Figure 2.15. Examples of Pt^{II} bis(acetylide)-bridged molecular dyads.

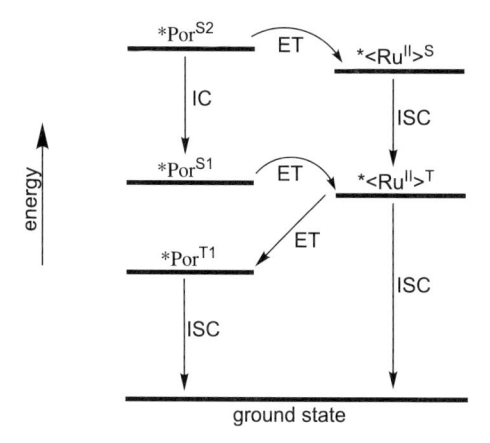

Figure 2.16. Scheme representing the various processes that follow from selective excitation into the various subunits in **26**: ET = energy transfer, ISC = intersystem crossing, IC = internal conversion.

radiative decay of excited states provided the reactants are positioned at the correct site and provided due attention is paid to the ensuing energetics. This realization, in turn, provides new opportunities to design interesting molecular dyads that operate by way of upper excited states.

The Pt^{II} bis-acetylide connector has also been used to link pyrene to ruthenium(II) tris(2,2'-bipyridine) in the form of complex **28** (Fig. 2.15) [116]. Here, the triplet energy of the pyrene unit lies slightly below that of the metal complex, although the two triplets are in thermal equilibrium at room temperature. Excitation into the metal complex results in rapid formation of the MLCT triplet state localized on this subunit. Intramolecular triplet energy transfer takes place with a rate constant of 3×10^8 s^{-1} to establish the thermodynamic distribution of triplet states (Fig. 2.17). This thermally equilibrated mixture decays with a lifetime of 17 μs in deoxygenated acetonitrile at room temperature. The net effect is to prolong the triplet lifetime of the metal complex by a factor of ca. 17. Since the latter triplet exhibits luminescence around 620 nm this makes for a promising biolabel. Because of the energy gap between the two triplets, the equilibrium mixture is dominated by the pyrene triplet, as evidenced by transient absorption spectroscopy. For the corresponding osmium(II) tris(2,2'-bipyridine) complex [131], the MLCT triplet lies at much lower energy than the pyrene-based triplet and there is no equilibrium distribution. In this case, the MLCT triplet decays with the expected lifetime of around 65 ns. Replacing 2,2'-bipyridine with 2,2':6',2"-terpyridine, thereby forming complex **29**, also pushes the energy of the triplet MLCT to be lower than that of the appended pyrene [116]. Thus, there is a fine balance in achieving the appropriate prolongation of the triplet lifetime.

The small series of compounds illustrated in Fig. 2.18 was synthesized in an ef-

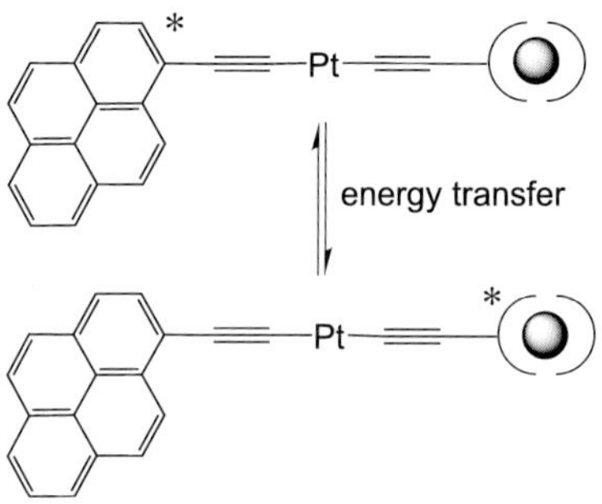

Figure 2.17. Pictorial representation of reversible triplet energy transfer within **28**.

fort to examine the relative importance of long-range vs. stepwise triplet energy transfer along a short bridge [132]. In these molecules, the photoactive terminals are ruthenium(II) and osmium(II) tris(2,2'-bipyridine) complexes and the bridges are diethynylated aromatic residues. The energetics are such that there is a reasonable thermodynamic driving force for intramolecular energy transfer from Ru^{II} to Os^{II} in each case. For the phenylene-bridged system, **30P**, triplet energy transfer occurs by way of long-range super-exchange interactions with a rate constant of 8×10^8 s^{-1}. The π,π^* triplet state localized on the bridge is situated at too high energy for this species to act as a real intermediate in the overall energy-transfer process (Fig. 2.19). As such, the observed rate reflects energy transfer over a distance of 18.3 Å and is characterized by an electronic coupling matrix element of ca. 2 cm^{-1}. Comparing this rate with the inherent triplet lifetime of the corresponding binuclear Ru^{II} complex indicates that triplet energy transfer is essentially quantitative at room temperature. Calculations indicate that Förster-type energy transfer contributes only about 1% to the total rate under these conditions.

Replacing phenylene with naphthalene [132], complex **31N**, has the effect of bringing the triplet energy of the bridge to a point slightly lower than that of the Ru^{II} donor but still well above that of the Os^{II} acceptor (Fig. 2.19). Triplet energy transfer now occurs via a stepwise process in which the naphthalene-based triplet is formed as a real intermediate. The first step is very fast, because of the short separation distance, but reversible. This step occurs with a rate constant of 1.1×10^{10} s^{-1} and is characterized by $V_{DA} = 11.7$ cm^{-1}. Energy transfer from the bridge-localized triplet to the terminal Os^{II}-based acceptor is also fast and takes place with a rate constant of 1.7×10^{10} s^{-1}. For this latter step, V_{DA} is 11.5 cm^{-1}. The net result is very fast triplet energy transfer along the molecular axis with the naphthalene-like triplet state appearing as a real intermediate [132].

Upon exchanging naphthalene for anthracene, complex **32A**, there is a major change in the energetics of the system (Fig. 2.19). Now, the triplet state localized on the bridge is at lower energy than those based on the terminals and the anthracene-like triplet acts as a bottle-neck [132]. Triplet energy transfer from the Ru^{II}-based donor to the bridge is somewhat restricted because this process takes place well within the Marcus inverted region. The rate constant is 1.2×10^{10} s^{-1}

30P
R_{MM} = 18.3 Å
k_{ET} = 8.0 10^8 s^{-1}

31N
R_{MM} = 18.3 Å
k_{ET1} = 1.1 10^{10} s^{-1}
k_{ET2} = 1.7 10^{10} s^{-1}

32A
R_{MM} = 18.3 Å
k_{ET1} = 1.2 10^{10} s^{-1}
k_{ET2} = 2.9 10^9 s^{-1}

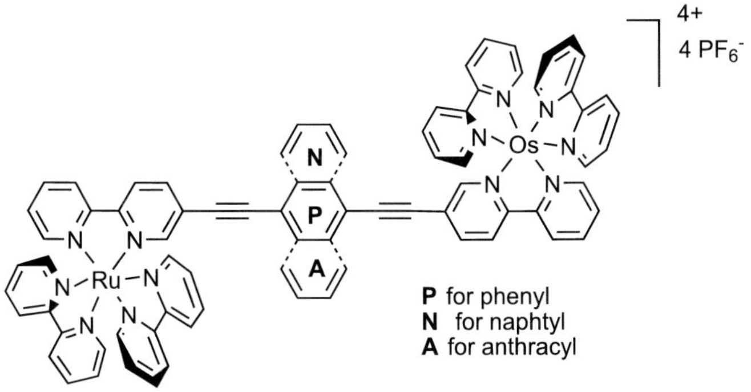
4+
4 PF6-
N
P
A
P for phenyl
N for naphtyl
A for anthracyl

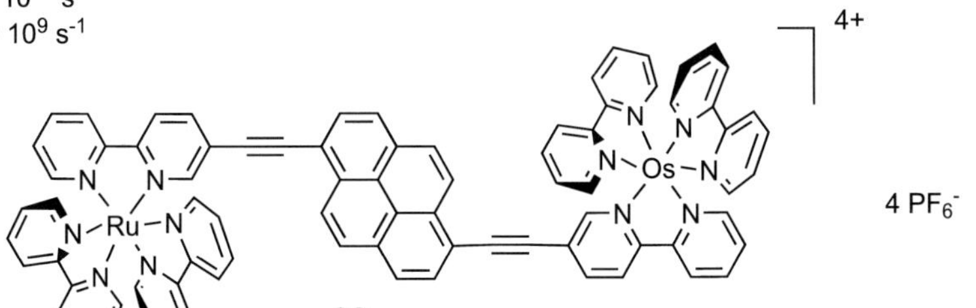
4+
4 PF6-

33

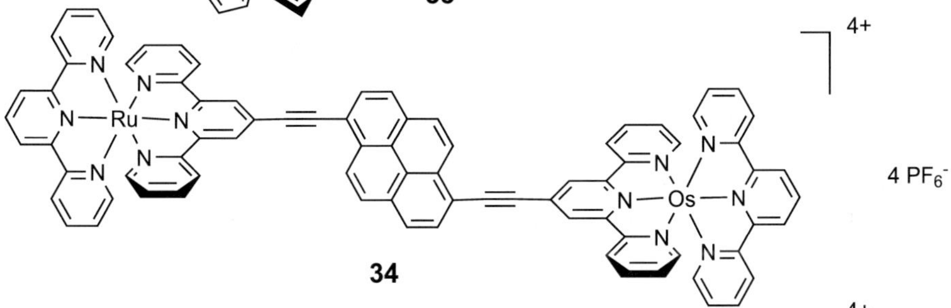
4+
4 PF6-

34

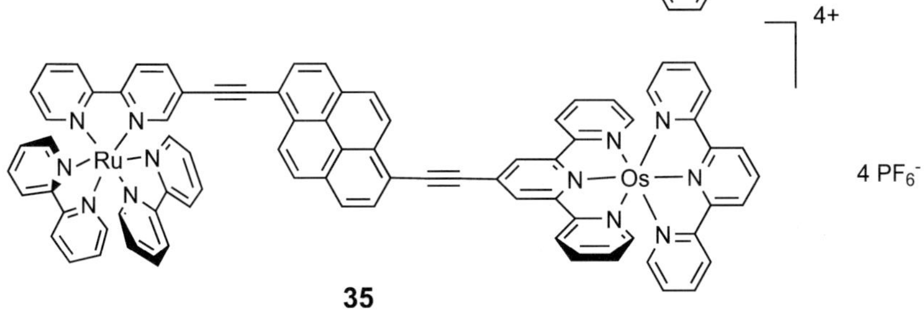
4+
4 PF6-

35

Figure 2.18. Examples of molecular dyads bearing central diethynylated aromatic groups as the bridge.

and V_{DA} is 7.3 cm^{-1}. Triplet energy transfer from the Os^{II}-based terminal to the bridge occurs with a rate constant of 2.9×10^{10} s^{-1} and with a V_{DA} of 11.8 cm^{-1}. Since these latter two triplets lie within thermal equilibrium at room temperature, there is a slight reverse transfer from the bridge triplet to the Os^{II}-based acceptor.

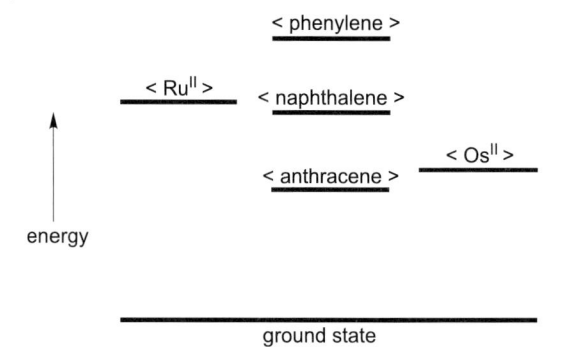

```
                    < phenylene >
                    ─────────────

          < Ru^II >      < naphthalene >
          ────────       ──────────────
                                              < Os^II >
                         < anthracene >       ────────
                         ─────────────
 ↑
 │
 │
energy

                    ─────────────────────
                         ground state
```

Figure 2.19. Energy level diagram for the Ru^{II}/Os^{II} molecular dyads having bridging phenylene-, naphthalene-, or anthracene-based units.

This process has the effect of prolonging the triplet lifetime of the Os^{II} complex [131].

Similar processes take place with the corresponding 2,2′:6′,2″-terpyridine complexes [133] and it is clear that there is a definite kinetic advantage to be gained from the stepwise process. Here, long-range triplet energy transfer from the Ru^{II} donor to the complementary Os^{II}-based acceptor occurs with a rate constant of 1.1×10^8 s^{-1} at room temperature. Interspersing a naphthalene residue in place of the bridging phenylene unit increases the rate constant to 1.1×10^9 s^{-1}. Again, the bridge-localized triplet state functions as a real intermediate in the overall process. The naphthalene spacers can be replaced by pyrene, forming complexes **33–35** (Fig. 2.18), with similar effects.

An interesting case of electronic energy transfer from the Ru-based donor to the Os-based acceptor takes place through the anthracene bridge in complex **36** (Fig. 2.20) with almost unitary efficiency and at a high rate [134]. The semi-flexible nature of the spacer does not facilitate detailed mechanistic understanding and there is no indication for successive energy transfer involving the triplet localized on the central anthracene unit. In principle, this latter triplet should possess an energy somewhere between those of the Ru^{II} complex and the corresponding Os^{II} complex. When the aromaticity of the anthracene core is destroyed by a self-photosensitized oxygenation reaction, leading to complex **37** in Fig. 2.20, both the rate and overall efficiency of the energy-transfer process are decreased by a value of about 80-fold relative to those values found for the reference complex **36**. It is surmised that the formation of an endoperoxide species **37** curtails electronic coupling across the anthracene moiety due to attack by singlet molecular oxygen [134]. For this latter system, it is possible that the semi-flexible spacer facilitates Förster-type interactions. The large disparity in rates of energy transfer found for **36** and **37** suggests that this chemical oxidation, which could be reversible, might form the basis of a switching mechanism. However, endoperoxide formation is too slow for

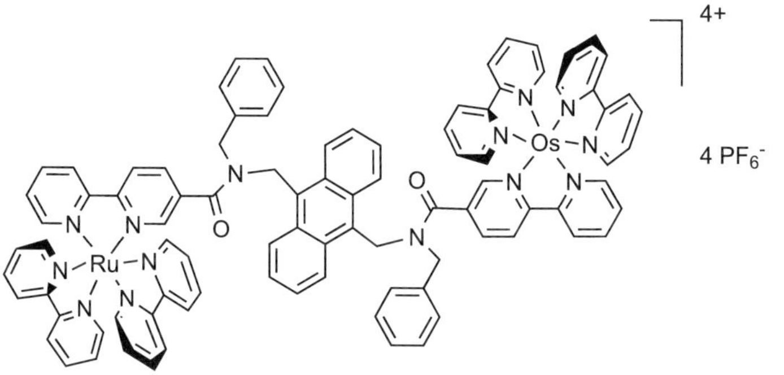

36, R_{MM} = 12.2 Å, k_{ET} = 4.7 10^8 s^{-1}

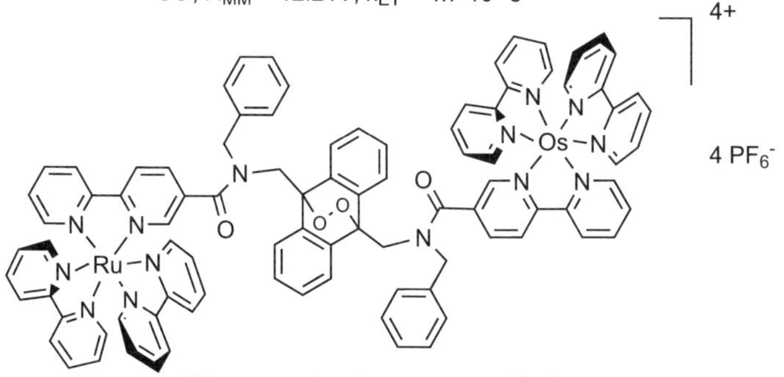

37, R_{MM} = 17.8 Å, k_{ET} = 0.6 10^7 s^{-1}

Figure 2.20. The anthracene-bridged binuclear Ru^{II}/Os^{II} complex and the corresponding oxygenated product.

practical purposes. Incidently, there now exists a rather thorough understanding of the kinetics and mechanism of singlet molecular oxygen production by sensitization with metal poly(pyridine) complexes [135–137].

Light-induced electron transfer has been observed in the homometallic complex **38**, in which the central anthraquinone unit acts as an electron acceptor for the triplet MLCT state localized on the Ru^{II} complex [138]. Presumably, the extended phenazine-based ligand is involved in formation of the lowest-energy MLCT triplet but the photophysical properties of these complexes remain poorly understood despite detailed investigation [139, 140]. Replacing the quinone with different diimides (Fig. 2.21) allows intramolecular electron transfer to continue. When the terminal Os^{II} unit is selectively oxidized to Os^{III}, forming complex **39**, electron transfer occurs from the reduced diimide to the Os^{III} center [141]. For the triad **40**, an interesting two-step process is seen in which the central diimide acts as an electron relay, allowing formation of the long-range charge-separated state in 75% yield. This latter species has a lifetime of 11 ns at room temperature. The rate of the photo-induced electron transfer is independent of the nature of the diimide core, probably because the flexible propane tethers favor conformational heterogeneity.

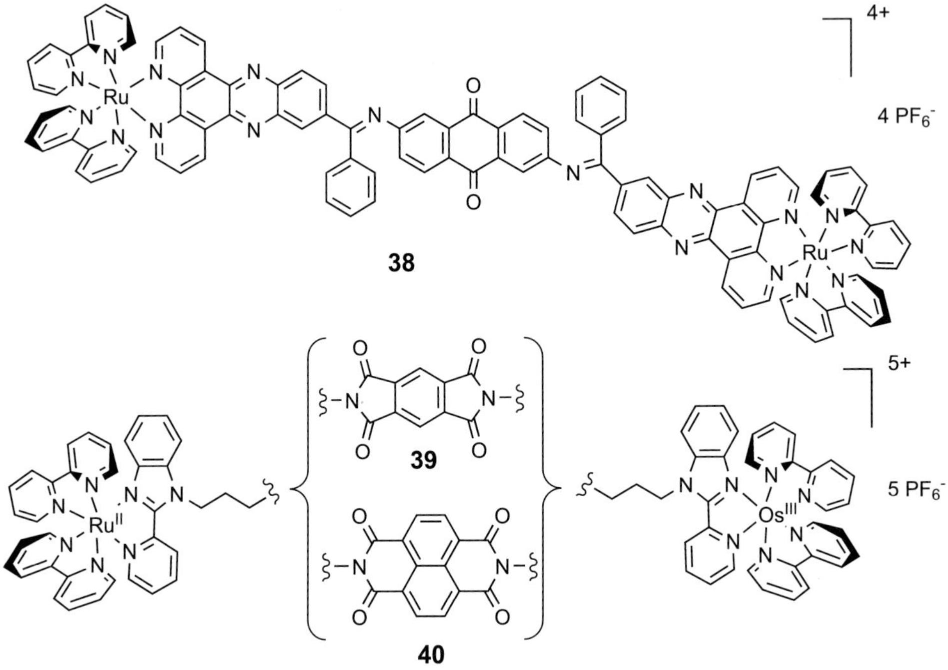

Figure 2.21. Additional Ru^{II}/Os^{II} molecular dyads used to study intramolecular triplet energy transfer.

2.3.3
Supramolecular Systems

Covalent bonding provides for well-defined structures but is expensive to employ. With the notable exception of trivial tasks such as building up a library of compounds by replacing the central metal in a porphyrin, each compound has to be synthesized separately. An alternative, favored by Nature, is to rely on noncovalent interactions to hold together the donor and acceptor functions in a supramolecular entity. Four approaches towards this ideal can be considered; these include the use of coordinative bonding, multi-point hydrogen bonding, π-stacking or electrostatic effects to self-assemble the required superstructure. In particular, the use of coordinative bonding to direct light-induced electron transfer in multi-porphyrin arrays has been pioneered by Sauvage [142]. Several donor–connector–acceptor triads have been engineered in which a metal complex plays a pivotal role [143–146], sometimes appearing as a real redox intermediate [147]. Such systems can be assembled and dismantled by treatment with appropriate precursors and competing ligands. The main motivation behind this research has been to better mimic the electron-transfer chemistry occurring within the photosynthetic reaction center.

Coordinative bonding has also been used to assemble polynuclear metallo-helicates where multitopic ligands wrap around cations of appropriate coordination geometry [148]. Many such systems have been reported [149–151], and the field has developed so as to produce molecular squares, grids, ladders and polymers [152]. It is interesting to note that certain binuclear metallo-helicates retain redox activity and undergo reversible electron-transfer reactions [153]. More relevant to the present discussion is the observation that intervalence charge-transfer character can be seen in mixed-valence Cu^{II}/Cu^{I} species [153]. This observation indicates that the helicate can promote charge transfer, at least over short distances. These systems have not yet been built into molecular-scale wires but related work has demonstrated the viability of forming metallo-mesogens by the same strategy [154].

Considerable attention has been given to the self-assembly of photo-active arrays by way of multiple hydrogen bonding. This field was developed primarily by Sessler [155] who has produced a series of porphyrin-based systems that display intra-assembly energy [156–160] or electron transfer [161, 162]. In most cases it is difficult to conclude if electron transfer proceeds through the hydrogen bonds, as opposed to diffusional contact between the main reactants, but in certain semi-rigid arrays this seems to be the case. Mostly, hydrogen bonding takes place by way of covalently attached nucleic acid bases in nonprotic solvents. This original work has now been extended to include the assembly of many other donor/acceptor pairs. Thus, Fig. 2.22 shows the self-assembly of donor-acceptor dyads by

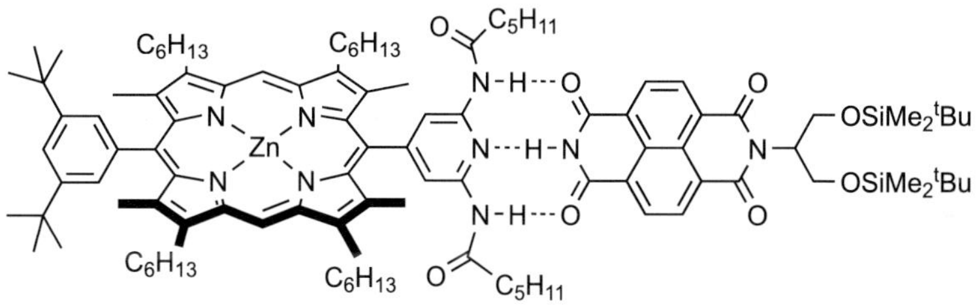

41, K_{BIND} = 280 M^{-1} (C_6D_6), k_{CS} = 4.0 10^{10} s^{-1}, k_{CR} = 3.7 10^9 s^{-1}

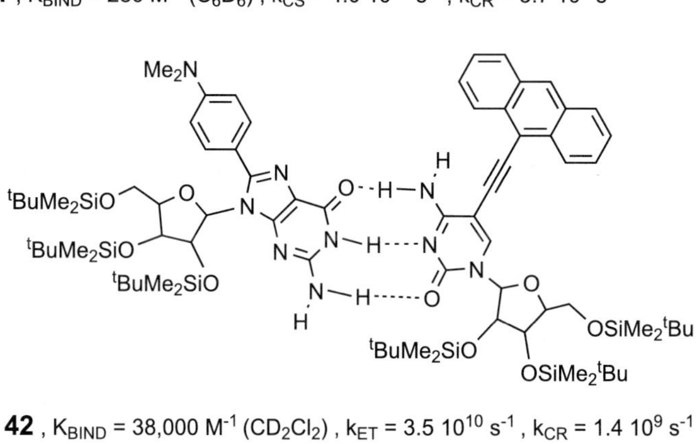

42, K_{BIND} = 38,000 M^{-1} (CD_2Cl_2), k_{ET} = 3.5 10^{10} s^{-1}, k_{CR} = 1.4 10^9 s^{-1}

Figure 2.22. Examples of hydrogen-bonded donor–acceptor dyads.

way of 3-point hydrogen bonding [163]. In the supramolecular arrangement **41** a 2,6-diacylaminopyridylporphyrin (electron donor) is tethered to a naphthalene diimide partner, as an electron acceptor [163]. Despite relatively weak association, the rate of forward electron transfer (k_{CS}) from the first singlet excited state of the porphyrin to the diimide fragment is very fast. The lifetime of the transient charge-separated species is 270 ps. In contrast, the binding constant for **42** is about 130-fold stronger, but the rate of photo-induced electron transfer from the dimethylaniline donor to the singlet excited state of the anthracene unit is similar to that in complex **41** whilst the lifetime of the resultant charge-separated state is 700 ps [164].

In a similar fashion, Ru^{II} and Os^{II} poly(pyridine) complexes have been preorganized via complementary base pairing (Fig. 2.23) [165]. In such systems, triplet energy transfer is relatively efficient, although in the case of **43** this is hampered by weak association of the bases. For this dyad, the guanine–cytosine couple

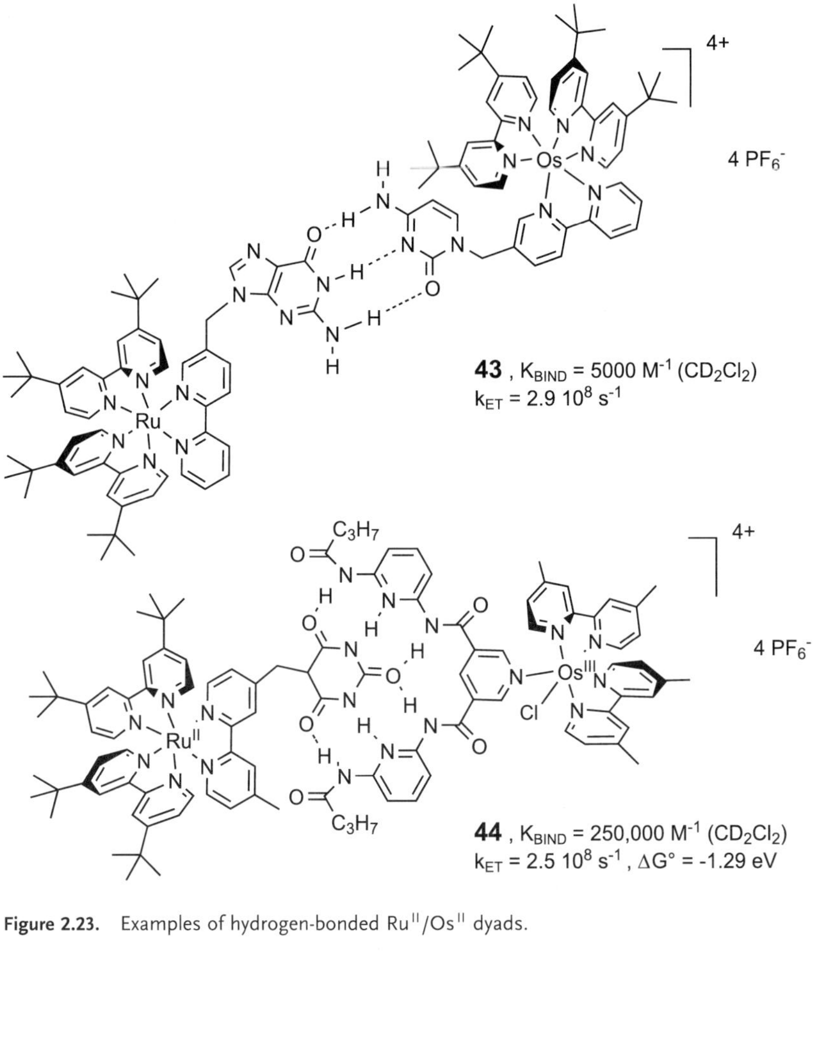

43, K_{BIND} = 5000 M^{-1} (CD$_2$Cl$_2$)
k_{ET} = 2.9 10^8 s^{-1}

44, K_{BIND} = 250,000 M^{-1} (CD$_2$Cl$_2$)
k_{ET} = 2.5 10^8 s^{-1}, $\Delta G°$ = -1.29 eV

Figure 2.23. Examples of hydrogen-bonded Ru^{II}/Os^{II} dyads.

merely brings the reactants into a smaller reaction volume and the energy-transfer step requires diffusional contact between donor and acceptor. The design of artificial systems with a greater number of hydrogen bonds offers the promise of higher association constants, such as in the supramolecular ensemble **44** (Fig. 2.23) [166]. The barbiturate moiety attached to the Ru^{II} portion facilitates a six-point H-bonding array with the bis-diacylaminopyridine fragment of the Os^{II} complex. The resulting stabilization is increased virtually 50-fold compared to **43**, keeping in mind that the electrostatic repulsion remains constant. Despite a higher-order of hydrogen bonding and a larger driving force, the rate of forward electron transfer from the Ru(II) excited state to the Os(III) center is modest and in a similar range to that found for complex **43**. These systems have been developed into light-harvesting dendrimers and polymers decorated with Ru^{II} and/or Os^{II} poly(pyridine) complexes [167–170].

Molecular dyads have been assembled by complementary electrostatic attractive forces and used to demonstrate the principle of light-induced electron transfer [171–173]. Such systems involve short-range effects and appear unlikely to offer the potential to construct extended arrays. A useful alternative is to self-assemble donor-acceptor systems around π-bonding [174]. Thus, light-active species are readily intercalated into DNA and can be used to promote long-range energy or electron transfer along the duplex [175]. Benniston et al. [176] developed this concept to produce the first light-activated molecular shuttle where illumination at one terminal caused a rotaxane to migrate along its thread. Related systems have been used to provide directed electron transfer along one branch of a dual-pathway system [177] whilst Stoddart and coworkers have described an impressive collection of molecular devices based on π-bonding interactions [178–180]. Such materials can be self-assembled into ordered stacks where the π-systems of adjacent molecules are aligned [181]. This is a promising route for the construction of electronic conduits that might eventually rival carbon nano-tubes and covalent arrays.

2.3.4
Prolonging the Excited State Lifetime

In order to devise molecular-scale photonic devices capable of very long-range electron transfer it is necessary to ensure that the rate of transfer can compete effectively with non-radiative decay of the excited state. This requirement precludes the use of ruthenium(II) bis(2,2′:6′,2″-terpyridine) as the photoactive donor because its triplet lifetime is only ca. 250 ps at room temperature [182]. The triplet state of the corresponding ruthenium(II) tris(2,2′-bipyridine) is significantly longer lived, with a typical lifetime of ca. 1 μs in deoxygenated solution at room temperature [96, 97]. It is interesting to note that a hybrid binuclear complex formed by linking Ru-terpy to Ru-bipy via a single ethynylene spacer, complex **45** (Fig. 2.24), displays photophysical properties between those found for the extreme cases [183]. Thus, the emission peak lies at 705 nm whilst the luminescence quantum yield is 0.006. The triplet lifetime measured in deoxygenated acetonitrile at room temperature is 350 ns. Replacing the Ru-terpy terminal with Os-terpy, forming complex **46**, has

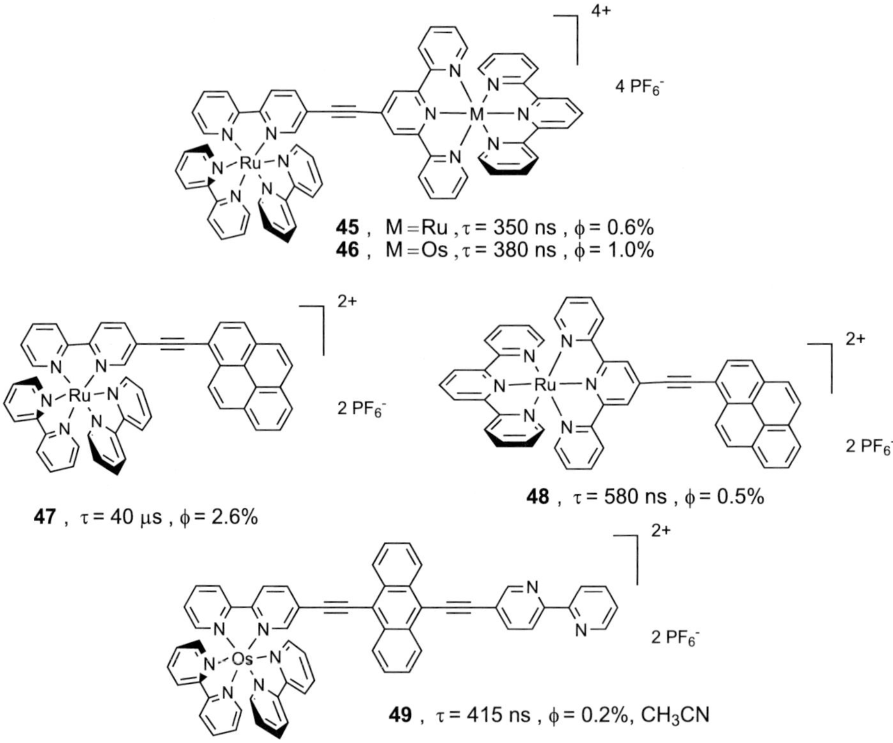

45, M = Ru, τ = 350 ns, φ = 0.6%
46, M = Os, τ = 380 ns, φ = 1.0%

47, τ = 40 μs, φ = 2.6%

48, τ = 580 ns, φ = 0.5%

49, τ = 415 ns, φ = 0.2%, CH_3CN

Figure 2.24. Examples of systems displaying prolonged triplet lifetimes for the MLCT state.

the effect of switching on intramolecular triplet energy transfer from the Ru-bipy terminal. For this system, the Os-terpy unit has a triplet lifetime of 380 ns.

A well-established approach for stabilizing the lowest-energy triplet state has involved the covalent attachment of an aromatic polycyclic hydrocarbon of comparable triplet energy. Here, the triplet lifetime of the metal complex is prolonged because of reversible triplet energy transfer with the appended polycycle. Thus, Ford and Rodgers [184] observed reversible triplet energy transfer between a ruthenium(II) tris(2,2′-bipyridine) subunit and pyrene linked to the metal complex via a flexible hydrocarbon chain. Equilibration between the two triplet states extended the phosphorescence lifetime of the metal complex from ca. 1 μs to 11.2 μs in deoxygenated solution at room temperature. Related work by Sasse et al. [185, 186] also described reversible triplet energy transfer between the metal complex and pyrene, although the phosphorescence lifetime of the ruthenium(II) tris(2,2′-bipyridine) complex was somewhat lower, being 5.2 μs in deoxygenated methanol at 20 °C.

Harriman et al. [187] subsequently described a series of photoactive dyads bearing pyrene and metal (M = Ru^{II} or Os^{II}) tris(2,2′-bipyridine) terminals bridged by an ethynylene group (Fig. 2.24). For the Os^{II}-based dyads, the triplet state remained essentially unperturbed by the presence of the appended pyrene fragment. However, the triplet state localized on the metal complex in the corresponding Ru^{II}-based dyads, complex **47**, was involved in reversible energy transfer with the triplet state associated with the pyrene unit, which is situated at slightly lower energy. When the terminal metal complex is a Ru-terpy fragment, complex **48**, the triplet levels are inverted such that the pyrene-like triplet state lies slightly above that of the metal complex. Even so, the triplet lifetime of the Ru-terpy unit is extended to ca. 580 ns. For the ethynylene-bridged systems, equilibration between the two triplet states was reached [116] within 10 ps while, in the case of the Ru-bipy terminal, the equilibrium mixture decayed with a lifetime of 42 μs in deoxygenated acetonitrile at room temperature [187]. The corresponding Os^{II}-based dyad having anthracene attached via an ethynylene group, complex **49**, has a triplet lifetime of ca. 400 ns, which is very long for such a chromophore [131].

Castellano et al. [188] have described the photophysical properties of ruthenium(II) tris(1,10-phenanthroline) derivatives having either one or three pyrene units attached at the 5-position, complex **50** (Fig. 2.25). These latter systems show phosphorescence only from the MLCT state of the metal complex. The triplet lifetimes are much enhanced relative to the parent complex, with the dyad and triad, respectively, displaying phosphorescence lifetimes of 24 and 148 μs in deoxygenated acetonitrile at room temperature. At the time, this was the longest lifetime to be assigned to the MLCT triplet state of a ruthenium(II) poly(pyridine) complex. The long triplet lifetimes are attributed to reversible triplet energy transfer with the appended pyrene moieties and, using ultrafast transient absorption spectroscopy, the rates of formation of the equilibrium distribution of triplet states were determined [188].

Other systems have been constructed that contain more than one pyrene unit attached to the metal fragment; for example, see complex **51** in Fig. 2.25 [189]. Thus, the excited triplet state lifetimes of a series of nine Ru-bipy derivatives have been found to increase linearly with the number of appended pyrene residues [190]. The lifetimes range from 0.8 to 18.1 μs. A supramolecular structure has been assembled around a central zinc(II) cation in such a way that the metal complex is held close to a pyrene unit [191]. Here, the triplet lifetime of the metal complex is somewhat enhanced because of reversible triplet energy transfer between the two chromophores. The pyrene unit can be replaced with other chromophores provided the triplet energy levels can be balanced. Thus, Castellano et al. [192] have reported prolonged triplet lifetimes for complexes **52** and **53**.

It is interesting to compare the triplet lifetimes measured for the three binuclear Ru-terpy derivatives illustrated in Fig. 2.26. Here, the phenylene-bridged complex, **54**, has a triplet lifetime of 110 ns in deoxygenated acetonitrile at room temperature [193]. For the naphthalene-bridged system **55** the triplet lifetime increases to 415 ns because of reversible triplet energy transfer to the bridge. For the corresponding anthracene-bridged complex **56** the triplet lifetime of the Ru-terpy terminal is only ca. 90 ps. This greatly shortened triplet lifetime arises because of irreversible triplet energy transfer to the bridge.

A set of ruthenium(II) poly(pyridine) complexes has been synthesized in which a

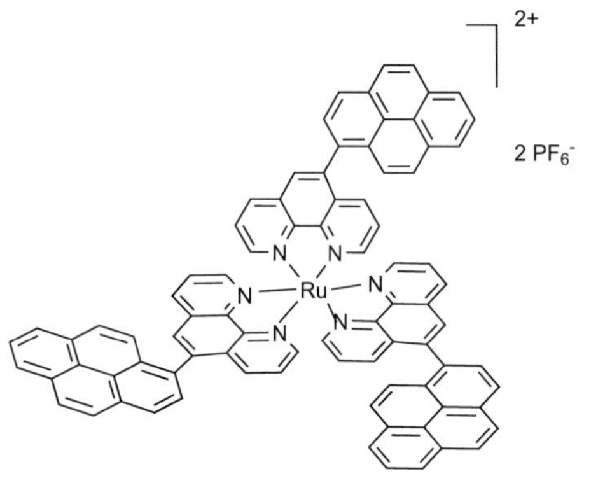

50, τ = 148 μs, CH$_3$CN

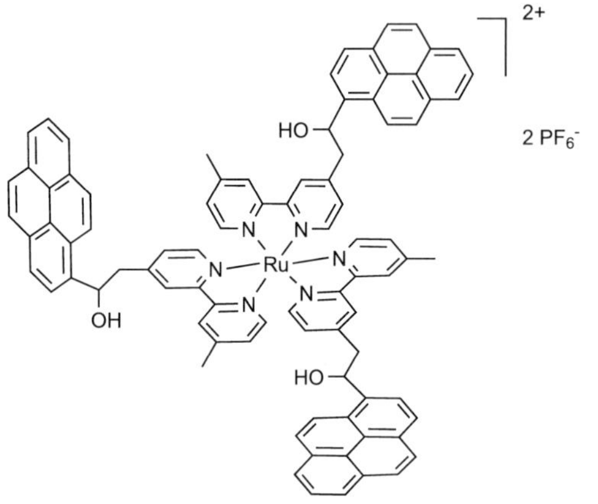

51, τ = 9 μs, φ = 6.4%, CH$_3$CN

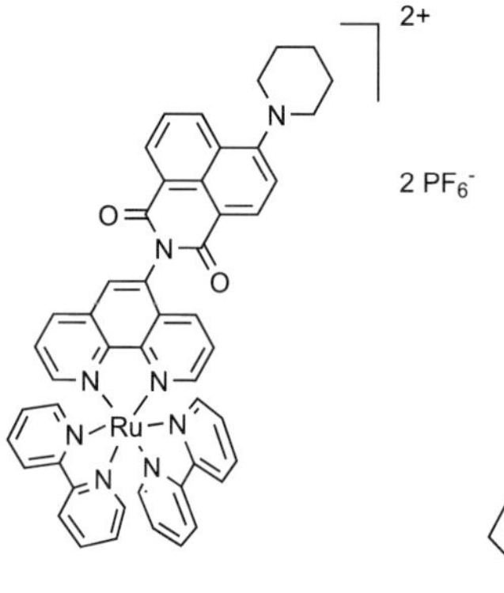

52, τ = 47 μs, φ = 3.3%, DMSO

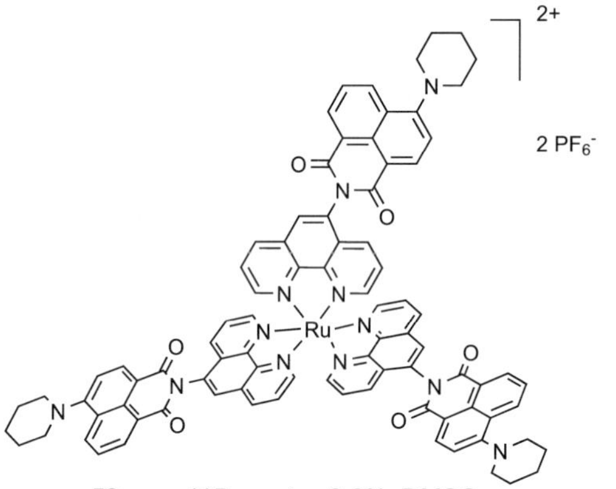

53, τ = 115 μs, φ = 3.2%, DMSO

Figure 2.25. Structures of Ru^{II} poly(pyridine) complexes bearing pyrene residues.

central diethynylated pyrene moiety separates the 2,2′-bipyridine- and 2,2′:6′,2″-terpyridine-based terminals (Fig. 2.27) [194]. The mononuclear complex, having only the 2,2′-bipyridine ligand coordinated with the metal cation, **57**, and the corresponding binuclear complex, **58**, show remarkably similar luminescence properties in deoxygenated acetonitrile solution at room temperature. Two emission bands are evident in the spectrum. These bands appear to be in thermal equilibrium over the temperature range 0–60 °C but only a single emitting species is seen in a frozen glass at 77 K. The phosphorescence lifetimes, being 140 and 165 μs respectively for mono- and binuclear complexes, are significantly longer than those associated with the parent complexes under the same conditions [194]. The analogous compound having two ruthenium(II) tris(2,2′-bipyridine)-based terminals, **59**, shows comparable behavior but the triplet lifetime at room temperature is only 48 μs. Allowing for all of the measured photophysical and electrochemical properties, it is concluded that the triplet manifold has the metal-to-ligand, charge-transfer

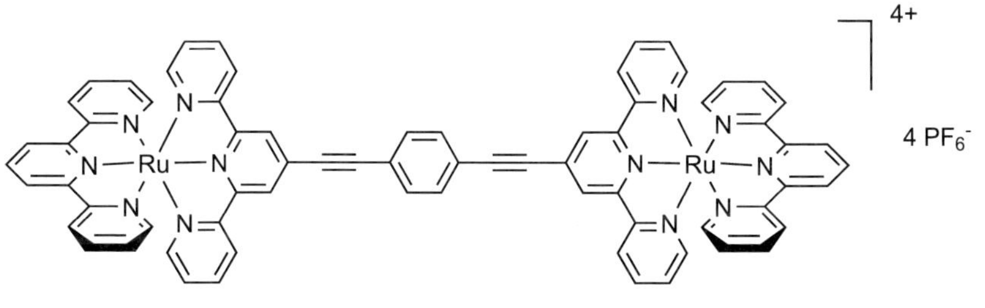

54, $\tau = 110$ ns, $\phi = 0.6\%$, CH_3CN

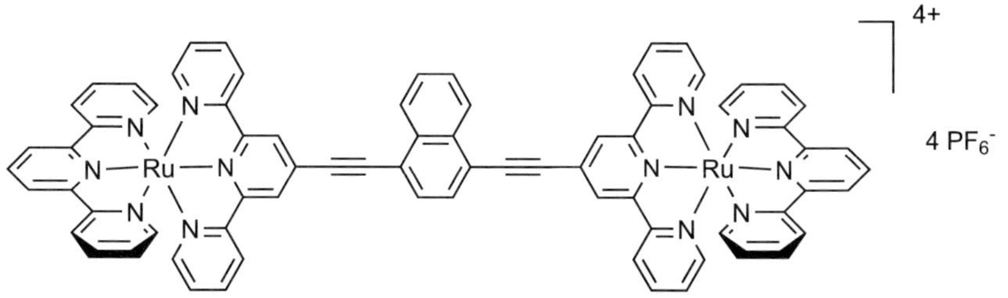

55, $\tau = 415$ ns, $\phi = 0.8\%$, CH_3CN

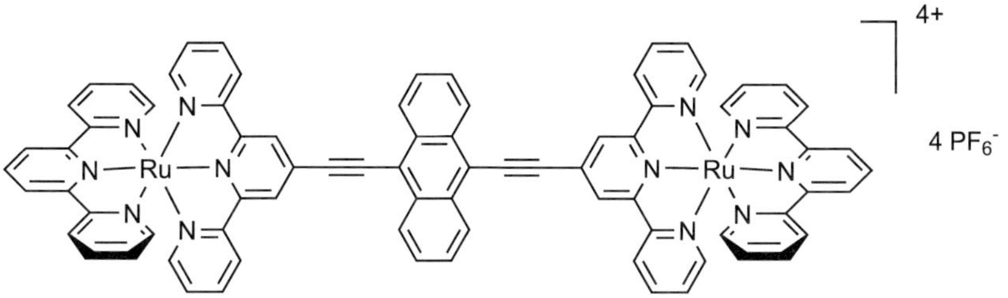

56, $\tau = 0.09$ ns, $\phi = 0.2\%$, CH_3CN

Figure 2.26. Examples of binuclear Ru[II] poly(pyridine) complexes built around ethynylated aromatic ligands.

state localized on the metal complex in equilibrium with an intramolecular charge-transfer (CT) state involving the pyrene and a coordinated poly(pyridine) group [194]. This latter state lies at lower energy in a polar solvent and controls the photophysics. At low temperature, only the MLCT triplet is observed. Other systems have been shown to involve an intramolecular CT triplet [195].

It is generally considered that the triplet lifetimes of ruthenium(II) poly-(pyridine) complexes are set by interaction between the lowest-energy MLCT triplet and a higher-lying metal-centered (MC) triplet state [96, 97]. This situation has been considered in detail for Ru-terpy [196]. A viable strategy for prolonging the triplet lifetime, therefore, involves raising the energy of the MC state, either by substitution [197] or by incorporating the compound in a zeolite framework [198]. The opposite approach of lowering the triplet energy of the metal complex has worked very well for Ru-terpy derivatives [199] and, by attaching ethynylene

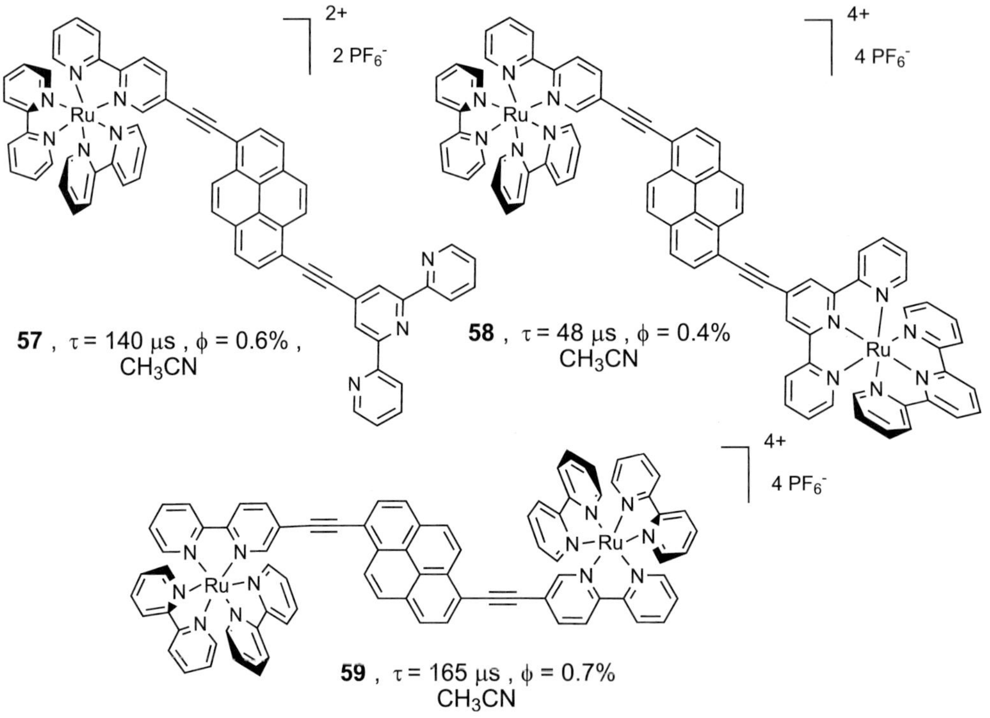

57, τ = 140 μs, ϕ = 0.6%, CH$_3$CN

58, τ = 48 μs, ϕ = 0.4% CH$_3$CN

59, τ = 165 μs, ϕ = 0.7% CH$_3$CN

Figure 2.27. Pyrene-based binuclear Ru^{II} poly(pyridine) complexes.

groups at the 4′-position (Fig. 2.28), it has been possible to isolate binuclear complexes, **60** and **61**, having triplet lifetimes approaching 1 μs at ambient temperature [200]. Replacing the ethynylene bridge with an ethenylene group, **61**, increases the triplet lifetime to 1.2 μs at room temperature but the corresponding ethyl-bridged binuclear complex, **62**, possesses a triplet lifetime of only 380 ps.

Considerable attention has been given to seeking an improved understanding of how unsaturated bridges help to stabilize the triplet state of Ru-terpy derivatives. It was shown that attaching phenylene groups to the 4′-position of the terpyridine ligand did not have a significant effect on the triplet lifetime. Thus, the series of complexes shown as **63** in Fig. 2.29 are seen to display triplet lifetimes not much longer than that of the parent Ru-terpy [199]. This is in marked contrast to those complexes where the acetylenic group is substituted directly onto the terpyridine, complexes **60** [199]. Likewise, the series of binuclear complexes reported by Sauvage et al. [201], compound **64** in Fig. 2.29, show only a modest prolongation of the triplet lifetime. There is a clear benefit from attaching a single phenylene unit but increasing the length of the bridge does not help to further increase the triplet lifetime. This situation should be contrasted to the general effect of the acetylenic group, as displayed by complexes **60** [199]. It is interesting to note that coordinat-

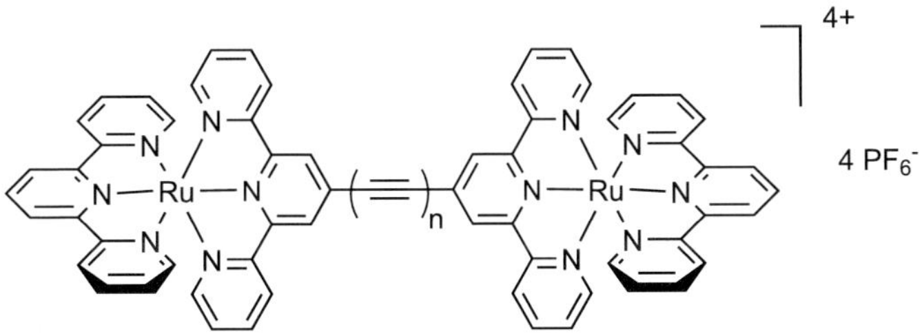

60₁, n = 1 ; λ_{LUM} = 722 nm ; τ_{LUM} = 570 ns
60₂, n = 2 ; λ_{LUM} = 735 nm ; τ_{LUM} = 720 ns

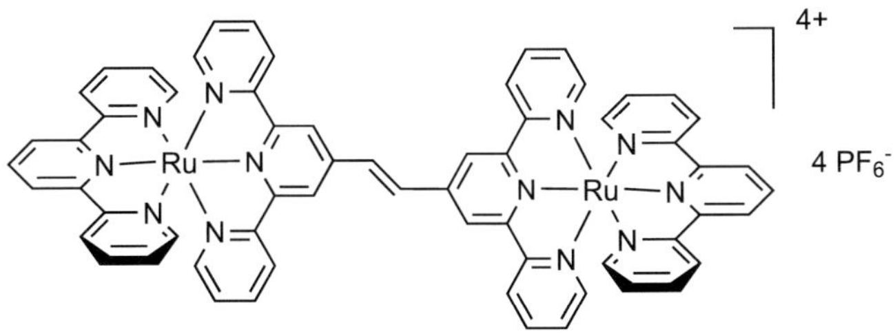

61 , λ_{LUM} = 810 nm , τ_{LUM} = 1220 ns

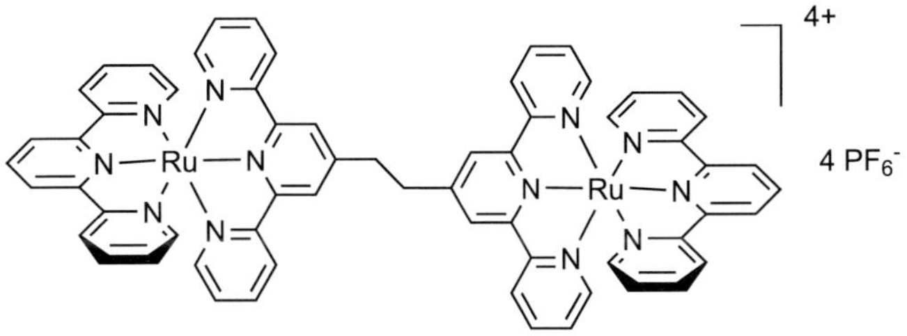

62 , λ_{LUM} = 650 nm , τ_{LUM} = 0.38 ns

Figure 2.28. Binuclear Ru^{II} poly(pyridine) complexes built around bridges of varying degree of hybridization.

ing a zinc(II) cation to the back-to-back terpyridine derivative **65** has a modest effect on the triplet lifetime [202].

It is clear from detailed spectral analyses that attaching an acetylenic group to the poly(pyridine) ligand lowers the triplet energy of the MLCT state. In part, this effect arises because the substituent makes the coordinated ligand easier to reduce [203]. As mentioned above, the decreased triplet energy has the effect of decreasing interaction between the lowest-energy MLCT triplet state and a higher-energy metal-centered (MC) triplet. By measuring the triplet lifetime as a function of temperature it is now apparent that the lowest-energy MLCT state also couples to a

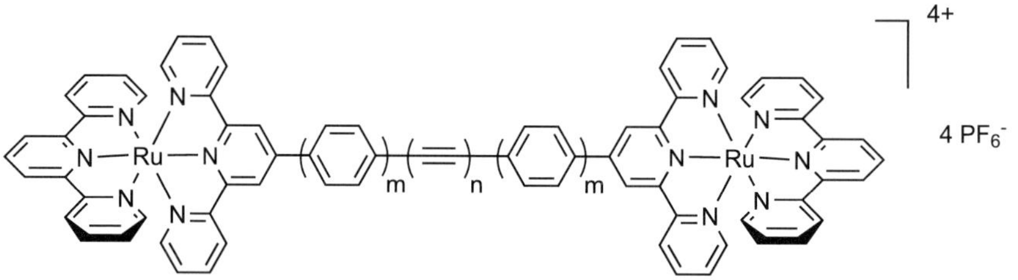

63₁, $n = 1$; $m = 0$, $\lambda_{LUM} = 722$ nm; $\tau_{LUM} = 565$ ns
63₂, $n = 2$; $m = 0$, $\lambda_{LUM} = 735$ nm; $\tau_{LUM} = 720$ ns
63₃, $n = 1$; $m = 1$, $\lambda_{LUM} = 670$ nm; $\tau_{LUM} = 3.2$ ns
63₄, $n = 2$; $m = 1$, $\lambda_{LUM} = 665$ nm; $\tau_{LUM} = 5.5$ ns

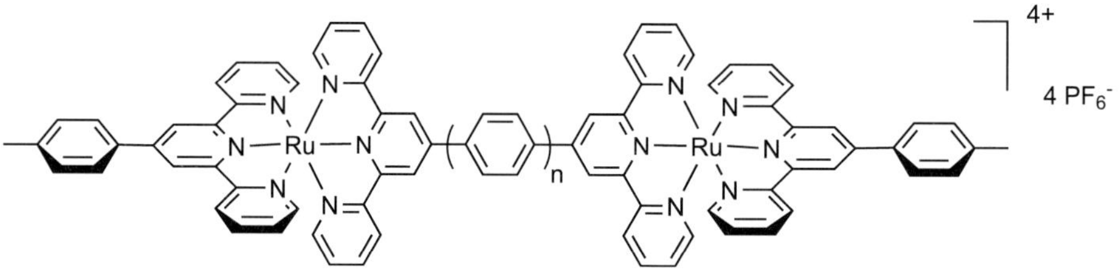

64₀, $n = 0$; no luminescence
64₁, $n = 1$; $\lambda_{LUM} = 720$ nm; $\tau_{LUM} = 565$ ns
64₂, $n = 2$; $\lambda_{LUM} = 735$ nm; $\tau_{LUM} = 720$ ns

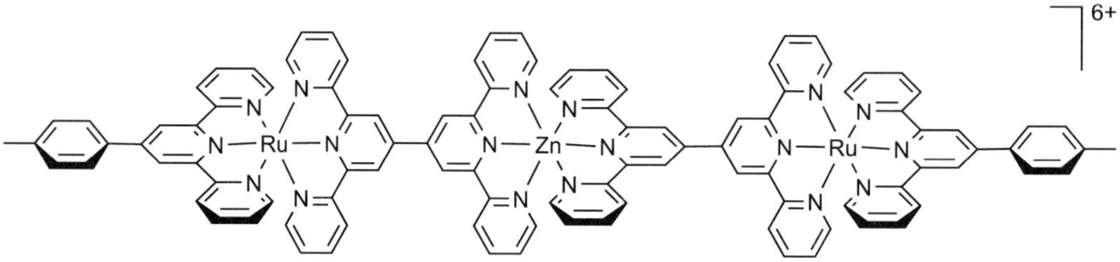

65, $\lambda_{LUM} = 700$ nm; $\tau_{LUM} = 75$ ns

Figure 2.29. Binuclear Ru^{II} poly(pyridine) complexes with elongated bridges.

second MLCT state that possesses increased singlet state character [204, 205]. Thus, the rate constant (k_T) for decay of the lowest-energy MLCT triplet can be expressed in terms of Eq. (6) [206].

$$k_T = \left(\frac{1}{\tau_{LUM}}\right) = \frac{k_0 + k_1 \exp\left(-\frac{E_A}{k_B T}\right) + k_2 \exp\left(-\frac{E_B}{k_B T}\right)}{1 + \exp\left(-\frac{E_A}{k_B T}\right) + \exp\left(-\frac{E_B}{k_B T}\right)} \tag{6}$$

Here, k_0 is the activationless rate constant for decay of the triplet state at low temperature. The term E_A represents the energy gap between the lowest-energy MLCT state and a second MLCT triplet situated at slightly higher energy. This second MLCT triplet decays with a rate constant of k_B. Typically, E_A has a value of a few hundred wavenumbers. The MC triplet is reached by passing over a substantial barrier, E_B. This latter barrier has a typical value of ca. 2000 cm^{-1}. The rate constant (k_B) for accessing the MC state tends to be very high, around 10^{11} s^{-1}, and is primarily responsible for the strong temperature effect found for many Ru^{II}-based MLCT triplet states. Because the corresponding Os^{II}-based MLCT triplets lie at much lower energy, they are less sensitive to changes in temperature. It has also been shown that the triplet lifetimes are affected by the nature of substituents attached to the parent ligands [207] whilst certain binuclear complexes form the triplet MLCT state with quantum yields much less than unity [208].

It is now recognized that unsaturated substituents, notably acetylenic groups, can provide a simple route for extended electron delocalization at the triplet level. This situation has been studied extensively [112] for the compounds shown in Fig. 2.27, while related work has examined electron delocalization in poly(phenylene)-bridged binuclear Ru^{II}-based complexes (Fig. 2.29) [201]. The general effect of electron delocalization is to prolong the triplet lifetime by decreasing the size of the reorganisation energy. This effect has been studied in detail for **66** (Fig. 2.30) [120], where a close comparison has been made with the photophysical properties of the parent Ru-terpy complex [196]. The single ethynylene group causes a 160-fold increase in triplet lifetime. There is a small increase in the radiative rate constant and a net decrease in the total reorganization energy accompanying deactivation of the triplet MLCT state. There is a more serious decrease in the magnitude of the electronic coupling matrix element that characterizes interaction between ground and excited states. It is this latter term that has most effect on the triplet lifetime [120].

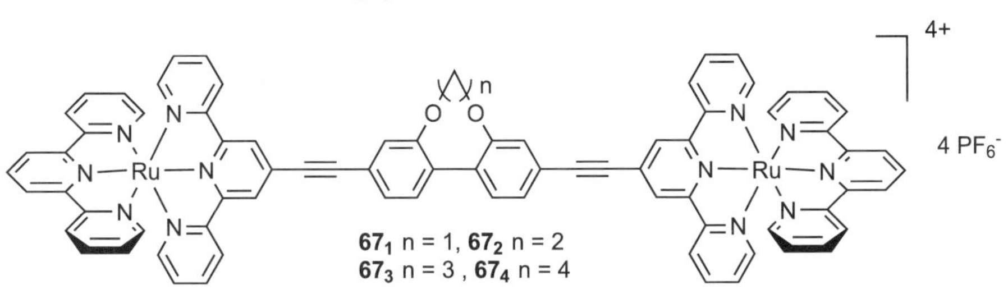

Figure 2.30. Binuclear Ru^{II} bis(2,2':6',2''-terpyridine) complexes built around a polytopic ligand with variable geometry.

Interestingly, the triplet lifetime is very sensitive to the geometry of the bridge in the series of compounds **67** illustrated in Fig. 2.30 [209]. Here, a biphenylene spacer is used as the polytopic ligand for a binuclear Ru-terpy complex. The spacer is equipped with a constraining tether such that the dihedral angle around the central biphenyl spacer can be altered systematically simply by changing the length of the tether. It was found that the activationless rate constant k_0 is strongly affected by the length of the strap.

Finally, it is noted that the triplet lifetime can be fine tuned by incorporating appropriate complexing units in the bridge, Fig. 2.31 [66]. This effect is well exemplified by complex **68** which has an additional 2,2′-bipyridine group inserted into the acetylenic connector. The energy of the bridge can now be tuned by coordination of different cations to the vacant ligand. Lowering of the bridge energy affords better blending of orbitals on the terpyridine ligands and on the bridge and this has the effect of increasing the extent of electron delocalization at the triplet level. Alkylation of the central bipyridine unit, forming **69**, switches on light-induced electron transfer since the resultant dication is a good electron acceptor [66]. The central bipyridine can be replaced with 1,10-phenanthroline units, complexes **70** and **71**, or bipyrimidine, as in complex **72** [210]. In each case, coordination of adventitious cations serves to prolong the triplet lifetime by increasing the degree of electron delocalization.

2.3.5
Long-range Triplet Energy Transfer

Considerable research effort has been expended on the search for molecular systems able to transfer excitation energy over long distances. Ideally, transfer has to occur with high efficacy over distances in excess of 100 Å, and maybe as far as 200 Å. This goal is easily accomplished with Förster-type energy transfer in certain cases, but is far outside the range of transfer distances attainable with electron exchange. Even Förster transfer is relatively inefficient with transition metal complexes because of their restricted spectral overlap integrals. For example, Fig. 2.32 shows two molecular dyads in which the terminal Ru^{II} and Os^{II} tris(1,10-phenantholine)-based terminals are connected via sterically constrained spiro linkages [211]. The photophysical properties of the corresponding mononuclear complexes show the expected temperature dependence arising from strong interactions with higher-lying excited states. Intramolecular triplet energy transfer occurs in both **73** and **74** at 77 K and at room temperature. In both complexes, energy transfer takes place exclusively via the Förster mechanism and it is clear that the spiro connector does not promote long-range super-exchange interactions. In **73**, the rate of energy transfer is some 15-fold higher than for **74** whilst the metal–metal separations are 9.6 Å and 13.5 Å respectively. In this latter case, the rate of Förster-type energy transfer is 5.5×10^7 s^{-1} at room temperature and 1.7×10^7 s^{-1} at 77 K [211]. Although the relative insensitivity to changes in temperature is a useful advantage of the Förster mechanism, it is obvious that this approach is not going to lead to the rational design of effective photo-active wires incorporating

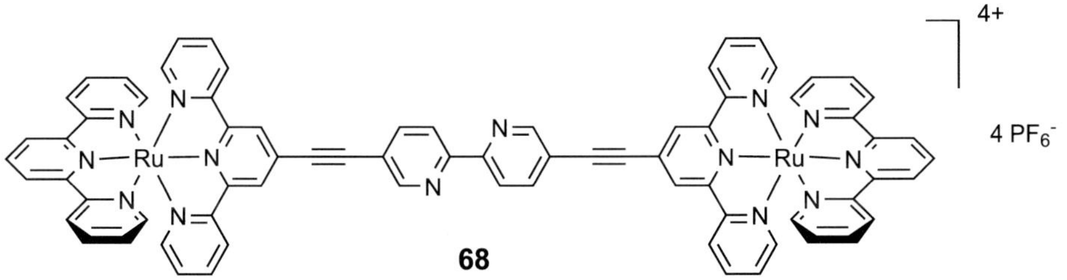

68

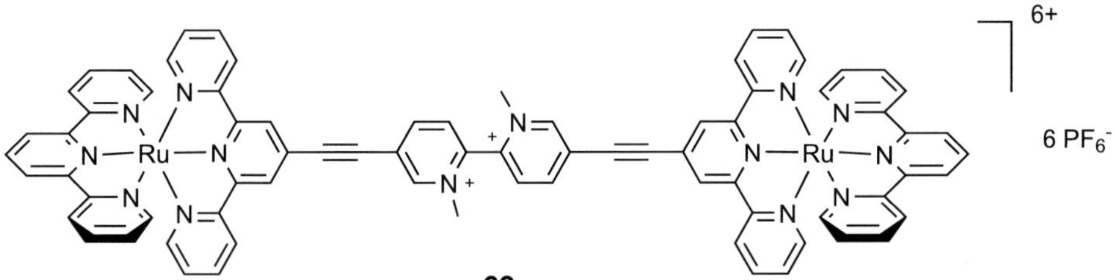

69

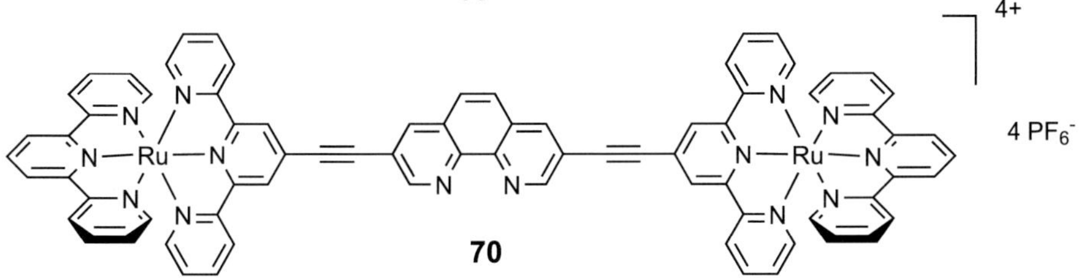

70

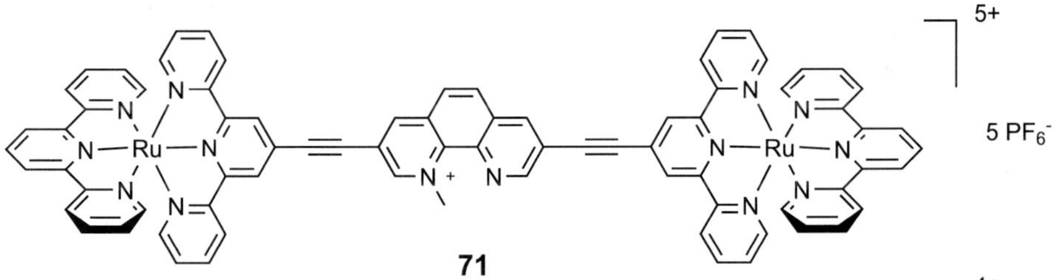

71

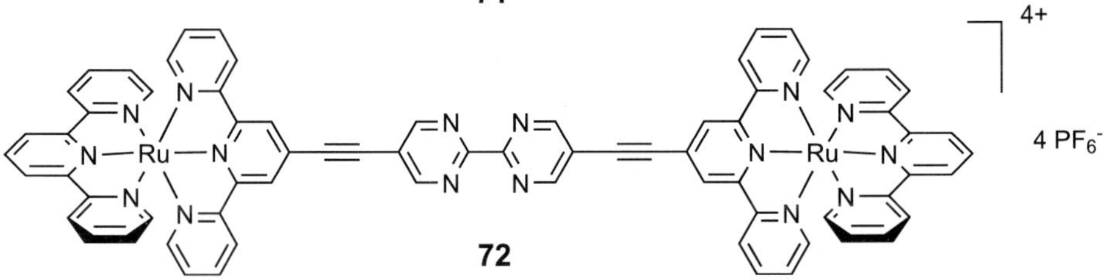

72

Figure 2.31. Binuclear Ru^{II} poly(pyridine) complexes equipped with central complexing units.

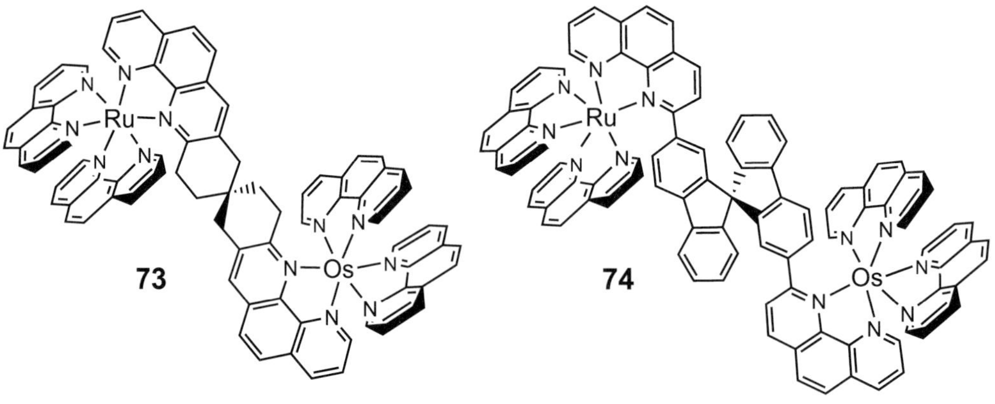

Figure 2.32. Structures of spiro-bridged binuclear complexes displaying efficient through-space energy transfer.

metallo-chromophores [212, 213]. Attention, therefore, has passed to the Dexter mechanism.

Shown in Fig. 2.33 are some of the prototypic Ru^{II}/Os^{II} bis(2,2':6',2''-terpyridine) complexes used to establish the viability of long-range electron exchange. Complexes 75_1 and 75_2 possess a carbon-rich acetylenic bridge and undergo extremely fast intramolecular triplet energy transfer at both room temperature and 77 K [199]. Adding an incremental acetylenic bond attenuates the rate of energy transfer by a factor of ca. 1.4 and, in both cases, the rate is not affected much by changes in temperature. For these binuclear complexes, energy transfer is solely by way of Dexter-type electron exchange and the approximate attenuation factor (β) is ca. 0.17 Å^{-1}. This is a relatively small attenuation factor that ensures electron exchange is quantitative in these systems. Note that the rate constant ($k_{ET} = 7.1 \times 10^{10}$ s^{-1}) is very high for 75_1. In principle, it should be possible to build long molecules capable of efficient energy transfer but there are two important considerations that tend to suggest this will not be so easy. Firstly, the carbon-rich bridges are poorly soluble and do not favor building long wires. Secondly, the triplet energy of the bridge will decrease with increasing length [68]. This is not a serious problem until quite long bridges are developed but the short bond length is such that many acetylenic groups are needed to accomplish separation distances of >100 Å. At this point, the triplet energy of the bridge becomes lower than that of the acceptor and energy transfer will no longer take place.

A possible solution to this problem is to insert insulating groups into the bridge in an effort to raise the triplet energy or to offset the effects of increased conjugation [68]. Thus, complexes **76** have a phenylene ring incorporated into the acetylenic wire. It has already been established that the corresponding Ru^{II} binuclear complexes do not possess unusually long-lived triplet lifetimes [199]. This suggests that the phenylene ring hinders electron delocalization at the triplet level. It was subsequently found that triplet energy transfer in compounds **76** is much slower

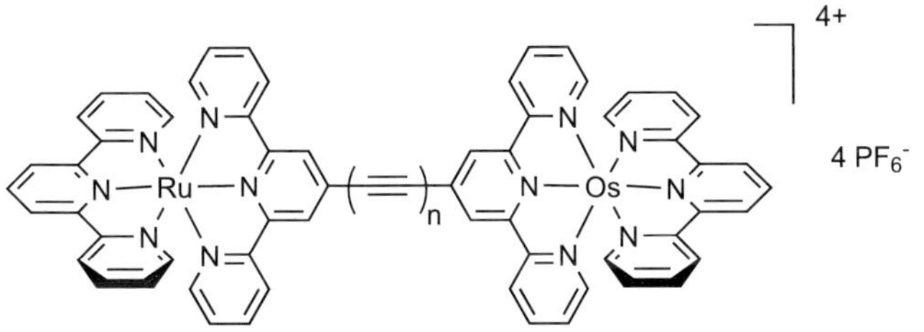

75₁, $n = 1$, $R_{MM} = 13.8$ Å, $k_{ET} = 7.1\ 10^{10}$ s^{-1}
75₂, $n = 2$, $R_{MM} = 16.2$ Å, $k_{ET} = 5.0\ 10^{10}$ s^{-1}

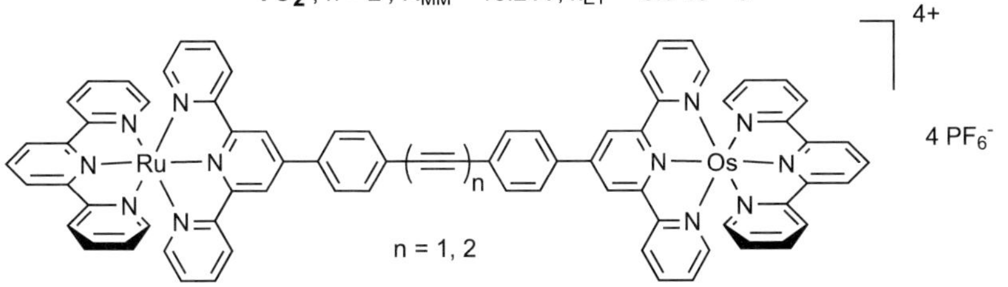

$n = 1, 2$

76₁, $n = 1$, $R_{MM} = 22.0$ Å, $k_{ET} = 8.2\ 10^{8}$ s^{-1}
76₂, $n = 2$, $R_{MM} = 24.7$ Å, $k_{ET} = 6.8\ 10^{8}$ s^{-1}

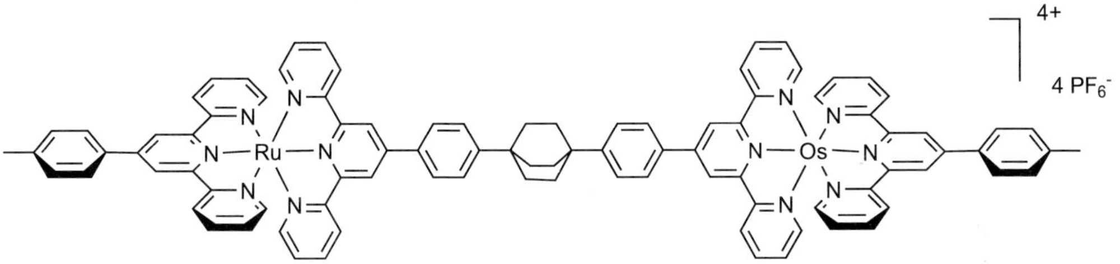

77, $R_{MM} = 23.5$ Å; $k_{ET} = 4.4\ 10^{6}$ s^{-1} (77K)

Figure 2.33. Examples of mixed-metal Ru^{II}/Os^{II} poly(pyridine) complexes built for long-range triplet energy transfer.

than in **75**. A similar situation is found with mixed-metal Ru^{II}/Os^{II} binuclear complexes. Thus, intramolecular triplet energy transfer is rather slow ($k_{ET} = 4.4 \times 10^6$ s^{-1} at 77 K) in the σ-bridged complex **77**. Here, the metal–metal separation distance is 23.5 Å but there is a reasonable thermodynamic driving force for triplet energy transfer. In marked contrast, the rate constant for intramolecular triplet energy transfer across 0, 1 or 2 phenylene rings exceeds 10^{10} s^{-1} at 293 K [214]. For these latter compounds, being the Ru^{II}/Os^{II} mixed-metal versions of **64** (Fig. 2.29), it is clear that energy transfer must occur by way of Dexter-type electron exchange. In the extreme case, the metal–metal separation distance is 19.6 Å. This work confirms the insulating nature of σ-bonded units.

It is noteworthy that triplet energy transfer in acetylene-bridged porphyrin dimers has been interpreted in terms of the conformational flexibility of the system [215]. Thus, rapid relaxation between triplet levels occurs in a series of meso-

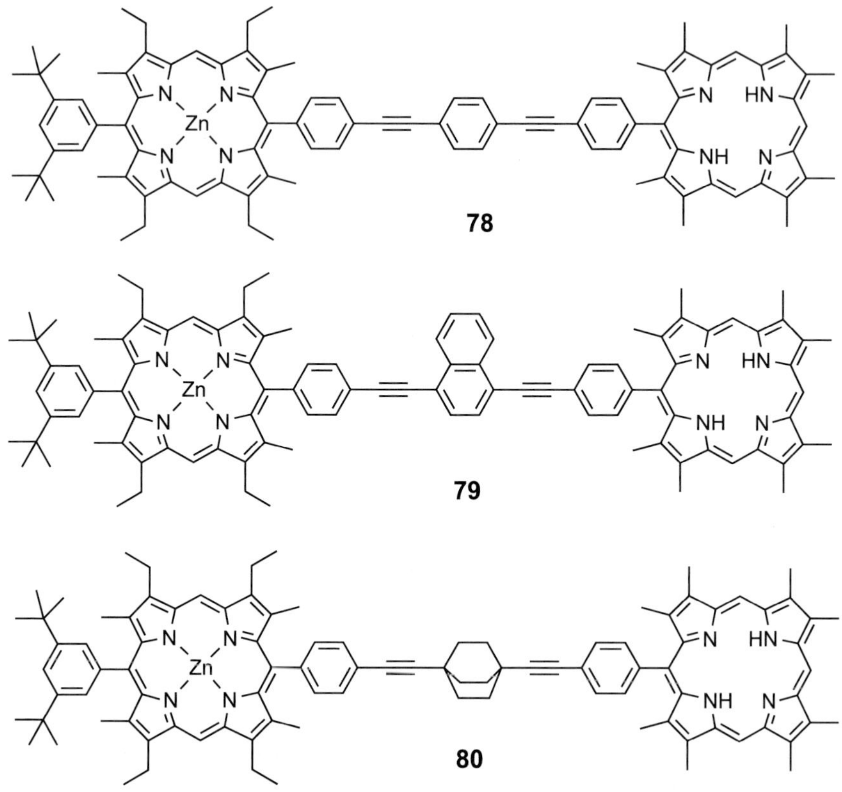

Figure 2.34. Porphyrin dimers built around acetylenic bridges.

diaryloctaalkylporphyrins over a wide temperature range. The two triplet states differ markedly in their photophysical properties and decay to the ground state is unusually fast. Even so, rapid intramolecular triplet energy transfer occurs within hybrid Zn^{II}/free-base porphyrin dimers linked by acetylenic groups. Rate constants for triplet energy transfer of ca. 10^6 s^{-1} are observed for **78** and **79** at 150 K across a separation distance of around 19 Å. Inserting a rigid σ-bonded structure into the bridge, forming complex **80** (Fig. 2.34), stops energy transfer, even though the geometry and energetics remain the same as for aryl-bridged dimers. In closely-related porphyrin dimers [216] it is shown that the structure of the porphyrin and the site of attachment both affect the rate of electronic energy transfer.

Intramolecular triplet energy transfer also occurs in the corresponding Ru^{II}/Os^{II} tris(2,2'-bipyridine)-based binuclear complexes (Fig. 2.8) [85]. Here, triplet energy transfer occurs over bridges comprising 3, 5 or 7 phenylene rings and appears to conform to the Dexter-type electron-exchange mechanism. The rate of energy transfer decreases exponentially with increasing separation distance and gives an attenuation factor of 0.32 Å^{-1}. This β value is inferior to that deduced for poly(acetylenes) but the use of poly(phenylene) bridges is more practical and allows

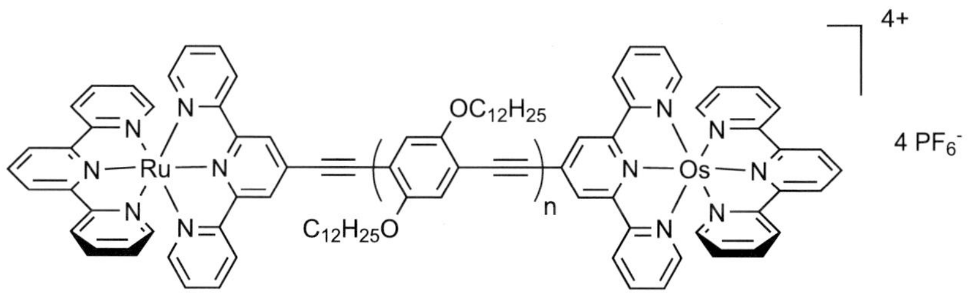

81₁, n = 1 ; R_{MM} = 19.8 Å ; k_{ET} = 11.6 10^7 s^{-1}
81₂, n = 2 ; R_{MM} = 27.9 Å ; k_{ET} = 3.2 10^7 s^{-1}
81₃, n = 3 ; R_{MM} = 36.0 Å ; k_{ET} = 1.3 10^7 s^{-1}
81₄, n = 4 ; R_{MM} = 44.1 Å ; k_{ET} = 0.77 10^7 s^{-1}
81₅, n = 5 ; R_{MM} = 52.3 Å ; k_{ET} = 0.22 10^7 s^{-1}

Figure 2.35. Mixed-metal Ru^{II}/Os^{II} poly(pyridine) complexes built around diethynylated benzene modules that display long-distance energy transfer.

for larger separations. Indeed, with seven phenylene rings the metal–metal separation distance is 42 Å and the rate constant is 1.0×10^7 s^{-1} at 293 K [85]. At the time, this was the longest distance over which electron exchange had been found to operate. It is notable that in these systems the rate of energy transfer seems to be slower for bipyridine-based bridges than for the analogous terpyridine-based systems.

A systematic study has shown that ethynylated phenylene units provide for an unusually shallow attenuation in the rate of through-bond electron exchange [217]. Thus, the Ru^{II}/Os^{II} binuclear complexes **81** depicted in Fig. 2.35 were found to undergo intramolecular triplet energy transfer at room temperature. The rate constant k_{ET} decreases exponentially with increasing separation distance according to the super-exchange mechanism (Fig. 2.36). Taken over the whole series, the decrease in rate represents an attenuation factor of only $\beta = 0.11$ Å^{-1}. This is the smallest β value yet determined for electron exchange, although it is still higher than recent reports for electron transfer. The maximum separation distance over which triplet energy transfer could be observed was 52 Å. In principle, it should be possible to build molecular dyads of increased length, but triplet energy transfer would then not compete with nonradiative decay of the excited state at room temperature. An additional issue with this system is that temperature-dependent studies are not available. This leaves open the question of mechanism and the possibility of a switch to a hopping mechanism at longer bridge lengths cannot be ruled out. Even so, this work represents a major step forward in the design of molecular-scale wires for electron exchange [217].

It has long been suggested that the extent of electronic coupling along poly(phenylene)-based wires should depend on the mutual orientation of adjacent rings. This problem has been difficult to study by experimental methods but recent

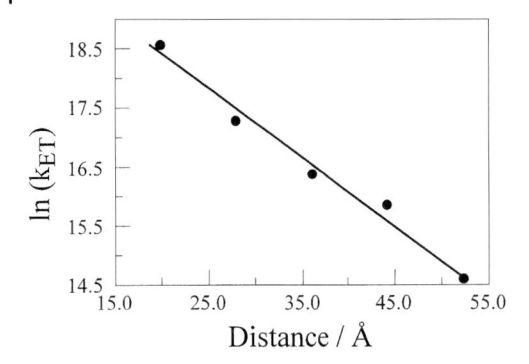

Figure 2.36. Effect of increased separation distance on the rate constant for intramolecular triplet energy transfer in **81**. The line drawn through the data points corresponds to an attenuation factor of $\beta = 0.11 \, \text{Å}^{-1}$.

work has shown that the electronic coupling matrix element for through-bond charge transfer depends crucially on the torsion angle of the central linker [218]. Electronic coupling is maximal for coplanar arrangements and at a minimum for orthogonal structures. The investigation of this effect relied on the synthesis of a set of biphenyl bridges having constrained geometry, as shown in Fig. 2.30 for compounds **67**.

Several studies have questioned whether poly(thiophene) might make a good conduit for long-range electron exchange. Thus, it is known that the polytopic bridge present in complex **82** (Fig. 2.37), favors extended electron delocalization at the triplet level [219]. The same bridge can be used to assemble trinuclear species of the type represented by complex **83**. Although it is realized that intramolecular triplet energy transfer does take place through a single thiophene ring, as in complex **84** [220, 221], it has been shown that longer ethynylated thiophene connectors introduce low-energy bridge-localized triplet states into the system [222]. These latter triplets lie close in energy to the MLCT triplet localized on the Ru-terpy donor and, therefore, might function as efficient energy relays for long-range transfer to the Os-terpy acceptor.

2.4
Molecular-scale Switches

The use of a single chemical bond as a switch would greatly reduce the power dissipation in electronic circuits, and a recent study [223] has shown that a molecular switch required only 47×10^{-21} J; some 10 000 times less than transistor switches used in current high-speed computers. The molecular switch in question consists of rotating one of the four phenyl legs attached to a porphyrin from one stable position to another by the use of an atomic force microscope. In other work it was

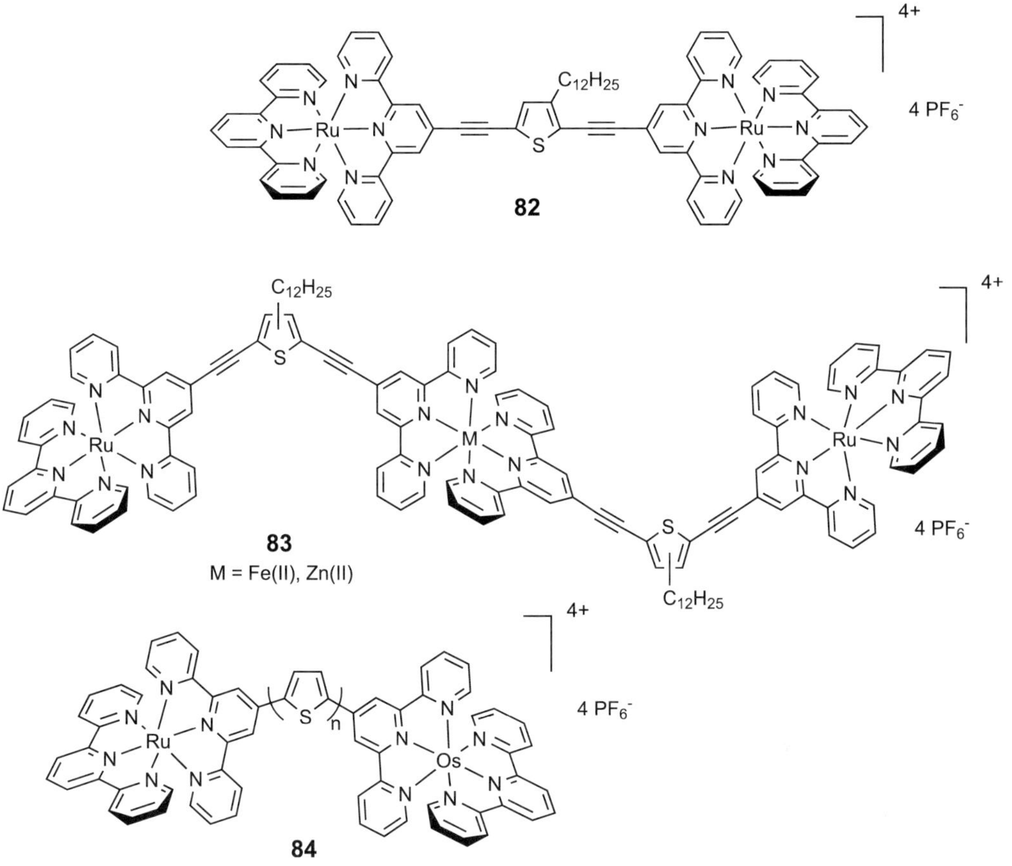

Figure 2.37. Examples of thiophene-bridged binuclear complexes.

found that the electrical resistance between a carbon nano-tube and a graphite substrate could be varied by a factor of 50 simply by rotating the nano-tube [224]. This simple strategy provides means by which to convert mechanical signals into electrical signals at the molecular level. A means to stabilize molecular switches based on chemical interactions with surrounding molecules using oligo(p-phenylene-ethynylene) molecules has been reported [225]. This effect has been achieved by embedding the conductive wire into a self-assembled monolayer of amide-containing alkanethiol molecules attached to a gold surface. The molecular-scale wires protrude beyond the monolayer and can be detected with a scanning tunneling microscope. Other molecular switches have been developed around catenanes [226] or the so-called chiropticene molecules [227].

Chiroptical molecular switches have been developed whereby discrimination between pseudo-enantiomers can be achieved simply by varying the frequency of the illumination source (Fig. 2.38) [228]. In these systems, the photo-active component is a sterically-crowded alkene which is forced into a helical shape to avoid undue

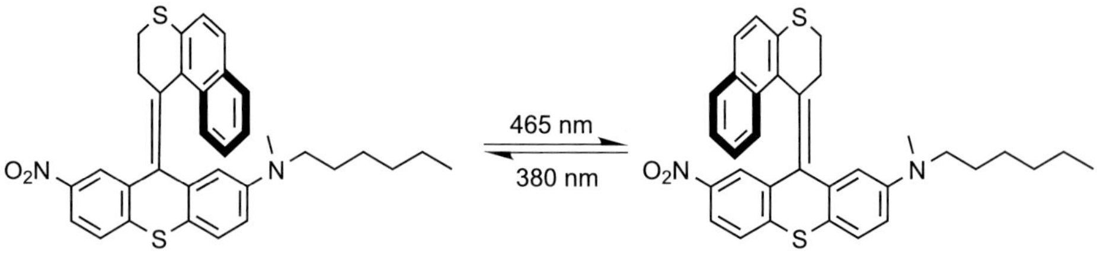

(M)-trans-85 ⇌ (465 nm / 380 nm) **(P)-cis-85**

Figure 2.38. A chiroptical photoswitch.

steric hindrance. Upon reversible photo-induced cis-trans isomerization, the helical shape of the molecule is lost. Because of the asymmetric substitution pattern there are small differences in the absorption profiles of the two forms that can be used as the basis for selective photochemistry. Under illumination at 380 nm a photostationary state is set up that comprises 70% (M)-trans-**85** and 30% (P)-cis-**85**. Illumination at 465 nm (where the cis-isomer barely absorbs) leads to a photostationary state consisting of 98% (P)-cis-**85** and 2% (M)-trans-**85** [229].

Over the past decade, considerable effort has been given to the study of photochromic materials and many such products are now commercially available. One particular class of photochromic materials, the so-called dithienylethenes [230], has shown particular promise for use in molecular photonic devices. Recent studies [231] have demonstrated that certain dithienylethenes can be incorporated into transition metal complexes in such a way that low-energy illumination can be used to change the conformation. Such realizations open the way to develop novel molecular-scale switches (Fig. 2.39). Thus, the known photochromic reagent dithienylperfluorocyclopentene has been functionalized with terminal Ru-bipy units linked through a phenylene ring, complex **86**. Illumination into the MLCT transition localized on the metal complex causes efficient ring closure, forming complex **87**. The reaction is believed to proceed by way of intramolecular triplet energy transfer from the Ru-bipy analogue to the open form of the bridge [231]. This process takes place on the nanosecond time scale and differs markedly from the chemistry that occurs upon direct illumination into the bridge. The photocyclization quantum yield is well above 0.5 and might reflect the consequences of severe steric repulsion between the bulky side-groups forcing the bridge into the reactive antiparallel conformation.

Photocyclization also takes place upon illumination into the MLCT band of the rhenium(I) complex **88** [232]. Here, the quantum yield for photocyclization reaches about 0.62. The closed form, complex **89**, can be converted back to the original open form by illumination at about 580 nm. The closed form also undergoes slow thermal opening but this occurs on a time scale far removed from that related to the photochemical processes. It seems that the ligand-localized triplet state lies at lower energy than the MLCT triplet in the closed form but the opposite is true in the open form.

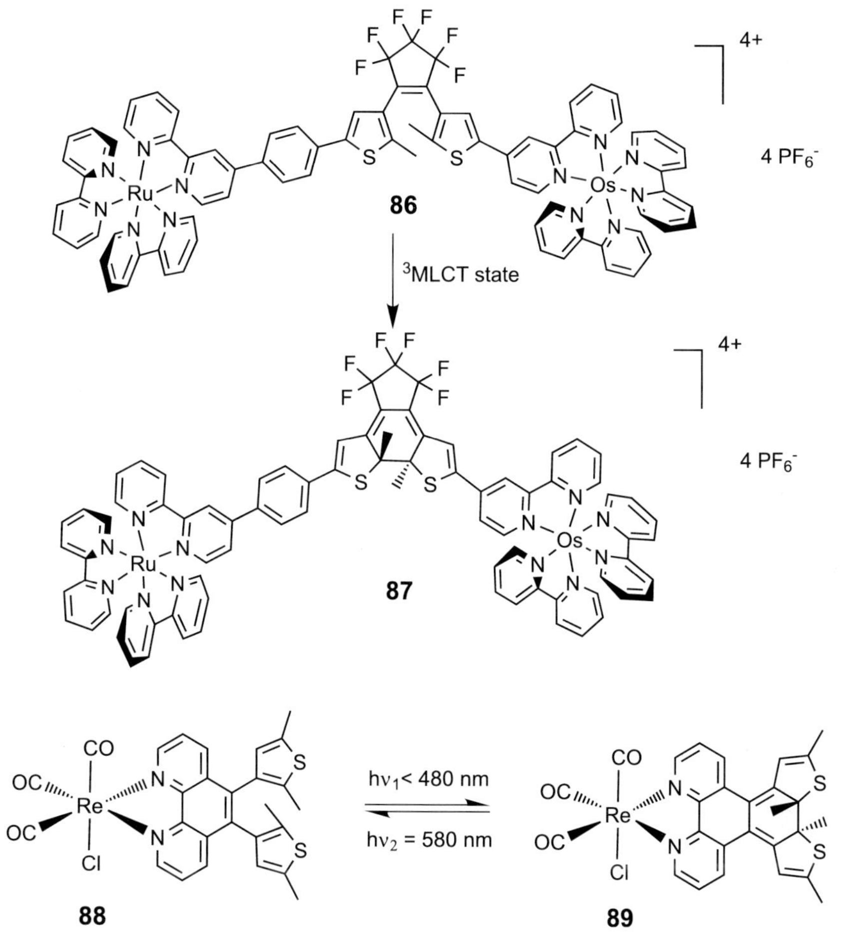

Figure 2.39. A dithienylethene-based optical photoswitch.

A different type of switch is displayed in Fig. 2.40. Here, a Ru-bipy complex is appended to a calixarene in which two of the phenoxyl groups have been oxidized to quinones, complex **90** [233, 234]. The calixarene also possesses a pendant bipyridine ligand. Upon illumination into the MLCT band localized on the metal complex, light-induced electron transfer takes place from the triplet state of Ru-bipy to one of the quinones. This process involves diffusive encounter between the reactants and it should be noted that ^{1}H NMR and molecular dynamics simulations indicate that the calixarene walls are highly mobile. Cations, such as Ba^{2+}, bind to the lower rim of the calixarene and are held in place by the additional bipyridine ligand. This has the effect of forcing the pendant Ru-bipy away from the calixarene in order to minimize electrostatic repulsion. The net effect is to curtail light-induced electron transfer. Thus, whereas **90** is nonluminescent the various cation-

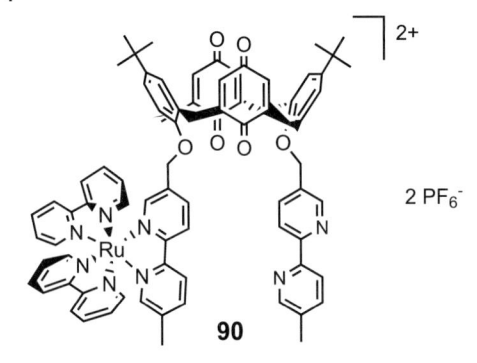

Figure 2.40. Structure of the calixquinone-based switch that responds to the presence of cations.

bound complexes emit in deoxygenated solution. The on–off level is set by the electronic charge of the bound cation [234].

A somewhat related system [235], shown in Fig. 2.41, involves a Ru^{II}/Os^{II} binuclear complex having the termini linked by a highly flexible poly(ethylene oxide)

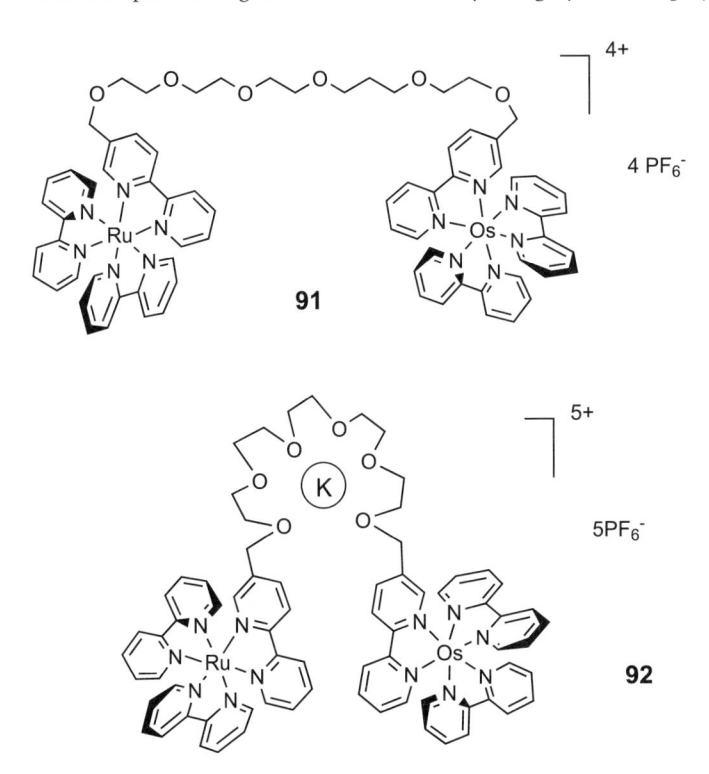

Figure 2.41. A poly(ethylene glycol)-based switch that changes conformation in the presence of cations.

connector, complex **91**. The termini are held too far apart for significant triplet energy transfer, although they can diffuse together at ambient temperature. However, addition of a cation causes the connector to fold into a closed form that brings the termini into closer proximity, as in complex **92**. This has the effect of switching-on triplet energy transfer.

2.5
Perspectives

The development of photoactive metallo-wires of nanometric dimension is still in its infancy. It has been shown that intramolecular triplet energy transfer by the electron-exchange mechanism offers a suitable analytical method by which to measure the efficacy of the wire at conducting electrons. A particularly appropriate system has Ru^{II}/Os^{II} poly(pyridine) complexes at either end of the wire. The best system produced to date displays triplet energy transfer over a distance of 52 Å [217]. This value, although impressive, is insufficient for the construction of viable molecular photonic wires. Considerably more success has been achieved with respect to the prolongation of the triplet lifetime of the Ru^{II}-based donor and it is now possible to prepare luminescent complexes with triplet lifetimes of the order of 150 µs in deoxygenated solution at room temperature. There still remains the problem, however, of engineering triplet energy transfer over distances in excess of 100 Å.

An alternative approach to the construction of photonic metallo-wires is to incorporate numerous identical chromophores into a one-dimensional array. Several such systems have been built in recent years, using conjugated aromatic residues [236] or fused porphyrins [237, 238]. Polyacetylenic wires have been synthesized that contain metal atoms in the conjugation pathway [239]. It has also proved possible to dope carbon nano-wires with transition metal ions [240]. Related systems are starting to appear that have many Ru^{II} poly(pyridine) complexes attached to a linear polytopic ligand. Such materials might operate as artificial light harvesters able to transport photons over long distances by way of a random walk. If so, it should be possible to cap the end of the wire with an Os^{II}-based acceptor, thereby creating an efficient photonic wire in which each chromophore lies in the same chemical and electronic environment (Fig. 2.42). Such ligands have recently been engineered [241].

An interesting strategy for constructing long, linear arrays has been developed [242]. This approach involves the synthesis of ethynylated ditopic ligands equipped with a metal complex at one end. Addition of a suitable metallo-synthon results in the self-assembly of the corresponding trinuclear complexes **93** and **94** (Fig. 2.43). Such materials display an interesting variety of intramolecular electron and/or energy transfer processes whilst the Zn^{II} complex exhibits extended electron delocalization at the triplet level. These systems could be adapted to produce arrays of higher nuclearity by careful selection of the bridging cation and reaction conditions. Indeed, it should be possible to use a bridging cation that introduces a metal bis(2,2':6',2"-terpyridine) complex with triplet energy between those of the primary

Figure 2.42. Molecular structures of new multitopic bipyridine-bridged diethynylthiophene ligands and schematic representation of energy hopping and end-trapping.

93_{1C} n = 1 ; M = Co^{2+}; 93_{1F} n = 1 ; M = Fe^{2+}; 93_{1Z} n = 1 ; M = Zn^{2+}
94_{2C} n = 2 ; M = Co^{2+}; 94_{1F} n = 2 ; M = Fe^{2+}; 94_{1Z} n = 2 ; M = Zn^{2+}

Figure 2.43. Molecular structures of trinuclear complexes built around an octahedral cobalt, iron or zinc cation.

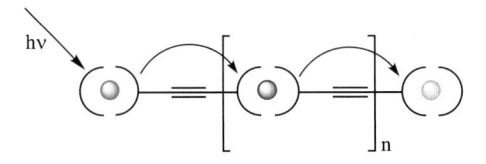

Figure 2.44. Design of self-assembled multinuclear complexes capable of long-distance vectorial energy transfer. The donor and acceptor units are Ru^{II} and Os^{II} bis(2,2′:6′,2″-terpyridine) complexes, respectively, while the central unit is a second Ru^{II} bis(2,2′:6′,2″-terpyridine) complex that has a triplet energy between those of the donor and acceptor.

donor and the ultimate acceptor. This would provide a photonic wire of the type shown in Fig. 2.44, where reversible energy migration allows for unusually long-distance excitation energy transfer. The main advantage of this type of approach is that it avoids the synthesis of multiple ligands of limited solubility.

2.6
Experimental: Selected Procedures

Representative synthetic protocols for selected compounds are presented in this section. These may be classified into two main categories: The first classical synthetic methodology starts from a preformed ligand bearing all vacant coordination sites, which are then saturated with appropriate metal centers. The second strategy starts from a preformed metal complex prepared from a monotopic ligand bearing reactive functions. These functionalities are further transformed using the *synthesis at the complex* approach to form the desired multitopic metallo-ligands. Finally, additional metal centers are added by selective complexation to vacant sites on the metallo-synthons. Preparation of the molecular-scale Ru-Os metallo-wires 81_1 to 81_5 falls in the first category and requires the prior synthesis of the terpyridine-based ditopic ligands [243]. With these ligands in hand, one of the terpy sites is complexed with a Ru center, usually end-capped with an unsubstituted terpy ligand, whilst, in a subsequent step, the vacant site is filled with an Os center, also end-capped with an unsubstituted terpy.

2.6.1
General Procedure for the Preparation of the Mononuclear [RuL$_n$]

In a Schlenk flask, a stirred solution of [Ru(terpy)(DMSO)Cl$_2$] (1 equiv.) and AgBF$_4$ (2.2 equiv.) in argon-degassed methanol solution was heated at 80 °C for 8 h. After cooling to room temperature, the deep-red solution was filtered over cotton-wool and transferred via cannula to an argon-degassed dichloromethane solution of the ditopic ligand (1 equiv.). During heating at 80 °C, the deep-red solution turned red-orange, showing slow ruthenium complexation. After complete consumption of starting material (determined by TLC), an aqueous solution (5 equiv.) of KPF$_6$ was added, the organic solvent was then removed under vacuum, and the precipitates were washed by centrifugation with water. The solid matter was purified by chromatography on alumina, eluting with dichloromethane using a gradient of methanol. The pure red-orange complexes were obtained by recrystallization from dichloromethane/hexane; **[RuL$_1$]** (Chart 2.1) 81% and **[RuL$_5$]** (Chart 2.2) 66%.

2.6.2
General Procedure for the Preparation of the Hetero-Dinuclear Complexes 81_1 and 81_5

To a stirred suspension of mononuclear **[RuL$_n$]** (1 equiv.) in tetrahydrofuran/water (1/1, v/v) was added dropwise a solution of [Os(terpy)(O)$_2$(OH)](NO$_3$)·2H$_2$O (1.2 equiv.) in water (3 mL). After heating at 45 °C for 15 min., three drops of hydrazine hydrate were added. Complexation was evidenced by slow development of a deep-

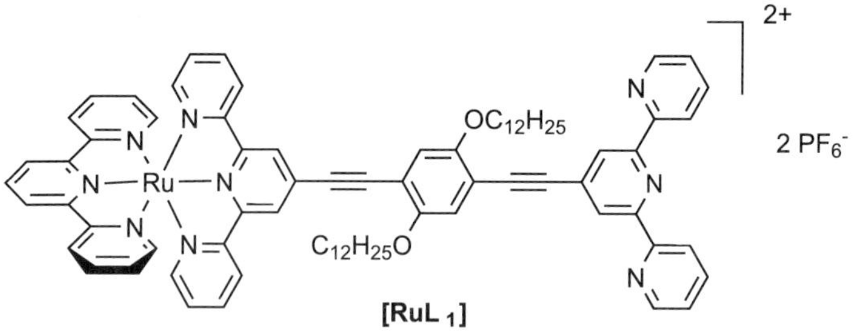

Chart 2.1

[RuL₁ structure shown]

brown color. After standing for a few days, an aqueous solution of KPF$_6$ (5 equiv.) was added, the organic solvent was removed under vacuum, the precipitates were washed by centrifugation with water and purified by chromatography on alumina eluting with CH$_2$Cl$_2$ using a gradient of methanol. The resulting red-brown complexes were recrystallized from acetone/hexane.

81$_1$ was prepared from 0.040 g (0.025 mmol) of **RuL$_1$** in 10 mL of THF/H$_2$O and 0.017 g (0.030 mmol) of [Os(terpy)(O)$_2$(OH)](NO$_3$)·2H$_2$O. After heating at 45 °C for 20 h, 0.023 g (0.126 mmol) of KPF$_6$ in 12 mL of H$_2$O was added, the organic solvent was removed under vacuum and the precipitate was washed by centrifugation with water (2 × 10 mL). Purification was carried out by chromatography on alumina using CH$_2$Cl$_2$/CH$_3$OH (0 to 10%) as eluant to give 0.030 g of **81$_1$** (52%).

81$_5$ was prepared from 0.015 g (0.004 mmol) of **RuL$_5$** in 10 mL of THF/H$_2$O and 0.003 g (0.005 mmol) of [Os(terpy)(O)$_2$(OH)](NO$_3$)·2H$_2$O. After heating at 45 °C for 4 days, 0.004 g (0.020 mmol) of KPF$_6$ in 4 mL of H$_2$O was added, the organic solvent was removed under vacuum and the precipitate was washed by centrifugation with water (2 × 10 mL). Purification was carried out by chromatography on alumina using CH$_2$Cl$_2$/CH$_3$OH (0 to 20%) as eluant to give 0.011 g of **81$_5$** (66%).

2.6.3
Synthesis at the Complex

The preparation of bipartite ligands bridged by e.g., aromatic polycycles, such as pyrene, falls within the second category – the so-called *synthesis at the complex*

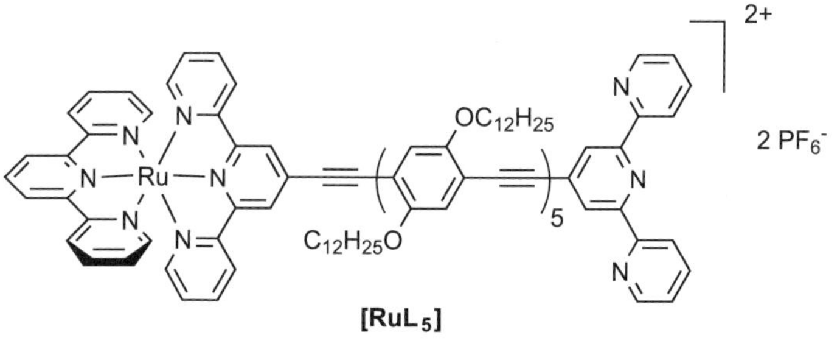

Chart 2.2

[RuL₅ structure shown]

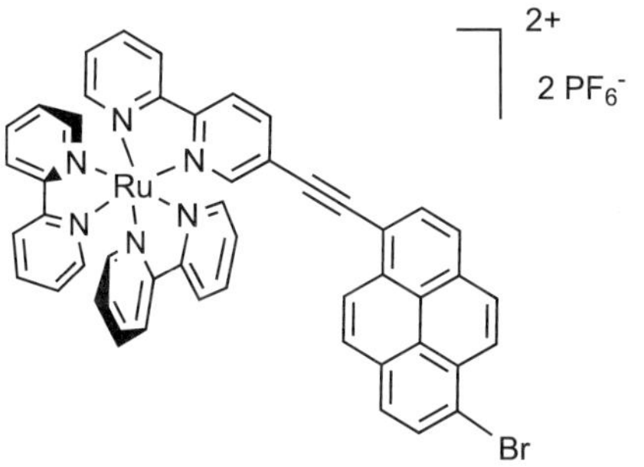

Chart 2.3

approach – and requires prior monofunctionalization of a pyrene building block that itself is difunctionalized. The quite selective mono-grafting of an ethynyl bipyridine fragment, followed by complexation with Ru salts, offers the possibility to further modify the complex with a different chelating fragment, such as terpyridine. This synthesis at the complex method is a very powerful tool to generate ligands, metallo-ligands and complexes displaying an asymmetry along the molecular axis.

[Ruthenium-bis-(2,2'-bipyridine){1-(5-ethynyl-2,2'-bipyridine)-6-bromopyrene}](PF$_6$)$_2$ (Chart 2.3) was prepared from 0.076 g (0.148 mmol) of cis-[Ru(bpy)$_2$Cl$_2$]·2H$_2$O, 0.017 g (0.068 mmol) of 1-(5-ethynyl-2,2'-bipyridine)-6-bromo-pyrene and 30 mL of C$_2$H$_5$OH. After complete consumption of the starting material, a solution of KPF$_6$ (5 equiv) in water (20 mL) was added, the organic solvent was then removed under vacuum and the precipitate was purified by chromatography on alumina, eluting with CH$_2$Cl$_2$ and using a gradient of CH$_3$OH (0 to 5%). The pure red-

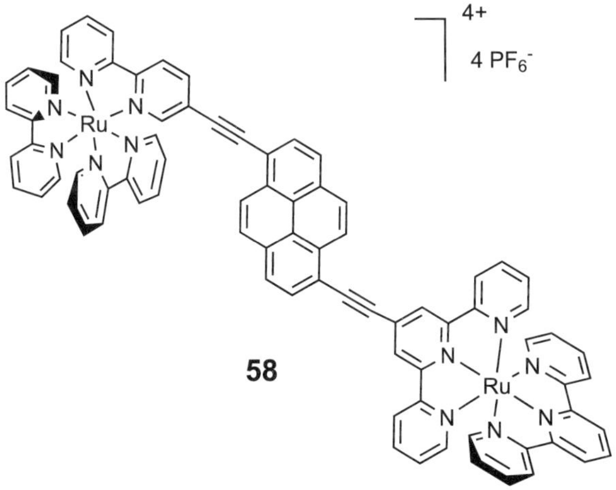

Chart 2.4

orange compound was obtained by recrystallization from CH_2Cl_2/hexane, to give 0.148 g of the target complex (86%).

57 was prepared from 0.120 g (0.104 mmol) of the above-described complex in 30 mL of THF, 0.0028 g (0.104 mmol) of 4′-ethynyl-2,2′:6′,2″-terpyridine, 0.008 g (0.008 mmol) of [Pd(PPh$_3$)$_4$] and 4 mL of diisopropylamine. Purification was performed by chromatography on alumina with $CH_3CN/H_2O/KNO_3$ (80/16/4) as eluant and afforded 0.114 g of **57** (81%).

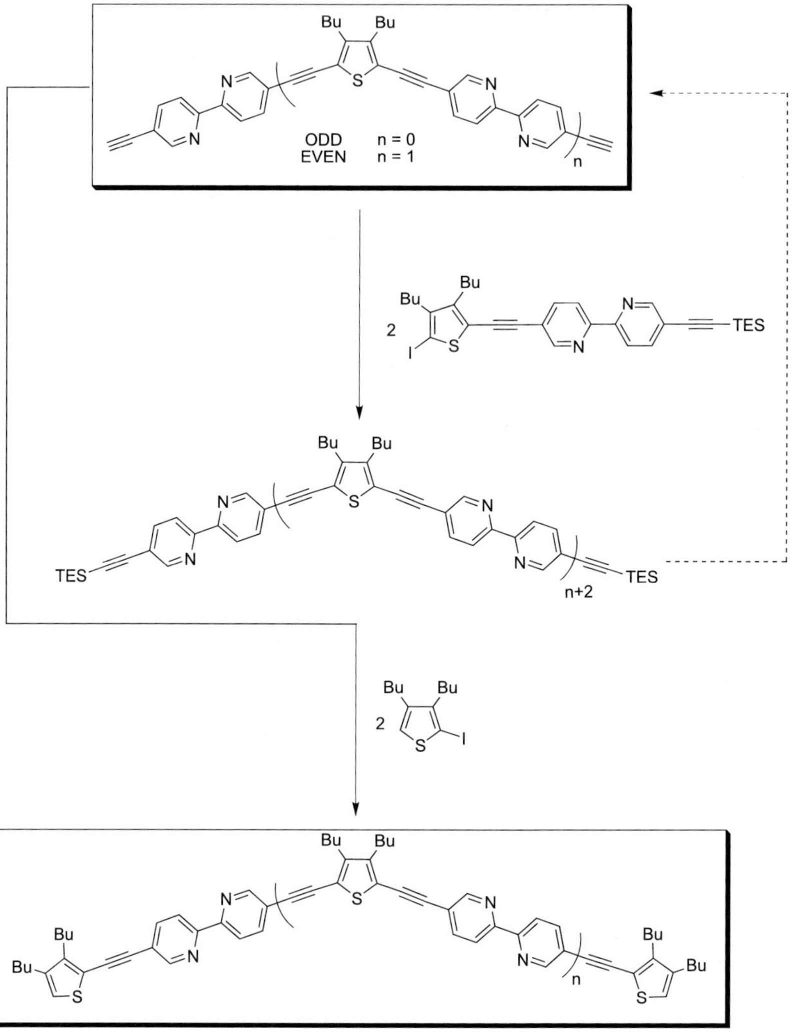

Figure 2.45. Schematic representation of the synthetic protocol required for the engineering of bipyridine based multi-site ligands.

58 was prepared from 0.090 g (0.104 mmol) of 57, following the procedure described above for preparation of the mononuclear [RuL$_n$] complexes. Purification was performed by chromatography on alumina with $CH_3CN/H_2O/KNO_3$ (80/16/4) as eluant and afforded 0.111 g of 58 (73%).

Preparation of multitopic bipyridine ligands bridged with diethynylthiophene is more complicated because two different pathways have to be devised, depending on the odd or even number of bipyridine subunits required. Figure 2.45 defines a single protocol applicable to synthesis of the multitopic ligands depicted in Fig. 2.42. Two main pathways are distinguished according to the pivotal building blocks constructed around a 5,5′-disubstituted-2,2′-bipyridine core and a thiophene moiety. In the top panel, coupling between a synthon bearing two terminal alkynes and a mono iodo-substituted hybrid bipy/thiophene molecule leads to the progressive addition of two bipy subunits to form the ligand precursor. The target ligands are obtained by cross-linking of these deprotected intermediates with 3,4-dibutyl-2-iodothiophene. The key concept here is the provision of the target ligands by an iterative sequence of reactions involving step-by-step introduction of each extending unit after a facile deprotection reaction. The choice of two solubilizing chains (butyl) was motivated by the relatively easy access to the 3,4-dibutylthiophene starting material. The presence of one protecting group (triethylsilyl, TES) and one reactive iodo function on the second starting material is critical for the success of the proposed reaction scheme. This approach allows an increase in the number of bipy units every two steps [244].

Acknowledgements

We gratefully acknowledge generous funding from the CNRS, the EPSRC, the University of Newcastle, the Université Louis Pasteur de Strasbourg and Johnson-Matthey Ltd for the loan of precious metal salts. The research efforts of many talented postgraduate and postdoctoral associates are acknowledged with our most sincere gratitude. The contributions of colleagues are credited by reference to original literature citations.

Abbreviations

LED	light emitting diode
MLCT	metal to ligand charge transfer state
CS	charge transfer
bipy	2,2′-bipyridine
terpy	2,2′:6′,6″-terpyridine
R_{MM}	metal to metal separation distance in Å
R_{CC}	center to center separation distance in Å
ET	energy transfer
ΔG	driving force for the energy of electron tranfer process in eV.

k_{ET} rate of triplet energy transfer process in s^{-1}
k_{CS} rate constant for the charge-transfer process in s^{-1}
k_{CR} rate constant for the recombination process in s^{-1}
ISC intersystem crossing
IC internal conversion
τ_{CS} lifetime for the charge-separated state.
MC metal-centered state
β attenuation factor, with units of Å^{-1}.

References

1 J. R. Heath, M. A. Ratner, *Phys. Today* **2003**, *56*, 43–49.
2 C. Joachim, J. K. Gimzewski, A. Aviram, *Nature* **2000**, *408*, 541–548.
3 P. A. Packan, *Science* **1999**, *285*, 2079–2081.
4 M. A. Ratner, *Mater. Today* **2002**, *5*, 20–27.
5 W. Liang, M. P. Shores, M. Bockrath, J. R. Long, H. Park, *Nature* **2002**, *417*, 725–729.
6 A. Nitzan, *Annu. Rev. Phys. Chem.* **2001**, *2*, 681–750.
7 H. Yu, Y. Luo, K. Beverly, J. F. Stoddart, H. R. Tseng, J. R. Heath, *Angew. Chem.* **2003**, *115*, 5884–5889; *Angew. Chem., Int. Ed.* **2003**, *42*, 5706–5711.
8 K. S. Kwok, J. C. Ellenbogen, *Mater. Today* **2002**, *5*, 28–37.
9 A. Amini, A. Harriman, *J. Photochem. Photobiol. C* **2003**, *4*, 155–177.
10 A. Amini, A. Harriman, *J. Phys. Chem. A* **2004**, *108*, 1242–1249.
11 A. Amini, A. Harriman, *Phys. Chem. Chem. Phys.* **2003**, *5*, 4556–4562.
12 A. Troisi, A. Nitzan, M. A. Ratner, *J. Chem. Phys.* **2003**, *119*, 5782–5788.
13 E. A. Weiss, L. E. Sinks, A. S. Lukas, E. T. Chernick, M. A. Ratner, M. R. Wasielewski, *J. Phys. Chem. B* **2004**, *108*, 10309–10316.
14 M. R. Wasielewski, G. L. Gaines, M. P. O'Neil, M. P. Niemczyk, W. A. Svec, *NATO Adv. Sci. Ser., Supramolecular Chem.* **1992**, *371*, 201–218.
15 A. Harriman, V. Heitz, M. Ebersole, H. Van Willigen, *J. Phys. Chem.* **1994**, *98*, 4982–4988.
16 H. Basch, M. A. Ratner, *J. Chem. Phys.* **2003**, *119*, 11943–11950.
17 J. Park, A. N. Pasupathy, J. I. Goldsmith, C. Chang, Y. Yaish, J. R. Petta, M. Rinkoski, J. P. Sethna, H. D. Abruña, P. L. McEuen, D. C. Ralph, *Nature* **2002**, *417*, 722–725.
18 A. M. Kuznetsov, J. Ulstrup, *J. Electroanal. Chem.* **2004**, *564*, 209–222.
19 A. Salomon, D. Cahen, S. Lindsay, J. Tomfohr, V. B. Engelkes, C. D. Frisbie, *Adv. Mater.* **2003**, *15*, 1881–1890.
20 M. W. Holman, R. C. Liu, D. M. Adams, *J. Am. Chem. Soc.* **2003**, *125*, 12649–12654.
21 Z. J. Donhauser, B. A. Mantooth, K. F. Kelly, L. A. Bumm, J. D. Monnell, J. J. Stapleton, D. W. Price Jr., A. M. Rawlett, D. L. Allara, J. M. Tour, P. S. Weiss, *Science*, **2001**, *292*, 2303–2307.
22 Y. Wada, *Curr. Appl. Phys.* **2002**, *2*, 331–334.
23 R. Niesner, R., W. Roth, K. H. Gericke, *Chem. Phys. Chem.* **2004**, *5*, 678–687.
24 M. Marrocco, *Appl. Opt.* **2004**, *43*, 5251–5262.
25 S. M. Lindsay, *Jpn. J. Appl. Phys.* **2002**, *41*, 4867–4870.
26 M. A. Reed, J. Chen, A. M. Rawlett, D. W. Price, J. M. Tour, *Appl. Phys. Lett.* **2001**, *78*, 3735–3737.
27 Y. Okawa, M. Aono, *Nature* **2001**, *409*, 683–684.
28 B. Grevin, P. Rannou, *Nat. Mater.* **2004**, *3*, 503–504.
29 L. Liang, J. Liu, C. F. Windisch, G. J.

Exarhos, Y. Liu, *Angew. Chem.* **2002**, *114*, 3817–3820; *Angew. Chem., Int. Ed.* **2002**, *41*, 3665–3668.

30 J. A. Massey, M. A. Winnik, I. Manners, V. Z. H. Chan, J. M. Ostermann, R. Enchelmaier, J. P. Spatz, M. Moller, *J. Am. Chem. Soc.* **2001**, *123*, 3147–3148.

31 A. Hatzor, P. S. Weiss, *Science* **2001**, *291*, 1019–1020.

32 P. G. Collins, M. S. Arnold, P. Avouris, *Science* **2001**, *292*, 706–709.

33 V. Derycke, R. Martel, J. Appenzeller, P. Avouris, *Nano Lett.* **2001**, *1*, 453–456.

34 H. W. C. Postma, T. Teepen, Z. Yao, M. Grifoni, C. Dekker, *Science* **2001**, *293*, 76–79.

35 R. J. Chen, Y. Zhang, D. Wang, H. Dai, *J. Am. Chem. Soc.* **2001**, *123*, 3838–3839.

36 A. Star, J. F. Stoddart, D. Steuerman, M. Diehl, A. Boukai, E. W. Wong, X. Yang, S. W. Chung, H. Choi, J. R. Heath, *Angew. Chem.* **2001**, *113*, 1771–1775; *Angew. Chem., Int. Ed.* **2001**, *40*, 1721–1725.

37 J. H. Schön, H. Meng, Z. Bao, *Nature* **2001**, *413*, 713–716.

38 X. Duan, Y. Huang, Y. Cui, J. Wang, C. M. Lieber, *Nature* **2001**, *409*, 66–69.

39 F. Cacialli, J. S. Wilson, J. J. Michels, C. Daniel, C. Silva, R. H. Friend, N. Severin, P. Samori, J. P. Rabe, M. J. O'Connell, P. N. Taylor, H. L. Anderson, *Nat. Mater.* **2002**, *1*, 160–164.

40 Y. Cui, Q. Wei, H. Park, C. M. Lieber, *Science* **2001**, *293*, 1289–1292.

41 A. C. Benniston, A. Harriman, D. J. Lawrie, A. Mayeux, K. Rafferty, O. D. Russell, *Dalton Trans.* **2003**, 4762–4769.

42 M. Heilemann, P. Tinnefeld, G. S. Mosteiro, M. G. Parajo, N. F. Van Hulst, M. Sauer, *J. Am. Chem. Soc.* **2004**, *126*, 6514–6515.

43 M. N. Paddon-Row, *Adv. Phys. Org. Chem.* **2003**, *38*, 1–85.

44 S. Blumstengel, F. Meinardi, R. Tubino, M. Gurioli, M. Jandke, P. Strohriegl, *J. Chem. Phys.* **2001**, *115*, 3249–3255.

45 R. Jin, Y. Cao, C. A. Mirkin, K. L. Kelly, G. C. Schatz, J. G. Zheng, *Science* **2001**, *294*, 1901–1903.

46 A. Harriman, R. Ziessel, *Chem. Commun.* **1996**, 1707–1716.

47 M. Heilemann, V. Buschmann, O. Piestert, P. Tinnefeld, K. D. Weston, M. Sauer, *Proc. SPIE-Int. Soc. Opt. Eng.* **2003**, *4962*, 38–46.

48 A. Osuka, H. Shimidzu, *Angew. Chem.* **1997**, *109*, 93–95; *Angew. Chem., Int. Ed.* **1997**, *36*, 135–140.

49 D. Kim, A. Osuka, *J. Phys. Chem. A* **2003**, *107*, 8791–8816.

50 A. Harriman, J.-P. Sauvage, *Chem. Soc. Rev.* **1996**, *25*, 41–50.

51 D. Gust, T. A. Moore, A. L. Moore, *Acc. Chem. Res.* **2001**, *34*, 40–48.

52 M. R. Wasielewski, *Chem. Rev.* **1992**, *92*, 435–461.

53 T. A. Moore, A. L. Moore, D. Gust, *Philos. Trans. R. Soc. London Ser. B* **2002**, *357*, 1481–1498.

54 M. P. O'Neil, M. P. Niemczyk, W. A. Svec, D. Gosztola, G. L. Gaines III, M. R. Wasielewski, *Science* **1992**, *257*, 63–65.

55 C. E. Olson, M. J. R. Previte, J. T. Fourkas, *Nat. Mater.* **2002**, *1*, 225–228.

56 R. W. Wagner, J. S. Lindsay, *J. Am. Chem. Soc.* **1994**, *116*, 9759–9760.

57 E. Hindin, R. A. Forties, R. S. Loewe, A. Ambroise, C. Kirmaier, D. F. Bocian, J. S. Lindsey, D. Holten, R. S. Knox, *J. Phys. Chem. B* **2004**, *108*, 12821–12832.

58 D. Holten, D. F. Bocian, J. S. Lindsey, *Acc. Chem. Res.* **2002**, *35*, 57–69.

59 R. K. Lammi, A. Ambroise, T. Balasubramanian, R. W. Wagner, D. F. Bocian, D. Holten, J. S. Lindsey, *J. Am. Chem. Soc.* **2000**, *122*, 7579–7591.

60 P. A. Liddell, D. Kuciauskas, J. P. Sumida, B. Nash, D. Nguyen, A. L. Moore, T. A. Moore, D. Gust, *J. Am. Chem. Soc.* **1997**, *119*, 1400–1405.

61 K. Li, P. J. Bracher, D. M. Guldi, M. A. Herranz, L. Echegoyen, D. I. Schuster, *J. Am. Chem. Soc.* **2004**, *126*, 9156–9157.

62 K. Ohkubo, H. Kotani, J. G. Shao,

Z. Ou, K. M. Kadish, L. Li, R. K. Pandey, M. Fujitsuka, O. Ito, H. Imahori, S. Fukuzumi, *Angew. Chem.* **2004**, *116*, 871–874; *Angew. Chem., Int. Ed.* **2004**, *43*, 853–856.

63 D. M. Guldi, H. Imahori, K. Tamaki, Y. Kashiwagi, H. Yamada, Y. Sakata, S. Fukuzumi, *J. Phys. Chem. A* **2004**, *108*, 541–548.

64 S. Fukuzumi, H. Kotani, K. Ohkubo, S. Ogo, N. V. Tkachenko, H. Lemmetyinen, *J. Am. Chem. Soc.* **2004**, *126*, 1600–1601.

65 J. Andreasson, G. Kodis, Y. Terazono, P. A. Liddell, S. Bandyopadhyay, R. H. Mitchell, T. A. Moore, A. L. Moore, D. Gust, *J. Am. Chem. Soc.* **2004**, *126*, 15926–15927.

66 M. Hissler, A. El-ghayoury, A. Harriman, R. Ziessel, *Angew. Chem.* **1998**, *110*, 1804–1807; *Angew. Chem., Int. Ed.* **1998**, *37*, 1717–1720.

67 A. Harriman, R. Ziessel, *Coord. Chem. Rev.* **1998**, *171*, 331–339.

68 R. Ziessel, M. Hissler, A. El-ghayoury, A. Harriman, *Coord. Chem. Rev.* **1998**, *178*, 1251–1298.

69 M. D. Ward, F. Barigelletti, *Coord. Chem. Rev.* **2001**, *216*, 127–154.

70 E. Baranoff, J.-P. Collin, L. Flamigni, J.-P. Sauvage, *Chem. Soc. Rev.* **2004**, *33*, 147–155.

71 A. M. Ramos, E. H. A. Beckers, T. Offermans, S. C. J. Meskers, R. A. J. Janssen, *J. Phys. Chem. A* **2004**, *108*, 8201–8211.

72 A. C. Benniston, V. Goulle, A. Harriman, J.-M. Lehn, B. Marczinke, *J. Phys. Chem.* **1994**, *98*, 7798–7804.

73 J.-P. Launay, *Chem. Soc. Rev.* **2001**, *30*, 386–397.

74 D. Gust, *Nature* **1994**, *372*, 133–134.

75 A. D. Joran, B. A. Leland, G. G. Geller, J. J. Hopfield, P. B. Dervan, *J. Am. Chem. Soc.* **1984**, *106*, 6090–6092.

76 M. N. Paddon-Row, M. J. Shephard, *J. Phys. Chem. A* **2002**, *106*, 2935–2944.

77 B. A. Leland, A. D. Joran, P. M. Felker, J. J. Hopfield, A. H. Zewail, P. B. Dervan, *J. Phys. Chem.* **1985**, *89*, 5571–5573.

78 A. M. Napper, I. Read, D. H. Waldeck, N. J. Head, A. M. Oliver, M. N. Paddon-Row, *J. Am. Chem. Soc.* **2000**, *122*, 5220–5221.

79 W. B. Davis, W. A. Svec, M. A. Ratner, M. R. Wasielewski, *Nature* **1998**, *396*, 60–63.

80 W. B. Davis, M. A. Ratner, M. R. Wasielewski, *J. Am. Chem. Soc.* **2001**, *123*, 7877–7886.

81 W. B. Davis, M. A. Ratner, M. R. Wasielewski, *Chem. Phys.* **2002**, *281*, 333–346.

82 F. Giacalone, J. L. Segura, N. Martín, D. M. Guldi, *J. Am. Chem. Soc.* **2004**, *126*, 5340–5341.

83 M. Furue, T. Yoshidzumi, S. Kinoshita, T. Kashida, S. Nozakura, M. Kamachi, *Bull. Chem. Soc. Jpn.* **1991**, *64*, 1632–1640.

84 B. Gholamkhass, K. Nozaki, T. Ohno, *J. Phys. Chem. B* **1997**, *101*, 9010–9021.

85 B. Schlicke, P. Belser, L. De Cola, E. Sabbioni, V. Balzani, *J. Am. Chem. Soc.* **1999**, *121*, 4207–4214.

86 F. Vögtle, M. Frank, M. Nieger, P. Belser, A. von Zelewsky, V. Balzani, F. Barigelletti, L. De Cola, L. Flamigni, *Angew. Chem.* **1993**, *105*, 1706–1709; *Angew. Chem., Int. Ed.* **1993**, *32*, 1643–1646.

87 L. De Cola, V. Balzani, F. Barigelletti, L. Flamigni, P. Belser, A. von Zelewsky, M. Frank, F. Vögtle, *Inorg. Chem.* **1993**, *32*, 5228–5238.

88 M. Frank, M. Nieger, F. Vögtle, P. Belser, A. von Zelewsky, L. De Cola, V. Balzani, F. Barigelletti, L. Flamigni, *Inorg. Chim. Acta* **1996**, *242*, 281–291.

89 C. Chiorboli, C. A. Bignozzi, F. Scandola, E. Ishow, A. Gourdon, J. P. Launay, *Inorg. Chem.* **1999**, *38*, 2402–2410.

90 V. Balzani, F. Barigelletti, P. Belser, S. Bernhard, L. De Cola, L. Flamigni, *J. Phys. Chem.* **1996**, *100*, 16786–16788.

91 S. Bernhard, P. Belser, *Synthesis* **1996**, 192–194.

92 L. Hammarström, F. Barigelletti, L. Flamigni, N. Armaroli, A. Sour, J. P.

Collin, J.-P. Sauvage, *J. Am. Chem. Soc.* **1996**, *118*, 11972–11973.

93 A. C. Benniston, G. M. Chapman, A. Harriman, M. Mehrabi, *J. Phys. Chem. A* **2004**, *108*, 9026–9036.

94 N. H. Damrauer, J. K. McCusker, *J. Phys. Chem. A* **1999**, *103*, 8440–8446.

95 A. C. Bhasikuttan, M. Suzuki, S. Nakashima, T. Okada, *J. Am. Chem. Soc.* **2002**, *124*, 8398–8405.

96 V. Balzani, A. Juris, M. Venturi, S. Campagna, S. Serroni, *Chem. Rev.* **1996**, *96*, 759–834.

97 J.-P. Sauvage, J.-P. Collin, J.-C. Chambron, S. Guillerez, C. Coudret, V. Balzani, F. Barigelletti, *Chem. Rev.* **1994**, *94*, 993–1019.

98 D. Lötscher, S. Rupprecht, P. Collomb, P. Belser, H. Viebrock, A. von Zelewsky, P. Burger, *Inorg. Chem.* **2001**, *40*, 5675–5681.

99 J.-P. Collin, A. Harriman, V. Heitz, F. Odobel, J.-P. Sauvage, *Coord. Chem. Rev.* **1996**, *148*, 63–69.

100 Th. Forster, *Discuss. Faraday Soc.* **1959**, *27*, 7–17.

101 D. L. Dexter, *J. Chem. Phys.* **1953**, *21*, 836–850.

102 G. L. Closs, P. Piotrowiak, J. M. MacInnis, G. R. Fleming, *J. Am. Chem. Soc.* **1988**, *110*, 2652–2653.

103 H. Oevering, J. W. Verhoeven, M. N. Paddon-Row, E. Cotsaris, N. S. Hush, *Chem. Phys. Lett.* **1988**, *143*, 488–495.

104 P. F. Barbara, T. J. Meyer, M. A. Ratner, *J. Phys. Chem.* **1996**, *100*, 13148–13168.

105 B. S. Brunschwig, N. Sutin, *Comments Inorg. Chem.* **1987**, *6*, 209–235.

106 Z. Murtaza, D. K. Graff, A. P. Zipp, L. A. Worl, W. E. Jones, W. D. Bates, T. J. Meyer, *J. Phys. Chem.* **1994**, *98*, 10504–10513.

107 P. Alborès, L. D. Slep, T. Weyhermuller, L. M. Baraldo, *Inorg. Chem.* **2004**, *43*, 6762–6773.

108 V. A. Mikhailova, A. I. Ivanov, E. Vauthey, *J. Chem. Phys.* **2004**, *121*, 6463–6469.

109 C. Joachim, M. A. Ratner, *Nanotechnology* **2004**, *15*, 1065–1075.

110 X.-Y. Zhu, *J. Phys. Chem. B* **2004**, *108*, 8778–8793.

111 E. G. Petros, V. I. Teslenko, V. May, *J. Chem. Phys.* **2004**, *121*, 5328–5338.

112 V. Grosshenny, A. Harriman, F. M. Romero, R. Ziessel, *J. Phys. Chem.* **1996**, *100*, 17472–17484.

113 V. Grosshenny, A. Harriman, J.-P. Gisselbrecht, R. Ziessel, *J. Am. Chem. Soc.* **1996**, *118*, 10315–10316.

114 V. Grosshenny, A. Harriman, M. Hissler, R. Ziessel, *J. Chem. Soc., Faraday Trans.* **1996**, *92*, 2223–2238.

115 A. Harriman, M. Hissler, O. Trompette, R. Ziessel, *J. Am. Chem. Soc.* **1999**, *121*, 2516–2525.

116 M. Hissler, A. Harriman, A. Khatyr, R. Ziessel, *Chem. Eur. J.* **1999**, *5*, 3366–3381.

117 A. Harriman, F. Odobel, J.-P. Sauvage, *J. Am. Chem. Soc.* **1994**, *116*, 5481–5482.

118 J.-P. Collin, A. Harriman, V. Heitz, F. Odobel, J.-P. Sauvage, *J. Am. Chem. Soc.* **1994**, *116*, 5679–5690.

119 I. M. Dixon, J.-P. Collin, J.-P. Sauvage, F. Barigelletti, L. Flamigni, *Angew. Chem.* **2000**, *112*, 1348–1351; *Angew. Chem., Int. Ed.* **2000**, *39*, 1292–1296.

120 A. C. Benniston, G. M. Chapman, A. Harriman, M. Mehrabi, *J. Phys. Chem. A* **2004**, *108*, 9026–9036.

121 A. Harriman, F. Odobel, J.-P. Sauvage, *J. Am. Chem. Soc.* **1995**, *117*, 9461–9472.

122 N. Mataga, S. Taniguchi, H. Chosrowjan, A. Osuka, N. Yoshida, *Chem. Phys.* **2003**, *295*, 215–228.

123 G. G. Gurzadyan, T.-H. Tran-Thi, T. Gustavsson, *J. Chem. Phys.* **1998**, *108*, 385–388.

124 G. A. Schick, *Thin Solid Films* **1989**, *179*, 521–527.

125 A. Nakano, Y. Yasuda, T. Yamazaki, S. Akimoto, I. Yamazaki, H. Miyasaka, A. Itaya, M. Murakami, A. Osuka, *J. Phys. Chem. A* **2001**, *105*, 4822–4833.

126 D. Gust, T. A. Moore, A. L. Moore, C. Devadoss, P. A. Liddell, R. Hermant, R. A. Nieman, L. J. Demanche, J. M. DeGraziano, I.

Gouni, *J. Am. Chem. Soc.* **1992**, *114*, 3590–3603.

127 H. Chosrowjan, S. Tanigichi, T. Okada, S. Takagi, T. Arai, K. Tokumaru, *Chem. Phys. Lett.* **1995**, *242*, 644–649.

128 N. Mataga, H. Chosrowjan, S. Taniguchi, Y. Shibata, N. Yoshida, A. Osuka, T. Kikuzawa, T. Okada, *J. Phys. Chem. A* **2002**, *106*, 12191–12201.

129 A. N. MacPherson, P. A. Liddell, D. Kuciauskas, D. Tatman, T. Gillbro, D. Gust, T. A. Moore, A. L. Moore. *J. Phys. Chem. B* **2002**, *106*, 9424–9433.

130 M. P. Debreczeny, M. R. Wasielewski, S. Shinoda, A. Osuka, *J. Am. Chem. Soc.* **1997**, *119*, 6407–6414.

131 A. El-ghayoury, A. Harriman, R. Ziessel, *Chem. Commun.* **1999**, 2027–2028.

132 A. El-ghayoury, A. Harriman, R. Ziessel, *J. Phys. Chem. A* **2000**, *104*, 7906–7915.

133 A. El-ghayoury, A. Harriman, A. Khatyr, R. Ziessel, *Angew. Chem.* **2000**, *112*, 4457–4460; *Angew. Chem., Int. Ed.* **2000**, *39*, 185–189.

134 P. Belser, R. Dux, M. Baak, L. De Cola, V. Balzani, *Angew. Chem.* **1995**, *107*, 634–637; *Angew. Chem., Int. Ed. Engl.* **1995**, *34*, 595–598.

135 M. C. DeRosa, R. J. Crutchley, *Coord. Chem. Rev.* **2002**, *233*, 351–371.

136 Q. C. Mulazzani, H. Sun, M. Z. Hoffman, W. E. Ford, M. A. J. Rodgers, *J. Phys. Chem.* **1994**, *98*, 1145–1150.

137 A. A. Abdel-Shafi, P. D. Beer, R. J. Mortimer, F. Wilkinson, *Helv. Chim. Acta* **2001**, *84*, 2784–2795.

138 L. Mishra, C.-S. Choi, K. Araki, *Chem. Lett.* **1997**, *26*, 447–448.

139 A. E. Friedman, J. C. Chambron, J.-P. Sauvage, N. J. Turro, J. K. Barton, *J. Am. Chem. Soc.* **1990**, *112*, 4960–4962.

140 E. J. C. Olson, D. Hu, A. Hörmann, A. M. Jonkman, M. R. Arkin, E. D. A. Stemp, J. K. Barton, P. F. Barbara, *J. Am. Chem. Soc.* **1997**, *119*, 11458–11467.

141 M. D. Hossain, M.-A. Haga, H. Monjushiro, G. Gholamkhass, K. Nozaki, T. Ohno, *Chem. Lett.* **1997**, *26*, 573–574.

142 P. Mobian, J.-M. Kern, J.-P. Sauvage, *Angew. Chem.* **2004**, *116*, 2446–2449; *Angew. Chem., Int. Ed.* **2004**, *43*, 2392–2395.

143 P. Mobian, J.-M. Kern, J.-P. Sauvage, *J. Am. Chem. Soc.* **2003**, *125*, 2016–2017.

144 J.-C. Chambron, J.-P. Collin, V. Heitz, D. Jouvenot, J.-M. Kern, P. Mobian, D. Pomeranc, J.-P. Sauvage, *Eur. J. Org. Chem.* **2004**, 1627–1638.

145 L. Flamigni, A. M. Talarico, J.-C. Chambron, V. Heitz, M. Linke, N. Fujita, J.-P. Sauvage, *Chem. Eur. J.* **2004**, *10*, 2689–2699.

146 E. Baranoff, K. Griffiths, J.-P. Collin, J.-P. Sauvage, B. Ventura, L. Flamigni, *New J. Chem.* **2004**, *28*, 1091–1095.

147 C. Hamann, J.-M. Kern, J.-P. Sauvage, *Inorg. Chem.* **2003**, *42*, 1877–1883.

148 A. Harriman, R. Ziessel, J.-C. Moutet, E. Saint-Aman, *Phys. Chem. Chem. Phys.* **2003**, *5*, 1593–1598.

149 E. C. Constable, F. Heirtzler, M. Neuburger, M. Zehnder, *J. Am. Chem. Soc.* **1997**, *119*, 5606–5617.

150 J. M. Lehn, *Chem. Eur. J.* **1999**, *5*, 2455–2463.

151 L. J. Charbonnière, A. F. Williams, C. Piguet, G. Bernardinelli, E. Rivara-Minten, *Chem. Eur. J.* **1998**, *4*, 485–493.

152 J. R. Nitschke, J.-M. Lehn, *Proc. Natl. Acad. Sci., USA* **2003**, *100*, 11970–11974.

153 A. El-ghayoury, A. Harriman, A. De Cian, J. Fischer, R. Ziessel, *J. Am. Chem. Soc.* **1998**, *120*, 9973–9974.

154 R. Ziessel, L. Douce, A. El-ghayoury, A. Harriman, A. Skoulios, *Angew. Chem.* **2000**, *112*, 1549–1553; *Angew. Chem., Int. Ed.* **2000**, *39*, 1489–1493.

155 A. Harriman, D. J. Magda, J. L. Sessler, *J. Phys. Chem.* **1991**, *95*, 1530–1532.

156 A. Harriman, D. J. Magda, J. L. Sessler, *J. Chem. Soc., Chem. Commun.* **1991**, 345–348.

157 J. L. Sessler, B. Wang, A. Harriman, *J. Am. Chem. Soc.* **1995**, *117*, 704–714.
158 J. L. Sessler, Y. Kubo, A. Harriman, *J. Phys. Org. Chem.* **1992**, *5*, 644–648.
159 V. Kràl, S. L. Springs, J. L. Sessler, *J. Am. Chem. Soc.* **1995**, *117*, 8881–8882.
160 S. L. Springs, D. Gosztola, M. R. Wasielewski, V. Kràl, A. Andrievsky, J. L. Sessler, *J. Am. Chem. Soc.* **1999**, *121*, 2281–2289.
161 J. L. Sessler, B. Wang, A. Harriman, *J. Am. Chem. Soc.* **1993**, *115*, 10418–10419.
162 A. Harriman, Y. Kubo, J. L. Sessler, *J. Am. Chem. Soc.* **1992**, *114*, 388–390.
163 A. Osuka, R. Yoneshima, H. Shiratori, T. Okada, S. Taniguchi, N. Mataga, *Chem. Commun.* **1998**, 1567–1568.
164 J. L. Sessler, M. Sathiosathan, C. T. Brown, A. Rhodes, G. Wiederrecht, *J. Am. Chem. Soc.* **2001**, *123*, 3655–3660.
165 N. Armaroli, F. Barigelletti, G. Calogero, L. Flamigni, C. M. White, M. D. Ward, *Chem. Commun.* **1997**, 2181–2182.
166 T. H. Ghaddar, E. W. Castner, S. S. Isied, *J. Am. Chem. Soc.* **2000**, *122*, 1233–1234.
167 C. Kim, H. Kim, *J. Organomet. Chem.* **2003**, *673*, 77–83.
168 V. Balzani, P. Ceroni, A. Juris, M. Venturi, S. Campagna, F. Puntoriero, S. Serroni, *Coord. Chem. Rev.* **2001**, *219*, 545–572.
169 M. Zhou, J. Roovers, *Macromolecules* **2001**, *34*, 244–252.
170 E. C. Constable, P. Harverson, *Inorg. Chim. Acta* **1996**, *252*, 9–11.
171 S. L. Springs, A. Andrievsky, V. Kràl, J. L. Sessler, *J. Porphy. Phthalocy.* **1998**, *2*, 315–325.
172 A. M. Brun, A. Harriman, S. M. Hubig, *J. Phys. Chem.* **1992**, *96*, 254–257.
173 D. M. Guldi, G. N. A. Rahman, J. Ramey, M. Marcaccio, D. Paolucci, S. Qin, W. T. Ford, D. Balbinot, N. Jux, N. Tagmatarchis, M. Prato, *Chem. Commun.* **2004**, 2034–2035.
174 B. Rybtchinski, L. E. Sinks, M. R. Wasielewski, *J. Phys. Chem. A* **2004**, *108*, 7497–7505.
175 A. M. Brun, A. Harriman, *J. Am. Chem. Soc.* **1994**, *116*, 10383–10393.
176 A. C. Benniston, A. Harriman, *Angew. Chem.* **1993**, *105*, 1553–1555; *Angew. Chem., Int. Ed. Engl.* **1993**, *32*, 1459–1461.
177 A. C. Benniston, P. R. Mackie, A. Harriman, *Angew. Chem.* **1998**, *110*, 376–378; *Angew. Chem., Int. Ed.* **1998**, *37*, 354–356.
178 A. H. Flood, R. J. A. Ramirez, W.-Q. Deng, R. P. Muller, W. A. Goddard III, J. F. Stoddart, *Aust. J. Chem.* **2004**, *57*, 301–322.
179 C. P. Collier, E. W. Wong, M. Behloradsky, F. M. Raymo, J. F. Stoddart, P. J. Kuekes, R. S. Williams, J. R. Heath, *Science* **1999**, *285*, 391–394.
180 N.-Q. Deng, R. P. Muller, W. A. Goddard III, *J. Am. Chem. Soc.* **2004**, *126*, 13562–13563.
181 J. R. Heath, J. F. Stoddart, R. S. Williams, *Science* **2004**, *303*, 1136–1136.
182 J. R. Winkler, T. L. Netzel, C. Creutz, N. Sutin, *J. Am. Chem. Soc.* **1987**, *109*, 2381–2392.
183 A. Harriman, M. Hissler, A. Khatyr, R. Ziessel, *Eur. J. Inorg. Chem.* **2003**, 955–959.
184 W. E. Ford, M. A. J. Rodgers, *J. Phys. Chem.* **1992**, *96*, 2917–2920.
185 G. J. Wilson, W. H. F. Sasse, A. W.-H. Mau, *Chem. Phys. Lett.* **1996**, *250*, 583–588.
186 G. J. Wilson, A. Launikonis, W. H. F. Sasse, A. W.-H. Mau, *J. Phys. Chem. A* **1997**, *101*, 4860–4866.
187 A. Harriman, M. Hissler, A. Khatyr, R. Ziessel, *Chem. Commun.* **1999**, 735–736.
188 D. S. Tyson, J. Bialecki, F. N. Castellano, *Chem. Commun.* **2000**, 2355–2356.
189 D. S. Tyson, F. N. Castellano, *J. Phys. Chem. A* **1999**, *103*, 10955–10960.
190 N. D. McClenaghan, F. Barigelletti, B. Maubert, S. Campagna, *Chem. Commun.* **2002**, 602–603.
191 A. Juris, L. Prodi, *New J. Chem.* **2001**, *25*, 1132–1135.

192 D. S. Tyson, C. R. Luman, X. Zhou, F. N. Castellano, *Inorg. Chem.* **2001**, *40*, 4063–4071.

193 M. Hissler, A. El-ghayoury, A. Harriman, R. Ziessel, *Angew. Chem.* **1998**, *110*, 1804–1807; *Angew. Chem., Int. Ed.* **1998**, *37*, 1717–1720.

194 A. Harriman, A. Khatyr, R. Ziessel, *Dalton Trans.* **2003**, 2061–2068.

195 J. F. Michalec, S. A. Bejune, D. C. Cuttell, G. C. Summerton, J. A. Gertenbach, J. S. Field, R. J. Haines, D. R. McMillin, *Inorg. Chem.* **2001**, *40*, 2193–2200.

196 A. Amini, A. Harriman, A. Mayeux, *Phys. Chem. Chem. Phys.* **2004**, *6*, 1157–1164.

197 M. T. Indelli, C. A. Bignozzi, F. Scandola, J.-P. Collin, *Inorg. Chem.* **1998**, *37*, 6084–6089.

198 A. A. Bhuiyan, J. R. Kincaid, *Inorg. Chem.* **1998**, *37*, 2525–2530.

199 A. C. Benniston, V. Grosshenny, A. Harriman, R. Ziessel, *Angew. Chem.* **1994**, *106*, 1956–1958; *Angew. Chem., Int. Ed. Engl.* **1994**, *33*, 1884–1886.

200 A. C. Benniston, A. Harriman, V. Grosshenny, R. Ziessel, *New J. Chem.* **1997**, *21*, 405–408.

201 L. Hammarström, F. Barigelletti, L. Flamigni, M. T. Indelli, N. Armaroli, G. Calogero, M. Guardigli, A. Sour, J.-P. Collin, J.-P. Sauvage, *J. Phys. Chem. A* **1997**, *101*, 9061–9069.

202 F. Barigelletti, L. Flamigni, G. Calogero, L. Hammarström, J.-P. Sauvage, J.-P. Collin, *Chem. Commun.* **1998**, 2333–2334.

203 M. Hissler, A. Harriman, A. Khatyr, R. Ziessel, *Chem. Eur. J.* **1999**, *5*, 3366–3381.

204 S. R. Allsop, A. Cox, S. H. Jenkins, T. J. Kemp, S. M. Tunstall, *Chem. Phys. Lett.* **1976**, *43*, 135–137.

205 G. H. Allen, R. P. White, D. P. Rillema, T. J. Meyer, *J. Am. Chem. Soc.* **1984**, *106*, 2613–2620.

206 R. S. Lumpkin, E. M. Kober, L. A. Worl, Z. Murtaza, T. J. Meyer, *J. Phys. Chem.* **1990**, *94*, 239–243.

207 A. C. Benniston, A. Harriman, F. M. Romero, R. Ziessel, *Dalton Trans.* **2004**, 1233–1238.

208 A. C. Benniston, V. Grosshenny, A. Harriman, R. Ziessel, *Dalton Trans.* **2004**, 1227–1232.

209 A. C. Benniston, A. Harriman, P. Li, C. A. Sams, *Phys. Chem. Chem. Phys.* **2004**, *6*, 875–877.

210 A. Harriman, A. Mayeux, C. Stroh, R. Ziessel, *J. Chem. Soc. Dalton* **2005**, 2925–2932.

211 A. Juris, L. Prodi, A. Harriman, R. Ziessel, M. Hissler, A. El-ghayoury, F. Wu, E. C. Riesgo, R. P. Thummel, *Inorg. Chem.* **2000**, *39*, 3590–3598.

212 A. Harriman, S. A. Roston, A. Khatyr, R. Ziessel, *Faraday Disc.* **2005**, in press.

213 T.-Y. Dong, M.-C. Lin, M. Y.-N. Chiang, J.-Y. Wu, *Organometallics* **2004**, *23*, 3921–3930.

214 F. Barigelletti, L. Flamigni, V. Balzani, J. P. Collin, J.-P. Sauvage, A. Sour, *New J. Chem.* **1995**, *19*, 793–798.

215 J. Andréasson, J. Kajanus, J. Mårtensson, B. Albinsson, *J. Am. Chem. Soc.* **2000**, *122*, 9844–9845.

216 J. Andréasson, H. Zetterqvist, J. Kajanus, J. Mårtensson, B. Albinsson, *J. Phys. Chem. A* **2000**, *104*, 9307–9314.

217 A. Harriman, A. Khatyr, R. Ziessel, A. C. Benniston, *Angew. Chem.* **2000**, *112*, 4457–4460H; *Angew. Chem., Int. Ed.* **2000**, *39*, 4287–4290.

218 A. C. Benniston, A. Harriman, P. Li, C. A. Sams, M. D. Ward, *J. Am. Chem. Soc.* **2004**, *126*, 13630–13631.

219 A. Harriman, A. Mayeux, A. De Nicola, R. Ziessel, *Phys. Chem. Chem. Phys.* **2002**, *4*, 2229–2235.

220 E. C. Constable, C. E. Housecroft, E. R. Schofield, S. Encinas, N. Armaroli, F. Barigelletti, L. Flamigni, E. Figgemeier, J. G. Vos, *Chem. Commun.* **1999**, 869–870.

221 Y. Liu, A. De Nicola, O. Reiff, R. Ziessel, F. S. Schanze, *J. Phys. Chem. A* **2003**, *107*, 3476–3485.

222 A. Barbiari, B. Ventura, F. Barigelletti, A. De Nicola, M. Quesada, R. Ziessel, *Inorg. Chem.* **2004**, *43*, 7359–7368.

223 F. Moresco, *Phys. Rep.* **2004**, *399*, 175–225.

224 J. Bernholc, D. Brenner, M. B. Nardelli, V. Meunier, C. Roland, *Annu. Rev. Mater. Sci.*, **2002**, *32*, 347–375.

225 P. A. Lewis, C. E. Inman, Y. Yao, J. M. Tour, J. E. Hutchison, P. S. Weiss, *J. Am. Chem. Soc.* **2004**, *126*, 12214–12215.

226 C. P. Collier, G. Mattersteig, E. W. Wong, Y. Luo, K. Beverly, J. Sampaio, F. M. Raymo, J. F. Stoddart, J. R. Heath, *Science* **2000**, *289*, 1172–1175.

227 K. A. Hutchison, J. P. Parakka, B. S. Kesler, R. R. Schumaker, *Proc. SPIE-Int. Soc. Opt. Eng.* **2000**, *3937*, 64–72.

228 M. Q. Sans, P. Belser, *Coord. Chem. Rev.* **2002**, *229*, 59–66.

229 B. L. Feringa, R. A. van Delden, N. Koumura, E. M. Geertsema, *Chem. Rev.* **2000**, *100*, 1789–1816.

230 K. Matsuda, M. Irie, *J. Photochem. Photobiol. C* **2004**, *5*, 169–182.

231 R. T. F. Jukes, V. Adamo, F. Hartl, P. Belser, L. De Cola, *Inorg. Chem.* **2004**, *43*, 2779–2792.

232 V. W.-W. Yam, C.-C. Ko, N. Zhu, *J. Am. Chem. Soc.* **2004**, *126*, 12734–12735.

233 A. Harriman, M. Hissler, P. Jost, G. Wipff, R. Ziessel, *J. Am. Chem. Soc.* **1999**, *121*, 14–27.

234 M. Hissler, A. Harriman, P. Jost, G. Wipff, R. Ziessel, *Angew. Chem.* **1998**, *110*, 3439–3443; *Angew. Chem., Int. Ed.* **1998**, *37*, 3249–3252.

235 N. C. Fletcher, M. D. Ward, S. Encinas, N. Armaroli, L. Flamigni, F. Barigelletti, *Chem. Commun.* **1999**, 2089–2090.

236 P. N. Taylor, A. J. Hagan, H. L. Anderson, *Org. Biomol. Chem.* **2003**, *1*, 3851–3856.

237 K. S. Suslick, N. A. Rakow, M. E. Kosal, J.-H. Chou, *J. Porphy. Phthalocyan.* **2000**, *4*, 407–413.

238 J. R. Reimers, T. X. Lü, M. J. Crossley, N. S. Hush, *Nanotechnology* **1996**, *7*, 424–429.

239 K.-T. Wong, J.-M. Lehn, S.-M. Peng, G.-S. Lee, *Chem. Commun.* **2000**, 2259–2260.

240 A. Leonhardt, M. Ritschel, R. Kozhuharova, A. Graff, T. Mühl, R. Huhle, I. Mönch, D. Elefant, C. M. Schneider, *Diamond Relat. Mater.* **2003**, *12*, 790–793.

241 A. De Nicola, S. Goeb, R. Ziessel, *Tetrahedron Lett.* **2004**, *45*, 7963–7967.

242 V. Grosshenny, A. Harriman, R. Ziessel, *Angew. Chem.* **1995**, *107*, 2921–2925; *Angew. Chem., Int. Ed. Engl.* **1995**, *34*, 2705–2708.

243 A. Khatyr, R. Ziessel, *J. Org. Chem.*, **2000**, *65*, 3126–3134.

244 S. Goeb, A. De Nicola, R. Ziessel, *J. Org. Chem.* **2005**, *70*, 1518–1529.

3
All-benzenoid Polycyclic Aromatic Hydrocarbons: Synthesis, Self-assembly and Applications in Organic Electronics

Jishan Wu and Klaus Müllen

3.1
A Brief Introduction to Polycyclic Aromatic Hydrocarbons

Polycyclic aromatic hydrocarbons (PAHs), which can be regarded as two-dimensional graphite segments, represent one of the most widely investigated classes of compounds in organic synthetic chemistry and materials science [1]. While PAHs are found in the residue of domestic and natural combustion of coal, wood and other organic materials, there are only a few examples of analytically pure and functionalized PAHs, especially for aromatics with high molecular weight. Pioneering contributions to the synthesis and characterization of polycyclic aromatics were made by Scholl and Clar in the first half of the 20th century [2]. At that time, they used harsh conditions such as high temperature and strong oxidizing reagents. As a result of progress in analytical techniques for the characterization of complicated compounds, synthetic breakthroughs were achieved which made the selective synthesis of various PAHs under mild conditions possible [3].

The investigations of polyarylene chemistry have been restricted to relatively few categories of polycyclic aromatic molecules of low molecular weight because the larger PAHs are usually insoluble and/or chemically unstable. The most familiar class of PAHs are the regular peri- and kata-condensed alternant hydrocarbons such as the "acenes" and the "phenes". The former are kata-annealated with the benzenoid rings fused in a linear arrangement. Examples of "acenes" are anthracene, tetracene, pentacene, hexacene, etc. (Fig. 3.1). With the increasing number of six-membered rings, the energy gaps of these "acenes" decrease rapidly and their chemical stability also decreases. For example, the violet pentacene and the green hexacene must be handled under an inert atmosphere, whereas the dark-green heptacene has never been obtained in a pure state because it is too readily oxidized by air [4]. The "phenes" consist of benzenoid rings fused in an angular arrangement, e.g., phenanthrene, tetraphene, pentaphene, hexaphene, etc. (Fig. 3.2). When the benzenoid rings are arranged in a helical mode, helicenes are obtained which deviate markedly from planarity as the number of rings exceeds five (Fig. 3.2) [5].

Carbon-Rich Compounds. Edited by Michael M. Haley and Rik R. Tykwinski
Copyright © 2006 WILEY-VCH Verlag GmbH & Co. KGaA, Weinheim
ISBN: 3-527-31224-2

3.1 A Brief Introduction to Polycyclic Aromatic Hydrocarbons

Anthracene

Tetracene (orange)

Pentacene (violet)

Hexacene (green)

Heptacene (dark-green)

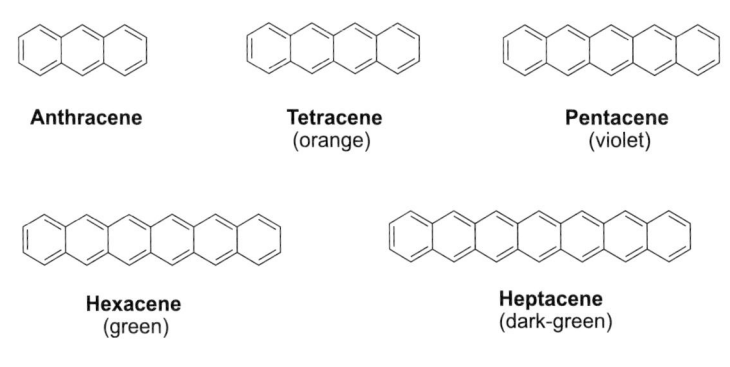

Figure 3.1. Examples of "acenes".

Phenanthrene

Tetraphene

Pentaphene

Hexaphene

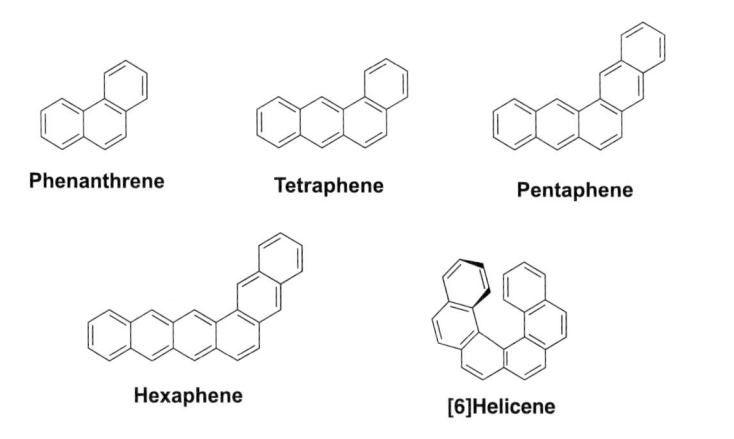

[6]Helicene

Figure 3.2. Examples of *kata*-condensed PAHs: "phenes" and "helicenes".

Another major class of PAHs consists of peri-condensed polyarenes, for example, pyrene, ovalene, hexa-*peri*-hexabenzocoronene (HBC), etc. (Fig. 3.3).

"Circulenes" represent an interesting class of PAHs which are characterized by fusion of rings in a macrocyclic arrangement, for example, corannulene, coronene, [7]circulene and kekulene (Fig. 3.4) [6]. The benzenoid rings may also be linked in a chain or macrocyclic arrangement, for example, the chain-linked oligophenyl-

Pyrene

Ovalene

Hexa-*peri*-hexabenzocoronene

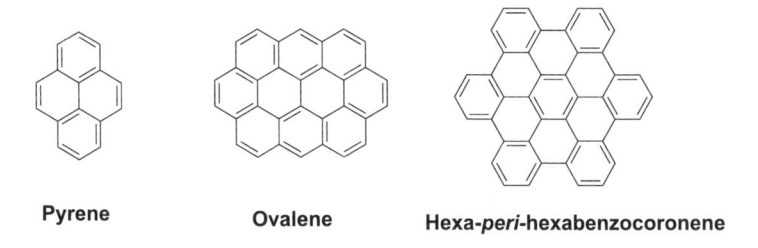

Figure 3.3. Examples of *peri*-condensed PAHs.

Corannulene   **Coronene**   **[7]Circulene**   **Kekulene**

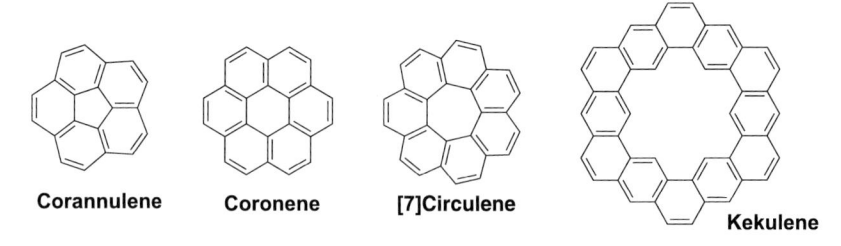

Figure 3.4. Examples of "circulenes".

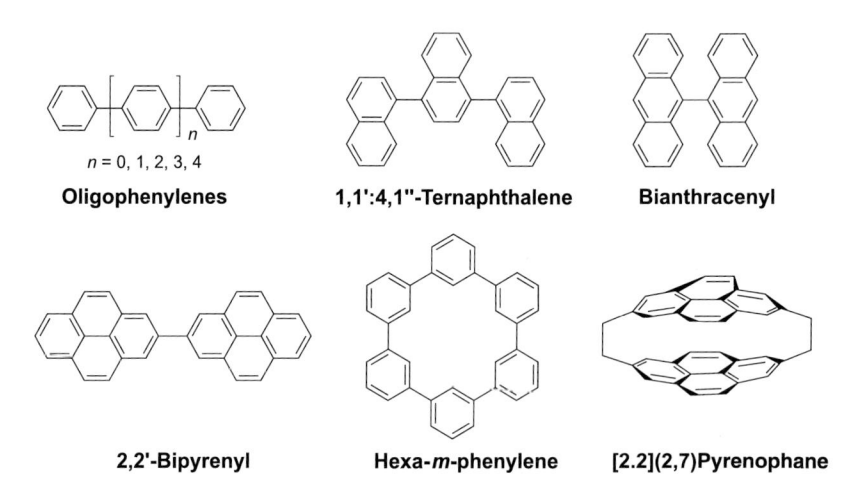

Oligophenylenes (n = 0, 1, 2, 3, 4)   **1,1':4,1''-Ternaphthalene**   **Bianthracenyl**

2,2'-Bipyrenyl   **Hexa-*m*-phenylene**   **[2.2](2,7)Pyrenophane**

Figure 3.5. Examples of chain-linked arylenes with linear or macrocyclic arrangements.

enes, bipyrenyl, bianthracenyl [7], meta-connected oligophenylene macrocycles [8] and cyclophanes (Fig. 3.5) [9].

The fusion of four-, five-, or six-membered rings with benzenoid rings usually produces nonalternant PAHs, examples of which are shown in Fig. 3.6 [10].

Various modern synthetic methods have been used to prepare PAHs with well-defined structures [1]. The classic reactions such as the Scholl reaction tend to involve relatively vigorous conditions and as a result the yields are usually low. Flash

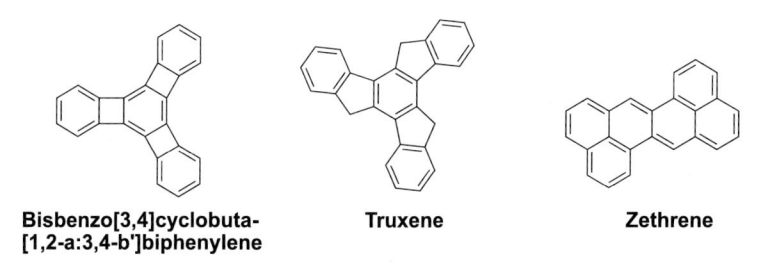

Bisbenzo[3,4]cyclobuta-[1,2-a:3,4-b']biphenylene   **Truxene**   **Zethrene**

Figure 3.6. Examples of four-, five- and six-membered rings fused arylenes.

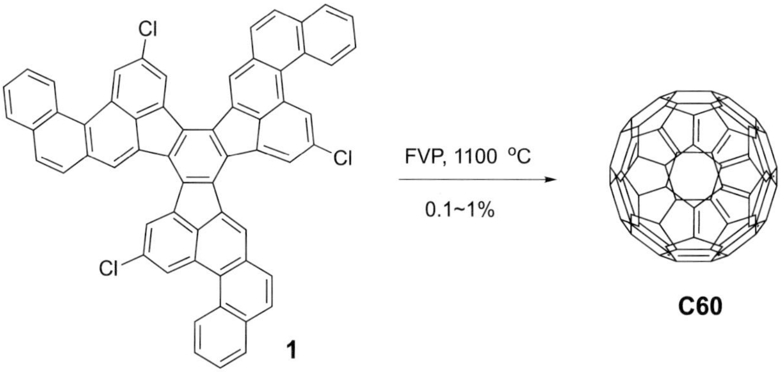

Scheme 3.1. Chemical synthesis of Buckministerfullerene C60 by flash vacuum pyrolysis.

vacuum pyrolysis (FVP) is also a powerful method to convert appropriate precursors into condensed PAHs at elevated temperature. Under FVP conditions, some planar hydrocarbons can even be transformed into nonplanar, bowl-shaped PAHs [11]. A remarkable example is the chemical synthesis of Buckministerfullerene C60 performed by Scott and coworkers [12]. As shown in Scheme 3.1, C60 can be chemically synthesized by FVP from the planar hydrocarbon precursor **1** containing 60 carbons and three cleavable C–Cl bonds. Other nonplanar PAHs such as corannulene, [7]circulene [13], and numerous fullerene fragments have also been successfully obtained by FVP [11], however, the yields are still low.

Modern synthesis of PAHs is usually performed under milder conditions. Various intramolecular cyclodehydrogenation methods have been developed. Among them, oxidative cyclodehydrogenation with strong oxidants such as $CuCl_2$–$AlCl_3$ and reductive cyclodehydrogenation with potassium have been widely used [14]. An example involving both processes is the synthesis of quaterrylene derivatives **2** (Scheme 3.2) [14a,b]. Intermolecular oxidative coupling is sometimes also an effec-

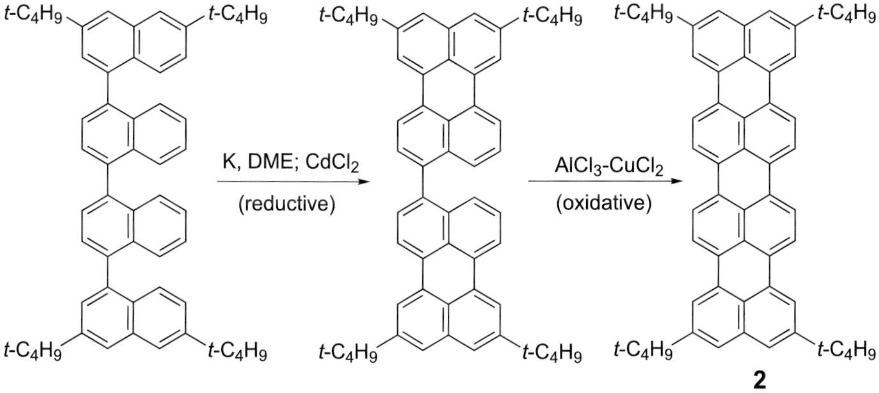

Scheme 3.2. Reductive and oxidative coupling reactions towards quaterrylenes.

tive method to make larger PAHs by doubling the size, for example, the oxidative coupling of corannulenes [15].

Photo-induced ring closure of stilbene-type compounds has been widely used in the preparation of condensed PAHs. One example is the synthesis of circumanthracene (**3**) in which a photocyclization and a subsequent oxidative cyclodehydrogenation were involved (Scheme 3.3) [16].

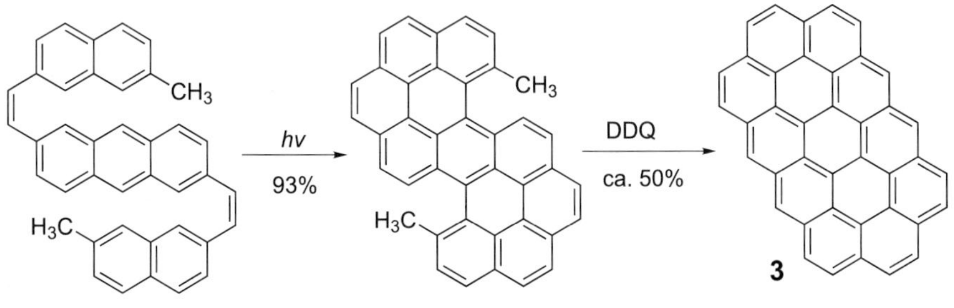

Scheme 3.3. Synthesis of circumanthracene by photocyclization and oxidative cyclodehydrogenation.

Cyclodehydration is another frequently employed method for making PAHs. The immediate synthetic precursor may be a carboxylic acid, a ketone, or an aldehyde. Acid-catalyzed cyclization of carboxylic acid substituted arylenes produces new rings, which are further converted to aromatic rings by dehydration and/or dehydrogenation. This method has been used for the synthesis of perylene **4** (Scheme 3.4) [17]. Similarly, ketones and aldehydes can be also used as precursors for the synthesis of both alternant and nonalternant PAHs such as dibenz[a,h]anthracene **5** via cyclodehydration (Scheme 3.4) [18]. This method is also a useful way to prepare polycyclic arylenes containing five-membered rings [1c].

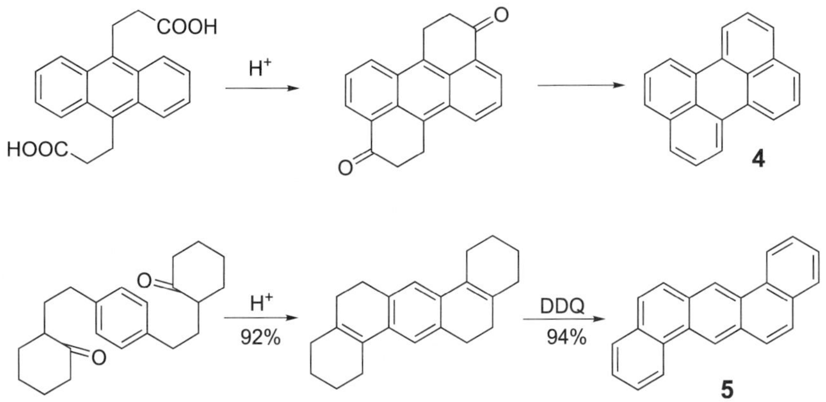

Scheme 3.4. Examples for the synthesis of PAHs via intramolecular cyclodehydration steps.

Diels-Alder cycloaddition provides one of the most useful methods for the construction of polycyclic aromatic ring systems. With the use of maleic anhydride as dienophile, two additional carbon atoms may be fused to the bay region sites of some polycyclic arylenes to generate additional rings. For example, the first step in the synthesis of benzo[*ghi*]perylene **6** involves a Diels-Alder cycloaddition between perylene and maleic anhydride (Scheme 3.5) [19]. Aryne intermediates generated *in situ* also serve as dienophiles for Diels-Alder cycloadditions; for example, reaction of 1,5-naphthadiyne (generated *in situ* from 2,6-dibromo-1,5-bis[(*p*-tolylsulfonyl)oxy]naphthalene with 2-methylisoindole forms an adduct which upon oxidation with *m*-chloroperbenzoic acid affords dibenzo[*b,k*]chrysene **7** (Scheme 3.5) [20].

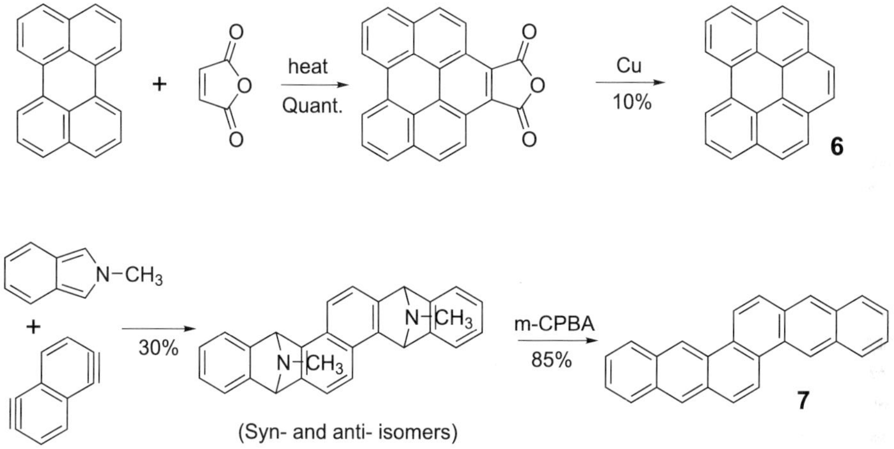

Scheme 3.5. Diels-Alder cycloaddition as key step for the synthesis of PAHs.

Other methods such as dimerization and trimerization of acetylenes and arynes [21], dehydrohalogenation [22], diaryl coupling using transition metal catalysts [23], Wittig-type reactions [24], and rearrangement reactions [25], have also been used in the synthesis of PAHs. A detailed description of the synthesis of different polycyclic arylene is available in the books by Clar and by Harvey et al. [1]. In this article, we will focus on the Scholl-type intramolecular cyclodehydrogenations for the synthesis of graphitic molecules, i.e., ring closure by treating appropriate precursors with Lewis acids such as $FeCl_3$.

The electronic properties and stability of PAHs depend not only on the size, but also markedly upon the arrangement of benzenoid rings. According to Clar's aromatic sextet rule, the stability of the isomers of certain polycyclic aromatics increases with the number of sextets. Thus, *all-benzenoid* polycyclic hydrocarbons (PBAHs) whose structure can be represented by a single Clar formula with no double bonds show extremely high stability, high melting point and low chemical reactivity. Some of them even exist in interstellar space [26]. A typical example is the hexa-*peri*-hexabenzocoronene (HBC, in Fig. 3.3) which has been studied since the middle of the 20th century and has recently attracted intense interest in our group for its unique materials properties.

A major breakthrough in semiconductor science was the discovery of conducting or semiconducting organic polymers in the 1970s, which has now led to promising applications in the field of organic semiconductors [27]. The electron delocalization within PBAHs and their chemical stability give rise to interesting electronic and optical properties and potential applications as promising semiconducting materials in organic devices such as light-emitting diodes (LEDs), field effect transistors (FETs), liquid crystal displays (LCDs) and solar cells [28]. Furthermore, two-dimensional all-benzenoid graphitic molecules are useful liquid crystalline materials because, if substituted with solubilizing flexible chains, they can form stable columnar mesophases, which are desirable for device processing due to their self-assembly and self-healing capabilities [29]. Encouraged by these potentials, a series of studies on the PBAHs has been made in our groups and in others. In this chapter, an overview of the recent progress on the synthesis, electronic properties, supramolecular self-assembly behavior, and device applications of all-benzenoid graphitic discs will be presented.

3.2
All-benzenoid PAHs – Synthesis, Structural Characterizations and Electronic Properties

3.2.1
Hexa-*peri*-hexabenzocoronene – An Old Story with New Discoveries

Among the all-benzenoid PAHs, hexa-*peri*-hexabenzocoronene (HBC) and its derivatives have attracted long-term interest due to their high symmetry, high stability, high charge carrier mobility and facile self-assembly. The first synthesis of the parent HBC was by Clar and coworkers in 1958 [30]. They found that the bromination of 2:3-7:8-dibenzo-*peri*-naphthene **8** in benzene gave a deep brown precipitate in which three bromine atoms were added (Scheme 3.6). The precipitate was heated at 153 °C and resulted in the formation of tetrabenzoperopyrene **9**. Compound **9**

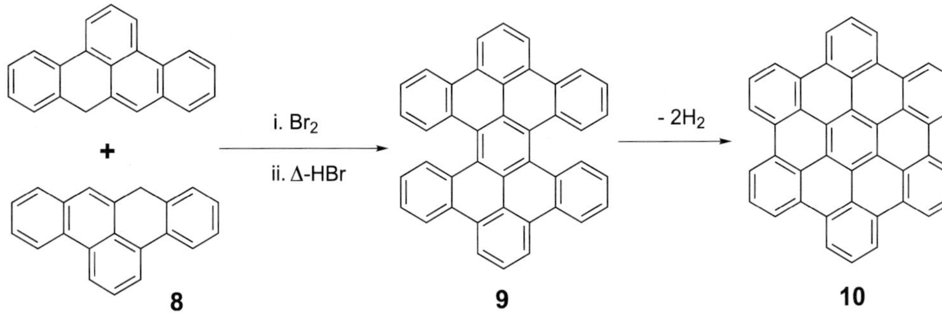

Scheme 3.6. Clar's route to HBC.

3.2 All-benzenoid PAHs – Synthesis, Structural Characterizations and Electronic Properties

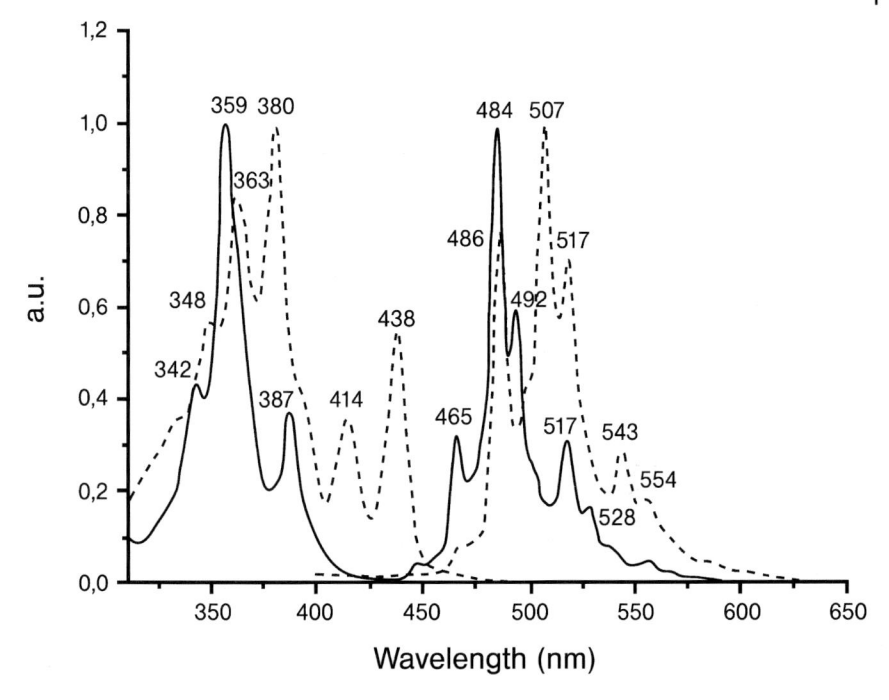

Figure 3.7. Combined UV–vis absorption and fluorescence spectra of HBC **10** (solid line) and HBC with additional fused conjugated ring **68a** (dotted line).

was further heated to 481 °C and HBC **10** was obtained as a pale yellow solid which did not melt even at 700 °C. HBC can also be obtained by melting compound **8** with sulfur or by sublimation of the residue from the preparation of dibenzo-*peri*-naphthene [31]. The UV–vis absorption spectrum of the dilute HBC solution in 1,2,4-trichlorobenzene revealed three typical bands: a low-energy α-band at around 440 nm, a β-band at 360 nm and a *p*-band at 387 nm (Fig. 3.7). They also observed orange-yellow phosphorescence with a very long lifetime in frozen solid solution in 1,2,4-trichlorobenzene at low temperature.

Halleux and coworkers later reported the synthesis of parent HBC using two different methods (Scheme 3.7) [32]. The cyclodehydrogenation of hexaphenylbenzene **11** by molten $AlCl_3/NaCl$ and the reaction of dibenz-1,9;2,3-anthrone **12** with $Zn/ZnCl_2$ both produced HBC. Two other hydrocarbons, tetrabenz-4,5;6,7;11,12;13,14-peropyrene and tetrabenz-1,2;3,4;8,9;10,11-bisanthene, have been isolated from the reaction of compound **12** with $Zn/ZnCl_2$.

A new synthesis of HBC was reported by Schmidt and coworkers in 1986 [33]. As shown in Scheme 3.8, reaction of the quinone **13** with phenyllithium afforded the diols **14** in 29%~46% yield. The diols were cyclized in an $AlCl_3/NaCl$ melt (3 min, 120 °C) and then subjected to aromatization with Cu at 400 °C to give a mixture of HBC (**10**) and dibenzo[*fg,ij*]phenanthro[9,10,1,2,3-*pqrst*]pentaphene in

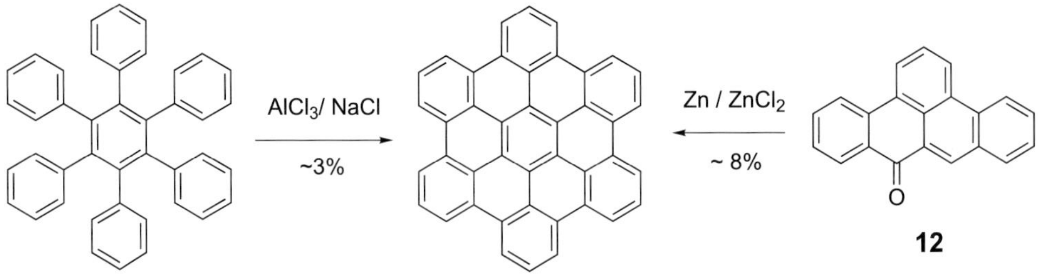

Scheme 3.7. Two different ways towards HBC by Halleux et al.

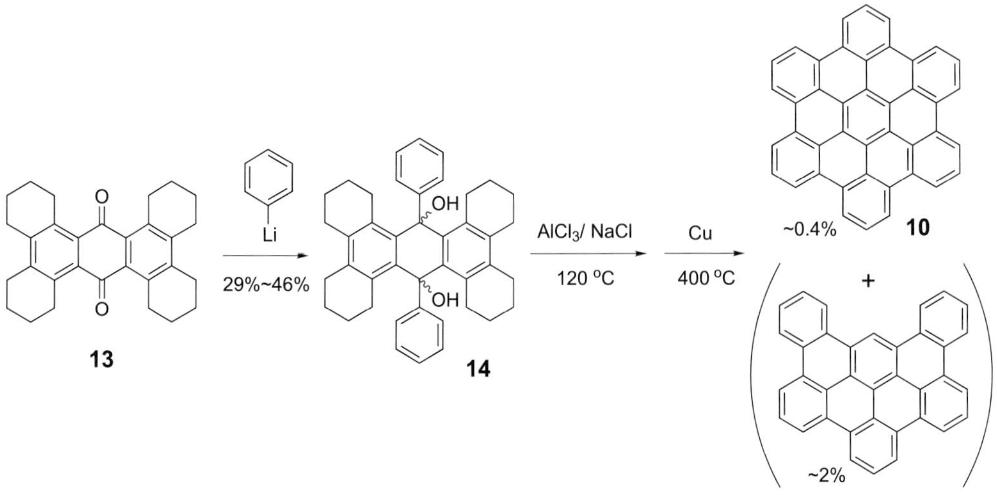

Scheme 3.8. Synthesis of HBC by Schmidt et al.

which one phenyl ring was lost. The mixture was separated by fractional sublimation and crystallization from trichlorobenzene.

All the above synthetic methods involve complicated experimental workup and only gave the desired compound in very low yield. Our group has developed an efficient way to prepare HBC and related PBAH structures by oxidative cyclodehydrogenation of branched oligophenylenes with Cu(II) salts such as $CuCl_2$ and $Cu(OTf)_2$ catalyzed by $AlCl_3$ [34]. Use of the weaker Lewis acid $FeCl_3$ renders the addition of an oxidant unnecessary because $FeCl_3$ possesses an oxidation potential sufficient for the C–C bond formation. The parent HBC was thus synthesized in nearly quantitative yield by treating hexaphenylbenzene with $CuCl_2/AlCl_3$ or $Cu(OTf)_2/AlCl_3$ in CS_2, or $FeCl_3$ dissolved in nitromethane (Scheme 3.9). A detailed study of the oxidative cyclodehydrogenation process by control of the quantity of oxidant and reaction time revealed that the reaction occurred exclusively in an intramolecular fashion without the formation of organic side products

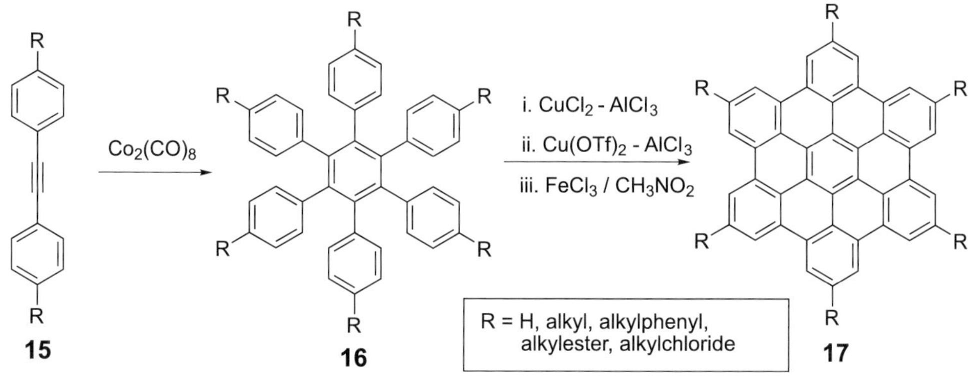

Scheme 3.9. High-yielding synthesis of HBCs by oxidative cyclohydrogenation.

and even the intermediate compound phenyldibenzo[*fg,ij*]phenanthro[9,10,1,2,3-*pqrst*]pentaphene was separated, indicating that the cyclohydrogenation proceeds in a step-wise process [35]. The introduction of solubilizing flexible side-chains onto HBC not only facilitates standard characterization by NMR spectroscopy and solution processing, but also permits the discotic molecules to form a thermotropic columnar liquid crystalline phase due to the melting of alkyl chains and the π–π stacking of the rigid cores. As shown in Scheme 3.9, the substituted hexaphenylbenzene **16** was synthesized by $Co_2(CO)_8$-catalyzed cyclotrimerization [36], and the cyclohydrogenation was usually conducted with iron (III) chloride instead of Cu(II)–$AlCl_3$ systems because Friedel-Crafts dealkylation sometimes occurred with the aluminum(III) chloride catalyst. By this method, the six-fold soluble alkyl [37], alkylphenyl [38], alkylester [39], and alkylchloride [40] substituted HBCs **17** were prepared in high yield.

The parent HBC as purified by sublimation is a yellow microcrystalline solid [35]. A single crystal of HBC has been grown from molten pyrene using mesitylene as a flux [41]. Single-crystal structure revealed that HBC crystallized in a γ motif with the planar discs stacked in a tilted column. The single crystal structure of hexa-*tert*-butyl substituted HBC, however, displays a sandwich-herringbone stacking due to the bulky substituents [42].

When the HBC is peripherally substituted with long alkyl chains, the disc-like molecules **17** self-assemble into one-dimensional columnar aggregates. Due to the strong π–π interactions between the HBC cores, a highly ordered columnar liquid crystalline phase is detected over a wide temperature range for the alkyl and alkylphenyl substituted HBCs [37]. In designing semiconducting materials with high charge carrier mobilities, the control of the order of the materials in the solid state is a key issue. The high order parameter found in HBC materials leads to high one-dimensional charge carrier mobility, as determined by pulse-radiolysis time-resolved microwave conductivity technique (PR-TRMC) [43]. For example, the one-dimensional charge carrier mobility ($\Sigma\mu_{1D}$) for the hexakis-dodecyl-*peri*-hexabenzocoronene (**HBC-C12**) in the liquid crystalline phase reached about

0.38 cm^2 V^{-1} s^{-1} at 110 °C and the hexakis-(4-dodecylphenyl)-*peri*-hexabenzocoronene (**HBC-PhC12**) showed a liquid crystalline phase at room-temperature with $\Sigma\mu_{1D} = 0.22$ cm^2 V^{-1} s^{-1}. The high charge carrier mobility and excellent mesophase properties qualify them as active semiconductors in organic electronics. The synthesis of HBC materials with different functions and symmetries was thus conducted to further control the order and thus their charge transporting properties.

The HBC derivatives with low symmetry were synthesized by an alternative route (Scheme 3.10). The key step was Diels-Alder cycloaddition between suitable tetraphenylcyclopentadienone derivatives **18** and substituted diphenylacetylenes **19** to afford a low symmetric hexaphenylbenzene derivative **20**, which was followed by the standard cyclodehydrogenation with FeCl$_3$ to give the fused HBC derivatives **21**.

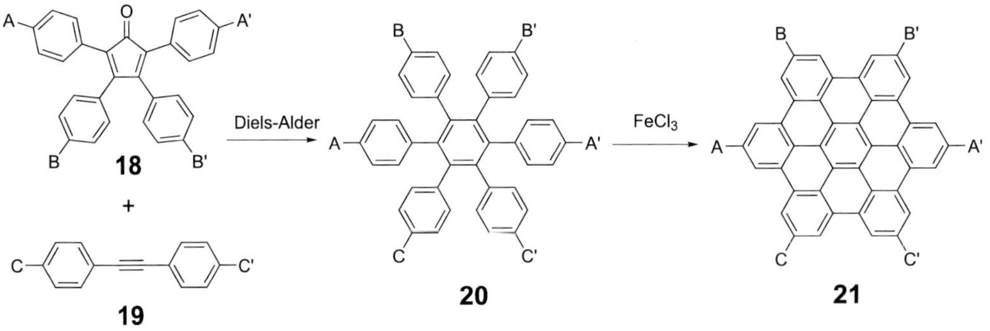

Scheme 3.10. Synthetic strategy of HBC derivatives with lower symmetry.

Based on the above synthetic strategy, HBCs **22–26** carrying solubilizing alkyl chains and one or multiple bromine-functionalities with different symmetries were synthesized (Fig. 3.8). Subsequent functionalization by transition metal-catalyzed coupling reactions such as Kumada, Suzuki, Hagihara, Negishi and Buchwald coupling reactions was possible [44]. The attachment of different functionalitites allows a more exact control of the order of HBCs in the bulk state, the alignment of the discs on the surface, the intramolecular binary energy/electron transport, and so on.

Whereas the normal alkylated HBCs have planar cores, the peralkoxylated HBCs such as **29a** and **29b** are nonplanar graphene molecules [45]. Following a similar synthetic strategy, the permethoxylated **29a** and perdodecoxylated HBCs **29b** were synthesized under standard cyclodehydrogenation conditions with FeCl$_3$ (Scheme 3.11). Due to the steric congestion in the bay position, the HBC molecules **29a** and **29b** are nonplanar, as revealed by the single-crystal analysis of molecule **29a**. It is noteworthy that the power of this cyclodehydrogenation method can even overcome such large steric congestion during the cyclization process. It was found that the outer phenyl rings flip up and down in an alternating manner with respect to the inner ring. A significant deviation from the planar geometry with a maximum

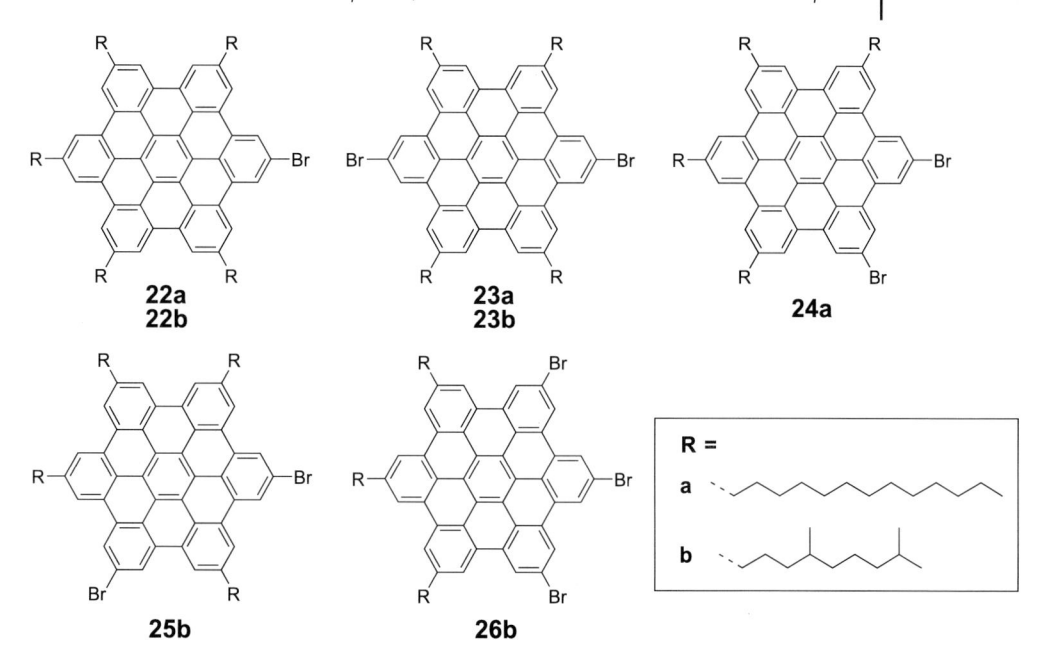

3.2 All-benzenoid PAHs – Synthesis, Structural Characterizations and Electronic Properties | 101

Figure 3.8. Bromine substituted HBC derivatives.

Scheme 3.11. Synthesis of peralkoxylated HBCs.

angle of $\beta = 16.8°$ is observed. The steric congestion actually leads to a "double-concave" conformation for molecule **29a**, which thus is a good supramolecular host for noplanar fullerene and electron-deficient perfluorobenzene. The complex of HBC **29a** and fullerene C60 was studied by single-crystal crystallography (Fig. 3.9). It was found that the C60 is perfectly included by the complementary double-concave of molecule **29a**, and further organized into a one-dimensional columnar structure. The strong interaction between the C60 and molecule **29a** can be

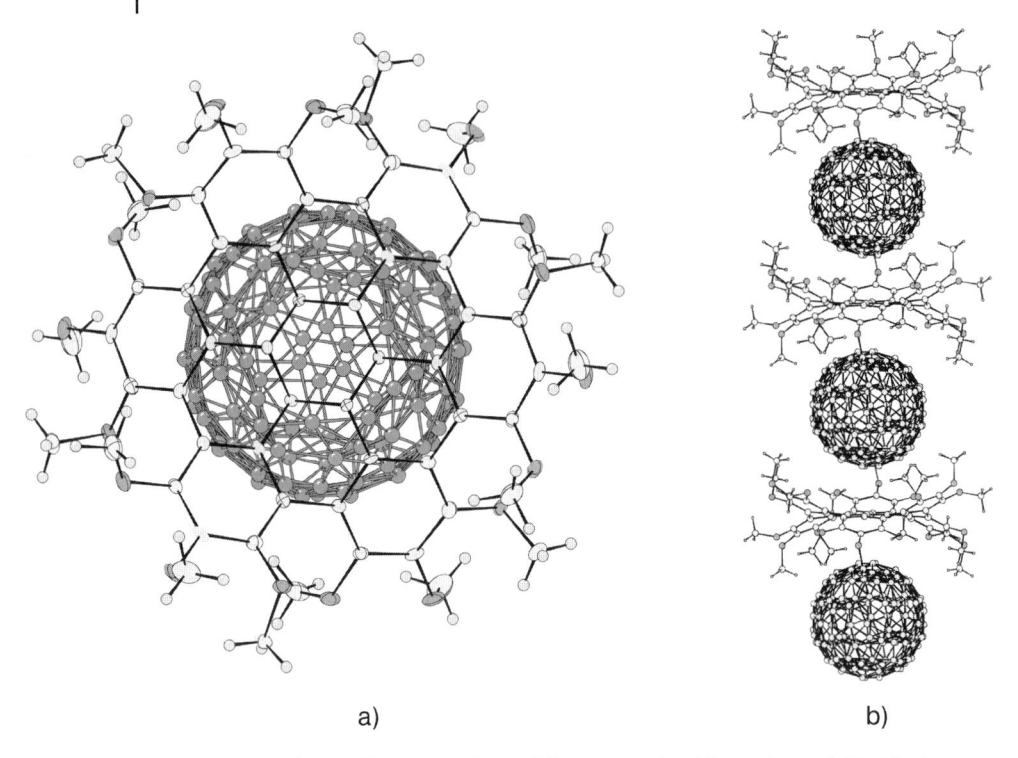

Figure 3.9. Single-crystal structure of HBC-fullerene complex: (a) top-view and (b) side-view.

explained in terms of van der Waals interactions and polar electrostatic interactions rather than of charge transfer interactions. On the other hand, molecule **29a** forms a 1:2 complex with hexafluorobenzene (HFB), i.e., two HFB molecules are accommodated into the cavity formed by the flexible surrounding methoxy groups. The interactions can be explained as arene–perfluoroarene interactions. The introduction of long alkyl chains drives the molecule **29b** into a room temperature columnar liquid crystalline phase [46].

While the above synthetic method for HBC materials is simple and flexible, and provides the desired compounds in high yields on a multigram scale, some drawbacks and limitations in its scope have been encountered. First, it requires multistep synthesis of the substituted diphenylacetylenes (**15** and **19**) and cyclopentadienones (**20**) with the desired functionality "R" (either solubilizing alkyl chains or functional groups such as alkyl esters etc.). Carrying these substituents through subsequent steps can sometimes lead to lower atom economy. Secondly, the identity of "R" is limited by the tolerance of the cyclotrimerization catalyst. For example, $Co_2(CO)_8$ catalyzed cyclotrimerization of bipyrimidylacetylene failed due to formation of a complex between the cobalt and nitrogen atoms. Finally, the "R" is limited by its compatibility with the final oxidation step. For example, six-fold geminal or vicinal ester and ketone groups (electron acceptors) substituted hexaphe-

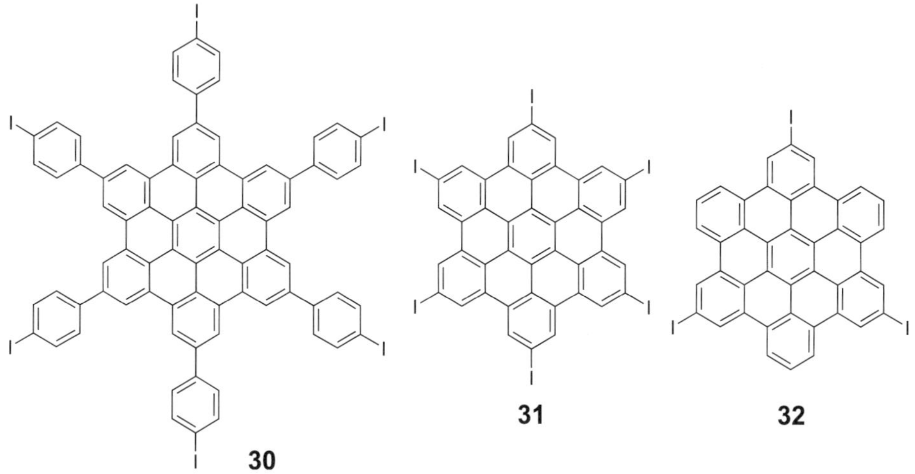

Figure 3.10. Three insoluble but highly reactive HBC building blocks.

nylbenzenes have so far precluded the application of the standard oxidation conditions. Direct attachment of heteroatoms like oxygen, nitrogen, and sulfur to the HBC core is expected to modify self-assembly and electronic properties; however, attempted cyclodehydrogenation of alkoxy and alkylthio-substituted hexaphenylbenzenes resulted in ether cleavage [46]. In addition, electro-active moieties such as triarylamine and pyridine are excluded presumably due to preferential localization of the charge as radical cations on nitrogens [47].

For reasons of atom economy and to broaden the scope of the synthesis of various HBC materials, several insoluble HBC building blocks carrying active iodine atoms, hexakis(4-iodophenyl)-*peri*-hexabenzocoronene **30**, hexakis-(4-iodo)-*peri*-hexabenzocoronene **31**, and a D_{3h}-symmetric triiodo HBC disc **32** (Fig. 3.10) were synthesized [48]. Although these building blocks are insoluble in normal solvents, they show a high reactivity in Pd catalyzed Hagihara-Sonogashira, Buchwald-Hartwig, and Kumada-Negishi cross coupling reactions, thus paving a way to versatile synthesis of functional HBC materials.

Despite the virtual insolubility of these building blocks, Hagihara-Sonogashira coupling reactions with solubilizing acetylenes work smoothly and give a series of soluble extended HBC derivatives **33** and **34a–d** in good yields. (Scheme 3.12) MALDI-TOF analysis of the crude products indicates quantitative conversions without any deiodination. The extension of the rigidity in molecule **34a–c** leads to highly ordered columnar liquid crystalline phases (vide infra) [48b]. The HBCs **33** and **34d** substituted with electroactive triarylamines adopt a coaxial columnar stacking and allow two paths for charge transport – along the stacks of the amines on the outside and along the HBC columns in the core – a so-called "double-cable" approach [48c]. Such a coaxial columnar stacking was supported by two-dimensional wide-angle X-ray scattering (2D WAXS) measurements on

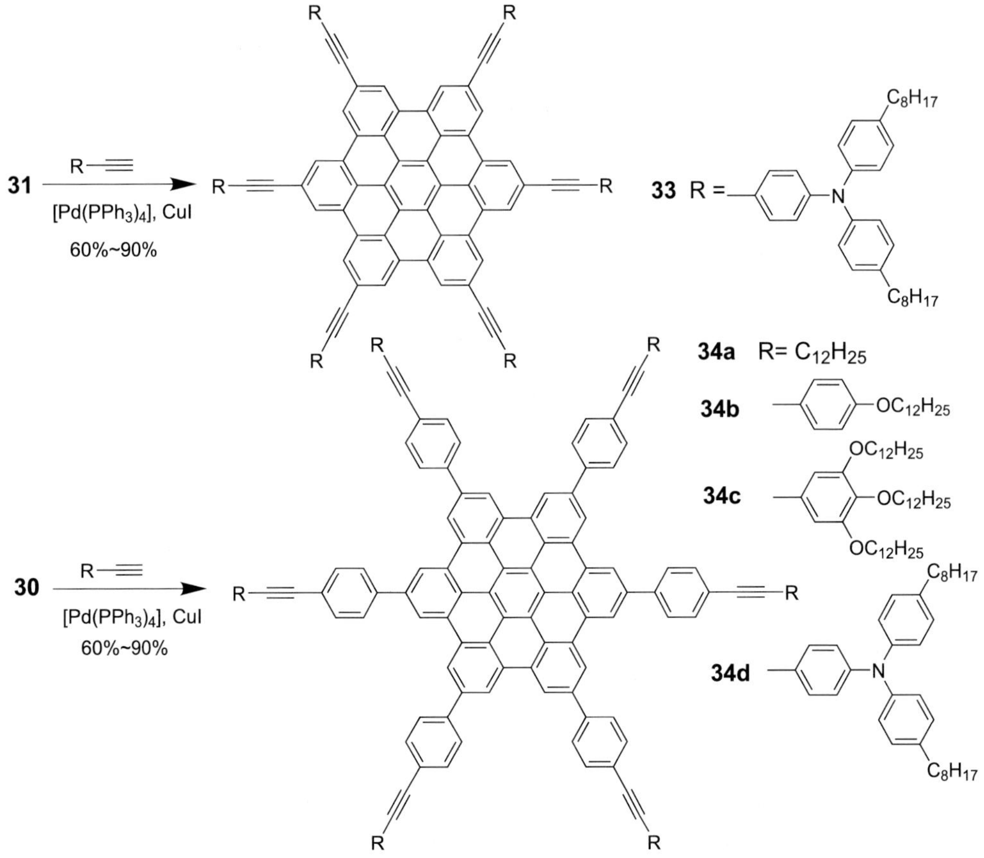

Scheme 3.12. Functionalization of the insoluble HBC building blocks **30** and **31** by Hagihara–Sonogashira coupling reactions.

extruded fibers. The charge carrier mobilities of these compounds detected by PR-TRMC techniques are much higher than for the normal amorphous triarylamine hole transport materials, thus, they are promising hole transporting materials for organic electronics applications.

Buchwald coupling reactions between the insoluble HBC building block **32** with di(4-octyl)phenylamine afford the soluble arylamine-substituted HBC **35** in modest yield. Similarly, the monoamino- and *para*-diamino-substituted HBCs **36** and **37** were synthesized from the corresponding bromine substituted HBCs **22a** and **23a**, respectively. Direct attachment of nitrogen atoms onto the HBC core dramatically lowers their oxidative potentials, as revealed by cyclic voltammetry measurements. Most interestingly, step-wise oxidation of compounds **35** and **37** leads to mixed-valence compounds, in which obvious intramolecular charge transfer between the nitrogen centers through the large HBC π-system was observed by UV–Vis–NIR spectroscopy and electronic paramagnetic resonance (EPR) measure-

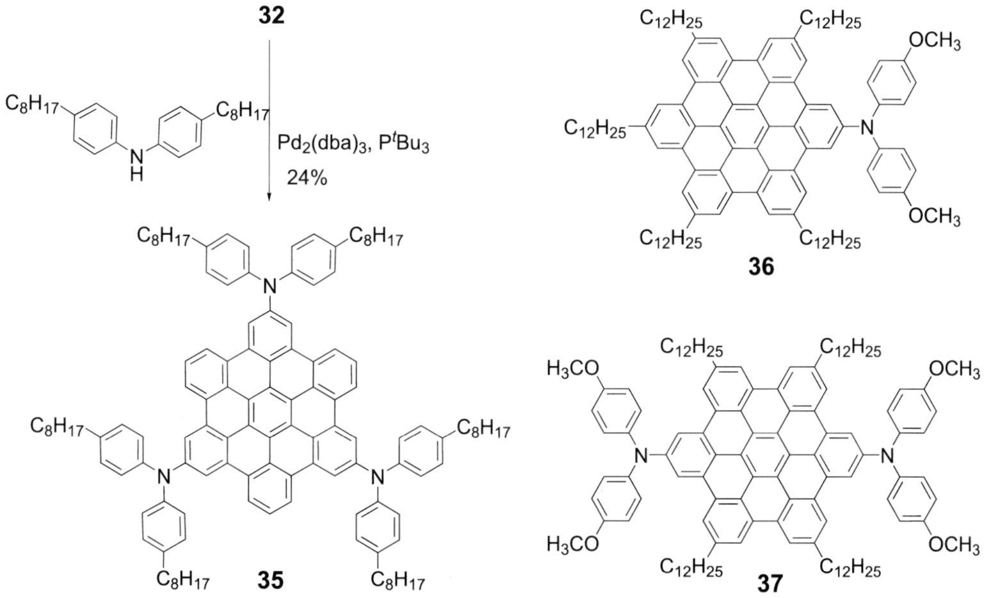

Scheme 3.13. Arylamine substituted HBCs prepared by Buchwald coupling reactions.

ments. At the same time, intermolecular association between the charged HBC species was also observed in solution by temperature dependent EPR spectroscopy [48c].

From an electronic point of view, HBC is regarded as an aromatic "superbenzene", which suggests the synthesis of the corresponding oligomers as "super-oligophenylenes". Dimers and trimers of hexa-*peri*-hexabenzocoronene (HBC) **38–41** (Fig. 3.11) with different geometries have been prepared. Their photophysical properties were studied by UV–vis and fluorescence spectra and their self-assembly has been studied by wide-angle X-ray diffraction (WAXD) in the bulk-state [49]. The dimer **38** was prepared by Yamamoto coupling of the corresponding mono-bromo-substituted HBC **22b**. The *para*- and *ortho*-connected trimers were synthesized by oxidative cyclodehydrogenation of the corresponding branched oligo(hexaphenylbenzenes) which were prepared by Diels-Alder cycloaddition of functionalised cyclopentadienones to hexaphenylbenzenes with diphenylethynyl-ene substituents at the *para*- and *ortho*-positions, respectively. Similarly the "superfluorene" **40** was made by cycloaddition of the diethynylfluorene with a cyclopentadienone and cyclodehydrogenation of the resulting compound. The dimer and trimers represent a new kind of electronically decoupled oligoarylenes and serve as model compounds for the corresponding poly-*para*-(*ortho*)-"superphenylenes". Large torsion angles between the HBC units and the small atomic orbital coefficients of the bridged-head carbons explain the observed weak electronic coupling. As a result their UV and PL spectra look essentially identical to those of the individual HBC units. By contrast the methylene bridge in **40** induces copla-

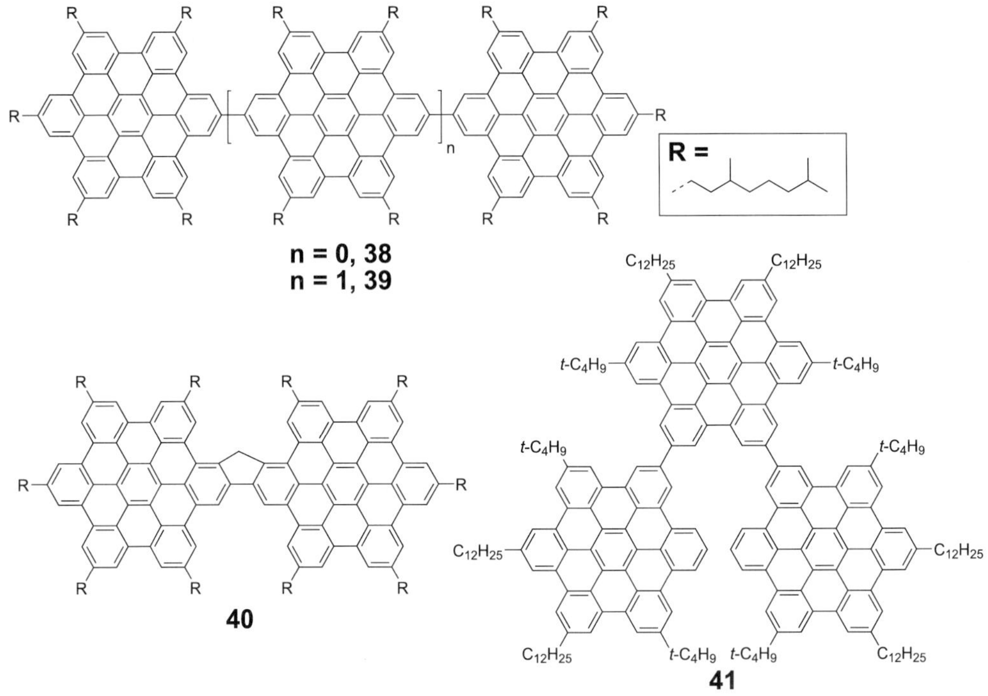

Figure 3.11. The oligomers of HBCs as oligo "superphenylenes".

narity of the two HBC units which improves the π-conjugation and suppresses the geometrical relaxation of the backbone upon electronic excitation, leading to a more prominent 0–0 transition band in the fluorescence spectrum than is seen for HBCs. Remarkably enough, an ordered columnar stacking has been detected for the dimer **38**, the *para*-trimer **39**, and the superfluorene **40**. Due to the strong out-of-plane twisting of the *ortho*-HBC units such stacking is not seen in the *ortho*-trimer **41**.

The high-yielding oxidative cyclodehydrogenation process with Cu(II)/AlCl$_3$ or FeCl$_3$ allowed the preparation of different HBC materials on a gram scale, and the novel synthetic strategy based on the insoluble HBC building blocks **30–32** dramatically broadened the scope of functions of HBC-based materials due to their highly ordered liquid crystalline phase and very good hole transporting properties. These new methods were therefore applied in the synthesis of larger graphitic aromatic hydrocarbons.

3.2.2
All-benzenoid Graphitic PAHs Larger than HBCs

Increasing the size of the core of discotic materials is predicted to improve the columnar stability and order due to the large overlap of the π-surface and thus to

3.2 All-benzenoid PAHs – Synthesis, Structural Characterizations and Electronic Properties

enhance their charge carrier mobility. We have accordingly investigated ways to synthesise soluble discotic PBAHs with cores significantly larger than in HBC. A general concept similar to the synthesis of HBCs is first to prepare appropriate branched oligophenylenes by Diels-Alder reactions, and then to fuse the phenyl rings by oxidative cyclodehydrogenation. As shown in Scheme 3.14 we developed a short, efficient route to superphenalenes **42a–f** containing 96 carbon atoms in their cores and different substitutents [50]. The oligophenylene precursors **44a–f** were prepared in nearly quantitative yield by Diels-Alder cycloaddition between the substituted cyclopentadienones **43a–f** and 1,3,5-triethynylbenzene in refluxing o-xylene. This was then followed by oxidative cyclodehydrogenation with a larger excess of $FeCl_3$ (90 equiv.) and longer reaction times. The dodecyl **42a**, phytyl **42b** and carbomethoxyphenyl **42e** substituted superphenalenes are readily soluble in normal organic solvents. In contrast, superphenalenes with 4-dodecylphenyl **42d** or 3,7-dimethyloctyl **42c** chains, which impart good solubility to HBCs, showed poor solubility. This is somewhat surprising and may be explained by the special space-filling and packing of the substitutents. The insoluble 4-iodophenyl substituted superphenalene **42f** was also synthesized and like its HBC counterpart **30**, can be functionalized by Pd catalyzed Hagihara-Sonogashira coupling reactions, even when the product is also virtually insoluble [51].

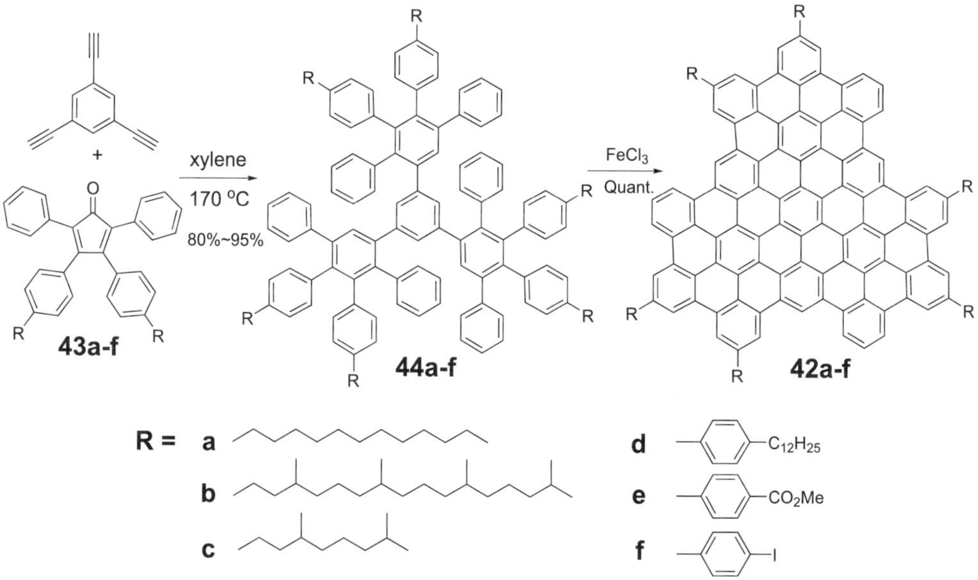

Scheme 3.14. Synthesis of "superphenalenes" containing 96 carbon atoms in the core.

Recently, two additional large three-fold symmetric graphitic discs, one containing 150 carbon atoms **45** and another containing 90 carbon atoms **46** at the core, were prepared (Scheme 3.15) [52]. The former is the largest disc with three-fold

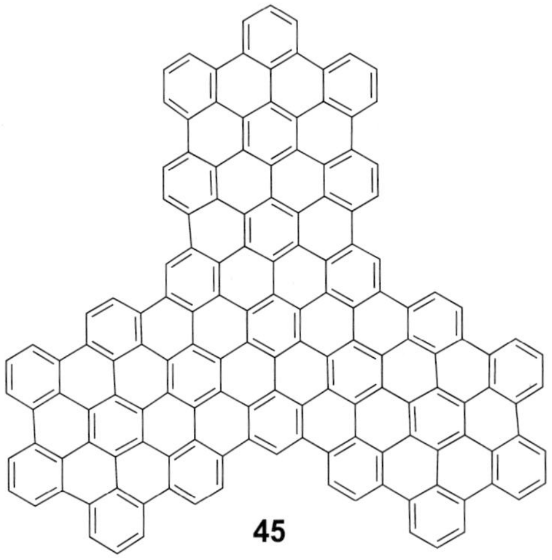

Scheme 3.15. Synthesis of two three-fold symmetric graphitic discs.

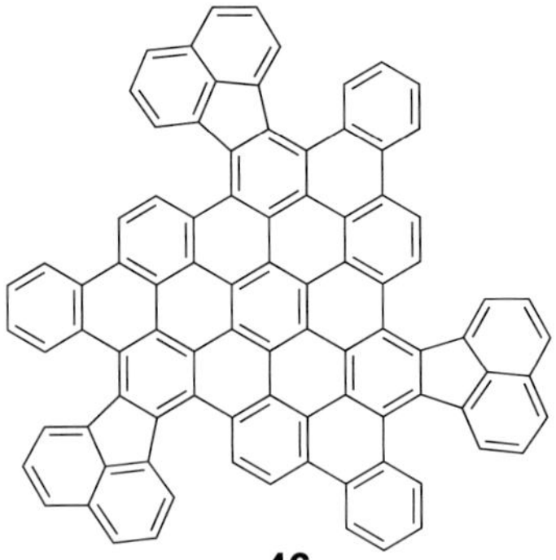

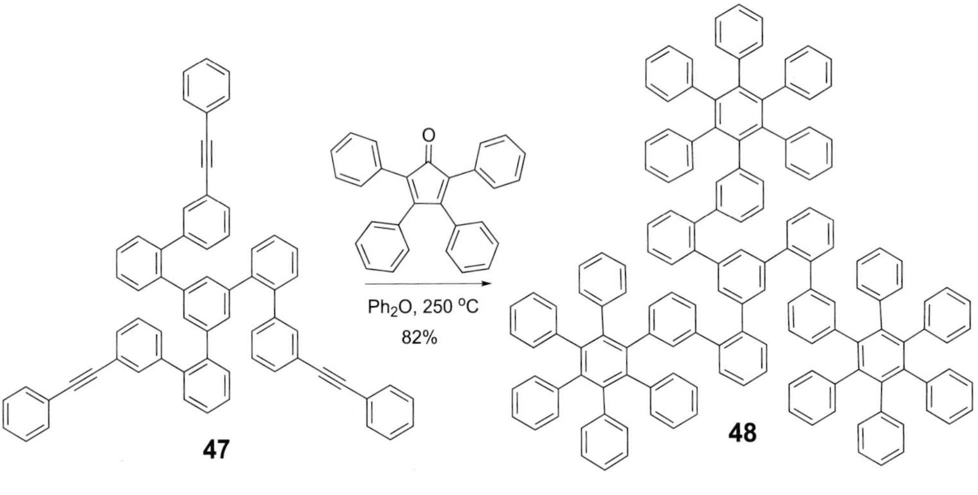

symmetry yet made, while the latter is nonplanar due to the five-membered rings, and so offers a possibility to make bowl-shaped molecules by flash vacuum pyrolysis. The synthesis of molecule **45** starts with a C_3 symmetric substituted trisbiphenylbezene **47**, which reacts with tetaphenylcyclopentadienone in refluxing diphenylether to give branched oligophenylene **48**. The single-crystal analysis of molecule **48** revealed a propeller conformation with three arms lying on the same side of the central benzene ring. Treating **48** with 100 equiv. $FeCl_3$ for 19 h in CH_2Cl_2 afforded the desired compound **45** as an insoluble powder, which was characterized by MALDI-TOF mass spectroscopy, solid state UV–vis absorption, and Raman spectroscopy. Compound **46**, a red fluorescent solid, was synthesized by a similar route to that for C96 discs but using 7,9-diphenyl-8H-cyclopenta[l]acenaphthylen-8-

3.2 All-benzenoid PAHs – Synthesis, Structural Characterizations and Electronic Properties

one instead of tetraphenylcyclopentadienone in the Diels-Alder cycloaddition step. Due to the steric congestion between periphery naphthyl groups, molecule **46** is not planar and thus is partially soluble in common organic solvents.

Using a similar synthetic strategy, other giant graphitic discs containing 132 **49** and 222 **50** carbon atoms at the core were synthesized (Scheme 3.16). The synthesis of the parent and phytyl substituted C132 discs **49** starts with a four-fold Diels-Alder cycloaddition between the tetraphenylcyclopentadienones and 3,3′,5,5′-tetraethynylbiphenyl **51**, followed by oxidative cyclodehydrogenation with $FeCl_3$ or

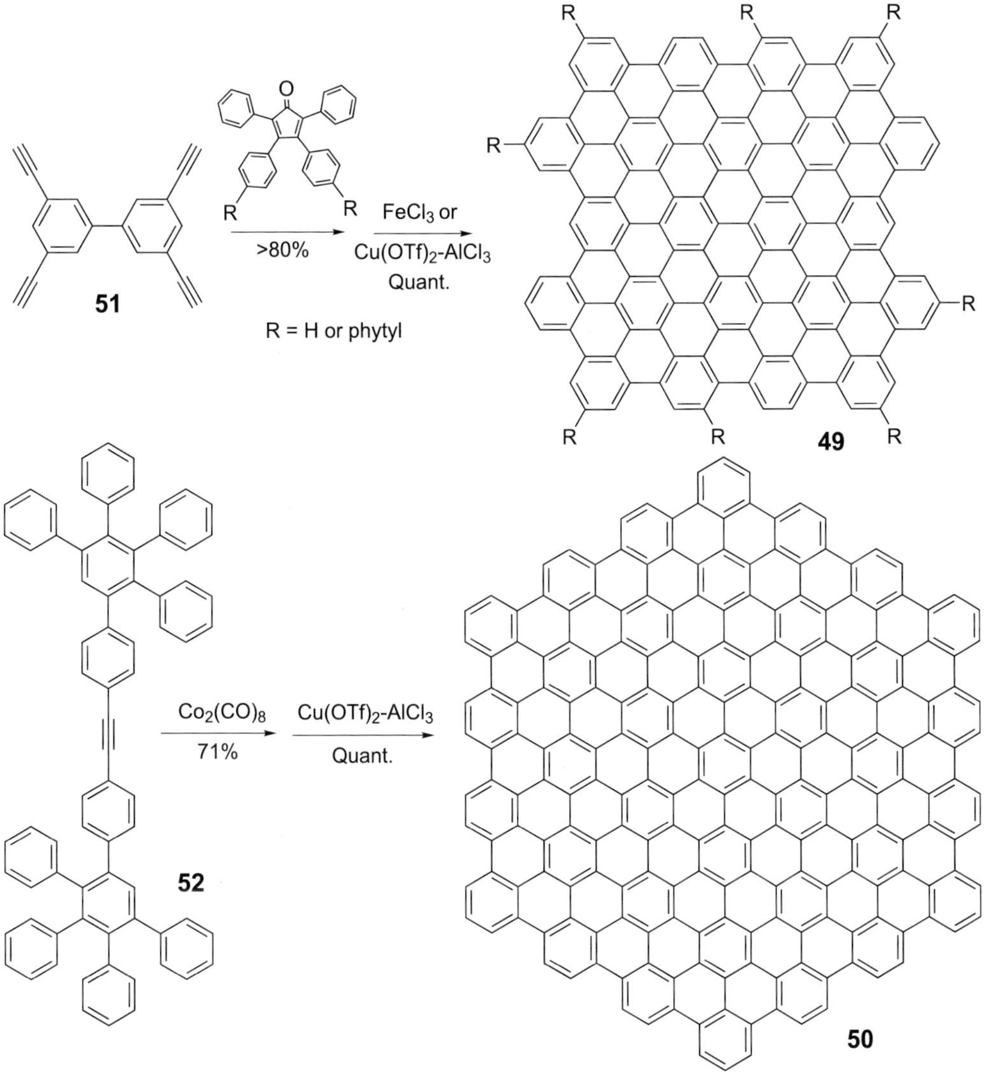

Scheme 3.16. Synthesis of giant graphitic discs containing 132 and 222 carbons.

Cu(OTf)$_2$/AlCl$_3$ [53]. The long flexible phytyl chains impart solubility in the normal organic solvents. To date, compound **50** is the largest polycyclic hydrocarbon with characterized structure. It was synthesized by a Co$_2$(CO)$_8$ catalyzed cyclotrimerization of compound **52** and the resulting precursor was planarized by Cu(OTf)$_2$ and AlCl$_3$ in CS$_2$ under carefully controlled conditions [54]. The insoluble product was characterized by isotope-resolved MALDI-TOF mass spectroscopy. The solid-state UV–Vis absorption spectrum measured for a film simply smeared onto quartz plates revealed a broad long-wavelength absorption band extended over the UV–Vis region from 250 nm to 1400 nm, indicating an extended conjugation.

The concept of completely planarizing large oligophenylenes reaches its limit when the branched oligophenylene becomes so large that the oxidative cyclodehydrogenation can only partially take place. It was found that the cyclodehydrogenation of the huge oligophenylene dendrimer containing 385 carbon atoms **53** preferentially started at the dendrimer arms and not in the interior, thus leading to a propeller structure with giant graphitic segments as "blades" (Scheme 3.17) [55]. The cyclodehydrogenation of **53** with various amounts of oxidant produces different propeller structures. The optimal conditions involved use of 378 equivalents of oxidant per mole of **53**, which yielded a product with a sharp main peak at 5830 Da, corresponding to loss of almost 200 hydrogens, and could be obtained reproducibly, together with a small peak at 5760 corresponding to the fully dehydrogenated product **54a**. Due to the insolubility of the product, characterization could be performed only by mass spectrometry and solid-state UV–vis absorbance spectroscopy. The product was treated with AlCl$_3$/NaCl in the melt, under which conditions PAHs are stable but single aryl–aryl bonds can break to produce new fragments with peaks at 1842 and 1914 Da corresponding to PAHs with 150 and 156 carbons. This suggested that the product contained the propeller-like structures **54b** and **54c** which contain three such fragments as blades around the central benzene. A C156 model compound was synthesized and its UV–vis spectrum was found to closely match that of the product with a maximum at about 560 nm. This further supported the proposed structure. Molecular modeling suggests that the graphene fragments in **54b** and **54c** are not coplanar and thus not conjugated to each other, so their UV–vis spectra should resemble those of a single blade. The proposed structures also suggest that dehydrogenation and accompanying planarization commence at the periphery of the dendrimer with the planarization at the core being the final stage of the process.

As structural analogues to naphthalene and triphenylene, molecules **55** and **56** can be regarded as "supernaphthalene" and "supertriphenylene", respectively (Fig. 3.12) [56]. Two or three HBCs annealed together improve the π-conjugation, in contrast to what was observed for the linear connected HBC oligomers **38** and **39**. Attachment of flexible alkyl chains onto the former leads to formation of an ordered columnar liquid crystalline phase [57].

The molecular design of larger graphitic molecules can either proceed via two-dimensional expansion to achieve larger and larger discs or through a one-dimensional extension to form ribbon-like graphite sheets. Thus a series of well-defined graphite ribbons **57–59** containing 60, 78, and 114 carbon atoms at the

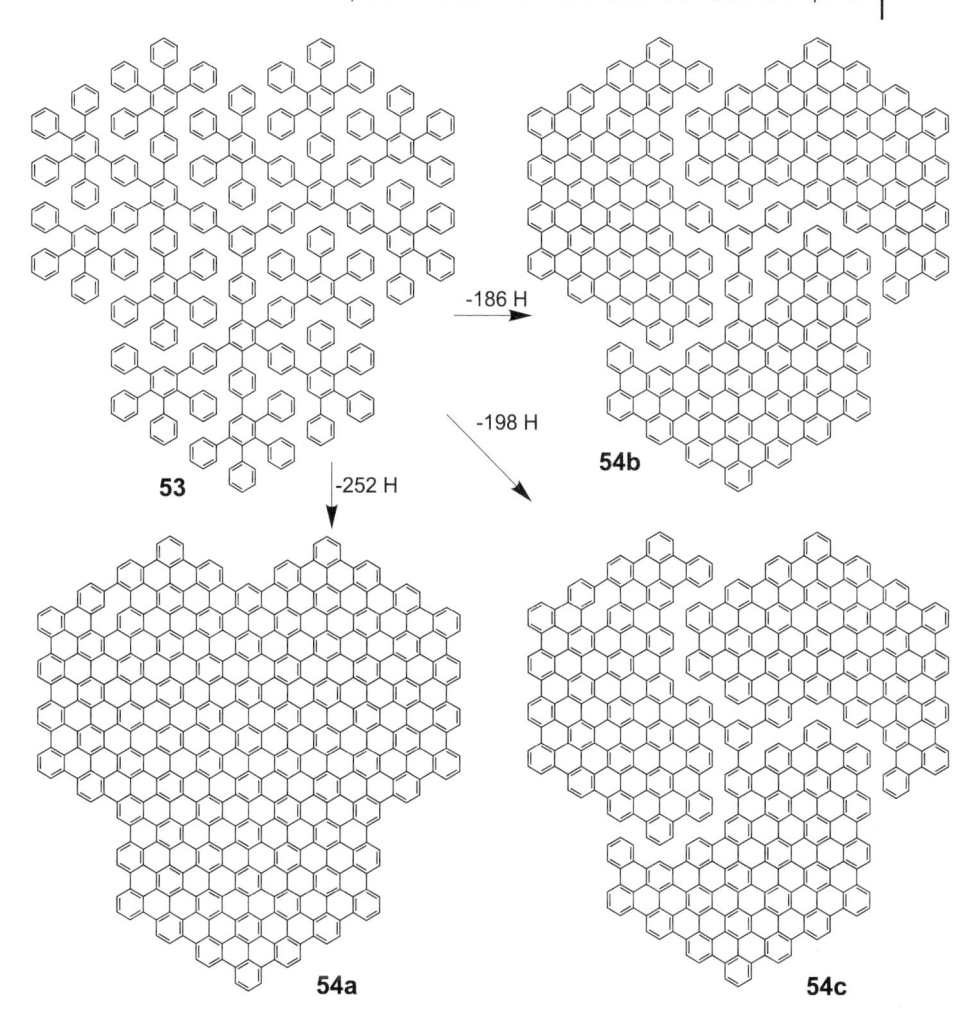

3.2 All-benzenoid PAHs – Synthesis, Structural Characterizations and Electronic Properties | 111

Scheme 3.17. Partial cyclodehydrogenation of huge oligophenylene dendrimers leads to three-dimensional graphite propellers.

core were prepared by a similar concept via a multistep procedure (Fig. 3.13) [58]. In analogy to the effective conjugation lengths in conjugated polymers, the electronic absorption maximum red-shifted with increasing molecular aspect ratio due to the ever-increasing delocalized π-surface. More recently [59], one-dimensional polymeric graphite ribbons **60** were synthesized by intramolecular oxidative cyclodehydrogenation of soluble branched polyphenylenes **61**, obtained by Diels-Alder polymerization of 1,4-bis(2,4,5-triphenylcyclopentadienone-3-yl)benzene **62** and diethynylterphenyl **63** in good yield (Scheme 3.18).

While the insolubility of the resulting graphite ribbons **60** precluded standard spectroscopic structure elucidation, their electronic and vibrational properties were

55 R = H, t-C$_4$H$_9$, C$_{12}$H$_{25}$

56 R = H, t-C$_4$H$_9$

Figure 3.12. Chemical structures of "supernaphthalene" and "supertriphenylene".

probed by solid-state UV–vis, Raman and infrared spectroscopy. A wide and unstructured absorption band covering the visible range of the electronic spectrum ($\lambda_{max} \sim 800$ nm) was observed, confirming the highly extended conjugated framework. The profile of the visible Raman spectrum of the material is characterized by two strong bands (at 1603 cm^{-1} and 1322 cm^{-1}), corresponding to the G and D

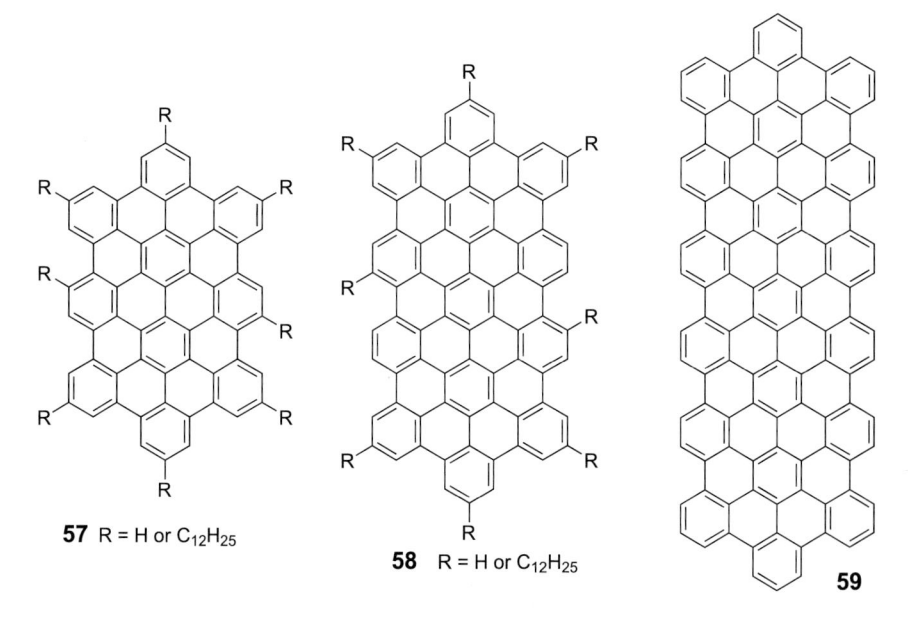

57 R = H or C$_{12}$H$_{25}$

58 R = H or C$_{12}$H$_{25}$

59

Figure 3.13. Chemical structures of ribbon-like graphite sheets.

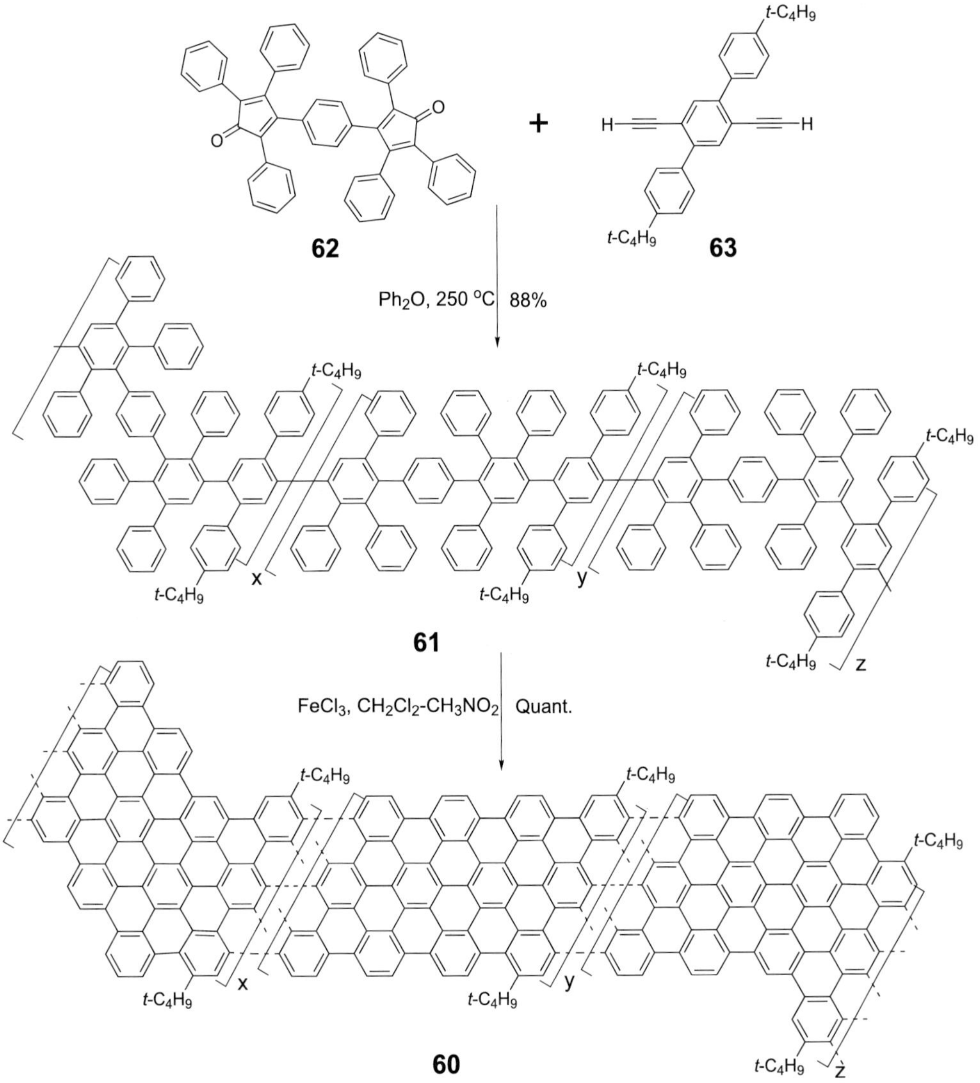

3.2 All-benzenoid PAHs – Synthesis, Structural Characterizations and Electronic Properties

Scheme 3.18. Synthesis of one-dimensional graphite ribbons.

bands of graphite. The obtained graphite ribbons are not linear, but rather contain "kinks" due to the isomeric substructures in the polyphenylene precursor. High resolution transmission electron miscroscopy (HRTEM) images of the black precipitate of **60** disclosed two different domain types: one contains an ordered graphite layer structure with a layer distance of 3.8 Å, while the other appears as a disordered area, as would be expected due to the nonlinear structure of the obtained graphite ribbons.

The optical absorption properties of graphitic discs with different sizes were studied and it was found that the maximum wavelength of the lowest energy absorption band (α-band) increases linearly with the number of "full" phenyl rings and the spectral features becomes increasingly broadened [29b]. This behavior is very similar to that of conjugated oligomers [60]. The charge transport properties of alkylchain-substituted graphite discs were studied by PR-TRMC techniques [61]. For compounds which are crystalline solids at room temperature, the one-dimensional charge carrier mobilities ($\Sigma\mu_{1D}$) lay in the range of 0.4~1.0 cm^2 V^{-1}S^{-1}; in the columnar mesophases, the $\Sigma\mu_{1D}$ was somewhat lower and varied from 0.08 to 0.38 cm^2 V^{-1}S^{-1}. The mobility values in both phases are considerably larger than the maximum found previously for discotic triphenylene derivatives. However, the charge carrier mobility does not follow a linear increase with the increase in molecular size, suggesting that besides the size, the supramolecular organization and order of the discs also play an important role in charge transport.

The above results show that the standard synthetic concept for HBCs can also be applied in the synthesis of larger graphitic discs and ribbons, thus opening the door to a broad range of versatile electronic materials with different functions.

3.2.3
PAHs with Varying Peripheries

Besides the shape, another important factor that affects the electronic properties and chemical reactivity of PAHs is the nature of the periphery. According to Clar's classification, the graphitic molecules with "armchair" and "cove" peripheries shown in Fig. 3.14 (A and B) are all-benzenoid PAHs. In addition to these linear topologies, Stein and Brown considered two other peripheral structures, i.e. "acene-like" (C) and "quinoidal" (D) structures, which lie in a higher energy state and thus show higher chemical reactivity [62].

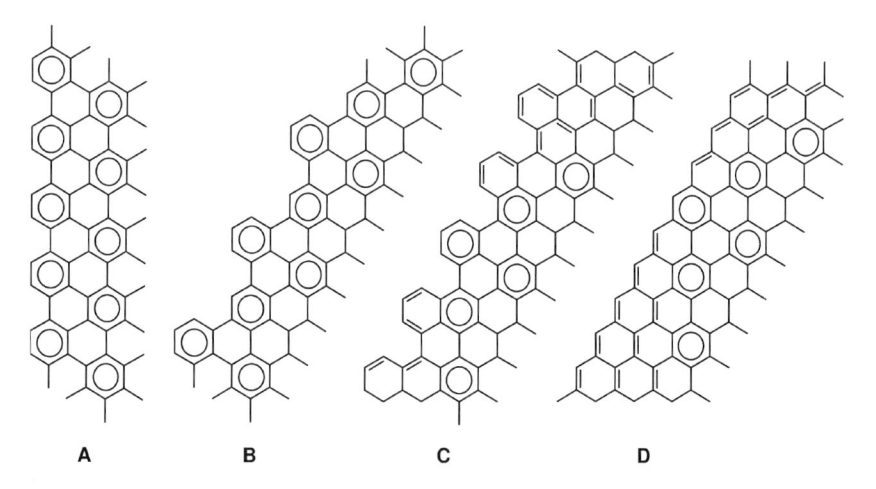

Figure 3.14. Peripheral topologies of large PAHs: (A) and (B) all-benzenoid, (C) acene-like and (D) quinoidal.

64

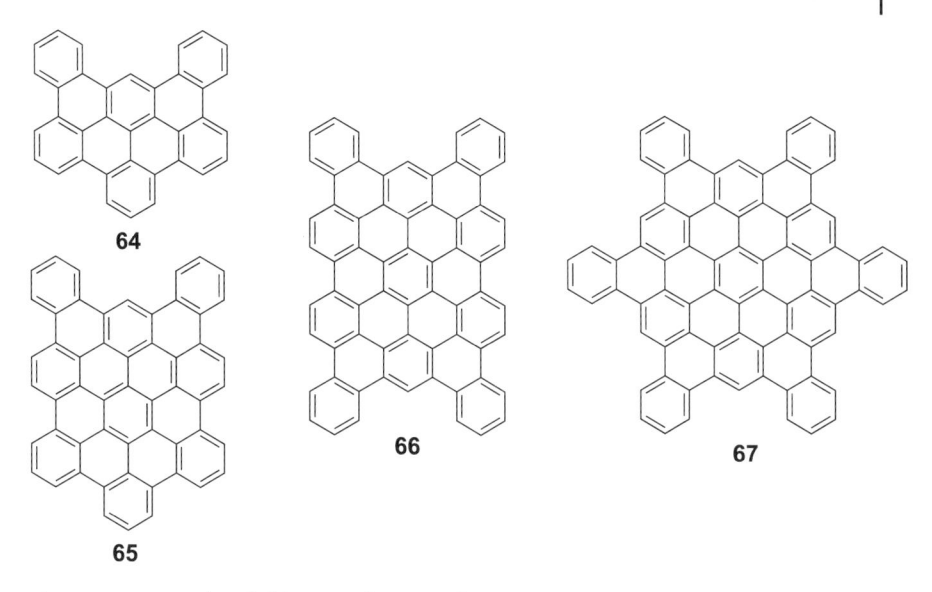

65

66

67

Figure 3.15. Examples of all-benzenoid PAHs with "cove"-type edge.

Several all-benzenoid hydrocarbons **64–67** with "cove"-type instead of "armchair" periphery have been prepared (Fig. 3.15) [35, 63]. The resonances of the protons in the "cove" position are strongly shifted to low-field in ^{1}H NMR spectra due to the deshielding of the aromatic surroundings. The UV–vis absorption spectra of these compounds in solution as well as in thin film show three types of bands (α, β and p) which are typical of polycyclic aromatic hydrocarbons. Similar to other all-benzenoid PAHs, the absorption maxima of these "cove"-types PAHs are progressively red shifted with increasing size of the π-system.

For the "acene" and "phene" series, the UV–vis and fluorescence spectra shift dramatically when the number of the phenyl rings increases. In contrast, the shifts for all-benzenoid PAHs with either "armchair" or "cove"-type edges are small and they show a high chemical stability. Very recently, graphitic molecules with partial "zig-zag" periphery such as **68a**, **68b** and **73** (Scheme 3.19) were synthesized. It was found that the introduction of two or six extra π-centers onto the all-benzenoid graphitic molecules dramatically influences their electronic properties, chemical reactivity and two- and three-dimensional self-assembly [64].

As shown in Scheme 3.19, the key building blocks, the cyclopentadienone derivatives **69a** and **69b**, were first prepared by Knoevenagel condensation between pyren-4,5-dione derivatives and benzil. Diels-Alder cycloadditions of **69a–b** with substituted diphenylacetylenes or 1,3,5-triethynylbenzene afford precursors **71a–b** and **72**, respectively. Oxidative cyclodehydrogenations using FeCl$_3$ or Cu(OTf)$_2$/AlCl$_3$ provide the desired graphitic molecules **68a–b** and **73** with partial "zig-zag" edges in high yield. The comparative UV–vis absorption and fluorescence spectra of **68a** and parent HBC (**10**) recorded in 1,2,4-trichlorobenzene are shown in Fig.

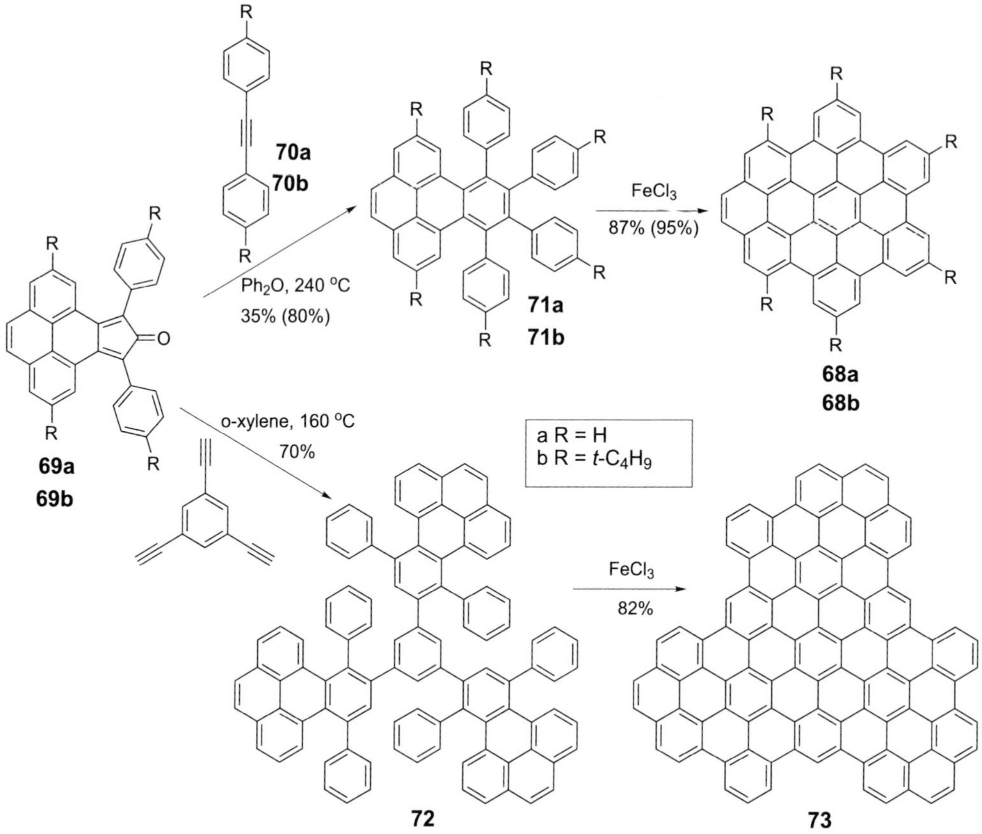

Scheme 3.19. Synthesis of graphitic molecules with partial "zig-zag" periphery.

3.7. The absorption maximum of **68a** ($\lambda_{max} = 380$ nm) shows a significant bathochromic shift with respect to the corresponding band of HBC **10** ($\lambda_{max} = 359$ nm) (0 → 0 band of the $S_0 \to \beta$ transition). Reducing the symmetry increases the intensity of 0 → 0 band of the α (or L_b) transition (486 nm) in the fluorescence spectrum of **68a**, which is otherwise very weak (symmetry forbidden) for the D_{6h}-symmetric unsubstituted HBC **10**. Interestingly, there is also a characteristic band at 438 nm that is not found in the all-benzenoid analog. ZINDO-CIS quantum chemical calculations were carried out to estimate the excitation energies and oscillator strengths for the B3LYP/3-21G equilibrium structure of this molecule. It was found that a remarkable net transition dipole moment is associated with the 0 → 0 band of the $S_0 \to L_a$ (or $S_0 \to p$) transition. It is clear that this unusually strong band originates from the strong electronic perturbation induced by the two additional π centers. The UV–vis absorption spectrum of **68b** is essentially unchanged with reference to the unsubstituted parent compound **68a**, in accord with the relatively weak donor effect of *tert*-butyl groups. A small bathochromic shift of about 6 nm with respect to **68a** is probably due to the expected nonplanarity caused by steric congestion. Crystals of **68b** suitable for X-ray structure analysis were obtained by slow evaporation of a hexane solution at room temperature. The molecule is bent owing to the steric interaction of the *tert*-butyl groups. The "outer rings" are alternatively bent up and down by approximately 15° each. The nonplanarity of **68b** hinders the formation of the stacking and thus provides cavities which are filled by the solvent. The graphitic molecule **73** with more extended "zig-zag" character was characterized by MALDI-TOF MS and solid UV–vis spectra. The absorption maximum of **73** ($\lambda_{max} = 535$ nm) shows a significant bathochromic shift (45 nm) with respect to its parent PAH without "zig-zag" edges. The additional two or six π-centers also provide reactive sites for further functionalization, e.g., preparation of its corresponding K-region oxide, thus opening up a wide range of "graphene" chemistry [65].

3.2.4
"Superbenzene" Chemistry and Others

Although HBC has a structural analogy to benzene, it displays a lower reactivity due to the low atomic orbit coefficients of the edge-carbon atoms. A series of reactions which are well-known in "benzene" chemistry such as halogenation, sulfonation, nitration only afford disappointing results with HBC due to the lack of regioselectivity and complexity of the products. However, recently we found that the catalytic region-specific hydrogenation of HBC and its derivatives gave peralkylated coronenes [66].

The HBC derivatives **74** carrying different substitutents are combined with Pd/C (10 wt.%) in dry THF, pressurized to 55~65 bar H_2, and heated to 60 °C (Scheme 3.20). The hydrogenation of n-alkyl chain-substituted HBCs gave the peralkylated coronenes **75** in nearly quantitative yield. The parent HBC (**10**), however, gave lower yield (10%) after a long reaction time, most probably due to its poor solubility. Increasing the steric bulk of the side chains was found to hinder the reaction. The structures of the peralkylated coronenes were confirmed by MALDI-TOF, NMR and UV–vis absorption spectroscopy. The puckered-ring periphery of these new discs does not prohibit self-assembly to columnar structures in a fashion sim-

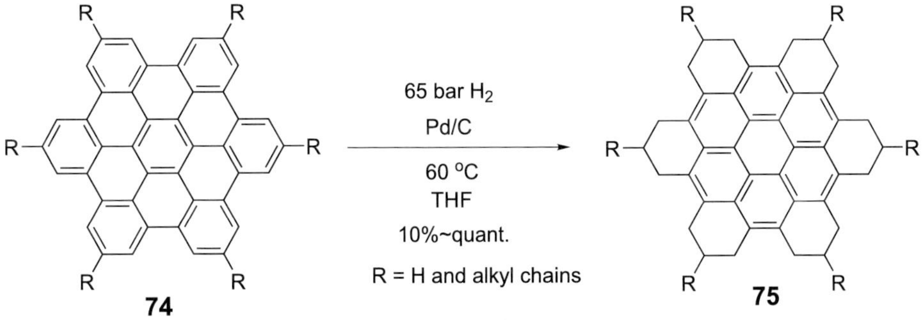

Scheme 3.20. Regiospecific hydrogenation of HBC derivatives.

ilar to that of planar HBC precursors but decreases the isotropic temperature by ~300 °C relative to the latter. The charge carrier mobility and lifetime within the bulk materials is still of the same order as those previously found for HBCs. A similar conversion of a homolog with 60 core carbons to the first peralkylated circumbiphenyl was also achieved [53].

HBC was also designed as a supramolecular building block in cyclophanes. Cyclophanes **76a–b** with two HBCs covalently linked by flexible alkyl chains were synthesized by intermolecular ring-closing olefin methathesis of dienes in good yield (Scheme 3.21) [67]. The HBC diene **77** was first prepared in high yield via a Pd-catalyzed Kumada coupling reaction between the dibromo-substituted HBC **23b** and 4-pentenylmagnesium bromide. Ring-closing olefin metathesis of diene **77** was effected in toluene at 0.4 mM concentration catalyzed by the Grubbs catalyst ($[(PCy_3)_2Cl_2Ru=CHPh]$). Cyclophane **76a** was obtained in 64% isolated yield. The $\pi-\pi$ stacking of the HBCs in solution was expected to promote the covalent capture of the two HBCs. The olefin linkages were quantitatively hydrogenated over Pd/C to produce cyclophane **76b**. Solution spectroscopic studies of these cyclophanes indicate a face-to-face intramolecular arrangement of the discs with a slight lateral displacement at room temperature. The new cyclophanes show well-defined, columnar liquid crystalline phases in the bulk state, which correspond to the arrangement of monomeric discs; that is, the covalent linkages do not cause a measurable variation in the inter- and intramolecular stacking. Self-assembly into two-dimensional crystals at a solid/liquid interface was visualized by STM, and the

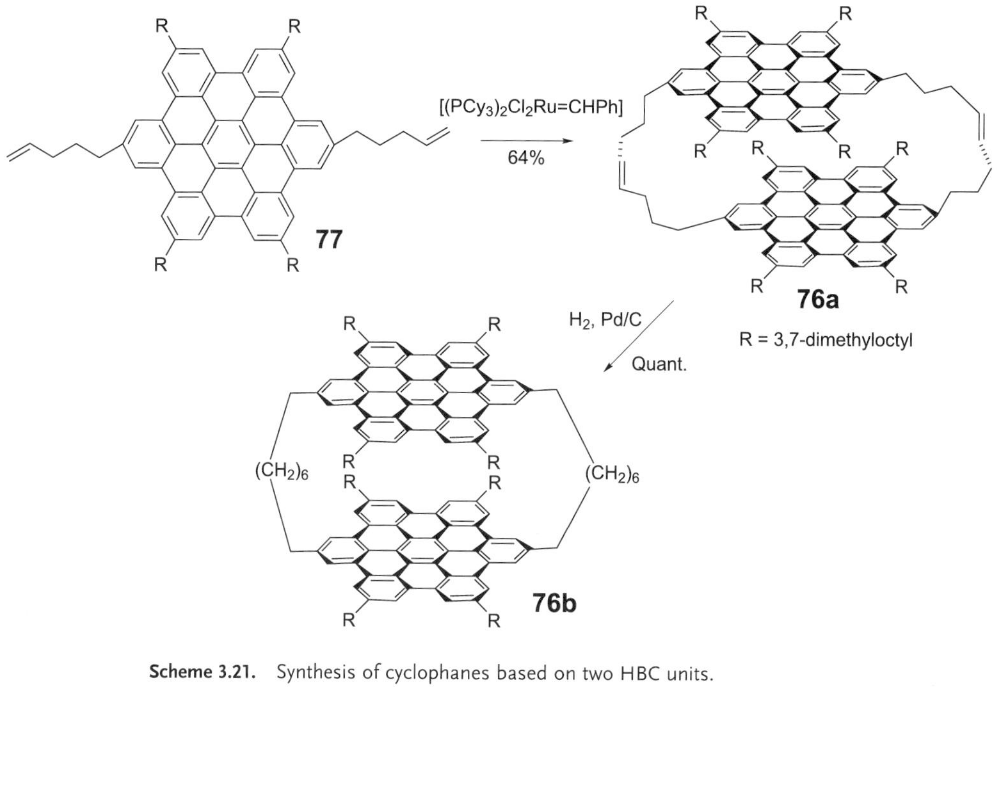

Scheme 3.21. Synthesis of cyclophanes based on two HBC units.

electronic properties of single molecules were assessed by scanning tunneling spectroscopy, revealing an unsymmetric diode-like behavior, i.e., the current tunneling through the HBC cores and the alkyl chains under positive and negative bias are different due to their different molecular orbital energy levels [34b].

The inclusion of heteroatoms such as nitrogen into the graphene structure is expected to dramatically change its electronic properties, as it does in N-doped carbon nanotubes [68]. Draper and coworkers prepared a "superbenzene" **78**, containing four nitrogens, by a similar process to the synthesis of the unsymmetric HBCs (Scheme 3.22) [69]. The precursor **81** was prepared by Diels-Alder cycloaddition between the bipyrimidylacetylene (**79**) and *tert*-butyl substituted tetraphenylcyclopentadienone (**80**). Cyclodehydrogenation of **81** using $CuCl_2$–$AlCl_3$ in CS_2 affords the first heterosuperbenzene **78**. The presence of the imine N atoms has increased the overall electron-accepting properties by comparison to the all-C analog. The strong green emission can be quenched by addition of acid which reacts with the imines. Metal coordination of **78** with Pd(II) and Ru(II) was done by adding suitable metal salts (Scheme 3.22) [70]. The Pd(II) coordination in complex **82**

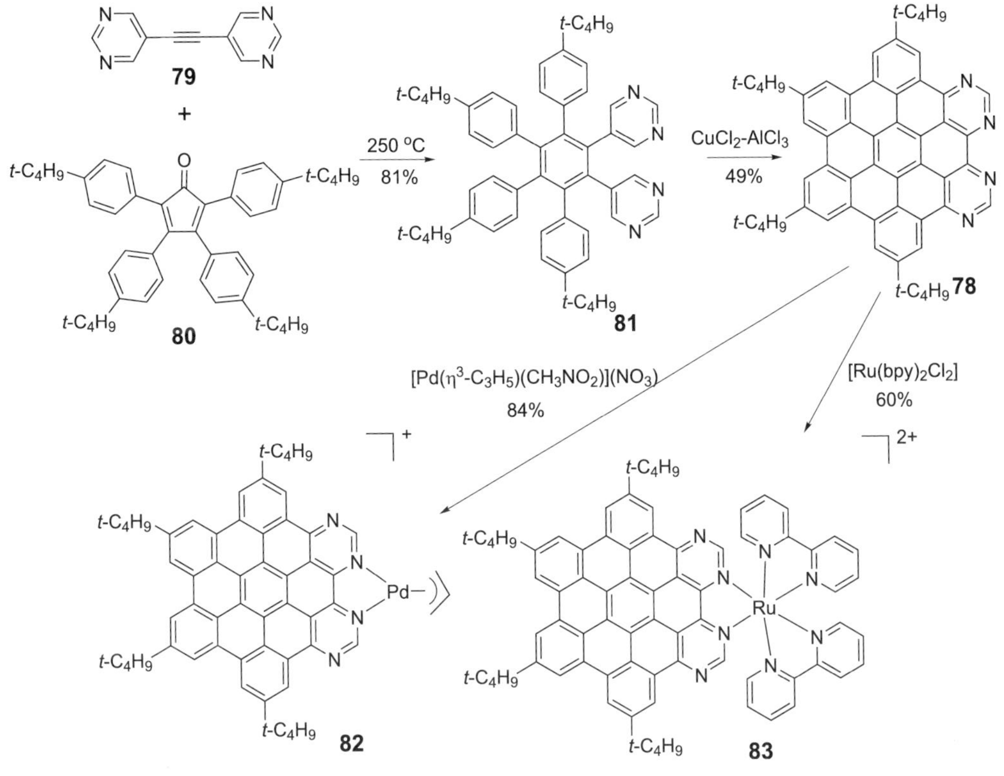

Scheme 3.22. Synthesis of nitrogen-containing "superbenzene" and its coordination with transition metal complexes.

causes a red-shift in the low-energy absorptions, a decrease in the intensity of the n–π* absorptions, and a quenching of the emission. The Ru(II) coordination complex 83 is a "black" MLCT absorber and a near-IR emitter, and has potential applications in solid-state electroluminescence devices.

In this section, we have presented the versatile synthesis of all-benzenoid polycyclic aromatic hydrocarbons with different sizes, shape and functionalities. Their excellent charge transporting properties indicate that they are suitable active semiconductor materials for organic electronic devices. To fully evaluate these molecules for such purposes, detailed studies on their self-assembly in the bulky state, in solution and even at the solid/liquid interface are necessary to understand the basic relations between the molecular structure, processing and self-assembly.

3.3
Self-assembly and Application of Columnar Liquid Crystals based on PBAHs

3.3.1
Columnar Superstructures in the Bulk State

One drawback of conjugated polymers as semiconductors in organic electronics is the formation of structural defects during synthesis or processing. The alkyl chain substituted all-benzenoid polycyclic aromatic hydrocarbon molecules can form self-assembled one-dimensional columns due to the strong π–π interactions between the aromatic cores and a natural nanophase separation. Under heating, these disc-like molecules form columnar liquid crystalline phases. The coin-like aromatics stack in a face-to-face motif, forming a 1D charge transport channel, thus mimicking that in the 1D conjugated polymers. Recent results showed that charge carrier mobilities as high as 1.0 cm^2 V^{-1} S^{-1} can be reached in these discotics [43]. There are several possible advantages of using alkyl chain-substituted discotics molecules: (i) They can be obtained as defect-free materials; (ii) they can be processed from solution or melts due to their low viscosity and (iii) they have some self-healing ability after columnar formation. One key point in using these materials in electronics is to control their supramolecular order to achieve high charge carrier mobility, and this can be tuned by (i) the size of the discs and the nature of their peripheries; (ii) the length and branching of the alkyl chains which allow the engineering of the phase transition and (iii) the functional substituents which bring additional noncovalent forces such as π–π interactions, hydrogen bonding etc..

Triphenylenes substituted with flexible chains have shown a rich variety of columnar phases varying from fully crystalline, through helical, plastic, nematic discotic, etc., which, however, generally require heteroatoms adjacent to the core [71]. Increasing the core size to larger discs such as HBCs induces stronger π–π interactions and enhances phase separation, thus leading to stable mesophases with wide phase widths. For the n-alkyl substituted HBCs (Fig. 3.16), the isotropic temperatures are higher than 450 °C [37]. Introduction of the bulkier branched substituents lowers not only the isotropic temperature to 46 °C, as found in the dove-tailed

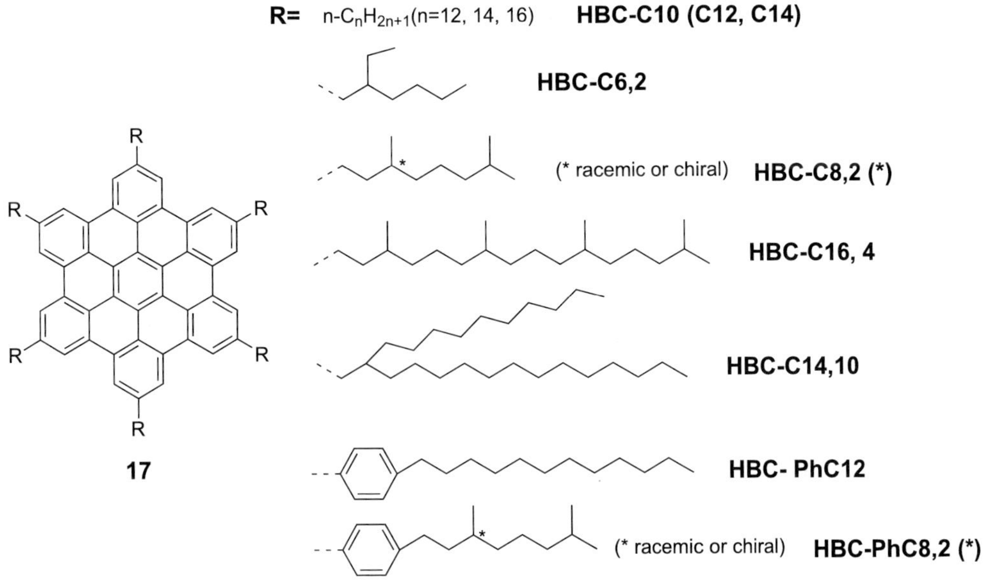

3.3 Self-assembly and Application of Columnar Liquid Crystals based on PBAHs | 121

Figure 3.16. HBCs with different alkyl and alkylphenyl substitutions.

HBC-C14,10, but also lowers the crystalline-to-liquid crystalline phase transition temperatures below room temperature, for example, in the compounds **HBC-C16,4** and **HBC-C14,10** [72, 73]. A wide-range columnar liquid crystalline phase from room temperature up to >400 °C was observed in the dodecylphenyl substituted **HBC-PhC12**, which is capable of uniform thin film formation at room temperature [38]. Further increasing the size of the core to 78 (**67**), 96 (**42**), and 132 (**49**) carbon atoms results in broader columnar liquid crystalline phases which do not become isotropic below 600 °C. Most interestingly, the six dodecyl substituted disc containing 96 core carbons **42a** displays good solubility, a highly ordered liquid crystalline phase at room temperature, good film formation capability and high charge carrier mobility; thus, it is a good candidate for FETs and photovoltaic devices [50]. The low isotropic temperatures (<250 °C) found in some highly branched alkyl chain substituted HBCs allow processing in the melt. The dovetailed alkyl chains in **HBC-C14,10** largely fill the free space around the aromatic core and dramatically influence the self-organization behavior of HBC. Extraordinary long-range self-assembly was observed by polarized optical microscopy (POM) during cooling from the melt [72].

The columnar superstructures in different phases determined by differential scanning calorimetry (DSC) are usually studied by two-dimensional wide-angle X-ray scattering (2D WAXS) assisted by solid-state NMR. The bulky discotic materials are extruded into fibers through a mini-extruder in the liquid crystalline phase and during this process, the discs are aligned along the extrusion direction. The X-ray beam is then perpendicular to the fibers and the 2D X-ray scattering patterns are

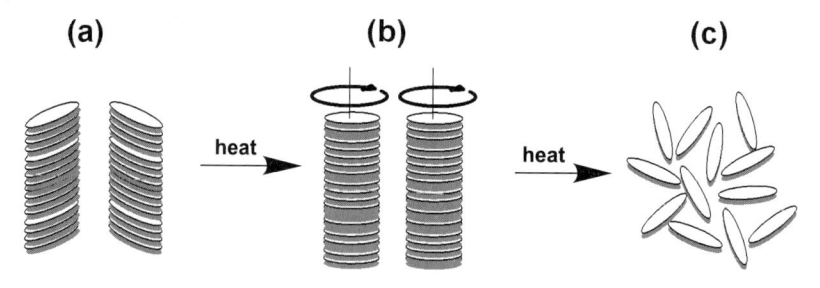

Figure 3.17. Schematic representation of crystalline phase-liquid crystalline phase-isotropic liquid transition for HBC.

recorded at different temperatures. The two- or three-dimensional positional or rotational order of the discs in the bulk state can then be derived from the 2D patterns [74]. The HBC based materials normally have a low temperature "microcrystalline" phase followed by at least one or more columnar liquid crystalline phases. For example, **HBC-C12** shows a crystalline phase below 107 °C and above a hexagonally ordered columnar liquid crystalline phase (D_h) until isotropic melting at about 417 °C. In the crystalline phase, the molecules are three-dimensionally ordered with a low mobility of the discs in the column and the discs are tilted with respect to the columnar axis (Fig. 3.17). In the D_h phase, rotation of HBC discs around the column axis is possible, and the discs are now perpendicular to the columnar axis [75]. In the isotropic phase, the molecules behave like a normal liquid, i.e. no anisotropy exists. Besides the hexagonal lattice, orthorhombic and monoclinic lattices are also observed in the HBCs materials, depending on the molecular symmetry and temperatures. For the larger discotics such as **67**, **42** and **49**, no tilting is observed in the higher temperature columnar liquid crystalline phase and enthalpic gain is achieved by a simple lateral offset of the graphene layer (Fig. 3.18). The strong reflections correlated to the intracolumnar $\pi-\pi$ stacking indicate high order parameters for these large discs [50, 53].

The supramolecular order of the columnar superstructures can be controlled by an additional weak interaction between the arms. For example, in the six-tolane substituted HBCs **34b** and **34c**, highly ordered helical columnar plastic crystalline phases were observed, i.e., the neighboring discs are rotated through 15° with respect to each other so that every fifth disc forms one helicate in a single column (Fig. 3.19) [48b]. The space between the arms is efficiently filled in a tightly packed helix with either edge- or face-to-face interactions of the exo tolanes. It is should be mentioned that this kind of helix is not induced by chiral substituents, but by weak interactions between the arms. A similar helical crystalline phase was observed in the hexa(hexylthio)triphenylenes which show high charge carrier mobility in this phase [76]. Atomic force microscopic measurements on spin-casting or drop-casting film of molecules **34b** and **34c** revealed single-column resolved nanowire and ribbon formation on highly oriented pyrolytic graphite (HOPG) surface. Further control by hydrogen bonding is possible to achieve highly ordered columnar superstructures.

HBC-(C$_{12}$H$_{25}$)$_6$

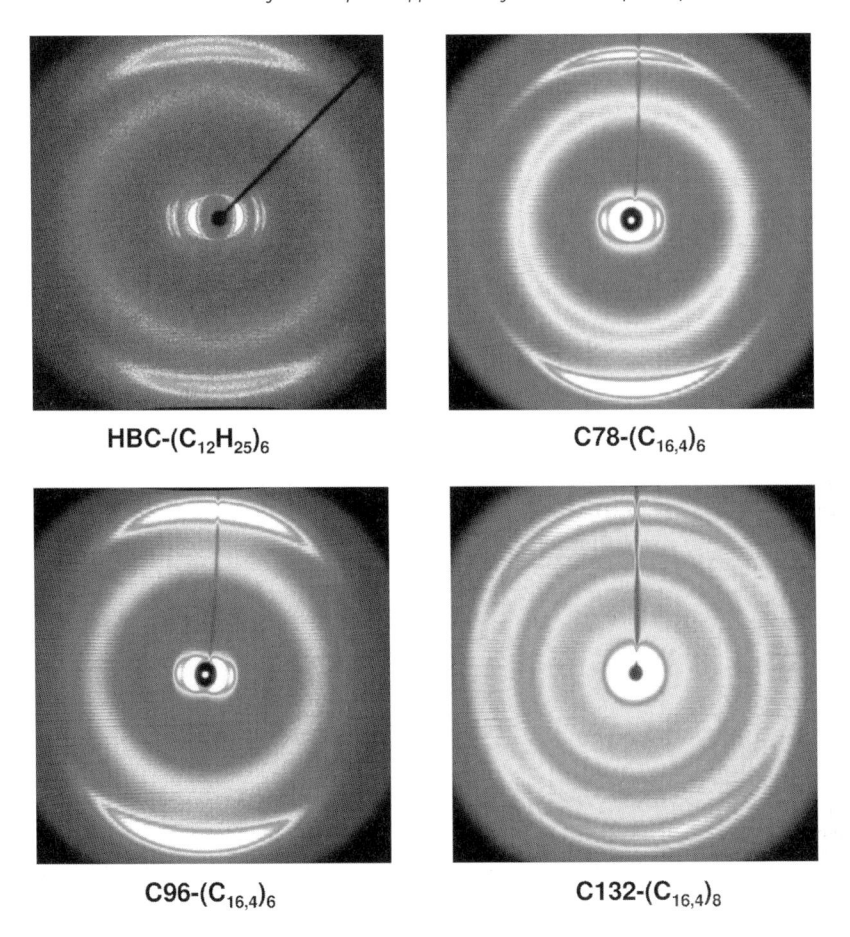

C78-(C$_{16,4}$)$_6$

C96-(C$_{16,4}$)$_6$

C132-(C$_{16,4}$)$_8$

Figure 3.18. 2D-WAXS diffractograms of C42 (**HBC-C12**), C78 (**67**), C96 (**42a**) and C132 (**49**) in the mesophase state showing the well-ordered, columnar structure. Reproduced by permission of The Royal Society of Chemistry.

3.3.2
Alignment on Substrates and Device Applications of Columnar Liquid Crystals

In devices, the control of the alignment of the discotic materials is the key issue for achieving good performance criteria such as high charge carrier mobility. Although the discotics may self-assemble into fibrillar or ribbon-like nanostructures from drop-casting or spin-casting films, the random orientation of assembles disappointingly leads to low performance in devices. Homogeneous edge-on alignment of the discs on substrates is required in the field-effect transistors, in which two electrodes are contacted with the aligned discs along the orientation direction (Fig. 3.20). In photovoltaic devices, however, the discs are desired to have a homeotropic alignment with a face-on arrangement. The donor columns are accompanied by home-

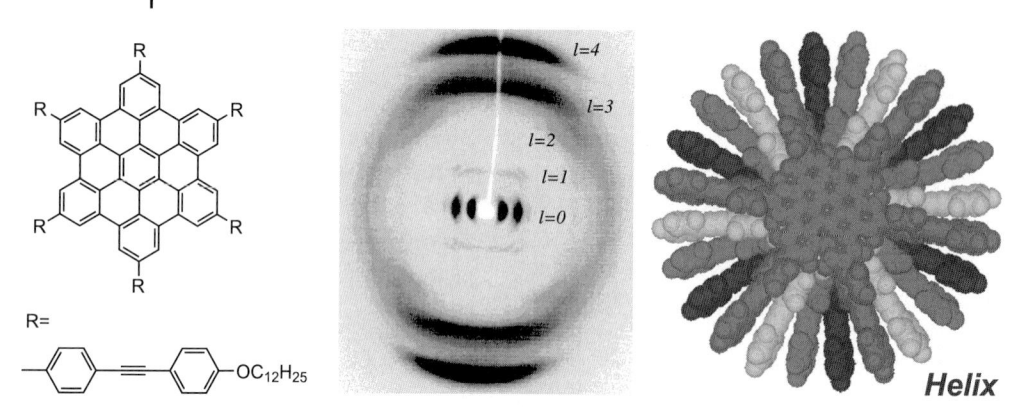

Figure 3.19. Chemical structure, 2D-WAXS diffractogram and proposed helical 3D-arrangement of HBC **34b**.

Field Effect Transistor: "edge-on"

vs.

Photovoltaic Device: "face-on"

Figure 3.20. Schematic representation of two different electronic device types and their desired arrangement of discotic molecules as electronically active components: edge-on for FETs versus face-on for photovoltaic devices. Reproduced by permission of The Royal Society of Chemistry.

otropically aligned acceptor columns. Various methods can be used to induce order in the macroscopically uniformly oriented thin films.

Simple mechanical shearing may induce orientation of the discotic materials, as shown in the 2D WAXS measurements; however, such techniques cannot induce uniform orientation of large domains. The normal alkyl chain substituted HBCs adopt a random edge-on arrangement on polar substrates such as glass and mica, as evidenced by the typical textures in POM measurements. When discotic materials with low isotropic temperature are sandwiched between two glass or ITO slides, thermal annealing around the isotropic point may result in partial formation of a thermodynamically stable homeotropic face-on arrangement. The photocurrent of the homeotropic areas is distinctly larger than that of the random areas [73]. Recently, it was found that introduction of oxygen atoms into the branched alkyl chains on the HBCs (**84a–d**) dramatically influenced the alignment of these discs after slow cooling from the isotropic state between two surfaces or on one surface as a thin film (Fig. 3.21) [77]. The affinity of the oxygen atoms with the surface may play an important role in controlling the thermodynamic process. Large mono-domain homeotropically arranged HBC thin films can be obtained which have obvious potential in photovoltaic applications.

Hydrophobic HBC-discs with six hydrophilic oligo(ethylene oxide) chains exhibit highly ordered columnar structures with enhanced molecular segregation in the bulk state [78]. The hydrophilic periphery forces a "face-on" orientation at the air/water interface and homeotropic alignment in cluster-separated thin layers after transfer by the LB-method.

High-efficiency organic photovoltaic devices derive their performance from two basic elements: (i) the dissociation within the active semiconductor layers of the photogenerated excited state, and (ii) the transport of the resulting separated charges to the electrodes. Discotic liquid crystals such as HBC-based materials benefit from their electronic donor properties, high one-dimensional hole-transport

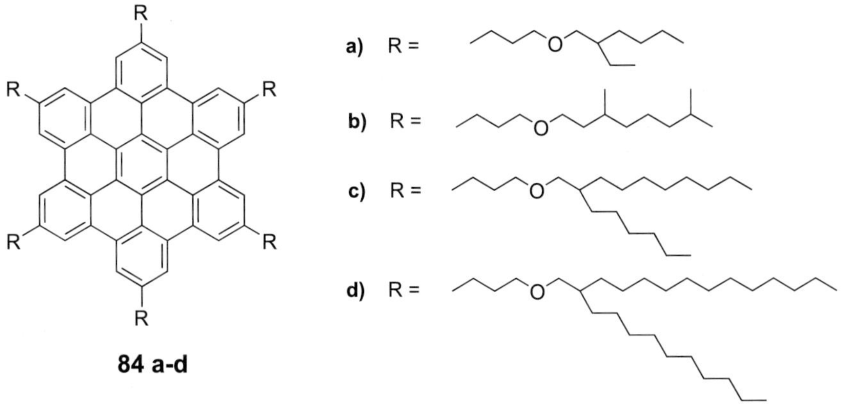

Figure 3.21. HBCs substituted with dove-tailed ether chains which can undergo homeotropic alignment after cooling from the melt.

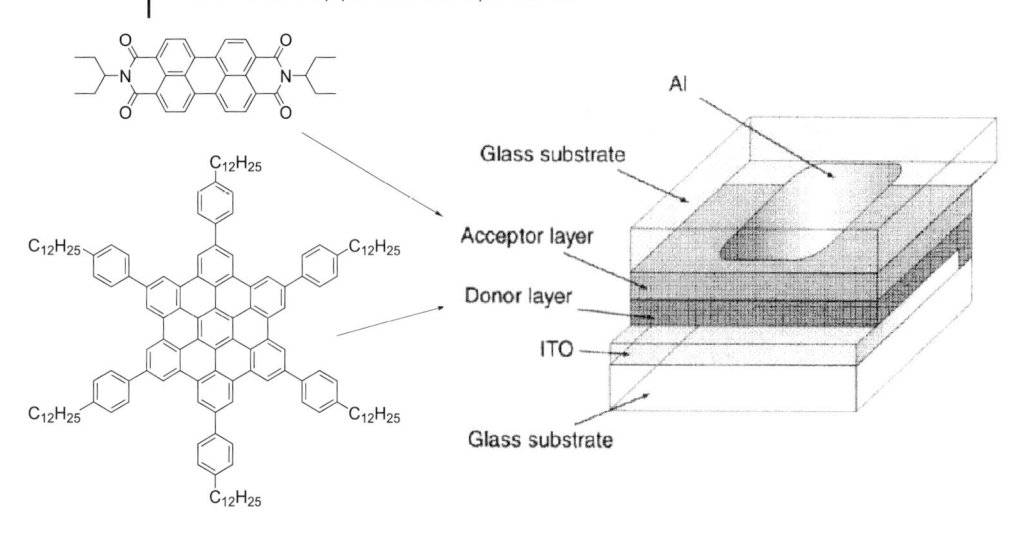

Figure 3.22. Highly efficient photodiodes based on discotic LC (**HBC-PhC12**) and crystalline (PDI) materials.

mobility and good solution processing. Major progress towards such devices was made by Schmidt-Mende et al. [79]. A mixture of the liquid crystalline **HBC-PhC12** and crystalline perylene diimide (PDI, electronic acceptor) in solution was spin-coated onto an ITO substrate (Fig. 3.22), and the obtained photodiodes exhibited extremely high external quantum efficiency (EQE = 34% at 490 nm). The facile phase separation between the HBCs and the crystalline acceptors, the large overlap of the donor and acceptor on the interface, the high charge carrier mobility along these discotic materials and the good molecular energy level gradient between the HBC and PDI are the crucial factors. Such simple but highly efficient methods are very promising for practical solar cell applications using discotic LCs. Use of the above described homeotropically arranged discs in such devices is under investigation.

The large area edge-on arrangement of discs can be obtained in two different ways. One is to utilize pre-oriented substrates. The nanoscale grooves in the surface of friction-transferred polytetrafluoroethene (PTFE) have been proven to be efficient templates for aligning discotic molecules such as triphenylenes and HBCs [80, 81]. The discotic materials were cast from solution onto the surface of rubbed PTFE and the solvents were evaporated slowly. An uniaxial aligned layer of discs along the rubbed direction was obtained, and uniform edge-on arrangement of the discs was proved by POM, surface X-ray diffraction, and electron diffraction techniques [81]. The perfect orientaion of discs on the PTFE surface is due to the dimensional matching of the discs with a certain number of PTFE chains [82].

A markedly simple and straightforward approach to alignment of the layer of discs is the so-called zone-casting technique. Deposition of an HBC solution from a stationary nozzle onto a moving substrate (glass or ITO), both of which are temperature-controlled, produces concentration and temperature gradients result-

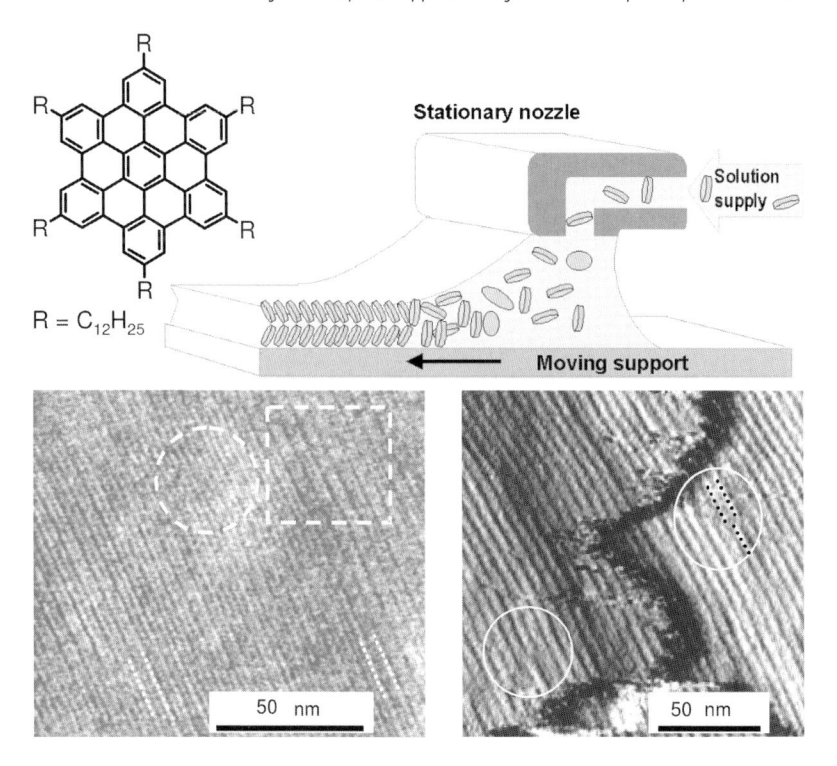

Figure 3.23. Schematic representation of zone-casting process and AFM images of the resulting films of **HBC-C12**, showing details of the long-ranged order columnar structures with some defects (in circle). Reproduced by permission of The Royal Society of Chemistry.

ing in uniaxial columnar growth (Fig. 3.23) [83, 84]. The uniform edge-on columnar arrangement of the large domain oriented layer was confirmed by AFM, transmission electron microscopy and surface X-ray diffraction techniques. The AFM images show extremely long columns with single crystalline order over several square centimeters. Some unavoidable structural defects were observed. Such a simple method is very promising for the alignment of other kind of discotics and conjugated polymers and may be used in fabrication of FET devices.

FET devices based on aligned HBC layers achieved by both methods were fabricated in a top-contact mode [81, 84]. An on-off ratio of 10^4 and a high field-effect mobility of 5×10^{-4} cm^2 V^{-1} s^{-1} were found for oriented HBC derivatives on an aligned PTFE layer on silicon. The oriented **HBC-C12** thin film prepared by the zone-casting technique, however, afforded much higher field-effect mobility up to 5×10^{-3} cm^2 V^{-1}s^{-1} in the saturated regime and a turn-on voltage of ca. −15 V. Thus, the zone-casting technique opens new pathways for fabricating large-scale thin layers of highly oriented semiconductors and enhancing their charge transport. This is an important step towards improving the performance of organic semiconductors in electronics.

Traditional Langmuir and Langmuir-Blodgett (LB) techniques are well-established methods for creating ordered monolayers and multilayers. Thus an amphiphilic HBC bearing five alkyl chains and one alkyl chain substituent terminated with a carboxylic acid group was used in LB film fabrication [85]. Two different packing arrangements at the air/water interface were observed as a function of applied pressure: under low pressure very well ordered π-stacked lamellae of HBC molecules tilted at about 45° relative to the surface normal were observed, while under higher pressure, the coherence of the π-stacks was lost. The LB-technique approach to ordered HBC films was further modified by using silicon wafers, functionalised with high molecular weight polyethylene imine (PEI), as substrates [86, 87]. The PEI serves as an anchor point for the first layer by forming an ionic interaction complex with the carboxylic acid function of the HBC. This resulted in highly ordered HBC films with a defined number of layers and thus thickness. The oriented layers achieved by LB techniques are also potential candidates for FET devices.

3.3.3
Controlled Self-assembly in Solution

Ordered thin film formation of discotic molecules in the bulk state strongly suggests a study of the self-assembly of the discs in solution. The aggregation of the discs has a great influence not only on their order, but also on the photophysical and electronic properties of each disc [88]. Like all other conjugated macrocycles or discs, the n-alkyl substituted HBCs show dynamic self-assembly in solution, i.e., a polydisperse mixture of aggregates such as dimers, trimers, and larger aggregates existed with the distribution dependent on the concentration and temperature. The high tendency of these alkylated HBCs, such as **HBC-C12**, to aggregate was demonstrated by concentration and temperature-dependent ^{1}H NMR spectroscopic measurements and nonlinear least squares analysis of the experimental data. To further control the self-assembly of HBCs in solution, bulky substituents such as oligophenylene dendrons are attached to the HBC core by Diels-Alder cycloaddition reactions from the ethynylene containing HBCs such as **34a** (Fig. 3.24) [89]. The rigid dendrons in molecule **85a** suppress the π–π interactions of the HBC cores to a certain extent and a slow (with respect to the NMR time scale) monomer–dimer equilibrium is observed. This unique equilibrium was further controlled by temperature, concentration, and solvent to afford discrete monomeric or dimeric species. Further structural modifications such as the replacement of dodecyl groups in **85a** with hydrogen atoms resulted in a stable dimer structure in **85b** due to diminished steric hindrance, as supported by quantum chemical calculations. Attaching the dendron arms more closely to the HBC core, as in molecule **86**, results in a material which exists only as a nonaggregated monomer. UV–vis absorption and fluorescence spectra of these discrete species revealed obvious differences in their electronic and optoelectronic properties which can be explained by the existence or absence of π–π interactions. This study of the controlled self-

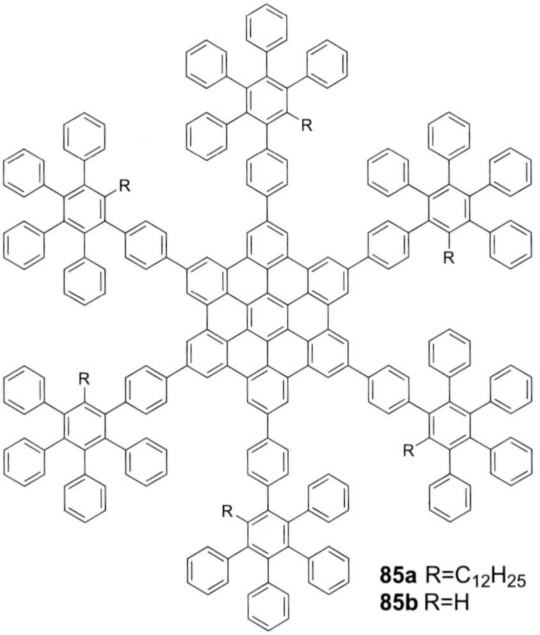

Figure 3.24. Oligophenylene dendrons substituted HBCs which show unique self-assembly in solution.

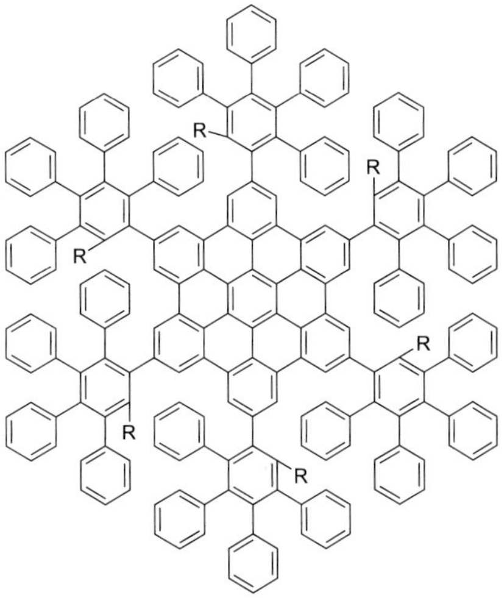

assembly of HBC molecules in solution shows how structural and environmental factors can affect the supramolecular behavior and electronic properties of disc-shaped π-systems. The extremely slow monomer–dimer equilibrium of **85a** is rarely observed in synthetic nonbiological systems. Further control of this dimerization should be possible, for example, by introducing additional weak interactions such as hydrogen bonding and static attraction/repulsion between the dendrons.

3.3.4
Two-dimensional Crystals at the Solid/Liquid Interface

The thermodynamically stable homeotropic arrangement of discotic materials on one or two glass surfaces indicates a special interaction of the aromatic discs with the surface. Vacuum deposition of parent HBC onto different substrates revealed that the HBC thin films showed an ordered face-on arrangement on HOPG, Au(111), MoS$_2$, and Cu(111) surfaces due to epitaxial growth [90, 91]. Highly ordered thin films of alkylated discs on MoS$_2$ were also obtained from solution processing. The electronic structures of these thin film were studied by angle-resolved X-ray photoelectron spectra in combination with low-energy electron diffraction (LEED) [92]. In the following, we will focus on the ordered growth of 2D crystals of the discs on the HOPG surface in solution, studied by scanning tunneling microscopy (STM) and scanning tunneling spectroscopy (STS) techniques. The

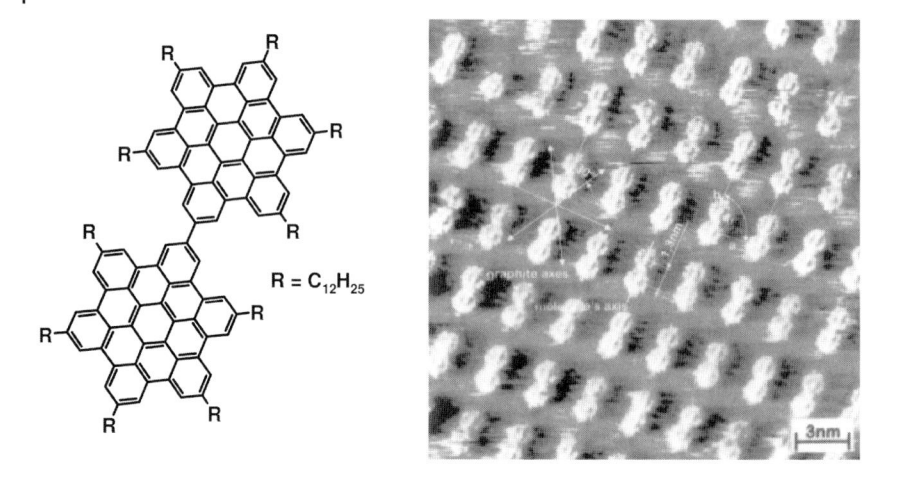

Figure 3.25. STM images of HBC dimer on the HOPG surface.

strong van der Waals interactions between the HOPG substrate and graphitic discs as well as alkyl chains induce ordered monolayer or bilayers at the solid/liquid interface.

Alkylated HBCs form regular monolayers on the HOPG surface in solution and asymmetric I/V characteristics were measured by STS [34b]. This diode-like effect has been attributed to positional asymmetry of either the vertical placement of the disc in the gap or of the frontier molecular orbitals relative to the electrodes. The alkylphenyl substituted HBCs self-assemble into 2D crystals of the discs with variable vertical displacements from the substrate [93]. When the alkylphenyl chains contain chiral centers, a regular staircase superstructure results. A submolecular visualization of a covalently linked HBC dimer on the HOPG surface revealed a contrast, which reflects the structure of the aromatic parts of the molecule, with the aromatic moieties being oriented like graphene layers in graphite (Fig. 3.25) [94].

Processing of large vanishingly soluble discs such as parent HBC and C132 **49** is very difficult. However, epitaxial layers of these unsubstituted discs with electron acceptors can be obtained by self-assembly from solution [95]. STM studies reveal that the interaction with the substrate induces a strong perturbation of the electronic structure of the pure donor in the first expitaxial monolayer. In the second epitaxial layer with a donor/acceptor stoichiometry 2:1 the molecules are unperturbed.

Other electronic components can be attached onto the HBC core and the self-assembled side-by-side monolayers provide opportunities to fabricate single molecular devices at the solid/liquid interface. The covalently tethered HBC and pyrene units co-assemble to form stable nano-phase-separated arrays [96]. A significantly different packing of the adlayer of the HBC-pyrene dyad **87** on the basal plane of the HOPG surface has been obtained, as shown in Fig. 3.26. This STM current

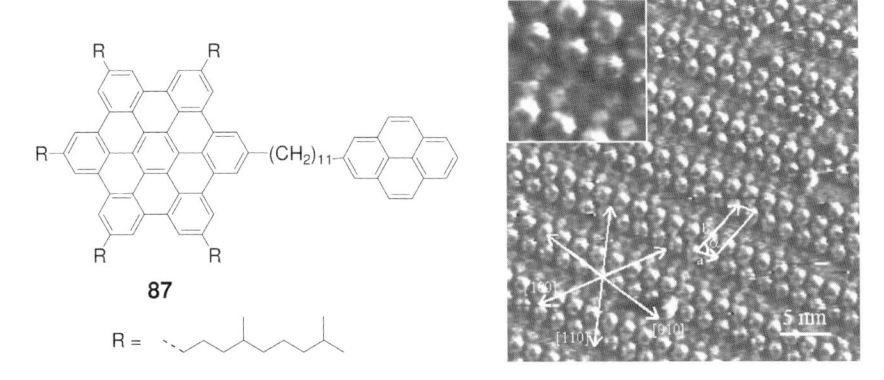

Figure 3.26. Molecular structure of HBC-Pyrene dyad and its nano-phase separated self-assembled monolayer on the HOPG surface.

image clearly reveals a 2D crystalline dimer structure and the bright spots can be attributed to the HBC π-conjugated cores. A careful inspection of Fig. 3.26 reveals some smaller bright spots tightly packed near the HBC cores. This feature becomes even more evident in the zoom-in image depicted in the top left corner and can be ascribed to the pyrene moieties. Other HBC-π-system dyads in which the tethered system is expected to contribute some electronic function, e.g., photo-excited charge separation and transport, will be the next targets. These could provide insight into the role of energy or electron transfer in such a system at surfaces.

3.4 Conclusion

Chemists have been pursuing new polycyclic aromatic hydrocarbons since the beginning of the last century and much knowledge about their synthesis, structural characterization and structure–property relationships has been obtained. The recent discovery of the novel electronic and optical properties of these π-conjugated molecules in electronic devices has promoted intensive studies on PAHs as semiconductors in organic electronic devices. All-benzenoid PAHs as chemically stable semiconducting molecules are extremely interesting considering the prolongation of the lifetime of organic devices. We have developed efficient routes to prepare all-benzenoid PAHs with different sizes, shapes, edge-structures, and substituents. Among them, the discotic liquid crystalline materials based on these mesogens have many merits such as high charge carrier mobility, easy solution processing and self-healing properties and thus are promising for practical device applications. The key issue is to control the supramolecular order and the alignment of the discotics in the bulk state. This can be achieved by both structural modification and the choice of processing techniques, e.g., using alignment layers, zone-casting and LB techniques. The field-effect transistor and photovoltaic devices based on these materials and techniques show promising performance, suggesting that the low-molecular weight discotic liquid crystalline materials could replace conjugated polymers, which usually contain structural defects acquired during chemical synthesis. Our targets in the future are discotic materials with even higher charge carrier mobility which are the product of rational supramolecular design. Related to the film formation from solution, the self-assembly of the discs in solution and thus the electronic properties can also be finely tuned by structural modification of the substituents. The disc-like molecules usually self-assemble into ordered monolayer and bilayers on the solution/graphite interface, which provides an opportunity to build up single-molecule electronic devices. The devices based on these discotics may be reasonably scaled down to one hundred nanometers by novel fabrication techniques to meet a goal of ultrahigh speed computer design.

3.5
Experimental: Selected Procedures

3.5.1
Synthesis of hexa-*peri*-hexabenzocoronene 10 by Cu(II)-mediated oxidative cyclodehydrogenation – a general procedure to prepare unsubstituted graphitic molecules [35]

$CuCl_2$ (9.68 g, 72.0 mmol) and $AlCl_3$ (9.60 g, 72.0 mmol) were stirred together with hexaphenylbenzene **11** (1.07 g, 2.00 mmol) in CS_2 (500 mL) under argon for about 2 d at room temperature. When no remaining reactant was observed by mass spectrometry the reaction was quenched by addition of EtOH (500 mL). The suspension was filtered and the residue washed with H_2O, conc. HCl, H_2O, conc. ammonia solution, H_2O, acetone, CS_2, CH_2Cl_2 (about 500 mL in small portions) and dried under vacuum. Yield: 1.0 g of HBC **10**.

3.5.2
Synthesis of hexakis(4-dodecylphenyl)-*peri*-hexabenzocoronene (HBC-PhC12) – a general synthetic method towards six-fold alkyl- and alkylphenyl-substituted HBCs [38]

The synthesis involves three steps:

(i) 4,4'-Bis(4-dodecylphenyl)diphenylacetylene:

In a 250 mL Schlenk flask, 18 mL of a 1 M solution of 4-dodecylphenylmagnesium bromide was added dropwise to 1 g (3 mmol) 4,4'-dibromodiphenylacetylene dissolved in 100 mL dry THF. 200 mg of $[PdCl_2(dppf)]$ catalyst was added to this solution. The resulting mixture was stirred under reflux in an inert atmosphere for 20 h. While cooling to room temperature, a white solid precipitated. The solid was then filtered and washed several times with petroleum ether and MeOH to yield 1.5 g (75%) of the title product.

(ii) Hexakis(4-dodecylbiphen-1-yl)benzene:

In a 100 mL round flask equipped with a reflux condenser, a suspension of 1 g (1.5 mmol) bis(4-dodecylphenyl)diphenylacetylene and 50 mL of dioxane was degassed several times; 78 mg (0.23 mmol) [$Co_2(CO)_8$] was then added and the resulting mixture was refluxed for 3 h. The solvent was evaporated under vacuum, and was purified using column chromatography on silica with petroleum ether/CH_2Cl_2 (8/2) as eluant, yielding 0.7 g (70%) of the title compound as an off-white solid.

(iii) Hexakis(4-dodecylphenyl)-*peri*-hexabenzocoronene:

A 250 mL two-necked round bottom flask was charged with 0.53 g (0.27 mmol) of hexakis(4-dodecylbiphen-1-yl)benzene and 70 mL of CH_2Cl_2. Using a glass capillary, a constant stream of argon was bubbled through the solution. Then 0.8 g (5 mmol) of $FeCl_3$ dissolved in CH_3NO_2 was added dropwise using a syringe. After 45 min, the mixture was quenched with MeOH and the precipitate was filtered. The resulting yellow solid was recrystallized from hot THF and dried under vacuum to yield 0.42 g HBC-PhC12 (80%).

3.5.3
Functionalization of insoluble HBC building blocks 30–32 by Sonogashira coupling reactions [48]. Synthesis of hexakis(1-dodecynylphenyl)-*peri*-hexabenzocoronene (34a) as a representative example

To a frame-dried 25 mL Shlenk flask was added 300 mg hexakis(4-iodophenyl)-*peri*-hexabenzocoronene **30** (0.173 mmol), 30 mg [$Pd(PPh_3)_4$] (2.5 mol% per iodo), 10 mg of CuI (5.0 mol% per I), and 20 mL piperidine. The mixture was degassed by bubbling argon for 15 min, then 403 mg dodecyne (2.076 mmol) was added. The suspension was heated to 50 °C overnight, cooled to RT, and 20 mL MeOH was added. The yellow precipitate was collected, followed by column chromatography (silica gel, PE/DCM = 3:2) to give 302 mg pure title compound as yellow powder (82%).

3.5.4
Synthesis of C96-C12 precursor 1,3,5-tris[3′,4′-di(4″-dodecylphenyl)-2′,5′-diphenylphenyl]benzene (44a) by Diels-Alder cycloaddition reaction – a representative procedure for the synthesis of branched oligophenylenes [50]

76 mg (0.51 mmol) of 1,3,5-triethynylbenzene and 1.31 g (1.82 mmol) of 3,4-bis(4-dodecylphenyl)-2,5-diphenylcyclopentadienone were dissolved in 4 mL of *o*-xylene under argon atmosphere and the resultant mixture was heated for 18 h at 170 °C.

After cooling, the solvent was removed under reduced pressure and the residue was purified by column chromatography (silica gel, petroleum ether/CH_2Cl_2 = 5:1) to afford 1.10 g title compound as a colorless solid (97%).

3.5.5
Hydrogenation of hexakis-dodecyl-*peri*-hexabenzocoronenes 74 [66]

Solid HBC-C12 sample (100~500 mg) was placed in a 300 mL, glass-lined autoclave along with an equivalent mass of Pd/C (10%). The autoclave was sealed and degassed via passage of an argon stream through a rubber septum. Dry THF was transferred via cannula under argon, the reaction vessel was sealed, pressurized to 50–60 bar, and heated to 65 °C for 12 h. After cooling to room temperature, the pressure was released to the atmosphere, the solution was passed through a 0.4 μm filter and then concentrated under reduced pressure. The peralkylated coronene 75 as yellow powder was obtained in quantitative yield.

3.5.6
Synthesis of HBC dimer 38 by Yamamoto coupling reaction [49]

To a deoxygenated mixture of 1,5-cyclooctadiene (COD, 3 mL) in toluene (25 mL), Ni(COD)$_2$ (116 mg, 0.42 mmol) and 2,2′-dipyridyl (68 mg, 0.42 mmol) were added. After the mixture was stirred at room temperature for 1 h, a deoxygenated solution of 2-bromo-5,8,11,14,17-penta(3,7-dimethyloctyl)-hexa-*peri*-hexabenzocoronene (0.45 g, 0.35 mmol) in toluene (50 mL) was added. The resulting mixture was heated at 60 °C for 20 h. The mixture was allowed to cool to room temperature; the precipitated product was filtered and washed with toluene, dibromoethane and methanol. The product was purified by recrystallization from toluene to give the title compound as a yellow solid in 87% yield.

Acknowledgments

This work was supported by the Bundesministerium für Bildung und Forschung (Zentrum für Multifunktionelle Werkstoffe und Miniaturisierte Funktionseinheiten, BMBF 03N 6500), the Fonds der Chemischen Industrie, the Deutsche Forschungsgemeinschaft (DFG, Schwerpunktprogramm organische Feldeffekttransistoren SFB 625), the Volkswagenstiftung (Elektronentransport durch konjugierte molekulare Scheiben und Ketten), the European Science Foundation (Project SMARTON) and the European Commission (Programs SISITOMAS, DISCEL, MAC-MES, NAIMO (NMP4-CT-2004-500355) and RADASA). The authors would like to thank Professors A. J. Bard (Austin), T. Björnholm (Kopenhagen), R. H. Friend (Cambridge), J. F. Nierengarten (Strasbourg), T. Pakula (Mainz), J. P. Rabe (Berlin), W. R. Salaneck (Linköping), L. Schlapbach (Fribourg), H. W. Spiess (Mainz), A. F. Thünemann (Potsdam), J. M. Warman (Delft), and their coworkers for the fruitful and stimulating collaboration of many years. We thank all the co-workers who contributed to this work.

Abbreviations

AFM	Atomic force microscopy
m-CPBA	meta-chloroperbenzoic acid
2D WAXS	two-dimensional wide-angle X-ray scattering
DDQ	2,3-dichloro-5,6-dicyano-1,4-benzoquinone
DME	1,2-dimethoxyethane
DSC	differential scanning calorimetry
EPR	electronic paramagnetic resonance
FETs	field effect transistors
FVP	flash vacuum pyrolysis
HBC	hexa-peri-hexabenzocoronene
HFB	hexafluorobenzene
HOPG	highly oriented pyrolytic graphite
HRTEM	high resolution transmission electron microscopy
ITO	Indium tin oxide
LB	Langmuir-Blodgett
LCDs	liquid crystal displays
LEDs	light-emitting diodes
LEED	low-energy electron diffraction
MALDI-TOF	matrix-assisted laser desorption ionization time-of-flight
NIR	near infrared
PAHs	polycyclic aromatic hydrocarbons
PBAHs	all-benzenoid polycyclic aromatic hydrocarbons
PEI	polyethylene imine
POM	polarized optical miscroscopy
PR-TRMC	pulse-radiolysis time-resolved microwave conductivity
PTFE	polytetrafluoroethene
STM	scanning tunneling microscopy
STS	scanning tunneling spectroscopy
Tf	trifluoromethanesulfonyl
WAXD	wide-angle X-ray diffraction
UV	ultraviolet

References

1 (a) *Polycyclic Hydrocarbons*, E. CLAR, Academic Press, New York, **1964**, Vol. I/II; (b) *The Aromatic Sextet*, E. CLAR, Wiley-VCH, London, **1972**; (c) *Polycyclic Aromatic Hydrocarbons*, R. G. HARVEY, Wiley-VCH, New York, **1997**.

2 (a) R. SCHOLL, C. SEER, R. WEITZEN-

bök, *Chem. Ber.* **1910**, *43*, 2202–2209; (b) R. Scholl, C. Seer, *Liebigs Ann. Chem.* **1912**, *394*, 111–123; (c) R. Scholl, C. Seer, *Chem. Ber.* **1922**, *55*, 330–341; (d) E. Clar, D. G. Stewart, *J. Am. Chem. Soc.* **1953**, *75*, 2667–2672; (e) E. Clar, W. Schmidt, *Tetrahedron* **1979**, *35*, 2673–2680.

3 See Ref. [1c] and *Carbon Rich Compounds I/II* in *Topics in Current Chemistry*, Springer, Berlin, **1998**, Vol. 196; **1999**, Vol. 201.

4 E. Clar, B. Boggiano, *J. Chem. Soc.* **1957**, 2681–2689.

5 W. H. Laarhoven, W. J. C. Prinsen, *Top. Curr. Chem.* **1984**, *125*, 63–130.

6 (a) W. E. Barth, R. G. Lawton, *J. Am. Chem. Soc.* **1971**, *93*, 1730–1745; (b) J. M. Robertson, J. G. White, *J. Chem. Soc.* **1945**, 607–617; (c) K. Yamamoto, T. Harada, Y. Okamoto, H. Chikamatsu, M. Nakazaki, Y. Kai, T. Nakao, M. Tanaka, S. Harada, N. Kasai, *J. Am. Chem. Soc.* **1988**, *110*, 3578–3584; (d) L. T. Scott, M. M. Hashemi, D. T. Meyer, H. B. Warren, *J. Am. Chem. Soc.* **1991**, *113*, 7082–7084; (e) R. Scholl, K. Meyer, *Chem. Ber.* **1932**, *65*, 902–915; (f) H. A. Staab, F. Diederich, *Chem. Ber.* **1983**, *116*, 3487–3528; (g) F. Vögtle, H. A. Staab, *Chem. Ber.* **1968**, *101*, 2709–2716.

7 *Electronic Materials: The Oligomer Approach*, K. Müllen, G. Wegner (eds.), Wiley-VCH, Weinheim, **1998**.

8 H. Irngartinger, L. Leiserowitz, G. M. J. Schmidt, *Chem. Ber.* **1970**, *103*, 1132–1156.

9 (a) *Cyclophanes*, J. D. Dunitz, K. Hafner, S. Ito, J. M. Lehn, K. N. Raymond, C. W. Rees, J. Thiem, F. Vögtle (eds.), Springer-Verlag, Berlin, **1994**; (b) *Modern Cyclophane Chemistry*, H. Hopf, R. Gleiter (eds.), Wiley-VCH, Weinheim, **2004**.

10 (a) R. Diercks, K. P. C. Vollhardt, *J. Am. Chem. Soc.* **1986**, *108*, 3150–3152; (b) B. Gómez-Lor, Ó. de Frutos, P. A. Ceballos, T. Granier, A. M. Echavarren, *Eur. J. Org. Chem.* **2001**, 2107–2114; (c) W. Kemp, I. T. Storie, C. D. Tulloch, *J. Chem. Soc.* **1980**, 2812–2817.

11 L. T. Scott, *Angew. Chem.* **2004**, *116*, 5102–5116; *Angew. Chem. Int. Ed.* **2004**, *43*, 4994–5007 and references therein.

12 (a) L. T. Scott, M. M. Boorum, B. McMahon, S. Hagen, J. Mack, J. Blank, H. Wegner, A. de Meijere, *Science* **2002**, *295*, 1500–1503; (b) M. M. Boorum, Y. V. Vasil'ev, T. Drewello, L. T. Scott, *Science* **2001**, *294*, 828–831.

13 (a) A. Borchardt, A. Fuchicello, K. V. Kilway, K. K. Baldridge, J. S. Siegel, *J. Am. Chem. Soc.* **1992**, *114*, 1921–1923; (b) K. Yamamoto, H. Sonobe, H. Matsubara, M. Sato, S. Okamoto, K. Kitaura, *Angew. Chem.* **1998**, *108*, 69–70; *Angew. Chem. Int. Ed.* **1996**, *35*, 69–70.

14 (a) K. H. Koch, K. Müllen, *Chem. Ber.* **1991**, *124*, 2091–2100; (b) M. Müller, H. Mauermann-Düll, M. Wagner, V. Enkelmann, K. Müllen, *Angew. Chem.* **1995**, *107*, 1751–1752; *Angew. Chem. Int. Ed.* **1995**, *34*, 1583–1585; (c) P. Kovacic, M. B. Jones, *Chem. Rev.* **1987**, *87*, 357–379.

15 (a) M. Zander, W. Franke, *Chem. Ber.* **1958**, *91*, 2794–2797; (b) R. Goddard, M. W. Haenel, W. C. Herndon, C. Küger, M. Zander, *J. Am. Chem. Soc.* **1995**, *117*, 30–41.

16 (a) R. D. Broene, F. Diederich, *Tetrahedron Lett.* **1991**, *32*, 5227–5230; (b) K. A. Muszkat, *Top. Curr. Chem.* **1981**, *88*, 89–143; (c) H. Meier, *Angew. Chem.* **1992**, *104*, 1425–1426; *Angew. Chem. Int. Ed.* **1992**, *31*, 1399–1420.

17 I. Postovoskii, N. P. Bednya, *Zh. Obsch. Khim. USSR* **1937**, *7*, 2919.

18 R. G. Harvey, J. Pataki, C. Cortez, P. DiRaddo, C. Yang, *J. Org. Chem.* **1991**, *56*, 1210–1217.

19 E. Clar, M. Zander, *J. Chem. Soc.* **1957**, 4616–4619.

20 C. S. LeHoullier, G. W. Gribble, *J. Org. Chem.* **1983**, *48*, 1682–1685.

21 (a) J. W. Barton, A. R. Grinham, *J. Chem. Soc. Perkin Trans. I* **1972**, 634–637; (b) L. Blanco, H. E. Helson, M. Hirthammer, H. Mestdagh, S. Spyroudis, K. P. C. Vollhardt, *Angew. Chem.* **1987**, *99*, 1276–1279;

Angew. Chem. Int. Ed. **1987**, *26*, 1246–1247.

22 S. AMIN, K. HUIE, N. HUSSAIN, G. BALANIKAS, S. S. HECHT, *J. Org. Chem.* **1985**, *50*, 1948–1954.

23 (a) X. Q. TANG, R. G. HARVEY, *J. Org. Chem.* **1995**, *60*, 3570; (b) J. E. RICE, Z. W. CAI, *J. Org. Chem.* **1993**, *58*, 1415–1424.

24 (a) A. MINSKY, M. RABINOVITZ, *Synthesis* **1983**, 497–498; (b) C. YANG, D. T. C. YANG, R. G. HARVEY, *Synlett* **1992**, 799–800.

25 (a) A. J. FLOYD, S. F. DYKE, S. E. WARD, *Chem. Rev.* **1976**, 509–562; (b) R. G. HARVEY, C. YANG, E. ABU-SHAQARA, *J. Org. Chem.* **1993**, *58*, 5866.

26 L. J. ALLAMANDOLA, A. G. TIELENS, J. R. BARKER, *Astrophys. J.* **1985**, *290*, L25–L28.

27 See 2000 Nobel Prize lectures: (a) H. SHIRAKAWA, *Angew. Chem.* **2001**, *113*, 2642–2648; *Angew. Chem. Int. Ed.* **2001**, *40*, 2574–2580; (b) A. G. MACDIARMID, *Angew. Chem.* **2001**, *113*, 2649–2659; *Angew. Chem. Int. Ed.* **2001**, *40*, 2581–2590; (c) A. J. HEEGER, *Angew. Chem.* **2001**, *113*, 2660–2682; *Angew. Chem. Int. Ed.* **2001**, *40*, 2591–2611.

28 (a) *Handbook of Conducting Polymers*, 2nd edn., T. A. SKOTHEIM, R. L. ELSENBAUMER, J. R. REYNOLDS (eds.), Marcel Dekker, New York, **1998**; (b) *Semiconducting Polymers: Chemistry, Physics, and Engineering*, G. HADZIIOANNOU, P. F. VAN HUTTEN (eds.), Wiley-VCH, Weinheim, **2000**.

29 (a) I. C. SAGE, in *Handbook of Liquid Crystals*, D. DEMUS, J. GOODBY, G. W. GRAY, H. W. SPIESS, V. VILL (eds.), Wiley-VCH, Weinheim, **1998**, Vol. 1, 731–762; (b) M. D. WATSON, A. FECHTENKÖTTER, K. MÜLLEN, *Chem. Rev.* **2001**, *101*, 1267–1300; (c) C. D. SIMPSON, J. WU, M. D. WATSON, K. MÜLLEN, *J. Mater. Chem.* **2004**, 494–504; (d) R. J. BUSHBY, O. R. LOZMAN, *Curr. Opin. Solid. State Mater. Sci.* **2002**, *6*, 569–578.

30 (a) E. CLAR, C. T. IRONSIDE, *Proc. Chem. Soc.* **1958**, 150; (b) E. CLAR, C. T. IRONSIDE, M. ZANDER, *J. Chem. Soc.* **1959**, 142–147.

31 E. CLAR, *Chem. Ber.* **1943**, *76*, 609–621.

32 A. HALLEUX, R. H. MARTIN, G. S. D. KING, *Helv. Chim. Acta* **1958**, *129*, 1177–1183.

33 W. HENDEL, Z. H. KHAN, W. SCHMIDT, *Tetrahedron* **1986**, *42*, 1127–1134.

34 (a) M. MÜLLER, C. KÜBEL, K. MÜLLEN, *Chem. Eur. J.* **1998**, *4*, 2099–2109; (b) A. STABEL, P. HERWIG, K. MÜLLEN, J. P. RABE, *Angew. Chem.* **1995**, *107*, 1768–1770; *Angew. Chem. Int. Ed.* **1995**, *34*, 1609–1611.

35 C. KÜBEL, K. ECKHARDT, V. ENKELMANN, G. WEGNER, K. MÜLLEN, *J. Mater. Chem.* **2000**, *10*, 879–886.

36 J. A. HYATT, *Org. Prep. Proced. Int.* **1991**, *23*, 460–463.

37 P. HERWIG, C. W. KAYSER, K. MÜLLEN, H. W. SPIESS, *Adv. Mater.* **1996**, *8*, 510–513.

38 A. FECHTENKÖTTER, K. SAALWÄCHTER, M. A. HARBISON, K. MÜLLEN, H. W. SPIESS, *Angew. Chem.* **1999**, *111*, 3224–3228; *Angew. Chem. Int. Ed.* **1999**, *38*, 3039–3042.

39 J. D. BRAND, C. KÜBEL, S. ITO, K. MÜLLEN, *Chem. Mater.* **2000**, *12*, 1638–1647.

40 Z. WANG, K. MÜLLEN, unpublished results.

41 R. GODDARD, M. W. HAENEL, W. C. HERNDON, C. KRÜGER, M. ZANDER, *J. Am. Chem. Soc.* **1995**, *117*, 30–41.

42 P. T. HERWIG, V. ENKELMANN, O. SCHMELZ, K. MÜLLEN, *Chem. Eur. J.* **2000**, *6*, 1834–1839.

43 (a) A. M. VAN DE CRAATS, J. M. WARMAN, K. MÜLLEN, Y. GEERTS, J. BRAND, *Adv. Mater.* **1998**, *10*, 36–38; (b) A. M. VAN DE CRAATS, J. M. WARMAN, A. FECHTENKÖTTER, J. D. BRAND, M. A. HARBISON, K. MÜLLEN, *Adv. Mater.* **1999**, *11*, 1469–1472.

44 (a) S. ITO, M. WEHMEIER, J. D. BRAND, C. KÜBEL, R. EPSCH, J. P. RABE, K. MÜLLEN, *Chem. Eur. J.* **2000**, *6*, 4327–4342; (b) A. FECHTENKÖTTER, N. TCHEBOTAREVA, M. D. WATSON, K. MÜLLEN, *Tetrahedron* **2001**, *57*, 3769–3783.

45 Z. WANG, F. DÖTZ, V. ENKELMANN, K. MÜLLEN, *Angew. Chem.* **2005**, *117*,

1273–1276; *Angew. Chem. Int. Ed.* **2005**, *44*, 1247–1250.

46 K. Weiss, G. Beernink, F. Dötz, A. Birkner, K. Müllen, C. H. Wöll, *Angew. Chem.* **1999**, *111*, 3974–3978; *Angew. Chem. Int. Ed.* **1999**, *38*, 3748–3752.

47 (a) C. Lambert, G. Nöll, *Angew. Chem.* **1998**, *110*, 2239–2242; *Angew. Chem. Int. Ed.* **1998**, *37*, 2107–2110; (b) C. Lambert, G. Nöll, *Chem. Eur. J.* **2002**, *8*, 3467–3477.

48 (a) J. Wu, M. D. Watson, K. Müllen, *Angew. Chem.* **2003**, *115*, 5487–5491; *Angew. Chem. Int. Ed.* **2003**, *42*, 5329–5333; (b) J. Wu, M. D. Watson, L. Zhang, Z. Wang, K. Müllen, *J. Am. Chem. Soc.* **2004**, *126*, 177; (c) J. Wu, M. Baumgarten, M. G. Debije, J. M. Warman, K. Müllen, *Angew. Chem.* **2004**, *116*, 5445–5449; *Angew. Chem. Int. Ed.* **2004**, *43*, 5331–5335.

49 J. Wu, M. D. Watson, N. Tchebotareva, Z. Wang, K. Müllen, *J. Org. Chem.* **2004**, *69*, 8194–8204.

50 Ž. Tomović, M. D. Watson, K. Müllen, *Angew. Chem.* **2004**, *116*, 733–777; *Angew. Chem. Int. Ed.* **2004**, *43*, 755–758.

51 Ž. Tomović, K. Müllen, unpublished results.

52 J. Wu, Ž. Tomović, V. Enkelmann, K. Müllen, *J. Org. Chem.* **2004**, *69*, 5179–5186.

53 C. D. Simpson, PhD Thesis, University of Mainz, **2003**.

54 C. D. Simpson, J. D. Brand, A. J. Berresheim, L. Przybilla, H. J. Räder, K. Müllen, *Chem. Eur. J.* **2000**, *8*, 1424–1429.

55 C. D. Simpson, G. Mattersteig, K. Martin, L. Gherghel, R. E. Bauer, H. J. Räder, K. Müllen, *J. Am. Chem. Soc.* **2004**, *126*, 3139–3147.

56 V. S. Iyer, M. Wehmeier, J. D. Brand, M. A. Keegstra, K. Müllen, *Angew. Chem.* **1997**, *109*, 1676–1679; *Angew. Chem. Int. Ed.* **1997**, *36*, 1603–1606.

57 D. Wasserfallen, K. Müllen, unpublished results.

58 (a) V. S. Iyer, K. Yoshimura, V. Enkelmann, R. Epsch, J. P. Rabe, K. Müllen, *Angew. Chem.* **1998**, *110*, 2843–2846; *Angew. Chem. Int. Ed.* **1998**, *37*, 2696–2699; (b) M. Müller, V. S. Iyer, C. Kübel, V. Enkelmann, K. Müllen, *Angew. Chem.* **1997**, *109*, 1679–1682; *Angew. Chem. Int. Ed.* **1997**, *36*, 1607–1610; (c) C. D. Simpson et al., unpublished results.

59 J. Wu, L. Gherghel, M. D. Watson, J. Li, Z. Wang, C. D. Simpson, U. Kolb, K. Müllen, *Macromolecules* **2003**, *36*, 7082–7089.

60 R. E. Martin, F. Diederich, *Angew. Chem. Int. Ed.* **1999**, *38*, 1350–1377.

61 M. G. Debije, J. Piris, M. P. de Haas, J. M. Warman, Z. Tomovic, C. D. Simpson, M. D. Watson, K. Müllen, *J. Am. Chem. Soc.* **2004**, *126*, 4641–4645.

62 S. E. Stein, R. L. Brown, *J. Am. Chem. Soc.* **1987**, *109*, 3721–3729.

63 F. Dötz, J. D. Brand, S. Ito, L. Gherghel, K. Müllen, *J. Am. Chem. Soc.* **2000**, *122*, 7707–7717.

64 Z. Wang, Z. Tomovic, M. Kastler, R. Pretsch, F. Negri, V. Enkelmann, K. Müllen, *J. Am. Chem. Soc.* **2004**, *126*, 7794–7795.

65 Z. Wang, K. Müllen, unpublished results.

66 M. D. Watson, M. G. Debije, J. M. Warman, K. Müllen, *J. Am. Chem. Soc.* **2004**, *126*, 766–771.

67 M. D. Watson, F. Jäckel, N. Severin, J. P. Rabe, K. Müllen, *J. Am. Chem. Soc.* **2004**, *126*, 1402–1407.

68 Y. Huang, X. Duan, Y. Cui, L. J. Lauhon, K. H. Kim, C. M. Lieber, *Science* **2001**, *294*, 1313–1317.

69 S. M. Draper, D. J. Gregg, R. Madathil, *J. Am. Chem. Soc.* **2002**, *124*, 3486–3487.

70 S. M. Draper, D. J. Gregg, E. R. Schofield, W. R. Browne, M. Duati, J. G. Vos, P. Passaniti, *J. Am. Chem. Soc.* **2004**, *126*, 8694–8701.

71 S. Chandrasekhar, S. Krishna, *Contemp. Phys.* **1999**, *40*, 237.

72 W. Pisula, M. Kastler, D. Wasserfallen, T. Pakula, K. Müllen, *J. Am. Chem. Soc.* **2004**, *126*, 8074–8075.

73 C. Y. Liu, A. Fechtenkötter, M. D. Watson, K. Müllen, A. J. Bard, *Chem. Mater.* **2003**, *15*, 124–130.

74 I. Fischbach, T. Pakula, P. Minkin, A. Fechtenkötter, K. Müllen, H. W. Spiess, K. Saalwachter, *J. Phys. Chem. B* **2002**, *106*, 6408–6418.
75 S. P. Brown, I. Schnell, J. D. Brand, K. Müllen, H. W. Spiess, *J. Am. Chem. Soc.* **1999**, *121*, 6712–6718.
76 D. Adam, P. Schuhmacher, J. Simmerer, L. Häussling, K. Siemensmeyer, K. H. Etzbach, H. Ringsdorf, D. Haarer, *Nature* **1994**, *371*, 141–143.
77 W. Pisula, Ž. Tomović, B. El Hamaoui, M. D. Watson, T. Pakula, K. Müllen, *Adv. Funct. Mater.* **2005**, *15*, 893–904.
78 M. Lee, J. W. Kim, S. Peleshanko, K. Larson, Y. S. Yoo, D. Vaknin, S. Markutsya, V. V. Tsukruk, *J. Am. Chem. Soc.* **2002**, *124*, 9121–9128.
79 L. Schmidt-Mende, A. Fechtenkötter, K. Müllen, E. Moons, R. H. Friend, J. D. MacKenzie, *Science* **2001**, *293*, 1119–1122.
80 S. Zimmermann, J. H. Wendorff, C. Weder, *Chem. Mater.* **2002**, *14*, 2218–2223.
81 A. M. van de Craats, N. Stutzmann, O. Bunk, M. M. Nielson, M. D. Watson, K. Müllen, H. D. Chanzy, H. Sirringhaus, R. H. Friend, *Adv. Mater.* **2003**, *15*, 495–499.
82 O. Bunk, M. M. Nielsen, T. I. Solling, A. M. van de Craats, N. Stutzmann, *J. Am. Chem. Soc.* **2003**, *125*, 2252–2258.
83 A. Tracz, J. K. Jeszka, M. D. Watson, W. Pisula, K. Müllen, T. Pakula, *J. Am. Chem. Soc.* **2003**, *125*, 1682–1683.
84 W. Pisula, A. Menon, M. Stepputat, I. Lieberwirth, U. Kolb, A. Tracz, H. Sirringhaus, T. Pakula, K. Müllen, *Adv. Mater.* **2005**, *17*, 684–689.
85 N. Reitzel, T. Hassenkam, K. Balashev, T. R. Jensen, P. B. Howes, K. Kjaer, A. Fechtenkötter, N. Tchebotareva, S. Ito, K. Müllen, T. Bjørnholm, *Chem. Eur. J.* **2001**, *7*, 4894–4901.
86 S. Kubowicz, U. Pietsch, M. D. Watson, N. Tchebotareva, K. Müllen, A. F. Thünemann, *Langmuir* **2003**, *19*, 5036–5041.
87 A. F. Thüneman, D. Ruppelt, S. Ito, K. Müllen, *J. Mater. Chem.* **1999**, *9*, 1055–1058.
88 (a) A. Bayer, J. Hübner, J. Kopitzke, M. Oestreich, W. Rühle, J. H. Wendorff, *J. Phys. Chem. B* **2001**, *105*, 4596–4602; (b) S. Marguet, D. Markovitsi, P. Millié, H. Sigal, S. Kumar, *J. Phys. Chem. B* **1998**, *102*, 4697–4710; (c) A. J. Fleming, J. N. Coleman, A. B. Dalton, A. Fechtenkötter, M. D. Watson, K. Müllen, H. J. Byrne, W. J. Blau, *J. Phys. Chem. B* **2003**, *107*, 37–43.
89 J. Wu, A. Fechtenkötter, J. Gauss, M. D. Watson, M. Kastler, C. Fechtenkötter, M. Wagner, K. Müllen, *J. Am. Chem. Soc.* **2004**, *126*, 11311–11321.
90 M. Keil, P. Samorí, D. A. dos Santos, T. Kugler, S. Stafström, J. D. Brand, K. Müllen, J. L. Brédas, J. P. Rabe, W. R. Salaneck, *J. Phys. Chem. B* **2000**, *104*, 3967–3975.
91 P. Ruffieux, O. Gröning, M. Bielmann, C. Simpson, K. Müllen, L. Schlapbach, P. Gröning, *Phys. Rev. B* **2002**, *66*, 073409.
92 R. Friedlein, X. Crispin, C. D. Simpson, M. D. Watson, F. Jäckel, W. Osikowicz, S. Marciniak, M. P. de Jong, P. Samori, S. K. M. Jönsson, M. Fahlman, K. Müllen, J. P. Rabe, W. R. Salaneck, *Phys. Rev. B* **2003**, *68*, 195414.
93 P. Samorí, A. Fechtenkötter, F. Jackel, T. Böhme, K. Müllen, J. P. Rabe, *J. Am. Chem. Soc.* **2001**, *123*, 11462–11467.
94 S. Ito, P. T. Herwig, T. Böhme, J. P. Rabe, W. Rettig, K. Müllen, *J. Am. Chem. Soc.* **2000**, *122*, 7698–7706.
95 P. Samorí, N. Severin, C. D. Simpson, K. Müllen, J. P. Rabe, *J. Am. Chem. Soc.* **2002**, *124*, 9454–9457.
96 N. Tchebotareva, X. Yin, M. D. Watson, P. Samorí, J. P. Rabe, K. Müllen, *J. Am. Chem. Soc.* **2003**, *125*, 9734–9739.

4
[N]Phenylenes: a Novel Class of Cyclohexatrienoid Hydrocarbons

Ognjen Š. Miljanić and K. Peter C. Vollhardt

4.1
Introduction

Aromaticity is one of the most frequently employed concepts in organic chemistry [1]. Despite the omnipresent use of the term, a unique definition is lacking to this day. Aromaticity is most commonly viewed through the prisms of structural [1a, 2], energetic [1a, 3], and magnetic [1a, 4] properties of the systems under study. Structurally, aromatic bond lengths lie between those of normal single and double bonds. Aromatic rings are more stable than their open-chain counterparts, and their unusual magnetic characteristics are reflected in the specific values of magnetic susceptibilities and ^{1}H-NMR chemical shifts. Experimentalists often use chemical reactivity as another qualitative measure of aromatic character. A unifying characteristic of aromatic compounds is the preference for substitution versus addition reactions, which is a manifestation of their tendency to retain the π-electronic skeleton; however, attempts to quantify this effect have met with limited success [5]. Krygowski and Cyrañski describe aromaticity as *an excess property, a deviation from an additive scheme* [2]. While there is a certain degree of correlation between the various criteria given above [6], the issues are sufficiently complex to have induced practitioners to treat aromaticity as a "multidimensional phenomenon" [6, 7].

Two simple hydrocarbons, benzene and cyclobutadiene, stand at opposite ends of the aromaticity continuum, regardless of the criterion chosen. All six C–C bonds in benzene are equal in length (1.398 Å) [8], in contrast to the distinctly single (1.526, 1.581 Å) and double (1.441, 1.359 Å) bonds in the crystallographically characterized peralkylated and persilylated cyclobutadienes, respectively [9]. The resonance energies of the two compounds are also drastically different: relative to an isolated double bond, benzene is stabilized by 32 kcal mol^{-1}, cyclobutadiene is destabilized by 48 kcal mol^{-1} [10]. The vastly different stabilities of the two molecules are reflected in the fact that benzene has been known since Faraday's times [11], whereas the first isolation of cyclobutadiene was reported only in 1973 [12]. This behavior, as well as the corresponding alternating properties of the higher annulenes [13], is in accord with Hückel's rule [14], which states that fully conjugated

Carbon-Rich Compounds. Edited by Michael M. Haley and Rik R. Tykwinski
Copyright © 2006 WILEY-VCH Verlag GmbH & Co. KGaA, Weinheim
ISBN: 3-527-31224-2

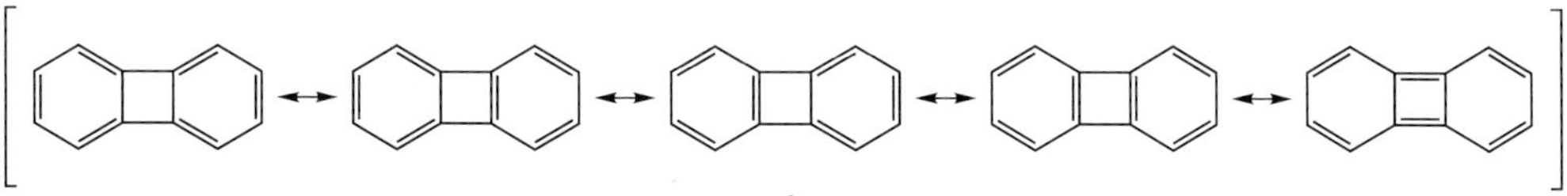

Figure 4.1. The resonance forms of biphenylene (**1**).

systems with $(4n+2)$ π-electrons should share the stabilization of benzene, whereas those with $(4n)$ π-electrons should be destabilized by cyclic conjugation.

In the light of this divergence, the juxtaposition of the benzene and cyclobutadiene structural motifs fused in a single molecule is an intriguing topology. The simplest stable system to have such a fusion is biphenylene **1** (Fig. 4.1), the five resonance forms of which range from "[12]annulenoid" to increasingly "cyclobutadienoid." Originally prepared by Lothrop in 1941 by reacting 2,2′-dibromobiphenyl with Cu$_2$O at 350 °C [15], biphenylene has since been synthesized in a multitude of ways [16] and is now commercially available [17]. Most biphenylene syntheses can be classified into three categories (Scheme 4.1, left): (i) dimerizations of arynes [16], (ii) oxidative dehalogenations of 2,2′-dihalobiaryls [15, 16, 18], and (iii) small molecule extrusions from bridged biaryls [16, 19].

Despite the presence of cyclobutadienoid circuits, the chemical reactivity of **1** (Scheme 4.1, right) reflects considerable aromatic character: biphenylene under-

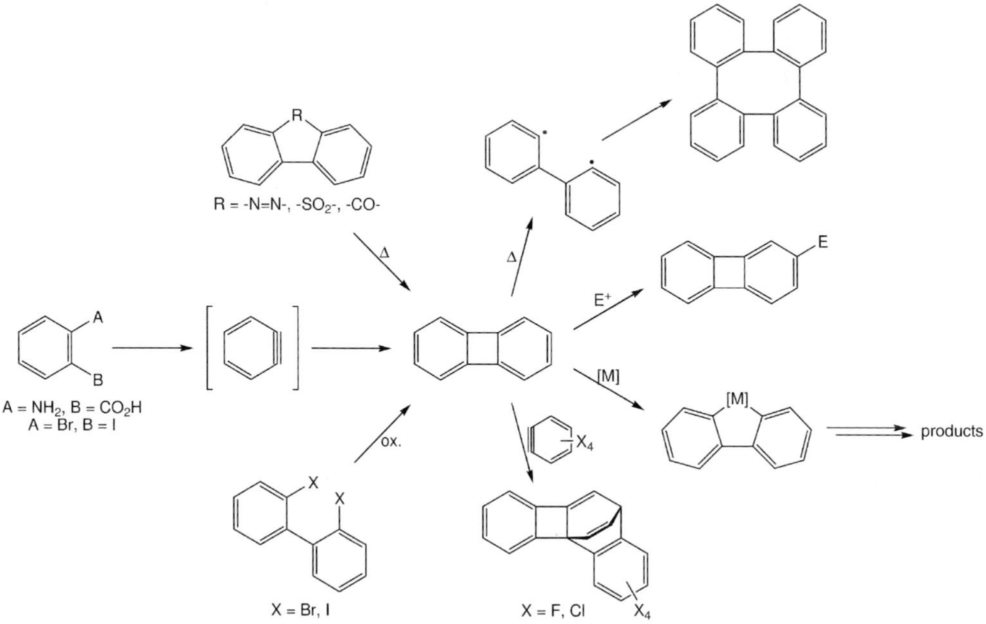

Scheme 4.1. General modes of biphenylene preparation (left) and reactivity (right).

goes electrophilic substitution, rather than addition, almost exclusively at the β-positions and at a rate that is comparable to that of naphthalene [16]. The four-membered ring is thermolyzed, most likely to the 2,2'-biphenyldiyl diradical, which dimerizes to tetrabenzocyclooctatetraene [20]. The aryl–aryl C–C bond in biphenylene is also readily attacked by a number of metal complexes, and the organometallic intermediates thus obtained can lead to a variety of ring-opened and insertion products [16, 21]. Biphenylene is relatively inert in the Diels-Alder reaction: it does not react with tetracyanoethene [22], benzyne [23], or maleic anhydride [16a]. However, it functions as a dienophile with respect to the more electron-deficient tetrachloro- and tetrafluorobenzynes, producing monoadducts [22].

The above reactivity notwithstanding, there are strong indications that the cyclobutadienoid ring has a profound influence on the properties of the system. Thus, a crystal structure [24] highlights the reluctance of **1** to allow conjugation between the two benzene nuclei, with relatively long aryl–aryl bonds (1.514 Å) and noticeably shorter fused bonds (1.426 Å). Conversely, the six-membered rings are distorted in such a fashion as to minimize cyclobutadienoid character in the center, exhibiting pronounced bond alternation (long bonds 1.426 and 1.423 Å, short bonds 1.372 and 1.385 Å). In short, the first resonance form in Fig. 4.1 is a strong contributor to the description of the molecule. Despite these distortions, the electronic spectrum of **1** [25] is distinctly different from that of biphenyl, with peaks that are strongly shifted bathochromically, signaling a substantial narrowing of the HOMO–LUMO gap. Cross-conjugation is also evidenced by substituent effects on reactivity and IR absorptions [16a]. Perhaps most informative, the ^{1}H-NMR spectrum of **1** exhibits relatively shielded resonances at $\delta = 6.60$ ppm (α-hydrogens) and 6.70 ppm (β-hydrogens) [26], ascribed to the presence of a paramagnetic ring current in the cyclobutadiene ring. ^{13}C-NMR spectroscopy is diagnostic of σ strain effects and reveals peaks at 117.8 ppm (α-carbon), 128.4 ppm (β-carbon) and 151.7 ppm (quaternary) [27].

The cumulative experimental data on **1** are to be viewed within the context of recent advances in the understanding of how both σ- and π-effects impinge on the aromaticity of benzene [28]. To what extent are these effects operational in **1**? Shaik, Hiberty and coworkers have suggested that the D_{6h} structure of benzene is the result of a σ–π balance: while π-electrons tend to distort the molecule into the D_{3h} symmetry of cyclohexatriene, the rigidity of the σ-framework acts to enforce higher symmetry [29]. Recently, Schaefer and Schleyer [30] showed that, as a general rule, π-distortivity overcomes σ-rigidity in higher annulenes – benzene is thus a fortuitous exception, rather than a prototype! In this context, **1** is not readily classified as aromatic, non-, or antiaromatic. Hückel's rule seemingly does not apply to it [1, 31], as it would predict a cyclically delocalized 12π-electron system to be unstable. In addition, the strain of the four-membered ring complicates the picture, consequently making **1** an excellent subject on which to study π- and σ-strain in polycyclic compounds.

Biphenylene is the simplest member of a novel class of polycyclic hydrocarbons in which benzene rings are fused to cyclobutadiene moieties in an alternating

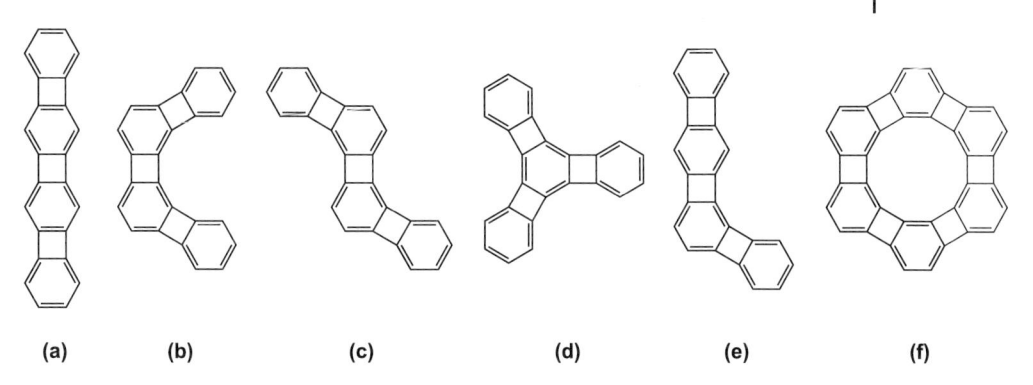

Figure 4.2. Simple phenylene topologies: (a) linear [4]-, (b) angular [4]-, (c) zigzag [4]-, (d) branched [4]-, (e) (mixed) bent [4]-, and (f) circular [6]phenylene.

manner. The name **[N]phenylenes** was coined for these molecules, in which N equals the number of benzene rings. Higher phenylenes exist as several isomers [32], due to the different modes of fusion between the individual rings. A phenylene can be linear, angular, zigzag, branched, or circular, based on the mode of fusion, and mixed topologies are possible. Fig. 4.2 exemplifies these designations.

The various topologies of the [N]phenylenes offer the opportunity to test the hypotheses advanced for the understanding of **1**, to significantly expand the range of available strained ring aromatics in a systematic manner, and to explore new avenues in the area of electronic materials. For example, appropriate design, as in angularly fused derivatives, should provide compounds in which benzene ring distortion is enhanced compared to **1**. Alternatively, linear fusion would enforce a different, bisallylic type deformation, due to symmetry constraints. Moreover, Trinajstić has suggested that the HOMO–LUMO gap along the linear series should drop rapidly [33], whereas the isomeric zigzag relatives should show much attenuated electronic activation. Apart from the anticipated unusual physical properties, the reactivity of the phenylenes is expected to be unique, due to the combination of electronic and ring-strain factors. Synthetically, these structures pose a challenge, in large part due to the presence of multiple cyclobutadiene rings, the cumulative ring strain of which (of the order of 50 kcal mol^{-1} per cyclobutadiene ring [34]) seems prohibitive.

Phenylenes are closely related to the much larger family of the polycyclic aromatic hydrocarbons (PAHs). The chemistry of PAHs has been studied comprehensively with respect to synthesis [35], theory [31], and material science [36]. Each phenylene is correlated to a unique PAH (its "hexagonal squeeze" [37]) by formal removal of the cyclobutadiene cycles through fusion of the attached benzene rings[1]. This topological connection (Fig. 4.3) is general, as it exists in one (linear

---

1) Alternatively, phenylenes can be treated as PAHs in which neighboring benzene rings are separated by a two-bond spacer.

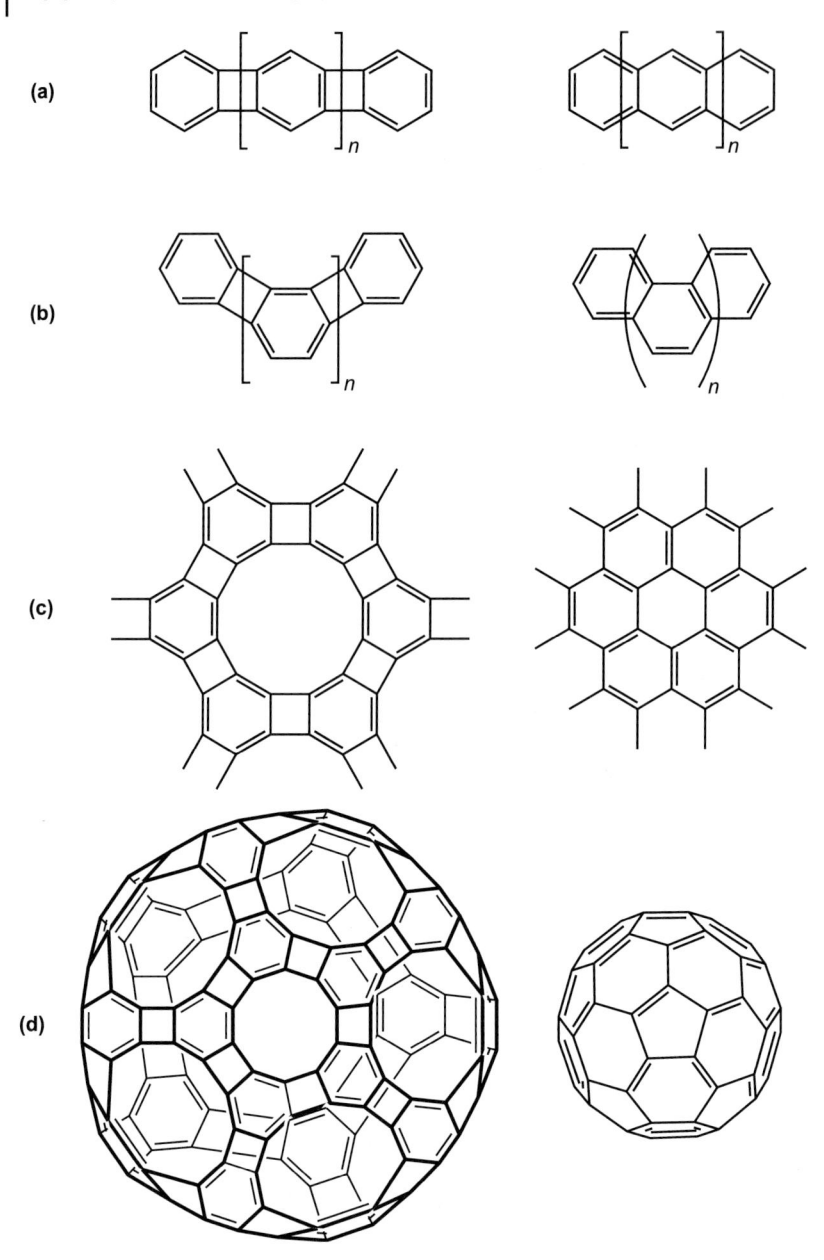

Figure 4.3. Phenylenes and topologically related PAHs: (a) linear [N]phenylenes and polyacenes; (b) angular/zigzag [N]phenylenes and polyphenanthrenes/helicenes; (c) "circular [6]phenylene sheet" and graphite; (d) archimedene (C_{120}) and fullerene (C_{60}).

phenylenes – acenes), two (e.g., circular [6]phenylene sheets – graphite), and three dimensions (e.g., archimedene – fullerene). There are important differences, however, starting with the incremental change in the number of π-electrons along the respective series. For example, PAHs increase this count in increments of four, thus maintaining their $(4n+2)$ π-character. Phenylenes, on the other hand, are homologated by the addition of a C_6 fragment and accordingly alternate between $(4n+2)$ and $(4n)$ π-electrons. Circular phenylenes preserve the π-electron count of their open counterparts, whereas PAHs lose two electrons in this formal transformation and switch from $(4n+2)$ to $(4n)$. Finally, both fullerenes [38] and the three-dimensional phenylenes alternate between $(4n+2)$ and $(4n)$ electron count.

Gutman associated several theoretical parameters of the phenylenes with those of the analogous PAHs [40]. He showed that the algebraic structure count (ASC) [40] of phenylenes equals the number of Kekulé structures (K) of their hexagonal squeezes [37]. ASC and K serve as measures of stability in nonbenzenoid and benzenoid hydrocarbons, respectively [31, 41]. The stability of phenylenes therefore appears to parallel that of their corresponding PAHs. The Wiener index, used to predict the boiling points of hydrocarbons based on their structures [42], correlates linearly between the two classes [43]. It has been proposed that six-membered rings in phenylenes follow the anti-Clar's rule: if a certain ring in phenylene is conjugated strongly, its analog in the hexagonal squeeze is conjugated weakly (i.e. is "empty" in Clar's terminology) and vice versa [44]; however, as later Sections will show, this is not a general trend. The list of analogies is not exhausted here [39, 44], and future research may reveal new ties between the two classes.

The following account will describe progress in the synthesis and the chemical and physical properties of the phenylenes, in that order. It is written with the aim of placing all presently known members of this class of hydrocarbons, including **1**, on a comparative footing [45].

4.2
Preparation of Phenylenes

4.2.1
Early Synthetic Strategies [45]

Although **1** had been constructed in a variety of ways [16], at the outset of this work, attempts to extend these methods to the synthesis of higher phenylenes either failed [46] or were limited. Nevertheless, Barton and coworkers managed to apply the extrusion of nitrogen from benzodicinnolines by flash vacuum pyrolysis (FVP) (precedented for biphenylene [19]) to the relatively low-yielding preparation of angular and linear [3]phenylene [47]. Application of this technique to the isolation of branched [4]phenylene was unsuccessful [48], possibly indicating the limits of this methodology.

The breakthrough that enabled the chemistry described in this account came through the discovery of a new versatile biphenylene synthesis based on the cyclotrimerization of alkynes catalyzed by [CpCo(CO)$_2$] [49]. Thus, a variety of substituted biphenylenes could be made by the cocyclization of 1,2-diethynylbenzene **2** with alkynes, in the case of bis(trimethylsilyl)acetylene (BTMSA) yielding **3** in a remarkable 96% yield (Scheme 4.2) [50]. Exploiting the silyl substituents as masked ethynyl groups and using tin instead of silicon, as appropriate [45], gave access to **4** and **6** and, hence, the linear homologs **5** [51] and **7** [52] by iterative sequences involving up to three separate co-oligomerization steps (for **7**). These linear [N]phenylenes were targeted first for synthesis, because they are distinct from their angular isomers, as this topology (in which cyclobutadienoid circuits cannot be completely avoided) imparts relative electronic activation [53].

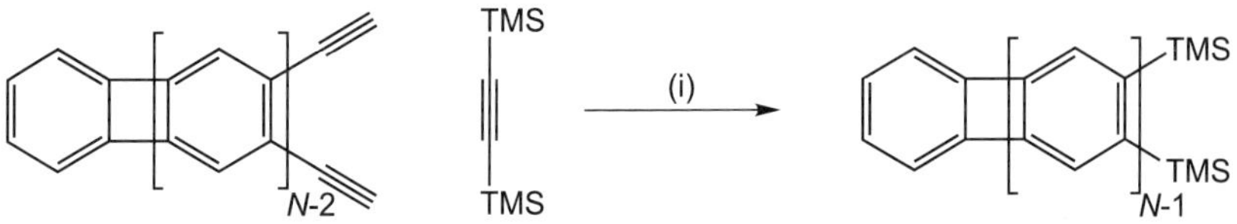

2, N = 2
4, N = 3
6, N = 4

3, N = 2
5, N = 3
7, N = 4

Scheme 4.2. The last step in the preparation of linear [N]phenylenes **3**, **5**, and **7** by an (iterative) single cocyclization strategy: N = 2, (i) [CpCo(CO)$_2$], hν, Δ, 96%; N = 3, (i) [CpCo(CO)$_2$], hν, Δ, 36%; N = 4, (i) [CpCo(CO)$_2$], THF, hν, Δ, 9 h, 30%, then CO (1 atm), 90 °C, 16 h, 100%.

The increasingly long linear sequences necessitated by the single cocyclization approach were significantly shortened by employing more convergent double cocyclizations (Scheme 4.3). In this variant, a tetraethynylated arene precursor undergoes biscycloadditions to generate four rings in a single operation, leading to **9** [51, 54], **11** [52], and **13** [55]. The power of the transition-metal-based approach is evident, when one recognizes that eight of the nine rings in **13** are made by [CpCo(CO)$_2$].

The second topology addressed in this early work was the angular frame. In contrast to their linear counterparts, angular [N]phenylenes possess one, presumably dominant, resonance form that completely avoids double bonds in the four-membered rings (Scheme 4.4). This simple representation should translate into increased bond localization and alkene-like reactivity of the internal nuclei. Retrosynthetically, the prototype angular [3]phenylene **15** can be unraveled by retrocyclization of the terminal or the internal rings (Scheme 4.4a). The former strategy, while successful for derivatives of **15** [56], is not readily extendable to the higher homologs of **15**; therefore only the latter is described. This approach is distinct, in as much it requires an intramolecular alkyne cyclotrimerization (a cycloisomeriza-

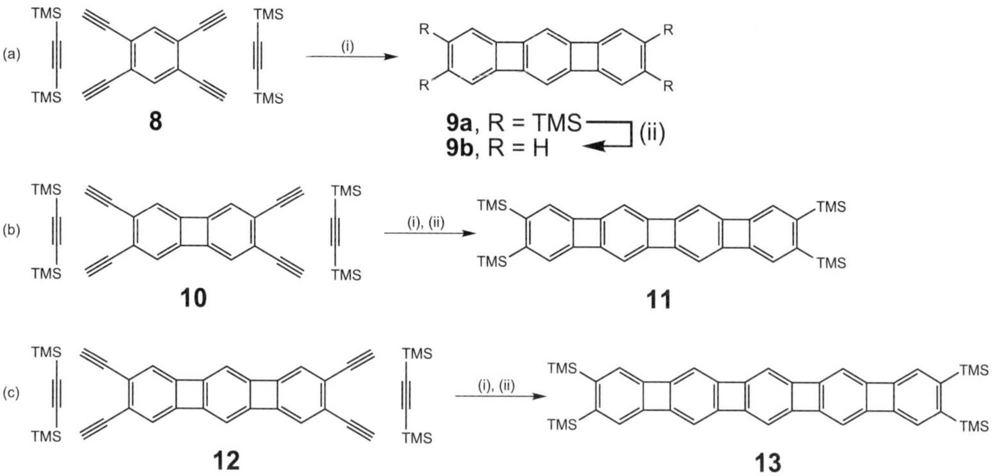

4.2 Preparation of Phenylenes

Scheme 4.3. The last step in the preparation of linear [N]phenylenes **9b**, **11**, and **13** by a double cocyclization strategy: (a) N = 3, (i) [CpCo(CO)₂], PhCH₃/DMF, hν, Δ, 6 h, 71%; (ii) t-BuOK, t-BuOH, THF/DMSO, 85 °C, 6 h, 79%; (b) N = 4, (i) [CpCo(CO)₂], THF, hν, Δ, 13 h, 30%; (ii) CO (1 atm), 120 °C, 72 h, 99%; (c) N = 5, (i) [CpCo(CO)₂], THF, hν, Δ, 16 h, 20%; (ii) CuCl₂·2H₂O (4 equiv), 1,2-diethoxyethane, H₂O/NEt₃, 0 °C, 3 h, 40%.

tion), initially deemed a dubious proposition considering the large amount of ring strain that is generated during the process. In the event, however, **15** could be made from **14** by [CpCo(CO)₂]-mediated cyclization in 30% yield [57]. The generality of this transformation was evident with the biphenylenyl substituted analogs of **14**, namely **16** and **18**, which isomerized successfully to angular [4]- (**17**) and [5]phenylene (**19**), in 30 and 5% yield, respectively [58].

Investigations since these early syntheses have brought about a marked improvement in yields through a stepwise protocol. Thus, exposure of **14** [59] or **16** [60] to [CpCo(eth)₂] [61] at low temperatures gave the corresponding cobaltacyclopentadiene(alkyne) complexes, which, when heated in the presence of a CpCo trap (e.g. 1,3-cyclohexadiene), furnished **15** and **17** in 70 and 51% yields, respectively. The reasons for these improvements may be the use of stoichiometric cobalt at low temperatures which serves to bind all the alkyne units, thus obviating adverse polymerization or other processes, and the subsequent isomerization-demetallation under conditions that bind CpCo irreversibly, thus avoiding strained ring opening by cobalt fragments (see Section 4.3.5).

The third topology to be targeted early was the branched frame of **21b** (Scheme 4.5). The central benzene ring of this system was expected to be maximally bond localized, perhaps representing the first example of a 1,3,5-cyclohexatriene – a long-sought experimental model for the estimation of the resonance energy in benzene. The preparation of **21b** was achieved via an ambitious triple cocyclization strategy, in which **20** [62] added three molecules of BMTSA to provide **21a** in 39%

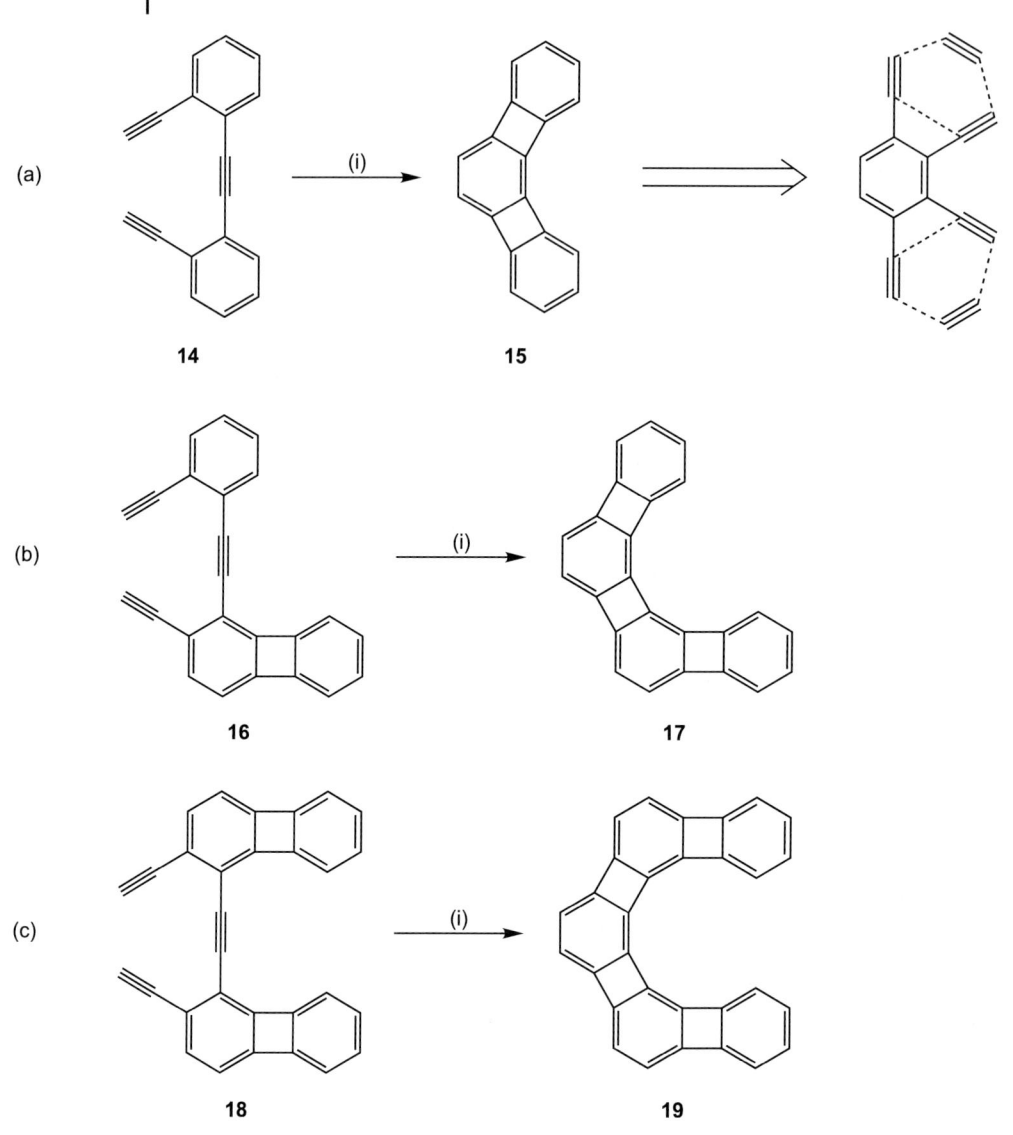

Scheme 4.4. Preparation of angular [N]phenylenes **15**, **17**, and **19**: (a) (i) [CpCo(CO)₂], hv, Δ, 30% or [CpCo(eth)₂], THF, −30 °C, followed by CO (8 atm), 100 °C, 70%; (b) [CpCo(CO)₂], m-xylene, hv, Δ, 30% or [CpCo(eth)₂], THF, −25 °C, 16 h, followed by 1,3-cyclohexadiene, THF, 100 °C, 2 h, 51%; (c) [CpCo(CO)₂], m-xylene, hv, Δ, 5%.

yield, which could be readily protodesilylated to the parent **21b** [63]. The construction of **21a** is remarkable, considering the explosive nature of **20**, the regioselectivity of the individual cotrimerizations, the fact that six rings are generated in one step, and, again, the strain in the product.

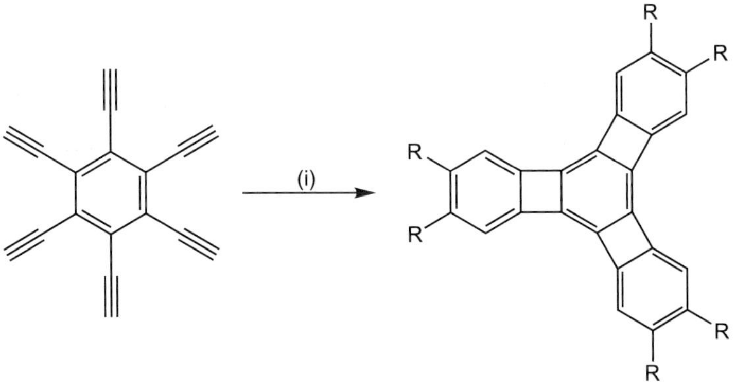

Scheme 4.5. Preparation of branched [4]phenylene **21b** by triple cocyclization: (i) BTMSA, [CpCo(CO)$_2$], hν, Δ, 39%; (ii) CF$_3$CO$_2$H, CHCl$_3$, 77%.

4.2.2
Syntheses of New Phenylenes

The previous section summarized the essence of what was known at the time of the last review of the subject [45a]. Since then, twelve new phenylenes of increasing size and topological complexity have been prepared. With the exception of the linear series, forays have been made into the assembly of all types of phenylenes depicted in Fig. 4.2. The following five subsections will describe, in order, the syntheses of angular [5]–[9]phenylene, also dubbed "heliphenes", because of their helical configuration [64]; the preparation of zigzag [4]- and [5]phenylene, through both intra- and intra/intermolecular cyclizations; the construction of three phenylenes with new mixed topologies; new branched phenylenes; and synthetic efforts towards the (still) elusive class of circular phenylenes.

4.2.2.1 Angular and Helical Phenylenes
Molecular models indicate that, starting with angular [6]phenylene, the two ends of the angular phenylenes suffer steric interactions that render them helical, an expectation that was quantified theoretically [65]. The hexagonal squeezes of these helical phenylenes (heliphenes) are helicenes, a class of PAHs that has received much scrutiny [66].

As described in Section 4.2.1, the key step in the preparation of angular [3]- to [5]phenylene employed a single cobalt-catalyzed cycloisomerization of the respective precursor triynes **14**, **16**, and **18**. Such a strategy was no longer feasible for the higher analogs, as suitable building blocks based on functionalized angular [3]phenylene derivatives are not (yet) readily available. Hence for the higher systems, multiple cycloisomerizations had to be designed using the same build-

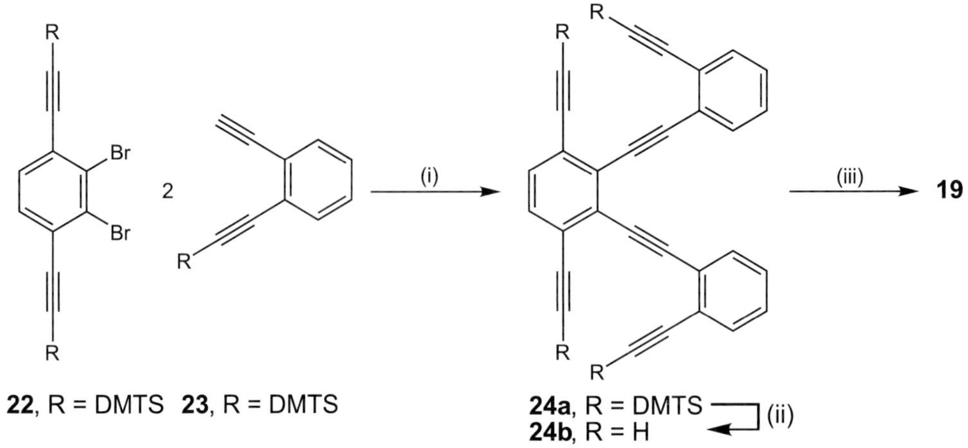

22, R = DMTS   **23**, R = DMTS   **24a**, R = DMTS
                                        24b, R = H

Scheme 4.6. Preparation of angular [5]phenylene **19** by double intramolecular cyclization: (i) [PdCl$_2$(PPh$_3$)$_2$], CuI, NEt$_3$, Δ, 57%; (ii) TBAF, THF, (95%); (iii) [CpCo(CO)$_2$], m-xylene, hν, Δ, 33%.

ing blocks. The already known angular [5]phenylene **19** was chosen as a testing ground for a double cyclization scheme (Scheme 4.6) [64]. Crucial for the success of the preparation of starting material **24** was the discovery that 1,2,3,4-tetrabromobenzene can be selectively alkynylated at the 1- and 4-positions to render **22** [67]. Sonogashira coupling of **22** to the previously reported **23** [58] produced **24a** (57%). The deprotected **24b** was cyclized to **19** in 33% yield [64].

Replacing the terminal benzene substituents in **24b** once and twice by biphenylenyl groups, in a manner analogous to that employed in the extension of the synthesis of **15** to **17** and **19** (Scheme 4.4), furnished hexaynes **25** and **27**, respectively, both of which underwent double cycloisomerization to [6]- and [7]heliphene **26** (12%) and **28** (8%), respectively (Scheme 4.7) [64].

With the synthesis of **28**, we have reached the limits of the double intramolecular cyclization approach, and access to the next higher homologs required the execution of even more ambitious triple cyclizations. The viability of such reactions was tested with **28** (Scheme 4.8) [68]. Thus, starting with tetrayne **29** [64], Sonogashira coupling with TMSA and selective deprotection provided **30**. This alkyne was reacted with another equivalent of **29** and the resulting nonayne completely desilylated to give **31**. Cobalt-catalyzed cyclization then afforded **28** in 2% yield. While this yield is low, one needs to remember that the reaction generates nine rings in one step, including six four-membered rings with an estimated strain of over 300 kcal mol^{-1}.

Having demonstrated the feasibility of triple cycloisomerizations, synthetic schemes were once again developed that replaced the terminal benzene moieties with biphenylene, giving rise to **32** and **34**, respectively. The former then provided

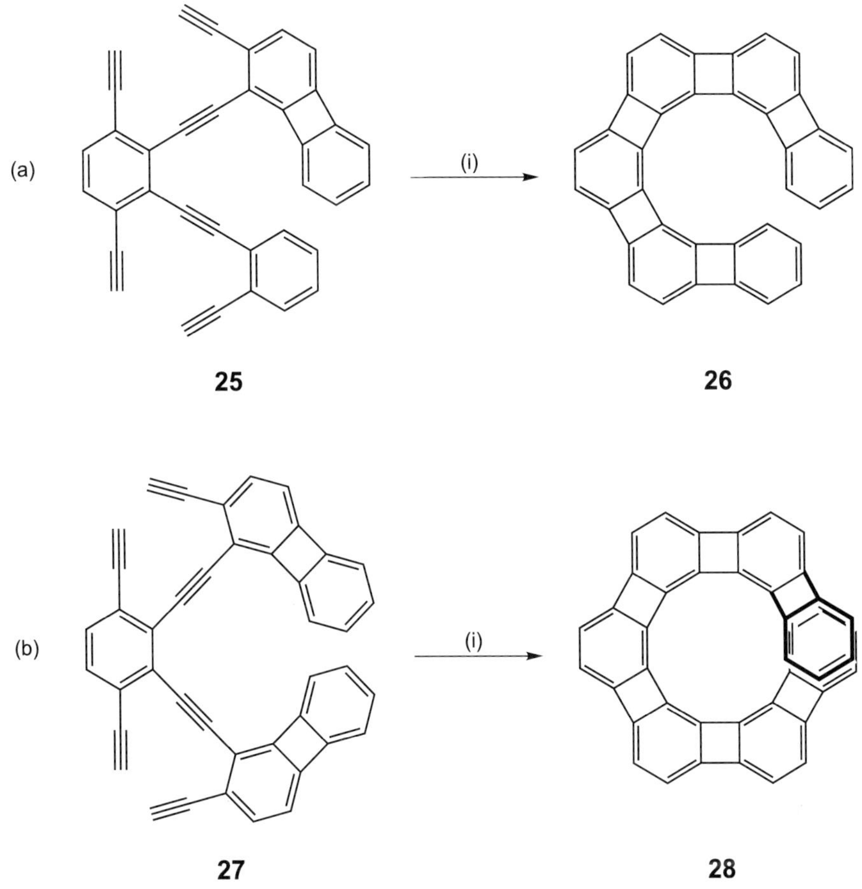

Scheme 4.7. Preparation of heliphenes **26** and **28** by double intramolecular cyclization: (a) (i) [CpCo(CO)$_2$], *m*-xylene, *hv*, Δ, 30 min, 12%; (b) (i) [CpCo(CO)$_2$], *m*-xylene, *hv*, Δ, 30 min, 8%.

33, the latter **35** (both in 2% yield; Scheme 4.9) [68]. These two compounds represent the largest phenylenes hitherto known.

4.2.2.2 Zigzag Phenylenes

The family of zigzag phenylenes is closely related to the angular isomers, in as much as it has the same repeating angular fusion of benzocyclobutadiene units, although "helical strain" is absent [65]. The electronic properties of its members are thus expected to be fairly similar. These phenylenes are also interesting as models for the one-dimensional zigzag-phenylene polymer, with properties different from the infinite linear [*N*]phenylene [69]. Finally, both archimedene (Fig. 4.2d) [53a, 70] and the octahedral C$_{48}$ [71] contain zigzag phenylene subunits.

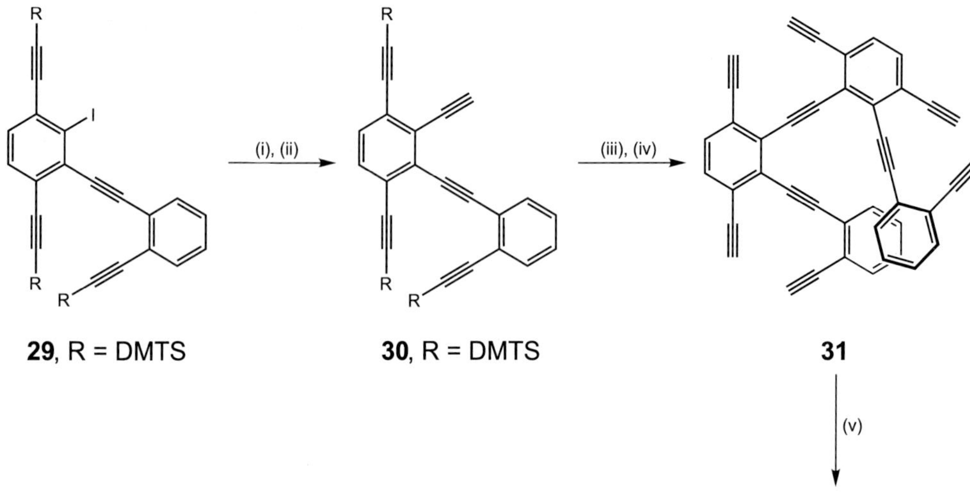

29, R = DMTS   **30**, R = DMTS   **31**

↓ (v)

28

Scheme 4.8. Preparation of helical [7]phenylene **28** by triple intramolecular cyclization: (i) TMSA, [PdCl$_2$(PPh$_3$)$_2$], CuI, NEt$_3$, 85 °C, 14 h 52%; (ii) K$_2$CO$_3$, THF/MeOH, 30 min, 92%; (iii) **29**, [PdCl$_2$(PPh$_3$)$_2$], CuI, NEt$_3$, 65 °C, 14 h, 41%; (iv) TBAF, THF, 23 °C, (95%); (v) [CpCo(CO)$_2$], m-xylene, hν, Δ, 1 h, 2%.

The topological analogy between the angular and the zigzag family of phenylenes is reflected in the resemblance of the synthetic strategies to the two classes. The parent zigzag [4]phenylene **38** was approached via **37**, a regioisomer of **16** (Scheme 4.4), in which the two alkynyl substituents on the biphenylene nucleus have traded places (Scheme 4.10a). Compound **37** was in turn made via a three-step elaboration of 1,2-diiodobiphenylene **36** [58]. Cobalt then converted **37** into **38** in 31% yield [72]. An alternative route (Scheme 4.10b) constituted the first example of a combination of intra- and intermolecular cyclizations in a single reaction step. It started with tetrabromobenzene **39**, which was elaborated with **23** (Scheme 4.6), followed by three-fold coupling with TMSA and full deprotection, ultimately giving **41**. This pentayne was cocyclized with TMSA (see Experimental Section) and subsequently protodesilylated to afford **38**. This method was extended to the synthesis of the bent [4]phenylenes (Section 4.2.2.3) and could, in principle, be used also on a simplified route to angular [4]phenylene **17**, a task yet to be tackled.

A variant of the double cycloisomerization route to angular [5]phenylene (Scheme 4.6) was used to prepare zigzag [5]phenylene **44** (Scheme 4.11) [72]. Starting once more with **39**, double alkynylation with **23** assembled tetrayne **42**, which was further substituted with TMSA. Removal of all the silyl protecting groups provided **43**, a regioisomer of **24b** (Scheme 4.6). Compound **43** was then cyclized to **44** in 2% yield [72].

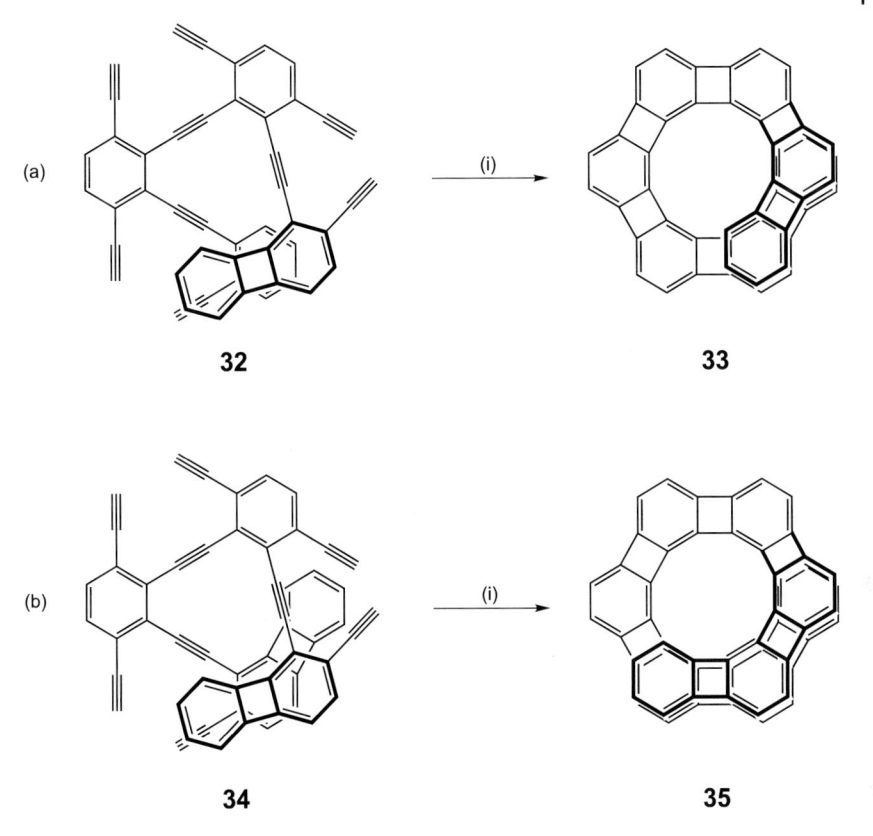

Scheme 4.9. Preparation of heliphenes **33** and **35** by triple intramolecular cyclization: (a) (i) [CpCo(CO)$_2$], m-xylene, hv, Δ, 30 min, 2%; (b) (i) [CpCo(CO)$_2$], m-xylene, hv, Δ, 20 min, 2%.

4.2.2.3 Phenylenes with Mixed Topology: the "Bent" Isomers

All the phenylene topologies discussed so far contained only one mode of repeating fusion: either linear or angular. The smallest molecule with mixed linear/angular connectivity is bent [4]phenylene **48** (Scheme 4.12), the last [4]phenylene isomer to be made [52, 58, 63, 72]. This isomer is intriguing, in particular because of the unusual nature of the two juxtaposed internal six-membered rings and their surroundings. Its synthesis entailed application of a regioisomeric variation of the intramolecular approach to **17**, through **46** (Scheme 4.12a), formed by reaction of 2,3-diiodobiphenylene **45** [51] with **23**. Further ethynylation eventually resulted in triyne **47**, which was cyclized to **48** in 33% yield [73]. The 9,10-bis(trimethylsilyl) derivative of **48**, **52**, was made by the combination of intramolecular cyclization and cocyclization with BTMSA, precedented for **38** (Scheme 12b) [72]. The starting 1,2,4,5-tetrabromobenzene **49** was desymmetrized into **50**. A sequence of two Sonogashira couplings, first with **23** and then with TMSA, was followed by the full deprotection to give **51**. Cyclization proceeded in 19% yield, producing **52** [73].

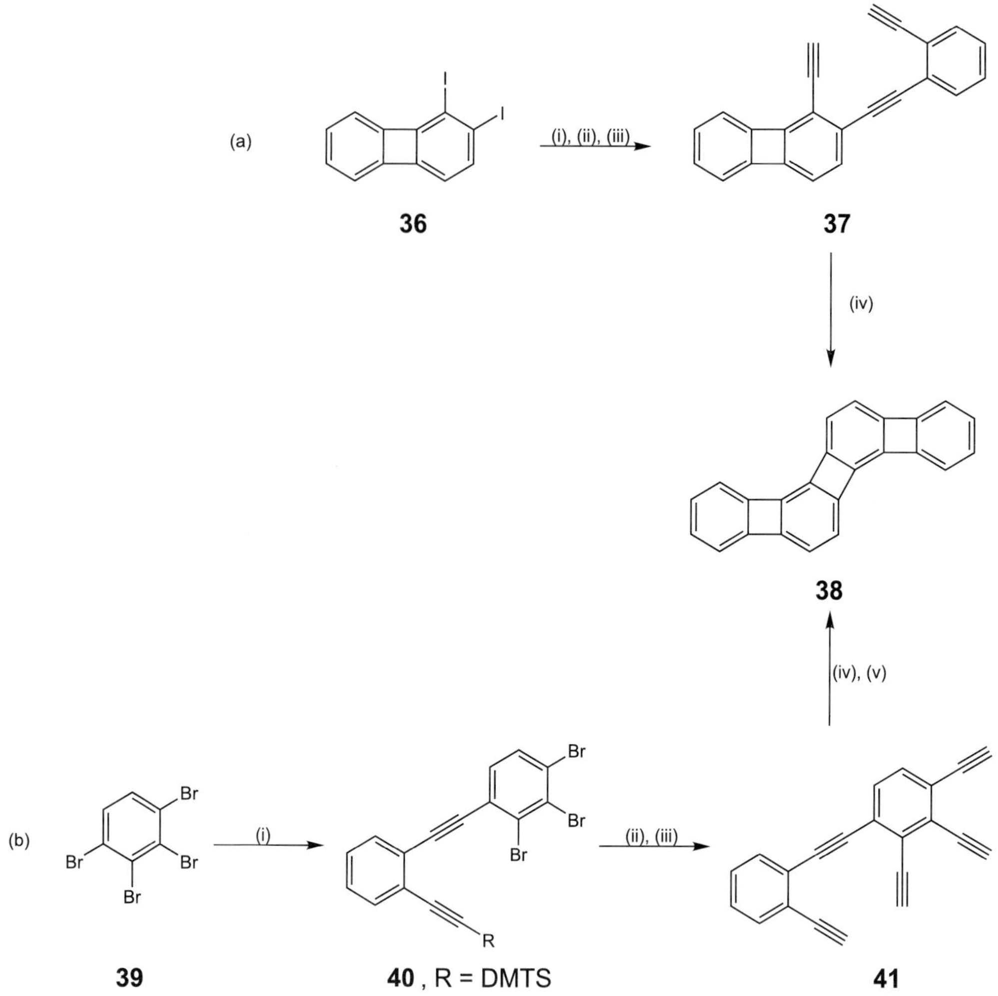

Scheme 4.10. Two syntheses of zigzag [4]phenylene **38**: (a) intramolecular approach, (i) **23**, [PdCl$_2$(PPh$_3$)$_2$], CuI, Et$_3$N, 23 °C, 15 h; (ii) TMSA, [PdCl$_2$(PPh$_3$)$_2$], CuI, Et$_3$N, 50 °C, 2 d, 62% (over 2 steps); (iii) TBAF, THF, 23 °C, 40 min; (iv) [CpCo(CO)$_2$], m-xylene, hν, Δ, 18 h, 29% (over 2 steps); (b) mixed intra/intermolecular approach, (i) **23**, [Pd(PPh$_3$)$_2$Cl$_2$], CuI, Et$_3$N, 50 °C, 24 h, 66%; (ii) TMSA, [PdCl$_2$(PPh$_3$)$_2$], CuI, piperidine, 100 °C, 7 d; (iii) TBAF, THF, 33% (over 2 steps); (iv) [CpCo(CO)$_2$], BTMSA, hν, Δ, 10 h, 15%; (v) CF$_3$CO$_2$H/CHCl$_3$, 23 °C, 12 h, 74%.

The success of Scheme 4.12 encouraged approaches to the higher homologs of **48**, *anti*- (**56**, Scheme 4.13), and *syn*-doublebent [5]phenylene (**60**, Scheme 4.14). These systems would allow an investigation of the effect of increasing bond localization of the termini of the linear [3]phenylene fragment on the properties of the center piece. Strategically, the approach to both systems was modeled

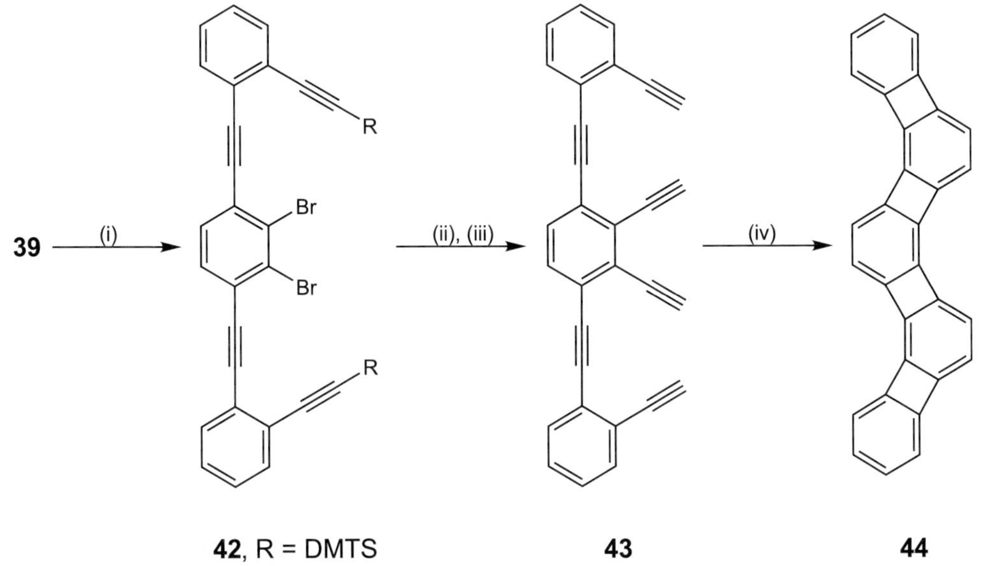

Scheme 4.11. Synthesis of zigzag [5]phenylene: (i) **23**, [PdCl$_2$(PPh$_3$)$_2$], CuI, Et$_3$N, 23 °C, 5 d, 66%; (ii) TMSA, [PdCl$_2$(PPh$_3$)$_2$], CuI, piperidine, 80 °C, 3 d; (iii) TBAF, THF, 23 °C, 2 h, 80% (over 2 steps); (iv) [CpCo(CO)$_2$], *m*-xylene, hν, Δ, 2 h, 2%.

after Schemes 4.6 (for **19**) and 4.12, utilizing regioisomeric double intramolecular cyclizations.

The synthesis of anti-doublebent [5]phenylene **56** [74] commenced with the tetrahalogenated C$_{2h}$-symmetric **53** [75]. Another use of the versatile building block **23** provided **54**. Subsequent coupling with TMSA and deprotection afforded **55**. The cyclization initially failed when attempted with [CpCo(CO)$_2$] as the catalyst, but was later rendered successful by the application of the milder [CpCo(eth)$_2$] conditions [74].

In an analogous (but slightly altered) manner, the synthesis of **60** (Scheme 4.14) started with 1,5-dibromo-2,4-diiodobenzene **57** [76] as a C$_{2v}$-symmetric template. Reaction with TMSA and deprotection gave 1,5-dibromo-2,4-diethynylbenzene. Another Sonogashira coupling, this time with 1-bromo-2-iodobenzene, provided the tetrabrominated **58**. This material underwent a four-fold exchange of bromides with TMSA and, after fluoride-assisted deprotection, yielded hexayne **59**. The cyclization to **60** proceeded smoothly under the conditions of [CpCo(eth)$_2$] catalysis (see Experimental Section).

4.2.2.4 Branched Phenylenes

Two other types of mixed topology are the branched/linear and branched/angular motifs. To what extent can the bond localization of the central cyclohexatriene in

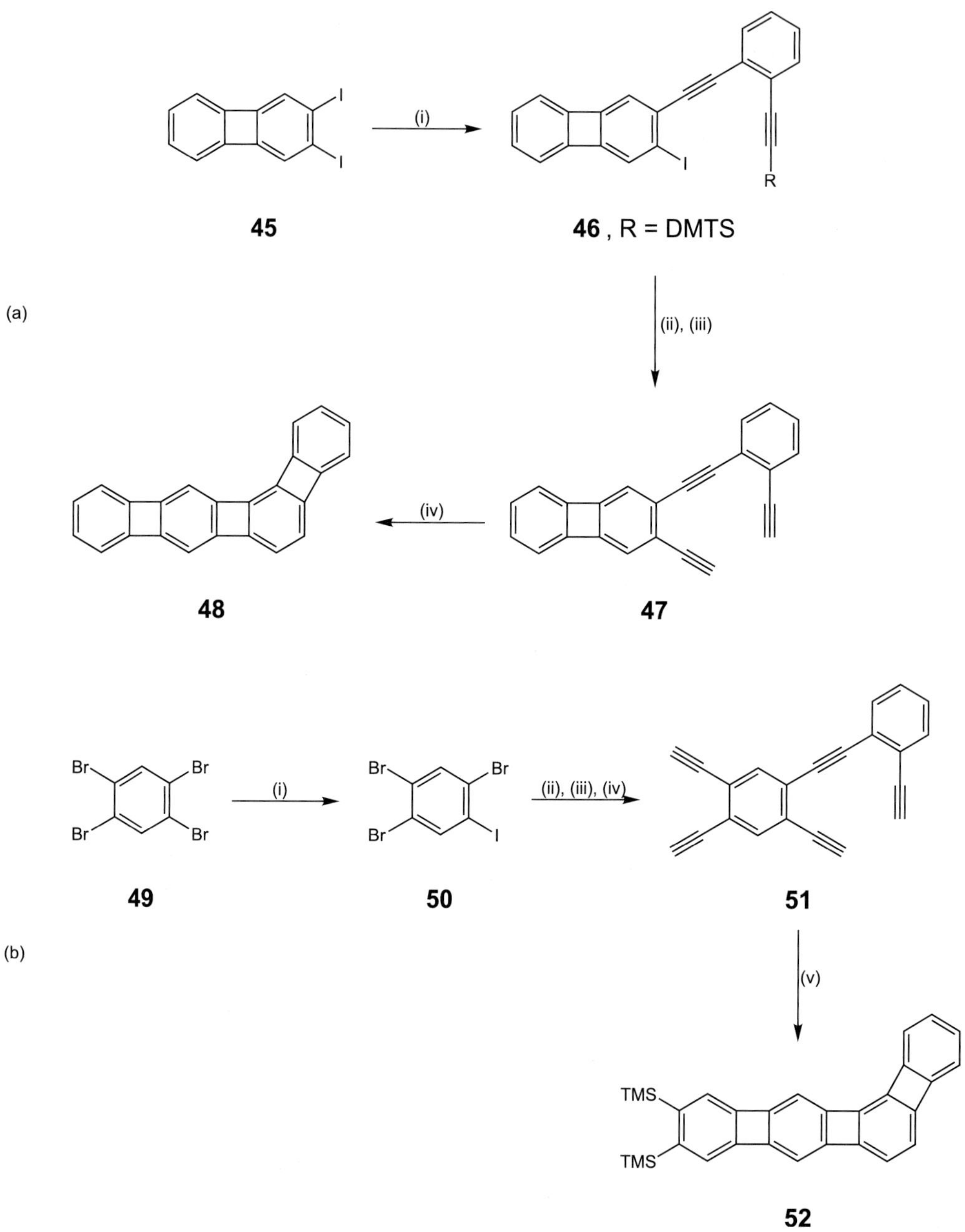

Scheme 4.12. The syntheses of bent [4]phenylenes **48** and **52**: (a) (i) **23**, [PdCl$_2$(MeCN)$_2$], CuI, PPh$_3$, piperidine, 90 °C, 40 h, 16%, (ii) TMSA, [PdCl$_2$(PPh$_3$)$_2$], CuI, piperidine, 44 h, 93%, (iii) TBAF, THF, 20 min, (95%), (iv) [CpCo(CO)$_2$], *m*-xylene, *hν*, Δ, 15 h, 33%; (b) (i) BuLi, Et$_2$O, −78 °C, followed by I$_2$, Et$_2$O, −78 °C, 93%, (ii) **23**, [PdCl$_2$(PPh$_3$)$_2$], CuI, PPh$_3$, Et$_3$N, 23 °C, 15 h, (iii) TMSA, [PdCl$_2$(PPh$_3$)$_2$], CuI, Et$_3$N, 120 °C, 2.5 d, 29% (over two steps), (iv) TBAF, THF, 2 h, (95%), (v) [CpCo(CO)$_2$], BTMSA, *hν*, Δ, 16 h, 19%.

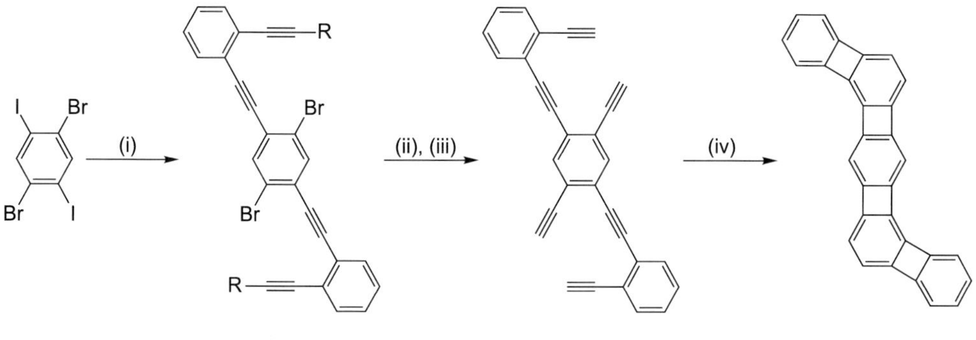

Scheme 4.13. The synthesis of anti-doublebent [5]phenylene **56**: (i) **23**, [PdCl$_2$(PPh$_3$)$_2$], CuI, Et$_3$N, 72%; (ii) TMSA, [PdCl$_2$(PPh$_3$)$_2$], CuI, Et$_3$N, 120 °C, 70%; (iii) TBAF, THF, 2 h, (95%); (iv) [CpCo(eth)$_2$], THF, −25 °C, 16 h, followed by 1,3-cyclohexadiene, THF, 110 °C, 2 h, 7%.

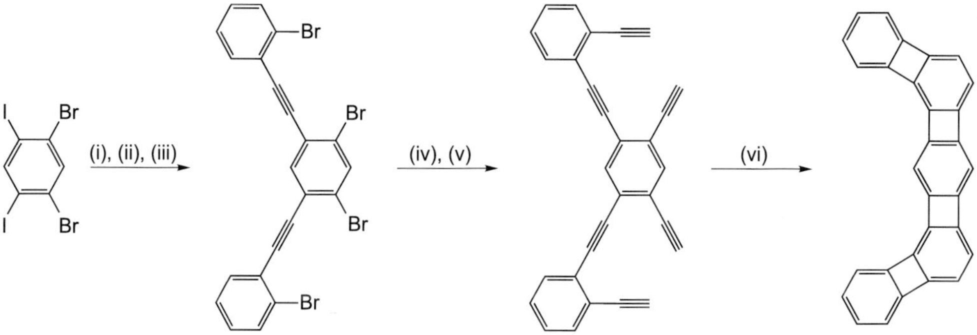

Scheme 4.14. The synthesis of syn-doublebent [5]phenylene **60**: (i) TMSA, [PdCl$_2$(PPh$_3$)$_2$], CuI, Et$_3$N, 23 °C, 2 h, 96%; (ii) KOH, Et$_2$O/EtOH, (iii) 1-bromo-2-iodobenzene, [PdCl$_2$(PPh$_3$)$_2$], CuI, Et$_3$N, 120 °C, 44% (over 2 steps); (iv) TMSA, [PdCl$_2$(PPh$_3$)$_2$], CuI, Et$_3$N, 120 °C, 47%; (v) TBAF, THF, 2 h, (95%); (vi) [CpCo(eth)$_2$], THF, −25 °C, 16 h, followed by 1,3-cyclohexadiene, THF, 110 °C, 2 h, 14%.

the branched [4]phenylene **21b** be manipulated by additional fusions? One might expect linear fusion to increase it, whereas angular fusion should cause the opposite. To validate this expectation, branched [5]phenylene **64b** (Scheme 4.15), C_{3h}-symmetric branched **66** (Scheme 4.16), and its D_{3h}-symmetric isomer **71** (Scheme 4.17) were constructed.

The preparation of **64b** relied on a modification of the iterative cocyclization strategy to linear [N]phenylenes (Section 4.2.1) [77]. Thus, diyne **61** [63] was cocyclized with bis(triisopropylsilyl)-1,3,5-hexatriyne **62** [55]. The resulting **63** was

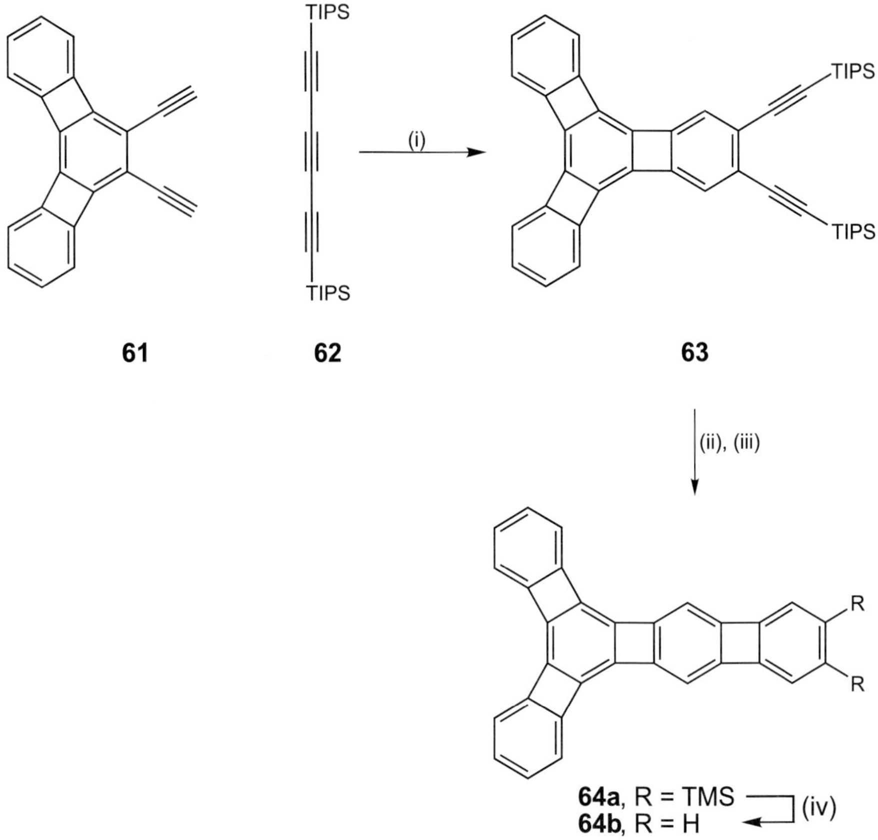

Scheme 4.15. The synthesis of branched [5]phenylene **64b**: (i) [CpCo(CO)$_2$], PhCH$_3$, hv, Δ, 16 h, 32%; (ii) TBAF, THF, 23 °C, 2 h, (95%); (iii) BTMSA, [CpCo(CO)$_2$], THF, hv, Δ, 16 h, 33%; (iv) CF$_3$CO$_2$H, CH$_2$Cl$_2$, 23 °C, 16 h, 65%.

deprotected and subjected to a second cocyclization, this time with BTMSA, providing the Y-shaped **64a** in 33% yield (over 2 steps). Acid-catalyzed removal of the silyl groups produced the parent branched [5]phenylene (**64b**, Scheme 4.15) [77].

The synthesis of C_{3h}-symmetric branched **66** (Scheme 4.16) represents an extension of Scheme 4.16. It starts with hexaethynylbenzene **20** [62], which was cocyclized with **62** in 38% yield. The resulting hexaalkynyl substituted **65a** was treated with TBAF to afford **65b**. This material was cocyclized with BTMSA in 37% yield (over 2 steps), producing the C_3-symmetric hexakis(trimethylsilyl)[7]phenylene **66** [77].

For the preparation of **71** (Scheme 4.17), a strategy was necessary that desymmetrized the sixfold symmetry of **20** to allow for the generation of angular fusion. It started with trialdehyde **67** [78], which was coupled with TMSA in 97% yield, to

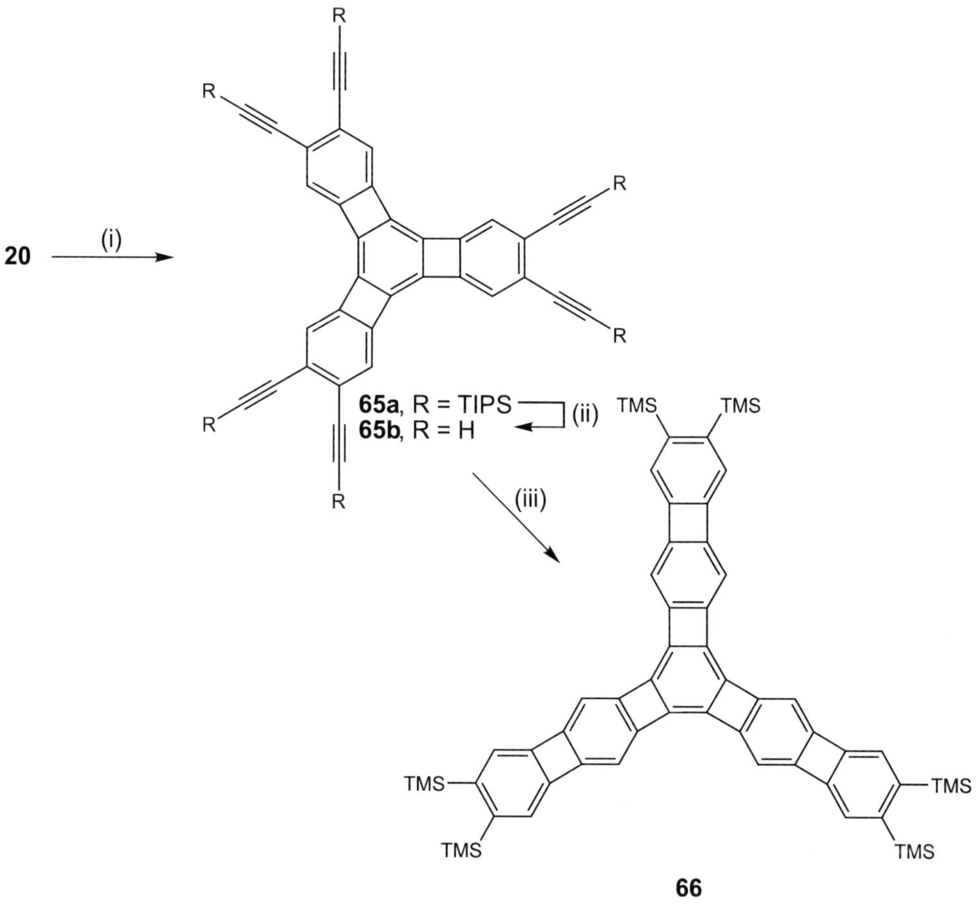

Scheme 4.16. The synthesis of C_{3h}-symmetric branched **66**: (i) **62** (7 equiv), [CpCo(CO)$_2$], PhCH$_3$, $h\nu$, Δ, 16 h, 38%; (ii) TBAF, THF, 23 °C, 30 min, (95%); (iii) BTMSA, [CpCo(CO)$_2$], THF, $h\nu$, Δ, 16 h, 37%.

give **68**. A Corey-Fuchs dibromoolefination, followed by treatment with LDA provided the hexayne **69**. The remaining three benzene rings of **70** were introduced by reacting **69** with 1-iodo-2-(TMSethynyl)benzene [79]. Base-catalyzed removal of all six TMS groups was followed by threefold [CpCo(CO)$_2$]-mediated cycloisomerization to **71** (2% yield) [80]. Compound **71** has the distinction of representing the largest synthesized subunit of the "Archimedean solid" archimedene (C$_{120}$, Fig. 4.2d) [53, 70]. The successful conversion of the nonayne precursor to **71** provides a valuable additional example of a triple intramolecular cyclization, differing topologically from those employed on route to **28**, **33**, and **35** by the fact that six (of nine) reacting triple bonds reside on a single benzene ring.

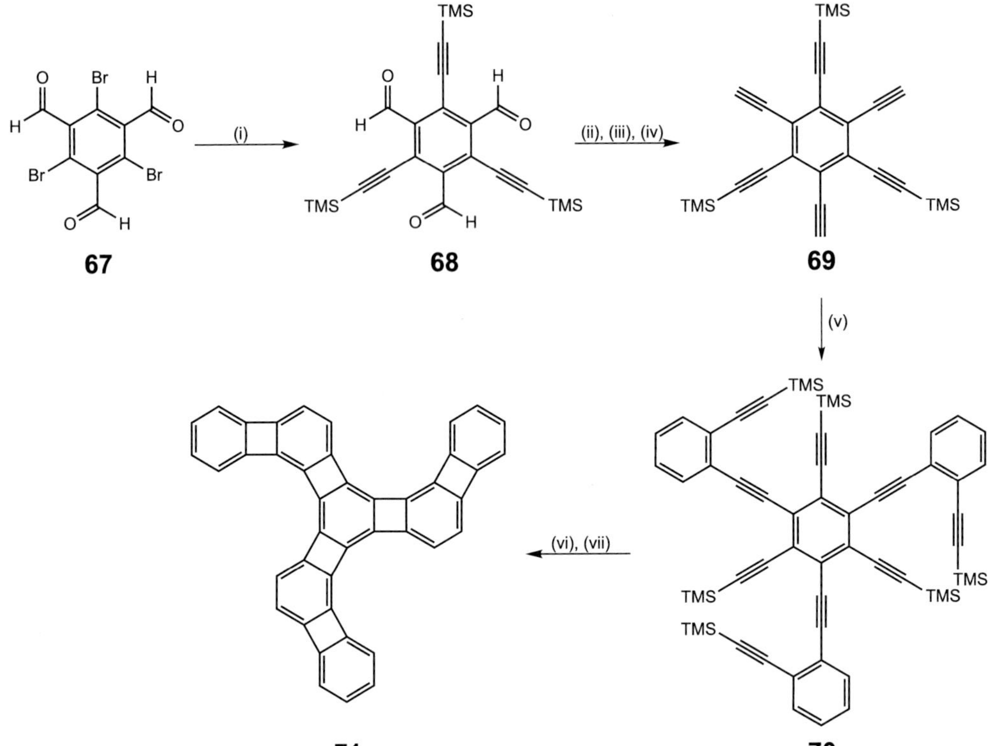

Scheme 4.17. The synthesis of **71**: (i) TMSA, [PdCl$_2$(PPh$_3$)$_2$], CuI, Et$_3$N, THF, 97%; (ii) CBr$_4$, Zn, PPh$_3$, CH$_2$Cl$_2$, 99%; (iii) LDA, THF, −78 °C; (iv) aq. NH$_4$Cl, 95% (over 2 steps); (v) 1-iodo-2-(TMSethynyl)benzene, [PdCl$_2$(PPh$_3$)$_2$], CuI, i-Pr$_2$NH, THF, 77%; (vi) K$_2$CO$_3$, MeOH/THF, 61%; (vii) [CpCo(CO)$_2$], m-xylene, hν, Δ, 1.2%.

4.2.2.5 Circular Phenylenes

Circular phenylenes have the distinguishing characteristic of a resonance picture that includes forms that encompass both the inner and outer peripheral loops, a phenomenon described as superdelocalization [81]. This class of circular phenylenes remains elusive [67, 82]. The simplest member of this series that does not suffer from additional "circular" strain is [6]phenylene **77d** (Scheme 4.18), also christened antikekulene [57] to highlight its relationship to kekulene, its all-benzenoid relative with an equal number of rings [83]. In antikekulene, avoidance of (benzo)cyclobutadienoid local circuits is expected to enhance the contribution of the potentially superdelocalized resonance form depicted for the structure in Scheme 4.18, albeit with the added and destabilizing feature that both inside and outside peripheries contain a (4n) electron count.

An oligoalkyne polycyclization route to any circular phenylene is conceptually different from those developed for the other topologies, as it requires the elaboration of a suitably functionalized dehydrobenzannulene, a significant synthetic

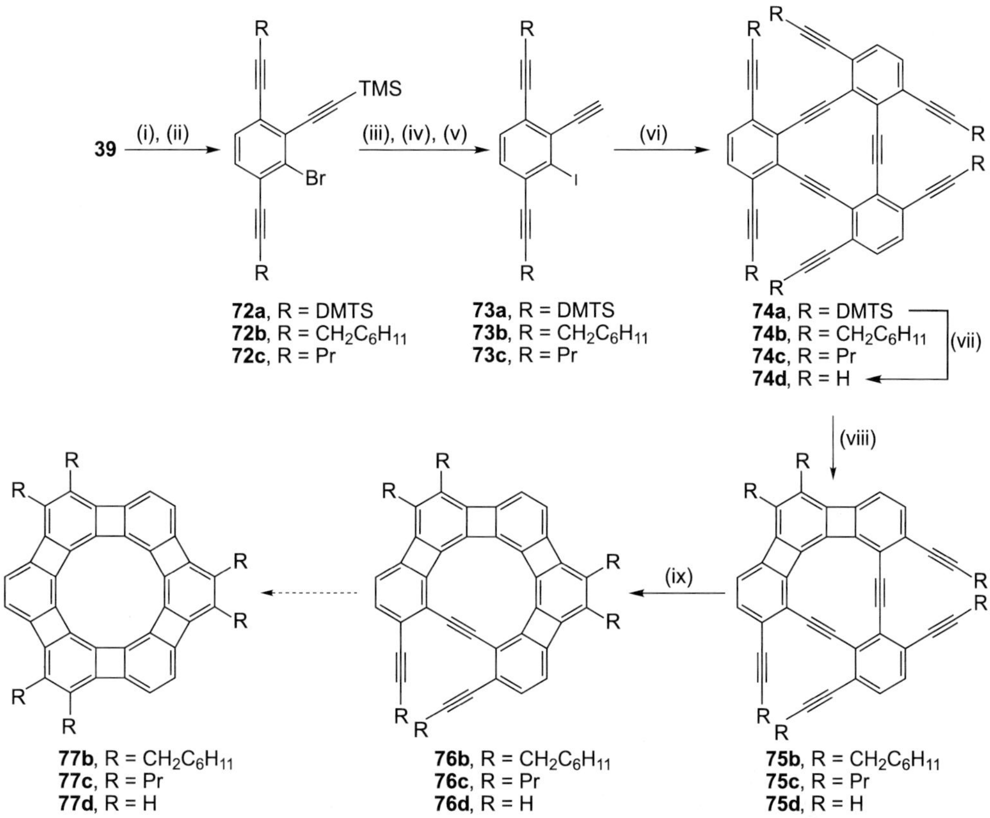

4.2 Preparation of Phenylenes

72a, R = DMTS
72b, R = CH$_2$C$_6$H$_{11}$
72c, R = Pr

73a, R = DMTS
73b, R = CH$_2$C$_6$H$_{11}$
73c, R = Pr

74a, R = DMTS
74b, R = CH$_2$C$_6$H$_{11}$
74c, R = Pr
74d, R = H

77b, R = CH$_2$C$_6$H$_{11}$
77c, R = Pr
77d, R = H

76b, R = CH$_2$C$_6$H$_{11}$
76c, R = Pr
76d, R = H

75b, R = CH$_2$C$_6$H$_{11}$
75c, R = Pr
75d, R = H

Scheme 4.18. Attempted syntheses of circular [6]phenylenes **77b–d**: (i) RC≡CH, [PdCl$_2$(PPh$_3$)$_2$], CuI, Et$_3$N, 23–60 °C, 3 d, 80% (**72a**), 58% (**72b**), 51% (**72c**); (ii) TMSA, [PdCl$_2$(PPh$_3$)$_2$], CuI, Et$_3$N, 100 °C, 4 h–2.5 d, 49% (**72a**), 31% (**72b**), 27% (**72c**); (iii) BuLi, Et$_2$O, −78 °C, 30 min; (iv) I$_2$, Et$_2$O, from −78 °C to 23 °C; (v) K$_2$CO$_3$, CH$_3$OH, 1 h, 91% (**73a**), 86% (**73b**), 73% (**73c**); (vi) CuCl, NH$_4$OH, EtOH, 1 h, followed by pyridine, Δ, 6 h, 20% (**74a**), 36% (**74b**), 32% (**74c**); (vii) TBAF, THF, CH$_3$CN, 5 h, 95%; (viii) [CpCo(CO)$_2$], m-xylene, hv, Δ, 20 min, 45% (**75b**), 14% (**75c**), 0% (**75d**); (ix) [CpCo(CO)$_2$], 1,2,4-trichlorobenzene, hv, Δ, 20 min, 40% (**76b**), 14% (**76c**).

enterprise in its own right. This is witnessed by the fact that even the preparation of the parent dehydrobenz[12]annulene (also known as tribenzocyclyne, TBC) remains a challenging task [82b, 84], almost 40 years after its original synthesis by Staab and Graf [85]. In the case of **77d**, the appropriate tribenzocyclyne is **74d** (Scheme 4.18). Its synthesis commenced with **39**, which was manipulated into bromide **72a**. Bromine–iodine exchange, followed by TMS group removal delivered **73a** in 91% overall yield. Attempted cyclocoupling under Sonogashira conditions was complicated by irreproducibility. Switching to the Stephens-Castro reaction gave better results, and cyclyne **74a** emerged in 20% yield (see Experimental Section). Deprotection with TBAF gave **74d** in 95% yield [67]. Compound **74d** is the largest synthesized substructure of the novel carbon allotrope [86] graphyne [87] and organizes into a remarkable supramolecular framework in the crystal [82a].

Unfortunately, attempted threefold cobalt-mediated cyclization of **74d** gave only insoluble dark brown materials. Suspecting that the insolubility of intermediates or **77d** itself might be the problem, the cyclohexylmethyl- and propyl-substituted materials were prepared (**74b** and **74c**, respectively; Scheme 4.18). Application of standard cyclization conditions to these derivatives furnished the singly cyclized **75b** and **c**, respectively. Resubjecting these materials to the reaction conditions in the higher-boiling 1,2,4-trichlorobenzene afforded the products of the double cyclization **76b** and **c**, respectively. Despite extensive efforts, the third cyclization did not take place even in sulfolane (reaction temperature ∼ 200 °C). This result is puzzling, especially in view of the ready metallacycle formation from triyne **14** and [CpCo(eth)$_2$] [59]. A possible explanation might be the increasing distance between the reacting triple bonds along the series **74b-75b-76b** (all of which were crystallographically characterized). The notion that the problems of the final cyclization are kinetic in nature is supported by the finding that the conversion of **76d** into **77d** is calculated to be exothermic by -45.50 kcal mol^{-1} (B3LYP/6-31G*) [67].

To summarize this section, to date 19 phenylenes have yielded to synthesis. They can be divided broadly into 5 families (# of examples): linear (3), angular/helical (7), zigzag (2), bent (3) and branched (4). Their topologies have been accessed through 26 different routes, 15 of which involved in the crucial step an all-intramolecular cobalt-catalyzed cyclization, 9 used intermolecular variants and 2 a combination of the two strategies.

4.3
Comparative Reactivity of the Phenylenes

The presence of strained cyclobutadiene moieties [34] and cyclohexatrienoid rings renders the phenylenes susceptible to various reactions; thus, hydrogenation, metal complexation, ring openings, and cycloadditions are all feasible. Early work focused on the chemistry of linear [3]- [51, 58] and branched [4]phenylene [63, 89] and has been reviewed [45a]. The following sections will concentrate on selected recent examples featuring the comparative reactivity of angularly fused cyclohexatrienoid rings.

4.3.1
Hydrogenation

With the caveat of the mechanistic complexities of heterogenous catalytic hydrogenations [90], the relative ease of hydrogenation of the cyclohexatrienoid rings in the phenylenes (Scheme 4.19) can be used as a qualitative measure of reactivity. Thus, while **9b** [51] and **21b** [89] could be hydrogenated readily (Pd/C, 1 atm H$_2$), **15** required more stringent conditions (Pd/C, 10 atm H$_2$) [57], and **1** was inert or underwent hydrogenolytic four-membered ring opening [16a].

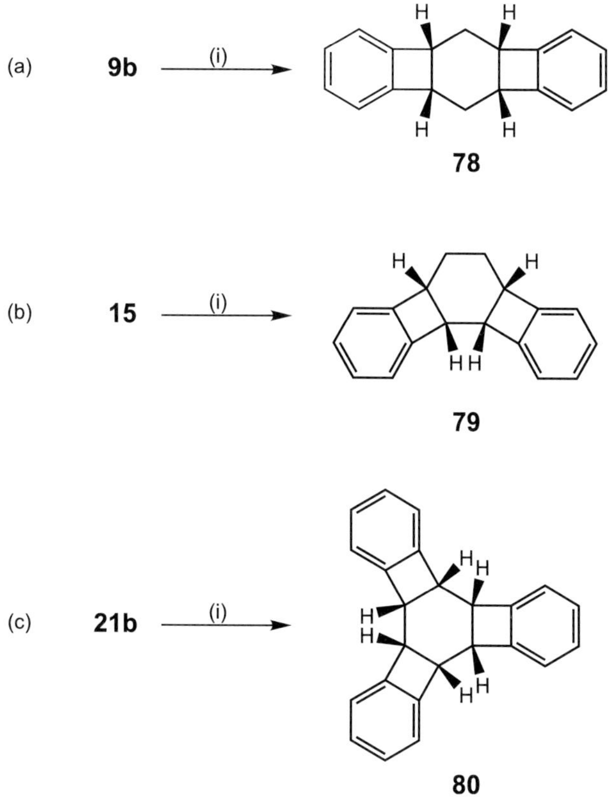

Scheme 4.19. Hydrogenation of phenylenes **9b**, **15**, and **21b**: (a) (i) 5% Pd/C, H$_2$ (1 atm), THF, 23 °C, 3 h, 74%; (b) (i) Pd/C, H$_2$ (10 atm), THF, 23 °C, 99%; (c) (i) Pd/C, H$_2$ (1 atm), THF, 23 °C, 18 h, 87%.

Preliminary observations thus suggested a reactivity order of **9b** ≥ **21b** > **15** > **1**. The ambiguities in the kinetics notwithstanding, thermodynamic measurements clearly point to the fact that the central ring in **21b** is more cyclohexatrienic than that in **15**. Thus, the measured heats of hydrogenation, corrected for the strain present in the respective all-cis-hexahydroderivatives **80** and **79** (Scheme 4.19), are −(83.0 to 84.2) kcal mol^{-1} and −(68.1 to 73.6) kcal mol^{-1}, respectively, revealing that the central ring in **15** enjoys more resonance stabilization than that in **21b** by at least ∼10 kcal mol^{-1}. Perhaps even more interestingly, the corrected ΔH_{hyd} of **21b** is remarkably close to that estimated for three cyclohexene double bonds (−84.8 kcal mol^{-1}), suggesting that the central ring is a true cyclohexatriene, possibly devoid of any resonance interaction between the π bonds. Such a picture has also been painted employing other methods [91]. Finally, the estimated ΔH_{hyd} of biphenylene (**1**), corrected for strain in the product, using a similar approach to that described for **15** and **21b**, has a value of −64.8 kcal mol^{-1}, attesting to its expected attenuated activation relative to the other two phenylenes, although still featuring benzene rings that are less aromatic than benzene itself ($\Delta H_{hyd} = -49.1$ kcal mol^{-1}) [34].

The relatively higher reactivity of **9b** compared to **15** made the hydrogenation of bent [4]phenylene **48** an interesting proposition: which one of the two internal rings is the more reactive? On the basis of simple resonance arguments, the fusion of an additional benzocyclobutadiene fragment should stabilize the linear and destabilize the angular component of **48**, and thus possibly invert the reactivity order observed for the parents **9b** and **15**. Because **48** was not available in sufficient quantities, the problem was addressed with its bis(trimethylsilyl) derivative **52** (Scheme 4.20) [73]. Upon subjecting **52** to the reaction conditions previously used on **9b** and **21b** (Pd/C, 1 atm H$_2$, see also Experimental Section), the **B** ring was hydrogenated cleanly to give **81**. This result was clearly in consonance with expectation, even though the effect of the presence of the remote silyl groups in **52** may have contributed to its outcome. More experimentation is in order to corroborate these findings.

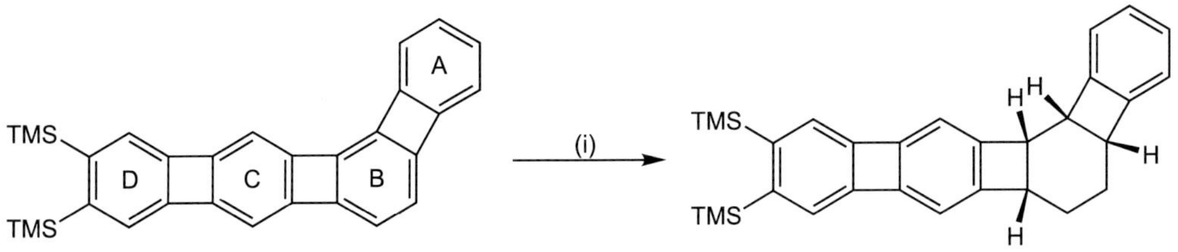

Scheme 4.20. The hydrogenation of **52**: (i) H$_2$ (1 atm), Pd/C, Et$_2$O, 10 min, 44%.

In syn-doublebent [5]phenylene **60**, the central linear moiety is stabilized even further compared to **48**, due to the presence of two angular fusions. The angular components, in turn, are still destabilized compared to the parent **15**, but to a lesser extent than in **48** (since they "share" the destabilization caused by the linear fusion). An overall decrease in reactivity of all rings, relative to **48** (or **52**) is thus expected. Preliminary results confirm this prediction, since, in contrast to **52**, **60** remains inert to hydrogenation (Pd/C, 1 atm H$_2$, 2 h) [92]. Similarly, dipropyl substituted zigzag [5]phenylene resisted hydrogenation even at increased pressures (5% Pd/C, 12.2 atm H$_2$) [72], in agreement with the notion that extension of the angular/zigzag phenylene frame causes an (at least initial) decrease in cyclohexatrienoid character of the internal rings [45a, 58].

4.3.2
Oxacyclopropanation and Cyclopropanation

In the light of the difficulty to attach meaning to the relative kinetic reactivities of the phenylenes in catalytic hydrogenations, it would be instructive to inspect their direct reactions with electrophilic species capable of attacking the activated six-membered rings. Indeed, and further corroborating the cyclohexatrienic character of the phenylenes, it was possible to effect oxacyclopropanations of **1**, **15**, and **21b**. Using dimethyldioxirane (DMDO) [93] as the oxidant, biphenylene (**1**) was converted sluggishly into the corresponding trisoxacyclopropane **82** (Scheme 4.21a). Its stereochemistry was assigned as trans on mechanistic grounds; however, a cis-geometry would also be consistent with the spectral data [94]. In contrast to the

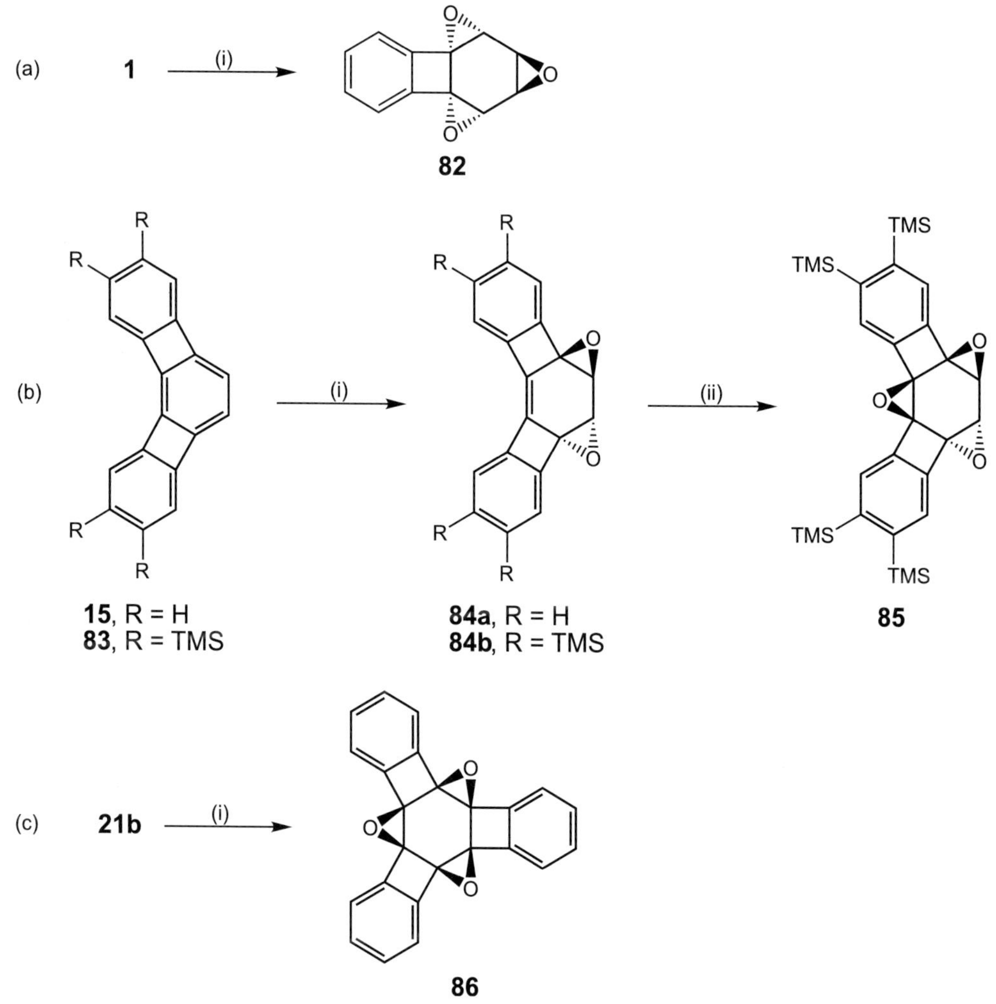

Scheme 4.21. Oxacyclopropanation of **1**, **15**, **83**, and **21b**: (a) (i) DMDO, acetone, 23 °C, 24 h, 30%; (b) (i) for **15** – DMDO, acetone, 23 °C, 30 min, (100%), for **83** – DMDO, acetone, 23 °C, 1 h, (100%); (ii) for **84b** only – DMDO, acetone, 23 °C, 6 h, 26%; (c) (i) DMDO, acetone, 23 °C, 84%.

slow conversion of **1**, angular [3]phenylene **15** was oxidized comparatively quickly under these conditions, but only to the moisture-sensitive (and hence difficult to completely characterize) bisoxacyclopropane **84a** (Scheme 4.21b). Switching to tetrakis(trimethylsilylated) **83** provided the more stable **84b**, the connectivity of which could be proven by (especially) NMR spectroscopy. Only on renewed oxidation of this compound was the trisoxacyclopropane **85** obtained in 26% yield [95]. In the latter, the asymmetry of the trans,trans,cis-arrangement manifests itself diagnostically in the ^{1}H-NMR spectrum. This stereochemical assignment also corroborates the proposed trans-geometries of **84a** and **b**, for which NMR data were not definitive [95], and possibly provides further support for the proposed structure of **82**. Finally, and to complete the series, **21b** underwent complete, but now all-cis, oxacyclopropanation to **86** during the course of just one hour (84% yield; Scheme 4.21c) [89]. Its structure was ascertained by an X-ray crystallographic analysis (Fig. 4.4).

The different stereochemical outcome of the oxidations of **1** and **15** compared to **21b** may be a consequence of the unique all-benzofusion in **86**, resulting in significant steric hindrance to trans attack due to the outside rings, even after the first oxacyclopropanation and pronouncedly so after the second.

Compared to the results of the above oxidations, the picture is less clear for the topologically seemingly analogous cyclopropanations. Thus, **1** transforms in the

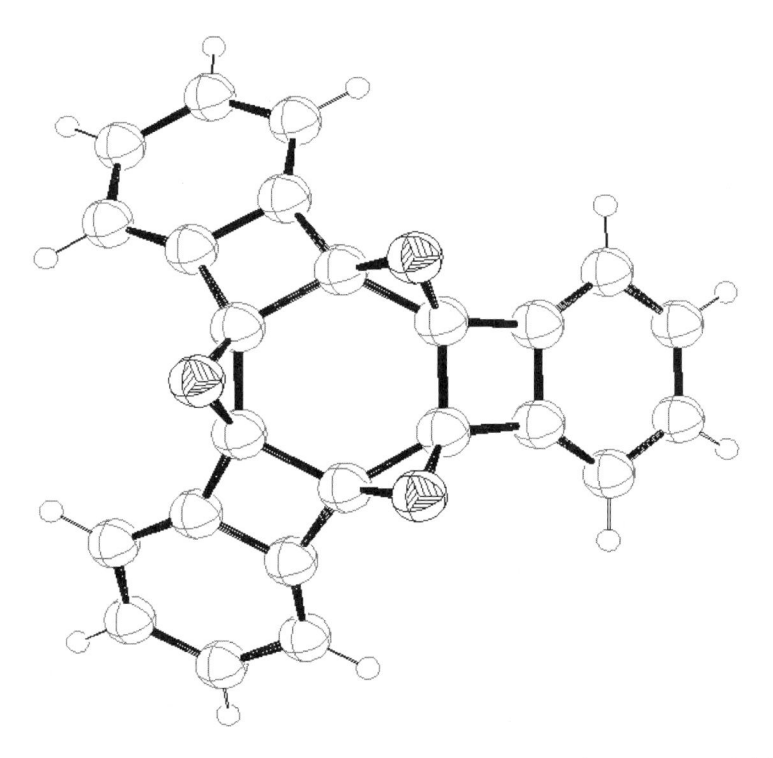

Figure 4.4. X-ray crystal structure of **86** (thermal ellipsoids are shown at 50% probability).

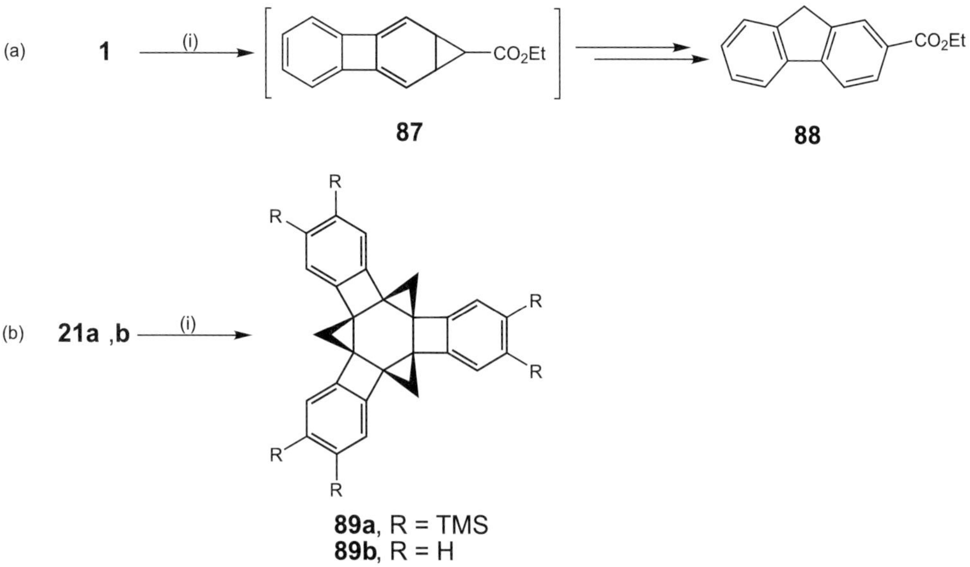

Scheme 4.22. Cyclopropanation of **1**, **21a**, and **21b**: (a) (i) ethyl diazoacetate (N$_2$CHCO$_2$Et), 165 °C, 15%; (b) (i) Et$_2$Zn, PhCH$_3$, 60 °C, 78% (**89a**), 97% (**89b**).

presence of ethyl diazocarboxylate to **88** only at elevated temperature (Scheme 4.22a), presumably through intermediate adduct **87** [16a, 96]. On the other hand, while angular [3]phenylene **15** was inert to modified Simmons-Smith conditions (Et$_2$Zn, PhCH$_3$, 60 °C) [95, 97], branched **21a,b** responded to this reagent by providing the triscyclopropanated **89a** and **b** in excellent yields (Scheme 4.22b, also Experimental Section) [89]. In analogy to the trisoxacyclopropanation of **21b** (Scheme 4.21c), carbene addition occurs all-cis, as rigorously ascertained by an X-ray crystal structure of **89a**.

4.3.3
[4+2]Cycloadditions

Another measure of the degree of diene character of phenylenes is their relative susceptibility to undergo [4+2]cycloadditions. Such reactions would lead to highly strained products, which might be expected to be labile. In addition, cycloadditions should be regiocontrolled by the desire to avoid ensuing cyclobutadienoid circuits. In this respect, singlet oxygen [98] has proven to be an interesting dienophile. For example, the oxidation of **1** with this species (Scheme 4.23a) [94] was proposed to generate intermediate endoperoxide **90**, which underwent ring-opening to **91**, followed by a series of skeletal rearrangement and an ene-reaction with the reagent, ultimately giving hemiacetal hydroperoxide **92** in 56% yield. Tetrasilylated linear [3]phenylene **9a** reacted with atmospheric oxygen through an analogous endoper-

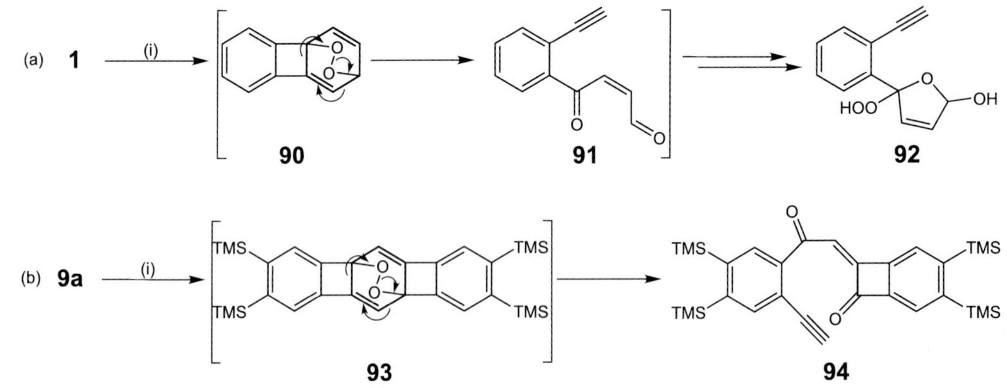

Scheme 4.23. Reactions of **1** and **9a** with singlet oxygen: (a) (i) O$_2$, tetraphenylporphyrin, hν, acetone, −40 °C, 5 d, 56%; (b) (i) O$_2$, hν, C$_6$H$_6$, 23 °C, 1–2 h, 80% (Z:E = 3:1, by NMR).

oxidation-ring opening sequence (interestingly without the necessity of irradiation or added sensitizer), giving the diketone **94**. Unlike the related **91**, this compound could be isolated and characterized (along with its E-isomer) [99]. In both cases, the regioalternative mode of initial cycloaddition, which would have generated one (for **1**) or two (for **9a**) benzocyclobutadiene subunits, was avoided.

On the basis of the above results, analogous endoperoxidation of the angular **15** was expected to be even more facile, as the subsequent skeletal rearrangement should allow the opening of both four-membered rings. This expectation was confirmed by the reaction of **15** with singlet oxygen (now requiring irradiation in the presence of a sensitizer), which produced the Z-dione **96** in 70% yield (Scheme 4.24a) [95]. The corresponding conversion of dipropyl-substituted zigzag [5]phenylene **97** (again without added sensitizer, Scheme 4.24b) provided **98**, the structure of which was confirmed crystallographically (Fig. 4.5) [72]. Unfortunately, no data are available that would allow for an estimate of the relative reactivity of **1**, **9a**, **15**, and **97**. However, it is interesting to note that the branched **21b**, although containing the most highly cyclohexatrienic ring, was recovered unchanged under these conditions. The reason must be that there is no pathway available that does not generate a benzocyclobutadiene derivative.

Considering the success of singlet oxygen cycloadditions, it seemed logical to extend this chemistry to carbon-based dienophiles. Indeed, biphenylene, while generally inert, even in the presence of o-benzyne, transforms to isolated Diels-Alder adducts with more reactive benzyne derivatives (Scheme 4.1) [16a, 22, 23]. Angular phenylene **15** appears to be more reactive, as expected, but undergoes further rearrangements driven by the release of ring strain in the cycloadducts [95]. Thus, on exposure to tetracyanoethene (TCNE), **15** formed a green charge-transfer complex, which, on heating, resulted in the dibenzodehydro[10]annulene **99** (Scheme 4.25a, also Experimental Section). Mechanistically, this transformation

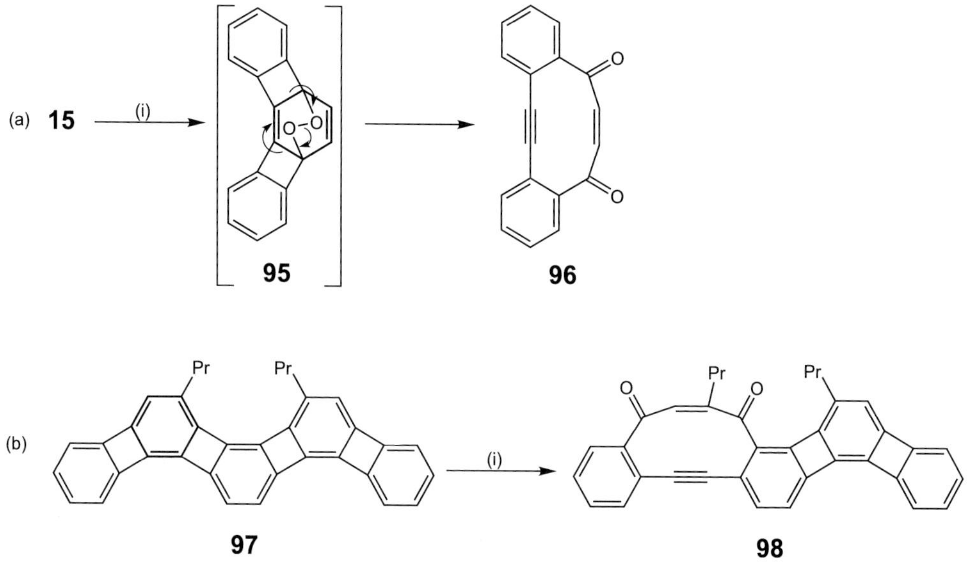

Scheme 4.24. Reactions of **15** and **97** with singlet oxygen: (a) (i) O_2, methylene blue, $h\nu$, CH_2Cl_2, 23 °C, 70%; (b) (i) O_2, 23 °C, 12 h, 9%.

can be envisaged to proceed by a process similar to that leading to **96**, except that double bond isomerization has occurred (possibly during work-up).

Remarkably, changing the dienophile to the alkyne dimethyl butynedioate (DMAD), activated by added $AlCl_3$, did not alter the course of the reaction, even though a highly strained product is generated via **100** in the form of **101** (Scheme

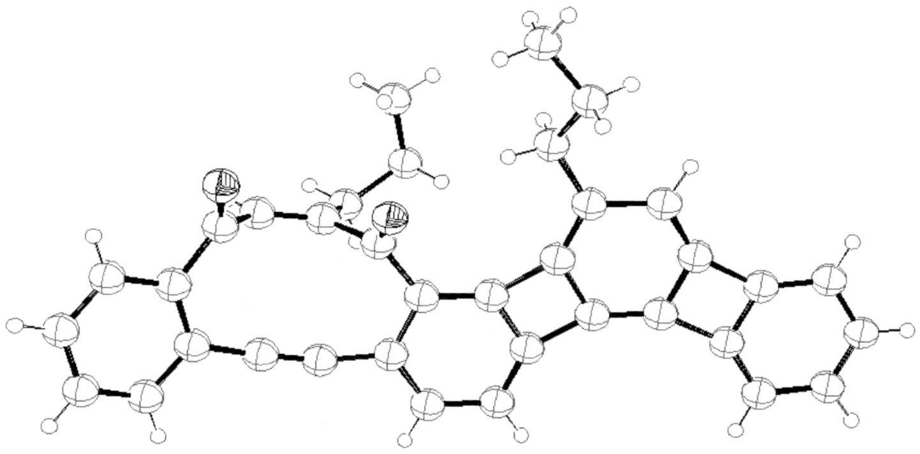

Figure 4.5. X-ray crystal structure of **98** (thermal ellipsoids are shown at 50% probability).

Scheme 4.25. Cycloaddition reactions of **15**: (a) (i) TCNE (1 equiv), CH_3CN, Δ, 8 h, 78%; (b) (i) DMAD (1.6 equiv), $AlCl_3$ (1 equiv), $PhCH_3$, 23 °C, 1 h, 74%.

4.25b). The extraordinary structure of **101**, the most distorted fully unsaturated [6]paracyclophane, was confirmed by X-ray crystallography (Fig. 4.6) [95]. In contrast, and again as expected, branched **21b** was unreactive to these reagents, with the exception of TCNE, which produces a charge-transfer complex.

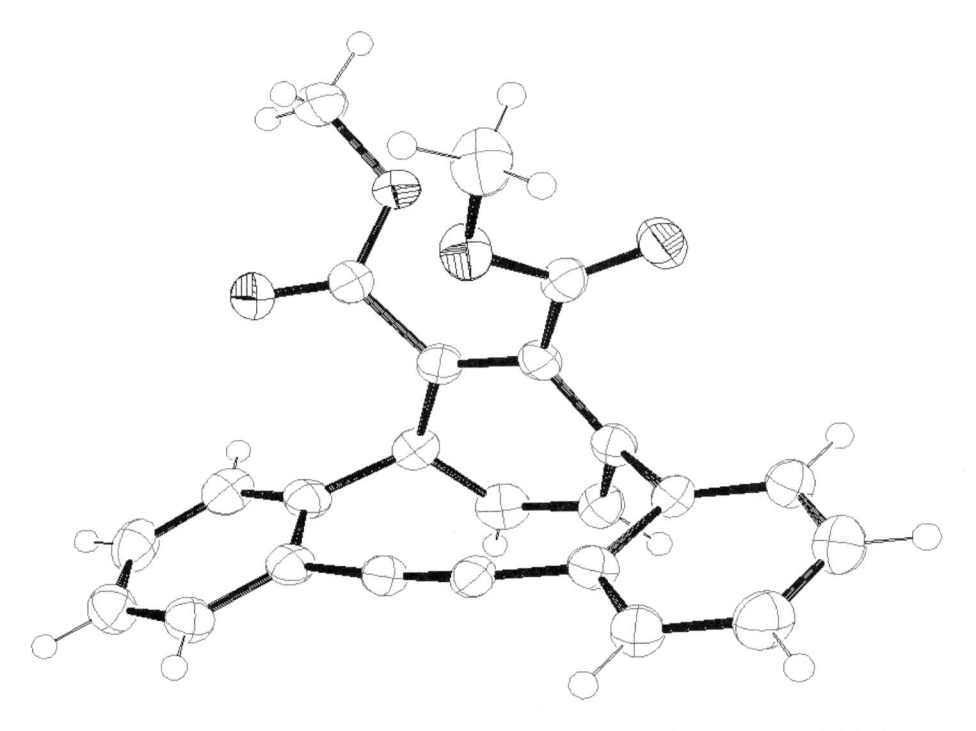

Figure 4.6. X-ray crystal structure of **101** (thermal ellipsoids are shown at 50% probability).

4.3.4
Flash Vacuum Pyrolysis

The remarkable ring opening reactions in the preceding section herald the phenylenes as "loaded springs", not surprising in the light of their ring strain and hence high heat of formation (Section 4.3.1). One might therefore anticipate that, much like other strained hydrocarbons [100], they would enter isomerization manifolds, ultimately ending in PAHs as thermodynamic minima. Indeed, under flash-vacuum pyrolysis (FVP) conditions, **1** had been shown to isomerize to acenaphthylene **104** as the major (Scheme 4.26a) and transient *as*-indacene **103** as a minor product, the existence of the latter inferred through the isolation of a Diels-Alder adduct to **104** (Scheme 4.26a) [101]. Isomers **103** and **104** are derived from a common intermediate, benzopentalene **102**, in this cascade, which is generated by a sequential hydrogen shift/ring contraction from **1**, as indicated summarily in its structural drawing. A second such process leads to **103**. Acenaphthylene **104**, in turn, is the result of a vinylidene carbene deinsertion/reinsertion sequence from benzopentalene (wavy lines) [102]. These results prompted an investigation of the behavior of the two isomeric [3]phenylenes **9b** and **15** under these conditions. Aside from probing the kinds of PAHs that might be formed, it was of interest to see whether the two compounds would interconvert prior to further conversion, a possibility that, if realized, would shed experimental light on their relative stability, a much debated issue [33, 53]. Recent calculations suggest that **15** is slightly more stable than **9b** [53].

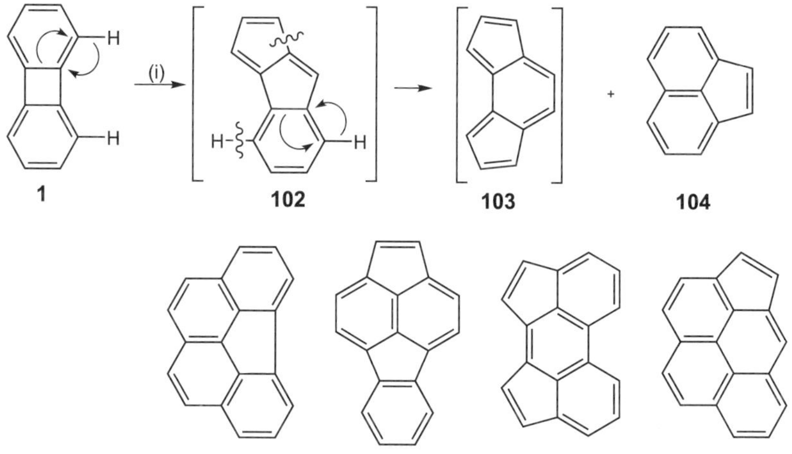

(a) **1** → [**102**] → [**103**] → **104**

(b) **9b** (i)→ **105** (10 %)   **106** (4 %)   **107** (1 %)   **108** (1 %)   **15** (1 %)

(c) **15** (i)→ **105** (24 %)   **106** (11 %)   **107** (3 %)   **108** (7 %)

Scheme 4.26. FVP of **1**, **9b**, and **15**: (a) (i) 900 °C, vacuum, 45% (**104**), 55% (adduct of **103** to **104**); (b) (i) 1000 °C, 5×10^{-7} Torr; (c) (i) 1000 °C, 5×10^{-7} Torr.

In fact, **9b** [103] and **15** [103, 104] gave not only the same mixture of PAHs on FVP, but the linear isomer could be shown to isomerize to its angular relative at 1000 °C (Scheme 4.26b and c). A mechanism for this isomerization is patterned after a related isomerization in the literature [105] and proposes a four-membered ring opening, followed by hydrogen shifts in the resulting biradical and ring closure. ^{13}C-Labeling experiments narrowed considerably the number of mechanistic pathways leading to the PAHs. Details, too lengthy to be presented here, are reported in the original publication [103], and all suggest initial hydrogen or carbon shift/ring contraction from **15**.

4.3.5
Interaction with Organometallic Fragments

As σ and π activated hydrocarbons, phenylenes should be susceptible to interaction with metal fragments. This notion is already borne out with **1**, which readily undergoes metal-promoted ring openings [16a, 21, 106] and π complexation [16a, 107]. A systematic comparison of the reactivity of higher phenylenes with transition metal complexes is yet to be executed; therefore, the following provides simply a summary of what has been done so far.

In the linear phenylenes, the increased cyclobutadienoid character of the four-membered rings manifests itself already in their synthesis, since both **11** and **13** were formed initially as CpCo complexes. The use of an external ligand [52] or oxidation of the cobalt center [55] was necessary to release the free phenylene (Scheme 4.3b and c). While not recorded for CpCo, the C(aryl)–C(aryl) bond in **9a** can be activated with [Fe$_2$(CO)$_9$] to give rise to dibenzoferroles **109** and **110**, as well as the bisallylic complex **111** (Scheme 4.27) [51].

The angular [3]phenylene **15** underwent double C–C activation by [CpCo(eth)$_2$] to afford **112** in 71% yield (Scheme 4.28a, also Experimental Section) [95]. The for-

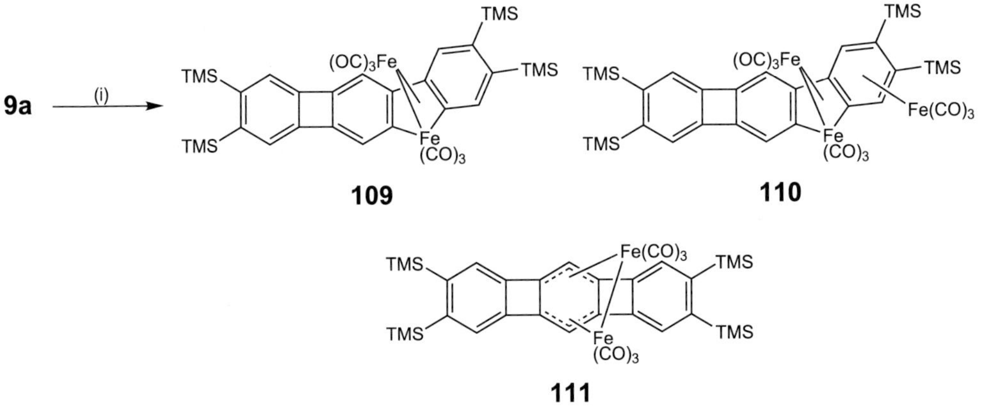

Scheme 4.27. Reaction of **9a** with [Fe$_2$(CO)$_9$]: (i) [Fe$_2$(CO)$_9$] (5.5 equiv), C$_6$H$_6$, Δ, 24 h, 67% (**109**), 18% (**110**), 14% (**111**).

4.3 Comparative Reactivity of the Phenylenes | 173

(a) **15** →(i)

112

(b) **14** →(i), (ii)

113

Scheme 4.28. (a) C(aryl)–C(aryl) activation in **15**: (i) [CpCo(eth)$_2$] (10 equiv), C$_6$H$_6$, 70 °C, 6 h, 71%. (b) Preparation of the η^4-complex of **15** to a Cp*Co-fragment: (i) [Cp*Co(eth)$_2$], THF, −20 °C, 16 h; (ii) Δ.

mation of an η^4-complex between the central benzene ring in **15** and CpCo was not observed in this reaction. This finding is surprising, considering that the Cp*Co complex **113** can be prepared via a stepwise sequence from the cyclization precursor **14** (Scheme 28b). In **113**, the cobalt is attached in such a way as to minimize cyclobutadienoid circuits (Scheme 4.28b, Fig. 4.7(a)) [59].

(a) (b)

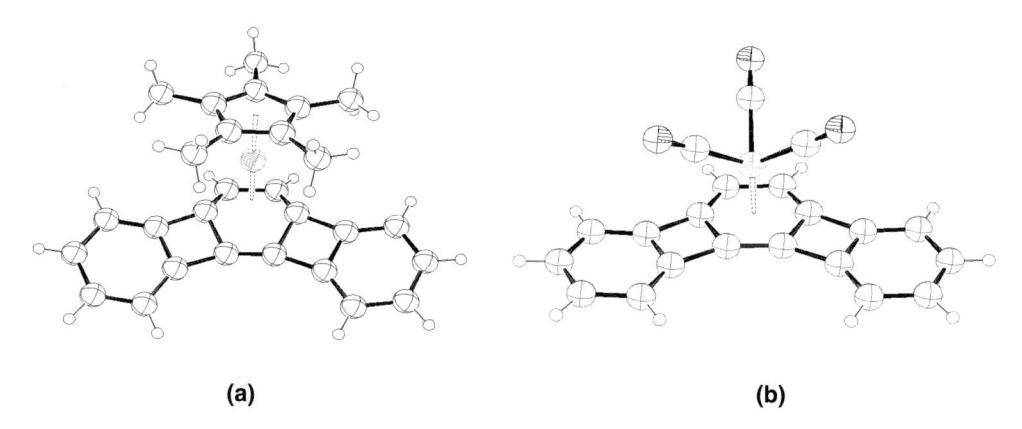

Figure 4.7. X-ray crystal structures of **113** (a) and **114** (b), thermal ellipsoids are shown at 50% probability.

4 [N]Phenylenes: a Novel Class of Cyclohexatrienoid Hydrocarbons

Finally, like biphenylene [107a, 108], phenylenes appear to be readily complexed by chromium tricarbonyl. For example, exposure of **15** to [Cr(CO)$_3$(NH$_3$)$_3$] produced complex **114** (Scheme 4.29a) [109]. The crystal structure of **114** (Fig. 4.7(b)) showed that the three Cr–C–O axes are perpendicular to those of the formal single bonds of **15** [110]. Similarly, the branched skeleton of **21a** is susceptible to metalation by Cr(CO)$_3$, however, here giving rise to two regioisomeric complexes

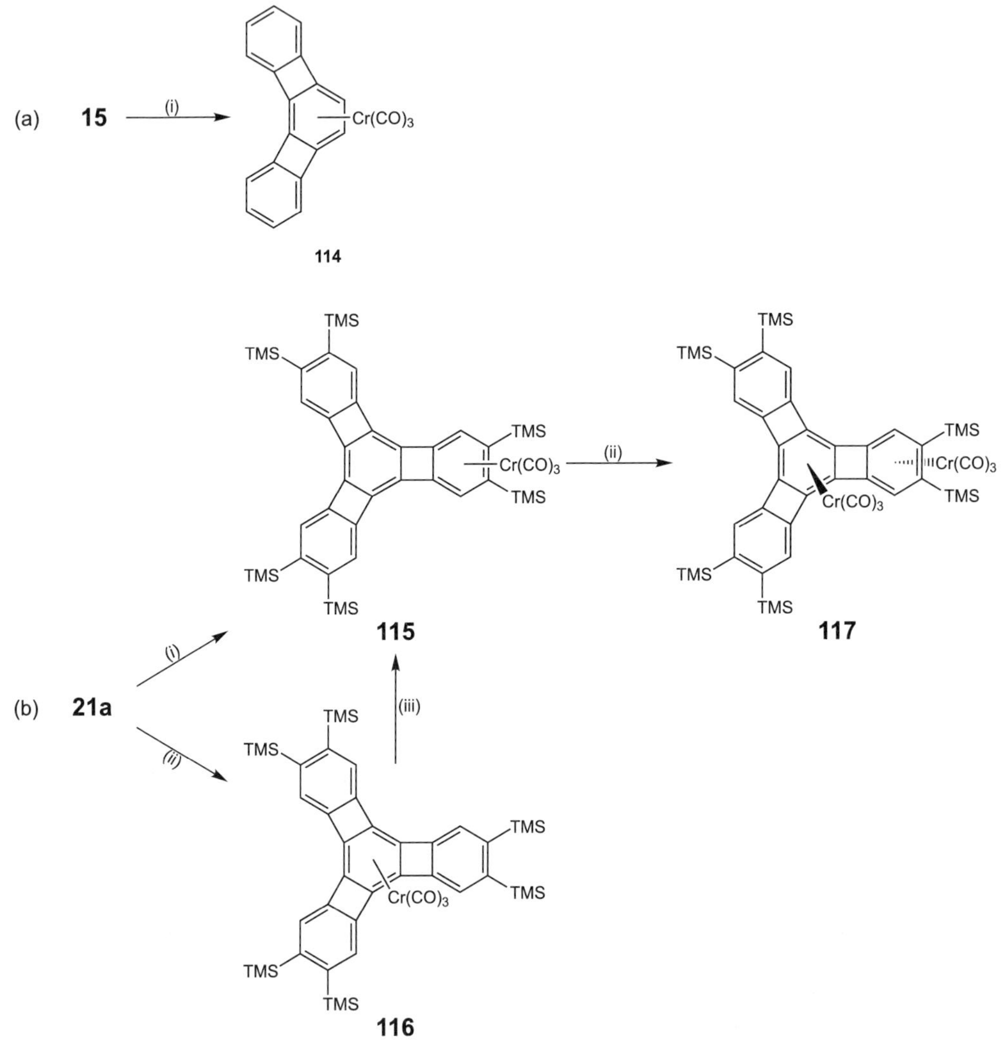

Scheme 4.29. Complexation of six-membered rings in **15** and **21**: (a) (i) [Cr(CO)$_3$(NH$_3$)$_3$], dioxane, 100 °C, 4–5 h; (b) (i) [Cr(CO)$_3$(NH$_3$)$_3$], dioxane, 100 °C, 14 h, 57%; (ii) naphthalene-Cr(CO)$_3$, THF/Et$_2$O, 60 °C, 14 h, 89% (**116**), 43% (**117**); (iii) 90 °C.

(Scheme 4.29b). Treatment with [Cr(CO)$_3$(NH$_3$)$_3$] resulted in the (so-called) exo-complex **115**, while naphthalene-Cr(CO)$_3$ provided the endo isomer **116**. The latter appears to be a kinetic product, as it could be converted thermally to **115**. Further complexation of **115** generated the bischromium complex **117** [111]. It is clear from these cursory experiments that phenylenes should be a rich source of new organometallic compounds.

4.4
Physical Properties of the Phenylenes

As a novel class of hydrocarbons, the most important aspect of the phenylenes lies in their physical properties. The following sections will compare (to the extent that it is possible) structural, spectroscopic, and calculated aspects of the known 19 phenylenes.

4.4.1
Structural Properties

As mentioned repeatedly in previous sections, the unique interplay of the π and σ frame in the phenylenes gives rise to unusually distorted benzene rings, a feature that manifests itself in experimental and calculated structural parameters. Generally, two types of distortion are observed. The first is typical of linearly annulated systems, in which, for reasons of symmetry, the inner six-membered rings cannot adopt a cyclohexatrienoid configuration. Rather, the effect of fusion is to impart bisallylic character, with long fused and shorter adjacent bonds. This is accompanied by a change in the fusion angles in the six-membered rings to more obtuse.

The second and, at this point, more frequently encountered distortion of inner rings is typical of angular and branched topologies and easier to understand, namely cyclohexatrienoid bond alternation. Here, a simple descriptor of average bond length alternation, i.e. (Σ 3 long bonds – Σ 3 short bonds)/3, can be employed for comparative purposes. This number can also be expressed as the degree of bond alternation (in %), by assigning a 0% value to benzene and choosing the exocyclic diene unit in 1,2-dimethylenecyclobutene as the 100% standard. The difference between the 1.497 Å long bond and 1.338 Å short bond in this reference equals 0.159 Å [112]. This model was chosen because of its appropriate geometry and the fact that the two exocyclic bonds show almost no interaction, thus minimizing cyclobutadienoid resonance.

Keeping in tune with the experimental tenor of this paper, Fig. 4.8 summarizes the available experimental ^{1}H-NMR and structural data (bond alternation percentages) for all known parent phenylenes. Exceptions are **11**, **13**, and **66**, for which the parent systems have not been made. In these cases, calculated bond lengths of the parent systems were employed in determining the extent of bond alternation. For **38**, **44**, and **48**, for which X-ray data could not be collected, the experimental geometries of substituted derivatives were used, in conjunction with calculated

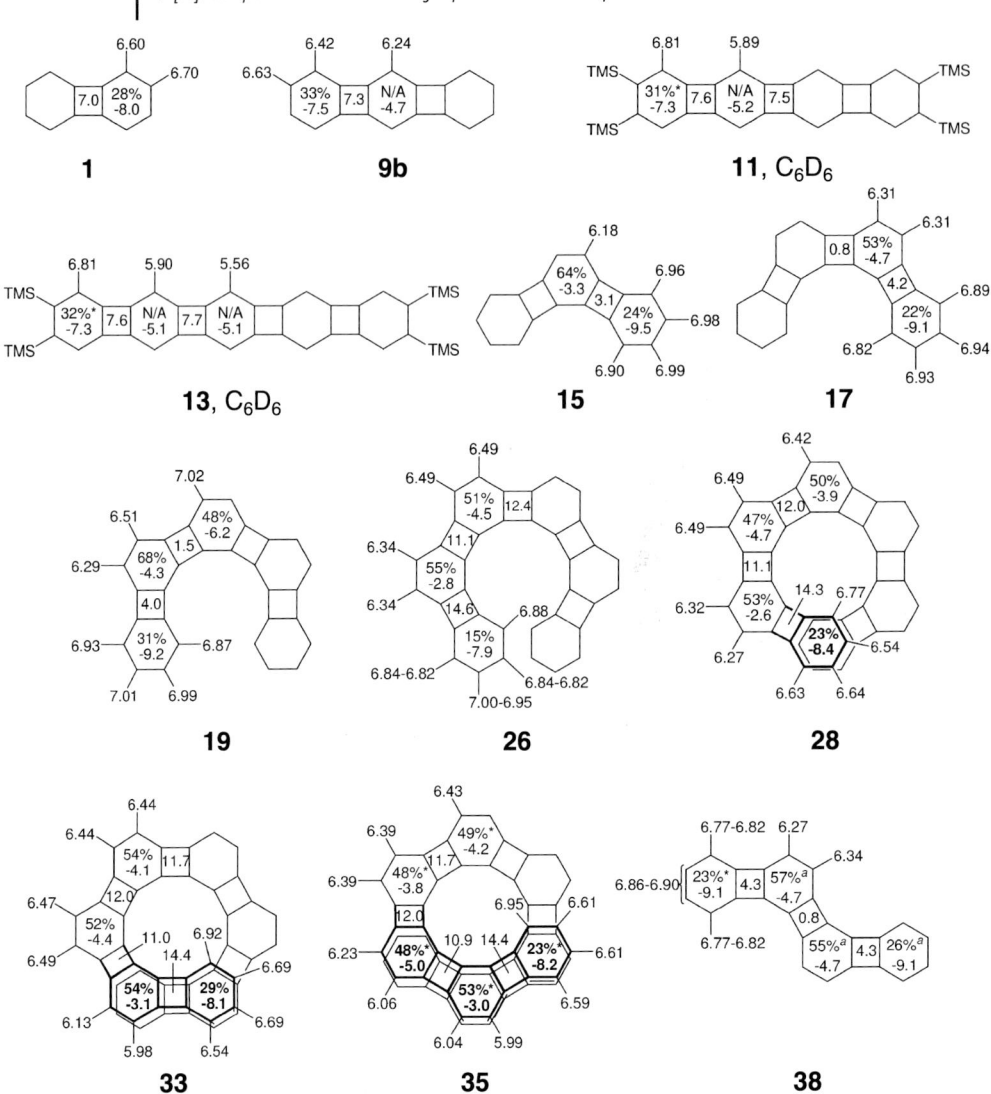

Figure 4.8. Experimental ^{1}H-NMR chemical shifts (CDCl$_3$, unless mentioned otherwise), measured bond localization percentages (from X-ray data), and calculated NICS(1) values [NICS(0) for heliphenes] of known phenylene topologies. All NICS values refer to the parent compounds. An asterisk denotes a calculated bond localization percentage for the parent system. Double bonds are omitted for clarity. ^{a} For the 2,3-bis(trimethylsilyl) derivative; ^{b} for **97**; ^{c} for **52**.

data (parent) for the substituted rings. Finally, the missing X-ray information for **60**, **64b**, and **71** has been replaced by calculated values. Justification for blending experimental with calculated information comes from the finding that the latter reproduces experimental trends perfectly, although it tends to underestimate slightly

4.4 *Physical Properties of the Phenylenes* | 177

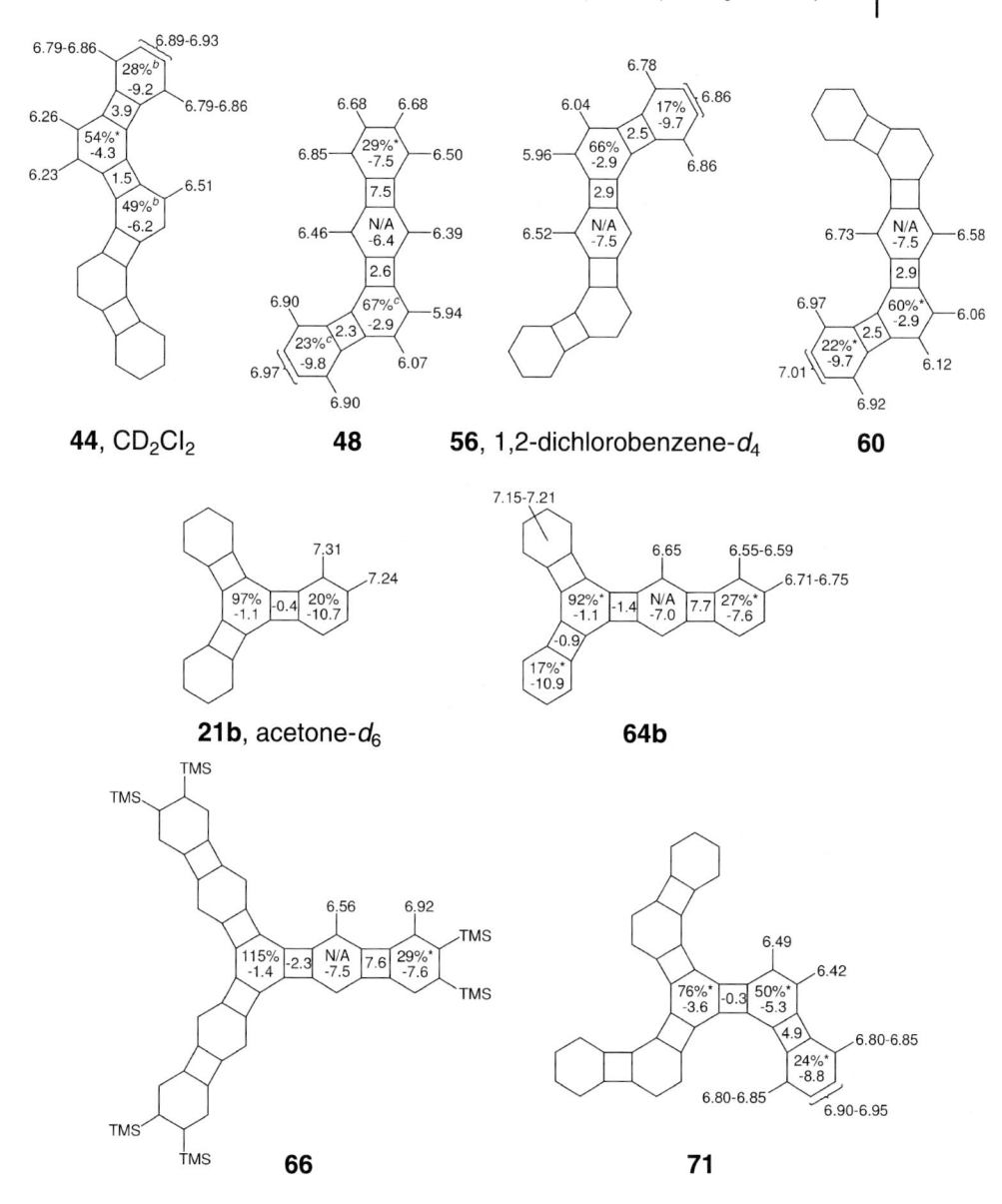

44, CD$_2$Cl$_2$ **48**   **56**, 1,2-dichlorobenzene-d$_4$ **60**

21b, acetone-d$_6$ **64b**

66   **71**

Figure 4.8. (*continued*)

bond alternation percentages. Figure 4.8 also lists NICS values (vide infra) for the parent systems. The following discussion attempts to place these data on a comparative footing, focusing on selected illustrative examples. As will be seen, a fairly consistent picture emerges.

Inspection of the terminal rings of all the phenylenes in Fig. 4.8 shows that they are the least localized (average value 26.8%), and thus, based on this structural criterion, the most aromatic. For comparison, **1** shows 28% bond-localization (Fig. 4.8) [24].

Formal linear fusion of a benzocyclobutadiene fragment to **1** gives rise to **9b**. On the basis of the simple bond localization picture developed for biphenylene (Section 4.1), this mode of fusion should increase the cyclobutadienoid character of the four-membered ring and, in turn, force the terminal benzene ring of **9b** to become more localized. This is indeed the case, albeit subtly so: the termini of **9b** are 33% localized [113]. The core in **9b** has D_{2h} symmetry, imposed upon it by the "conflicting" fusion of the adjacent cyclobutadiene rings. In such cases, the bond localization analysis is not applicable, since it assumes pseudo-three fold symmetry of the system. The central ring instead exhibits the expected bisallylic character (Fig. 4.9): two symmetric C_3 fragments (bond lengths 1.392 Å) are connected by two long bonds (1.417 Å) [113]. Skeletal elongation in the linear series apparently has little effect on bond alternation in the terminal rings, judged by the calculated values of 31% (**11**) and 33% (**13**).

In contrast to **9b**, **15** evolves from **1** by formal angular fusion of a benzocyclobutadiene unit. This alteration acts to dramatically increase bond localization in the center (64%) [57], in turn decreasing the cyclobutadienoid character of the four-membered rings and hence increasing delocalization of the terminal benzene nuclei (24%). The effect of further angular fusions on the termini is quite small, as corresponding values average 23.9% (Fig. 4.8). However, such elongation imposes a distinct pattern of oscillating bond alternation values on the internal rings in the angular and, to the extent of available members, also the zigzag series [72]. A simple rationale for this phenomenon is that the most delocalized terminus enforces the highest degree of bond localization in the first internal ring, which then allows for some "relaxation" of the second internal ring, which in turn increases relative localization in the third and so on. This pattern seems to be attenuating with size, as judged by the heliphene series [68], perhaps approaching a limiting value of ~50% in the corresponding polyheliphene. The fusion of three benzocyclobutadienes to benzene, as in branched **21b**, maximizes its cyclohexatriene character (97% bond alternation), thus allowing the termini to be maximally delocalized (20%) [56].

The bond-alternation approach can be used to evaluate the effects of mixed fusions. For example, the skeleton of **48** can be built formally from **15** by linear benzocyclobutadieno fusion. This change, by relay, appears to lead to increased localization of the angularly fused six-membered ring (67% vs. 64% in **15**). Alternatively, regarding **48** as built by angular terminal benzocyclobutadieno fusion to **9b**, the effect of increasing bond alternation in one of the termini of **9b** "relaxes" the other (29% of the linear end in **48** vs. 33% in **9b**). Similar effects are observed in **56** [74]. While the numbers are small, the trends are consistent.

In the mixed branched examples, the elongation of **21b** either linearly (as in **64b** or **66**) or angularly (as in **71**) has the expected localizing and delocalizing effects, respectively, on the centers of branching. For example (experimentally), whereas the central ring of **21b** exhibits 97% localization, in **66**, this value is increased to 115% – larger than the reference 1,2-dimethylenecyclobutene! Taking recourse to

4.4 Physical Properties of the Phenylenes | 179

9b

52

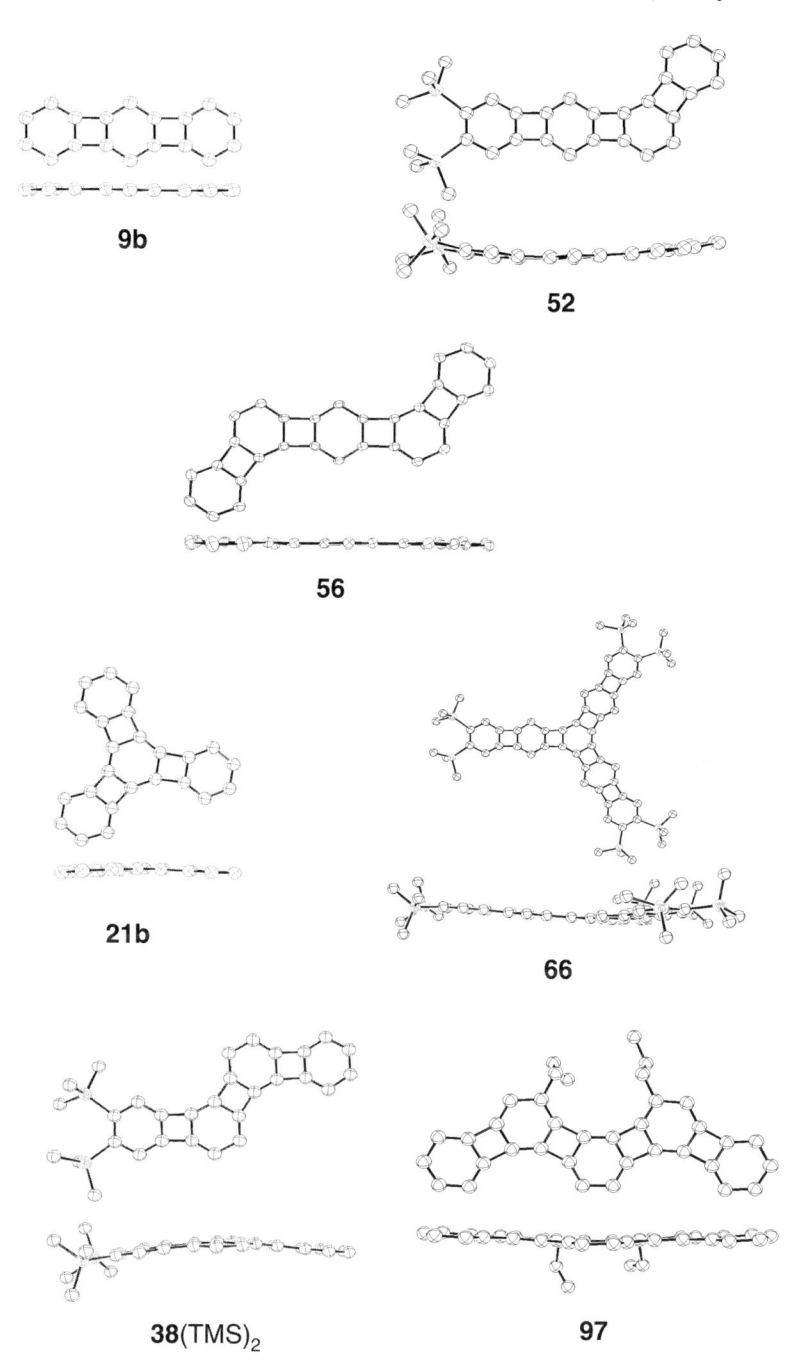

56

21b

66

38(TMS)₂

97

Figure 4.9. X-ray structures of selected [N]phenylenes – top and side views. Hydrogen atoms omitted for clarity, thermal ellipsoids shown at 50% probability.

calculated structures for the other members in the branched family (and, for the sake of consistency, using calculated numbers also for **21b** and the parent of **66**) confirms the experimental trend: 89% (**21b**) – 92% (**64b**) – 92% (parent of **66**) – 76% (**71**) [74, 80]. One notes that the juxtaposition of branched and angular fusion in **71** delocalizes not only the center (vs. **21b**), but also the penultimate six-membered rings (50%, vs. 64% in **15**) [80].

Figure 4.9 shows the X-ray structures of some of the phenylenes discussed. While **9b** and **56** are essentially planar, **21b** and the substituted **52**, **66**, **38**(TMS)$_2$, and **97** show noticeable deplanarization. Such deplanarization is general and independent of topology, size, and substitution pattern and is thought to be due to crystal-packing effects that are comparable in energy to deformation energies (several kcal mol^{-1}) [56]. VT-NMR experiments on a derivative of **21b** bearing pro-chiral substituents revealed the absence of decoalescence at the experimental low temperature limit of −93 °C, indicating either a very low barrier to planarization, or a planar structure in solution, as indeed also calculated for **21b**. In addition, calculations showed that the phenylenes are more deformable than their hexagonal squeezes [37]. This flexibility was ascribed to ready pyramidalization of the four-membered ring carbons as a result of two phenomena: hyperconjugation of the low lying σ antibonding orbitals of the strained bonds with the HOMO of the π system [114] and minimization of antiaromatic overlap in the cyclobutadiene nuclei [115]. The observation of such facile deplanarization is encouraging in view of projected syntheses of archimedene [53, 70] and circular [5]phenylene (the phenylene analog of corannulene) [53].

In the light of the preceding discussion, it is instructive to view the compilation of X-ray structures of the heliphenes (Fig. 4.10). Even in the absence of nonbonded interactions, angular [5]phenylene **19** already shows a small, prehelical deviation from planarity. The higher angular [N]phenylenes ($N > 5$) can no longer adopt planar structures and are helical [64, 68]. The "helical strain" is not large, as determined by calculations, for example, only 3.2, 5.4, and 7.0 kcal mol^{-1} for **26**, **28**, and **33**, respectively [65]. Table 4.1 summarizes some of the structural parameters of the heliphenes, highlighting the steady increase in the helix climb and in-plane turn in the series. The angle between the planes of the terminal benzene rings is the highest in [7]heliphene. The inner helix of the [6]- and [7]helicene (hexagonal squeezes of **26** and **28**) climbs more steeply than that in their heliphene counterparts, due to the smaller diameters of the PAH systems [116, 117].

As the heliphenes are chiral, the possibility of enantiomer separation is intriguing. Because of their extensive delocalization, the heliphenes should show remarkable chiroptical properties [66c,d, 131]. In an attempt to probe the feasibility of their resolution, configurational stability was probed by NMR experiments. For this purpose, a series of heliphene derivatives bearing potentially diastereotopic substituents (isopropyl and methoxymethyl, respectively) was prepared [64, 68]. In accord with the calculated low barrier to enantiomerization of **26** [65], decoalescence of the methyl group signals for an isopropyl derivative was not evident at the limiting temperature of −75 °C. Turning to methoxymethyl [7]heliphene, methylene decoa-

4.4 Physical Properties of the Phenylenes | 181

15

17

19

26

28

33

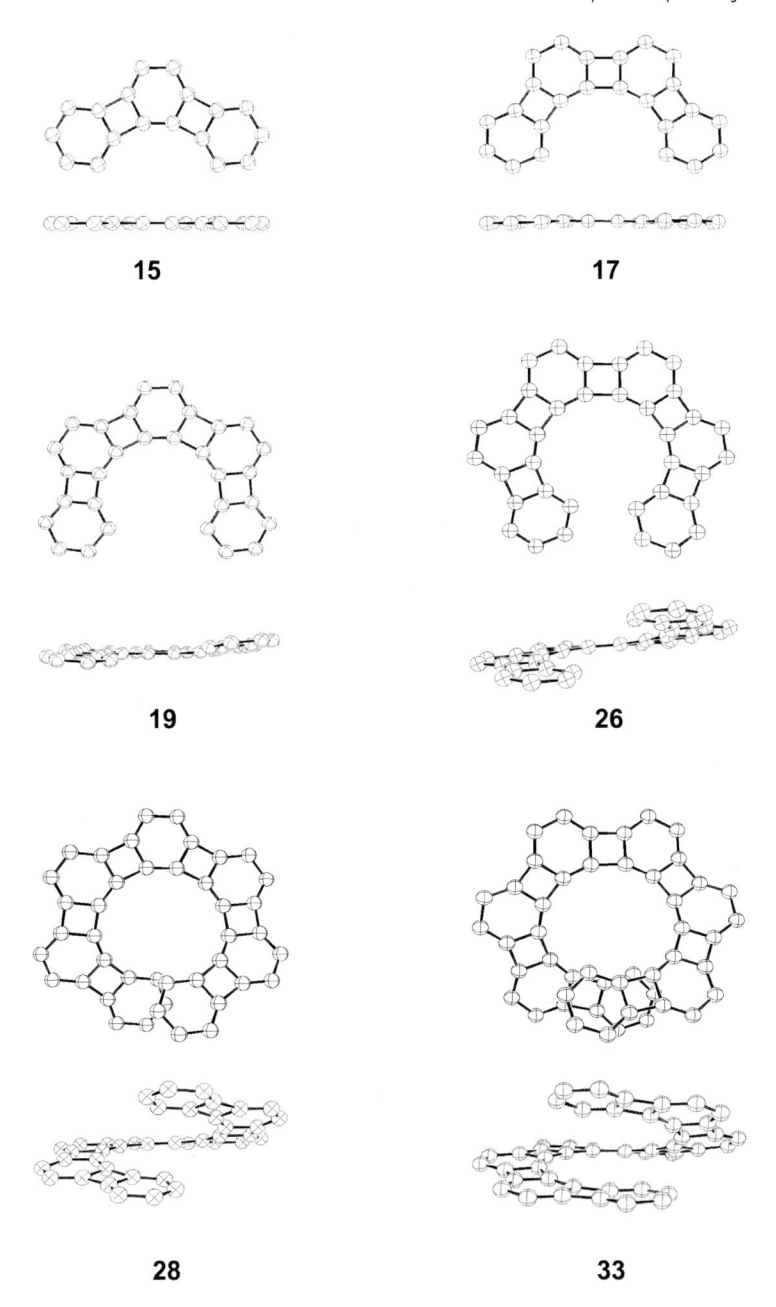

Figure 4.10. X-ray structures of [N]heliphenes – top and side views. Hydrogen atoms omitted for clarity, thermal ellipsoids shown at 50% probability.

Table 4.1. Comparison of the helix parameters of heliphenes and helicenes (calculated values given in parentheses).

Molecule	Terminal ring centroid distance (Å)	Terminal ring interplanar angle (°)	Inner helix climb (Å)	Inner helix in-plane turn (°)	Racemization barrier (kcal mol^{-1})	Helical strain (kcal mol^{-1})
26	5.62 (5.87)[a]	22.8 (27.2)[a]	2.16 (2.30)[a]	337.3 (332.4)[a]	N/A (3.6)[b]	(3.2)[b]
28	4.07 (4.54)[a]	30.1 (40.6)[a]	3.29 (3.64)[a]	361.3 (361.6)[a]	12.6 ± 0.5[a] (17.0)[a]	(5.4)[b]
33	4.41 (5.49)[c]	23.6 (41.1)[c]	3.35 (4.24)[c]	393.0 (389.1)[c]	13.4 ± 0.4[c]	(7.0)[b]
35	N/A (7.48)[c]	N/A (33.2)[c]	N/A (5.07)[c]	N/A (415.7)[c]	<12.0[c]	
[6]helicene	4.44[d]	58.5[d]	3.20[d]	314.3[d]	36.2[e]	
[7]helicene	3.83[f]	32.3[f]	3.75[f]	380.8[f]	41.7[e]	

a Ref. [64]; b Ref. [65]; c Ref. [68]; d Ref. [116]; e Ref. [66c,d, 131]; f Ref. [117].

lescence was recorded at −27 °C, indicating a barrier of 12.6 ± 0.4 kcal mol^{-1} for helix flipping [64] – less than a third of the value for the corresponding helicene [66c,d, 131]. Surprisingly, the analogous barrier for methoxymethyl [8]heliphene was only slightly higher, 13.4 ± 0.4 kcal mol^{-1} and methoxymethyl substituted **35** showed no signal splitting for the methylene hydrogens on cooling to its solubility limit at −45 °C [64, 68]. The flexibility of these systems is therefore extraordinary, a consequence of both ready in- and out-of-plane deformations (vide supra and infra).

4.4.2
Magnetic Properties

^{1}H-NMR chemical shifts are highly diagnostic of whether a compound is aromatic or not [4]. Hydrogens located on the inside of aromatic rings exhibit relatively high field chemical shifts, while those on the outside are relatively deshielded. Antiaromatic circuits have the opposite effect. In the phenylenes, the generally observed shielding of all protons (relative to alkylbenzenes) is the result of the simultaneous operation of two effects: the decreased diatropism of the six-membered rings and the shielding influence of the cyclobutadiene nuclei on the protons in their vicinity.

To better understand the relative contributions of the component rings to the observed chemical shifts in the phenylenes, recourse was taken to nucleus-independent chemical shift (NICS) [118] calculations, which provide such data (in ppm) for a point nucleus at any given position in a molecule. For cyclic polyenes

4.4 Physical Properties of the Phenylenes

this is typically 1.0 Å above the center of the ring, chosen to minimize local perturbations [119]). Negative NICS values denote an aromatic ring (NICS(1)$_{benzene}$ = −12.5), whereas positive values indicate an antiaromatic circuit (NICS(1)$_{cyclobutadiene}$ = 15.1) [120]. Used together, NICS and NMR are useful tools in the following analysis of the corresponding entries in Fig. 4.8. As will be seen, in the phenylenes, the magnetic data correlate well with the structural criterion of (anti)aromaticity.

As mentioned previously, the α- and β-protons in **1** resonate at δ 6.60 ppm and 6.70 ppm, respectively. The relative shielding of the α-proton is a consequence of the residual paratropicity of the neighboring four-membered ring, applicable to all phenylenes, with the exception of the branched isomers. The corresponding NICS values are −8.0 and 7.0 for the six- and four-membered rings, respectively. In general, the terminal rings exhibit the most negative NICS values and, correspondingly, the highest NMR chemical shifts, in consonance with the occurrence of the smallest extent of bond alternation.

In going from **1** to **9b**, the paratropism (antiaromaticity) of the cyclobutadienes (NICS = 7.3) is subtly increased, while the diatropism (aromaticity) of the terminal benzene rings decreases, as evidenced by the lower NMR chemical shifts (δ = 6.42 and 6.63 ppm) and less negative NICS value (−7.5). The diatropic character is lowest in the central ring (NICS = −4.7), which, in conjunction with the paratropism of the two adjacent cyclobutadiene fragments, leads to strong shielding of its hydrogen (δ = 6.24 ppm) [120]. While the NICS values for the internal six-membered rings fluctuate somewhat along the linear series **9b–11–13**, the cyclobutadienes appear to become increasingly paratropic, providing an explanation for the observation of incremental shielding of the central hydrogens [120].

The angular mode of fusion in **15** further reduces the diatropism of the center (NICS = −3.3; δ = 6.18 ppm), in conjunction with decreased paratropism of the cyclobutadienes (NICS = 3.1). As a consequence, the termini are more diatropic than in biphenylene (NICS = −9.5 vs. −8.0) [120]. In the remainder of the angular series, the arguments advanced previously for the rationalization of the trends in bond-localizations are clearly augmented by the magnetic data. Thus, **19**, as an example, shows the alternation of diatropism of the six-membered rings: NICS = −9.2, −4.3, −6.2. The paratropism of the cyclobutadiene moieties also oscillates: it is high in the outer ring (NICS = 4.0) and less so in the inner one (NICS = 1.5), reflecting the interplay with the neighboring six-membered rings and their respective aromaticity. For the heliphenes, the use of NICS(1) values was abandoned, since the areas above and below the rings are now inequivalent. Instead, NICS(0) data were computed, which, although numerically not directly comparable with NICS(1) numbers, showed the same alternating trends, in agreement with experimental NMR chemical shifts and bond localization numbers [64, 68]. The NMR and the NICS values of angular [4]- and [5]phenylene are essentially identical to those of their zigzag counterparts, highlighting the similarity between the two topologies.

In the branched **21b**, the central six-membered ring becomes essentially atropic (NICS = −1.1), as do the adjacent cyclobutadienes (NICS = −0.4), allowing for maximum diatropism of the three terminal cycles (NICS = −10.7; δ = 7.24, 7.31 ppm). As such, the system can be described essentially as an extended stilbene [120]. In support of this view, the signal for H$_\alpha$ (δ = 7.31 ppm) and H$_\beta$ (δ = 7.24 ppm) have traded their "normal" places, appearing in the order observed for ordinary benzocycloalkanes, such as indane: H$_\alpha$ δ = 7.06 ppm, H$_\beta$ δ = 6.99 ppm [121].

For the mixed topologies realized in **48** and **56/60**, a component analysis is in accord with the calculated magnetic behavior. Thus, if these systems are viewed as perturbed linear [3]phenylenes, the added angular fusions serve to increase the cyclohexatrienoid character of one or both termini of the linear substructure, consequently reducing their diatropism, as observed. At the same time, the paratropism of the four-membered rings (of the linear substructure) is also reduced, thus rendering the center ring more diatropic (aromatic). The trends in the NICS values of the latter agree with this analysis: −4.7 (**9b**), −6.4 (**48**), −7.5 (**56/60**), as do NMR chemical shifts: δ = 6.24 (**9b**), 6.39 (**48**), 6.58 (**60**) ppm. Along the same lines, completely removing the diatropism of one terminus of **9b**, as it occurs in **64b** and **66**, should have a similar pronounced effect on the diatropism of the central benzene, as reflected by the corresponding numbers for **64b** (δ = 6.65 ppm, NICS = −7.0) and **66** (δ = 6.56 ppm, NICS = −7.5). Turning to the mixed system **71** and viewing it as a perturbed **15**, the reduced diatropism of the core (the "perturbed" end of **15**) goes with reduced paratropism of the adjacent four-membered rings and increased diatropism of the next six-membered cycle (δ$_{average}$ = 6.58 ppm, NICS = −3.6). Alternatively, viewing **71** as a perturbed **21b**, the effect of reduced diatropism of the terminal rings of the substructure of branched [4]phenylene is to increase the diatropism of the core (NICS = −3.6 vs. −1.1).

Carbon-13 NMR spectroscopy is not usefully diagnostic of ring currents, and a typical ^{13}C-NMR spectrum of a phenylene exhibits four groups of signals. At the highest chemical shifts (δ ∼ 145–155) are the signals corresponding to the four-membered ring carbon atoms without adjacent other four-membered rings. Those that are adjacent to a second four-membered ring are relatively shielded [122] and absorb at δ ∼ 133–140 ppm. The remaining carbons, namely the nonquaternary nuclei of the six-membered rings, are also split into two categories – those adjacent to a cyclobutadiene (δ ∼ 113–120 ppm) and those distant from it (δ ∼ 125–130 ppm). The averaged (over all six carbons) for the internal six-membered rings in angular [3]-, [4]- and [5]phenylene changes very little and lies between 132.5 and 134.5 ppm.

4.4.3
Energetic Properties

The determination of the ground-state energies of the [N]phenylenes is of crucial importance in the evaluation of their aromaticity [3] and strain. On the other hand, their frontier orbital separation constitutes a measure of their kinetic stability [123] and is central to organic conductor applications [124]. The excited states of the phenylenes are also of interest for probing the changes in aromaticity that occur upon excitation and for identifying the nature of radiative relaxation pathways (fluorescence and/or phosphorescence).

Experimental enthalpies of formation for members of the series have been obtained only for **1** [125], **15**, and **21b**, and the agreement between the calculated and observed $\Delta H°_f$ data is remarkable [34]. For other phenylenes, only calculated data are available and the following will highlight some key findings.

Although the conjugated-circuit model [33] suggested that the linear [N]phenylenes are more stable than their angular isomers, the application of *ab initio* methods proved the opposite [126]. Schulman and Disch's examination of the problem by modern DFT methods placed the stabilization of **15** vs. **9b** at 2.4 kcal mol^{-1} [53]. Branched [4]phenylene **21b** is the most stable of the five [4]phenylenes, followed by **38** (relative energies: +4.1 kcal mol^{-1}), **17** (+4.3 kcal mol^{-1}), **48** (+4.8 kcal mol^{-1}), and linear [4]phenylene (+8.5 kcal mol^{-1}). The energies of zigzag [4]- (**38**) and [5]phenylene (**44**) are almost identical to those of their angular isomers [53]. A comparison of the relative energies of the twelve [5]phenylenes reveals the same general trends: the linear isomer is the least (+10.7 kcal mol^{-1}), the branched **64b** the most stable (0.0 kcal mol^{-1}), surprisingly more stable than the other branched isomer (+0.9 kcal mol^{-1}), which is devoid of a linear [3]phenylene substructure. Similarly, among the nonbranched [5]phenylenes, the two doublebent isomers (**56** and **60**) have the lowest energies (+3.8 kcal mol^{-1}), despite the presence of the linear fragment [65]. This disagreement with expectation, albeit associated with small numbers, was attributed to the opposing energetic contributions of the σ- (stabilizing) and π- (destabilizing) components of the linear frame [74].

The "helical strain" in the smaller heliphenes is relatively small (Table 4.1), but becomes substantial for larger systems. Thus, in order to model the limiting properties of larger phenylenes, calculations have been executed on various illustrative topologies of [19]phenylene [69]. The results indicate that the helical topology is 26.9 kcal mol^{-1} (1.4 kcal mol^{-1} per ring) less stable than its zigzag counterpart. Linear [19]phenylene is the least stable, 40.4 kcal mol^{-1} (2.1 kcal mol^{-1} per ring) more energetic than the zigzag isomer.

The electronic spectra of the phenylenes feature two typical sets of bands: one at lower wavelengths, with relatively large ε values, and a second at higher wavelengths, with lower extinction coefficients. The exact position of these absorptions, however, depends on the phenylene topology (Table 4.2). In the linear series, the λ_{max} value increases dramatically in going from **1** to **9a** ($\Delta\lambda_{max} = 75$ nm), **11** ($\Delta\lambda_{max} = 54$ nm), and then **13** ($\Delta\lambda_{max} = 38$ nm) [45a, 51, 52, 55]. The extrapolation of this limited set of experimental data to infinite N provides $\lambda_{max} = 662$ nm, corresponding to a small band gap of 1.87 eV for the linear polymer, boding well for potential electronic applications.

In contrast to the linear frame, the λ_{max} values of angular phenylenes attenuate more rapidly ($N = \infty$, $\lambda_{max} = 578$ nm, band gap = 2.14 eV) [45a, 57, 58, 64, 68]. The same seems to be true for the zigzag isomers, again with the caveat that only four experimental values are available ($N = \infty$, $\lambda_{max} = 587$ nm, band gap =

Table 4.2. HOMO–LUMO gaps and λ_{max} values in angular, zigzag and linear phenylenes.

N	λ_{max} (nm)			HOMO–LUMO gap (eV)		
	Angular	Zigzag	Linear	Angular	Zigzag	Linear
2	363[a] (isooctane)	363[a]	363[a]	3.42	3.42	3.42
3	428[b] (THF)	428[b]	438[b,d] (THF)	2.90	2.90	2.83
4	448[b] (THF)	465[c] (THF)	492[b,d] (THF)	2.77	2.67	2.52
5	470[b] (THF)	484[c] (CH$_2$Cl$_2$)	530[b,d] (THF)	2.64	2.56	2.34
6	491[e] (CH$_2$Cl$_2$)	–	–	2.53	–	–
7	503[e] (CH$_2$Cl$_2$)	–	–	2.47	–	–
8	515[f] (CH$_2$Cl$_2$)	–	–	2.41	–	–
9	524[f] (CH$_2$Cl$_2$)	–	–	2.37	–	–
∞	578[f]	587[g]	662[g]	2.14	2.11	1.87

a Ref. [45a]; b Ref. [58]; c Ref. [72]; d for tetrasilylated compounds; e Ref. [64]; f Ref. [68]; g Ref. [132].

2.11 V) [72]. The so-estimated band gaps for the linear, angular, and zigzag family conform with theoretical predictions [33, 69].

In agreement with the virtual absence of antiaromatic circuits, the UV spectrum of branched **21b**, while still exhibiting the diagnostic phenylene pattern, is hypsochromically shifted, with a highest wavelength absorption at 379 nm, reflecting a HOMO–LUMO gap of 3.28 eV – the highest among the [4]phenylenes [63].

The electronic spectra of the "mixed" phenylenes are strongly influenced by the presence of the linear substructures, which cause strong bathochromic shifts. Thus, **48** exhibits a λ_{max} of 486 nm, which is almost exactly equal to that of its linear relative and higher than that of the remaining isomers [73]. Syn-doublebent [5]phenylenes **56** and **60** absorb at 505 and 507 nm, respectively, at energies significantly lower than their isomers **19**, **44**, and **64b**, but higher than the all-linear **13** (Table 4.1) [74]. Finally, in the series of branched phenylenes, the effect of the presence of linear annelation is highlighted by the changes in λ_{max} when going from **21b** to **64b** ($\Delta\lambda_{max} = 107$ nm), **64b** to **66** ($\Delta\lambda_{max} = 37$ nm [77]), and, particularly, **71** to **66** ($\Delta\lambda_{max} = 67$ nm [81]).

The exploration of the photophysics of the phenylenes, still in its infancy, has focused on the smaller linear, angular, and branched systems, as well as **1**. On the basis of their rates of internal conversion of the first excited state (S$_1$), the systems studied have been labeled as either "fast IC compounds", with $k_{IC} > 10^9$ s^{-1}, or "slow IC compounds", with $k_{IC} \approx 10^7$ s^{-1}. Fast IC compounds, biphenylene and **9b**, relax their S$_1$ state predominantly through internal conversion ($\Phi_{IC} > 99\%$), since the rates of this reaction are significantly greater than those of the competing intersystem crossing and fluorescence processes. In angularly fused, slow IC compounds **15**, **38**, and **44**, other relaxation mechanisms gain in importance and become dominant in zigzag [5]phenylene **44**, which releases 21% of its excited state energy through fluorescence and crosses over efficiently ($\Phi_{ISC} = 45\%$) to the triplet state (T_1). The corresponding quantum yields for the branched **21b** are $\Phi_F = 15\%$ and $\Phi_{ISC} = 3\%$ [127].

Shpol'skii spectroscopy [128] of the phenylenes, together with DFT calculations, have provided significant insights into the vibrational characteristics of their excited states. The resemblance between the fluorescence emission and excitation spectra, observed for **15** and **44**, has indicated that, in phenylenes with angular substructures, the S_0 and S_1 states have similar geometries [129]. Studies performed on deuterated derivatives of **21b** allowed for the distinction between C–C and C–H vibrational modes in the S_1 state and demonstrated that, while the terminal rings of **21b** are essentially unperturbed in the excited state, the central cyclohexatrienoid ring partly delocalizes, suggesting rearomatization [130]. This is an important result, in view of the fact that the opposite effect – dearomatization of benzene in the first excited state – is well known [1a]. Photophysical studies confirmed the presence of a theoretically invoked [53, 65] low-frequency ($\sim$35 cm^{-1}) out-of-plane vibration and also revealed a new in-plane vibration in angular phenylenes ($\sim$100 cm^{-1}), notably absent from the spectra of corresponding PAHs. The first vibration deplanarizes the angular fragment by moving the terminal rings out of the molecular plane in opposite directions. The second increases the angle between the centroids of the three successive six-membered rings of an angular fragment, in turn pushing the terminal rings away from each other [129]. Both of these distortions have been proposed as operational in the transition state for the racemization of [6]heliphene [65] and their magnitude provides a convincing rationale for the ease by which this process occurs for the other heliphenes studied so far.

4.5
Conclusion and Outlook

The last decade has seen a significant increase in the number of synthesized members of the class of phenylenes. The experimental physical properties of these molecules, augmented by details derived from calculations, are painting a picture that allows for some clear trends to be recognized within the various topologies. These trends are pointing the way to further extensions. The preliminary chemical behavior of these novel hydrocarbons reflects the embedded σ- and π-strain, suggesting a host of new applications. Synthetic efforts will focus on the preparation of the higher members of the linear and zigzag topologies, the resolution of suitably modified heliphenes, the completion of the synthesis of the family of [5]phenylenes, the synthesis of the first circular, and the first belt-shaped phenylenes and the construction of ever larger subunits of archimedene. One would hope that in this way a fundamental contribution to the understanding of aromaticity in general and the chemistry of novel benzenoid hydrocarbons in particular can be made. Similarly, one might anticipate new applications in material science and the discovery of new transformations and products.

4.6
Experimental: Selected Procedures

4.6.1
Preparation of Bis(trimethylsilyl) Zigzag [4]Phenylene: A One-pot Intra- and Intermolecular [CpCo(CO)$_2$]-Mediated Cyclization

BTMSA (31 mL) was degassed in a 100 mL round-bottom flask, fitted with a reflux condenser. In a separate flask, **27** (213 mg, 0.778 mmol) was dissolved in THF (4 mL) and degassed. Each of the two solutions was treated with [CpCo(CO)$_2$] (104 µL). The THF solution was taken into a syringe and injected into the refluxing BTMSA over 10 h, while irradiating with a 300 W projector lamp (at a 5 cm distance). After the addition was complete, irradiation and reflux were maintained for an additional 3 h. The solvent was removed in vacuo and the reaction mixture preadsorbed onto silica gel. Column chromatography on silica (hexanes/CH$_2$Cl$_2$ 90:10) gave the product as brown oil. The oil was washed with pentane and the precipitated orange powder filtered off. This procedure was repeated, until no further precipitate formed. The combined yield of **28** was 38 mg (11%).

4.6.2
Preparation of *syn*-Doublebent [5]Phenylene: A [CpCo(eth)$_2$]-Mediated Cyclization

A solution of **58** (39 mg, 0.059 mmol) in THF (10 mL) was treated with of 1 M TBAF in THF (0.2 mL, 0.200 mmol). The mixture turned brown immediately and was stirred at room temperature for 1 h. Water (0.2 mL) was then added. After further stirring for 20 min, the mixture was filtered under N$_2$ through a short plug of silica into a 50 mL Schlenk tube. The silica was washed with two additional portions of dry THF and the combined filtrates concentrated to a 25 mL volume in vacuo.

The resulting solution of **59** was cooled to −25 °C and [CpCo(eth)$_2$] (23.4 mg, 0.130 mmol) in ether (10 mL) was added. The mixture was kept at −25 °C for 18 h and then allowed to warm up slowly. As soon as the temperature reached −10 °C, 1,3-cyclohexadiene (0.50 mL, 0.42 g, 5.25 mmol) was added in one portion. The Schlenk tube was closed and heated at 100 °C for 90 min. The mixture was then preadsorbed on silica and purified by column chromatography (hexanes/CH$_2$Cl$_2$ 85:15) to elute **60** as the first, green fluorescent fraction (3.0 mg, 14%).

4.6.3
Preparation of Dehydrobenz[12]annulene 74a: A Stephens-Castro Coupling

Triyne **73a** (603 mg, 1.07 mmol) was dissolved in absolute ethanol (40 mL) and added to a degassed solution of CuCl (107 mg, 1.08 mmol) in NH$_4$OH (40 mL). After stirring for 2 h, the resulting yellow-orange precipitate was filtered off and washed with water five times, followed by five ethanol washings. The dry precipitate was dissolved in dry, degassed pyridine (40 mL) and the solution was refluxed for 7 h. The crude product obtained after the solvent was evaporated was chromatographed on silica twice (hexanes/ether 97:3), finally yielding **74a** (93 mg, 20%) as a yellow solid.

4.6.4
Hydrogenation of Bis(trimethylsilyl) Bent [4]Phenylene

A solution of bis(trimethylsilyl)bent [4]phenylene **52** (80 mg, 0.18 mmol) in ether (60 mL) was charged with palladium on carbon (5 wt%, 120 mg). Subsequently, hydrogen gas was bubbled through the reaction mixture for 10 min. The solids were removed by filtration, the solvent evaporated and the resulting residue was purified by column chromatography (hexanes/CH_2Cl_2 98:2), giving **81** as light yellow oil (35 mg, 44%).

4.6.5
Cyclopropanation of Triangular [4]Phenylene

Branched [4]phenylene **21b** (42 mg, 0.139 mmol) was dissolved in toluene (25 mL) in a 50 mL three-neck round-bottom flask fitted with a thermometer, condenser and a septum. After flushing the system with argon, diethylzinc (0.90 mL of a 1.1 M solution in toluene, 1.0 mmol) and diiodomethane (220 mg, 0.821 mmol) were respectively added via syringe and the reaction mixture maintained at 60 °C under an argon atmosphere for 4 d. The solution was allowed to cool to room temperature and the solvent removed in vacuo. Chromatography of the residue on silica (hexanes/CH_2Cl_2 85:15), followed by recrystallization from hexanes gave 46 mg (97%) of the triscyclopropanated **89b** as a white solid.

4.6.6
Reaction of Angular [3]Phenylene with TCNE

Tetracyanoethene (22.4 mg, 0.175 mmol) was dissolved in acetonitrile (20 mL) and the resulting solution was added to a degassed solution of **15** (40 mg, 0.177 mmol) in acetonitrile (20 mL). The color changed instantly from yellow to green. The solution was degassed for an additional 5 min and subsequently kept at reflux for 8 h. The green color changed slowly to reddish-brown. The mixture was then cooled to room temperature and the solvent removed under reduced pressure. Purification by column chromatography on silica (CH_2Cl_2) was followed by recrystallization from CH_2Cl_2 to give **99** as a green microcrystalline solid (49 mg, 78%).

4.6.7
Reaction of Angular [3]Phenylene with [CpCo(eth)$_2$]

Angular [3]phenylene **15** (10 mg, 0.044 mmol) was dissolved in an NMR tube in benzene-d_6 (0.5 mL). To this solution was added [CpCo(eth)$_2$] (11 mg, 0.044 mmol), the tube was sealed and the mixture heated at 70 °C. An additional portion of [CpCo(eth)$_2$] (7 mg, 0.039 mmol) was added after 2 h, followed by the third one (32 mg, 0.178 mmol) after the total reaction time of 4 h. After all **15** reacted, NMR analysis indicated a yield of 71%. The solution was subsequently diluted with pentane (1 mL) and the resulting solid was filtered. Crystallization from benzene/pentane gave 7 mg (22%) of **112** as dark brown-black crystals.

Acknowledgments

One of the authors (P.V.) is greatly indebted to a group of superbly gifted co-workers and to a number of collaborators who developed the chemistry described in the preceding pages. Their names are listed in the corresponding references. The National Science Foundation (CHE-0071887) funded this work. The Center for New Directions in Organic Synthesis is supported by Bristol-Myers Squibb as a Sponsoring Member and Novartis as a Supporting Member.

Abbreviations

ASC	algebraic structure count
BTMSA	bis(trimethylsilyl)acetylene
Bu	butyl
Cp	cyclopentadienyl
Cp*	pentamethylcyclopentadienyl
DFT	density functional theory
DMF	dimethylformamide
DMAD	dimethyl butynedioate
DMDO	dimethyldioxirane
DMSO	dimethylsulfoxide
DMTS	dimethylthexylsilyl
Et	ethyl
eth	ethene
FVP	flash vacuum pyrolysis
HOMO	highest occupied molecular orbital
IC	internal conversion
IR	infrared
ISC	intersystem crossing
LDA	lithium diisopropylamide
LUMO	lowest unoccupied molecular orbital
Me	methyl
NICS	nucleus independent chemical shift
NMR	nuclear magnetic resonance
PAH	polycyclic aromatic hydrocarbon
Pr	propyl
Ph	phenyl

TBAF	tetrabutylammonium fluoride
TBC	tribenzocyclyne
TCNE	tetracyanoethene
THF	tetrahydrofuran
TIPS	trisisopropylsilyl
TMS	trimethylsilyl
TMSA	trimethylsilylacetylene
UV	ultraviolet
VT	variable temperature

References

1 (a) V. I. Minkin, M. N. Glukhovtsev, B. Ya. Simkin, *Aromaticity and Antiaromaticity: Electronic and Structural Aspects*, Wiley, New York, **1994**; (b) P. J. Garratt, *Aromaticity*, Wiley, New York, **1986**; (c) *Chem. Rev.* **2001**, *101*, 1115–1566, Special Issue: Aromaticity (some of the reviews from this issue are also cited separately).

2 T. M. Krygowski, M. K. Cyrański, *Chem. Rev.* **2001**, *101*, 1385–1419 and references cited therein.

3 (a) L. J. Schaad, B. A. Hess, Jr., *Chem. Rev.* **2001**, *101*, 1465–1476 and references cited therein; (b) S. W. Slayden, J. F. Liebman, *Chem. Rev.* **2001**, *101*, 1541–1566 and references cited therein.

4 (a) R. H. Mitchell, *Chem. Rev.* **2001**, *101*, 1301–1315 and references cited therein; (b) J. A. N. F. Gomes, R. B. Mallion, *Chem. Rev.* **2001**, *101*, 1349–1383 and references cited therein.

5 (a) J. W. Armit, R. Robinson, *J. Chem. Soc., (Abstracts)* **1925**, *127*, 1604–1618; (b) J. Kruszewski, T. M. Krygowski, *Tetrahedron Lett.* **1970**, *11*, 319–324; (c) T. M. Krygowski, *Tetrahedron Lett.* **1970**, *11*, 1311–1312; (d) W. T. Dixon, *J. Chem. Soc. B* **1970**, 612–616.

6 M. K. Cyrański, T. M. Krygowski, A. R. Katrizky, P. v. R. Schleyer, *J. Org. Chem.* **2002**, *67*, 1333–1338.

7 (a) K. Jug, A. M. Köster, *J. Phys. Org. Chem.* **1991**, *4*, 163–169; (b) A. R. Katrizky, P. Barczynski, G. Masumarra, D. Pisano, M. Szafan, *J. Am. Chem. Soc.* **1989**, *111*, 7–15.

8 G. A. Jeffrey, J. R. Ruble, R. K. McMullan, J. A. Pople, *Proc. R. Soc. London, Ser. A* **1987**, *414*, 47–57.

9 (a) H. Irngartinger, M. Nixford, *Angew. Chem.* **1983**, *95*, 415–416; *Angew. Chem., Int. Ed. Engl.* **1983**, *22*, 403–404; (b) A. Sekiguchi, M. Tanaka, T. Matsuo, H. Watanabe, *Angew. Chem.* **2001**, *113*, 1721–1723; *Angew. Chem. Int. Ed.* **2001**, *40*, 1675–1677.

10 A. A. Deniz, K. S. Peters, G. J. Snyder, *Science* **1999**, *286*, 1119–1122.

11 M. Faraday, *Philos. Trans. R. Soc. London* **1825**, *115*, 440–466.

12 (a) O. L. Chapman, C. L. McIntosh, J. Pacansky, *J. Am. Chem. Soc.* **1973**, *95*, 614–617; (b) A. Krantz, C. Y. Lin, M. D. Newton, *J. Am. Chem. Soc.* **1973**, *95*, 2744–2746.

13 R. D. Kennedy, D. Lloyd, H. McNab, *J. Chem. Soc., Perkin Trans 1* **2002**, 1601–1621 and references cited therein.

14 (a) E. Hückel, *Grundzüge der Theorie ungesättigter und aromatischer Verbindungen*, Verlag Berlin, **1938**; (b) E. Hückel, *Z. Phys.* **1931**, *70*, 204–286.

15 W. C. Lothrop, *J. Am. Chem. Soc.* **1941**, *63*, 1187–1191.

16 (a) M. K. Shepherd, *Cyclobutarenes: The Chemistry of Benzocyclobutene, Biphenylene, and Related Compounds*, Elsevier, New York, **1991**; (b) F. Toda, P. J. Garratt, *Chem. Rev.* **1992**, *92*, 1685–1707.

17 Sigma-Aldrich Fine Chemicals Catalog, **2003**–2004.
18 M. IYODA, S. M. H. KABIR, A. VORASINGHA, Y. KUWATANI, M. YOSHIDA, *Tetrahedron Lett.* **1998**, *39*, 5393–5396.
19 (a) S. KANOKTANAPORA, J. A. H. MACBRIDE, *J. Chem. Res. (S)* **1980**, 203–207; (b) J. A. MACBRIDE, *J. Chem. Soc., Chem. Commun.* **1972**, 1219–1220.
20 (a) J. P. DROSKE, J. K. STILLE, *Macromolecules* **1984**, *17*, 1–10; (b) L. FRIEDMAN, D. F. LINDOW, *J. Am. Chem. Soc.* **1968**, *90*, 2324–2328; (c) D. F. LINDOW, L. FRIEDMAN, *J. Am. Chem. Soc.* **1967**, *89*, 1271–1272.
21 C. PERTHUISOT, B. L. EDELBACH, D. L. ZUBRIS, N. SIMHAI, C. N. IVERSON, C. MÜLLER, T. SATOH, W. D. JONES, *J. Mol. Catal. A* **2002**, *189*, 157–168 and references cited therein.
22 H. JEANY, K. G. MASON, J. M. SKETCHLEY, *Tetrahedron Lett.* **1970**, *11*, 485–488.
23 H. HEANEY, P. LEES, *Tetrahedron Lett.* **1964**, *5*, 3049–3052.
24 (a) A. YOKOZEKI, C. F. WILCOX, S. H. BAUER, *J. Am. Chem. Soc.* **1974**, *96*, 1026–1032; (b) J. K. FAWCETT, J. TROTTER, *Acta Crystallogr.* **1966**, *20*, 87–93; (c) T. C. W. MAK, J. TROTTER, *J. Chem. Soc.* **1962**, 1–8.
25 H. YAMAGUCHI, M. ATA, J. F. W. MCOMIE, J. W. BARTON, H. BAUMANN, *J. Chem. Soc., Faraday Trans. 2* **1983**, *79*, 599–609.
26 H. P. FIGEYS, N. DEFAY, R. H. MARTIN, J. F. W. MCOMIE, B. E. AYERS, J. B. CHADWICK, *Tetrahedron* **1976**, *32*, 2571–2578.
27 A. J. JONES, P. J. GARRATT, K. P. C. VOLLHARDT, *Angew. Chem., Int. Ed. Engl.* **1973**, *12*, 241–243; *Angew. Chem.* **1973**, *85*, 260–261.
28 (a) Z. B. MAKSIĆ, M. ECKERT-MAKSIĆ, O. MÓ, M. YÁÑEZ, in *Pauling's Legacy: Modern Modelling of the Chemical Bond*, Z. B. MAKSIĆ, W. J. ORVILLE-THOMAS (eds.). Elsevier: Amsterdam, **1999**, 47–101 and references cited therein; (b) A. STANGER, N. BEN-MERGUI, S. PERL, *Eur. J. Org. Chem.* **2003**, 2709–2712; (c) A. STANGER, E. TKACHENKO, *J. Comput. Chem.* **2001**, *22*, 1377–1386; (d) A. STANGER, *J. Am. Chem. Soc.* **1998**, *120*, 12034–12040; (e) N. L. FRANK, J. S. SIEGEL, in *Advances in Theoretically Interesting Molecules*, R. P. THUMMEL (ed.), JAI: London, **1995**, Vol. 3, pp. 209–261; (f) J. S. SIEGEL, *Angew. Chem., Int. Ed. Engl.* **1994**, *33*, 1721–1723; *Angew. Chem.* **1994**, *106*, 1808–1810 references cited therein; (g) K. JUG, P. C. HIBERTY, S. SHAIK, *Chem. Rev.* **2001**, *101*, 1477–1500 and references cited therein.
29 (a) S. SHAIK, A. SHURKI, D. DANOVICH, P. C. HIBERTY, *Chem. Rev.* **2001**, *101*, 1501–1539 and references cited therein; (b) P. C. HIBERTY, D. DANOVICH, A. SHURKI, S. SHAIK, *J. Am. Chem. Soc.* **1995**, *117*, 7760–7768 and references cited therein. For a popular account: (c) G. NARAHARI SASTRY, *Curr. Sci.* **2001**, *81*, 1288–1290.
30 C. S. WANNERE, K. W. SATTELMEYER, H. F. SCHAEFER III, P. v. R. SCHLEYER, *Angew. Chem.* **2004**, *116*, 4296–4302; *Angew. Chem. Int. Ed.* **2004**, *43*, 4200–4206.
31 M. RANDIĆ, *Chem. Rev.* **2003**, *103*, 3449–3605 and numerous references cited therein.
32 I. GUTMAN, S. J. CYVIN, J. BRUNVOLL, *Monatsh. Chem.* **1994**, *125*, 887–894. For illustration, there are 12 [5]phenylenes, 122 [7]phenylenes, and 6387 [10]phenylenes.
33 N. TRINAJSTIĆ, T. G. SCHMALZ, T. P. ŽIVKOVIĆ, S. NIKOLIĆ, G. E. HITE, D. J. KLEIN, W. A. SEITZ, *New J. Chem.* **1991**, *15*, 27–31.
34 H.-D. BECKHAUS, R. FAUST, A. J. MATZGER, D. L. MOHLER, D. W. ROGERS, C. RÜCHARDT, A. K. SAWHEY, S. P. VEREVKIN, K. P. C. VOLLHARDT, S. WOLFF, *J. Am. Chem. Soc.* **2000**, *122*, 7819–7820.
35 (a) M. D. WATSON, A. FECHTENKÖTTER, K. MÜLLEN, *Chem. Rev.* **2001**, *101*, 1267–1300 and references cited therein; (b) R. G. HARVEY, *Polycyclic Aromatic Hydrocarbons*, Wiley-VCH, New York, **1995**; (c) E. CLAR, *The Aromatic Sextet*, Wiley, London, **1972**.
36 Recent examples: (a) M. G. DEBIJE, J.

Piris, M. P. de Haas, J. M. Warman, Ž. Tomović, C. D. Simpson, M. D. Watson, K. Müllen, *J. Am. Chem. Soc.* **2004**, *126*, 4641–4645; (b) J. S. Wu, L. Gherghel, M. D. Watson, J. Li, Z. Wang, C. D. Simpson, U. Kolb, K. Müllen, *Macromolecules* **2003**, *36*, 7082–7089; (c) H. Meng, M. Bendikov, G. Mitchell, R. Helgeson, F. Wudl, Z. Bao, T. Siegrist, C. Kloc, C.-H. Chem, *Adv. Mater.* **2003**, *15*, 1090–1093; (d) Y. Zhang, J. R. Petta, S. Ambily, Y. Shen, D. C. Ralph, G. C. Malliaras, *Adv. Mater.* **2003**, *15*, 1632–1635; (e) C. D. Dimitrakopoulos, I. Kymissis, S. Purushothaman, D. A. Neumayer, P. R. Duncombe, R. B. Laibowitz, *Adv. Mater.* **1999**, *11*, 1372–1375.

37 I. Gutman, *J. Chem. Soc., Faraday Trans.* **1993**, *89*, 2413–2416.

38 M. Bühl, A. Hirsch, *Chem. Rev.* **2001**, *101*, 1153–1183 and references cited therein.

39 Ž. Tomović, I. Gutman, *Hemijski Pregled*, **2001**, *42*, 29–37 and references cited therein.

40 (a) C. F. Wilcox, *Tetrahedron Lett.* **1968**, *9*, 795–800; (b) C. F. Wilcox, *J. Am. Chem. Soc.* **1969**, *91*, 2732–2736.

41 (a) I. Gutman, N. Trinajstić, C. F. Wilcox, *Tetrahedron* **1975**, *31*, 143–146; (b) C. F. Wilcox, I. Gutman, N. Trinajstić, *Tetrahedron* **1975**, *31*, 147–152.

42 (a) S. Nikolić, N. Trinajstić, Z. Mihalić, *Croat. Chem. Acta.* **1995**, *68*, 105–129; (b) I. Gutman, Y. N. Yeh, S. L. Lee, Y. L. Luo, *Indian J. Chem.* **1993**, *32A*, 651–661; (c) H. Wiener, *J. Am. Chem. Soc.* **1947**, *69*, 17–20.

43 Lj. Pavlović, I. Gutman, *J. Chem. Inf. Comput. Sci.* **1997**, *37*, 355–358.

44 I. Gutman, V. Ivanov-Petrović, *J. Mol. Struct. (THEOCHEM)* **1997**, *389*, 227–232.

45 For the two earlier summaries, see: (a) K. P. C. Vollhardt, D. L. Mohler, in *Advances in Strain in Organic Chemistry*, Vol. 5, B. Halton (ed.), JAI, London, **1996**, 121–160; (b) K. P. C. Vollhardt, *Pure. Appl. Chem.* **1993**, *65*, 153–156.

46 E. N. Losey, E. LeGoff, *J. Org. Chem.* **1973**, *38*, 3812.

47 (a) J. W. Barton, R. B. Walker, *Tetrahedron Lett.* **1978**, *19*, 1005–1008; (b) N. C. Jamieson, G. E. Lewis, *Aust. J. Chem.* **1967**, *20*, 321–338.

48 J. W. Barton, M. K. Shepard, *Tetrahedron Lett.* **1984**, *25*, 4967–4970.

49 (a) S. Saito, Y. Yamamoto, *Chem. Rev.* **2000**, *100*, 2901–2915; (b) C. Aubert, O. Buisine, M. Petit, F. Slowinski, M. Malacria, *Pure. Appl. Chem.* **1999**, *71*, 1463–1470; (c) D. B. Grotjahn, in *Comprehensive Organometallic Chemistry II*, Vol. 12, E. W. Abel, F. G. A. Stone, G. Wilkinson (eds.), L. S. Hegedus (vol. ed.), Pergamon, Oxford, **1995**, pp. 741–770; (d) K. P. C. Vollhardt, *Angew. Chem., Int. Ed. Engl.* **1984**, *23*, 539–556; *Angew. Chem.* **1984**, *96*, 525–541; (e) N. Schore, *Chem. Rev.* **1988**, *88*, 1081–1119.

50 B. C. Berris, Y.-H. Lai, K. P. C. Vollhardt, *J. Chem. Soc., Chem. Commun.* **1982**, 953–854.

51 B. C. Berris, G. H. Hovakeemian, Y.-H. Lai, H. Mestdagh, K. P. C. Vollhardt, *J. Am. Chem. Soc.* **1985**, *107*, 5670–5687.

52 M. Hirthammer, K. P. C. Vollhardt, *J. Am. Chem. Soc.* **1986**, *108*, 2481–2482.

53 J. M. Schulman, R. L. Disch, *J. Am. Chem. Soc.* **1996**, *118*, 8470–8474 and references cited therein.

54 B. C. Berris, G. H. Hovakeemian, K. P. C. Vollhardt, *J. Chem. Soc., Chem. Commun.* **1983**, 502–503.

55 L. Blanco, H. E. Helson, M. Hirthammer, H. Mestdagh, S. Spyroudis, K. P. C. Vollhardt, *Angew. Chem.* **1987**, *99*, 1276–1277; *Angew. Chem., Int. Ed. Engl.* **1987**, *26*, 1246–1247.

56 D. Holmes, S. Kumaraswamy, A. J. Matzger, K. P. C. Vollhardt, *Chem. Eur. J.* **1999**, *5*, 3399–3412 and unpublished results.

57 R. Diercks, K. P. C. Vollhardt, *Angew. Chem.* **1986**, *98*, 268–270; *Angew. Chem., Int. Ed. Engl.* **1986**, *25*, 266–268.

58 R. H. Schmidt-Radde, K. P. C.

Vollhardt, *J. Am. Chem. Soc.* **1992**, *114*, 9713–9715.

59 R. Diercks, B. E. Eaton, S. Gürtzgen, S. Jalisatgi, A. J. Matzger, R. H. Radde, K. P. C. Vollhardt, *J. Am. Chem. Soc.* **1998**, *120*, 8247–8248.

60 P. I. Dosa, G. D. Whitener, K. P. C. Vollhardt, A. D. Bond, S. J. Teat, *Org. Lett.* **2002**, *4*, 2075–2078.

61 K. Jonas, E. Deffense, D. Habermann, *Angew. Chem.* **1983**, *95*, 729; *Angew. Chem., Int. Ed. Engl.* **1983**, *22*, 716–717.

62 R. Diercks, J. C. Armstrong, R. Boese, K. P. C. Vollhardt, *Angew. Chem.* **1986**, *98*, 270–271; *Angew. Chem., Int. Ed. Engl.* **1986**, *25*, 268–269.

63 R. Diercks, K. P. C. Vollhardt, *J. Am. Chem. Soc.* **1986**, *108*, 3150–3152.

64 S. Han, A. D. Bond, R. L. Disch, D. Holmes, J. M. Schulman, S. J. Teat, K. P. C. Vollhardt, G. D. Whitener, *Angew. Chem.* **2002**, *114*, 3357–3361; *Angew. Chem. Int. Ed.* **2002**, *41*, 3223–3227.

65 J. M. Schulman, R. L. Disch, *J. Phys. Chem. A* **1997**, *101*, 5596–5599.

66 (a) H. Hopf, *Classics in Hydrocarbon Chemistry*, Wiley-VCH, Weinheim, **2000**, pp. 321–330; (b) F. Vögtle, *Fascinating Molecules in Organic Chemistry*, Wiley, New York, **1992**, pp. 156–180; (c) K. P. Meurer, F. Vögtle, *Top. Curr. Chem.* **1985**, *127*, 1–76; (d) W. H. Laarhoven, W. J. C. Prinsen, *Top. Curr. Chem.* **1984**, *125*, 63–130.

67 C. Eickmeier, H. Junga, A. J. Matzger, F. Scherhag, M. Shim, K. P. C. Vollhardt, *Angew. Chem.* **1997**, *109*, 2194–2199; *Angew. Chem., Int. Ed. Engl.* **1997**, *36*, 2103–2108.

68 S. Han, D. R. Anderson, A. D. Bond, H. V. Chu, R. L. Disch, D. Holmes, J. M. Schulman, S. J. Teat, K. P. C. Vollhardt, G. D. Whitener, *Angew. Chem.* **2002**, *114*, 3361–3364; *Angew. Chem. Int. Ed.* **2002**, *41*, 3227–3230.

69 J. M. Schulman, R. L. Disch, *J. Phys. Chem. A* **2003**, *107*, 5223–5227.

70 (a) J. M. Schulman, R. L. Disch, *Chem. Phys. Lett.* **1996**, *262*, 813–816;

(b) A. D. J. Haymet, *Chem. Phys. Lett.* **1985**, *122*, 421–424.

71 B. T. Dunlap, R. Taylor, *J. Phys. Chem.* **1994**, *98*, 11018–11019.

72 C. Eickmeier, D. Holmes, H. Junga, A. J. Matzger, F. Scherhag, M. Shim, K. P. C. Vollhardt, *Angew. Chem.* **1999**, *111*, 856–860; *Angew. Chem., Int. Ed. Engl.* **1999**, *38*, 800–804.

73 D. T.-Y. Bong, L. Gentric, D. Holmes, A. J. Matzger, F. Scherhag, K. P. C. Vollhardt, *Chem. Commun.* **2002**, 278–279.

74 D. T.-Y. Bong, E. W. L. Chan, R. Diercks, P. I. Dosa, M. M. Haley, A. J. Matzger, O. Š. Miljanić, K. P. C. Vollhardt, A. D. Bond, S. J. Teat, A. Stanger, *Org. Lett.* **2004**, *6*, 2249–2252.

75 H. Hart, K. Katsumasa, C.-J. F. Du, *J. Org. Chem.* **1985**, *50*, 3104–3110.

76 M. B. Goldfinger, K. B. Crawford, T. M. Swager, *J. Am. Chem. Soc.* **1997**, *119*, 4578–4593.

77 R. Boese, A. J. Matzger, D. L. Mohler, K. P. C. Vollhardt, *Angew. Chem.* **1995**, *107*, 1630–1633; *Angew. Chem., Int. Ed. Engl.* **1995**, *34*, 1478–1481.

78 J. E. Anthony, S. I. Khan, Y. Rubin, *Tetrahedron Lett.* **1997**, *38*, 3499–3502.

79 P. N. Baxter, *J. Org. Chem.* **2001**, *66*, 4170–4179.

80 D. Bruns, H. Miura, K. P. C. Vollhardt, A. Stanger, *Org. Lett.* **2003**, *5*, 549–552.

81 (a) P. v. R. Schleyer, H. Jiao, *Angew. Chem.* **1996**, *108*, 2548–2551; *Angew. Chem., Int. Ed. Engl.* **1996**, *35*, 2383–2386; (b) J. Aihara, *J. Chem. Soc., Faraday Trans.* **1995**, *91*, 237–239; (c) D. Babić, N. Trinajstić, *J. Mol. Struct. (THEOCHEM)* **1994**, *314*, 321–327; (d) I. Gutman, S. J. Cyvin, *J. Mol. Struct. (THEOCHEM)* **1993**, *288*, 85–91; (e) J. Aihara, *Bull. Chem. Soc. Jpn.* **1993**, *66*, 57–60; (f) J. Aihara, *J. Am. Chem. Soc.* **1992**, *114*, 865–868; (g) J. Cioslowski, P. B. O'Connor, E. D. Fleischmann, *J. Am. Chem. Soc.* **1991**, *113*, 1086–1089.

82 (a) A. J. Matzger, M. Shim, K. P. C. Vollhardt, *Chem. Commun.* **1999**,

1871–1872; (b) O. Š. Miljanić, K. P. C. Vollhardt, G. D. Whitener, *Synlett* **2003**, 29–34.
83 (a) H. A. Staab, F. Diederich, *Chem. Ber.* **1983**, *116*, 3487–3503; (b) H. A. Staab, F. Diederich, C. Krieger, D. Schweitzer, *Chem. Ber.* **1983**, *116*, 3504–3512.
84 (a) I. D. Campbell, G. Eglinton, W. Henderson, R. A. Raphael, *Chem. Commun.* **1966**, 87–89; (b) D. Solooki, J. D. Ferrara, D. Malaba, J. D. Bradshaw, C. A. Tessier, W. J. Youngs, *Inorg. Synth.* **1997**, *31*, 122–128; (c) C. Huynh, G. Linstrumelle, *Tetrahedron* **1988**, *44*, 6337–6344; (d) M. Iyoda, A. Vorasingha, Y. Kuwatani, M. Yoshida, *Tetrahedron Lett.* **1998**, *39*, 4701–4704.
85 (a) H. A. Staab, F. Graf, *Tetrahedron Lett.* **1966**, *7*, 751–757; (b) H. A. Staab, F. Graf, *Chem. Ber.* **1970**, *103*, 1107–1118.
86 (a) U. H. F. Bunz, Y. Rubin, Y. Tobe, *Chem. Soc. Rev.* **1999**, *28*, 107–119; (b) F. Diederich, *Nature* **1994**, *369*, 199–207.
87 (a) V. R. Coluci, D. S. Galvão, R. H. Baughman, *J. Chem. Phys.* **2004**, *121*, 3228–3237; (b) N. Narita, S. Nagai, S. Suzuki, *Phys. Rev. B* **2001**, *64*, 245408/1–245408/7; (c) N. Narita, S. Nagai, S. Suzuki, K. Nakao, *Phys. Rev. B* **2000**, *62*, 11146–11151; (d) N. Narita, S. Nagai, S. Suzuki, K. Nakao, *Phys. Rev. B* **1998**, *58*, 11009–11014; (e) M. M. Haley, *Synlett* **1998**, 557–565.
88 G. H. Hovakeemian, K. P. C. Vollhardt, *Angew. Chem.* **1983**, *96*, 1001–1002; *Angew. Chem., Int. Ed. Engl.* **1983**, *22*, 994–996.
89 D. L. Mohler, K. P. C. Vollhardt, S. Wolff, *Angew. Chem.* **1990**, *102*, 1200–1202; *Angew. Chem., Int. Ed. Engl.* **1990**, *29*, 1151–1154.
90 R. L. Augustine, *Heterogeneous Catalysis for the Synthetic Chemist*, Marcel Dekker, New York/Basel/Hong Kong, **1996**.
91 (a) A. Soncini, R. W. A. Havenith, P. W. Fowler, L. W. Jenneskens, E. Steiner, *J. Org. Chem.* **2002**, *67*, 4753–4758; (b) P. F. Fowler, R. W. A. Havenith, L. W. Jenneskens, A. Soncini, E. Steiner, *Chem. Commun.* **2001**, 2386–2387.
92 O. Š. Miljanić, K. P. C. Vollhardt, unpublished results.
93 (a) R. Jeyaraman, R. W. Murray, *J. Am. Chem. Soc.* **1984**, *106*, 2462–2463; (b) R. Murray, R. Jeyaraman, *J. Org. Chem.* **1985**, *50*, 2847–2853; (c) W. Adam, Y.-Y. Chan, D. Cremer, J. Gauss, D. Scheutzow, M. Schindler, *J. Org. Chem.* **1987**, *52*, 2800–2803; (d) W. Adam, R. Curci, J. O. Edwards, *Acc. Chem. Res.* **1989**, *22*, 205–211.
94 W. Adam, M. Balci, H. Kılıç, *J. Org. Chem.* **1998**, *63*, 8544–8546.
95 S. Kumaraswamy, S. S. Jalisatgi, A. J. Matzger, O. Š. Miljanić, K. P. C. Vollhardt, *Angew. Chem.* **2004**, *116*, 3797–3801; *Angew. Chem. Int. Ed.* **2004**, *43*, 3711–3715.
96 A. S. Kende, P. T. MacGregor, *J. Am. Chem. Soc.* **1964**, *86*, 2088–2089.
97 J. Furukawa, N. Kawabata, J. Nishimura, *Tetrahedron* **1968**, *24*, 53–58.
98 F. Sevin, M. L. McKee, *J. Am. Chem. Soc.* **2001**, *123*, 4591–4600 and references cited therein.
99 H. Mestdagh, K. P. C. Vollhardt, *J. Chem. Soc., Chem. Commun.* **1986**, 281–282.
100 (a) J. J. Gajewski, *Hydrocarbon Thermal Isomerizations*, Academic Press, New York, **1981**; (b) R. G. Harvey, *Curr. Org. Chem.* **2004**, *8*, 303–323; (c) L. T. Scott, *Angew. Chem.* **2004**, *116*, 5102–5116; *Angew. Chem. Int. Ed.* **2004**, *43*, 4994–5007.
101 U. Wiersum, L. W. Jenneskens, *Tetrahedron Lett.* **1993**, *34*, 6615–6618.
102 For recent mechanistic reviews, see: (a) A. Necula, L. T. Scott, *J. Anal. Appl. Pyrol.* **2000**, *54*, 65–87; (b) R. F. C. Brown, *Eur. J. Org. Chem.* **1999**, 3211–3222.
103 P. I. Dosa, A. S. Schleifenbaum, K. P. C. Vollhardt, *Org. Lett.* **2001**, *3*, 1017–1020.
104 A. J. Matzger, K. P. C. Vollhardt, *Chem. Commun.* **1997**, 1415–1416.
105 D. V. Preda, L. T. Scott, *Org. Lett.* **2000**, *2*, 1489–1492.

106 (a) M. Oh, K. Yu, H. Li, E. J. Watson, G. B. Carpenter, D. A. Sweigart, *Adv. Synth. Catal.* **2003**, *345*, 1053–1060 and references cited therein; (b) C. Perthuisot, B. L. Edelbach, D. L. Zubris, W. D. Jones, *Organometallics* **1997**, *16*, 2016–2023.

107 (a) Y. Oprunenko, I. Gloriozov, K. Lyssenko, S. Malyugina, D. Mityk, V. Mstislavsky, H. Günther, G. von Firks, M. Ebener, *J. Organomet. Chem.* **2002**, *656*, 27–42; (b) C. A. Dullaghan, G. B. Carpenter, D. A. Sweigart, *Chem. Eur. J.* **1997**, *3*, 75–78; (c) C. Elschenbroich, J. Schneider, W. Massa, G. Baum, H. Mellinghoff, *J. Organomet. Chem.* **1988**, *355*, 163–176.

108 A. Ceccon, A. Gambaro, A. M. Romanin, A. Venzo, *J. Organomet. Chem.* **1982**, *239*, 345–351.

109 M. Nambu, J. S. Siegel, *J. Am. Chem. Soc.* **1988**, *110*, 3675–3676.

110 M. Nambu, K. Hardcastle, K. K. Baldridge, J. S. Siegel, *J. Am. Chem. Soc.* **1992**, *114*, 369–371.

111 M. Nambu, D. L. Mohler, K. Hardcastle, K. K. Baldridge, J. S. Siegel, *J. Am. Chem. Soc.* **1993**, *115*, 6138–6142.

112 R. D. Brown, P. D. Godfrey, B. T. Hart, A. L. Ottrey, M. Onda, M. Woodruff, *Aust. J. Chem.* **1983**, *36*, 639–648.

113 A. Schleifenbaum, N. Feeder, K. P. C. Vollhardt, *Tetrahedron Lett.* **2001**, *42*, 7329–7332.

114 (a) W. V. Volland, E. R. Davidson, W. T. Borden, *J. Am. Chem. Soc.* **1979**, *101*, 533–537; (b) N. L. Allinger, *J. Am. Chem. Soc.* **1958**, *80*, 1953–1955; (c) N. L. Allinger, J. T. Sprague, *J. Am. Chem. Soc.* **1972**, *94*, 5734–5747; (d) D. W. Rogers, H. V. Voitkenberg, N. L. Allinger, *J. Org. Chem.* **1978**, *43*, 360–361; (e) W. L. Mock, *Tetrahedron Lett.* **1972**, 475–478; (f) L. Radom, J. A. Pople, W. L. Mock, *Tetrahedron Lett.* **1972**, 479–482.

115 (a) E. J. Petersson, J. C. Fanuele, M. R. Nimlos, D. M. Lemal, G. B. Ellison, J. G. Radziszewski, *J. Am. Chem. Soc.* **1997**, *119*, 11122–11123; (b) M. Bremer, P. v. R. Schleyer, U. Fleischer, *J. Am. Chem. Soc.* **1989**, *111*, 1147–1148.

116 C. de Rango, G. Tsoucaris, J. P. Declerq, G. Germain, J. P. Putzeys, *Acta Crystallogr. Sect. C* **1973**, *2*, 189–192.

117 (a) T. E. M. van den Hark, P. T. Beurskens, *Acta Crystallogr. Sect. C* **1976**, *5*, 247–252; (b) P. T. Beurskens, G. Beurskens, T. E. M. van den Hark, *Acta Crystallogr. Sect. C* **1976**, *5*, 241–246.

118 P. v. R. Schleyer, C. Maerker, A. Dransfeld, H. Jiao, N. J. R. van Eikema Hommes, *J. Am. Chem. Soc.* **1996**, *118*, 6317–6318.

119 P. v. R. Schleyer, M. Manoharan, Z.-X. Wang, B. Kiran, H. Jiao, R. Puchta, N. J. R. van Eikema Hommes, *Org. Lett.* **2001**, *3*, 2465–2468.

120 J. M. Schulman, R. L. Disch, H. Jiao, P. v. R. Schleyer, *J. Phys. Chem. A* **1998**, *102*, 8051–8055.

121 W. Adcock, B. D. Gupta, T. C. Khor, D. Doddrell, W. Kitching, *J. Org. Chem.* **1976**, *41*, 751–759.

122 R. P. Thummel, W. Nutakul, *J. Org. Chem.* **1978**, *43*, 3170–3173.

123 K. Fukui, *Orientation and Stereoselection*, Springer, Berlin, **1970**.

124 A. J. Heeger, *Angew. Chem.* **2001**, *113*, 2660–2682; *Angew. Chem. Int. Ed.* **2001**, *40*, 2591–2611 (Nobel lecture).

125 *NIST Chemistry WebBook, NIST Standard Reference Database Number 69*, November 1998, National Institute of Standards and Technology, Gaithersburg, MD 20899 (http://webbook.nist.gov).

126 Z. B. Maksić, D. Kovaček, M. Maksić-Eckert, M. Böckmann, M. Klessinger, *J. Phys. Chem.* **1995**, *99*, 6410–6416.

127 C. Dosche, H.-G. Löhmannsröben, A. Bieser, P. I. Dosa, S. Han, M. Iwamoto, A. Schleifenbaum, K. P. C. Vollhardt, *Phys. Chem. Chem. Phys.* **2002**, *4*, 2156–2161.

128 E. V. Shpol'skii, A. A. Il'ina, L. A. Klimova, *Dokl. Akad. Nauk SSSR* **1952**, *87*, 935–938.

129 C. Dosche, M. U. Kumke, F. Ariese,

A. N. Bader, C. Gooijer, P. I. Dosa, S. Han, O. Š. Miljanić, K. P. C. Vollhardt, R. Puchta, N. J. R. van Eikema Hommes, *Phys. Chem. Chem. Phys.* **2003**, *5*, 4563–4569.

130 C. Dosche, M. U. Kumke, H.-G. Löhmannsröben, F. Ariese, A. N. Bader, C. Gooijer, O. Š. Miljanić, M. Iwamoto, K. P. C. Vollhardt, R. Puchta, N. J. R. van Eikema Hommes, *Phys. Chem. Chem. Phys.* **2004**, *6*, 5476–5483.

131 S. Grimme, J. Harren, A. Sobanski, F. Vögtle, *Eur. J. Org. Chem.* **1998**, 1491–1509.

132 (a) G. Wenz, M. A. Müller, M. Schmidt, G. Wegner, *Macromolecules* **1984**, *17*, 837–850; (b) R. E. Martin, F. Diederich, *Angew. Chem.* **1999**, *111*, 1440–1469; *Angew. Chem. Int. Ed.* **1999**, *38*, 1350–1377; (c) Y. Geerts, G. Klärner, K. Müllen, in *Electronic Materials: The Oligomer Approach*, Wiley-VCH, Weinheim, **1998**.

5
A Solid State Strategy for the Preparation of Carbon-rich Polymers

Frank W. Fowler and Joseph W. Lauher

5.1
Introduction

The ultimate carbon rich polymer, with only carbon atoms as the constitutional repeat unit, is often referred to as carbyne. Being an allotrope of carbon, the one-dimensional polymer carbyne is analogous to the two-dimensional polymer graphite and the three-dimensional polymer diamond. In contrast to graphite and diamond, there is no general agreement on a material that is carbyne and it remains a controversial as well as confusing topic in the literature [1].

Carbyne is the most carbon-rich polymer in a series of linear conjugated polymers that begins with polyacetylene and its derivatives [2], Figure 5.1. There are numerous other members of this series containing a combination of double and triple bonds as the constitutional repeat unit such as the known linearly conjugated polydiacetylenes [3] (one double, one triple bond) and polytriacetylenes [4] (one double, two triple bonds). In addition to these linear conjugated polymers there are also the possibilities of cross conjugated polymers such as iso-polyacetylene and iso-polydiacetylene derivatives [5]. Because carbon rich polymers contain a high concentration of carbon–carbon multiple bonds they possess useful electrical and optical properties for advanced materials development with applications for the preparation of light emitting diodes, solar cells, transistors, chemical sensors, etc. [6].

The preparation of new polydiacetylenes and polytriacetylenes is complicated by the fact that no one has demonstrated a direct 1,4-diacetylene or a 1,6-triacetylene polymerization in solution, 1,2-polymerizations being more favorable. However, the polymers can be prepared in the solid state as the result of a topochemical polymerization. Topochemical reactions are solid state reactions in which the product and the regio- and stereochemistry of a reaction are directly controlled by the preorganization of the reactants.

In this report we will focus on the most ideal cases of topochemical polymerizations, transformations of single crystals of a monomer to single crystals of a polymer. Advantages of polymer single crystals are that the polymer chains are highly

Carbon-Rich Compounds. Edited by Michael M. Haley and Rik R. Tykwinski
Copyright © 2006 WILEY-VCH Verlag GmbH & Co. KGaA, Weinheim
ISBN: 3-527-31224-2

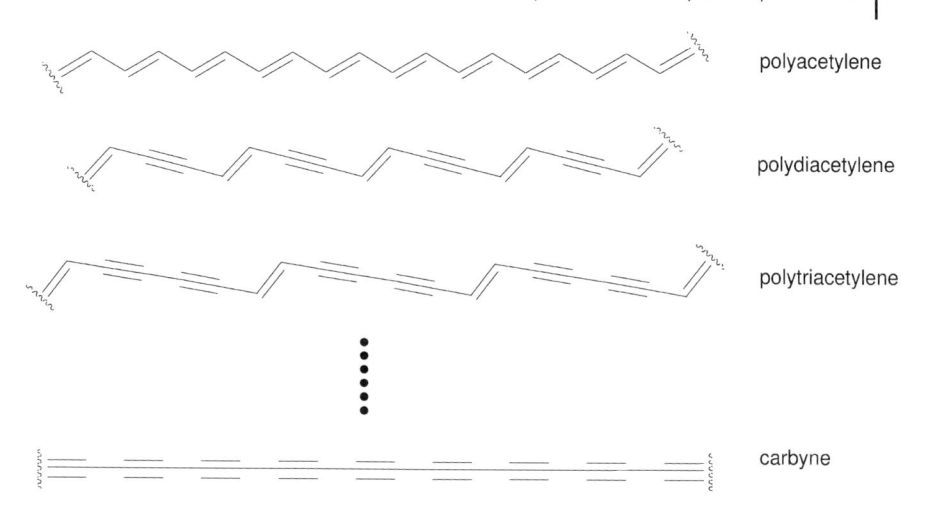

Figure 5.1. Conjugated polymers with increasing carbon content. Polyacetylene is the first member of this series that ends with carbyne.

organized, very pure and their molecular as well as their supramolecular structures are precisely known. A disadvantage is that the transformation of single crystal of monomer to a single crystal of polymer requires a precise organization of reactants. As a consequence single crystal polymers are extremely rare compared to traditional polymers. However, the rapid development of supramolecular synthesis holds great promise to change this situation making topochemical polymerizations producing single crystal polymers more common.

5.2
Topochemical Diacetylene Polymerizations

Topochemical polymerizations have their origins in the known instability of crystalline diacetylenes. In 1969 Wegner [7] reported that this instability was the result of the a 1,4-polymerization of the diacetylene [8]. This was a remarkable discovery. Not only has this polymerization been reported to produce a single crystal of a polymer but it also produced a linearly conjugated polymer analogous to polyacetylene. In contrast to polyacetylene, these crystalline polydiacetylenes are completely organized maximizing their electrical and optical properties [9]. Polydiacetylenes hold promise as a class of conjugated polymers for the development of advanced materials.

There are two steps involved in the preparation of single crystal polydiacetylenes. First, an appropriate monomeric diacetylene is prepared and, second, appropriate conditions are used for the assembly of this diacetylene in the solid state. These

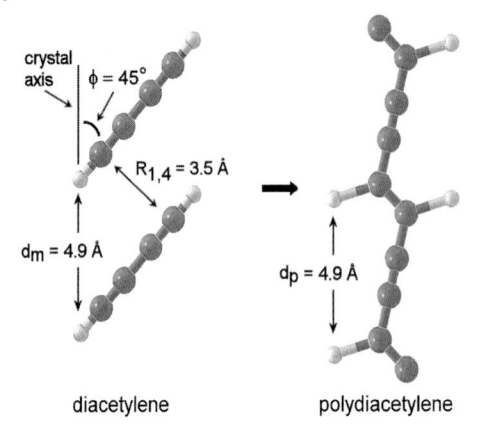

Scheme 5.1. Supramolecular structural parameters for a topochemical diacetylene polymerization. If the reaction is to proceed there must be a good match between the monomer repeat distance, d_m, and the known 4.90 Å polymer repeat distance d_p. A close approach, $R_{1,4}$, between the reacting C1–C4' carbon atoms is also necessary.

two steps, molecular and supramolecular synthesis, are required for the successful preparation of crystalline polydiacetylenes.

The structural parameters needed for a topochemical diacetylene polymerization are well understood [8]. If one examines known polydiacetylene structures (Scheme 5.1) one finds that the average distance, d_p, between equivalent atoms along the polymer backbone is 4.90 Å[1]. In order for a topochemical reaction to succeed it is absolutely essential for the diacetylene monomers to be aligned in a one-dimensional array with an intermolecular spacing, d_m, as near as possible to this 4.9 Å spacing. If the monomer spacing, d_m, is not commensurate with the polymer repeat, d_p, then large molecular displacements must occur during the polymerization process and the crystal lattice can be destroyed, inhibiting further the topochemical 1,4-diacetylene polymerization.

Although d_m is the most important parameter, the reacting atoms, C1–C4', should also be in close approach if the reaction is to proceed. The closest reasonable distance, $R_{1,4}$, is a van der Waals contact of about 3.5 Å. Along with the inherent length of the diacetylene unit itself, three distances are defined and one can calculate the angles. This calculation indicates that the diacetylenes should form an angle of 45° (ϕ), with respect to the stacking axis.

These two distances, d_m and $R_{1,4}$, must both be incorporated into any designed supramolecular structure if one wishes to achieve a topochemical 1,4-

---

1) This distance was determined from 14 known crystal structures of polydiacetylenes reported in the Cambridge Structural Database.

polymerization[2]. Having a close C1–C4' approach without the proper one-dimensional array of monomers aligned at the desired repeat distance, d_m, is insufficient. Indeed one can find cases where there is a close C1–C4' approach, but neighboring diacetylenes are perpendicular to the repeat axis, meaning that d_m and $R_{1,4}$ are identical [10], clearly polymerization will not occur. The repeat distance, d_m, is the most critical of the two distances, past experience has shown us that it should not deviate more than about 0.20 Å from the ideal value of 4.90 Å. The $R_{1,4}$ distance is less critical and crystal-to-crystal reactions may occur with $R_{1,4}$ values as large as 4 Å, if the d_m value is on target [11].

Inspection of the known crystal structures of polydiacetylenes reveals that two common intermolecular interactions have been observed to establish the important spacing of the monomers. These are hydrogen bonding and, more commonly, $\pi-\pi$ stacking [12] particularly of aromatic sulfonates.

Clearly the topochemical polymerization of diacetylenes is an extremely powerful method for the preparation of polymers. The crystal lattice environment suppresses unwanted side reactions and a highly ordered product is formed[3]. Both of these properties, a very pure and ordered product, are valuable for the development of advanced materials. Unfortunately, of the thousands of diacetylenes that are known[4] less than twenty have been reported[5] to undergo the very desirable transformation of a single crystal of the monomer to a single crystal of the polymer (SCSC) [13]. The reason for this situation is that there is a very low probability that a given diacetylene will adopt the necessary supramolecular structure suitable for a SCSC polymerization. The strategy of preparing a new diacetylene and hoping it will adopt the necessary supramolecular structure features for a SCSC polymerization is extremely unlikely to be successful for the preparation of new polydiacetylenes. The topochemical polymerization of diacetylenes to produce single crystal polydiacetylenes for advanced materials is a dramatic demonstration of the need to develop rational strategies for supramolecular synthesis.

2) Although these supramolecular structural features are a requirement for a successful topochemical reaction, the nature of the diacetylene substituents can also have an impact on the polymerization. During the polymerization process the diacetylenes experience a change in their atomic coordinates and their relationship to the bonded substituents. For example, the bond angle from the diacetylene carbon atoms to the substituents must change from approximately 180 to 120°. The attached substituents must be able to accommodate this structural change within the crystal lattice. This is often accomplished by rotations about series of single bonds. Thus, successful SCSC diacetylene polymerizations have not been observed with aryl groups as the only substituent connected to the diacetylene functionality.

3) With respect to organizing reactants for a specific reaction, the crystal lattice is analogous, if not superior, to an enzyme.

4) If one considers only diacetylenes not in a ring, with a molecular weight less than 700 and conforming to the general structure $RCH_2CCCCCH_2R$, a Scifinder search (October, 2004) reveals 3565 have been reported in the literature.

5) For inclusion in this list of single crystal polymers, X-ray determined atomic coordinates of polymer must have been reported.

5.2.1
Diacetylene Polymerizations

We began our studies on the development of supramolecular strategies with a strategy for the preparation of layered materials [14]. At that time the urea functional group played an important role for the formation of layered materials and we realized that it possessed a translational repeat distance (ca. 4.7 Å) that might be successful for spacing diacetylenes suitable for a topochemical polymerization. This led to the establishment of one of our early synthetic goals; the preparation of layered diacetylenes. Attainment of this goal requires success with both molecular and supramolecular synthesis.

The first step of our original strategy for the preparation of layered materials was to identify two functional groups that assemble in one dimension. The second step was to use molecular synthesis to incorporate these *two* functional groups into a single molecule. If, at the supramolecular level, the original one-dimensional assemblies persist, then a two-dimensional material will result. One of the early examples of this strategy was the ureylene dicarboxylic acid **1**, Fig. 5.2 [14]. The functional groups capable of assembling in one dimension are the urea and dicarboxylic acid. Single crystal X-ray diffractometry of this material demonstrates that it forms a layered structure. That is, within a layer, each ureylene dicarboxylic acid molecule forms eight hydrogen bonds to its nearest neighbors. However, the same ureylene dicarboxylic acid only forms weak van der Waals and dipolar interactions with molecules in adjacent layers.

Using the above strategy for the preparation of layered compounds the next step for the preparation of a layered diacetylene was to design a second molecular structure containing the diacetylene and necessary functionality for its incorporation into the layered hydrogen bonded network [15]. A reasonable choice is diacetylene **2** which contains the pyridine functionality, a functionality well known to hydrogen bond to carboxylic acids forming supramolecular networks [16].

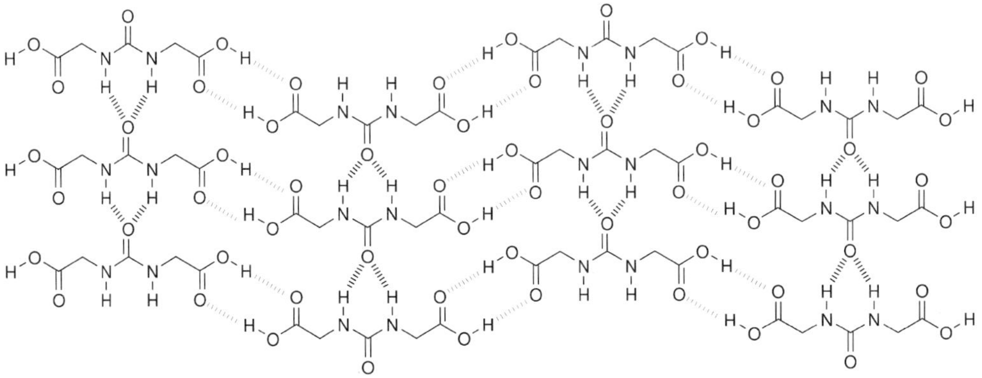

1

Figure 5.2. An example of a designed two dimensional layered compound using ureylene dicarboxylic acid **1**.

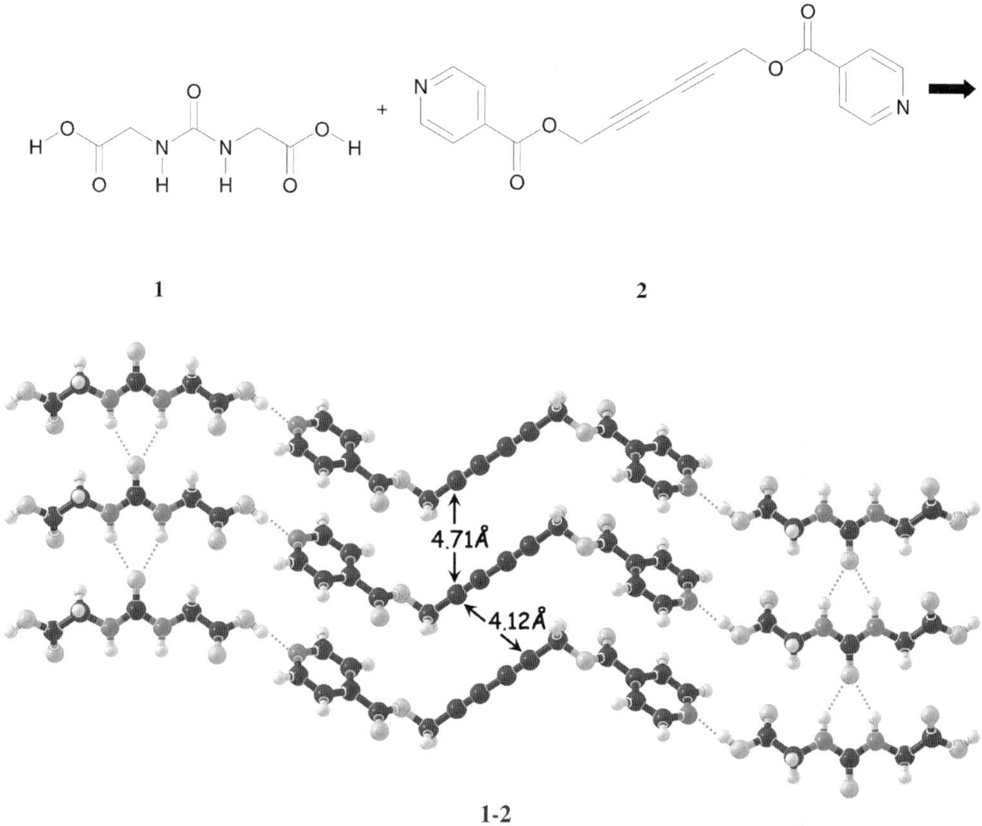

5.2 Topochemical Diacetylene Polymerizations

Scheme 5.2. The preparation of the host–guest compound
1–2 ($d_m = 4.71$ Å, $R_{1,4} = 4.12$ Å and $\phi = 57°$).

Slow evaporation from a methanol solution of a 1:1 molar mixture of host **1** and guest **2** produced a 1:1 host–guest complex whose crystal structure is shown in Scheme 5.2. The crystal structure of co-crystal **1–2** demonstrates that the hydrogen bonded host–guest complex was formed as designed. Although heating this complex induces a color change indicative of diacetylene polymerization, further investigation revealed that the crystal lattice was destroyed during the polymerization and only a few percent of an insoluble polymer was produced. Inspection of the crystal structure of **1–2** reveals two difficulties with respect to a topochemical polymerization. The distance between the reacting carbon atoms ($R_{1,4}$) of adjacent diacetylene functionalities is 4.12 Å with a translation spacing (d_m) of 4.71 Å. The change of $R_{1,4}$ from 4.12 Å to the expected 1.34 Å of a C=C bond of the product would require a considerable structural change and the translational spacing of the diacetylenes is short of the 4.9 Å anticipated to be ideal for the polydiacetylene product.

Because supramolecular structural features, such as the close contact of the diacetylenes, are extremely difficult to predict from a given molecular structure, they are beyond our current ability to control in the development of supramolecular synthetic strategies. However, the translation spacing of the diacetylenes issue can be addressed. In principle, this distance can be increased by replacing the urea functionality with a functionality that has a longer translational repeat distance. One possibility is the oxalamide which commonly self-assembles with a spacing of about 5 Å. Slow evaporation of a methanol solution of oxalamide **3** and diacetylene **2** produced a 1:1 host–guest complex. Unfortunately, the anticipated hydrogen bond motif of the oxalamide was not formed and the organization of the diacetylenes was not favorable for polymerization [17].

Interestingly, slow evaporation of a methanol solution of the isomeric diacetylene did produce the designed supramolecular structure **3–4** (Scheme 5.3). The

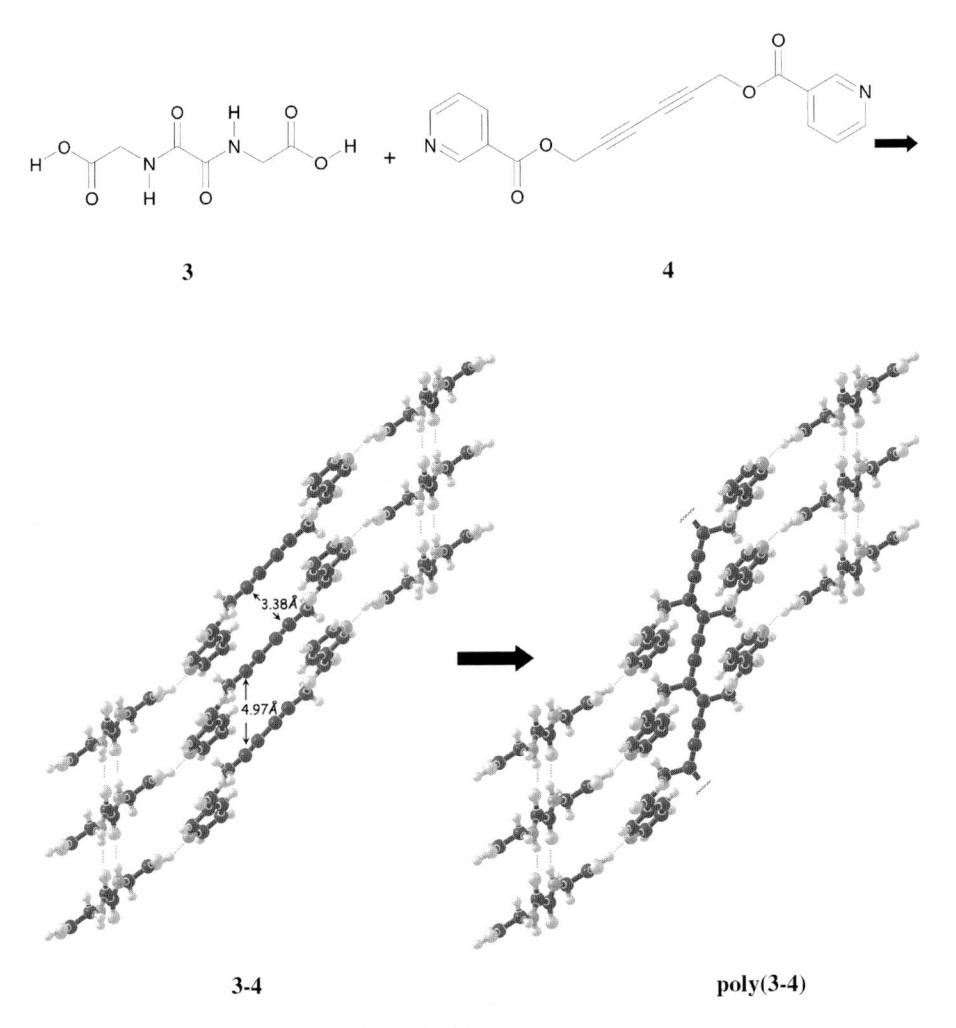

Scheme 5.3. Crystal structures illustrating the conversion of the host–guest compound **3–4** ($d = 4.97$ Å, $R_{1,4} = 3.38$ Å and $\phi = 43°$) to the corresponding polydiacetylene.

(a) The 001 face of poly(3-4).

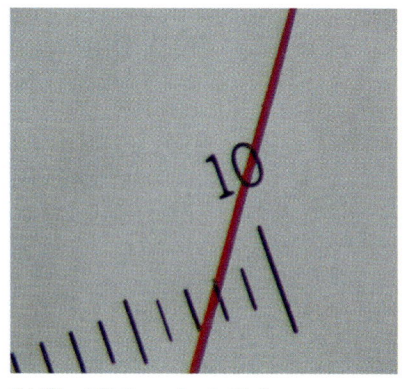

(b) The 010 face of poly(3-4)

(c) Poly(3-4) in ambient light showing its metallic gold luster.

Figure 5.3. Photographs of PDA crystals showing their anisotropy and metallic luster.

4.97 Å spacing of the diacetylenes is much closer to that anticipated for the polymer (4.9 Å) and the distance between the reacting carbon atoms of adjacent diacetylenes is within van der Waals contact (3.38 Å). On standing at room temperature or by thermal annealing over night at 110 °C the crystals of the diacetylene were converted to the polydiacetylene [17].

The polydiacetylene crystals (Figure 5.3) are remarkable for organic compounds in that their color depends upon their orientation. The crystals are colorless if viewed along the 001 face but deep red if viewed along the 010 face. When the polydiacetylene crystals are viewed in ambient light they assume a metallic gold luster.

Major features of the optical reflectance spectra [18] are the characteristic PDA absorption band strongly polarized along the chain direction and a near infrared charge transfer band involving the polymer chain and the electron deficient pyridine rings. A similar charge transfer band has recently been reported for a different PDA [19].

The preparation of the above single crystal polydiacetylene **3–4** is a success of the host–guest strategy for the preparation of designed supramolecular structures. Using the same supramolecular synthons, the one-dimensional oxalamide hydrogen bonded network for the translation repeat of the host and the pyridine-carboxylic acid hydrogen bond for association of the guest with the host, other supramolecular structures can be designed for a topochemical diacetylene polymerization. A simple example is the host–guest complex of oxalamide **5** and diacetylene **6** [20].

Although **5** and **6** readily formed co-crystals from a number of different solvents such as 50% *tert*-butanol in water, great difficulty was encountered obtaining crystals of suitable quality for a single crystal X-ray structure determination of the monomer. There were two reasons for this difficulty. The crystals were extremely small and polymerization appeared to be occurring at room temperature, which decreased the crystal quality. Thermal annealing at 120 °C for 12 h resulted in complete polymerization and a single crystal suitable for X-ray structure determination could be found (Scheme 5.4).

In order to further explore the scope of this strategy for the preparation of polydiacetylenes, attempts were made to produce the host–guest complex of other diacetylene diacids with oxalamide **5** [15, 20]. All diacetylene diacids chosen for study were symmetrical with a different alkyl spacer $-(CH_2)_n-$ separating the diacetylene from the acid functionalities. All of these diacetylene diacids formed co-crystals with **5** whose supramolecular structures are according to the original design. However, none of these readily produced a single crystal polymer upon the thermal annealing that was successful for the preparation of polymer **5–6** (Table 5.1).

The failure of **5–7** through **5–9** to undergo polymerization can be due to one or a combination of many factors. The hydrogen bonded network that is so critical for the formation of the designed supramolecular structure is also relatively rigid and difficult to distort during the polymerization process. The translational spacings (d_m) are all greater than the targeted value of 4.9 Å. The **5–9** cocrystal has the shortest spacing, but the closest approach ($R_{1,4}$) of the diacetylenes is unfavorable for polymerization. Presumably a crystal structure of the **5–6** monomer structure would show more ideal parameters, but that system polymerizes so readily we were never able to obtain a structure of the monomer.

5.2.1.1 Dipyridyldiacetylenes

Polydiacetylenes with conjugating substituents, such as aromatic rings, directly attached to the polymer backbone are of interest for materials development. It has often been proposed that these substituents will significantly enhance the nonlinear optical properties of the polymer [21–23]. However, until recently, few of these polydiacetylenes were known to exist [23–25]. No single crystal polydiactylenes with aromatic rings directly attached to the polymer backbone are known. The difficulty that these compounds encounter in undergoing a topochemical polymerization to produce a single crystal polymer is that the aromatic ring, which is directly connected to the reacting carbon atom, must undergo considerable motion in the crystal lattice during the reaction. The reacting carbon atom of one diacetylene

5.2 Topochemical Diacetylene Polymerizations

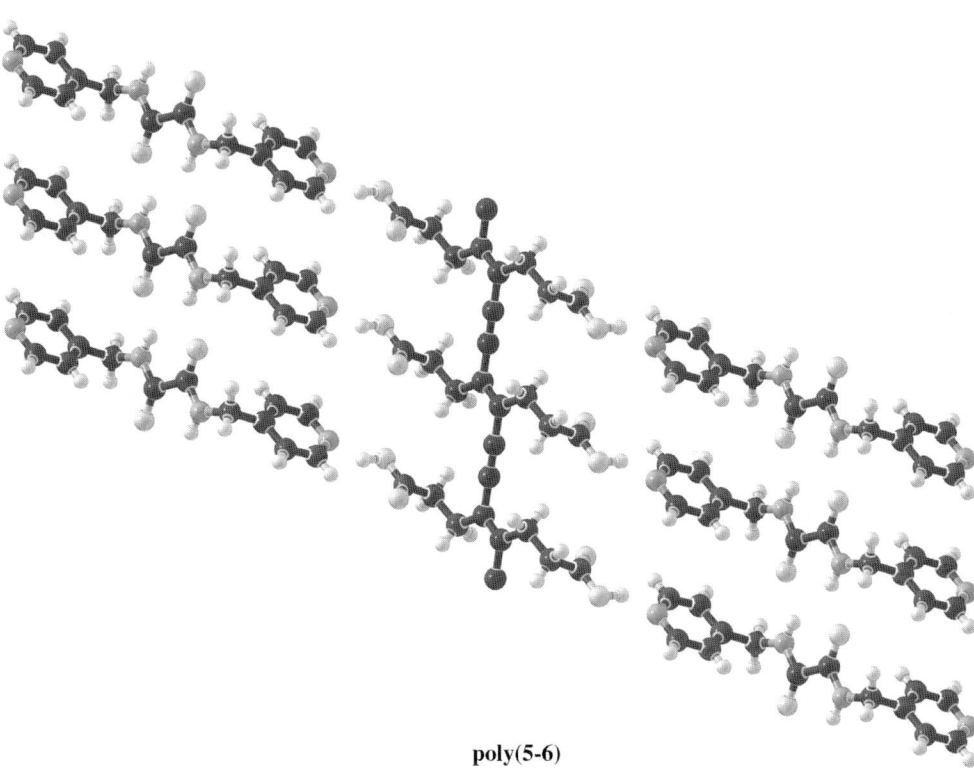

Scheme 5.4. The conversion of **5–6** to the host–guest compound and the co-crystal of the diacetylene polymer.

Table 5.1. Structural parameters for host–guest complexes derived from oxalamide **5** and dicarboxylic acids **7–9**.

	$-(CH_2)_2-$ (polymer) poly(5-6)	$-(CH_2)_3-$ 7	$-(CH_2)_4-$ 8	$-(CH_2)_8-$ 9
d_m (Å)	d_p 4.92	5.16	5.10	5.07
$R_{1,4}$ (Å)	(na)	3.55	3.85	4.12
ϕ (°)	(na)	43.5	48.7	53.3

must move over 1 Å and the bond angles of this carbon atom change from about 180 to 120°. With the known single crystal polymers these changes were accommodated by small displacements and conformational changes of the connecting chain.

Eight host–guest compounds (Table 5.2) were produced from the dipyridyl diacetylenes **14–15** and the oxamides **10–13** [26]. X-ray diffraction showed that all of these host–guest compounds possessed the designed supramolecular structure. An examination of the structural parameters, Table 5.2, shows that three of the co-crystals, **10–14**, **10–15**, and **11–15** have a repeat distance, d_m, close to the ideal d_p value of 4.9 Å. Of these three co-crystals only **10–14** and **11–15** have close $R_{1,4}$ contacts. We subjected all eight co-crystals to various polymerization conditions, but only **10–14**, the one with the best set of parameters, showed any tendency to undergo topochemical polymerization. Both ultraviolet and γ irradiation produced a substance that did not melt at a temperature below 300 °C. A unit cell was determined after γ irradiation and demonstrated that a change in the crystal structure had occurred but the quality of the crystal did not allow for a molecular structure determination. After the crystals were exposed to ultraviolet irradiation, a methanol extraction of the host and unreacted guest gave a residue whose weight suggested the polymerization proceeded to the extent of 61%.

These observations are consistent with the difficulty that aromatic substituted diacetylenes have in undergoing a topochemical polymerization. The preparation of single crystal polydiacetylenes with aromatic substituents directly attached to the polymer backbone remains a challenge for supramolecular synthesis.

5.2.1.2 Monosubstituted Diacetylenes

Until 2003 [11] no single crystal to single crystal transformation of a monosubstituted diacetylene to polydiacetylenes had been reported [27]. The Cambridge Structural Database contains only four entries for terminal diacetylenes and none of these have the supramolecular structural features necessary for a topochemical 1,4-polymerization as outlined in Scheme 5.1. There are many reasons for the lack of information on the topochemical polymerization of unsymmetrical monosubstituted diacetylenes. One is that the oxidative coupling procedure, readily applied for the preparation of symmetrical diacetylenes, is not easily applied to the preparation of unsymmetrical diacetylenes. Another factor is that unsymmetrical diacetylenes lack a center of symmetry and are less likely to pack with simple translational symmetry, a structural feature commonly observed for diacetylenes that undergo a topochemical polymerization (see Scheme 5.1).

The application of the host–guest strategy to the polymerization of layered monosubstituted diacetylenes requires a careful design. Because of the previous success with the oxalamide it is an obvious choice for achieving the necessary spacing of the monosubstituted diacetylenes for polymerization. One approach is to have the only substituent on the diacetylene contain *two* functionalities capable of associating with a suitable oxalamide host. Because phenol hydroxy groups are known to hydrogen bond to pyridines one possibility is the catachol diacetylene shown below.

Dissolution of **5** and **16** in aqueous methanol followed by slow evaporation gave

Table 5.2. The supramolecular structural parameters for all eight host–guest compounds derived from diacetylenes **14–15** and oxalamides **10–13**.

	10	**11**	**12**	**13**
14	$d_m = 4.93$ Å $R_{1,4} = 3.62$ Å $\phi = 47.7°$	$d_m = 5.23$ Å $R_{1,4} = 5.66$ Å $\phi = 76.0°$	$d_m = 5.21$ Å $R_{1,4} = 5.53$ Å $\phi = 74.2°$	$d_m = 5.23$ Å $R_{1,4} = 4.91$ Å $\phi = 64.0°$
15	$d_m = 4.90$ Å $R_{1,4} = 4.17$ Å $\phi = 55.1°$	$d_m = 5.04$ Å $R_{1,4} = 3.84$ Å $\phi = 48.8°$	$d_m = 5.20$ Å $R_{1,4} = 3.70$ Å $\phi = 45.4°$	$d_m = 5.22$ Å $R_{1,4} = 3.96$ Å $\phi = 49.1°$

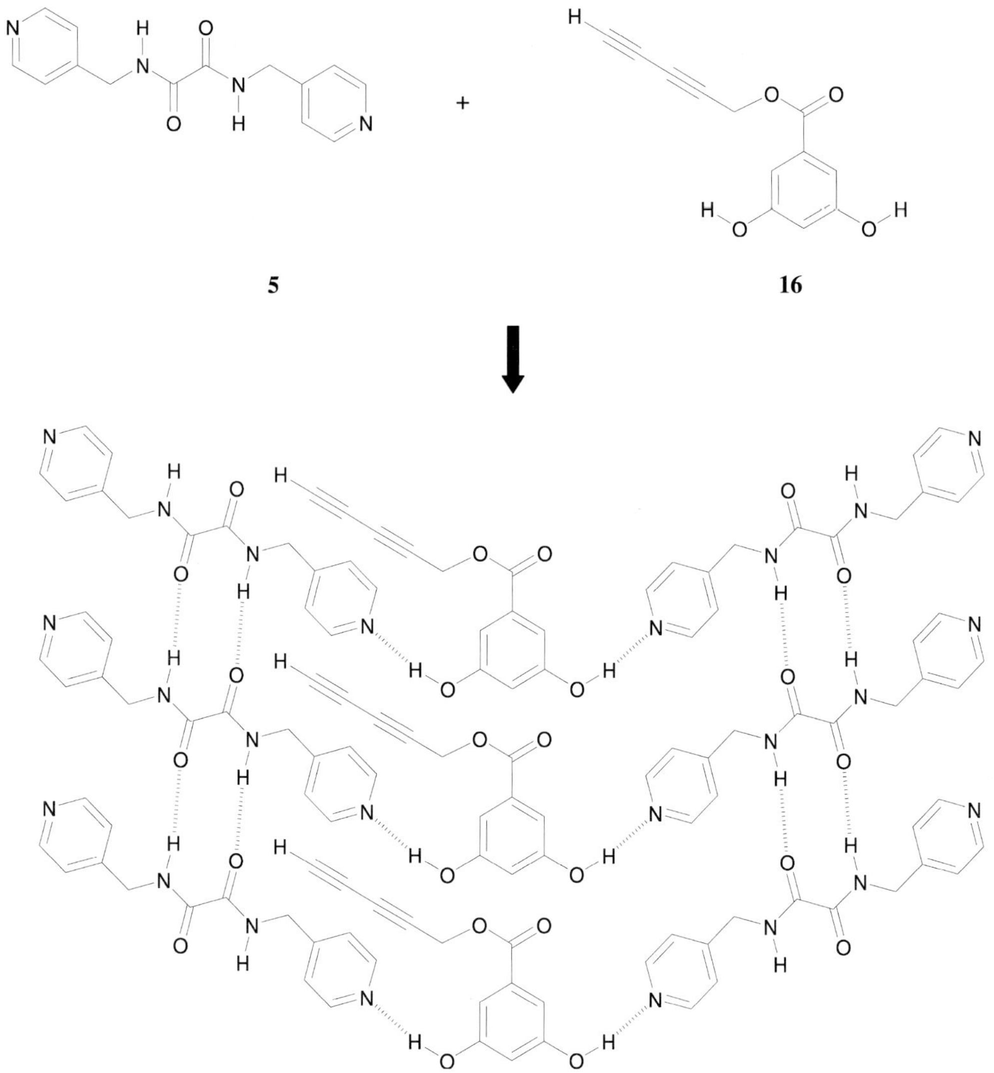

Scheme 5.5. The design of a supramolecular synthetic method for the preparation of a supramolecular structure with structural parameters suitable for a topochemical polymerization of a monosubstituted diacetylene.

a 1:1 co-crystal whose X-ray structure demonstrated the formation of the hydrogen bonding motif shown in Scheme 5.5. However, the supramolecular structure of **5–16** was not layered. Instead, the hydrogen bonded array shown in Scheme 5.5 consisted of three arrays of **5–16** formed a triple helix with two oxalamides per

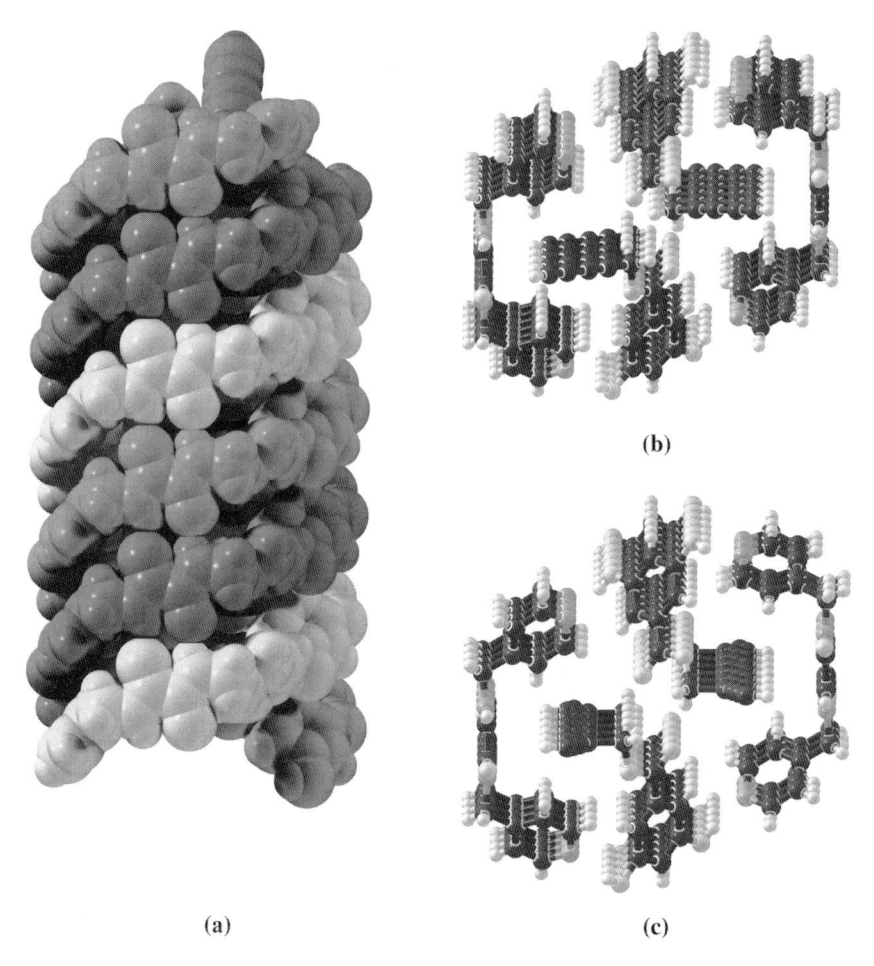

Figure 5.4. (a) The crystal structure of **5–16** illustrating the hydrogen bonded networks. (b) A view down the helical network of **5–16** ($d_m = 5.09$ Å, $R_{1,4} = 3.61$ Å and $\phi = 45°$). (c) A view down the helical network of poly(**5–16**).

turn. Two diacetylenes per turn occupy the internal space of the triple helix (Fig. 5.4).

The diacetylenes in this structure have the relationship suitable for a topochemical polymerization. The spacing of diacetylenes is 5.09 Å forming an angle of 45° with respect to the stacking axis. The distance separating the reacting carbon atoms (C1–C4′) of adjacent diacetylenes is 3.64 Å. These are close to ideal parameters for a topochemical polymerization and thermal annealing of **5–16** at 150 °C for 24 h gave the polydiacetylene (Fig. 5.4). The translational repeat distance along the polymer direction shrank to 4.925 Å from 5.090 Å. The two polydiacetylene chains within the triple helix represent potential insulated molecular wires.

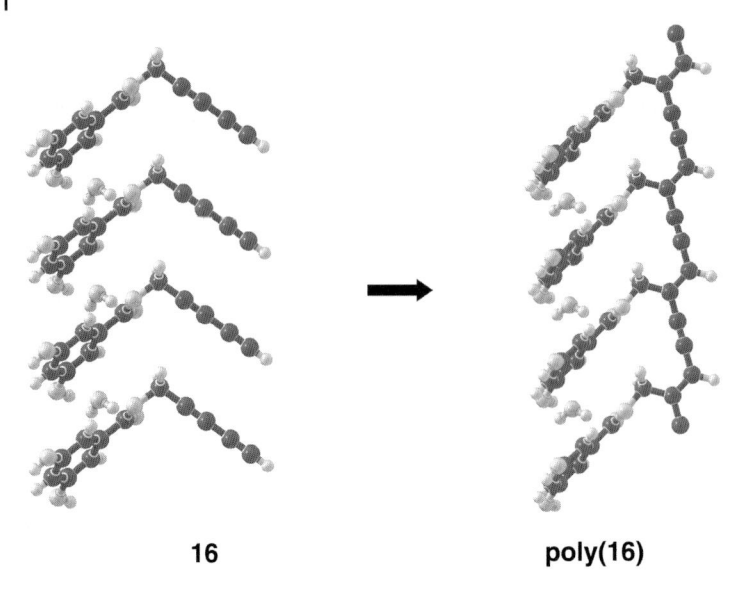

Scheme 5.6. Crystal structures illustrating the conversion of **16** H_2O ($d_m = 4.73$ Å, $R_{1,4} = 3.96$ Å and $\phi = 54°$) to the corresponding polydiacetylene.

Interestingly, the guest diacetylene **16**, by itself, crystallizes as a hydrate from aqueous methanol with supramolecular structural parameters for a possible topochemical polymerization (Scheme 5.6). The spacing, d_m, of the diacetylenes is 4.725 Å forming an angle of 54° with respect to the stacking axis. The distance, $R_{1,4}$, separating the reacting carbon atoms (C1–C4′) is 3.95 Å. Because these parameters are further from the ideal parameters derived from the model (Scheme 5.1) than **5–16**, we were pleased to observe that polymerization was complete on heating to 50 °C for 20 h, milder conditions than were required for the host–guest compound **5–16**.

Another interesting aspect of the polymerization of **16** compared to **5–16** is the nature of the molecular motion that occurs in the crystal lattice during the reaction. If C1 of the aromatic ring in both **16** and **5–16** is defined as an origin of rotation then both C1 and C4 of the diacetylene in **5–16** move about the same distance but in opposite directions (turnstile) when proceeding from the reactant to the product (Fig. 5.5–5.7). That is, the rotation of the diacetylene functionality is about a point midway between C2 and C3. Since all other known SCSC diacetylene polymerizations involve disubstituted diacetylenes with most being symmetrical, they undergo a similar motion proceeding from the reactant to the product. However, the motion of the monosubstituted diacetylene in **16** is not about the center of the functional group but much nearer the substituent (swinging gate).

Because the spatial requirements of the terminal hydrogen are small it can tolerate considerably more molecular motion than the larger substituent on the diacetylene. This larger motion can tolerate a larger mismatch of the diacetylene supra-

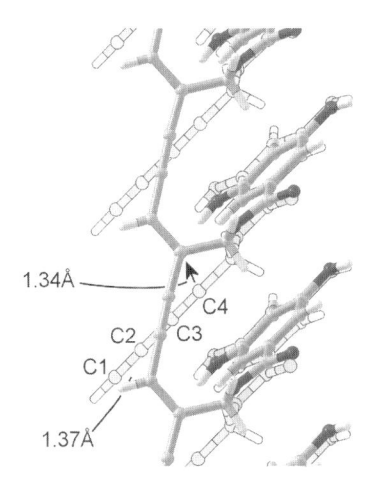

Figure 5.5. The pale background drawing shows the structure of the cocrystal diacetylene monomers (**5–16**). The bold foreground drawing shows the structure of the resulting polydiacetylene. The main motion is a rotation of the diacetylene accompanied by a pivot of the methylene group. The C1 and C4 atom movements are similar, 1.37 and 1.34 Å.

molecular structural parameters compared to the polymer. A consequence of this hypothesis is that terminal diacetylenes may more readily undergo a single crystal to single crystal transformation. This is interesting since **16** is the only known monosubstituted polydiacetylene.

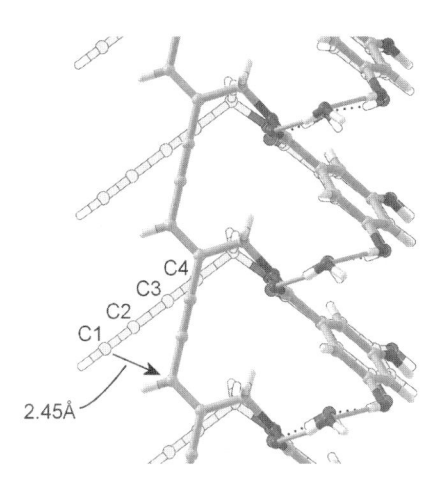

Figure 5.6. The pale background drawing shows the structure of the **16** · H_2O terminal diacetylene monomer structure. The bold foreground drawing shows the structure of the resulting polydiacetylene. The main motion is an inward swing of the diacetylene unit with a bend at C4. Atom C1 moves inward 2.45 Å. This inward motion of the diacetylene causes a shrinkage of the unit cell c axis and a 9% increase in crystal density.

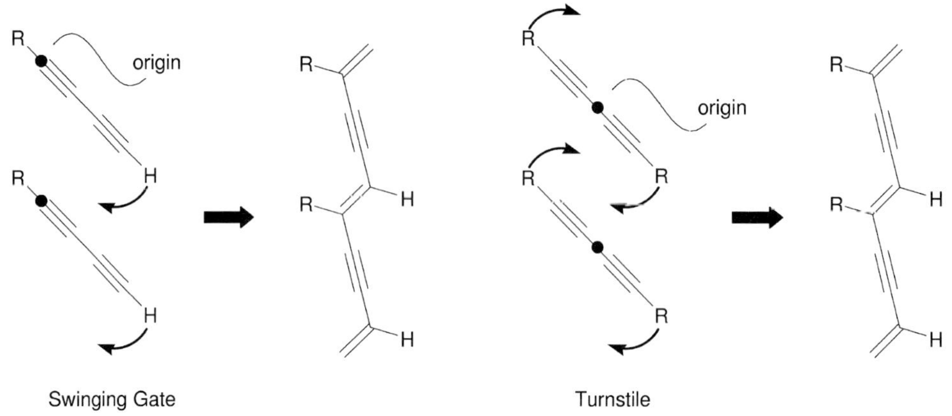

Swinging Gate

Turnstile

Figure 5.7. Two extreme mechanisms for the diacetylene polymerization. In the swinging gate mechanism the diacetylene pivots about the terminal carbon of the diacetylene. This mechanism is a possibility for unsymmetrical reactants such as monosubstituted diacetylenes. In the turnstile mechanism the diacetylene pivots about a point between the two diacetylenes. This mechanism is possible for symmetrical diacetylenes.

5.2.1.3 Summary of Diacetylene Polymerization

An important aspect of the host–guest strategy for the supramolecular synthesis of functional materials is the identification of persistent supramolecular synthons such as hydrogen bonding networks. Examples in the current work are the urea and oxalamide groups. Both of the one-dimensional networks resulting from these units, particularly the urea, are persistent because they are self-complementary. The persistency of the carboxylic acid-pyridine hydrogen bond is based upon the general rule that the best hydrogen bond donor will usually combine with the best hydrogen bond acceptor.

A second aspect of the host–guest strategy is to use two molecules to prepare a functional supramolecular structure. The host is used primarily to control supramolecular structure and the guest contains the functional group of interest. Both of these molecules must also contain functional groups for their assembly into the designed supramolecular structure. An advantage of this strategy is that it is efficient. Because the functionalities necessary for the designed supramolecular structure are distributed among two molecules, the molecular synthesis is simplified. The overall strategy is convergent in the last step. The host–guest strategy is also readily adaptable to a combinatorial approach to the preparation of functional materials. For example, three hosts and three compatible guests can give rise to nine different host–guests complexes for study.

It is also important to recognize that the application of supramolecular synthesis[6] to the preparation of new materials does not usually require complete control of supramolecular structure [28]. The goal of supramolecular synthesis is often to control a few supramolecular structural elements required for a given property. Examples of useful supramolecular structural elements are the organizing of molecules into layers, with their dipoles oriented in the same direction, or, as with diacetylene polymerizations, with a specific translational spacing.

Finally, it should not be anticipated that the host–guest approach will always produce a designed supramolecular structure. No synthetic method is 100% successful. The goal of the host–guest is to greatly improve the probability of producing a designed supramolecular structure. For example, in our early work we prepared four host–guest complexes using the two hosts, urea **1** and oxalamide **3**, with the two diacetylenes, **2** and **4**. Only two of these, **1–2** and **3–4**, gave the designed supramolecular networks. The other two combinations, **1–4** and **3–2**, formed co-crystals but the hydrogen bonded networks were different than the design. That is, the supramolecular synthons were not completely successful in these latter two cases.

A greater degree of successful supramolecular synthesis was experienced using the oxalamide **5** with diacetylene diacids **6–9**. Oxalamide **5** formed the designed, hydrogen bonded host–guest complexes with all of these diacids. However, only one of these host–guest complexes (**5–6**) was successfully converted into a crystalline polydiacetylene upon thermal annealing. Table 5.1 demonstrates that in these different host–guest complexes the diacetylene functionality has a slightly different spatial relationship to its nearest translationally related neighbor. These small differences in supramolecular structure are difficult to predict and are even more difficult to control.

5.3
Topochemical Triacetylene Polymerization

Unsuccessful attempts to perform a topochemical single crystal to single crystal polymerization of triacetylene have been reported over the years [29]. The difficulty of this reaction is that previous studies and molecular modeling [30, 31] of the polymer suggest a molecular repeat of about 7.4 Å will be required for a successful 1,6-polymerization of a triacetylene **17** (Scheme 5.7). A close approach, $R_{1,6}$, of the reacting C1–C6′ carbon atoms will also be required. It is rather unlikely that any triacetylene will fortuitously organize itself with *both* of these supramolecular structural parameters; the necessary angle ϕ is a very acute 28.5°. The direct

---

6) Supramolecular synthesis is the process, using noncovalent bonds, to organize molecules with certain defined supramolecular structural features. Because of the difficulty of determining the supramolecular structure of noncrystalline materials supramolecular synthesis is often applied to prepare crystalline materials. X-ray crystallography is without peer for the ability to determine supramolecular structure of crystalline materials. For this reason, supramolecular synthesis and crystal engineering are often used interchangeably. In a technical sense, supramolecular synthesis is more general since it could be applied to the preparation of noncrystalline materials.

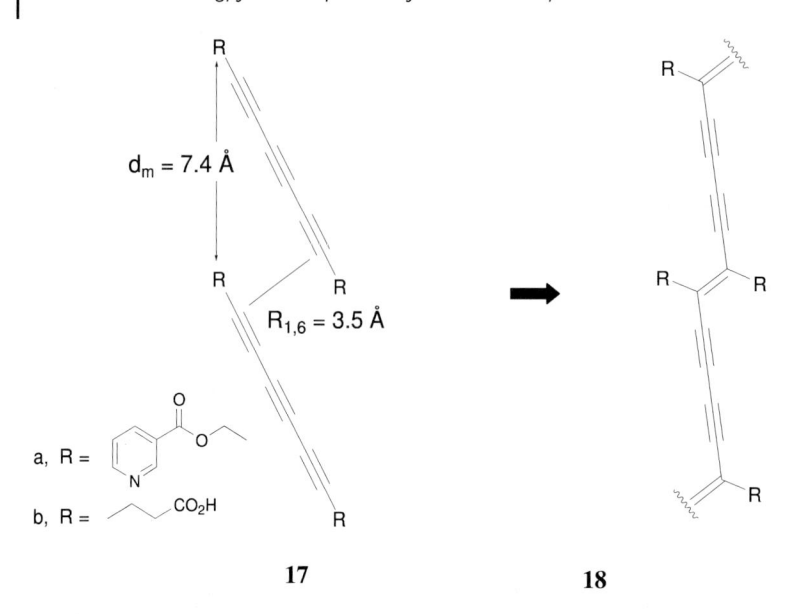

Scheme 5.7. Proposed supramolecular structure features[8] required for a topochemical triacetylene polymerization.

polymerization of a triacetylene to produce the polytriacetylene **18**[7] is thus an interesting challenge to the host–guest strategy. An approach to this problem first requires the identification of a functional group that self-assembles with a repeat distance of approximately 7.4 Å[8].

In initial studies the 2,5-diaminoquinone system **19** was prepared [33]. The X-ray crystal structure of this compound demonstrated that the 2,5-diaminoquinone assembled with a molecular repeat distance of 7.46 Å, which is close to the d_m value of 7.4 Å believed to be suitable for a topochemical polymerization (Fig. 5.8). Unfortunately, these simple 2,5-diaminoquinones proved to be very insoluble and not suitable for host–guest studies. However, the sub-structural feature responsible for the repeat distance is the vinylogous amide. Simple vinylogous amides such as **20** should have better solubility properties although the persistency of this functional group to assemble into the required motif is more problematical [35][9]. Among the

---

7) Diederich and co-workers have successfully prepared the conjugated poly(triacetylene) backbone, not by triacetylene polymerization but by the coupling of enediynes. Their considerable studies on these conjugated polymers have revealed some intriguing properties with possible application to advanced materials. For a review see Ref. 32.

8) If the molecular repeat distance (d_m) is 7.4 Å and the closest van der Waals approach of atoms C1–C6 of adjacent triacetylenes is 3.5 Å then the angle (φ) of the triacetylene chains with respect to the molecular repeat direction will be 28.2°.

9) All of the six vinylogous amides with structure **20** reported in the CSD have the supramolecular chemistry shown in Fig. 5.8. However, only one (ZAMVEO) assembles using the necessary simple translational motif. The remaining five examples assemble using a glide plane or screw axis. These types of symmetry double the crystallographic repeat distance to an undesirable 14–15 Å [34].

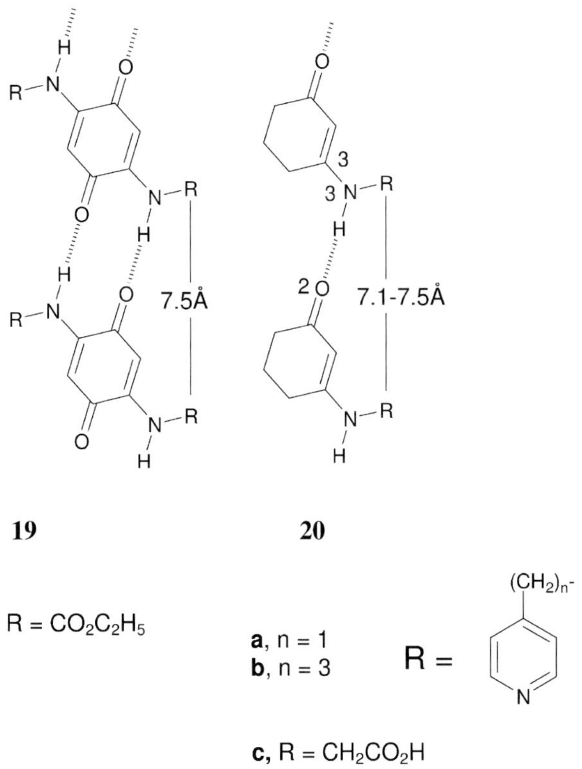

Figure 5.8. Possible hosts to organize triacetylenes with 7.4 Å spacing along the stacking direction for a topochemical polymerization.

six simple derivatives of the vinylogous amide reported in the Cambridge Structural Database (CSD), only one self-assembled with the required simple translational symmetry. However, we were pleased to observe that the repeat distance of this vinylogous amide[9] was 7.31 Å, a value in the range we were seeking.

After identifying a suitable host functionality to organize the triacetylene guest, the next step is to choose suitable functionalities on both the host and guest to facilitate the formation of the designed supramolecular network. Because of our earlier success with the diacetylene polymerization [4] we explored using analogous nicotinate and isonicotinate derivatives of triacetylene **17** as hydrogen bond acceptors. This choice required a hydrogen bond donor on the host. One possible host, carboxylic acid **20c** was readily prepared from 1,3-cyclohexanedione and glycine. We were unsuccessful at obtaining a host–guest compound using a hydrogen bond donor on the host and a hydrogen bond acceptor on the guest.

Fortunately, the above strategy for the preparation of supramolecular structures has flexibility. Another approach to the desired supramolecular structure is to attach the hydrogen bond donor (carboxylic acid) to the guest triacetylene and the

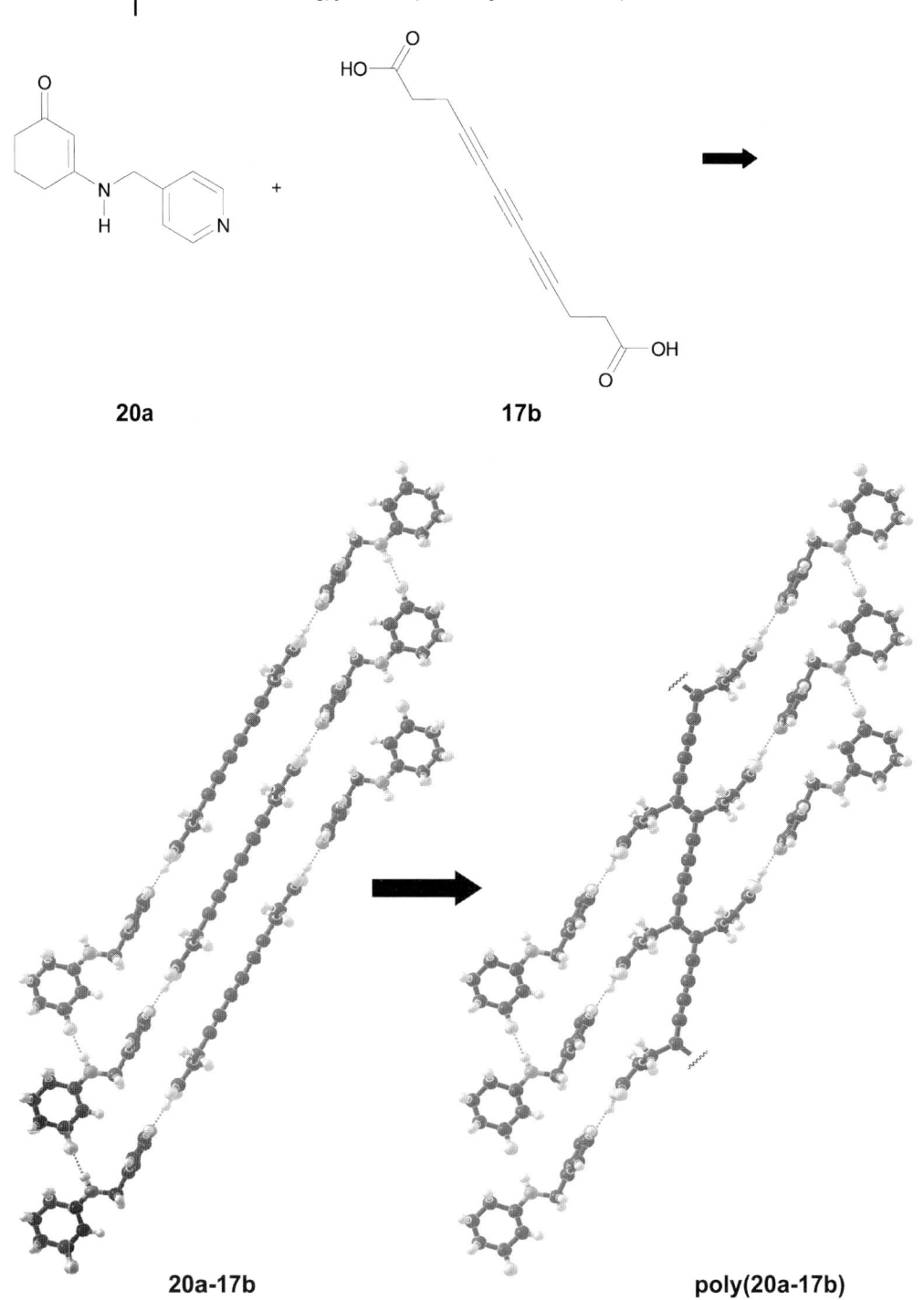

Scheme 5.8. The X-ray crystal structure of co-crystal **20a–17b**, ($d_m = 7.14$ Å, $R_{1,6} = 3.48$ Å and $\phi = 29°$) and its polymer.

hydrogen bond acceptor to the host (vinylogous amide **20a**). For this study we prepared triacetylenedicarboxylic acid guests **17b**. The host compound, pyridine **20a**, was readily prepared from condensation of cyclohexanedione with the corresponding amine.

Slow evaporation of a 2:1 methanol/ethyl acetate solution of pyridine host **20a** and the triacetylene guest **17b** produced a host–guest compound as pale red crystals with a melting point 144–145 °C which is between the melting point of the vinylogous amide host **20a** and the triacetylene guest **17b** (Scheme 5.8). A single crystal X-ray structure of the host–guest complex **20a–17b**, clearly shows that the molecules have self-assembled in accordance with the designed supramolecular structure, Scheme 5.7. The monomers are spaced at a d_m value of 7.143 Å with the triacetylene tilted at a ϕ angle of 29.2° with respect to the translational direction. The triacetylene chains are in van der Waals contact with the crucial nonbonded C1 to C6 distance, $R_{1,6}$, between adjacent triacetylene molecules of 3.487 Å. Although the repeat distance of 7.143 Å is shorter than the targeted value of 7.4 Å, overall these parameters are in good agreement with the numbers illustrated in Scheme 5.7.

Unfortunately, several attempts to induce polymerization of the triacetylene host–guest compound using either heat or ultraviolet irradiation were unsuccessful[10]. This has sometimes been a problem with diacetylene polymerization and gamma irradiation has proven successful. In fact, subjecting host–guest **20a–17b** to ^{60}Co γ-radiation[11] resulted in the crystals becoming dark red, a color transformation consistent with polymerization. The topochemical polymerization was followed by X-ray crystallography. Individual single crystals were mounted and the structure of each individual crystal was determined repeatedly after radiation doses of about 8–10 Mrad of γ-radiation [4].

The analysis of the diffraction data showed that the irradiated crystals consist of a solid solution of the monomer and its polymer. After 40 Mrad of radiation, the crystals polymerize to about 70%, the resulting polytriacetylene crystal structure is shown in Scheme 5.8. Further irradiation and polymerization brings about a sudden phase change in the crystals to an amorphous "glassy" state that no longer gives useful diffraction data.

It is curious to note that the conversion of the long van der Waal's contact to a covalent bond causes the lattice to increase (7.14 Å to 7.21 Å) along the *a*-axis. This is the direction of the polymer backbone which, ideally, should be about 7.4 Å.

The polytriacetylene was isolated by first extracting the irradiated crystals with concentrated aqueous HCl to remove the pyridine host and then by extracting the residue with methanol to remove any unreacted triacetylene. A red, amorphous solid, that does not melt up to 300 °C, was isolated in about 70% yield. It is insoluble in common organic solvents such as methanol, acetone, ethyl acetate and methylene chloride. The carboxylic acid polymer slowly dissolved to form a red

---

10) No significant polymerization was observed when the crystals were heated to 109 °C for 72 h or irradiated for 72 h with a Hanovia 550 W medium pressure mercury lamp.

11) The γ-ray source used had an output of 0.8 Mrad h^{-1} and was located at Brookhaven National Laboratories, Upton, NY.

solution in concentrated 1 M NaOH or alternatively a yellow solution in dilute 0.1 M NaOH [37]. The UV/VIS spectrum (0.01 g L^{-1} in 0.1 M NaOH) of the polymer shows a λ_{max} = 446 nm.

Raman spectroscopy has proven to be an excellent method for characterizing conjugated polymers [38]. The nonpolar multiple bonds usually display intense Raman bands. The polytriacetylene described here is no exception. The Raman spectrum of the polytriacetylene (0.011 g L^{-1} in 0.1 M NaOH) shows only two intense bands in the 2500–500 cm^{-1} region at 2148 and 1552 cm^{1}. These can assigned to the C≡C and C=C bonds respectively [4].

5.4
Topochemical Diene and Triene Polymerization

Although the polymerization of dienes and trienes does not give a conjugated polymer the similarity of these reactions to diyne and triyne polymerizations warrants discussion. The single crystal to single crystal polymerization of dienes has been elegantly developed in recent years by Matsumoto and co-workers [39, 40]. Their first example of this transformation was the polymerization of diethyl (Z,Z)-muconate 21 to give the stereoregular polymer poly(21) (see Scheme 5.9) [41]. Recently, other ester derivatives of muconic acid have been reported to undergo similar crystal to crystal transformations [42, 43]. It is interesting to note that diethyl (Z,Z)-muconate and other esters have been known for some time and yet it was not until 1994 that Matsumoto discovered their topochemical polymerization. In contrast, the topochemical polymerization of diacetylenes was identified about 25 years earlier.

Successful single crystal to single crystal topochemical diene polymerizations are analogous to diacetylene polymerizations. Important factors are the spacing of dienes, their angle with respect to the stacking direction and the distance between reacting atoms. The values for these parameters are similar to those for diynes as shown in Scheme 5.9. An additional factor for dienes, not necessary for diacetylenes, is the alignment of the π-system of one diene with respect to the π-systems of its reacting neighbors. Simply stated, the best σ-overlap of adjacent p-orbitals is achieved when the line connecting reacting atoms is at 90° with respect to the plane of the reacting π-systems.

Related to the topochemical diene polymerization is the topochemical triene polymerization. Because dienes successfully undergo a topochemical polymerization with supramolecular structural features similar to topochemical diyne polymerization it would be anticipated that trienes would require similar supramolecular parameters as for triynes.

In order to investigate the host–guest approach for the topochemical polymerization of trienes, the vinylogous amides 22 and 23 were prepared as guests. A simple hydrogen bonded translational assembly of the vinylogous amide generally gives a spacing (7.1–7.5 Å). This translational distance is anticipated to be necessary for a topochemical triacetylene polymerization. Unfortunately, all attempts to prepare host–guest compounds 22–24, 22–25, 23–24 and 23–25 for a topochemical triene polymerization were unsuccessful (Table 5.3) [44].

21 → **poly(21)**

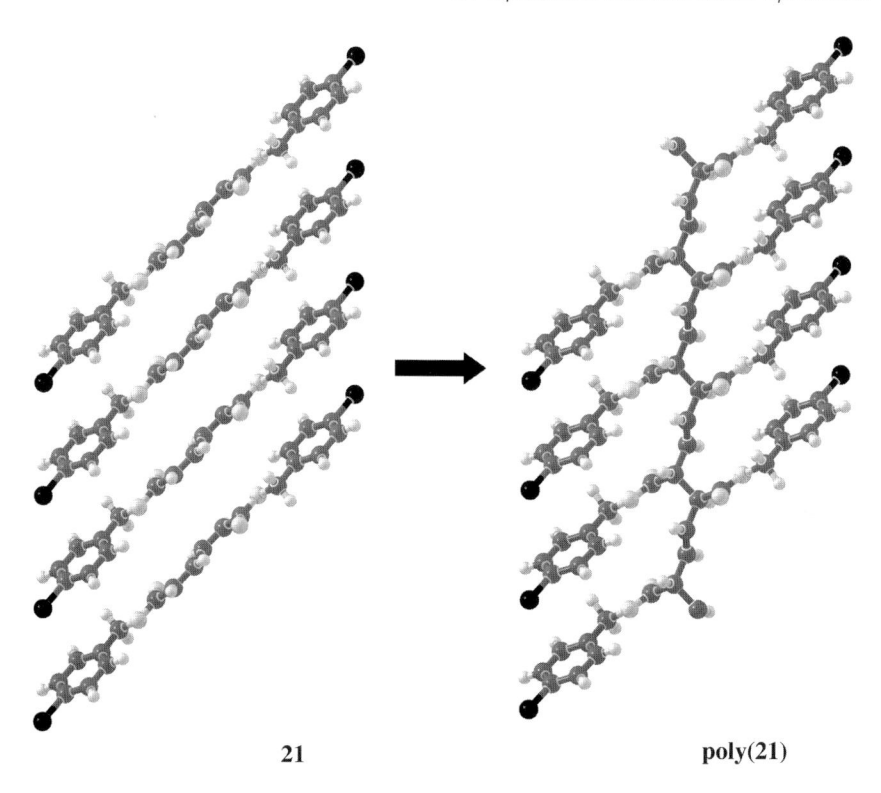

Scheme 5.9. The X-ray crystal structure of p-bromobenzyl muconate **21** ($d_m = 5.12$ Å, $R_{1,4} = 4.24$ Å and $\phi = 44°$) and its polymer [43].

However, the crystal structure of triene **24** demonstrated that it possessed a supramolecular structure suitable for a topochemical triene polymerization (Scheme 5.10) [45]. The spacing of trienes along the stacking direction is 7.252 Å with an angle of 34° with respect to the stacking direction. Molecular modeling suggests the ideal spacing along the stacking direction would be 7.2 Å. If the reacting carbon atoms of one triene approach those of its neighbors with a close contact of 3.5 Å then the trienes will be at an angle of 29° with respect to the stacking direction. Heating crystals of **24** to 110 °C for 8 h resulted in a smooth single crystal to single crystal transformation to poly(**24**). This represents the first reported example in the literature of a single crystal polytriene.

This supramolecular structure arises as a result of the isonicotine rings stacking by π–π interactions. The question is whether or not this is a persistent supramolecular synthon that can be applied to the polymerization of trienes as well as triynes. Janiak has recently analyzed π–π stacking of pyridines [46]. It was observed in the Cambridge Structural Database that a common assembly of pyridines is for them to be oriented in opposite directions with a near-zero angle between the planes of the rings and a close contact distance of 3.57 Å (Fig. 5.9). Although there was some

Table 5.3. Application of the host–guest strategy to the organization of trienes for a topochemical polymerization.

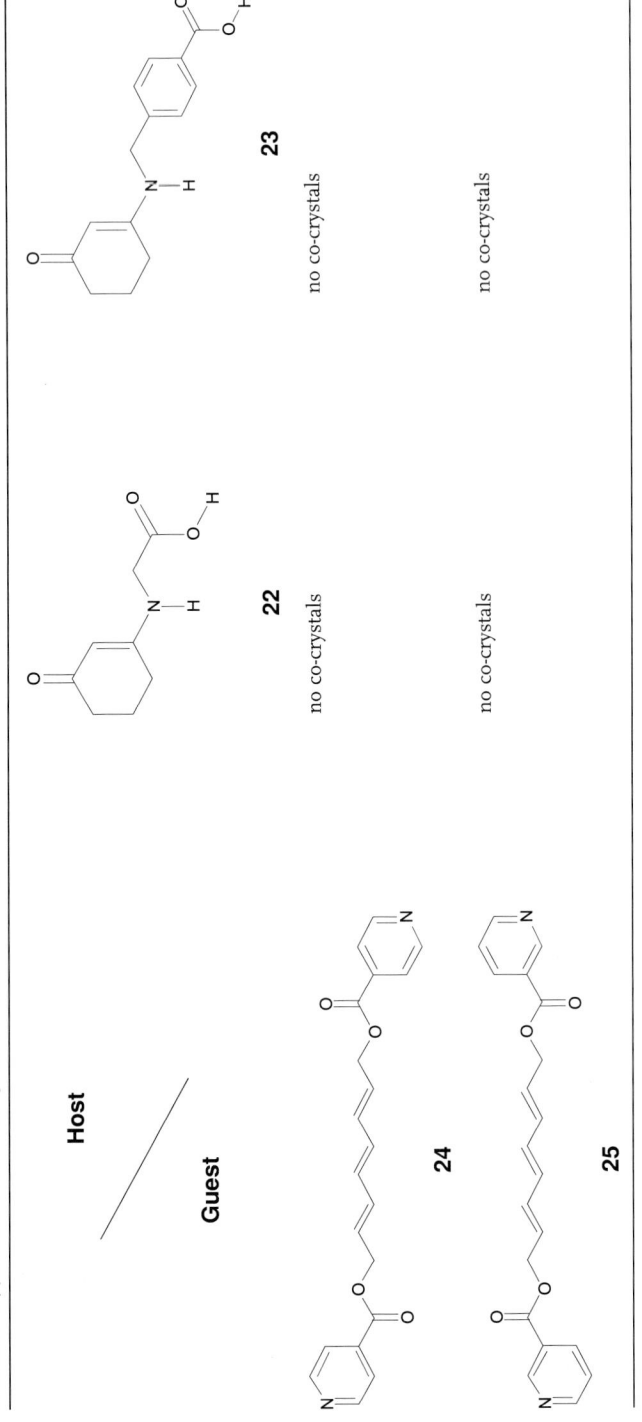

Host	22	23
Guest		
24	no co-crystals	no co-crystals
25	no co-crystals	no co-crystals

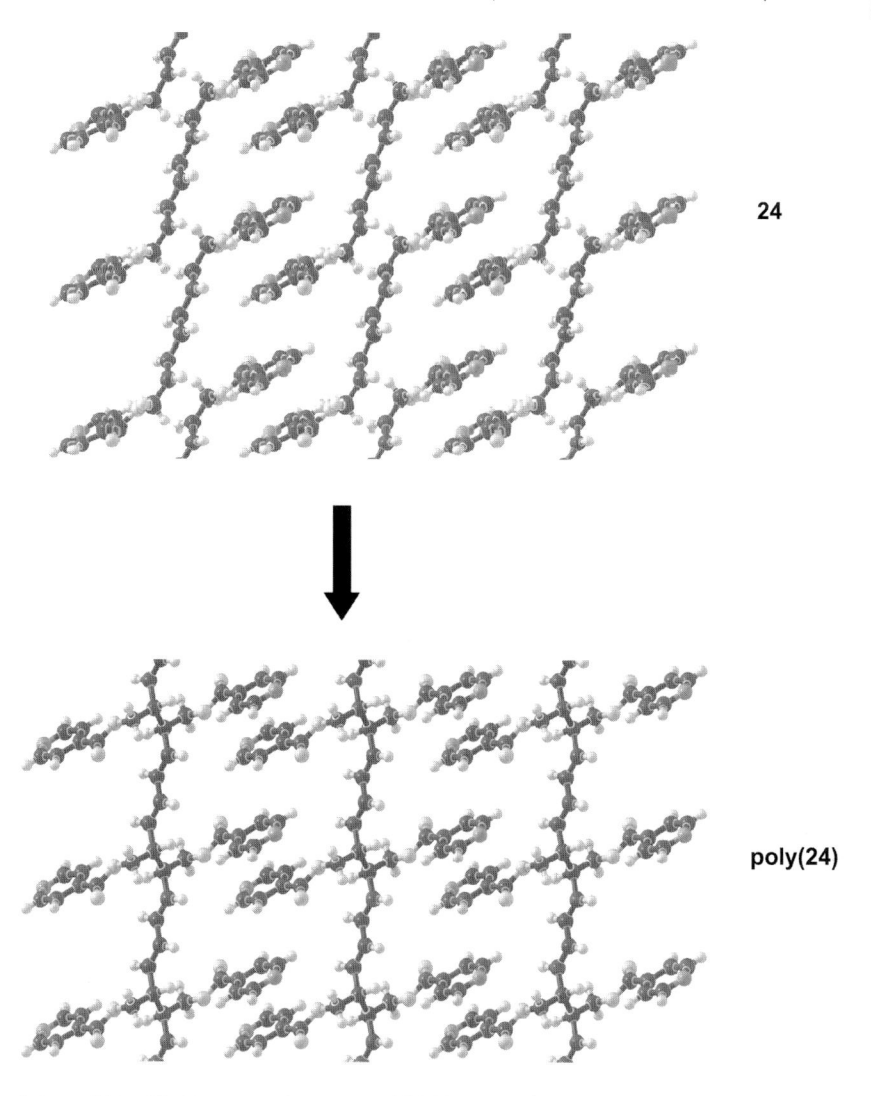

Scheme 5.10. The X-ray crystal structure of **24** ($d_m = 7.22$ Å, $R_{1,6} = 4.24$ Å and $\phi = 44°$) and its polymer [45].

variability in stacking, adjacent pyridine rings were commonly offset with a distance between the center of the rings being 3.80 Å. Neither of these dimensions are correct for triene polymerization, however, a simple translation of this assembly using π–π interactions (Fig. 5.9a) produces an array with a distance of 7.14 Å between equivalent molecules. It is also interesting to note that the stacking distance only depends upon the close contact of the pyridine rings and not the offset of adjacent pyridine rings. It remains to be seen whether or not this type of π–π stacking will be useful for topochemical triene and triyne polymerization.

(a)   (b)

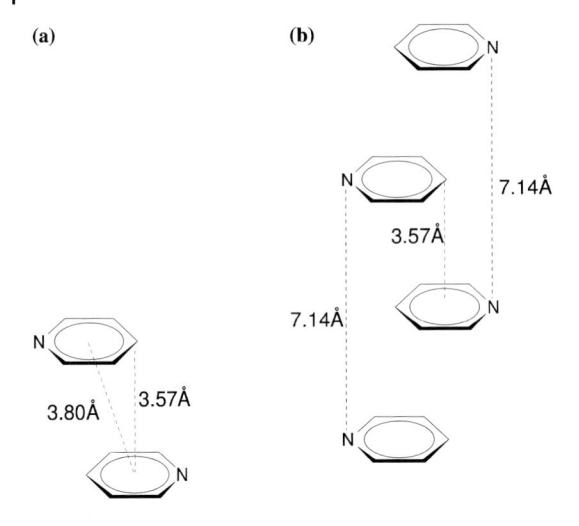

Figure 5.9. The π–π stacking of pyridine rings illustrating how interweaving of pyridines can achieve a suitable spacing for triene polymerization.

5.5
Summary

Single crystal carbon rich polymers such as the polydiacetylenes, polytriacetylenes, polydienes and polytrienes are now known. Although the early examples of these single crystal polymers were serendipitous, recent preparations have had a rational basis. For example, the host–guest strategy has successfully prepared disubstituted polydiacetylenes. It has also resulted in the preparation of the first example of a monosubstituted polydiacetylene and the first example of a polytriacetylene. This method is not always successful. Sometimes the proper conditions for preparing the designed co-crystals cannot be found. Sometimes the designed co-crystals form but the supramolecular features necessary for a topochemical polymerization are not present. The important point is that the host–guest approach is more successful than serendipity. Currently, supramolecular synthesis is just in its early stages of development. As we learn the principles of supramolecular synthesis we will be able to prepare, with a higher degree of success, new examples of single crystal polymers and discover new reactions that can be controlled by supramolecular chemistry.

5.6
Experimental: Selected Procedures

5.6.1
3-[(Pyridin-4-ylmethyl)-amino]-cyclohex-2-enone (20a)

Cyclohexanedione (0.56 g, 5 mmol) and 4-aminomethyl pyridine (0.54 g, 5 mmol) were dissolved in a minimum amount of benzene. The reaction mixture was refluxed and the water produced was removed with a Dean-Stark trap. Removal of the benzene *in vacuo* gave a yellow solid which was recrystallized from acetone to give 1.0 g (91%) of the vinylogous amide, mp = 138–139 °C.

5.6.2
Dodeca-4,6,8-triynedioic acid (17b)

To 4-pentynoic acid (0.24 g, 2.5 mmol), 0.1 eq. of CuCl and 0.3 eq. of $NH_2OH \cdot HCl$ in MeOH were added 3 eq. of propylamine while the mixture was stirred under N_2 in an ice bath. To this reaction mixture was slowly added, over 1 h using a syringe pump, bromoethynyltriethylsilane (0.54 g, 2.5 mmol). The reaction mixture was stirred overnight and the solvent was removed *in vacuo*. Water was added to the residue and the product mixture made acidic using concentrated HCl. The acidic product mixture was extracted with diethyl ether. This organic layer was washed with saturated NH_4Cl aqueous solution and dried with $MgSO_4$. Removal of the solvent *in vacuo* gave 0.34 g (59%) of the diyne as a pale yellow oil. The triethylsilyl group was removed by stirring the above product in 10 mL of diethyl ether containing 1.43 mL of a 1.0 M solution of tetra-n-butylammonium fluoride in tetrahydrofuran. The precipitate formed was dissolved in water and the solution made acidic using concentrated HCl. The mixture was extracted with ether, washed with brine and dried with $MgSO_4$. The solvent was removed *in vacuo* and the crude product recrystallized from petroleum ether giving 0.18 g (95%) of hepta-4,6-diynoic acid, mp = 83–84 °C.

To hepta-4,6-diynoic acid (0.1 g, 0.82 mmol), 0.1 eq. of CuCl and 0.3 eq. of $NH_2OH \cdot HCl$ in MeOH were added 3 eq. of propylamine while the mixture was stirred under N_2 in an ice bath. Using a syringe pump, 7-bromo-hepta-4,6-diynoic acid (0.15 g, 0.85 mmol) was added to the mixture. The reaction solution was stirred overnight and the solvent was removed *in vacuo*. Water was added to the residue and the product mixture made acidic using concentrated HCl. The product mixture was extracted with diethyl ether. This organic layer was washed with saturated NH_4Cl aqueous solution and dried with $MgSO_4$. Removal of the solvent *in vacuo* gave 0.17 g (95%) of dodeca-4,6,8-triynedioic acid (**17b**) which slowly turned red at room temperature and had no melting point.

5.6.3
Poly(20a–17b)

Dodeca-4,6,8-triynedioic acid **17b** (21.8 mg) was dissolved in 8 mL EtOAc. After dodeca-4,6,8-triynedioic acid was completely dissolved 3-[(pyridin-4-ylmethyl)-amino]-cyclohex-2-enone (**20a**) was added followed by 1.0 mL of methanol. The reaction was heated and stirred until the reactants dissolved. The mixture was then allowed to slowly evaporate at room temperature during which time red crystals precipitated, mp 144–145 °C. X-ray crystallography demonstrated these to be the host–guest complex **20a–17b**. Irradiation of this complex with a ^{60}Co γ-ray source (0.8 Mrad h^{-1}) for 80 h resulted in a slow conversion to the poly(triacetylene).

Acknowledgments

We wish to thank the National Science Foundation for financial support, CHE 0300008. Most of the work described in this review has been carried out by a group of excellent Stony Brook undergraduate and graduate students. It has been a pleasure to work with these students and we owe them our thanks.

Abbreviations and Acronyms

d_m crystallographic repeat distance between equivalent atoms in a stack of monomers
d_p crystallographic repeat distance between equivalent atoms in a polymer
ϕ angle between a diacetylene or triacetylene monomer and the principle crystallographic axis
PDA polydiacetylene
PTA polytriacetylene
$R_{1,4}$ distance between reacting atoms of neighboring diacetylene monomers
$R_{1,6}$ distance between reacting atoms of neighboring triacetylene monomers
SCSC single crystal to single crystal transformation

References

1 (a) R. B. Heimann, S. E. Evsyukov, L. Kavan, *Carbyne and Carbynoid Structures, Physics and Chemistry of Materials with Low-Dimensional Structures*, Vol. 21, Kluwer, Dordrecht, 1999; (b) S. Szafert, J. A. Gladysz, *Chem. Rev.* **2003**, *103*, 4175–4205.

2 A. G. MacDiarmid, *Rev. Modern Phys.* **2001**, *73*, 701–712.

3 For reviews of diacetylenes see: (a) A. Sarkar, S. Okada, H. Matsuzawa, H. Matsuda, H. Nakanishi, *J. Mater. Chem.* **2000**, *10*, 819–828; (b) *Nonlinear Optical Properties of Organic Molecules and Crystals*, Vol. 2, D. S. Chemla, J. Zyss (eds.), Academic Press, Orlando, 1987, p. 3; (c) R. R. Chance, M. W. Washabaugh, D. J. Hupe in *Polydiacetylenes*, D. Bloor, R. R. Chance (eds.), Martinus Nijhoff, Dordrecht, Netherlands, 1985, pp. 239–256; (d) H. Shirakawa, T. Masuda, K. Takeda in *The Chemistry of Triple-Bonded Functional Groups*, Supplement C2, S. Patai (ed.), John Wiley & Sons: New York, 1994, pp. 946–1007; (e) *Advances in Polymer Science*, Vol. 63, H.-J. Cantow (ed.), Springer-Verlag, Berlin, 1984.

4 J. Xiao, M. Yang, J. W. Lauher, F. W. Fowler, *Angew. Chem.* **2000**, *112*, 2216–2219; *Angew. Chem. Int. Ed. Engl.* **2000**, *39*, 2132–2135.

5 R. R. Tykwinski, Y. Zhao, *Synlett* **2002**, 1939–1953.

6 R. Kiebooms, R. Menon, K. Lee, Synthesis, in *The Handbook of Advanced Electronic and Photonic Materials and Devices*, Vol. 8, H. S. Halwa (ed.), Academic Press, San Diego, 2001, Ch. 1.

7 G. Wegner, *Z. Naturforsch. B* **1969**, *24*, 824–832.

8 For reviews on polydiacetylenes see: (a) V. Enkelmann, *Adv. Polym. Sci.* **1984**, *63*, 91–136; (b) *Polydiacetylenes: Synthesis, Structure and Electronic Properties*, D. Bloor, R. R. Chance (eds.), NATO ASI Series E 102, Martinus Nijhoff, Dordrecht, 1985.

9. A. J. Heeger, *Angew. Chem.* **2001**, *113*, 2660–2682; *Angew. Chem. Int. Ed. Engl.* **2001**, *40*, 2591–2611.
10. In the crystal structure of diacetylenedicarboxylic acid, dihydrate, both d_m and $R_{1,4}$ have a value of 3.74 Å. J. D. Dunitz, J. M. Robertson, *J. Chem. Soc.* **1947**, 1145–1156.
11. X. Ouyang, F. W. Fowler, J. W. Lauher, *J. Am. Chem. Soc.* **2003**, *125*, 12400–12401.
12. J. W. Steed, J. L. Atwood, *Supramolecular Chemistry*, John Wiley & Sons, Chichester, 2000.
13. G. S. Papaefstathiou, Z. Zhong, L. Geng, L. R. MacGillivray, *J. Am. Chem. Soc.* **2004**, *126*, 9158–9159.
14. Y.-L. Chang, M.-A. West, F. W. Fowler, J. W. Lauher, *J. Am. Chem. Soc.* **1993**, *115*, 5991–6000.
15. J. J. Kane, R.-F. Liao, J. W. Lauher, F. W. Fowler, *J. Am. Chem. Soc.* **1995**, *117*, 12003–12004.
16. C. M. Paleos, D. Tsiourvas, *Angew. Chem* **1995**, *107*, 1837–1855; *Angew. Chem., Int. Ed. Engl.* **1995**, *34*, 1696–1711.
17. T. L. Nguyen, *The Design and Preparation of Two-Dimensional Hydrogen Bonded Networks*, Ph.D. Thesis, SUNY Stony Brook, 1998.
18. J. M. Pigos, Z. Zhu, J. L. Musfeldt, *Chem. Mater.* **1999**, *11*, 3275–3278.
19. J. L. Foley, L. Li, D. J. Sandman, M. J. Vela, B. M. Foxman, R. Albro, C. J. Eckhardt, *J. Am. Chem. Soc.* **1999**, *121*, 7262–7263.
20. T. Tran, *The Preparation of Polydiacetylenes Using Supramolecular Chemistry*, MSc Thesis, SUNY Stony Brook, 2001.
21. O. J. Orchard, S. K. Tripathy, *Macromolecules* **1986**, *19*, 1844–1850.
22. S. E. Zutaut, S. P. McManus, M. Jalali-Heravi, *Chem. Mater.* **1996**, *8*, 2223–2231.
23. A. Sarkar, S. Okada, H. Nakanishi, H. Matsuda, *Macromolecules* **1998**, *31*, 9174–9180.
24. A. Sarkar, N. B. Kodali, M. B. Kamath, L. P. Bhagwat, S. S. Talwar, *J. Macromol. Sci., Pure Appl. Chem.* **1999**, *36*, 211–224.
25. A. Masaki, H. Matsuda, S. Okada, S. Shimada, H. Takeda, H. Shinbo, K. Hayamizu, F. Nakanisi, H. Nakanishi, *Mol. Cryst. Liq. Cryst. A* **1995**, *267*, 15–20.
26. (a) S. Curtis, *Designed Supramolecular Systems for the Polymerization of Diacetylenes*, Ph.D. Thesis, SUNY Stony Brook, 2003; (b) N. Le, *The Design and Preparation of Polydiacetylenes and Open Molecular Networks through a Convergent Host-Guest Strategy*, MSc Thesis, SUNY Stony Brook, 2000.
27. There have been attempts to polymerize terminal diacetylenes. For example: K. Inoue, N. Koga, H. Iwamua, *J. Am. Chem. Soc.* **1991**, *113*, 9803–9810.
28. For an excellent discussion of a comparison between crystal engineering and crystal structure prediction see: B. Moulton, M. J. Zaworotko, *Chem. Rev.* **2001**, *101*, 1629–1658.
29. For previous attempts at polytriacetylene polymerization see: (a) R. H. Baughman, K. C. Yee, *J. Polym. Sci. Macromol. Rev.* **1978**, *13*, 219–239; (b) A. Sarkar, S. Okada, H. Matsuzawa, H. Matsuda, H. Nakanishi, *J. Mater. Chem.* **2000**, *10*, 819–828; (c) J. Kiji, J. Kaiser, G. Wegner, R. C. Schulz, *Polymer* **1973**, *14*, 433–439; (d) Ref. 30.
30. V. Enkelmann, *Chem. Mater.* **1994**, *6*, 1337–1340.
31. The crystal structure of a small polytriacetylene oligomer is known (CSD reference code PIRROX). This structure suggests a molecular repeat distance of 7.39 Å of the monomer would be commensurate with the polymer. (J. Anthony, C. Boudon, F. Diederich, J.-P. Gisselbrecht, V. Gramlich, M. Gross, M. Hobi, P. Seiler, *Angew. Chem.* **1994**, *106*, 794–798; *Angew. Chem., Int. Ed. Engl.* **1994**, *33*, 763–766). Macromodel version 6.2 using the MM2 force field was used in modeling calculations. F. Mohamadi, N. B. G. Richards, W. C. Guida, R. Liskamp, M. Lipton, C. Caufield, G. Chang, T. Hendrickson, W. C. Still, *J. Comput. Chem.* **1990**, *11*, 440–467. Also see Ref. 30.

32 M. J. Edelmann, S. Odermatt, F. Diederich, *Chimia* **2001**, *55*, 132–138.

33 P. A. Cranwell, R. D. Haworth, *Tetrahedron* **1971**, *27*, 1831–1837.

34 K.-S. Huang, D. Britton, M. C. Etter, *Acta Crystallogr. Sect. C* **1995**, *51*, 1661–1663.

35 For a general discussion of vinylogous amide self-assembly see: V. Bertolasi, P. Gilli, V. Ferretti, G. Gilli, *Acta Cryst. B* **1998**, *54*, 50–65.

36 F. W. Fowler, J. W. Lauher, *J. Phys. Org. Chem.* **2000**, *13*, 850–857.

37 H. R. Bhattacharjee, A. F. Preziosi, G. N. Patel, *J. Polym. Sci. Polym. Symp.* **1984**, *71*, 259–270.

38 *Conjugated Polymers and Related Materials: The Interconnection of Chemical and Electronic Structure*, W. R. Salaneck, I. Lundström, B. Rånby (eds.), Oxford University Press, Oxford, 1993.

39 For a review see: A. Matsumoto, T. Odani, *Macromol. Rapid Commun.* **2001**, *22*, 1195–1215.

40 A. Matsumoto, *Top. Curr. Chem.* **2005**, *254*, 263–306.

41 K. Tashiro, A. N. Zadorin, S. Saragai, T. Kamae, A. Matsumoto, K. Yokoi, S. Aoki, *Macromolecules* **1999**, *32*, 7946–7950.

42 A. Matsumoto, H. Nakazawa, *Macromolecules* **2004**, *37*, 8538–8547.

43 T. Tanaka, A. Matsumoto, *J. Am. Chem. Soc.* **2002**, *124*, 9676–9677.

44 T. Hoang, *A Host-Guest Strategy to Supramolecular Synthesis of Linear 1,6-Polytrienes*, MSc Thesis, SUNY Stony Brook, 2002.

45 T. Hoang, J. W. Lauher, F. W. Fowler, *J. Am. Chem. Soc.* **2002**, *124*, 10656–10657.

46 C. Janiak, *J. Chem. Soc., Dalton Trans.* **2000**, 3885–3896.

6
Chiral Carbon-rich Macrocycles and Cyclophanes

Katie Campbell and Rik R. Tykwinski

6.1
Introduction

Over the last three decades, chiral carbon-rich compounds have evolved into an interesting and useful class of materials due to the many unique properties that result from the installation of conjugated chromophores within a chiral framework. Chiral carbon-rich materials possess distinct chiroptical and electronic properties that may prove useful for a number of intriguing applications ranging from optically active liquid crystals to nonlinear optical materials. Chiral carbon-rich compounds also represent a potentially valuable scaffold for use in asymmetric reactions and catalysis.

A particularly fascinating subset of this broad class of materials is that of chiral carbon-rich macrocycles and cyclophanes. The presence of a preorganized, chiral cavity that may be functionalized through synthetic manipulations allows these cyclic macromolecules to be utilized as materials for enantioselective sensing, chiral guest recognition, chiral separations, as well as ligands for asymmetric catalysis. In addition to their myriad of potential applications, chiral macrocycles are also attractive targets due to the synthetic challenge that they represent. Finally, many also demonstrate aesthetically pleasing structures, which naturally adds to their allure. Shape-persistent macrocycles in general are widely recognized as demanding synthetic targets [1]. Chirality, and the resulting issues of enantio- and diastereoselectivity, often introduces a greater layer of complexity to the synthesis and characterization of these fascinating molecules.

6.2
Strategies for Achieving Chiral Macrocycles or Cyclophanes

The chiral macrocycles and cyclophanes that have been included in this Chapter have loosely been classified with respect to the synthetic strategy employed to provide the asymmetric, cyclic product. One of the simplest and most economical methods for introducing chirality involves the use of only achiral building blocks to

Carbon-Rich Compounds. Edited by Michael M. Haley and Rik R. Tykwinski
Copyright © 2006 WILEY-VCH Verlag GmbH & Co. KGaA, Weinheim
ISBN: 3-527-31224-2

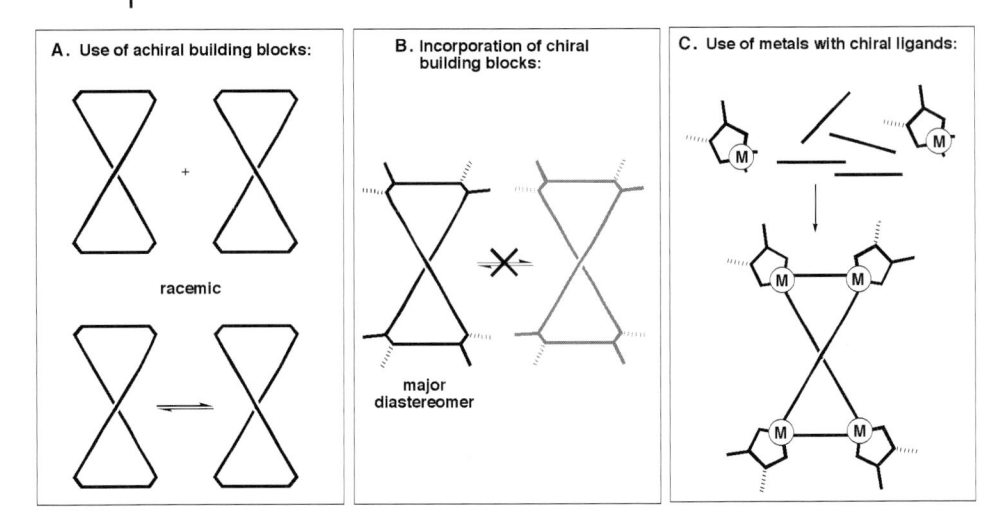

Figure 6.1. Schematic representation of possible synthetic approaches to chiral carbon-rich macrocycles.

construct a helically chiral species. This approach takes advantage of geometrical constraints and/or steric interactions between the building blocks that form the core of the macrocycle to provide a conformational twist, resulting in helical chirality (Fig. 6.1A). While this strategy is attractive in that it precludes the use of expensive, chiral building blocks, it nevertheless provides mainly racemic products that are often in rapid equilibrium at room temperature. Homochiral macrocycles can, however, be achieved in an enantio- or diastereoselective fashion through the application of a second strategy, which relies upon the use of chiral, and in many cases enantiopure, building blocks (Fig. 6.1B). This approach depends on the incorporation of chiral components (e.g., binaphthyl) to establish a chiral macrocyclic backbone. In many cases, the presence of chiral functional groups results in chirality transfer to the macrocyclic core, amplifying helical chirality. The final synthetic strategy under consideration in this Chapter involves the use of metal–ligand interactions. Chiral ligands may, through the application of self-assembly processes or through the formation of covalent metal–ligand bonds, be used to efficiently form chiral, metal-containing macrocycles.

The following discussion is intended to provide an introduction to the latest innovations in the design and synthesis of chiral carbon-rich macrocycles while providing some historical context of their development and applications. The systems under review are limited primarily to chiral macrocycles that are conjugated or that possess highly conjugated substructures. Excluded from the discussion are, for example, porphyrins [2, 3], cyclodextrins [4] and calixarenes [5], either because they do not fit within the context of our discussion or because they have been extensively reviewed elsewhere.

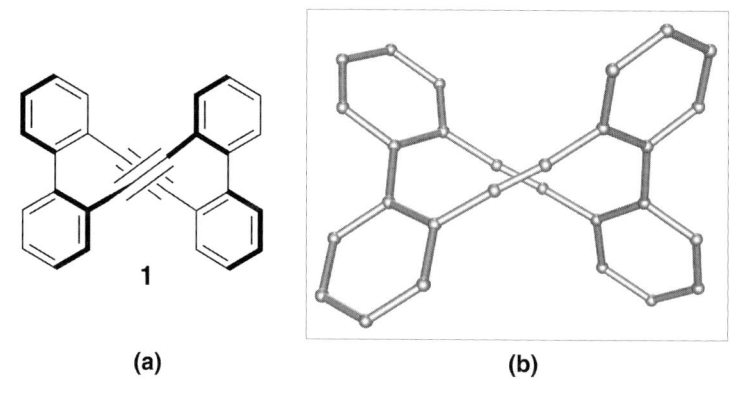

Figure 6.2. Molecular structure (a) and solid-state structure (b) of **1**.†

6.2.1
Helical Chirality Using Achiral Building Blocks

One of the simplest ways to access chiral macrocycles involves the use of achiral building blocks and takes advantage of 'molecular entanglements' to bias the cyclic core such that a helical conformation is preferred. An early example of a carbon-rich macrocycle possessing helical chirality is racemic compound **1**, prepared by Staab and coworkers in 1972 (Fig. 6.2) [6]. The synthesis of **1** was inspired by an interest in the intramolecular interactions of transannular triple bonds. Crystallographic analysis later confirmed the helical structure of **1** (Fig. 6.2) and the distance between the centers of the triple bonds was measured to be 2.851 Å [7].

Youngs and coworkers reported the synthesis of a hexa-benzannulene in the early 1990s [8]. Annulene **2** was prepared in low yield via the Castro–Stephens coupling of *ortho*-iodophenylacetylene (Fig. 6.3). X-ray crystallographic analysis confirmed the helical conformation of **2** in the solid-state and showed that two of the diphenylacetylene moieties are nearly eclipsed in this somewhat strained, distorted-boat conformation.

In 1997 Vollhardt and coworkers reported the synthesis of dehydroannulenes **3** and **4** (Fig. 6.4) [9]. The authors showed that racemic **3** adopts a nonplanar, chiral conformation in the solid-state. Interestingly, crystallographic analysis indicated that annulene **3** was essentially free from strain. Calculations suggested a barrier to enantiomerization of **3** through a planar intermediate of 7.5 kcal mol^{-1}. The tetrakis-isopropyl analog **4** was prepared, and the presence of diastereotopic methylene protons allowed for the direct measurement, via variable temperature NMR spectroscopy, of this energy barrier. The measured value of 9.3 kcal mol^{-1} was in close agreement with the calculated value. The authors also reported that the explosive decomposition of **3** upon heating to 245 °C led to the formation of carbon rich "bucky-onions".

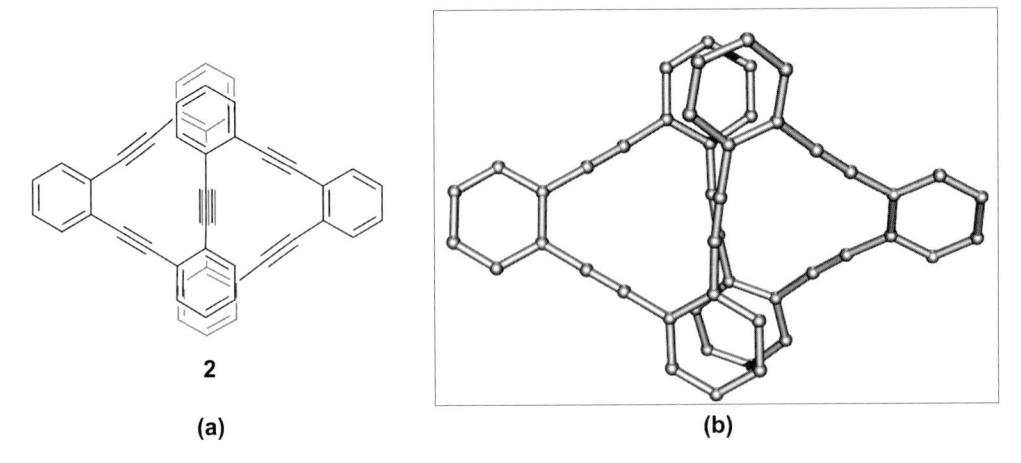

Figure 6.3. Molecular structure (a) and solid-state structure[†] (b) of benzannulene **2**.

Haley and coworkers have prepared the analogous expanded dehydrobenzannulenes as outlined in Scheme 6.1 [10]. Triynes **5** and **6** were oxidatively homocoupled using a unique, one-pot desilylation/dimerization procedure to provide hexaynes **7** and **8**, respectively. Dehydrobenzannulene **9** and the solubilized derivative **10** were achieved in very good yields via the Cu-mediated intermolecular oxidative homocoupling of the appropriate hexayne **7** or **8**, subsequent to desilylation using TBAF. The authors noted the remarkable stability of these [32]annulene derivatives.

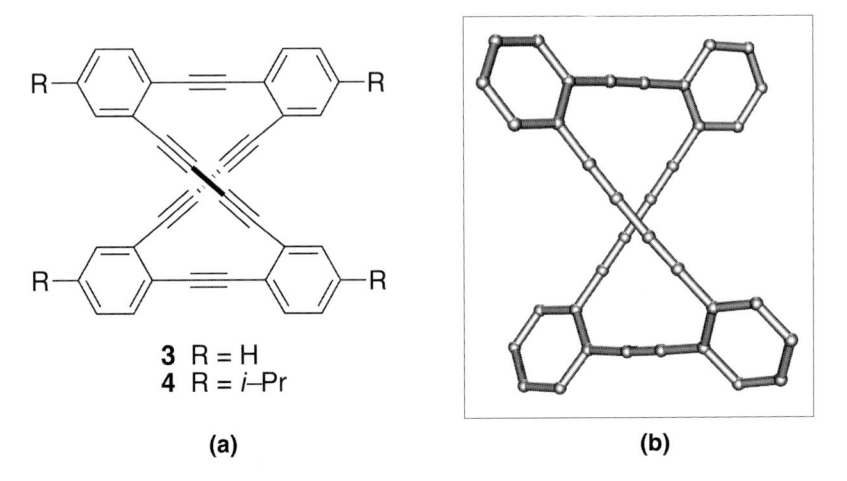

3 R = H
4 R = i–Pr

Figure 6.4. Molecular structure of helical annulenes **3** and **4** (a) and solid-state structure[†] of compound **3** (b).

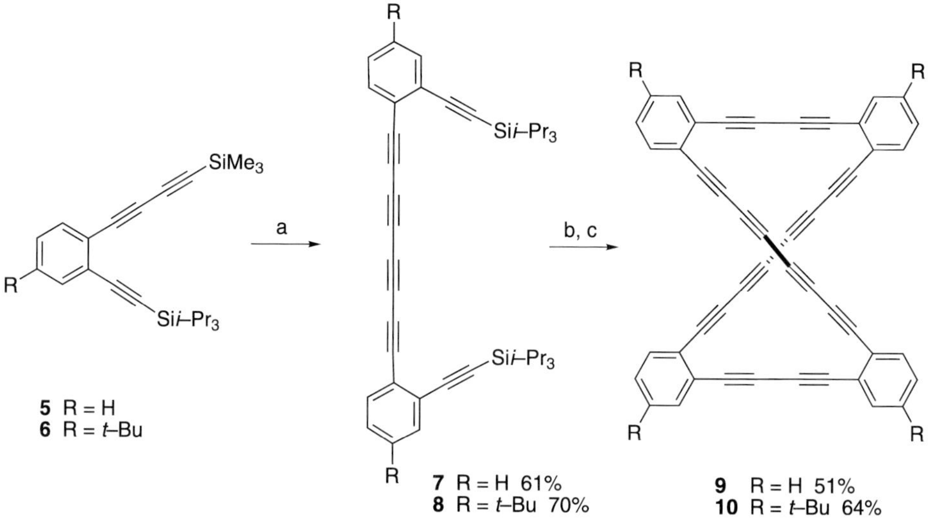

Scheme 6.1. Synthesis of annulenes **9** and **10**. Reagents and conditions: (a) K$_2$CO$_3$, Cu(OAc)$_2$ H$_2$O, py, MeOH; (b) TBAF, EtOH, THF; (c) CuCl, Cu(OAc)$_2$, py.

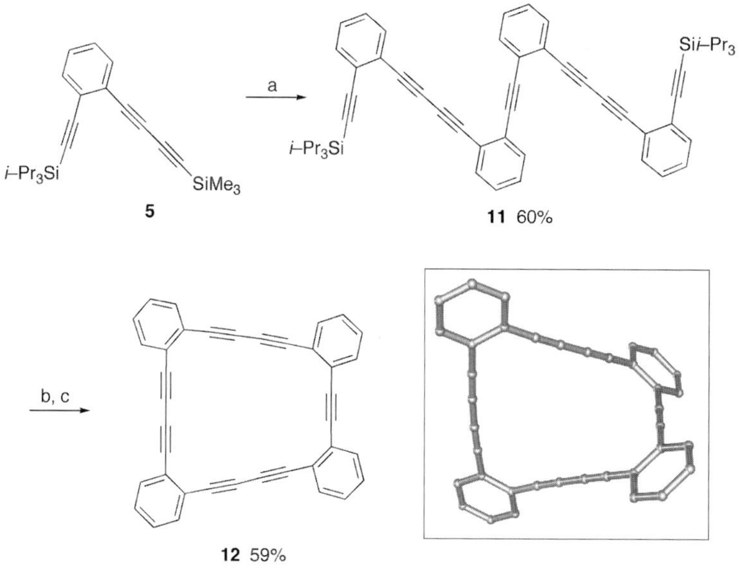

Scheme 6.2. Synthesis and solid-state structure[†] (inset) of annulene **12**. Reagents and conditions: (a) bis(2-iodophenyl)ethyne, [PdCl$_2$PPh$_3$)$_2$], CuI, KOH, H$_2$O, Et$_3$N; (b) TBAF, MeOH, THF; (c) CuCl, Cu(OAc)$_2$, py, 60 °C.

Utilizing the same achiral building block **5**, Haley and coworkers have also synthesized the less-symmetrical [22]annulene (Scheme 6.2) [11]. Triyne **5** was converted to oligomer **11** via a one-pot desilylation/alkynylation reaction with bis(2-iodophenyl)ethyne. Compound **11** was subsequently desilylated and subjected to intramolecular oxidative homocoupling conditions to provide the chiral annulene **12** in 59% yield. X-ray crystallographic analysis of **12** showed that the helical macrocycle adopts a distorted saddle-shaped conformation in the solid-state that is relatively free from strain.

Two particularly interesting examples, also by the Haley group, are the mixed paracyclophane/dehydrobenzannulenes (Scheme 6.3) [12]. The parent paracyclophane **13** was converted to dialdehyde **14**, which was subsequently alkynylated to

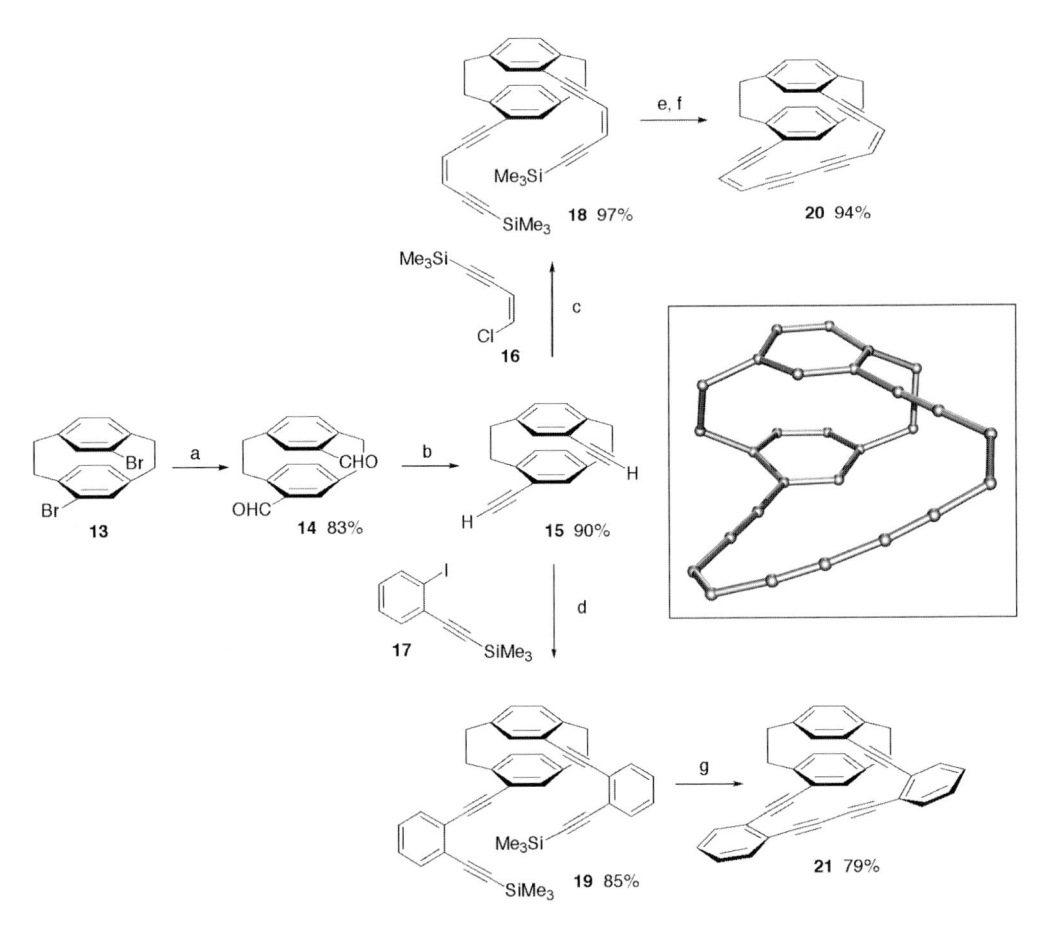

Scheme 6.3. Synthesis of paracyclophane/dehydrobenzannulene hybrids **20** and **21** and solid-state structure† of **20** (inset). Reagents and conditions: (a) (i) BuLi, THF, (ii) N-formylpyrrolidine; (b) dimethyl 1-diazo-2-oxopropylphosphonate, K_2CO_3, MeOH; (c) [Pd(PPh$_3$)$_4$], CuI, PrNH$_2$, THF; (d) [Pd(PPh$_3$)$_4$], CuI, Et$_3$N, THF; (e) K_2CO_3, MeOH, THF; (f) Cu(OAc)$_2$, MeCN; (g) K_2CO_3, CuCl, Cu(OAc)$_2$, py, MeOH.

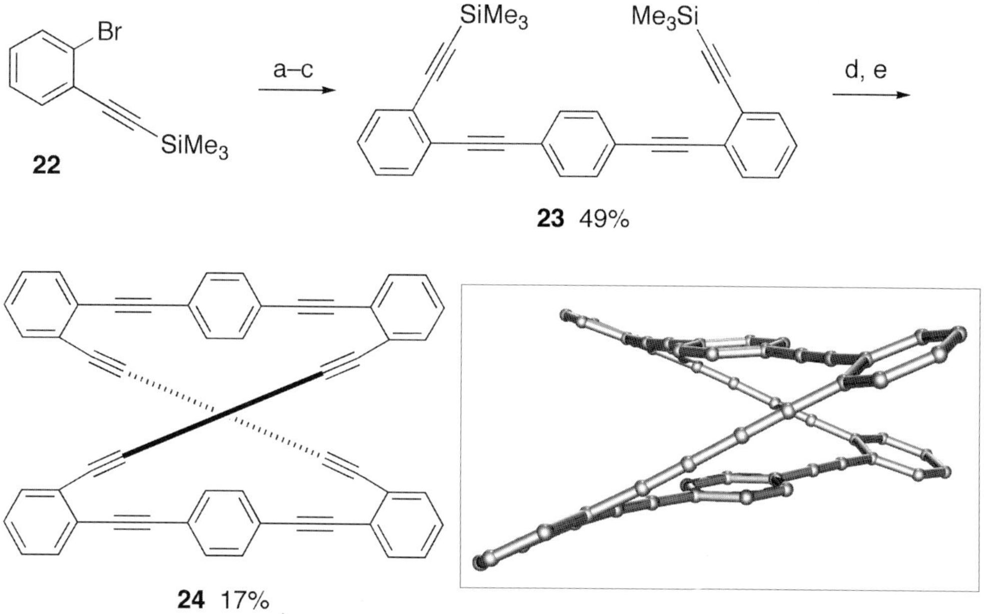

6.2 Strategies for Achieving Chiral Macrocycles or Cyclophanes | 235

Scheme 6.4. Synthesis and solid-state structure† (inset) of cyclophane **24**. Reagents and conditions: (a) BuLi, THF, −78 °C; (b) ZnBr$_2$, THF, 0 °C; (c) 1,4-bis(bromoethynyl)-benzene, [Pd(PPh$_3$)$_4$], reflux; (d) K$_2$CO$_3$, THF, MeOH, H$_2$O; (e) Cu(OAc)$_2$, py/Et$_2$O (3:1).

give cyclophane **15**. Reaction with the appropriate vinyl or aryl halide (**16** or **17**), using Pd-catalysis, led to the formation of compounds **18** and **19**, respectively. The synthesis of **20** required stepwise deprotection, followed by ring-closing using Cu-catalysis, providing the hybrid cyclophane in a remarkable 94% yield. For the other cyclophane **21**, compound **19** was deprotected and cyclized using a one-pot procedure affording the cyclic product in 79% yield. X-ray crystallographic analysis of racemic **20** confirmed the helical structure and indicated only a small distortion from the normal parent paracyclophane geometry.

Fallis and coworkers have reported the synthesis of an acetylenic cyclophane that also possesses helical chirality as established by crystallographic analysis (Scheme 6.4) [13]. The bromoethynylbenzene derivative **22** was elaborated via lithium-halogen exchange, treatment with ZnBr$_2$, and coupling of the resultant organozincate with 1,4-bis(bromoethynyl)benzene using Pd-catalysis to afford oligomer **23**. Desilylation of **23** and Eglinton homocoupling led to the formation of cyclophane **24**, which was isolated in a modest 17% yield over the two steps.

Using a similar synthetic protocol, Fallis and coworkers also prepared cyclophane **25** (Fig. 6.5) [14]. Crystallographic analysis of this macrocycle confirmed that **25** adopted an essentially unstrained, helical conformation in the solid-state.

25
(a)   (b)

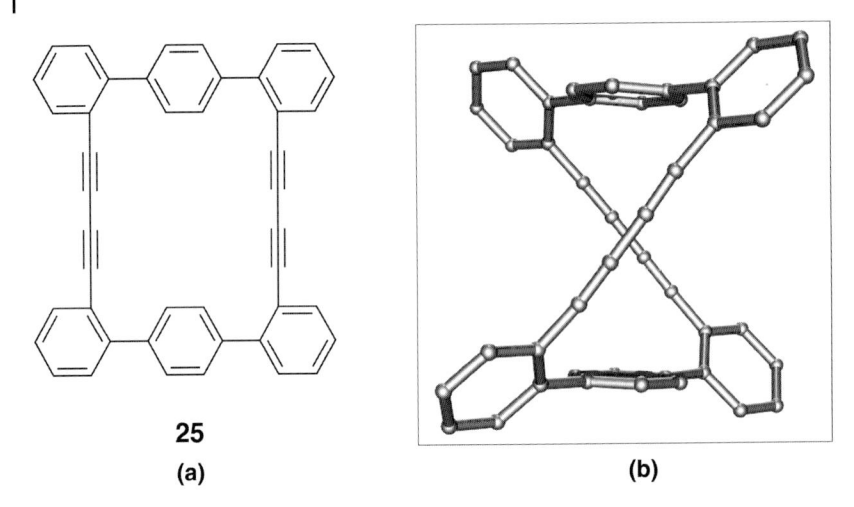

Figure 6.5. Molecular structure (a) and solid-state structure[†] of helical cyclophane **25** (b).

The same authors also targeted the synthesis of cyclophanes **26–29** that, due to the presence of long alkyl side chains, were expected to possess liquid crystalline properties [13]. Macrocycles **26–29** were isolated in slightly higher yields than **24** and **25**, presumably the result of the introduction of the solubilizing side chains. The helical racemic cyclophanes **26–29** were evaluated for liquid crystalline behavior, but these analyses were inconclusive due to the fact that all compounds were a mixture of regioisomers.

26 R = CH$_3$
27 R = CO$_2$Me

28 R = CH$_3$
29 R = CO$_2$Me

The presence of pendant side chains can, in some cases, be the actual source of chirality in a conformationally rigid macrocycle. For example, Tsuji and coworkers

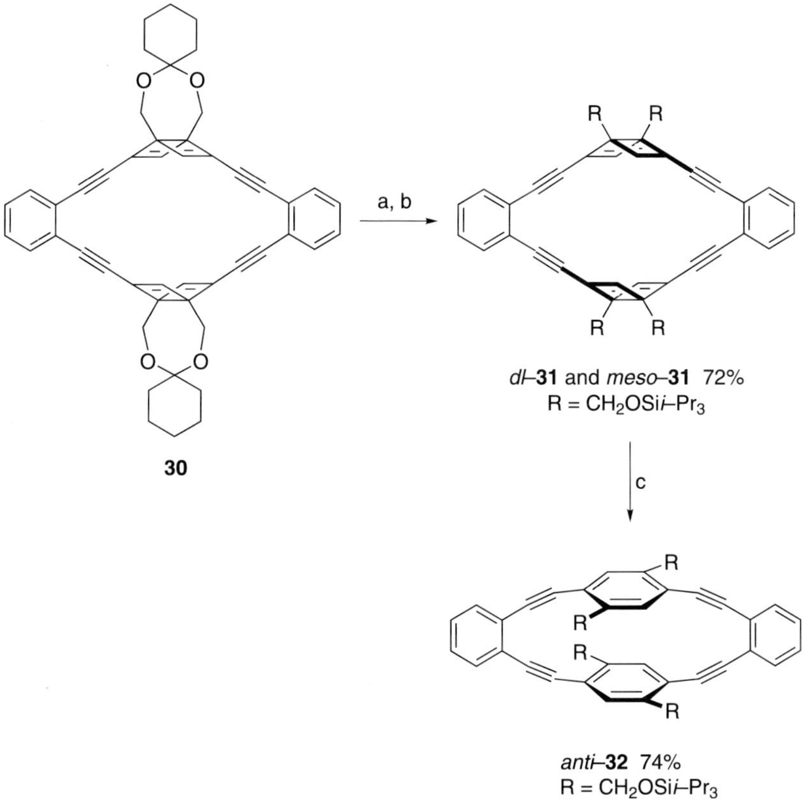

Scheme 6.5. Synthesis of oligophenylene macrocycles **31** and **32**. Reagents and conditions: (a) 1 M HCl, THF; (b) i-Pr$_3$SiOTf, Et$_3$N; (c) $h\nu$, CH$_2$Cl$_2$, 12 °C.

have recently reported the synthesis of phenylene macrocycles that are chiral, not due to a helical twist in the macrocyclic backbone but as a result of the regiochemistry of the pendant sides chains (Scheme 6.5) [15]. Macrocycle **30**, present as a mixture of *meso-* and *dl-*isomers, was converted to compound **31** via acid hydrolysis followed by silylation. The *meso-* and *dl-*isomers of macrocycle **31** were separated and *dl-***31** was subjected to irradiation to provide the chiral macrocycle *anti-***32**. Interestingly, under the same reaction conditions, *meso-***31** provided only decomposition products. The structure of *anti-***32** was confirmed crystallographically.

Due to an interest in molecules that could lead to the generation of new fullerenes, Fallis and coworkers have prepared helical cyclophanes that feature 60 carbon atoms [16]. The authors relied upon a combination of Negishi and Sonogashira cross-coupling strategies in combination with acetylenic homocoupling as outlined in Scheme 6.6. Reaction of **16** with BuLi at low temperature followed by trapping with ZnBr$_2$ gave the organozincate **33**, which was then

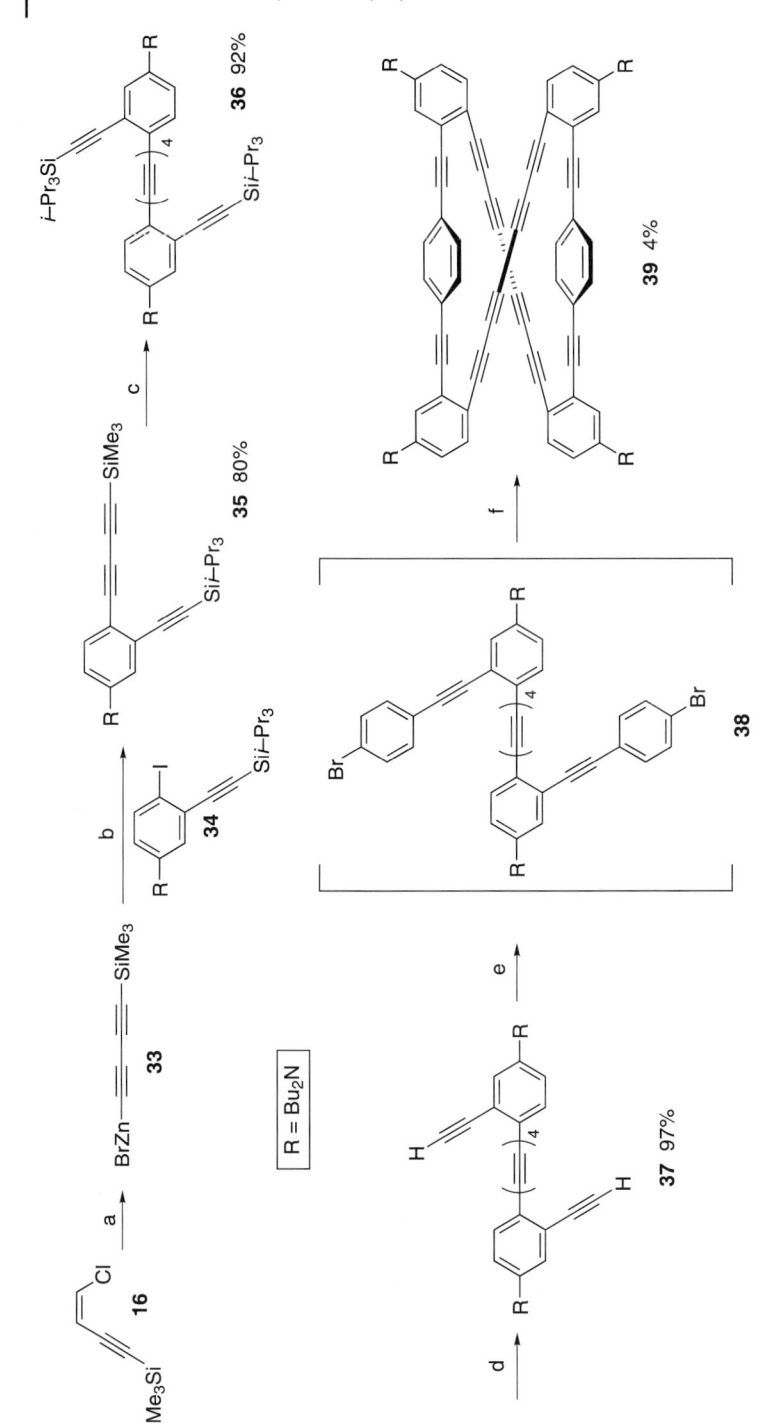

Scheme 6.6. Synthesis of C_{60} helical cyclophane **39** (R = Bu_2N). Reagents and conditions: (a) (i) BuLi (2 equiv.), THF, −78 °C, (ii) $ZnBr_2$, 0 °C; (b) **34**, [$Pd(PPh_3)_4$], reflux; (c) K_2CO_3, $Cu(OAc)_2$, MeOH, py; (d) TBAF, THF; (e) 1-bromo-4-iodobenzene, [$PdCl_2(PPh_3)_2$], CuI, Et_3N, THF, reflux; (f) **37**, [$PdCl_2(PhCN)_2$], t-Bu_3P.

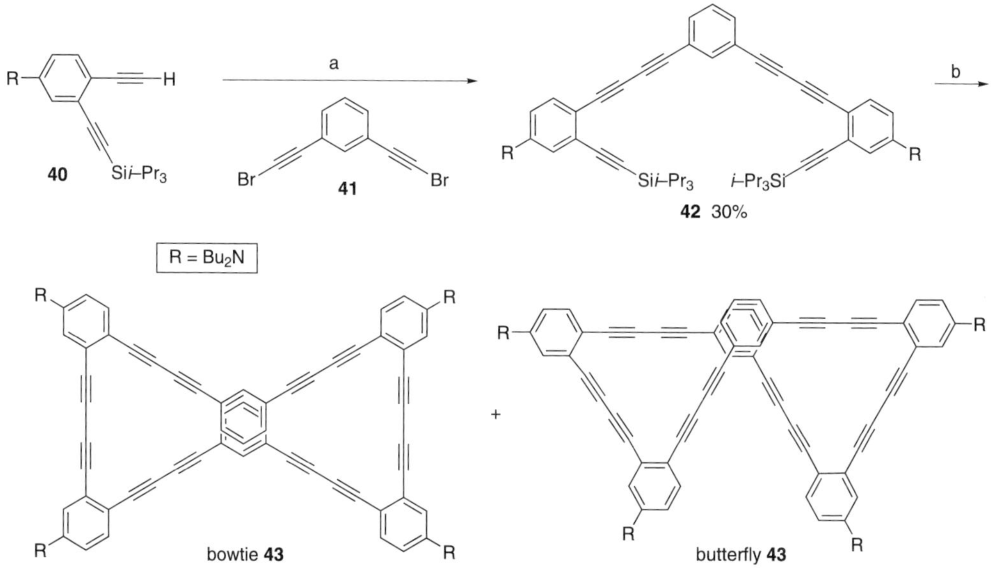

6.2 Strategies for Achieving Chiral Macrocycles or Cyclophanes | 239

Scheme 6.7. Synthesis of helical C$_{60}$ cyclophanes **43** (R = Bu$_2$N). Reagents and conditions: (a) [Pd$_2$(dba)$_3$], CuI, 1,2,2,6,6-pentamethylpiperidine, THF; (b) TBAF, Cu(OAc)$_2$, Et$_2$O, py.

cross-coupled with **34** to give triyne **35**. Dimerization of **35** was achieved via the *in situ* desilylation/acetylenic homocoupling protocol developed by Haley [10]. Desilylation of **36** gave the oligomer **37**, and Sonogashira cross-coupling reaction with 1-bromo-4-iodobenzene gave oligomer **38**, which was not isolated due to instability. The helicene **39** was achieved in low yield via a two-fold cross-coupling reaction between **38** and the desilylated oligomer **37**. The authors noted that the use of the more effective ligand system [PdCl$_2$(PhCN)$_2$] and Pt-Bu$_3$ was necessary to promote this final, ring-closing transformation. Molecular modeling was consistent with the proposed helical conformation of cyclophane **39**.

The same authors prepared the C$_{60}$ acetylenic cyclophanes detailed in Scheme 6.7, which possesses *meta*-substituted benzene rings at the core [16]. A Cadiot-Chodkiewicz type cross-coupling reaction of terminal acetylene **40** with dibromide **41** led to the formation of compound **42**. Subsequent to desilylation, the oligomer **42** was cyclodimerized to provide the cyclophanes **43** in 15% combined yield. The C$_{60}$ cyclophane **43** was achieved as two atropisomers, the "bowtie" and "butterfly" conformations as a 3:1 mixture, respectively, which were separable via size-exclusion chromatography. The authors stated that there was a significant barrier to interconversion between the two isomers, as evidenced by conformational stability at temperatures up to 100 °C.

Clearly the installation of rigid *ortho*-substituted benzene rings into the backbone of a carbon-rich macrocycle can provide the conformational rigidity re-

44 H
45 R = O(CH$_2$)$_9$CH$_3$
46 R = O(CH$_2$)$_{13}$CH$_3$

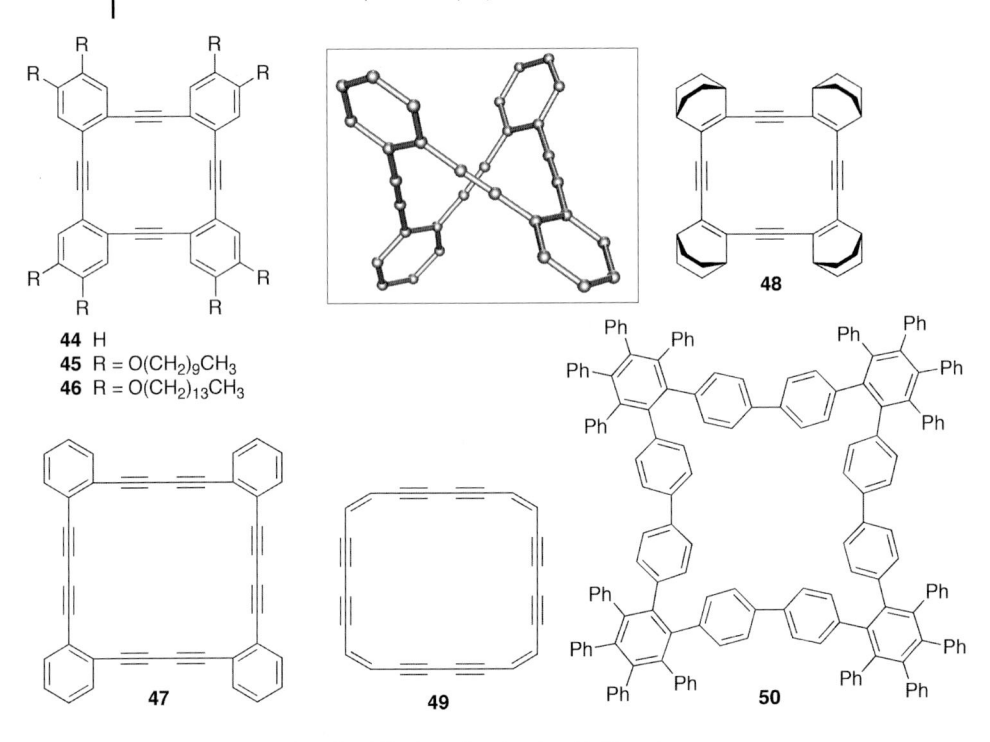

Figure 6.6. Examples of saddle shaped molecules **44–50** and solid-state structure[†] of compound **44** (inset).

quired to enforce helical chirality. There are also numerous examples of similar macrocycles possessing this substitution pattern that may be highly symmetric and/or conformationally mobile such that they are not intrinsically chiral. Nevertheless, many of these macrocycles can achieve asymmetric conformations in either the solid or solution state. For example, macrocycles **44–46** and **47**, prepared by the groups of Youngs [8, 17] and Haley [11, 18], respectively, each possess *ortho*-diethynylbenzene substitution and can presumably adopt a low-energy, saddle-type (but achiral) conformation (Fig. 6.6). A slight deformation of this saddle-conformation (the result of crystal packing effects, for example) would, however, result in a helically chiral conformation, as shown for **44** (Fig. 6.6, inset). Macrocycle **48**, synthesized by Komatsu and coworkers, possesses the same geometric motif, and therefore the same conformational possibility [19]. Other interesting examples, among many others, that fall into this category include the [24]dehydroannulene **49** prepared by Sondheimer and coworkers in 1970 [20], and the polyphenylene macrocycle **50**, comprised entirely of aromatic rings, that was synthesized by the group of Pascal [21].

While there is an abundance of helical cyclophanes based on the incorporation of *ortho*-substituted aromatic rings, there are far fewer examples of helically chiral

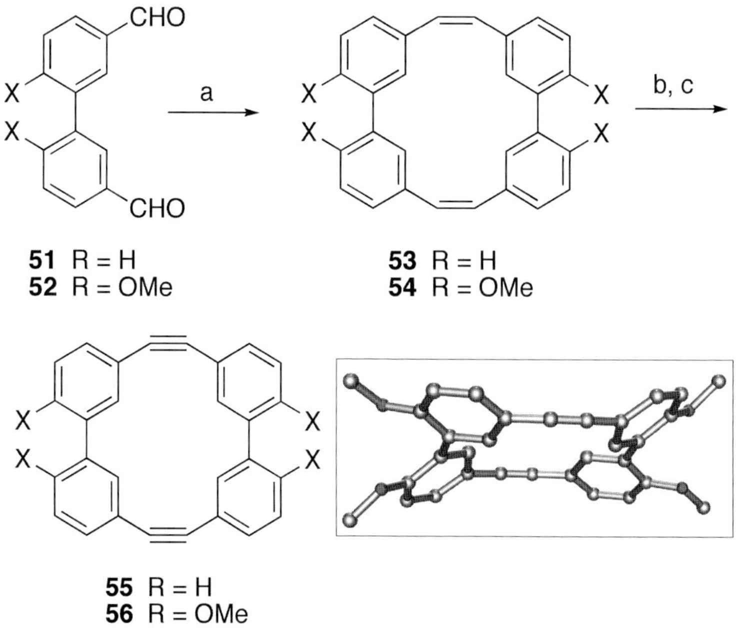

Scheme 6.8. Synthesis of metacyclophanes **55** and **56** and solid-state structure† of **56** (inset). Reagents and conditions: (a) TiCl$_4$, Zn–Cu, DME; (b) Br$_2$, benzene; (c) t-BuOK, THF.

macrocycles that feature purely *meta*-linkages. Recent examples are the metacyclophanes prepared by Oda and Kawase (Scheme 6.8) [22]. Their syntheses relied upon a McMurray coupling of dialdehydes **51** and **52** that yielded dienes **53** and **54**, respectively. These macrocycles were then converted to the desired cyclophanes **55** and **56** via bromination followed by dehydrohalogenation. Both cyclophanes have two diastereomeric conformers, the *meso* and *dl*. Crystallographic analysis of **56** showed that the *meso*-conformation is adopted in the solid-state, but variable temperature NMR spectroscopy revealed that isomerization between the two conformers occurs at temperatures as low as −40 °C.

Diene macrocycle **53** had, in fact, been previously synthesized in 1977 by Wennerström and Thulin via a Wittig reaction [23, 24]. It was subsequently irradiated in the presence of iodine to give the bi-4,5-phenanthrene cyclophane **57** (Scheme 6.9). Oxidation of cyclophane **57** by heating to 140 °C in a melt composed of NaCl and AlCl$_3$ provided the fused helicene **58** [23]. Attempts to further oxidize **58** to the fully closed derivative, however, were not successful.

Tsuji and coworkers have also synthesized a chiral macrocycle **59** constructed from *meta*-substituted aryl rings [15]. Only one rotamer, *anti*-**59**, is chiral, however, and the authors showed that the *para*-phenylene units can freely rotate, leading to fast interconversion.

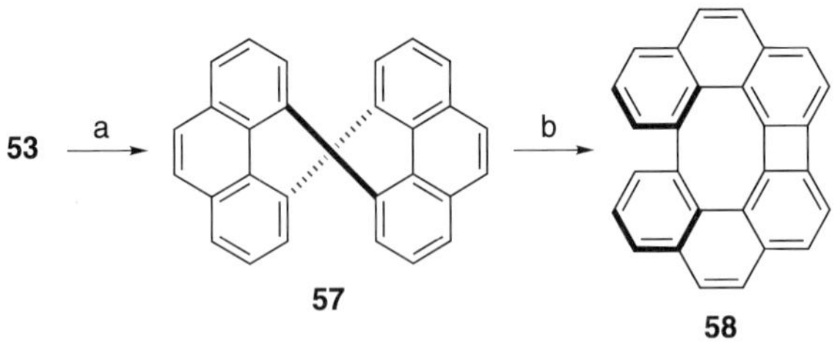

Scheme 6.9. Synthesis of phenanthrene macrocycle **57** and fused **58**. Reagents and conditions: (a) 254 nm, I$_2$, cyclohexane; (b) NaCl, AlCl$_3$, 140 °C.

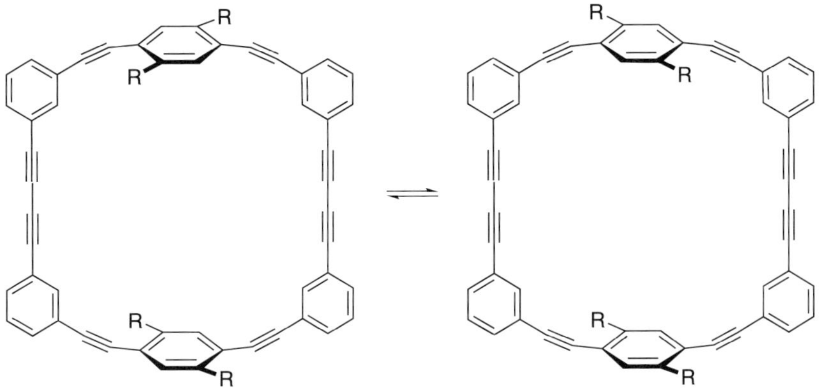

anti–**59**
R = CH$_2$OSi*t*–BuMe$_2$

syn–**59**
R = CH$_2$OSi*t*–BuMe$_2$

One of the key components employed in the assembly of helical macrocycles has been the aryl-acetylene moiety, with the rigid geometry of this linkage used to provide predictable access to helical structures. There are also several interesting examples, of helical macrocycles that achieve conformational rigidity due to the incorporation of *cis*-alkenes in place of the typical *ortho*-substituted benzene building block. A good example of this is the paracyclophane **60**, synthesized by Cram and Dewhirst in 1959 [25]. Subsequent to this initial report, numerous studies have expanded on this general framework through the insertion of spacers between the two *cis*-alkenes. For example, Wennerström and coworkers have reported the synthesis of a series of helical cyclophanes based on *cis*-alkenes as building blocks and using Wittig chemistry to effect cyclization [26]. These cyclophanes (**61–64**) possess varying degrees of conformational flexibility dependent upon the tether employed.

Fallis and coworkers have also explored this concept through the preparation of larger paracyclophanes containing the *cis*-alkenyl linkage, christening them revolveneynes (Scheme 6.10) [27]. Their synthesis also relied upon Wittig chemistry to convert terephthalaldehyde to dibromide **65**. Compound **65** was then elaborated to

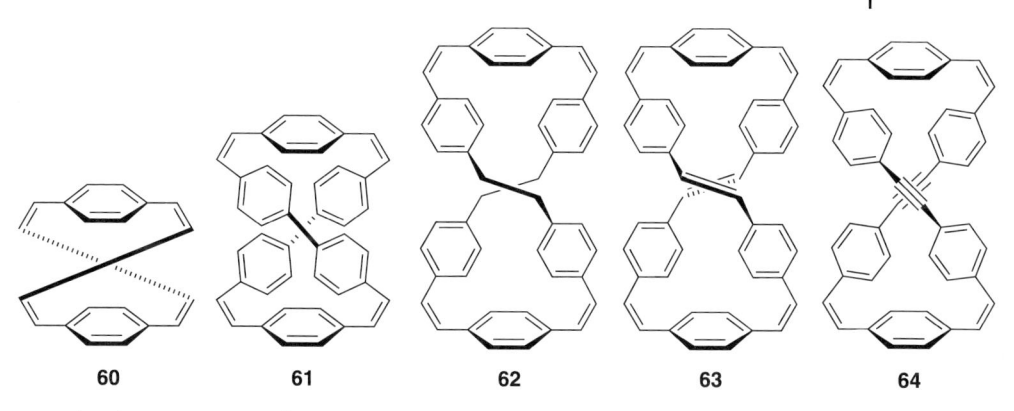

6.2 Strategies for Achieving Chiral Macrocycles or Cyclophanes | 243

60   61   62   63   64

give the diyne **66** in good yield. Pd-catalyzed cross-coupling of diyne **66**, subsequent to desilylation, with compound **65** gave the cyclophane **67** in low yield. Alternatively, the deprotected acetylene could be oxidatively homocoupled to provide the expanded cyclophane **68**. ^{1}H NMR spectroscopy was used to establish that free rotation of the phenyl rings of macrocycles **67** and **68** was occurring even at temperatures as low as −60 °C.

In addition to two-dimensional macrocycles, a number of research groups are actively pursuing the preparation of three-dimensional 'cage' structures due to tremendous current interest in the generation of C_{60} fullerene precursors. In 1996, Rubin and coworkers reported the synthesis of cyclophane **69** (Fig. 6.7) [28]. Crys-

Scheme 6.10. Synthesis of revolveneynes **67** and **68**. Reagents and conditions: (a) $Ph_3P^+CH_2Br\ Br^-$, t-BuOK, THF, −78 °C; (b) TMSA, [Pd(PPh$_3$)$_4$], CuI, BuNH$_2$; (c) K$_2$CO$_3$, MeOH; (d) **65**, [Pd(PPh$_3$)$_4$], CuI, BuNH$_2$; (e) Cu(OAc)$_2$, py, Et$_2$O, 35 °C.

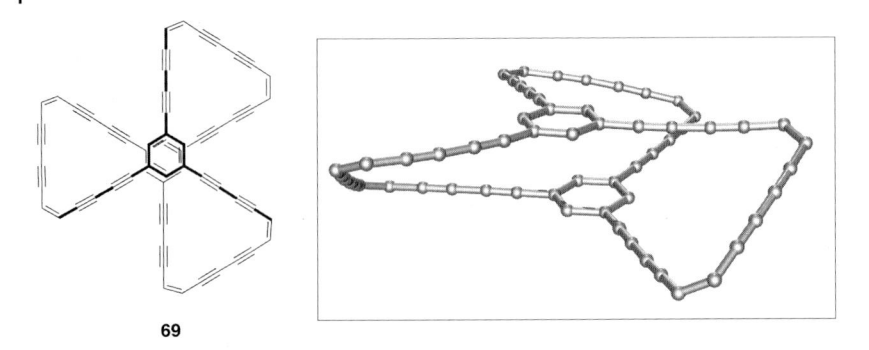

Figure 6.7. Molecular structure (left) and solid-state structure[†] (right) of helical cyclophane **69**.

tallographic analysis of **69** revealed that the two benzene rings are not stacked directly on top of one another, as in the molecular structure shown, but are offset by approximately 1.3 Å. The solid-state structure also showed that cyclophane **69** exists as two helical enantiomers. The authors determined, using computational analysis, that racemization of compound **69** occurs through an intermediate of D_{3h} symmetry where each of the three 'arms' pendant to the benzene rings occupy a single plane. The energy of this transition state was determined to be only 3–6 kcal mol^{-1} higher in energy than the chiral conformation, suggesting that racemization occurs rapidly at room temperature. The authors also explored the possibility of inducing cyclophane **69** to rearrange to form C_{60} in the gas phase. Their experiments showed that compound **69** underwent dehydrogenation to form $C_{60}H_{14}$, but no less-saturated peaks were observed.

Other cyclophanes sharing this motif have also been realized, such as compounds **70–71** and **72–76**, synthesized by the groups of Rubin [29] and Tobe [30], respectively. While crystallographic analysis was not performed on any of these derivatives, it is reasonable to assume that due to structural similarities to cyclophane **69**, they would also adopt helically chiral conformations. As with compound **69**, the three-dimensional cage compounds **70–71** and **72–75** were targeted as precursors to fullerene and diazafullerene structures [31]. The most recent members to join this class of compounds are the cyclophanes **76**, introduced by Tobe and coworkers as a precursor to C_{78} fullerene [32].

Wennerström and coworkers have synthesized numerous cage-type derivatives [33]. In 1982 they reported the synthesis of bicyclophanes **77–80**, and showed that the cyclophanes were conformationally mobile such that twisted, helical conformations could be achieved, and, in most cases, were even favored.

Herges and Okamoto have also reported the synthesis of a three-dimensional macrocyclic structure [34]. The "chiral tube" **81** was achieved in 15% yield and the two enantiomers were separated using chiral reverse-phase HPLC. CD spectroscopic measurements were acquired for both enantiomers, confirming the chiral separation as well as the structure assignment.

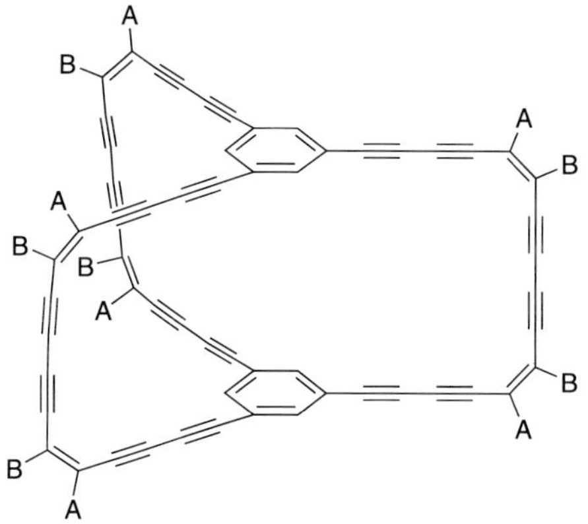

6.2 Strategies for Achieving Chiral Macrocycles or Cyclophanes | 245

70 A,B = (benzene ring)

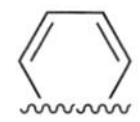

71 A,B = (cyclobutanedione)

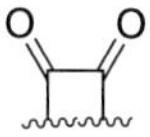

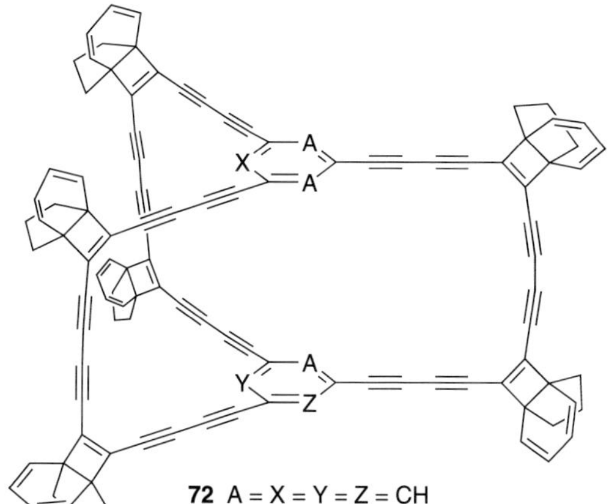

72 A = X = Y = Z = CH
73 X = Y = N, A = Z = CH
74 X = Z = N, A = Y = CH
75 A = X = Y = Z = CCl

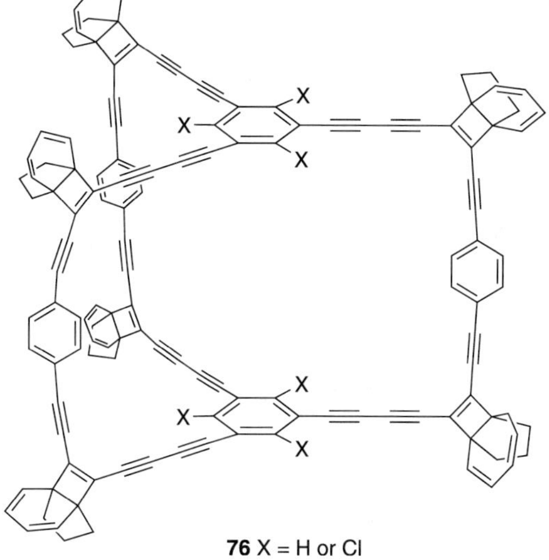

76 X = H or Cl

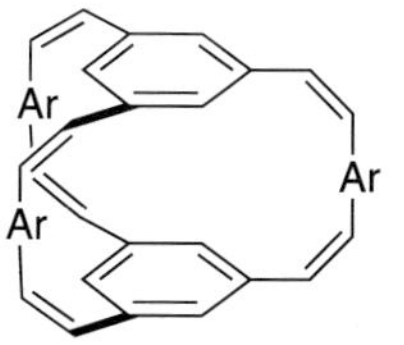

77 Ar = (1,3-phenylene)

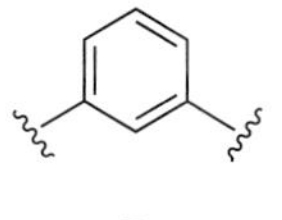

78 Ar = (5-bromo-1,3-phenylene)

79 Ar = (1,4-phenylene)

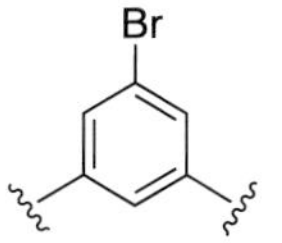

80 Ar = (2,5-thienylene)

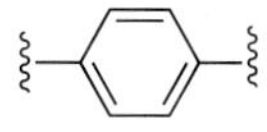

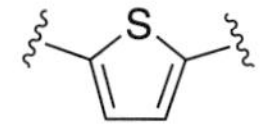

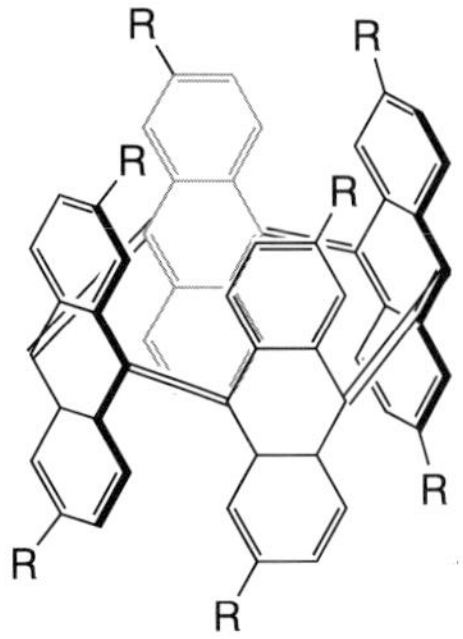

81 R = *t*-Bu

Several examples of helical cyclophanes bearing functional groups suitable for interaction with metal ions and/or salts have recently appeared in the literature. In 2001, Baxter reported the synthesis of a "twistophane" ion sensor, as shown in Scheme 6.11 [35]. Cross-coupling of iodobenzene **82** with the terminal acetylene **83** was achieved using Pd-catalysis. The resulting aryl halide **84** was stannylated with hexamethylditin and then homocoupled using [PdCl$_2$(dppf)]. This led to the formation of **85** in a remarkable 73% yield. Following deprotection of the acetylene groups, oxidative Hay homocoupling [36] gave twistophane **86** in 34% yield. Variable temperature ^{1}H NMR studies showed that the macrocycle **86** adopts a chiral twisted conformation where the two 2,2′-bipyridine moieties are in very close proximity. Upon heating, interconversion between enantiomeric conformers oc-

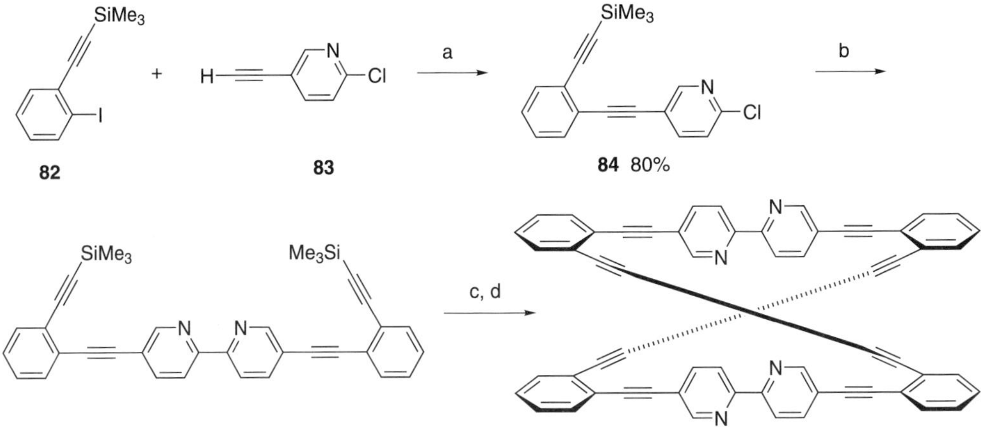

Scheme 6.11. Synthesis of twistophane **86**. Reagents and conditions: (a) [PdCl$_2$(PPh$_3$)$_2$], CuI, Et$_3$N; (b) Sn$_2$Me$_6$, [PdCl$_2$(dppf)·CH$_2$Cl$_2$], toluene, 120 °C; (c) TBAF, THF, H$_2$O; (d) CuCl, O$_2$, py.

curs, via an open, planar intermediate. Furthermore, the four nitrogens of the two bipyridine units may form a tetrahedral arrangement suitable for complexation in the presence of an appropriate metal ion. The resulting conformational changes could be sensed using UV–vis and/or fluorescence spectroscopy. Macrocycle **86** was found to bind specifically to Cu(II), Zn(II), Ni(II), and Co(II) salts as determined through rigorous UV–vis absorption and fluorescence emission analysis.

Baxter has also prepared, via an analogous synthetic protocol, the twistophane macrocycle **87**, where the 2,2′-bipyridine moieties are incorporated into the macrocyclic backbone at the 6,6′-position [37]. The selective ion sensory properties of this macrocycle were also established through UV–vis and fluorescence measurements. Calculations showed that twistophane **87** has three low-energy, chiral conformations.

87

The phenanthroline subunit may also be incorporated into the backbone of helical cyclophanes, as recently demonstrated by the Fallis group (Scheme 6.12) [38]. In this templated reaction, compound **88** was first treated with 0.5 equiv. of Cu(OAc)$_2$ in pyridine/Et$_2$O for a period of 2 h to generate the intermediate complex **89**, followed by the addition of a further 5.5 equiv. of the copper salt. Helical macrocycle **90** was then produced in a remarkable 70% yield upon treatment with KCN, which removes the Cu(II) and effects the release of the phenanthroline ligands. In addition to observed color changes that suggested the presence of complex **89**, the authors provided confirmation of this intermediate by performing the analogous coupling reaction in a single step, by the direct addition of all 6 equiv. of the Cu-catalyst. These conditions provided cyclophane **90** in only 15% yield, suggesting that templation via intermediate **89** was crucial in the step-wise synthesis.

The same authors also prepared the solubilized derivative **91** and its Cu-complex [(**91**)CuPF$_6$] using an analogous synthetic strategy [38]. Variable temperature ^{13}C NMR experiments showed that the energy barrier to enantiomerization for the

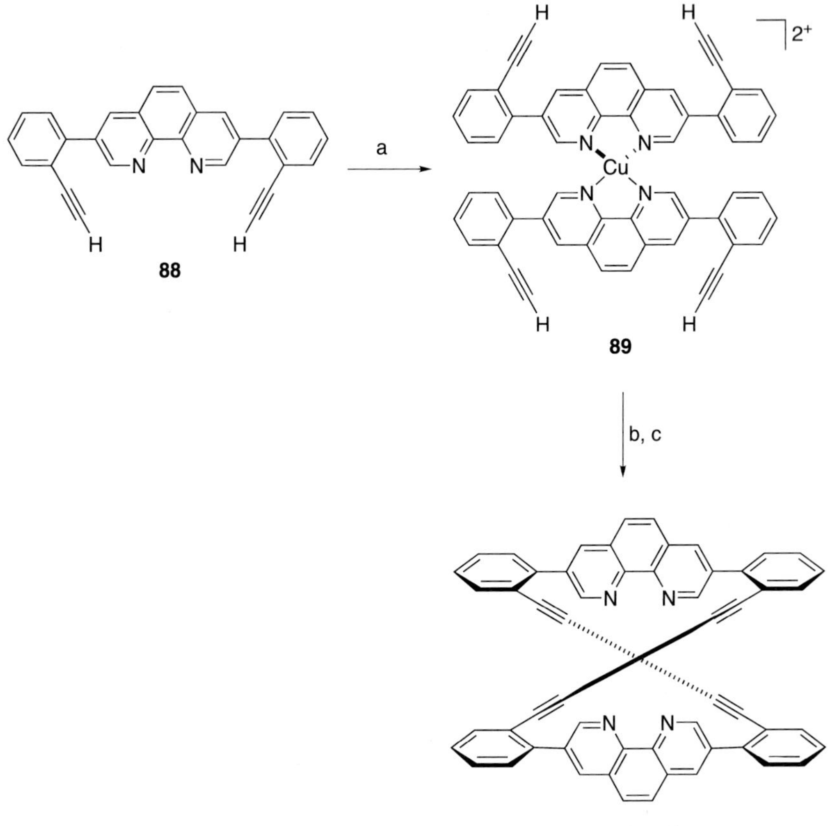

Scheme 6.12. Synthesis of cyclophane **90**. Reagents and conditions: (a) Cu(OAc)$_2$ (0.5 equiv.), Et$_2$O, py; (b) Cu(OAc)$_2$ (5.5 equiv.), Et$_2$O, py; (c) KCN (aq), CH$_2$Cl$_2$.

Cu-complex was approximately 4 kcal mol^{-1} greater in energy than for the uncomplexed helicene **91**.

Helical cyclophanes featuring enforced exotopic binding sites have been reported by Baxter [39]. As shown in Scheme 6.13, a standard sequence of Pd-catalyzed cross-coupling between an iodoarene and 1,4-diethynylbenzene leads to the formation of tetrayne **92**. Desilylation and intermolecular oxidative cyclodimerization using Hay conditions [36] gave macrocycle **93** in a remarkable 63% yield. Calculations showed that the energy-minimized structure of cyclophane **93** is twisted such that a helically chiral conformation is established. The ability of the fully conjugated macrocycle **93** to coordinate to transition metal ions was evaluated, and it was found that the fluorescence emission was quenched upon coordination to Pb(II) or Hg(II) salts. The ability of macrocycle **93** to act as a sensor for metal ions was further established by the observation that the selective precipitation of coordination polymers occurs when compound **93** is treated with Fe(II), Co(II), Ni(II), or Ag(I) salts.

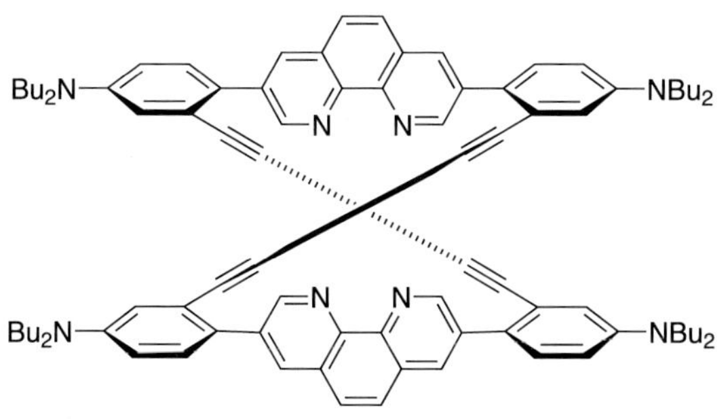

91

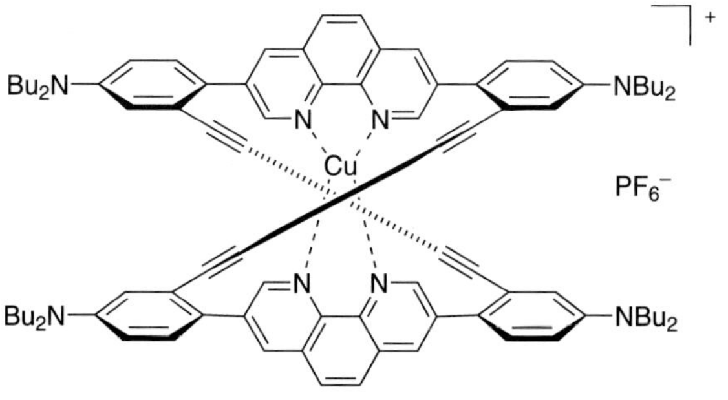

[(**91**)CuPF$_6$]

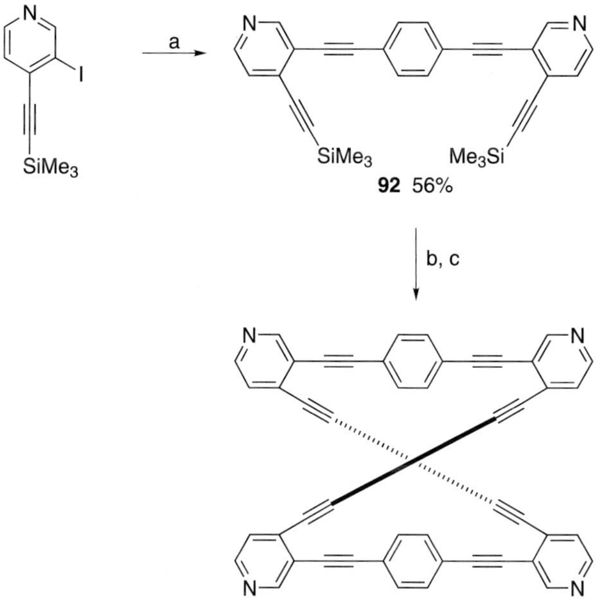

93 63%

Scheme 6.13. Synthesis of cyclophane **93**. Reagents and conditions: (a) 1,4-diethynylbenzene, [$PdCl_2(PPh_3)_2$], CuI, THF, Et_3N; (b) TBAF, THF, H_2O; (c) $Cu_2(OAc)_4$, py.

In addition to these pyridine-containing helical macrocycles, there are numerous reports of polyaza-macrocyclic species that are chiral due to the adaptation of a nonplanar conformation. Some current examples included the 'ruffled' benzimidazole-based ligands and the distorted phthalocyanines recently prepared by the groups of Chan [40] and Kobayshi [41], respectively. These systems do not fall within the context of this Chapter due to their resemblance to porphyrinic systems but they nevertheless represent an interesting class of chiral macrocycles based upon the incorporation of heterocyclic rings.

Baxter has employed a 2,5-diethynylthiophene building block leading to the formation of the mixed heterocyclic macrocycle, **94** [42], targeted due to the presence of both electron rich and electron poor heterocyclic rings. Compound **94** was synthesized in relatively high yield (46%) via a Cu-mediated acetylenic cyclodimerization. The author established, through extensive UV–vis absorption and fluorescence binding studies, that macrocycle **94** coordinates selectively to the following metals ions: Ag(I), Co(II), Ni(II), and Pd(II). Significant differences in the absorption and emission spectra were observed for **94** in the presence of these metal ions, compared to that of the free ligand, demonstrating the potential utility of these species for sensory applications. Intriguingly, in the presence of high concentrations (4 equiv.) of Ag(I), Ni(II), and Co(II) ions, the precipitation of what were likely high molecular weight coordination polymers was observed.

Using the analogous Wittig protocol that was successful for the phenyl based helical cyclophanes **61–64**, Wennerström and coworkers have also synthesized a number of mixed heterocyclic cyclophanes. Examples such as **95–97**, combining phenyl, furyl and thienyl building blocks, should be helical in nature. [26a, 43]. These cyclophanes were unfortunately all produced in low yield (<11%), due, at least in part, to their inherent instability.

Marsella and coworkers exploited perfluorophenyl–phenyl quadrupole interactions to template the synthesis of a thiophene-containing cyclophane [44]. As shown in Scheme 6.14, when the oligomeric precursors **98–101** were treated with Sonogashira cross-coupling conditions, the mixed perfluorophenyl–phenyl cyclophane **102** was reproducibly achieved in greater yield than either of the symmetrical analogs **103** and **104**. Other mixed macrocyclic derivatives were also prepared as part of this study, and their findings with these model systems supported

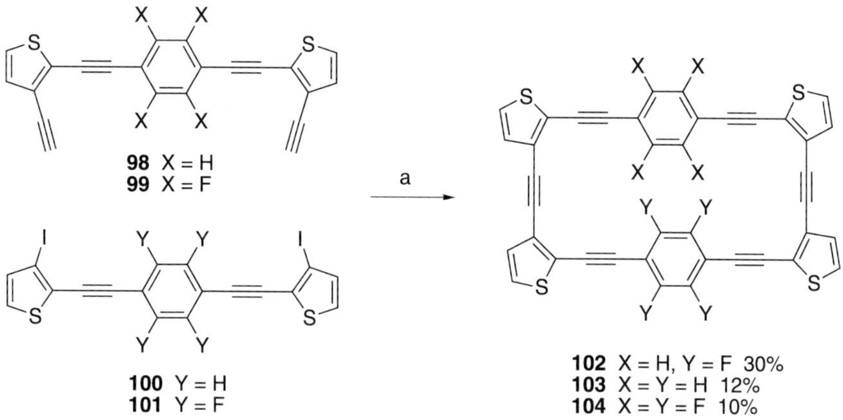

6.2 Strategies for Achieving Chiral Macrocycles or Cyclophanes | 251

98 X = H
99 X = F

100 Y = H
101 Y = F

102 X = H, Y = F 30%
103 X = Y = H 12%
104 X = Y = F 10%

Scheme 6.14. Synthesis of thiophene-containing cyclophanes **102**, **103**, and **104**. Reagents and conditions: (a) [PdCl$_2$(PPh$_3$)$_2$], CuI, THF, i-Pr$_2$NH, 80 °C.

the argument that electrostatic interactions were responsible for the self-assembly of the two coupling partners, leading to higher yields of the nonsymmetrical product.

Marsella and coworkers have also synthesized several double helical annulenes [45], based on the parent cyclooctatetrathiophene originally prepared by Kauffmann et al. [46]. Beginning with bromoarene **105**, *in situ* preparation of the thienyl Grignard followed by Ni-catalyzed cross-coupling led to the formation of the homodimer (Scheme 6.15). Regiospecific bromination gave compound **106**, and a

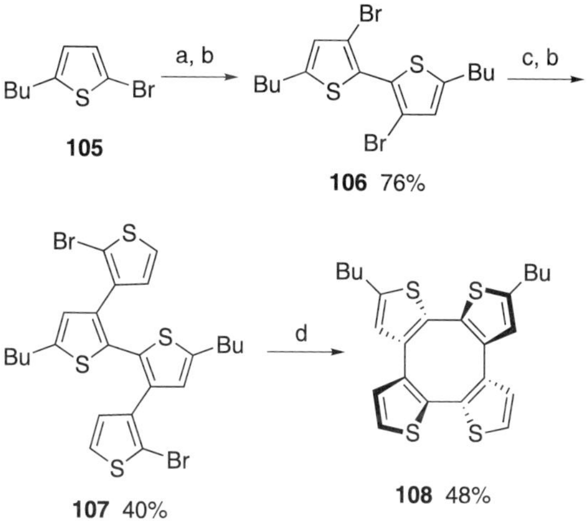

105

106 76%

107 40%

108 48%

Scheme 6.15. Synthesis of annulene **108**. Reagents and conditions: (a) BuLi (0.5 equiv.), MgBr$_2$, [NiCl$_2$(dppp)]; (b) NBS; (c) 3-thienylmagnesium bromide, [NiCl$_2$(dppp)]; (d) BuLi, FeCl$_3$.

Scheme 6.16. Synthesis of tetraiodo-annulene **111**. Reagents and conditions: (a) Br$_2$; (b) BuLi, then I$_2$.

second iteration of the cross-coupling/bromination sequence led to the formation of oligomer **107**. Subsequent to lithium–halogen exchange, Fe-catalyzed coupling of the dianion gave annulene **108**. The authors later polymerized compound **108** due to their interest in the development of polymeric electromechanical actuator devices. They hoped to take advantage of the helical conformation of the racemic annulene **108** by promoting redox-induced conformational changes, which could result in the desired expansion and contraction of the polymeric backbone, i.e., molecular actuation [47].

Marsella and coworkers later elaborated on the double helical scaffold of **108** [48]. Regiospecific bromination of the fully butylated derivative **109** (Scheme 6.16) takes place exclusively at the β-positions to provide compound **110** in near quantitative yield. The tetrabromide **110** was then converted to the tetraiodide **111**, which is suitable for further elaboration.

Helicenes **112–114** were all synthesized from **111** through application of the Sonogashira cross-coupling reaction, demonstrating the versatility of this building block (Fig. 6.8). ^{1}H NMR spectroscopy of double helical compounds **112** and **113**

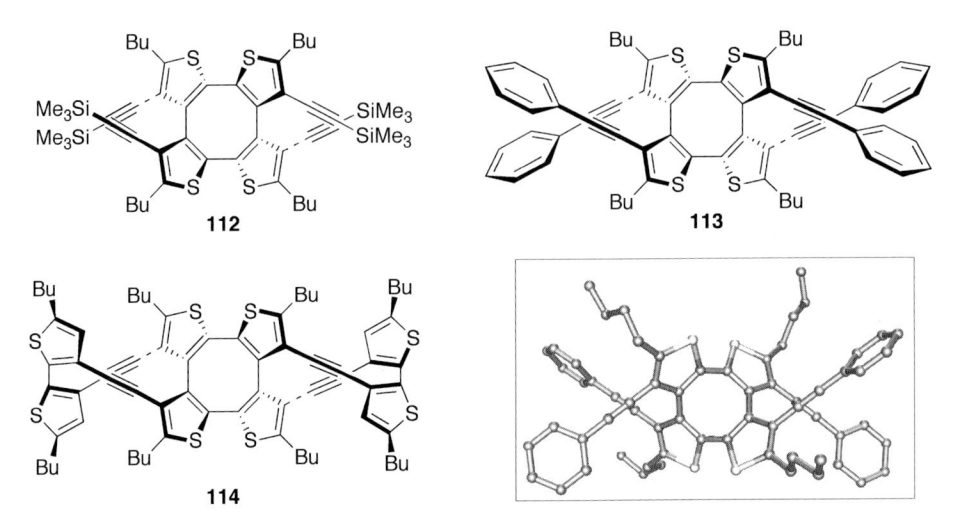

Figure 6.8. Marsella's helicenes **112–114** and solid-state structure[†] of **113** (inset).

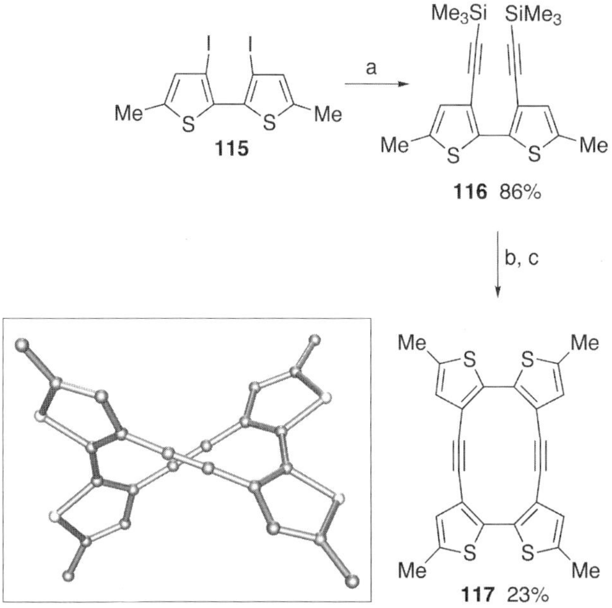

Scheme 6.17. Synthesis and solid-state structure† of expanded annulene **117** (inset). Reagents and conditions: (a) TMSA, [Pd(PPh₃)₄], CuI, i-Pr₂NH; (b) K₂CO₃, THF, MeOH; (c) **115**, [Pd₂(dba)₃], PPh₃, CuI, i-Pr₂NH.

revealed that these (racemic) compounds are conformationally stable at room temperature. X-ray crystallographic analysis of compound **113** confirmed the helical conformation. The authors also observed edge-to-face π-stacking interactions between phenyl groups of adjacent helicenes, resulting in the formation of a solid-state supramolecular polymer.

An "expanded" analog of the parent cyclooctatetrathiophene based on a [12]annulene core was then prepared in an effort to relieve steric congestion at the macrocyclic core (Scheme 6.17) [49]. The expanded analogue was synthesized by converting diiodo-bithiophene **115** to the diacetylene **116**, which was then cross-coupled with **115** to give macrocycle **117**. X-ray crystallographic analysis confirmed the helical conformation of **117**. The barrier to inversion (enantiomerization) was calculated by density functional theory (DFT) to be 9.7 kcal mol⁻¹, which is 14.3 kcal mol⁻¹ less than that of the parent cyclooctatetrathiophene.

Marsella's work in this area with polythiophenes coincided with that of Rajca and coworkers which focused on polyphenylenes [50]. Beginning with dibromide **118**, lithium–halogen exchange followed by oxidative coupling led to the formation of helical monomer **119** in 69% yield (Scheme 6.18). Alternatively, oxidative cross-coupling of **118** with **120** led to the formation of the dibromide **121** in 19% yield.

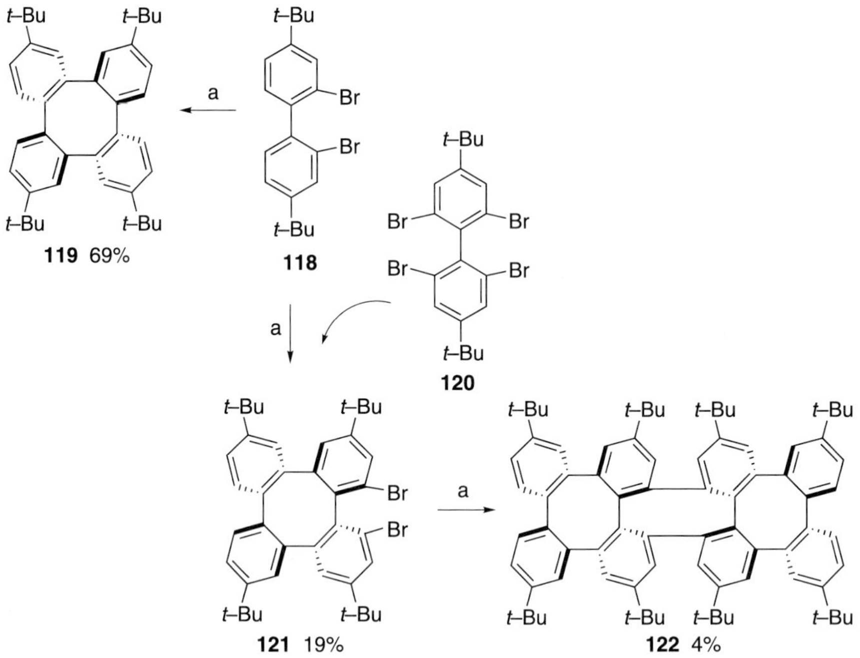

Scheme 6.18. Synthesis of helical monomer **119** and dimer **122**. Reagents and conditions: (a) t-BuLi, Et$_2$O, −78 °C then CuBr$_2$.

Formation of the dianion and subsequent oxidative homocoupling led to the formation of **122** in a modest 4% yield. The authors attribute the low yield to the fact that there are several competitive coupling pathways.

Rajca and coworkers have elaborated on the core structure of the cyclic *ortho*-tetraphenylene via the synthesis of a helical 1,1′-binaphthyl dimer as outlined in Scheme 6.19 [51]. This work represents the first example of a non-racemic macrocycle provided in this discussion. (*R*)–2,2′-Dibromo–1,1′-binaphthyl **123** was fully lithiated and subsequently oxidatively coupled using CuBr$_2$ to form the double helical **124** in >95% ee. The authors noted that similar reactions performed with racemic **123** led to complex mixtures of products and low isolated yields of the desired cyclic species **124**.

X-ray crystallographic analysis of **124** confirmed the helical structure and remarkably showed that the naphthalene rings remain essentially planar. The authors also prepared the dianion of **124** by treatment with Li and/or Na and showed through several NMR spectroscopic experiments that the barrier to racemization of the dianion is increased by approximately 25 to >40 kcal mol^{-1} versus the neutral species. This is in large part, although not exclusively, due to aromaticity in the central cyclooctatetraene ring upon formation of the dianion.

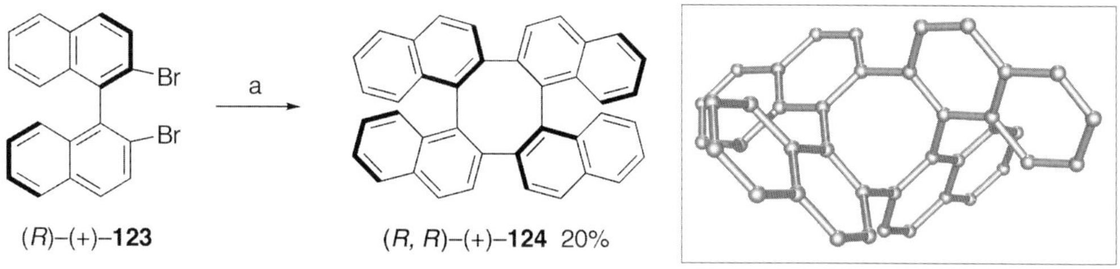

(R)-(+)-123

(R, R)-(+)-124 20%

Scheme 6.19. Synthesis and solid-state structure† of helical binaphthyl dimer **124** (inset). Reagents and conditions: (a) BuLi, −35 °C, Et$_2$O, then CuBr$_2$, −78 °C.

6.2.2
Chiral Macrocycles Using Chiral Carbon-rich Building Blocks

In addition to designing systems that will be chiral due solely to the presence of a conformational twist, the recent synthetic accessibility of building blocks of high enantiopurity has led to the realization of chiral macrocycles which derive their chirality from the direct presence of such chiral building blocks within the macrocyclic backbone. An advantage of this approach is that in some, but not all, cases these systems are nonracemic or a single stereoisomer has been achieved, as shown for (R,R)-**124** at the end of the last section. While syntheses that rely on the use of an expensive, enantiopure building block as their starting point may have certain drawbacks, this is nevertheless an effective strategy for accessing *homochiral* macrocyclic products. It has also been shown that, in some cases, the presence of chiral building blocks within a macrocyclic core can lead to chirality transfer resulting in amplified helical chirality.

In 1995 Diederich and coworkers reported the synthesis of a family of chiral binaphthyl-derived cyclophane receptors (Scheme 6.20) [52]. Oxidative Glaser-Hay coupling of (R)-**125** led to the formation of a mixture of cyclic oligomers, including trimer **126**, tetramer **127**, and pentamer **128**. Subsequent hydrolysis of the benzoyl ester provided the receptors **129** and **130** in excellent yield. It is worth noting that while the macrocycles were achieved in excellent yields (for a macrocyclization reaction), realization of the chiral building block (R)-**125** required several onerous synthetic steps. The authors demonstrated via ^{1}H NMR titration studies that trimeric **129** was capable of selective, and in some cases modestly enantioselective, carbohydrate recognition with several glucopyranosides. The same experiments with acyclic monomer **125** (after removal of the benzoyl esters) led to no significant complexation of the carbohydrate guests, demonstrating the need for multiple, cooperative, binding sites.

Diederich and coworkers later reported the synthesis of several modified cyclophane receptors such as **131–139**, as well as their larger analogs [53]. The D_3 symmetric receptors **129**, **131**, **135**, and **137** were found via computer modeling studies to adopt quite planar conformations. By contrast, the C_2 symmetric macrocycles

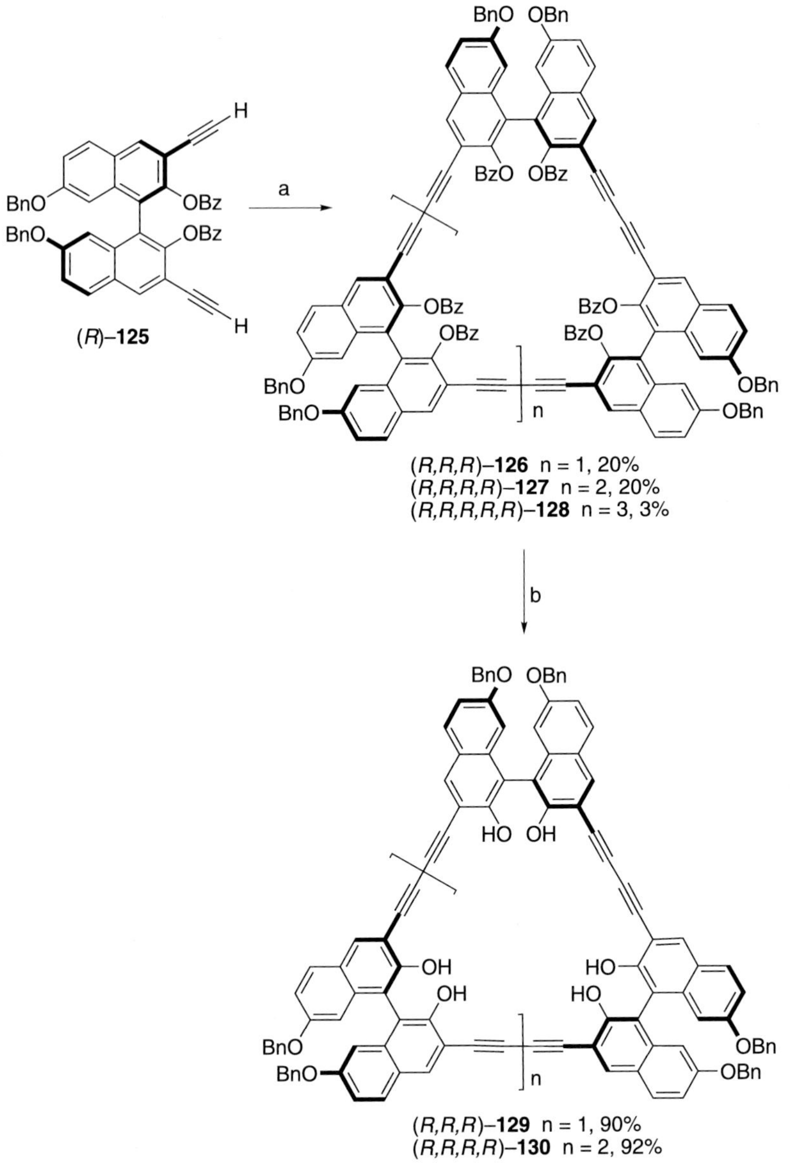

Scheme 6.20. Synthesis of cyclophane receptors **129** and **130**. Reagents and conditions: (a) CuCl, TMEDA, CH$_2$Cl$_2$; (b) KOH, MeOH, THF.

136 and **139** were predicted to adopt less planar, more flexible conformations. As a result of the less planar conformation, the less symmetrical receptors possessed much higher binding affinities for the carbohydrate guests and also demonstrated a greater degree of enantio- and diastereoselectivity.

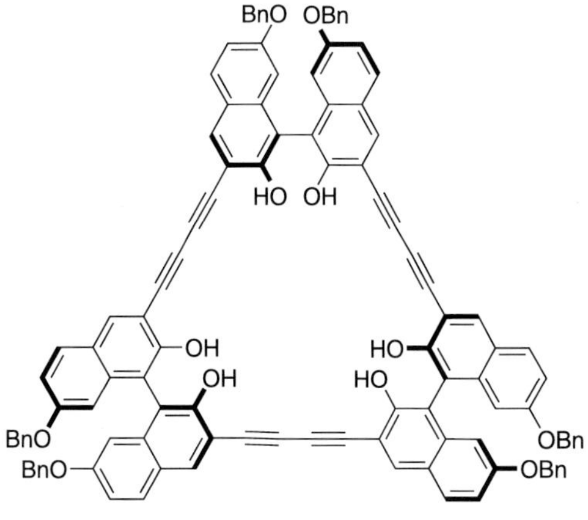

(S,S,S)-131

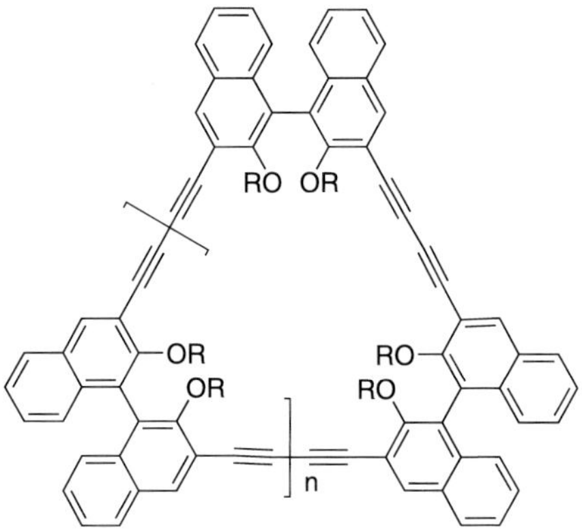

(S,S,S)-132 n = 1, R = MOM
(S,S,S,S)-133 n = 2, R = MOM
(S,S,S,S,S)-134 n = 3, R = MOM
(S,S,S)-135 n = 1, R = H
(R,R,S)-136 n = 1, R = H

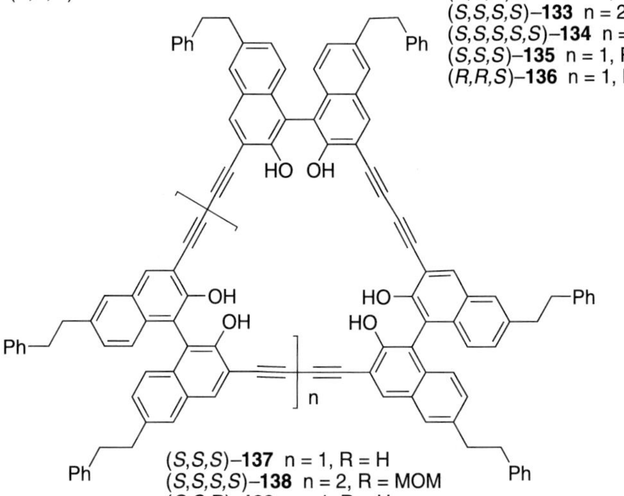

(S,S,S)-137 n = 1, R = H
(S,S,S,S)-138 n = 2, R = MOM
(S,S,R)-139 n = 1, R = H

Chiral tetraanionic analogs of these binaphthyl-derived receptors have also been reported by the same authors [52]. Receptor **140** was synthesized by treating macrocycle **130** with $POCl_3$ in the presence of Et_3N. Preliminary binding studies of this macrocycle with a glucopyranoside suggested dramatically increased complexation strength, relative to the neutral analogs.

Diederich and coworker have synthesized chiral receptors that possess additional carboxylate residues that are suitable for the complexation of disaccharides [54]. Cyclophanes **141–143** were prepared via a stepwise approach involving chain elongation steps followed by a single intramolecular cyclization reaction. The complexation of these receptors with two disaccharides was investigated by ^{1}H NMR spectroscopy and showed that the incorporation of the methyl carboxylate groups in **142** increased the association relative to the unfunctionalized derivative **141**. A subsequent publication reported on the detailed binding affinities of square cyclo-

(R,R,R,R)-**140**

phane **140** and tetrameric **141–143** [55] with mono- and disaccharide guests. In general, their findings showed that the smaller and more size complementary cyclophane **140** preferentially bound monosaccharide guests. The larger cyclophanes were ideal for selective disaccharide binding and the strength of the guest complexation increased concurrent with an increase in the number of host-guest interactions.

(R,R,R,R)-**141** R = H
(R,R,R,R)-**142** R = CO_2Me
(R,R,R,R)-**143** R = CO_2^- K^+

Other authors have also targeted chiral cyclophanes derived from a binaphthyl core, due in part to the accessibility of this building block in enantiopure form.

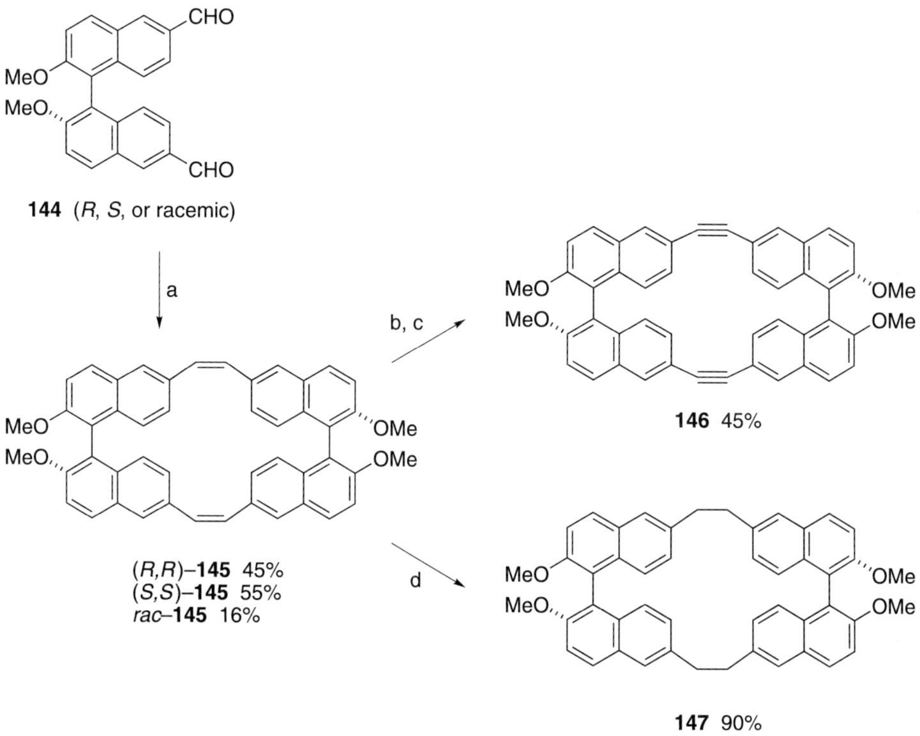

144 (R, S, or racemic)

(R,R)-**145** 45%
(S,S)-**145** 55%
rac-**145** 16%

146 45%

147 90%

Scheme 6.21. Synthesis of cyclophanes **145–147**. Reagents and conditions: (a) TiCl$_4$/Zn–Cu, DME; (b) Br$_2$, benzene; (c) t-BuOK, THF; (d) Pd–C/H$_2$.

Kawase and Oda have described the synthesis of doubly folded cyclophanes based on the elaboration of binaphthol into macrocyclic species [56]. Beginning with dialdehyde **144** (R, S, or racemic) the synthesis relied upon McMurry coupling for the formation of dimeric macrocycle **145** (Scheme 6.21). Interestingly, the use of racemic **144** results in the diastereoselective formation of racemic **145** and shows no evidence of the *meso*-coupling adducts. Elaboration of macrocycle **145** using either bromination/dehydrobromination or hydrogenation resulted in the formation of cyclophanes **146** and **147**, respectively. X-ray crystallographic analysis was performed on the diyne **146** and confirms the presence of a strained and twisted conjugated system.

Otera and coworkers have communicated the synthesis of a double helical alkynyl cyclophane (Scheme 6.22) [57]. Beginning with enantiopure **148** (either (R)- or (S)-forms), Pd-catalyzed cross-coupling with **149** or **150** led to the formation of **151** and **152**, respectively. Using a divergent strategy, **151** and **152** were carried on to the 'unmasked' iodides **153** and **154**, as well as the desilylated compounds **155** and **156**. The aryl iodides were then cross-coupled with the appropriate terminal acetylenes to give **157** and **158**. Conversion of the triazenes to the aryl iodides,

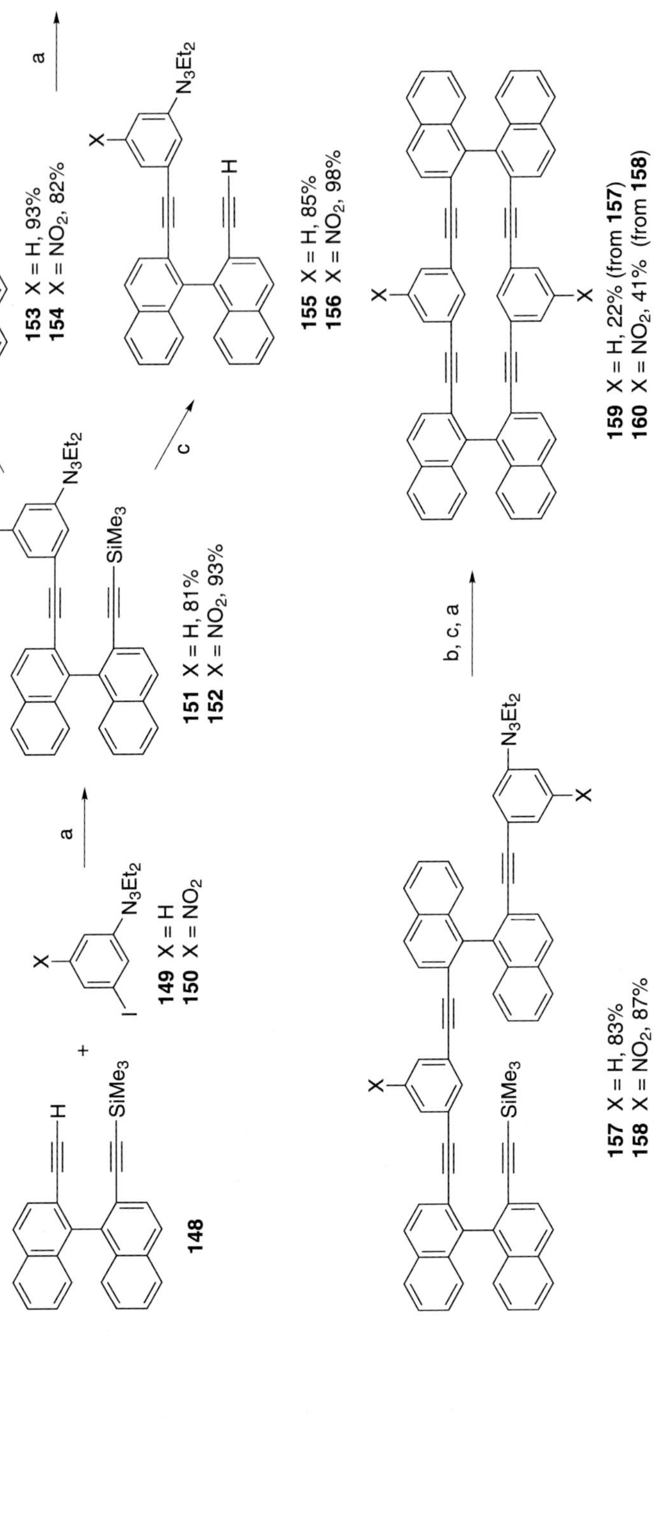

Scheme 6.22. Synthesis of double helical cyclophanes **159** and **160**. Reagents and conditions: (a) [Pd(PPh₃)₄], CuI, *i*-Pr₂NH, toluene; (b) MeI, reflux; (c) K₂CO₃, MeOH.

desilylation, and Sonogashira coupling effected the ring-closure, leading to chiral cyclophanes **159** and **160** (in both (R,P)- and (S,M)-forms). Single crystals suitable for X-ray crystallography could not be obtained from either **159** or **160** in enantiopure form. Single crystals grown from an equimolar mixture of (R,P)- and (S,M)-**160** were achieved and X-ray analysis showed a significant helical twist in the solid-state. Agreement between CD spectroscopy experiments performed in both the solution and solid-state suggested that the same structure is achieved in solution as that observed in the solid. Furthermore, the authors demonstrated a significant degree of chirality transfer by contrasting the CD spectra of **159** and **160** with that of an acyclic analog.

Pu and coworkers have synthesized and studied numerous binaphthyl-derived macrocycles and polymers. In 1998, they reported the synthesis of multipolar macrocycles **161–163** [58]. The macrocycles were produced as a separable mixture via a one-pot synthesis utilizing a Sonogashira cross-coupling reaction.

(R,R)–**161** n = 1
(R,R)–**162** n = 2
(R,R)–**163** n = 3

In a later publication, Pu and coworkers described the synthesis of a salophen-containing macrocycle that also features a binaphthyl-derived core (Scheme 6.23) [59]. The condensation of dialdehyde **164** with aryl diamine **165** led to the formation of the enantiopure macrocycle **166** in a remarkable 97% yield. The authors later prepared the Ni(II) complex of **166** by treating with two equivalents of [Ni(OAc)$_2$ · 4H$_2$O] and determined that the resulting complex was paramagnetic, indicating that one of the two nickel centers was coordinated to the chiral macrocyclic ligand in a non-planar, tetrahedral fashion. This is in contrast to previously described [Ni(salophen)] complexes that were square planar, and therefore diamagnetic [59].

Pu and coworkers have also reported the synthesis of numerous binaphthyl-containing macrocycles that are not fully unsaturated and thus, will not be extensively discussed within the context of this Chapter. An example, however, is the

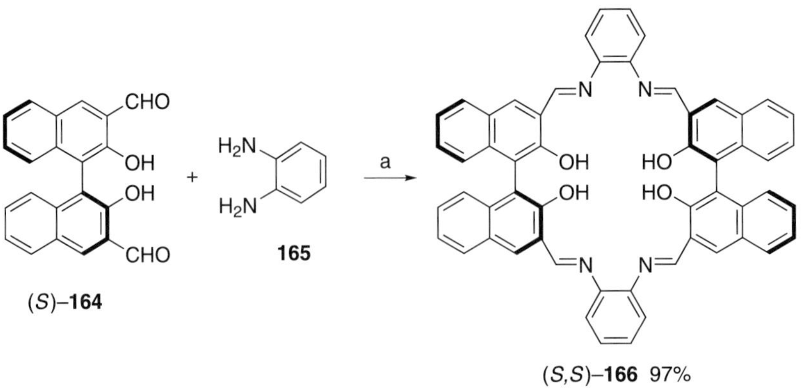

Scheme 6.23. Synthesis of salophen-cyclophane **166**. Reagents and conditions: (a) EtOH, reflux.

chiral macrocycle **167** which has been shown to be an effective enantioselective sensor for carboxylic acids including amino acids [60].

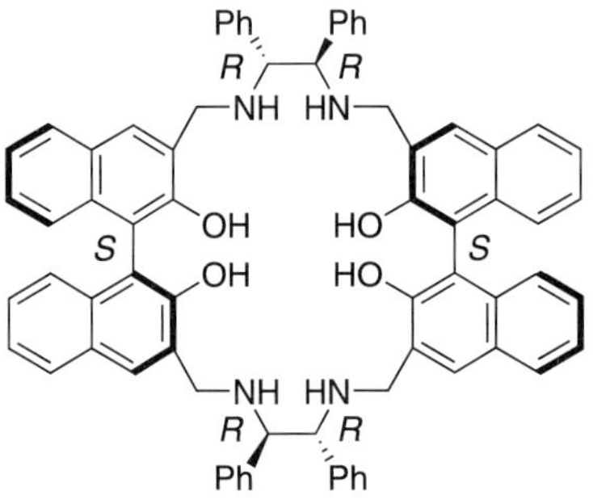

Binaphthyl-derived macrocycles featuring the incorporation of phenanthroline units have been described. In 1993, Cram and coworkers reported the synthesis of helically chiral (racemic) macrocycle **168** [61]. The four nitrogens of the phenanthroline units are arranged in a nearly tetrahedral array, suitable for the coordination of small metal ions. The authors explored the cation binding ability of the macrocycle versus that of an acyclic analogue, and their findings suggest that the increased preorganization of **168** relative to the acyclic species resulted in an increased in binding energy of 7.7–9.3 kcal mol^{-1} for small alkali metals.

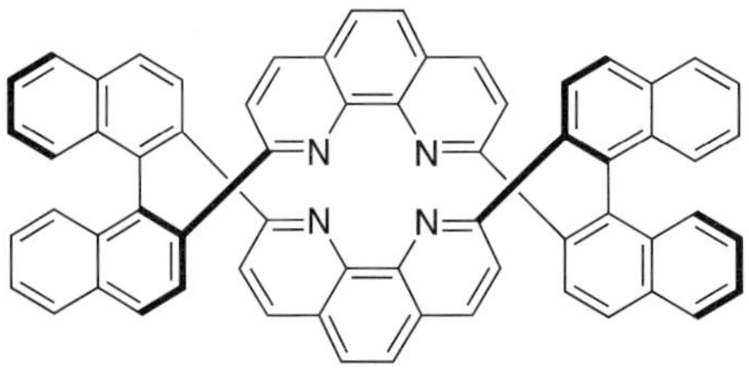

6.2 Strategies for Achieving Chiral Macrocycles or Cyclophanes | 263

Binaphthyl based structures are not the only chiral groups that have been incorporated into the backbone of chiral macrocycles. Recently, there has been an increasing number of reports of macrocycles that are chiral due to the incorporation of a helicene moiety [62]. One of the earliest structures to be reported was from Wennerström and coworkers [63] and linked directly two [5]helicenes to give macrocycle **169**, which came to be known as propellicene.

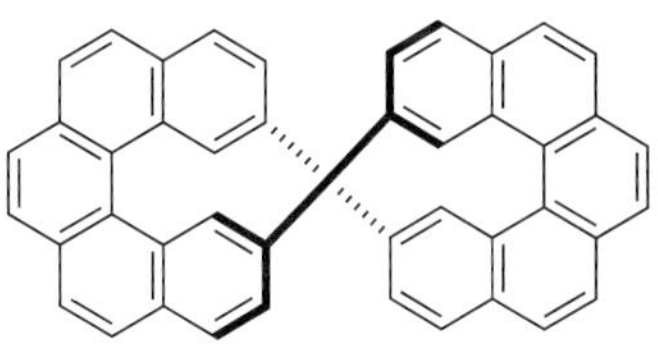

169

More recently, Yamaguchi and coworkers have prepared the helicene-containing analog of a *meta*-phenyleneethynylene macrocycle (Scheme 6.24) [64]. The diethynyl helicene **170** was cross-coupled with compound **171** and subsequently desilylated to provide oligomer **172** in 83% yield. Coupling of **172** with a diiodo-*meta*-

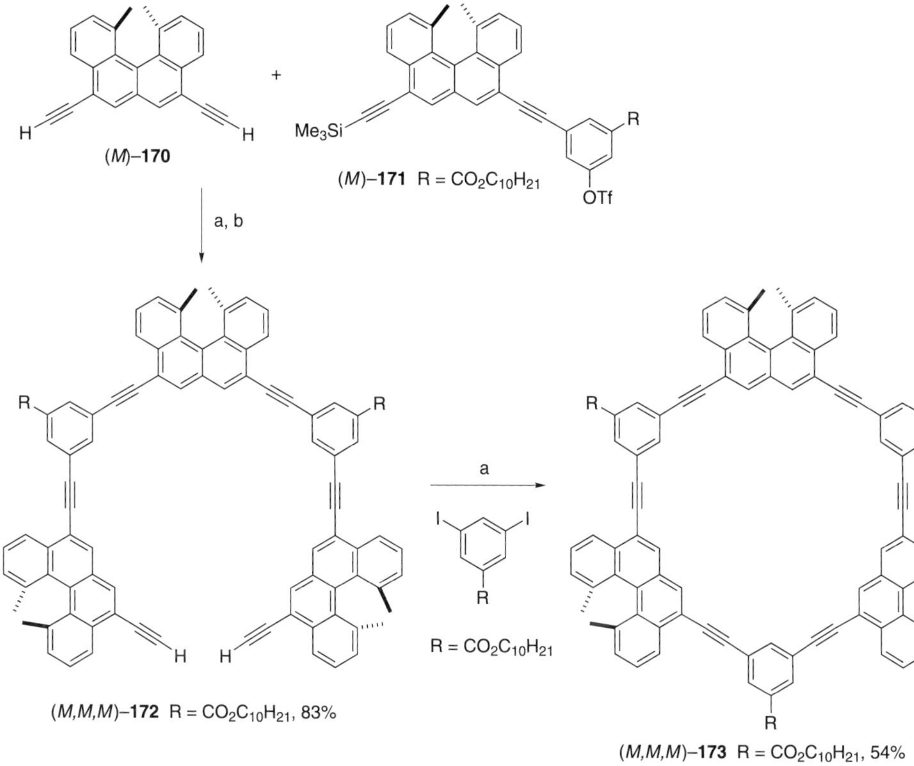

(*M*)–**170**

(*M*)–**171** R = $CO_2C_{10}H_{21}$

a, b

(*M*,*M*,*M*)–**172** R = $CO_2C_{10}H_{21}$, 83%

R = $CO_2C_{10}H_{21}$

a

(*M*,*M*,*M*)–**173** R = $CO_2C_{10}H_{21}$, 54%

Scheme 6.24. Synthesis of helicene-containing macrocycle **173**. Reagents and conditions: (a) [$Pd_2(dba)_3 \cdot CHCl_3$], CuI, $PMes_3$, PPh_3, Bu_4NI, *i*-Pr_2NEt, DMF; (b) TBAF, THF.

phenylene under high dilution conditions gave the desired helicene-containing macrocycle (*M,M,M*)-**173**. The authors also prepared the diastereomeric (*M,P,M*)-**173** macrocycle as well as the (*P,P,P*)-**173** and (*P,M,P*)-**173** isomers. The aggregation properties of the chiral macrocycles were evaluated, showing that self-aggregation was preferred in all cases.

Yamaguchi and coworkers have expanded on the structure of the chiral macrocycle **173**, using it as a building block to generate even larger chiral scaffolds, such as dimeric **174** and **175** and trimeric **176** [65]. The intra- and intermolecular aggregation properties of these cyclic species were evaluated via variable concentration ^{1}H NMR spectroscopy and it was shown that the assembly of these species was strongly dependent on the structure of linker (X) used. The use of a flexible linker such as that present in **174** leads to the formation of a very strong intramolecular aggregate. A rigid linker as in **175**, on the other hand, leads to the formation of a bimolecular aggregate. The trimeric species **176** forms a very strong bimolecular aggregate and shows no evidence of the monomeric species in solution. Interestingly, all of the compounds studied showed no evidence of higher aggregates, even at very high concentrations.

(*P,P,P*)/(*P,P,P*)–**174** R = $CO_2C_{10}H_{21}$

X = $-(CO_2(CH_2)_4-C\equiv C)_2-$

(*P,P,P*)/(*P,P,P*)–**175** R = $CO_2C_{10}H_{21}$

X = $-C\equiv C-C\equiv C-$

The same group has also prepared chiral dimeric cycloalkynes that are linked by an azo group [66]. A separate synthesis of both the *trans*-**177** and *cis*-**177** derivatives was required due to an inability to isomerize the macrocyclic species using either heat or irradiation. Variable temperature CD and ^{1}H NMR spectroscopy indicated that both isomers formed aggregates in solution. Vapor pressure osmometry and GPC were used to provide an estimate of the molecular weights of the aggregates. Interestingly, it was found that while the *trans*-isomer forms strictly bimolecular

$(P,P,P)/(P,P,P)/(P,P,P)$–**176** R = $CO_2C_{10}H_{21}$

aggregates, the *cis*-isomer forms trimeric and (at high concentration) polymeric aggregates.

Fox and coworkers have synthesized helicene-containing macrocycles, with an interest in preparing fully conjugated, chiral scaffolds [67]. Helical macrocycles **178** were produced as a mixture, but the monomeric ($n = 1$) and dimeric ($n = 2$) species were isolated in low yield by preparative TLC. UV–vis and CD spectroscopy suggested that the macrocyclic dimer is twisted such that conjugation between the respective helicene units is not possible. This conclusion was confirmed through computer modeling which revealed a highly twisted structure incapable of delocalization.

While not fully conjugated itself, the spiroindane moiety has been used by Scherf and coworkers to link carbon-rich arene segments, providing chiral macrocycles **179** and **180**. Both macrocycles inherit chirality from the spiro-centers that

266 | *6 Chiral Carbon-rich Macrocycles and Cyclophanes*

trans–(*M,M,M*)/(*M,M,M*)–**177**
cis–(*M,M,M*)/(*M,M,M*)–**177**

178 (n = 1–5)

are present in the framework [68]. The ability of **179** and **180** to function as chiral dopants for achiral, nematic liquid crystalline materials was explored.

The axial chirality of allenes makes them attractive building blocks for chiral macrocycles, yet very little has been done in this area. Krause and coworkers reported the first example of this new class of molecule, the allenophanes, in 1999 [69]. Their multistep synthesis of **181** was not stereoselective and ultimately afforded a mixture (presumably) of four diastereomers. All attempts to establish the presence of the four diastereomers via HPLC were thwarted by the low solubility of the compound.

Fallis and coworkers have recently reported the second member of this class of compounds [70]. The acetylenic allenophane **182** was synthesized as a single enantiomer, with the requisite chirality incorporated early on via a Sharpless asymmetric epoxidation. While crystallographic characterization of this intriguing molecule

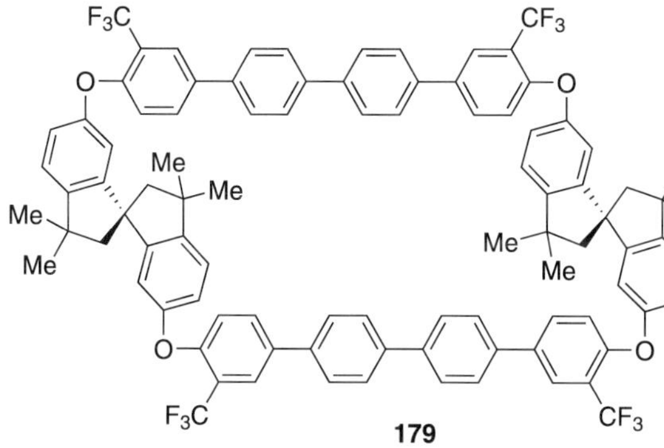

179

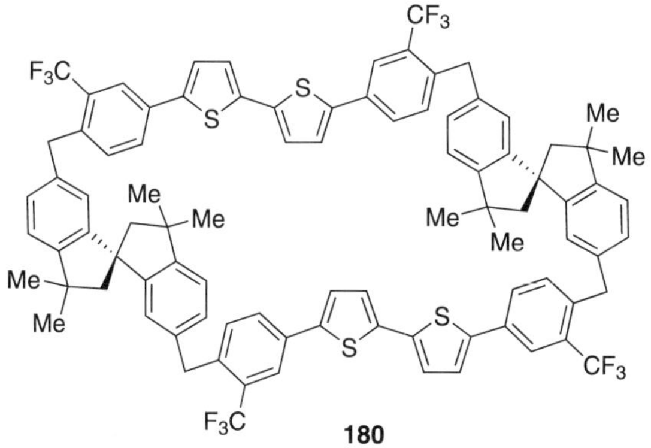

180

has thus far been elusive, CD spectroscopy and molecular modeling were used to probe its chiral structure.

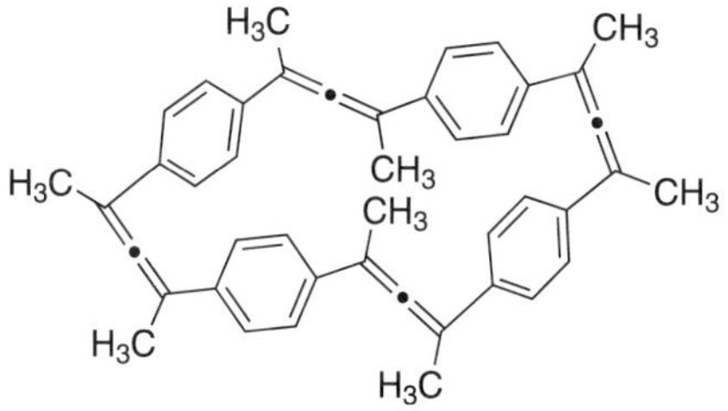

181

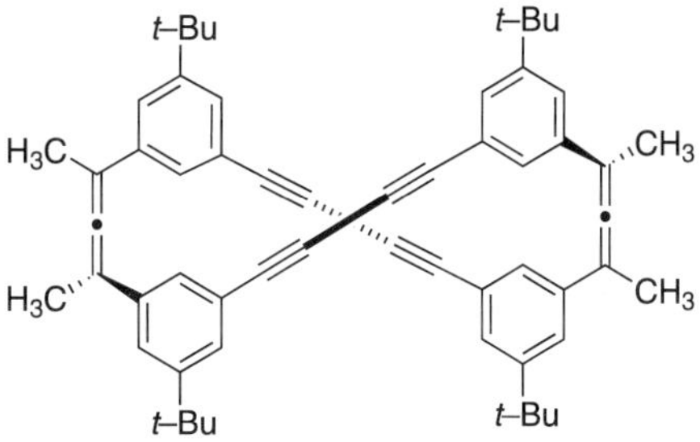

182

Chiral diamines have also been used to introduce asymmetry into the backbone of macrocyclic products. Villani and coworkers have reported the synthesis of a macrocyclic minireceptor **183** (Scheme 6.25) [71] via acylation of a commercially available, chiral diamine in the presence of Hunig's base. The enantiopure macrocycle **183** was immobilized on silica particles via further synthetic manipulations and was found to bind a wide range of amino acid derivatives with some of the highest degrees of enantioselection reported for a synthetic receptor.

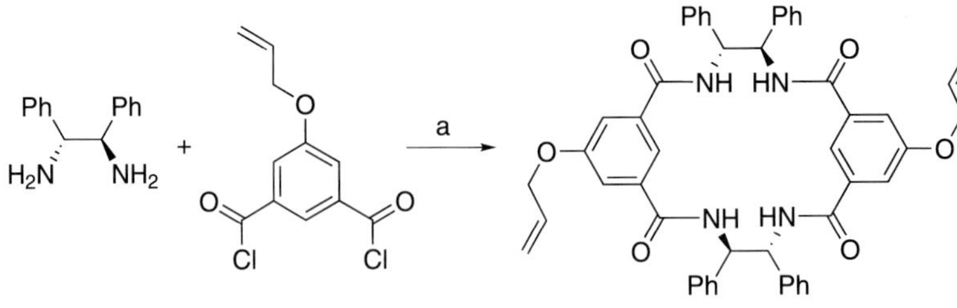

Scheme 6.25. Synthesis of macrocyclic receptor **183**. Reagents and conditions: (a) *i*-Pr$_2$NEt, THF.

Chiral macrocycles have also been achieved through the attachment of chiral groups to the periphery of a carbon-rich macrocyclic core. Tobe and coworkers have prepared chiral diethynylbenzene macrocycles **184** by decorating the core of the cyclophane with chiral side chains, and they then studied their self-assembly properties [72]. Given the wide range of macrocycles with this general framework, i.e., *meta*-diethynylaryl [1], it is surprising that this strategy has not been pursued more intensely.

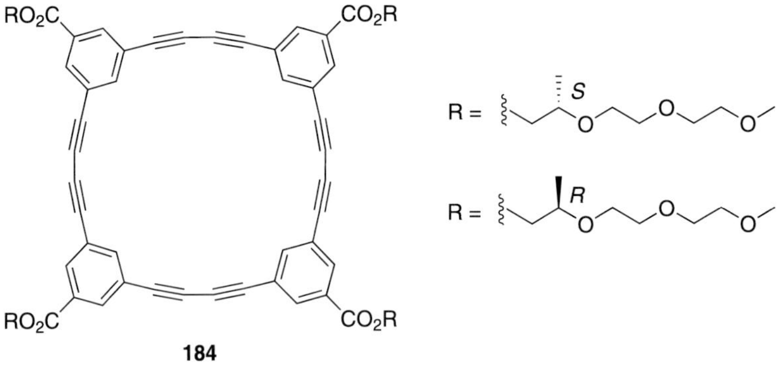

6.2.3
Metal-containing Chiral Macrocycles

Chiral, metal-containing macrocycles can be readily accessed through the self-assembly of coordinative building blocks with transition metals bearing chiral ancillary ligands. While this approach to chiral, carbon-rich macrocycles also relies upon the use of expensive, enantiopure building blocks, this disadvantage is offset to a large degree by the ease of assembly of these macromolecular products. This is best exemplified by the research described by Stang and coworkers in the late 1990s. In 1996, Stang's group reported the synthesis of chiral macrocyclic squares **185–188** derived from (*R*)-BINAP [73]. Macrocyclic squares **185** and **186** were both

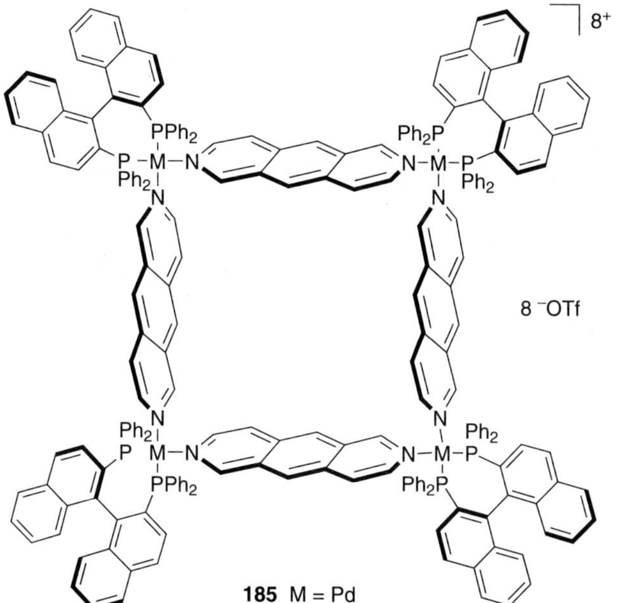

185 M = Pd
186 M = Pt

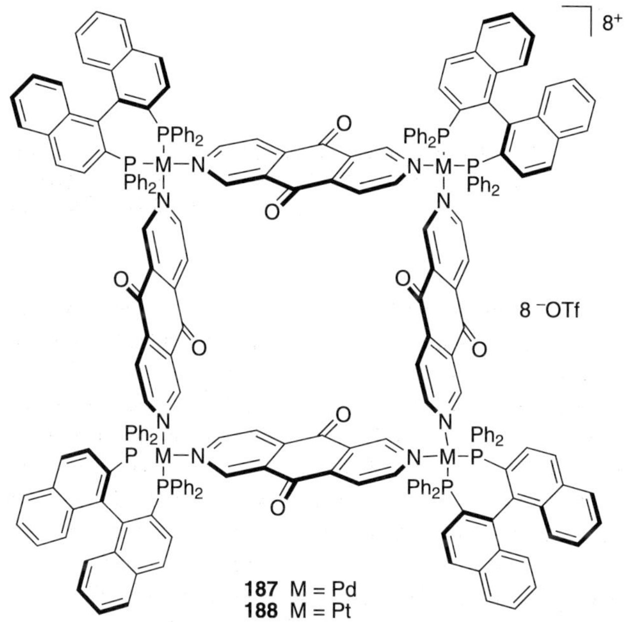

187 M = Pd
188 M = Pt

formed as a single stereoisomer under the self-assembly conditions. By contrast, compounds **187** and **188** were formed in a diastereomeric excess of 81 and 72%, respectively. The (S)-derivatives of compounds **187** and **188** were also prepared and CD spectroscopy confirmed the enantiomeric relationship. All of the squares **185–188** possess a helical twist in addition to the asymmetry introduced by the chiral BINAP ligand. The lack of rotational symmetry about the coordination axis of the diaza ligands results in a diastereoselective twist upon self-assembly with the transition metal center. The authors demonstrated that the diastereoselectivity is the result of chiral induction from the chiral diphosphine ligands by preparing the analogous squares using an achiral transition metal complex. The authors have also shown that this strategy can be extended to the synthesis of chiral three-dimensional polyhedra [74].

Stang and coworkers have also reported the synthesis of the related mixed iodonium-transition metal squares **189–192** [74]. Unlike the previous examples, the pyridyl ligands used possess rotational symmetry and thus, the squares **189–192** are chiral solely due to the presence of the chiral (R)-BINAP ligands.

189 M = Pd
190 M = Pt

191 M = Pd
192 M = Pt

In a subsequent study, Stang and coworkers reported the synthesis, in excellent yields, of macrocyclic squares **193–202** using both the (R)- and (S)-DIOP ligand in combination with (R)- and (S)-BINAP [75]. The Ag-complexes of **193**, **195**, **196**, **201**, and **202** were subsequently prepared by treating the chiral squares with two equivalents of AgOTf, and the resulting complexes were then evaluated for their ability to bind small, neutral, heteroatom-containing guests. CD spectroscopy was used to monitor the binding of these small organic guests, including phenazine, and tetramethylpyrazine.

Fujita and coworkers have utilized metal-mediated self-assembly reactions to form chiral, carbon-rich cyclic structures (Scheme 6.26) [76]. The authors showed

193 (R)–DIOP, M = Pd, L = PEt$_3$
194 (S)–DIOP, M = Pd, L = PEt$_3$
195 (R)–DIOP, M = Pt, L$_2$ = (R)–BINAP
196 (R)–DIOP, M = Pt, L$_2$ = (S)–BINAP
197 (S)–DIOP, M = Pt, L$_2$ = (R)–BINAP
198 (S)–DIOP, M = Pt, L$_2$ = (S)–BINAP
199 (R)–DIOP, M = Pd, L$_2$ = (R)–BINAP
200 (R)–DIOP, M = Pd, L$_2$ = (S)–BINAP
201 (S)–DIOP, M = Pd, L$_2$ = (R)–BINAP
202 (S)–DIOP, M = Pd, L$_2$ = (S)–BINAP

that ligand **203** self-assembled with Pd-complex **204** to form chiral complex **205**. They determined that complex **205** was in equilibrium with another species identified as catenane **206**. Hydrophobic media favored the formation of the catenated species **206**, and thus the two species **205** and **206** could be isolated from organic and aqueous media, respectively. Crystallographic analysis of **205** showed that the macrocycle adopts a nearly planar conformation in the solid-state and is thus achiral apart from the presence of the chiral diamine ligand. CD spectroscopic measurements of **205** showed little signal, indicating that an achiral conformation was also preferred in solution, while the CD spectrum of the catenated species **206** showed a very large response, indicating a high degree of helical induction.

Another common approach to the formation of chiral metallamacrocycles involves the introduction of coordinative functionality such as pyridine or bipyridine groups onto a chiral building block such as binaphthyl. The resulting chiral ligand can then be used as a component in metal-mediated self-assembly reactions. Hong and coworkers have reported the synthesis of chiral dimeric complexes using the chiral ligand **207** (Scheme 6.27) [77]. ^{1}H NMR titration experiments using either racemic or enantiopure **207** in reaction with **208** or **209** indicated a high degree of self-discrimination in the self-assembly process that led to the formation of only achiral complex [{(S)-**207**}{(R)-**207**}M(dppp)], with no formation of the homochiral complex. X-ray crystallographic analysis confirmed the achiral structure of **211**.

Larger ligands featuring multiple pyridine groups tethered to a binaphthyl core have been prepared by Otera and coworkers (Scheme 6.28) [78]. Diethynyl binaphthalene was elaborated via Sonogashira cross-coupling to form oligomer **212** which was then coupled with dipyridine compound **213** to give **214**. The authors noted that this specific synthetic sequence was required to achieve ligand **214**. The enantiopure chiral helicates (R,P)-**215** and (R,P)-**216** were created through simple metal complexation with AgPF$_6$ or [Cu(CH$_3$CN)$_4$]PF$_6$, respectively. The CD spectra of the helical complexes (R,P)-**215** and (R,P)-**216** revealed that chirality is transferred from the binaphthyl core to the ligand 'strands' as evidenced by the strong positive absorption that is observed in the spectrum of **215**; the CD spectrum of the

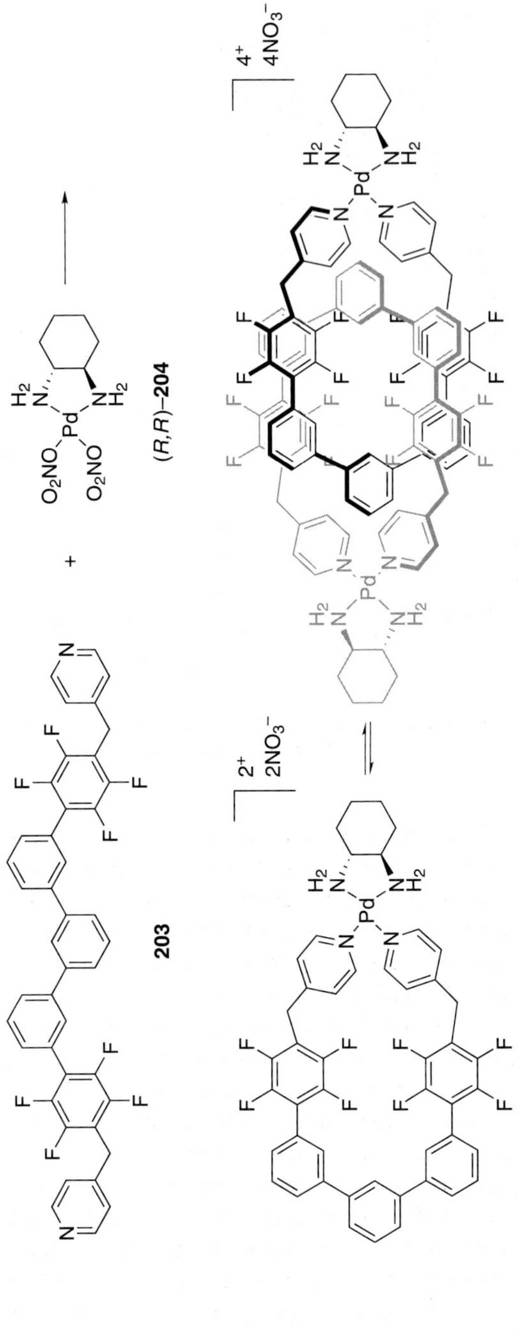

272 | *6 Chiral Carbon-rich Macrocycles and Cyclophanes*

Scheme 6.26. Synthesis of macrocycle **205** and catenane **206**.

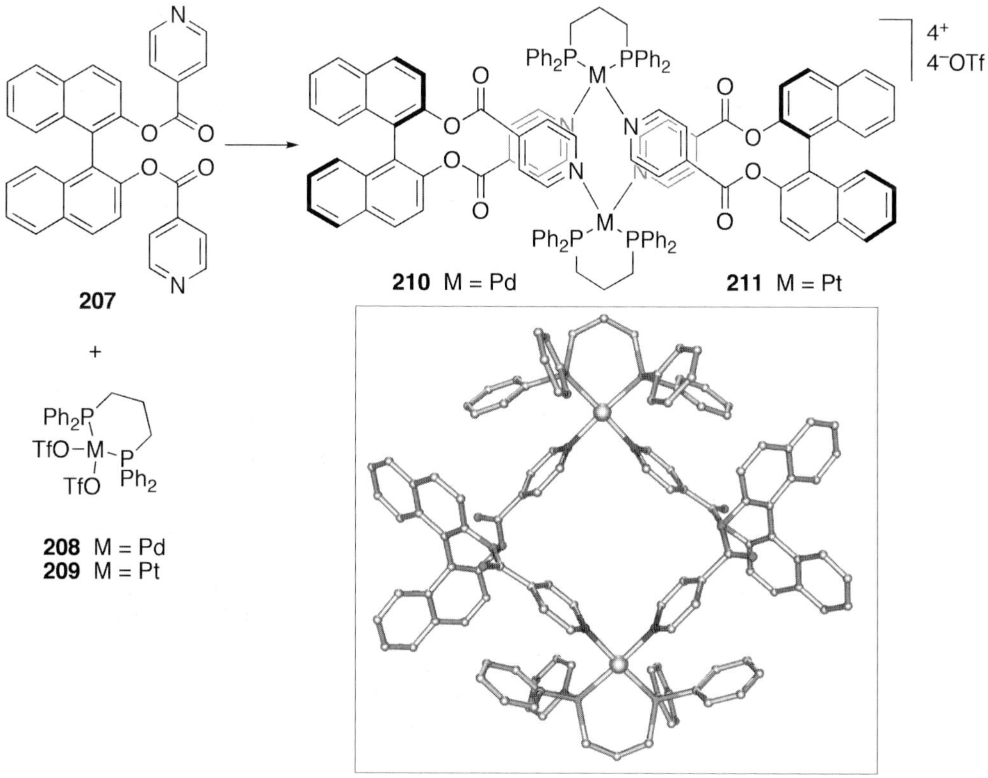

Scheme 6.27. Self-assembly reactions to form metalla-macrocycles **210** and **211** and solid-state structure[†] of **211** (inset).

uncoordinated ligand **214** shows a weaker, negative absorption. The other enantiomers, (*S,M*)-**215** and (*S,M*)-**216** (not shown) were assembled in a similar fashion.

To add support to several of their CD spectroscopic assignments, the same authors also prepared the chiral helicates **217–220**. Helicates **217** and **218**, which lack the binaphthyl–arylacetylene chromophore, were prepared to allow for the evaluation of the contribution of the helical strands at lower energy. In the absence of the low energy absorption that results from the presence of the binaphthyl-acetylene linkage, the chirality transfer to the ligand strands could be clearly determined. The incorporation of the anthracenyl moieties in **219** and **220** allowed the observation of negative CD exciton chirality at low energy wavelengths, confirming the proximity of the helical strands.

Lin and coworkers have reacted pyridine-containing binaphthyl-based ligands with [ClRe(CO)$_5$], resulting in the formation of the chiral molecular squares **221–224** (Scheme 6.29) [79]. ^{1}H NMR spectra of the squares showed well-resolved signals that indicated a single ligand environment and suggested that cyclization

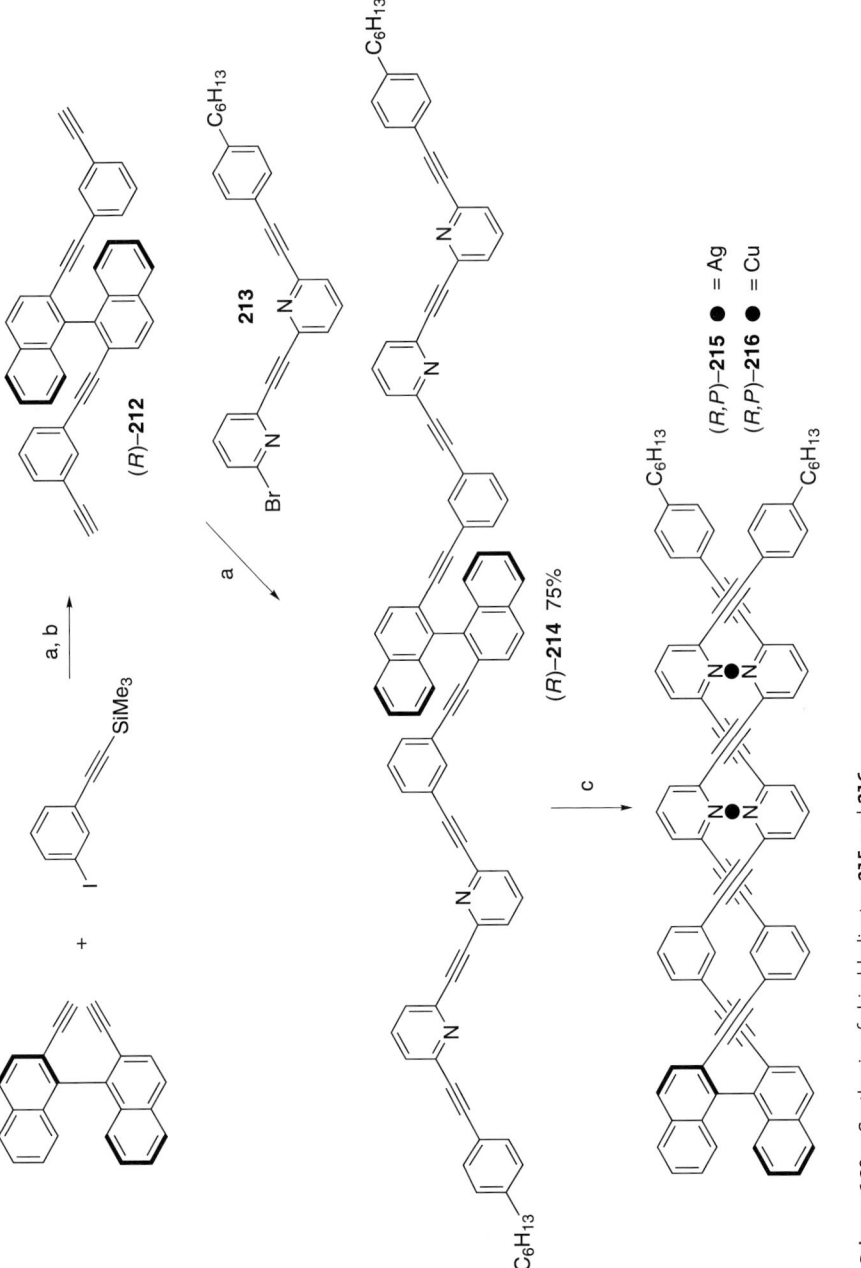

Scheme 6.28. Synthesis of chiral helicates **215** and **216**. Reagents and conditions: (a) [Pd(PPh$_3$)$_4$], CuI, i-Pr$_2$NH, toluene; (b) K$_2$CO$_3$, THF, MeOH; (c) AgPF$_6$, THF or [Cu(CH$_3$CN)$_4$]PF$_6$, CH$_2$Cl$_2$.

6.2 Strategies for Achieving Chiral Macrocycles or Cyclophanes | 275

(R,P)-217 ● = Ag
(R,P)-218 ● = Cu

(R,P)-219 ● = Ag
(R,P)-220 ● = Cu

occurs enantioselectively. The authors studied the luminescence quenching of **224** in the presence of chiral amino alcohols and found that the luminescence of enantiopure **224** is quenched at different rates by the two enantiomers of 2-amino-1-propanol. The degree of enantioselective binding observed for **224** was significantly greater than that observed when the same experiment was performed with the free ligand.

Changing the position of the pyridyl component on the binaphthalene core provided yet another geometric variation of this theme, and Lin and coworkers have shown how such a strategy can lead to a range of self-assembled macrocycles of variable size and substitution such as **225–230** [80]. All macrocycles could be produced with either Pd(II) or Pt(II) linkages in good to excellent yield (>74%, in all cases) and all are highly luminescent in solution.

Chiral binaphthyl ligands containing bipyridine groups have also been realized. In 2002 Lützen and coworkers reported the synthesis of ligands **231–233** (both (S)- and (R)-enantiomers) [81]. In addition to bipyridine groups, ligands **231–233** also possess additional, inwardly directed alkoxy groups (Fig. 6.9). Ligand **231** was treated with numerous metal salts and it was found to form discrete dinuclear metal complexes with [Ag(CH$_3$CN)$_2$(BF$_4$)] (Fig. 6.9) and Zn(BF$_4$)$_2$. X-ray crystallographic analysis was performed on the zinc complex [Zn$_2$(**231**)$_3$](BF$_4$)$_2$ and this, in combination with one- and two-dimensional NMR spectroscopic experiments, showed that both the Ag- and Zn-complexes are formed as a single diastereomer.

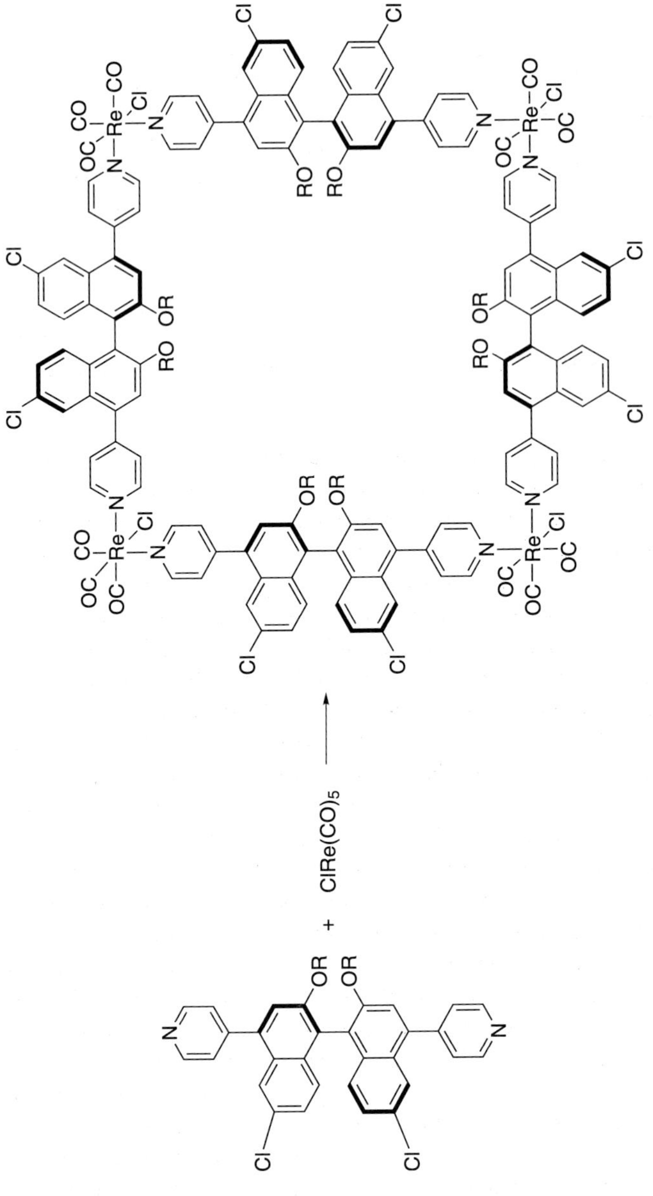

6 Chiral Carbon-rich Macrocycles and Cyclophanes

221 R = Et
222 R = t-BuMe$_2$Si
223 R = CH$_2$Ph
224 R = H

Scheme 6.29. Synthesis of chiral squares **221–224**.

228 R = Et, M = Pd or Pt
229 R = t-BuMe$_2$Si, M = Pd or Pt
230 R = C$_2$H$_4$OC$_2$H$_4$OCH$_2$, M = Pd or Pt

225 R = Et, M = Pd or Pt
226 R = t-BuMe$_2$Si, M = Pd or Pt
227 R = C$_2$H$_4$OC$_2$H$_4$OCH$_2$, M = Pd or Pt

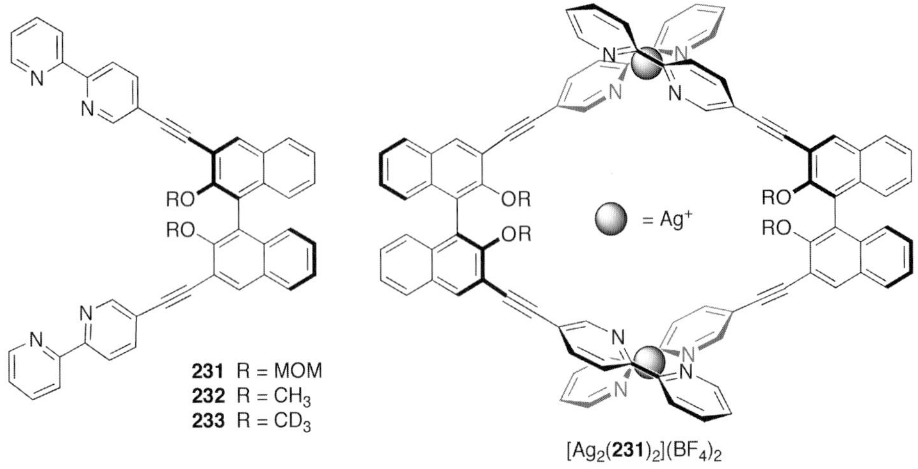

Figure 6.9. Ligands **231–233** and dimeric complex of **231** with silver.

Rather than relying upon self-assembly processes, it is also possible to synthesize metal-containing macrocycles by the direct formation of metal–carbon bonds. Tilley and Schafer have reported a unique, diastereoselective synthesis of zirconocene-containing metallamacrocycles [82]. As outlined in Scheme 6.30, treatment of racemic BINOL-based diyne **234** with Negishi-type zirconocene coupling conditions led to the diastereoselective formation of only two enantiomeric stereoisomers of metallamacrocycle (R,R)- and (S,S)-**235** in an excellent 91% yield. Use of (R)-**234** gave the expected homochiral enantiopure macrocycle (R,R)-**235**, confirming the diastereoselectivity of this process.

X-ray crystallographic analysis of racemic **235** confirmed the helical structure and showed a large dihedral angle of 100° between the planes of the naphthyl rings (Fig. 6.10 (a)). The metal-free analog **236** was produced by treating **235** with benzoic acid, and it was also characterized crystallographically (Fig. 6.10 (b)).

The authors surmised that a reduction in the allowed dihedral angle between the naphthyl rings could result in the formation of even larger metallamacrocycles. As shown in Scheme 6.31, use of the tethered BINAP derivative rac-**237** results in the diastereoselective formation of the homochiral trimers **238** in an excellent 80% yield. X-ray crystallographic analysis of **238** indicated a smaller average dihedral angle for the naphthyl rings of 60° and confirmed the cyclic structure. In the solid-state, compound **238** is highly folded into a C_2-symmetric conformation, and NMR spectroscopic analysis indicated that this conformation is maintained in solution. As with compound **235**, metallamacrocycle **238** could be demetallated to form the macrocycle **239**.

One of the most common methods used to install transition metal centers within carbon-rich structures takes advantage of the easily constructed metal–acetylide bond. In addition to their facile preparation, insertion of a metal center into an unsaturated material can introduce a range of exciting properties not pres-

6.2 Strategies for Achieving Chiral Macrocycles or Cyclophanes | 279

Scheme 6.30. Synthesis of macrocycles **235** and **236**. Reagents and conditions: (a) [Cp$_2$ZrCl$_2$], BuLi, THF, −78 °C to rt; (b) PhCO$_2$H, toluene.

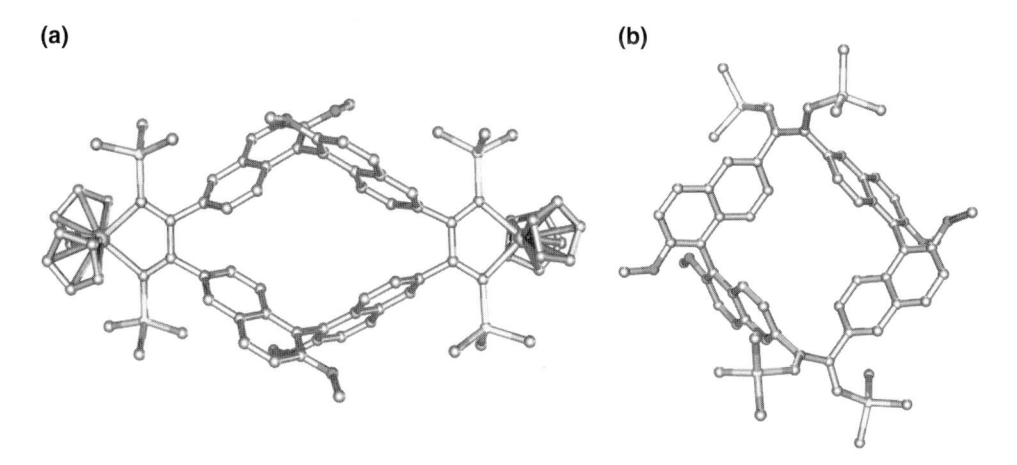

Figure 6.10. Solid-state structure† of macrocycles **235** (a) and **236** (b).

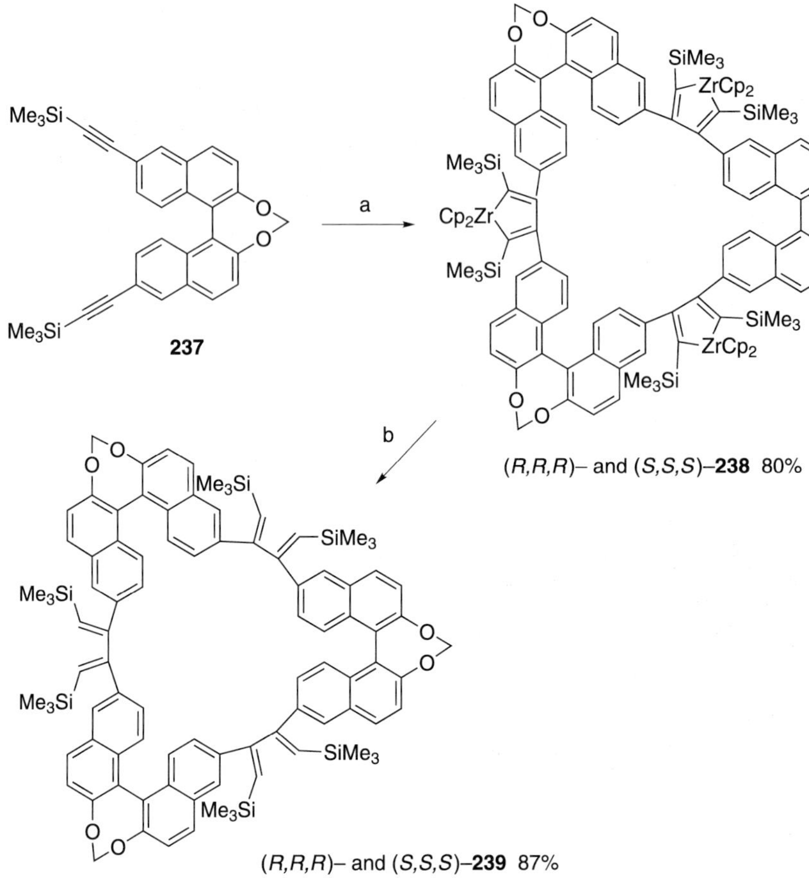

Scheme 6.31. Synthesis of macrocycles **238** and **239**. Reagents and conditions: (a) [Cp$_2$ZrCl$_2$], BuLi, THF, −78 °C to rt; (b) PhCO$_2$H, toluene.

ent within the purely organic structure. While numerous metal–acetylide based metallamacrocycles have been achieved [83], there are still very few examples of chiral derivatives.

Lin and coworkers have prepared macrocyclic structures based on the incorporation of metal–acetylide linkages. When the acetylide ligand is chiral, this approach can easily lead to the formation of chiral superstructures. In 2002 this group reported the synthesis of chiral organometallic triangles for use in asymmetric catalysis [84]. As detailed in Scheme 6.32, treatment of a precursor diyne with [cis-Pt(PEt$_3$)$_2$Cl$_2$] using Sonogashira's conditions [85] leads to the formation of triangles **240–242** in moderate yields. Formation of **243** under identical conditions was not successful, so this species was instead generated by desilylation of compound **242**. The authors exploited the presence of the hydroxy groups in macro-

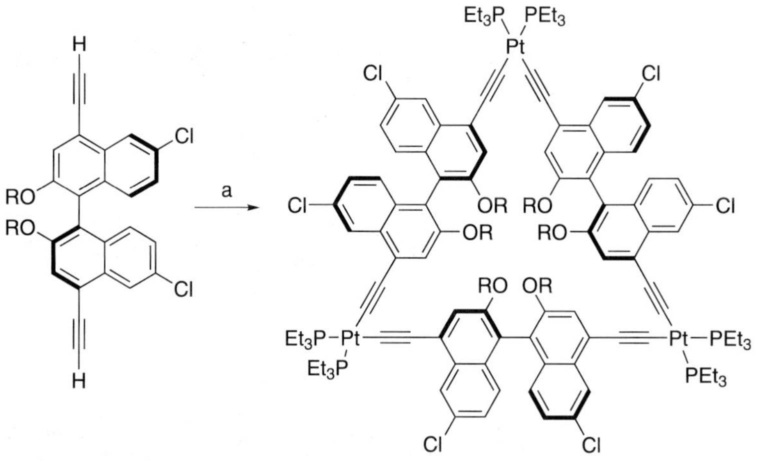

Scheme 6.32. Synthesis of chiral organometallic triangles 240–243. Reagents and conditions: (a) [cis-Pt(PEt$_3$)$_2$Cl$_2$], CuI, Et$_2$NH, CH$_2$Cl$_2$; (b) TBAF, THF.

240 R = Et, 38%
241 R = Me, 32%
242 R = t-BuMe$_2$Si, 45%
243 R = H, 91%

cycle 243 by using this species for enantioselective catalytic diethylzinc additions to aromatic aldehydes. They found that Ti(IV) complexes of ligand 243 catalyzed the addition of diethylzinc to aromatic aldehydes, providing chiral secondary alcohols in approximately 90% ee. The acyclic ligand, by contrast, provided only 80% ee under the same reaction conditions.

This same general protocol using Pt(II) diimine complex 244 as a precursor afforded chiral molecular triangle 245 and squares 246–247 (Scheme 6.33) [86]. Whereas both triangle 245 and square 246 were formed and isolated from the self-assembly reaction of an acetyl protected BINOL precursor, triangle 245 cleanly converts to square 246 when left in solution over time at room temperature. Conversely, square 247 was the only product isolated from the reaction of 244 with an ethylether protected BINOL precursor. The photophysical properties of these and related species were evaluated in solution, and molecular square 247 was also incorporated into a multilayer LED. Severe aggregation was observed for 247 in thin films, however, resulting in poor device performance.

Using analogous synthetic conditions, Lin and coworkers have also reported the synthesis of dimeric metallacyclophanes 248–250 in which the acetylenes are installed at the 6,6'-positions (Fig. 6.11) [87]. As with compound 243, the free hydroxy-containing macrocycle 250 was not directly accessible, but was instead formed via the hydrolysis of compound 249. Macrocycle 250 was evaluated as a ligand for the asymmetric addition of diethylzinc to aromatic aldehydes. It was

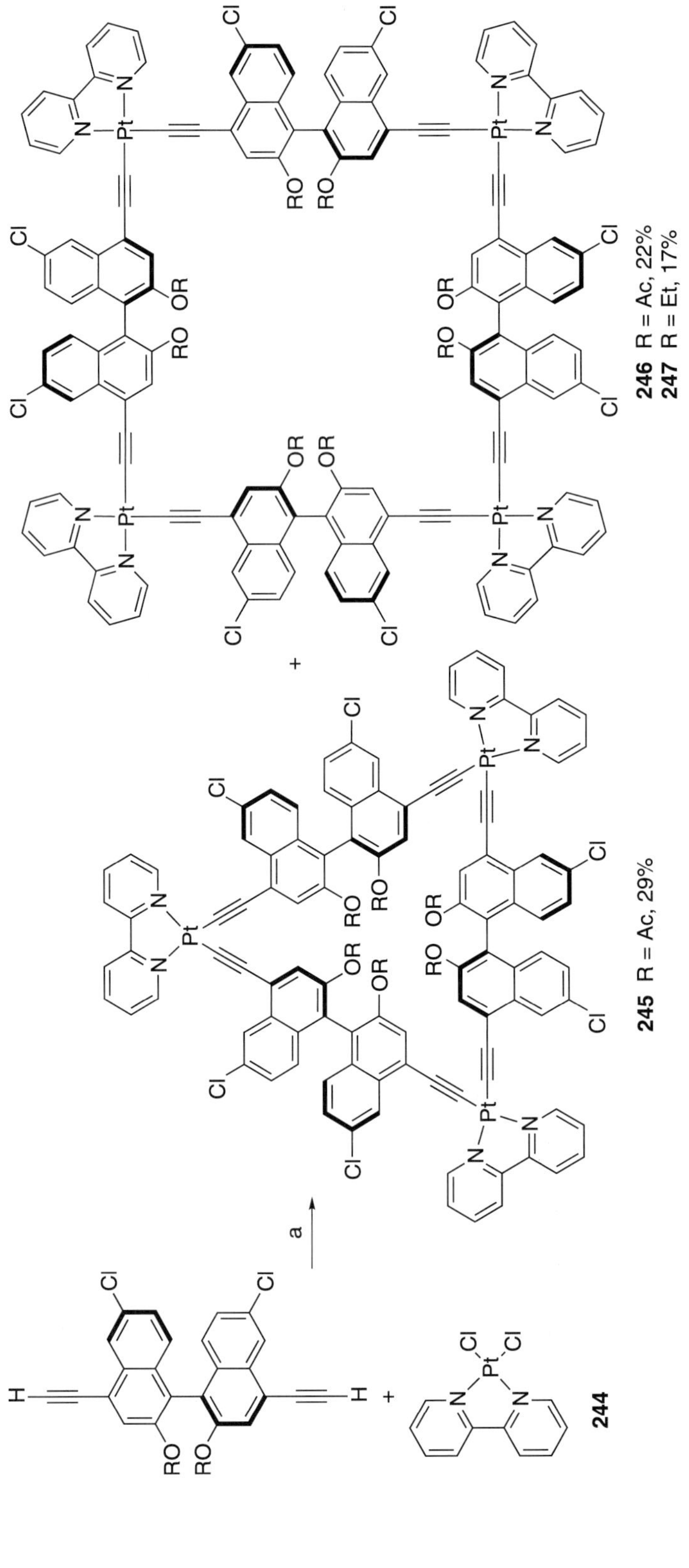

Scheme 6.33. Synthesis of chiral organometallic macrocycles **245–247**. Reagents and conditions: (a) CuI, Et$_2$NH, CH$_2$Cl$_2$.

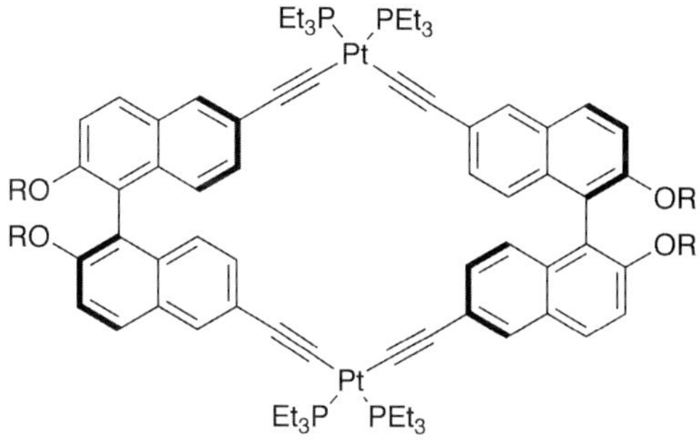

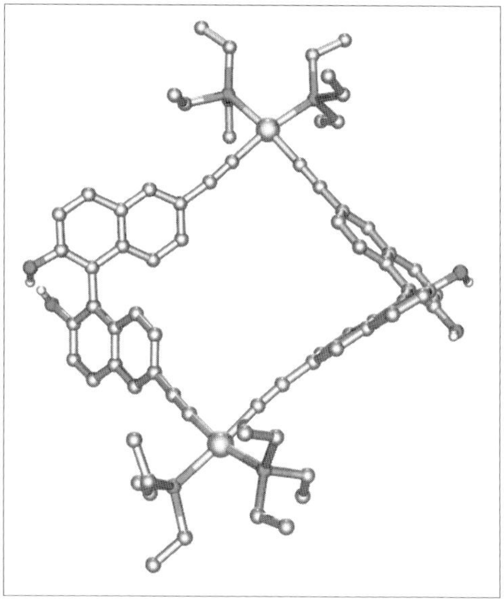

Figure 6.11. Chiral platinacycles **248–250** and solid-state structure[†] of **250** (inset).

248 R = Et
249 R = Ac
250 R = H

found that this macrocyclic ligand, which contains only exotopic hydroxy groups, does catalyze this transformation, but provides lower enantioselectivities than **243**.

Lin achieved higher enantioselectivities (79–91% ee) for the asymmetric addition of diethylzinc to aromatic aldehydes upon the installation of ligand binding sites in an endotopic fashion, as in metallamacrocycles **251–253** [88]. All three of these macrocycles were synthesized in good yield, and compounds **251** and **253** were both characterized crystallographically.

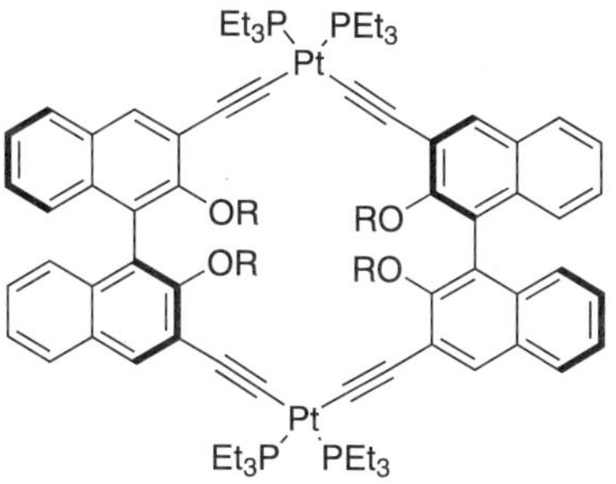

251 R = Me
252 R = Ac
253 R = H

Chiral molecular polygons have been prepared via treatment of either (*R*)- or (*S*)-**254** with [*trans*-PtCl$_2$(PEt$_3$)$_2$] (Scheme 6.34) [89]. This one-pot reaction affords a mixture of polygons **255–260**, with highest yields observed for the tetramer **256**

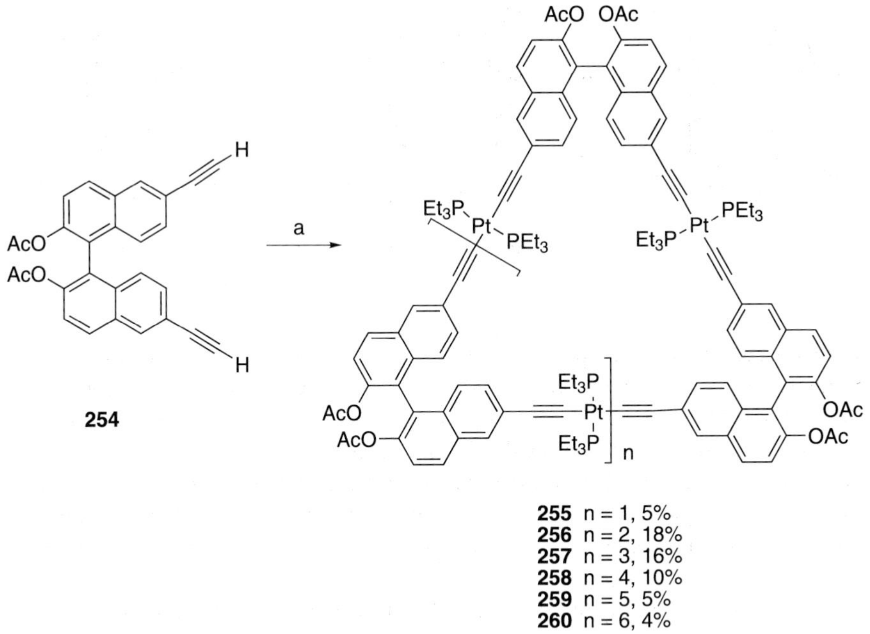

Scheme 6.34. Synthesis of chiral polygons **255–260**. Reagents and conditions: (a) [*trans*-PtCl$_2$(PEt$_3$)$_2$], CuCl, Et$_2$NH, CH$_2$Cl$_2$.

255 n = 1, 5%
256 n = 2, 18%
257 n = 3, 16%
258 n = 4, 10%
259 n = 5, 5%
260 n = 6, 4%

and pentamer **257**. The chiral polygons could be separated via silica gel chromatography and were all fully characterized. This same group has also recently reported the synthesis of even larger, nanoscopic and mesoscopic metallacycles by applying an analogous synthetic approach [90].

Tykwinski and coworkers have recently reported a novel synthetic method for the introduction of chirality into metal-acetylide based macrocycles [91]. As shown in Scheme 6.35, oligomer **261** was desilylated and subsequently treated with [PtCl$_2$(PPh$_3$)$_2$] using the reaction conditions described by Sonogashira and coworkers to provide the *trans*-Pt-acetylide macrocycle **262**. Treatment of the achiral macrocycle **262** with a chiral, chelating diphosphine ligand, (*S,S*)-chiraphos ((*S,S*)-**263**) results in the displacement of the PPh$_3$ groups via ligand exchange [92] providing the *cis*-chelated chiral macrocycle **264** in good yield.

The authors demonstrated the scope of this synthetic methodology via the preparation of chiral macrocycles **265–268** and also several acyclic species. This strategy was equally effective for carbocyclic systems such as **264**, **265**, and **267**, as well as pyridine-containing macrocycles **266** and **268**. Both enantiomers of macrocycle **268** were prepared by simply utilizing the opposite enantiomer of the chelating diphosphine ligand, (*R,R*)-**263**. This highlighted the economy of this methodology, which allows for the divergent and high yielding formation of either enantiomeric macrocycle from a common achiral precursor.

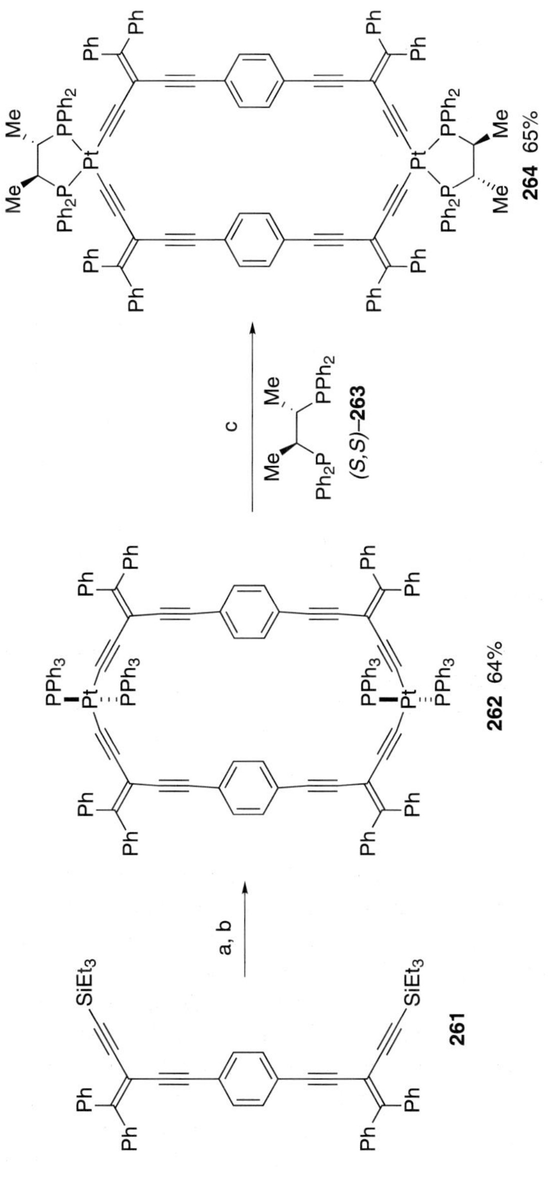

Scheme 6.35. Synthesis of chiral metallamacrocycle **264**. Reagents and conditions: (a) NaOH, THF/MeOH; (b) [PtCl$_2$(PPh$_3$)$_2$], CuI, Et$_2$NH, 50 °C; (c) CH$_2$Cl$_2$ or CD$_2$Cl$_2$.

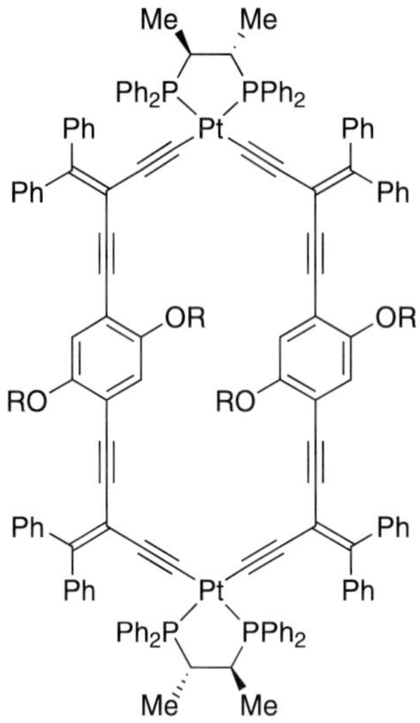

265 R = (CH$_2$)$_7$CH$_3$, 98%

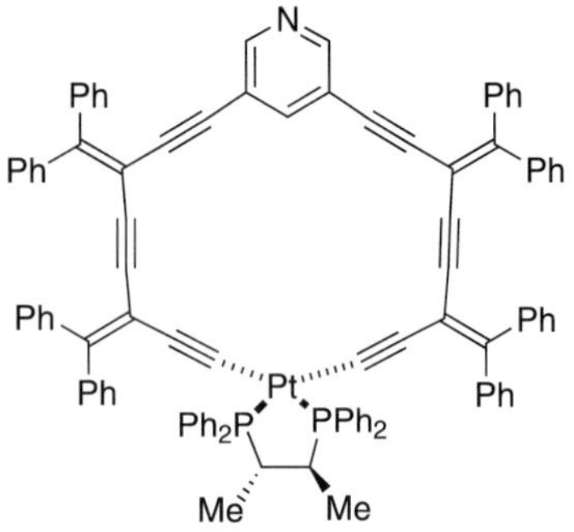

266 73%

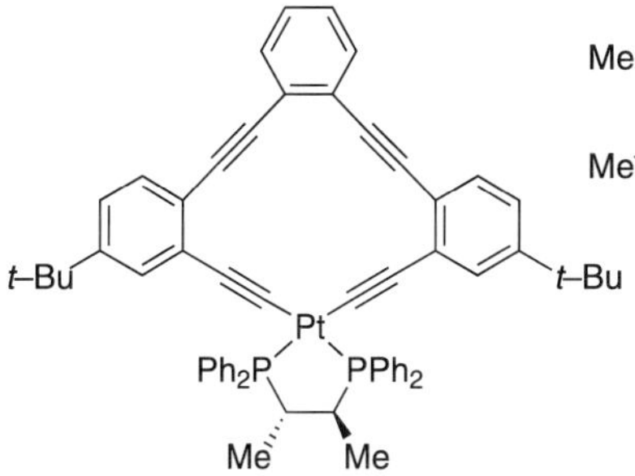

267 83%

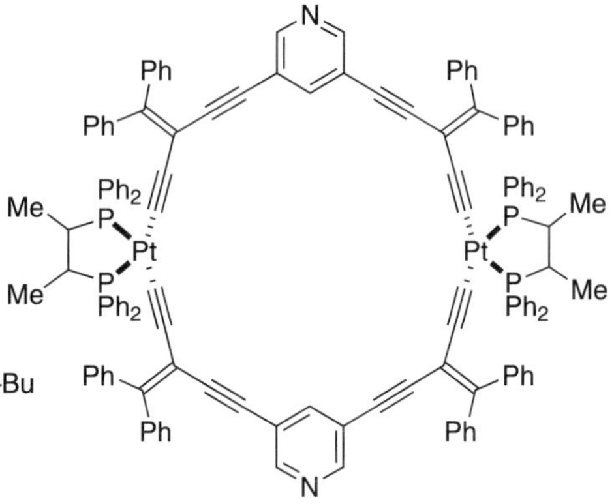

(S,S)/(S,S)–**268** 96%
(R,R)/(R,R)–**268** 92%

X-ray crystallographic analysis of platinacycle **266** confirmed its chiral structure and, interestingly, showed that the (S,S)-chiraphos ligand had adopted an unusual λ-conformation (Fig. 6.12). CD spectroscopy was used to demonstrate that in several cases, a significant degree of chirality transfer from the chelated chiral ligand to the macrocyclic species occurs.

6.3
Conclusions

Regardless of the synthetic strategy employed for their formation, carbon-rich chiral macrocycles and cyclophanes can be realized with both diverse structure and function. Modern synthetic techniques, in combination with more readily available

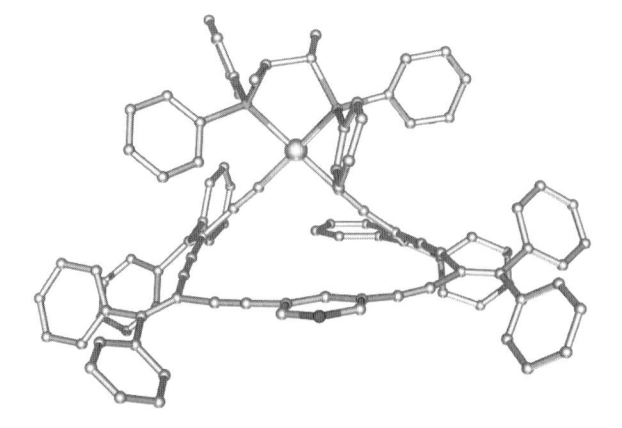

Figure 6.12. Solid-state structure† of chiral platinacycle **266**.

chiral building blocks, have greatly expanded the boundaries of this class of chiral macromolecules. Many chiral macrocycles and cyclophanes are still produced as racemic mixtures that are difficult or impossible to resolve. An increasing number, however, have been synthesized as a single enantiomer. Such achievements usher in a new era for this class of compounds: they have evolved from the realm of fundamentally interesting molecules into that of technologically useful materials.

6.4
Experimental: Selected Procedures

6.4.1
Cyclophane 12

Bis(2-iodophenyl)ethyne (323 mg, 0.75 mmol) was dissolved in mixture of H_2O/THF/Et_3N (0.01:1:5, v/v) charged with [$PdCl_2(PPh_3)_2$] (0.03 equiv.), CuI (0.06 equiv.), and KOH (10 equiv). In a separate vessel, **5** (663 mg, 1.75 mmol) was dissolved in Et_3N (1 M solution). Both solutions were degassed vigorously by either bubbling of N_2 or by three freeze–pump–thaw cycles. The solution of **5** was added via syringe pump to the bis(2-iodopenyl)ethyne solution over 16 h under nitrogen atmosphere at 50 °C. Upon completion, the mixture was concentrated in vacuo. The residue was dissolved in CH_2Cl_2 and filtered through a thick cake of silica gel using CH_2Cl_2. Solvent removal and chromatography on silica gel (4:1 hexanes/CH_2Cl_2) gave polyyne **11** (354 mg, 60% yield) as a dark orange gum.

A 0.01 M solution of **11** (100 mg, 0.127 mmol) in THF/MeOH (10:1) was treated with TBAF (1 M in THF, 0.32 mL, 0.32 mmol, 2.5 equiv.) at rt. The reaction was monitored by TLC and was complete within 30 min. The reaction mixture was diluted with Et$_2$O, washed with water (3×), brine (2×), and dried over MgSO$_4$. The solution was concentrated in vacuo and the resultant oil was dissolved in pyridine (ca. 10 mL). The pyridine solution was slowly added via syringe pump to a round bottom flask charged with Cu(OAc)$_2$ (25 equiv.), CuCl (20 equiv.), and pyridine (0.005 M solution based on **11**) at 60 °C. The addition was done under an ambient atmosphere over ca. 16 h. Upon completion of the reaction, the mixture was concentrated in vacuo. The residue was dissolved in CH$_2$Cl$_2$ and filtered through a thick cake of silica gel using CH$_2$Cl$_2$. Solvent removal and chromatography on silica gel (3:1 hexanes/CH$_2$Cl$_2$) gave cyclophane **12** (35 mg, 59% yield) as a cream-colored solid [11].

6.4.2
Cyclophane 43

A solution of TBAF (1 M in THF, 1.17 mL, 1.17 mmol, 2.2 equiv.) in pyridine/ether (3:1, 15 mL) was added over 5 h by syringe pump to a stirred solution of Cu(OAc)$_2$ (579 mg, 3.19 mmol, 6 equiv) and **42** (500 mg, 0.53 mmol, 1 equiv.) in pyridine/ether (3:1, 350 mL). The reaction was stirred for 18 h at rt. HCl (10% aq) and CH$_2$Cl$_2$ were added to the reaction, the organic phase separated, washed sequentially with HCl (10% aq), brine, dried over MgSO$_4$. Concentration to a dark yellow oil and chromatography on silica gel (petroleum ether/CH$_2$Cl$_2$, 9:1) gave a yellow glass (50 mg, 15%) as a mixture of isomers. Semipreparative size exclusion chromatography (SEC) (Jordi Gel DVB 100A, CH$_2$Cl$_2$) gave pure samples of butterfly and bowtie **43** [16].

6.4.3
Cyclophane 86

To a stirred solution of **85** (0.250 g, 0.456 mmol) in THF (15 mL) and distilled water (0.5 mL) was added TBAF (1 M in THF, 1.10 mL, 1.10 mmol, 2.4 equiv.) and the reaction was stirred at rt for 20 h. The solvent was removed in vacuo, distilled water (20 mL) was added to the residue and the mixture was briefly ultrasonicated. The suspended solid was isolated by filtration, washed with excess distilled water, and air dried. The crude product was purified by flash chromatography on silica gel (1% MeOH/CH$_2$Cl$_2$) and then washed with Et$_2$O and air dried to give the deprotected product (0.177 g, 96%) as white flakes (mp 201.0–201.5 °C dec). To a solution of deprotected **85** (0.135 g, 0.334 mmol) in pyridine (230 mL) was added a solution of CuCl (0.022 g, 0.22 mmol) in pyridine (3 mL). Oxygen was bubbled through the olive green reaction mixture, and then the solution was vigorously stirred for 120 h. (pyridine was added during the course of the reaction to keep the volume to ca. 230 mL). The reaction solution was then concentrated to ca. 4 mL, 20 mL of saturated aqueous KCN was added, and the suspension stirred for 24 h at rt. The solid was isolated by filtration under vacuum, washed with excess distilled water, and air dried. The resulting solid was suspended in 200 mL of CHCl$_3$, the mixture boiled and then filtered to remove the insoluble byproducts.

The solution was concentrated, the residue dissolved in boiling toluene (10–15 mL) and chromatographed on alumina (basic, activity III) with toluene as the eluant. The resulting product was finally purified by extraction with boiling acetone (2 × 150 mL) to give **86** (0.045 g, 34%) as a light-sensitive amorphous white solid [35].

6.4.4
(R)-(+)-124

BuLi (9.7 mL of a 2.5 M solution in hexane, 24.3 mmol) was added to a solution of (R)-(+)-2,2′-dibromo-1,1′-binaphthyl (4.00 g, 9.71 mmol, 80% ee) in Et_2O (400 mL) at −35 °C. After stirring for 2 h, the reaction mixture was cooled with a −78 °C bath, and $CuBr_2$ (13.0 g, 58.2 mmol) was added. The reaction mixture was allowed to warm to rt over 12 h, and cold water was then added. The organic layer was separated, washed with HCl, water, aqueous Na_2CO_3, and then dried over $MgSO_4$. Concentration in vacuo afforded 3.17 g of the crude product. A second iteration of the reaction, starting from (R)-(+)-2,2′-dibromo-1,1′-binaphthyl (3.34 g, 8.11 mmol, 85% ee), afforded 2.65 g of the crude product. The combined crude products were then purified three times by column chromatography (silica gel, TLC grade, hexane/$CHCl_3$ 3:1) to give (R,R)-(+)-**124** as a white powder (1.00 g, 22% yield, >95% ee) [51].

6.4.5
Complex 262

Oligomer **261** (58 mg, 0.076 mmol) was dissolved in wet THF/MeOH (1:1, 25 mL), NaOH (ca. 2.2 equiv.) was added, and the resulting solution was stirred, in the presence of air at room temperature, until TLC analysis indicated complete conversion to the desilylated product. Ether (25 mL) was added, and the resulting solution was washed with saturated NH_4Cl (2 × 25 mL) and dried. The solvent was reduced to ca. 1 mL and the resulting deprotected alkyne was added, with no further purification, to a degassed solution of $[PtCl_2(PPh_3)_2]$ (61 mg, 0.077 mmol) in Et_2NH (225 mL). A catalytic amount of CuI (8 mg, 0.04 mmol) was added and the mixture was stirred at 50 °C for 14 h. Ether and CH_2Cl_2 (75 mL each) were added, and the resulting solution was washed with water (2 × 50 mL), saturated NH_4Cl (6 × 100 mL) and dried over $MgSO_4$. Solvent removal in vacuo (*without heating*) and purification by gradient column chromatography on silica gel (hexanes/CH_2Cl_2 2:1 → neat CH_2Cl_2) afforded **262** (61 mg, 64%) as a bright yellow solid [91, 92].

6.4.6
Complex 264

(2S,3S)-(−)-Bis(diphenylphosphino)butane **263** (3.5 mg, 0.0082 mmol) was added to a solution of **262** (10 mg, 0.0040 mmol) in CD_2Cl_2 (1 mL), and the mixture was stirred at rt for 3 d. Solvent removal followed by purification on a gradient silica gel column (hexanes/CH$_2$Cl$_2$ 1:3 → CH$_2$Cl$_2$/acetone 3:1) afforded **264** (6 mg, 65%) as a bright yellow solid [91].

Acknowledgments

Financial support as provided by the University of Alberta and the Natural Sciences and Engineering Research Council of Canada (NSERC) is gratefully acknowledged. K.C. thanks NSERC for a post-graduate scholarship and a post-doctoral fellowship. We thank Dr. Bob McDonald for his expertise and generous help with generating the solid-state structures.

Abbreviations

Ac	acetyl
BINAP	2,2'-bis(diphenylphosphino)-1,1'-binaphthalene
Bn	benzyl
Bz	benzoyl
CD	circular dichroism
Cp	cyclopentadienyl
dba	dibenzylideneacetone
DFT	density functional theory
DIOP	isopropylidene-2,3-dihydroxy-1,4-bis(diphenylphosphino)butane
DME	1,2-dimethoxyethane
DMF	*N*,*N*-dimethylformamide
dppf	bis(diphenylphosphino)ferrocene
dppp	bis(diphenylphosphino)propane
GPC	gel permeation chromatography
HPLC	high performance liquid chromatography
LED	light emitting diode
Mes	mesityl
MOM	methoxymethyl
NBS	*N*-bromosuccinimide
NMR	nuclear magnetic resonance
py	pyridine
TBAF	tetrabutylammonium fluoride
Tf	trifluoromethanesulfonyl
THF	tetrahydrofuran
TLC	thin layer chromatography
TMEDA	*N*,*N*,*N*',*N*'-tetramethylethylenediamine
TMSA	trimethylsilylacetylene
UV–vis	ultraviolet visible

References

† Solid-state structure depicted in ball and stick format using POV-Ray 3.6.0 (http://www.povray.org) and based on X-ray crystallog-raphic analysis and coordinates obtained from the Cambridge Crystallographic Data Centre (CCDC).

1 Synthesis of macrocycles and cyclophanes, see: (a) C. S. JONES, M. J. O'CONNOR, M. M. HALEY, in *Acetylene Chemistry: Chemistry, Biology, and Material Science* (F. DIEDERICH, P. J. STANG, R. R. TYKWINSKI (eds.)), Wiley-VCH, Weinheim, **2005**, Ch. 8; (b) Y. TOBE, M. SONODA, in *Modern Cyclophane Chemistry* (R. GLEITER, H. HOPF (eds.)), Wiley-VCH, Weinheim, **2004**, Ch. 1; (c) S. HÖGER, *Chem. Eur. J.* **2004**, *10*, 1320–1329; (d) Y. YAMAGUCHI, Z. YOSHIDA, *Chem. Eur. J.* **2003**, *9*, 5430–5440; (e) D. ZHAO, J. S. MOORE, *Chem. Commun.* **2003**, 807–818; (f) C. GRAVE, A. D. SCHLÜTER, *Eur. J. Org. Chem.* **2002**, 3075–3098; (g) A. DE MEIJERE, S. I. KOZHUSHKOV, *Top. Curr. Chem.* **1999**, *201*, 1–42.

2 (a) G. SIMONNEAUX, P. LE MAUX, *Coord. Chem. Rev.* **2002**, *228*, 43–60; (b) N. KOBAYASHI, *Coord. Chem. Rev.* **2001**, *219*, 99–123.

3 T. FEKNER, J. GALLUCCI, M. K. CHAN, *J. Am. Chem. Soc.* **2004**, *126*, 223–236, and references therein.

4 E. ENGELDINGER, D. ARMSPACH, D. MATT, *Chem. Rev.* **2003**, *103*, 4147–4173.

5 T. W. BELL, N. M. HEXT, *Chem. Soc. Rev.* **2004**, *33*, 589–598, and references therein.

6 H. A. STAAB, E. WEHINGER, W. THORWART, *Chem. Ber.* **1972**, *105*, 2290–2309.

7 H. IRNGARTINGER, *Chem. Ber.* **1973**, *106*, 761–772.

8 D. SOLOOKI, J. D. BRADSHAW, C. A. TESSIER, W. J. YOUNGS, R. F. SEE, M. CHURCHILL, J. D. FERRARA, *J. Organomet. Chem.* **1994**, *470*, 231–236.

9 R. BOESE, A. J. MATZGER, K. P. C. VOLLHARDT, *J. Am. Chem. Soc.* **1997**, *119*, 2052–2053.

10 M. M. HALEY, M. L. BELL, S. C. BRAND, D. B. KIMBALL, J. J. PAK, W. B. WAN, *Tetrahedron Lett.* **1997**, *38*, 7483–7486.

11 M. L. BELL, R. C. CHIECHI, C. A. JOHNSON, D. B. KIMBALL, A. J. MATZGER, W. B. WAN, T. J. R. WEAKLEY, M. M. HALEY, *Tetrahedron* **2001**, *57*, 3507–3520.

12 A. J. BOYDSTON, L. BONDARENKO, I. DIX, T. J. R. WEAKLEY, H. HOPF, M. M. HALEY, *Angew. Chem.* **2001**, *113*, 3074–3077; *Angew. Chem. Int. Ed.* **2001**, *40*, 2986–2989.

13 S. K. COLLINS, G. P. A. YAP, A. G. FALLIS, *Org. Lett.* **2000**, *2*, 3189–3192.

14 S. K. COLLINS, G. P. A. YAP, A. G. FALLIS, *Angew. Chem.* **2000**, *112*, 393–396; *Angew. Chem. Int. Ed.* **2000**, *39*, 385–388.

15 M. OHKITA, K. ANDO, T. SUZUKI, T. TSUJI, *J. Org. Chem.* **2000**, *65*, 4385–4390.

16 M. A. HEUFT, S. K. COLLINS, A. G. FALLIS, *Org. Lett.* **2003**, *5*, 1911–1914.

17 D. ZHANG, C. A. TESSIER, W. J. YOUNGS, *Chem. Mater.* **1999**, *11*, 3050–3057.

18 W. B. WAN, S. C. BRAND, J. J. PAK, M. M. HALEY, *Chem. Eur. J.* **2000**, *6*, 2044–2052.

19 (a) T. NISHINAGA, T. KAWAMURA, K. KOMATSU, *J. Org. Chem.* **1997**, *62*, 5354–5362; (b) T. NISHINAGA, T. KAWAMURA, K. KOMATSU, *Chem. Commun.* **1998**, 2263–2264.

20 R. M. MCQUILKIN, P. J. GARRATT, F. SONDHEIMER, *J. Am. Chem. Soc.* **1970**, *92*, 6682–6683.

21 X. SHEN, D. M. HO, R. A. PASCAL JR., *Org. Lett.* **2003**, *5*, 369–371.

22 K. UTSUMI, T. KAWASE, M. ODA, *Chem. Lett.* **2003**, *32*, 412–413.

23 B. THULIN, O. WENNERSTRÖM, *Acta Chem. Scand. B* **1983**, *37*, 297–301.

24 B. THULIN, O. WENNERSTRÖM, *Tetrahedron Lett.* **1977**, *18*, 929–930.

25 D. J. CRAM, K. C. DEWHIRST, *J. Am Chem. Soc.* **1959**, *81*, 5963–5971.

26 (a) B. THULIN, O. WENNERSTRÖM, I. SOMFAI, B. CHMIELARZ, *Acta Chem. Scand. B* **1977**, *31*, 135–140; (b) I.

Raston, O. Wennerström, *Acta Chem. Scand. B* **1982**, *36*, 655–660; (c) D. Tanner, B. Thulin, O. Wennerström, *Acta Chem. Scand. B* **1979**, *33*, 443–448.

27 M. A. Romero, A. G. Fallis, *Tetrahedron Lett.* **1994**, *35*, 4711–4714.

28 Y. Rubin, T. C. Parker, S. I. Khan, C. L. Holliman, S. W. McElvany, *J. Am. Chem. Soc.* **1996**, *118*, 5308–5309.

29 (a) Y. Rubin, T. C. Parker, S. J. Pastor, S. Jalisatgi, C. Boulle, C. L. Wilkins, *Angew. Chem.* **1998**, *110*, 1353–1356; *Angew. Chem. Int. Ed.* **1998**, *37*, 1226–1229; (b) Y. Rubin, *Chem. Eur. J.* **1997**, *3*, 1009–1016.

30 (a) Y. Tobe, N. Nakagawa, K. Naemura, T. Wakabayashi, T. Shida, Y. Achiba, *J. Am. Chem. Soc.* **1998**, *120*, 4544–4545; (b) Y. Tobe, H. Nakanishi, M. Sonoda, T. Wakabayashi, Y. Achiba, *Chem. Commun.* **1999**, 1625–1626.

31 (a) U. H. F. Bunz, Y. Rubin, Y. Tobe, *Chem. Soc. Rev.* **1999**, *28*, 107–119; (b) Y. Tobe, N. Nakagawa, J.-y. Kishi, M. Sonoda, K. Naemura, T. Wakabayashi, T. Shida, Y. Achiba, *Tetrahedron* **2001**, *57*, 3629–3636.

32 Y. Tobe, R. Umeda, M. Sonoda, T. Wakabayashi, *Chem. Eur. J.* **2005**, *11*, 1603–1609.

33 H.-E. Högberg, O. Wennerström, *Acta Chem. Scand. B,* **1982**, *36*, 661–667.

34 R. Herges, M. Deichmann, T. Wakita, Y. Okamoto, *Angew. Chem.* **2003**, *115*, 1202–1204; *Angew. Chem. Int. Ed.* **2003**, *42*, 1170–1172.

35 P. N. W. Baxter, *J. Org. Chem.* **2001**, *66*, 4170–4179.

36 A. S. Hay, *J. Org. Chem.* **1962**, *27*, 3320–3321.

37 P. N. W. Baxter, *Chem. Eur. J.* **2002**, *8*, 5250–5264.

38 M. A. Heuft, A. G. Fallis, *Angew. Chem.* **2002**, *114*, 4702–4705; *Angew. Chem. Int. Ed.* **2002**, *41*, 4520–4523.

39 P. N. W. Baxter, *Chem. Eur. J.* **2003**, *9*, 2531–2541.

40 (a) T. Fekner, J. Gallucci, M. K. Chan, *Org. Lett.* **2004**, *6*, 989–992; (b) T. Fekner, J. Gallucci, M. K. Chan, *J. Am. Chem. Soc.* **2004**, *126*, 223–236.

41 N. Kobayashi, T. Fukuda, K. Ueno, H. Ogino, *J. Am. Chem. Soc.* **2001**, *123*, 10740–10741.

42 P. N. W. Baxter, *J. Org. Chem.* **2004**, *69*, 1813–1821.

43 A. Strand, B. Thulin, O. Wennerström, *Acta Chem. Scand. B* **1977**, *31*, 521–523.

44 M. J. Marsella, Z.-Q. Wang, R. J. Reid, K. Yoon, *Org. Lett.* **2001**, *3*, 885–887.

45 M. J. Marsella, R. J. Reid, *Macromolecules* **1999**, *32*, 5982–5984.

46 (a) T. Kauffmann, B. Greving, R. Kriegesmann, A. Mitschker, A. Woltermann, *Chem. Ber.* **1978**, *111*, 1330–1336; (b) T. Kauffmann, H. P. Mackowiak, *Chem. Ber.* **1985**, *118*, 2343–2352.

47 M. J. Marsella, *Acc. Chem. Res.* **2002**, *35*, 944–951.

48 M. J. Marsella, I. T. Kim, F. Tham, *J. Am. Chem. Soc.* **2000**, *122*, 974–975.

49 M. J. Marsella, G. Piao, F. S. Tham, *Synthesis* **2002**, 1133–1135.

50 A. Rajca, A. Safronov, S. Rajca, R. Shoemaker, *Angew. Chem.* **1997**, *109*, 504–507; *Angew. Chem., Int. Ed. Engl.* **1997**, *36*, 488–491.

51 A. Rajca, A. Safronov, S. Rajca, J. Wongsriratanakul, *J. Am. Chem. Soc.* **2000**, *122*, 3351–3357.

52 S. Anderson, U. Neidlein, V. Gramlich, F. Diederich, *Angew. Chem.* **1995**, *107*, 1722–1724; *Angew. Chem., Int. Ed. Engl.* **1995**, *34*, 1596–1600.

53 A. Bähr, A. S. Droz, M. Püntener, U. Neidlein, S. Anderson, P. Seiler, F. Diederich, *Helv. Chim. Acta* **1998**, *81*, 1931–1963.

54 A. S. Droz, F. Diederich, *J. Chem. Soc., Perkin Trans. 1* **2000**, 4224–4226.

55 A. S. Droz, U. Neidlein, S. Anderson, P. Seiler, F. Diederich, *Helv. Chim. Acta* **2001**, *84*, 2243–2289.

56 K. Utsumi, T. Nakamura, T. Kawase, M. Oda, submitted.

57 D. L. An, T. Nakano, A. Orita, J. Otera, *Angew. Chem.* **2002**, *114*, 179–181; *Angew. Chem. Int. Ed.* **2002**, *41*, 171–173.

58 L. Ma, L. Pu, Macromol. Chem. Phys. **1998**, *199*, 2395–2401.

59 H.-C. Zhang, W.-S. Huang, L. Pu, *J. Org. Chem.* **2001**, *66*, 481–487.

60 (a) J. Lin, H.-C. Zhang, L. Pu, *Org. Lett.* **2002**, *4*, 3297–3300; (b) Z.-B. Li, J. Lin, H.-C. Zhang, M. Sabat, M. Hyacinth, L. Pu, *J. Org. Chem.* **2004**, *69*, 6284–6293; (c) J. Lin, Z.-B. Li, H.-C. Zhang, L. Pu, *Tetrahedron Lett.* **2004**, *45*, 103–106.

61 J. K. Judice, S. J. Keipert, D. J. Cram, *J. Chem. Soc., Chem. Commun.* **1993**, 1323–1325.

62 For early examples of helicene based macrocycles, see: (a) M. Joly, N. Defay, R. H. Martin, J. P. Declerq, G. Germain, B. Soubrier-Payen, M. Van Meerssche, *Helv. Chim. Acta* **1977**, *60*, 537–560; (b) P. J. Jessup, J. A. Reiss, *Aust. J. Chem.* **1976**, *29*, 173–178; (c) H. Numan, H. Wynberg, *Tetrahedron Lett.* **1975**, *16*, 1097–1100.

63 B. Thulin, O. Wennerström, *Acta Chem. Scand. B* **1976**, *30*, 688–690.

64 K. Nakamura, H. Okubo, M. Yamaguchi, *Org. Lett.* **2001**, *3*, 1097–1099.

65 Y. Saiki, K. Nakamura, Y. Nigorikawa, M. Yamaguchi, *Angew. Chem.* **2003**, *115*, 5348–5350; *Angew. Chem. Int. Ed.* **2003**, *42*, 5190–5192.

66 Y. Saiki, H. Sugiura, K. Nakamura, M. Yamaguchi, T. Hoshi, J.-i. Anzai, *J. Am. Chem. Soc.* **2003**, *125*, 9268–9269.

67 J. M. Fox, D. Lin, Y. Itagaki, T. Fujita, *J. Org. Chem.* **1998**, *63*, 2031–2038.

68 B. Köhler, V. Enkelmann, M. Oda, S. Pieraccini, G. P. Spada, U. Scherf, *Chem. Eur. J.* **2001**, *7*, 3000–3004.

69 S. Thorand, F. Vögtle, N. Krause, *Angew. Chem.* **1999**, *111*, 3929–3931; *Angew. Chem. Int. Ed.* **1999**, *38*, 3721–3723.

70 (a) M. D. Clay, A. G. Fallis, *Angew. Chem.* **2005**, *117*, 4107–4110. *Angew. Chem. Int. Ed* **2005**, *44*, 4039–4042. (b) For a second, more recent example see: S. Odermatt, J. L. Alonso-Gómez, P. Seiler, M. M. Cid, F. Diederich, *Angew. Chem.* **2005**, *117*, 5203–5207; *Angew. Chem. Int. Ed.* **2005**, *44*, 5074–5078.

71 F. Gasparrini, D. Misiti, M. Pierini, C. Villani, *Org. Lett.* **2002**, *4*, 3993–3996.

72 Y. Tobe, T. Nagai, S. Araki, T. Ichikawa, A. Nomoto, M. Sonoda, K. Hirose, submitted.

73 B. Olenyuk, J. A. Whiteford, P. J. Stang, *J. Am. Chem. Soc.* **1996**, *118*, 8221–8230.

74 P. J. Stang, B. Olenyuk, D. C. Muddiman, R. D. Smith, *Organometallics* **1997**, *16*, 3094–3096.

75 C. Müller, J. A. Whiteford, P. J. Stang, *J. Am. Chem. Soc.* **1998**, *120*, 9827–9837.

76 (a) A. Hori, A. Akasaka, K. Biradha, S. Sakamoto, K. Yamaguchi, M. Fujita, *Angew. Chem.* **2002**, *114*, 3403–3406; *Angew. Chem. Int. Ed.* **2002**, *41*, 3269–3272; (b) A. Hori, H. Kataoka, A. Akasaka, T. Okano, M. Fujita, *J. Polym. Sci., Part A: Polym. Chem.* **2003**, *41*, 3478–3485.

77 T. W. Kim, M. S. Lah, J.-I. Hong, *Chem. Commun.* **2001**, 743–744.

78 A. Orita, T. Nakano, D. L. An, K. Tanikawa, K. Wakamatsu, J. Otera, *J. Am. Chem. Soc.* **2004**, *126*, 10389–10396.

79 S. J. Lee, W. Lin, *J. Am. Chem. Soc.* **2002**, *124*, 4554–4555.

80 S. J. Lee, J. S. Kim, W. Lin, *Inorg. Chem.* **2004**, *43*, 6579–6588.

81 A. Lützen, M. Hapke, J. Griep-Raming, D. Haase, W. Saak, *Angew. Chem.* **2002**, *114*, 2190–2194; *Angew. Chem. Int. Ed.* **2002**, *41*, 2086–2089.

82 L. L. Schafer, T. D. Tilley, *J. Am. Chem. Soc.* **2001**, *123*, 2683–2684.

83 For selected examples see: (a) R. Faust, F. Diederich, V. Gramlich, P. Seiler, *Chem. Eur. J.* **1995**, *1*, 111–117; (b) W. J. Youngs, C. A. Tessier, J. D. Bradshaw, *Chem. Rev.* **1999**, *99*, 3153–3180; (c) E. Bosch, C. L. Barnes, *Organometallics*, **2000**, *19*, 5522–5524; (d) D. Xu, B. Hong, *Angew. Chem.* **2000**, *112*, 1896–1899; *Angew. Chem. Int. Ed.* **2000**, *39*, 1826–1829; (e) P. Siemsen, U. Gubler, C. Bosshard, P. Günter, F. Diederich, *Chem. Eur. J.* **2001**, *7*, 1333–1341; (f)

H. Yao, M. Sabat, R. N. Grimes, F. Fabrizi de Biani, P. Zanello, *Angew. Chem.* **2003**, *115*, 1032–1035; *Angew. Chem. Int. Ed.* **2003**, *42*, 1002–1005; (g) K. Campbell, R. McDonald, M. J. Ferguson, R. R. Tykwinski, *Organometallics* **2003**, *22*, 1353–1355; (h) C. A. Johnson II, M. M. Haley, E. Rather, F. Han, T. J. R. Weakley, *Organometallics* **2005**, *24*, 1161–1172.

84 S. J. Lee, A. Hu, W. Lin, *J. Am. Chem. Soc.* **2002**, *124*, 12948–12949.

85 K. Sonogashira, T. Yatake, Y. Tohda, S. Takahashi, N. Hagihara, *J. Chem. Soc., Chem. Commun.* **1977**, 291–292.

86 L. Zhang, Y.-H. Niu, A. K.-Y. Jen, W. Lin, *Chem. Commun.* **2005**, 1002–1004.

87 H. Jiang, A. Hu, W. Lin, *Chem. Commun.* **2003**, 96–97.

88 J. Hua, W. Lin, *Org. Lett.* **2004**, *6*, 861–864.

89 H. Jiang, W. Lin, *J. Am. Chem. Soc.* **2003**, *125*, 8084–8085.

90 H. Jiang, W. Lin, *J. Am. Chem. Soc.* **2004**, *126*, 7426–7427.

91 K. Campbell, C. A. Johnson II, R. McDonald, M. J. Ferguson, M. M. Haley, R. R. Tykwinski, *Angew. Chem.* **2004**, *116*, 6093–6097; *Angew. Chem. Int. Ed.* **2004**, *43*, 5967–5971.

92 K. Campbell, R. McDonald, M. J. Ferguson, R. R. Tykwinski, *J. Organomet. Chem.* **2003**, *683*, 379–387.

7
Carbon-rich Cycles with Two and More 1,3-Butadiyne Units – Syntheses, Structures and Reactivities

Rolf Gleiter and Daniel B. Werz

7.1
Introduction

Alkyne units in a molecule provide rigid spacers between other fragments, two perpendicular π bonds for conjugation with high energy, and sp centers with hydrogens. The π conjugation attracted synthetic chemists in the 1960s for construction of conjugated systems by coupling of α,ω-polyacetylenes to annulenes [1]. The aromatic character of various dehydro[$4n+2$]annulenes demonstrated by Sondheimer [1] and others [2] shows that alkyne units can replace olefin units in aromatic rings. Furthermore, the insertion of an acetylene moiety in an annulene increases the conformational stability of the dehydroannulene system. With the discovery, of C_{60} as a new allotrope of carbon [3], in 1985, the alkyne units came again into the focus of chemists. The search for new all-carbon molecules [4] and new materials from carbon-rich skeletons [5] was enhanced by the development of novel procedures to assemble alkyne units such as novel metal-catalyzed cross-coupling methologies [6] and new techniques to prepare alkyne moieties [7].

In view of this very broad development of alkyne chemistry we have had to limit our review in this chapter and have focused mainly on our own recent work which was concerned with ring systems containing 1,3-butadiyne units. We do not intend to discuss here the many [$4n+2$]π- and [$4n$]π-dehydroannulenes containing 1,3-butadiyne units [8], the [N]pericyclynes [9], or shape-persistent macrocycles consisting of phenyl rings connected with alkyne units [10, 11].

Our interest in the topic of this chapter was stimulated by earlier spectroscopic investigations [12] on cyclic diynes with two parallel triple bonds in close proximity and transannular reactions between the triple bonds [13, 14]. We were further inspired by the results of Wegner [15] who found out that parallel aligned 1,3-butadiyne units (Scheme 7.1) yielded, upon irradiation or heating, a 1,4-addition in a topochemical reaction [15, 16].

In the resulting conjugated polymer the enediyne fragments adopt the trans configuration when the angle γ between the molecular axis and the stacking axis is approximately 45° and the distance d is less than 430 pm [16] (Scheme 7.1).

These results led us to concentrate on cyclic systems in which two or more 1,3-

Carbon-Rich Compounds. Edited by Michael M. Haley and Rik R. Tykwinski
Copyright © 2006 WILEY-VCH Verlag GmbH & Co. KGaA, Weinheim
ISBN: 3-527-31224-2

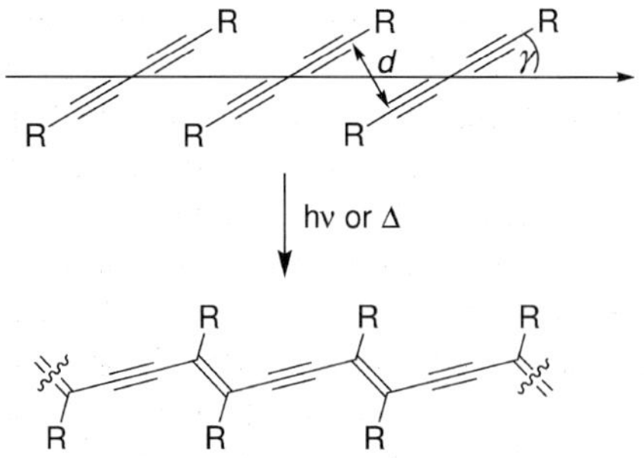

Scheme 7.1. Topochemical reaction of 1,3-butadiynes initiated by light or heat.

butadiyne units are kept in close proximity by nonconjugated bridges of at least two main group atoms.

7.2
Syntheses

The 1,3-butadiyne unit is usually generated by the oxidative coupling of two terminal alkynes. This protocol goes back to 1869 when Glaser prepared 1,4-diphenylbutadiyne from phenylacetylene, CuCl, and NH_3 in the presence of air [18]. Since that time many variations of this procedure and new approaches have been published. The most frequently used procedures are the Cu-mediated coupling reactions according to Eglinton [19] with $Cu(OAc)_2$ in MeOH/pyridine or the protocol proposed by Hay [20] using CuCl, TMEDA and O_2. These procedures are summarized in excellent reviews [21, 22]. As an example we show the oxidative coupling of the α,ω-diynes **1(n)** (Scheme 7.2).

Treatment of **1(3)–1(5)** [19, 23–26] with $Cu(OAc)_2$ in pyridine under high-dilution conditions produced as the main products the dimers **2(3)–2(5)** together with higher oligomers **3(n)–6(n)**. The severely strained **2(2)** was not found under Eglinton nor under regular Glaser conditions. However, a modified Glaser protocol with 1,5-hexadiyne (**1(2)**) in the presence of a large amount of benzene yielded cyclododeca-1,3,7,9-tetrayne (**2(2)**) in solution. All attempts to isolate **2(2)** failed due to its instability [27].

The above mentioned coupling procedures were also used to prepare a large variety of nonconjugated and conjugated systems with 1,3-butadiyne units as building blocks that are summarized in various reviews [1, 2, 28, 29].

For the synthesis of nonsymmetrically substituted 1,3-butadiynes the Cadiot-Chodkiewicz coupling is usually applied [30, 31]. In this protocol a terminal alkyne is reacted with a terminal bromoalkyne derivative in the presence of a Cu(I) salt and an amine. As examples we show in Scheme 7.3 the synthesis of the 1-azacyclotetradeca-3,5,10,12-tetrayne derivatives **10(3)a** to **10(3)f** [32], 1-isopropyl-1-azacyclopentadeca-3,5,11,13-tetrayne (**10(4)(c)**) [33] and 1-isopropyl-1-azacyclohexadeca-3,5,12,14-tetrayne (**10(5)(c)**) [33].

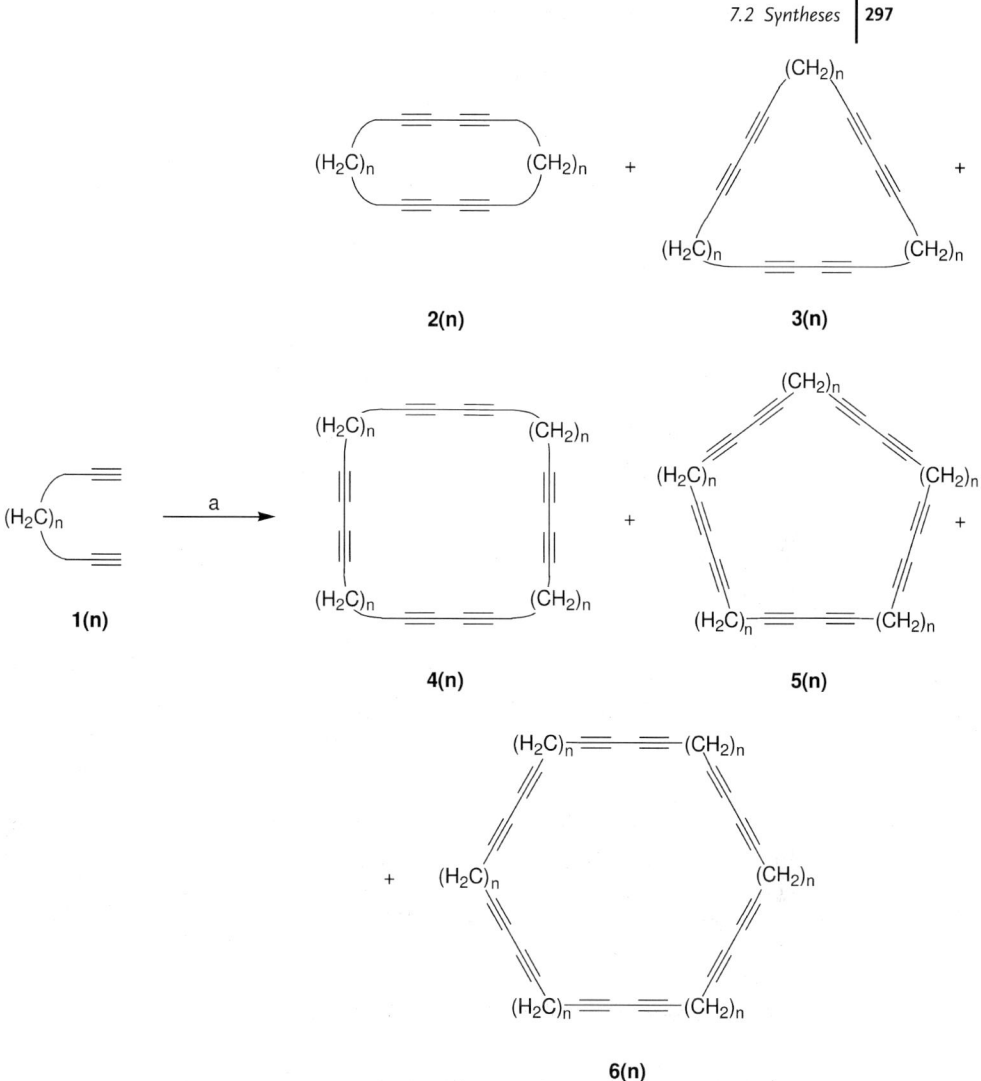

Scheme 7.2. Survey of cycles **2(n)–6(n)** obtained from α,ω-diynes **1(n)** via Hay coupling. (a) Cu(OAc)$_2$, pyridine/MeOH.

The key step in this sequence is the Cadiot-Chodkiewicz coupling between the corresponding terminal diyne **1(n)** and two equivalents of the THP-protected 1-bromoprop-1-yne-3-ol (**7**). The ring closure to **10(n)** was achieved by reacting **9(n)** with various primary amines in the presence of K$_2$CO$_3$ as base. The yield was highest (70%) for **10(5)c**.

Herein we mention that Pd-catalyzed coupling reactions between alkynes are also available to generate symmetrical 1,3-butadiyne units [21b, 22]. So far they have not been applied to the preparation of cyclic systems with 1,3-butadiyne units.

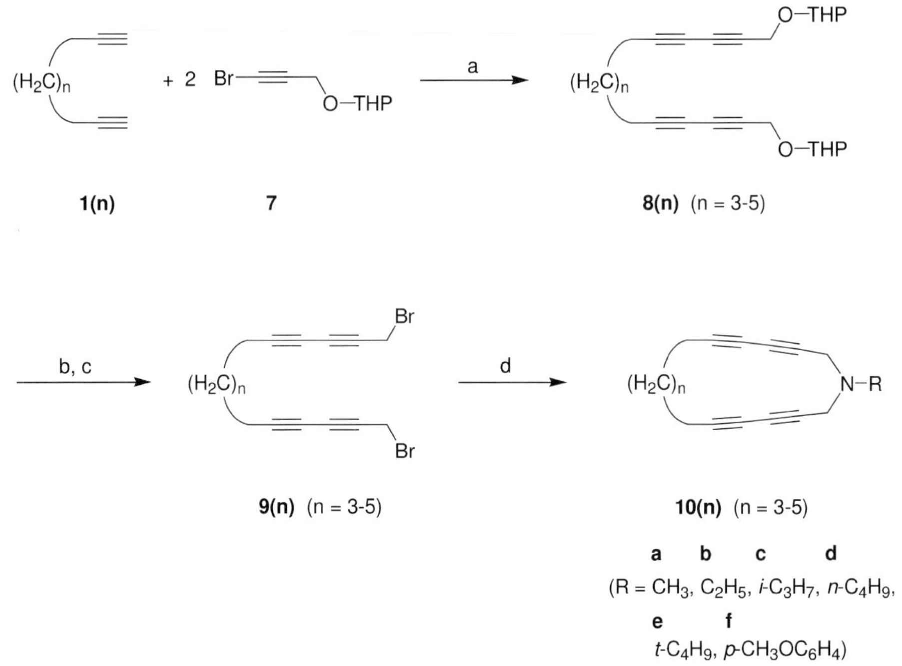

Scheme 7.3. Synthesis of nonsymmetrically substituted cyclic butadiynes **10(n)** using Cadiot-Chodkiewicz coupling as the key step. (a) CuI, pyrrolidine (35–66%), (b) H$_2$SO$_4$, H$_2$O/MeOH (67–84%), (c) PBr$_3$, pyridine/Et$_2$O (38–62%), (d) K$_2$CO$_3$, CH$_3$CN (5–70%).

Bäuerle et al. have recently published a noncatalytic procedure allowing the isolation of a bis-platinum macrocycle which was transferred to a conjugated system with two 1,3-butadiyne units by a reductive elimination process [34]. This technique was used to prepare cyclotetracosa-1,3-11,13-tetrayne (**15**) (Scheme 7.4) [35]. The reaction of dichloro-1,3-bis(diphenylphosphinopropane)platinum(II) (**12**) with 1,11-dodecadiyne yielded **13(11)**. This species was reacted again with **12** to afford the cyclic system **14(11)** which was oxidized with iodine to obtain cyclotetrayne **15**.

Another procedure to yield ethano-bridged 1,3-butadiynes was used by Scott et al. to prepare octamethylcyclododeca-1,3,7,9-tetrayne (**17**) (Scheme 7.5) [36]. This group found that *in situ* generated tetramethylhexapentaene (**16**) could be dimerized to **17** by Cu(I) catalysis. Similarly the related hexapentaene derivative **18** yielded in such a cycloaddition process the head-to-head dimer **19** [37a].

Cyclo[(disilanylene)(buta-1,3-diyne)]s were prepared by the reaction of 1,2-dichloro-1,1,2,2-tetra*iso*propyldisilane (**20**) and 1,4-dilithiobuta-1,3-diyne (**21**) yielding as the main product (11%) the 12-membered ring **22** and as side products the tri- and tetramers **23** and **24**, respectively (Scheme 7.6) [38].

Another example which uses a 1,3-butadiyne unit is given in Scheme 7.7 with the synthesis of 1,8-diazacyclotetradeca-3,5,10,12-tetraynes (**27**) [39]. In this proto-

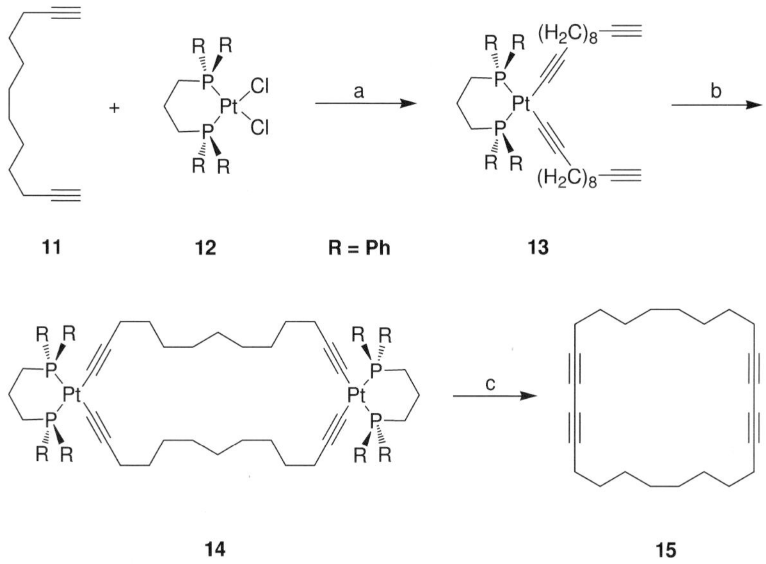

Scheme 7.4. Pd mediated coupling to prepare carbocyclic tetrayne **15**. (a) CuI, Et$_2$NH, DMF/THF (49%), (b) **12**, CuI, Et$_2$NH, DMF/THF (20%), (c) I$_2$, THF (62%).

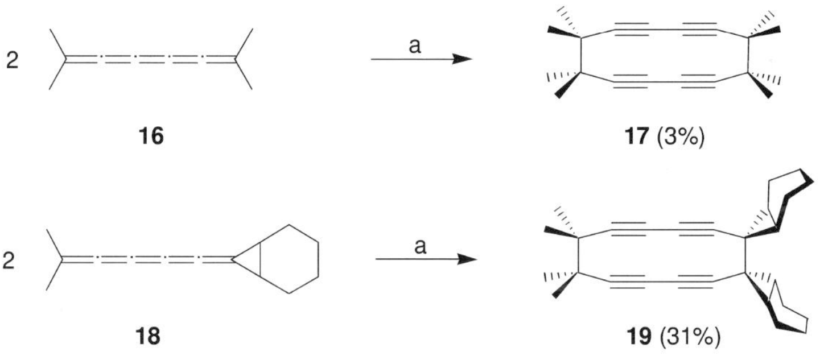

Scheme 7.5. Cu(I) catalyzed dimerization of *in situ* generated cumulenes **16** and **18** to afford the carbocycles **17** and **19**. (a) CuI.

col the 14-membered ring system is prepared by a four-component one-pot synthesis between 1,6-dibromohexa-2,4-diyne (**25**) and the corresponding primary amine (**26**) in presence of K$_2$CO$_3$ as base [39].

To prepare cyclic tetraynes consisting of 1,4-donor-substituted butadiyne units one can use the Glaser type coupling of terminal dithiadiynes **28(3)** or **28(4)** (Scheme 7.8). This method produces the 18- and 20-membered rings **29(3)** and **29(4)**, respectively, but no trimerization products were found [40].

To prepare macrocycles with three 1,4-donor-substituted butadiyne units the synthesis commences with the stable, commercially available bis-trimethylsilyl pro-

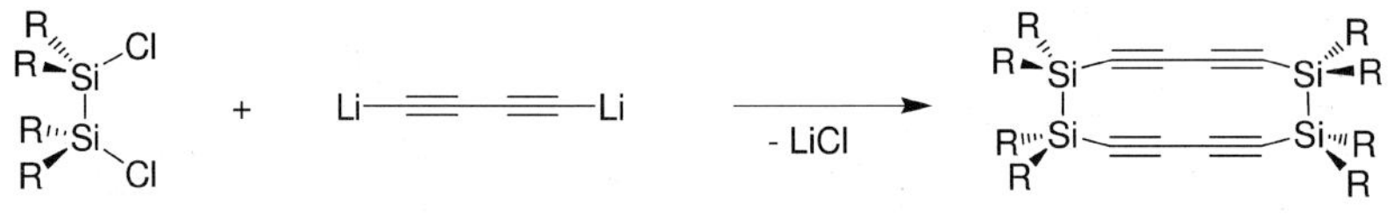

300 | *7 Carbon-rich Cycles with Two and More 1,3-Butadiyne Units – Syntheses, Structures and Reactivities*

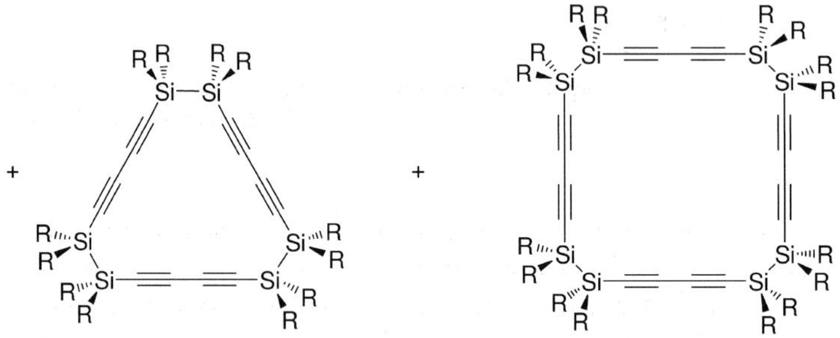

Scheme 7.6. Cyclization of lithiated butadiyne **21** with dichlorodisilane **20** to afford cyclic di-, tri- and tetramers **22–24**.

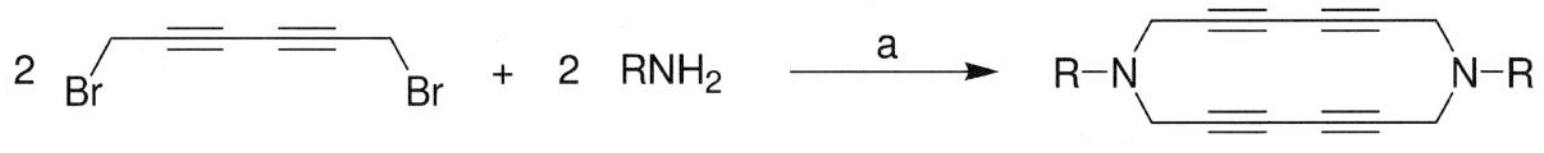

Scheme 7.7. Four-component synthesis of diazacyclotetraynes **27a–e**. (a) K_2CO_3, THF (1–17%).

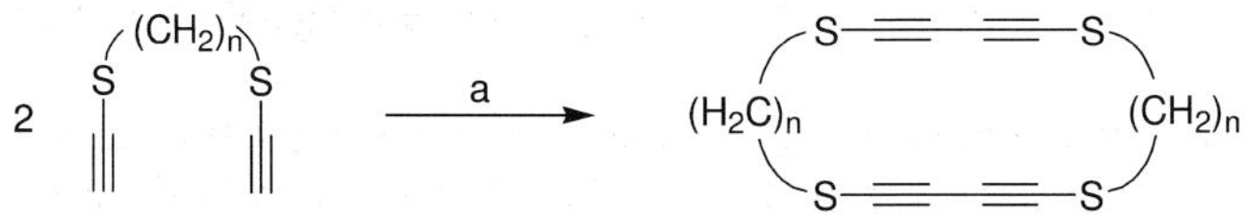

(R = CH_3, C_2H_5, i-C_3H_7, C_3H_5, $CH_2C_6H_5$)

Scheme 7.8. Glaser type coupling of dithiadiynes to yield tetrathiacyclotetraynes **29(3)** and **29(4)**. (a) CuCl, NH_3, MeOH, O_2, r.t.

29(n) (X = S, n = 2-5)   4-18%
33(n) (X = Se, n = 2-5) 2-17%

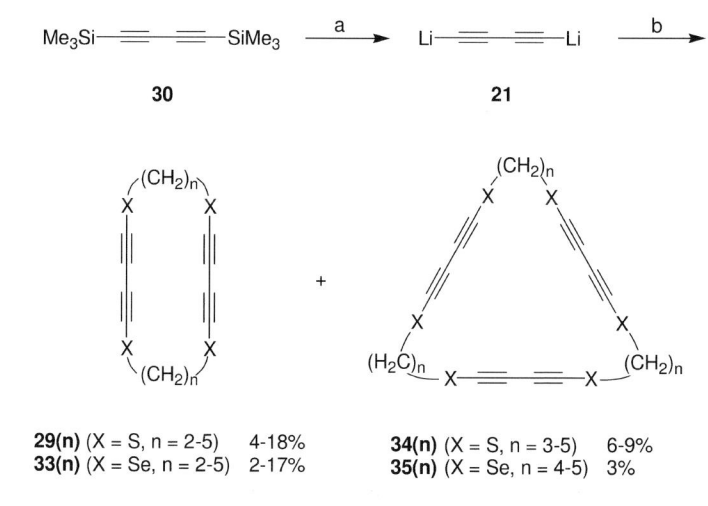

34(n) (X = S, n = 3-5)   6-9%
35(n) (X = Se, n = 4-5) 3%

Scheme 7.9. One-pot synthesis of macrocycles with 1,4-donor substituted butadiyne units. (a) MeLi/LiBr, THF, −78 °C to rt, (b) NCX–(CH$_2$)$_n$–XCN (**31(n)** (X = S, $n = 2$–5); **32(n)** (X = Se, $n = 2$–5)), THF, −40 °C to rt.

tected 1,3-butadiyne **30** (Scheme 7.9). The treatment of (**30**) with MeLi in the presence of LiBr in THF at −78 °C afforded bislithiated 1,3-butadiyne (**21**) which was reacted *in situ* with dithio- and diselenocyanatoalkanes **31(n)** and **32(n)**, respectively, to yield the cyclic species **29(n)** and **33(n)**–**35(n)** [40]. The trimers were obtained only for longer chains ($n = 3$–5) and in lower yields than the dimers **29** and **33**.

7.3
1,3-Butadiyne Moieties as Spacers and Framework

The 1,3-butadiyne units in **29** and **33–35** act as rigid framework which spans a cavity for guests (see below). This function of a polyalkyne unit was also used to construct cavities with receptor properties. Among monocyclic receptors were **36** and **37** with two and three π units and two and three butadiyne spacers [41, 42]. Both were constructed by Eglinton coupling. Other cyclic systems which form intracavity complexes are the cyclophanes **38** and **39** [43, 44] (Chart 7.1).

The latter system is soluble in water and acts as an acetyl acceptor. Water-soluble properties are also reported for the tetramer **40** obtained by homocoupling of a 1,4-*cis*-ethynyl substituted glucopyranose [45] (Chart 7.2). The tetramer **40** contains five water molecules in the crystal [45].

Rigid macrocycles with two, three and four butadiyne units (**40–43**) [45–47] are listed in Chart 7.2. All three isomers of the [8.8]cyclophane **41** (*ortho*, *meta* and *para*) [46] have been structurally characterized, but the hollow core proved to be too small for the inclusion of guest molecules. In the case of **42** the cubane units

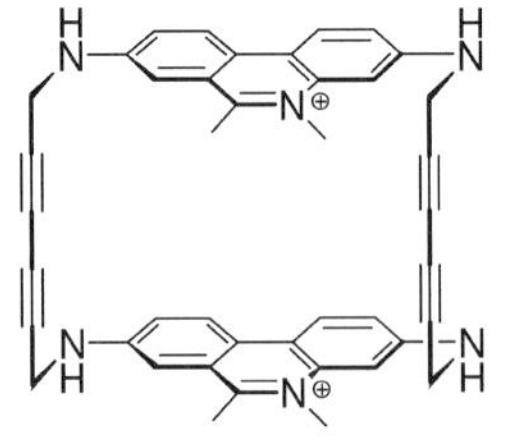

Chart 7.1. Examples of cyclophanes with butadiyne units.

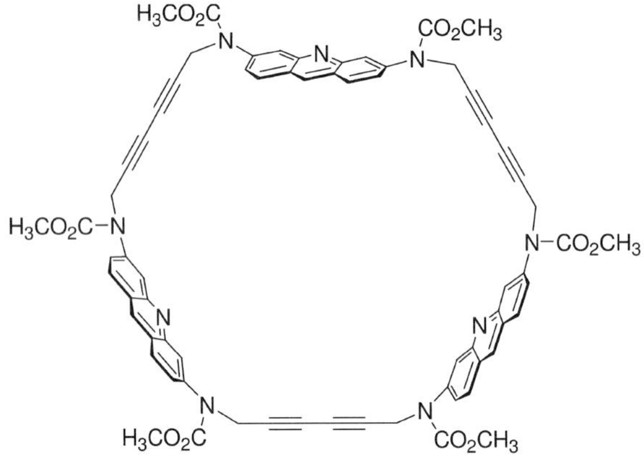

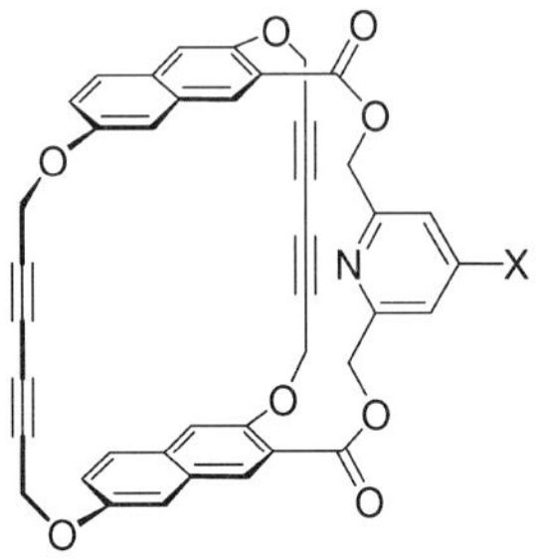

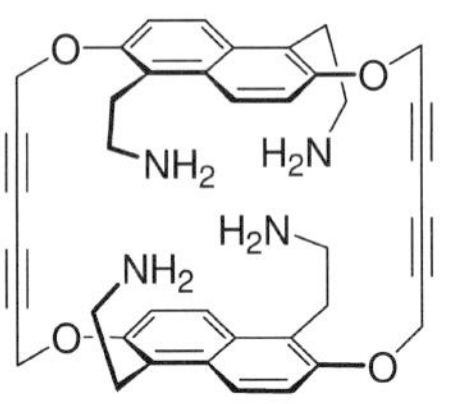

were rearranged to cyclooctatetraene [47]. The large cavity of **43** encloses dichloromethane in the solid state [48]. Bicyclic ring systems with three 1,3-butadiyne units are also available. The first hydrophobic cage system with three 1,3-butadiyne spacers was reported by Breslow et al. [49]. Building units were two 1,1,1-tris-(4-hydroxyphenyl)ethane units connected by three hexane-2,4-diyne units. The dimeric coupling reaction of the trispropargylphenylether **44** under standard high-dilution conditions (Eglinton or Hay procedure) failed. However, treatment of **44** with O_2-free pyridine and anhydrous $CuCl/CuCl_2$ at 0 °C gave the cage system **45** in 38% yield [49] (Scheme 7.10). This compound is able to enclose a benzene molecule in its cavity.

This concept was extended by Vögtle et al. who synthesized a series of cages capped by triphenylmethane [50] and triphenylmethyl dication species [51–53]. The Breslow protocol of joining the two fragments (e.g. **44**) was changed by using $Cu(OAc)_2 \cdot H_2O$ in CH_3CN. For the orange colored triphenylmethane dyes

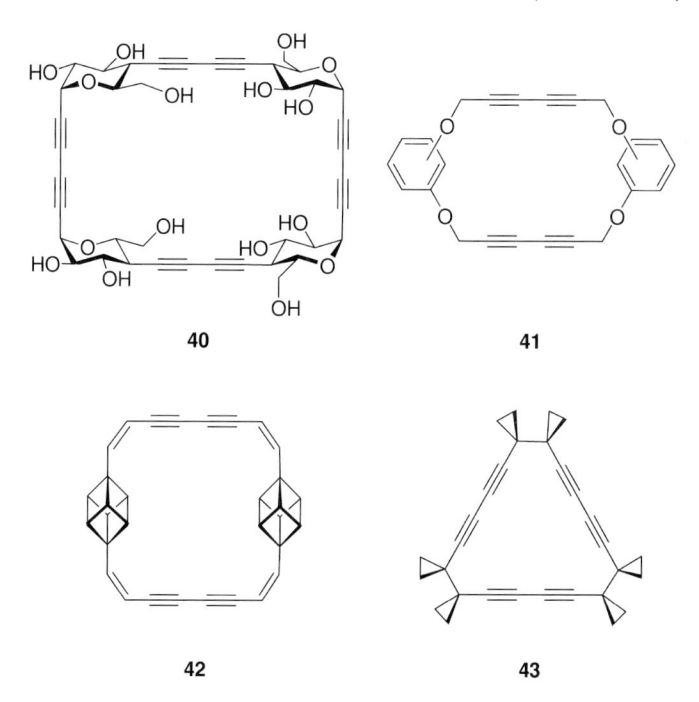

Chart 7.2. Rigid macrocycles **40–43** with two, three and four butadiyne units.

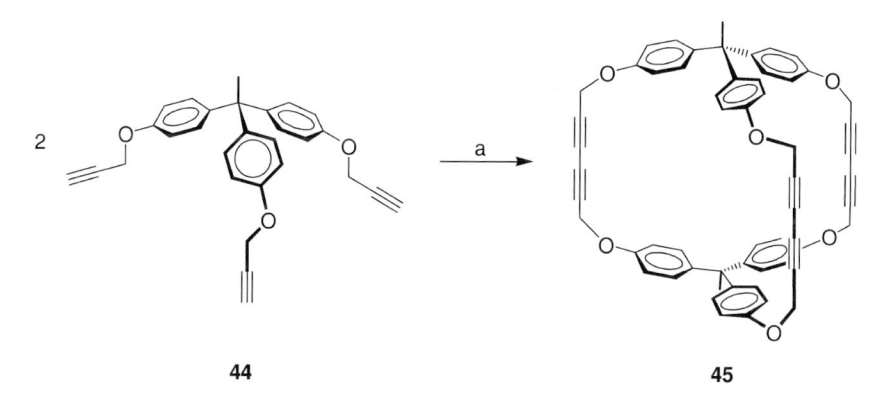

Scheme 7.10. Dimerization of **44** to afford cage compound **45**. (a) CuCl, CuCl$_2$, pyridine (35%).

solvatochromic and halochromic effects are reported which are different from those of nonmacrocyclic reference substances [52]. The triphenylmethane dyes can be reversibly transformed into the corresponding uncolored marocycles by changing the pH value (Scheme 7.11).

Similar cyclic systems with other chromophores are reported [52]. Related to the cage systems **45** and **46** are the phosphine oxide bifunctional macrocycles such as **48a** and **48b** [54] (Chart 7.3). These cyclophanes were prepared from tris-

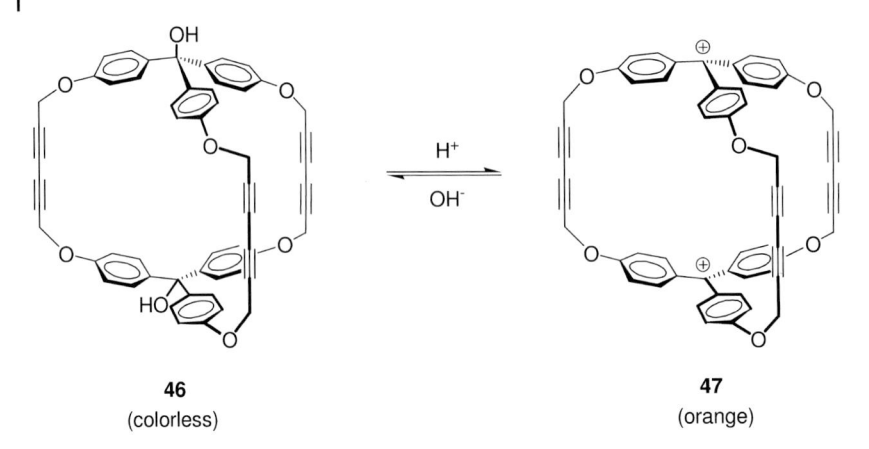

Scheme 7.11. pH dependent equilibrium between diol **46** and dication **47**.

(4-hydroxyphenyl)phosphine oxide analogous to **45** and **46**. The key reaction was an Eglinton-type coupling ($Cu(OAc)_2 \cdot H_2O$ in pyridine) at 60 °C. ^1H and ^{31}P NMR investigations revealed complexation of organic molecules such as pentafluorophenol, 2,6-dimethyl-4-nitrophenol and AcOH inside the cavity [54]. A single aromatic moiety, such as the phloroglucin unit in **49**, is also able to act as a terminus between the butadiyne units [50].

A rather long bicyclic system (1.47 nm) was prepared by Hay coupling of tris-(2-(5-ethynyl)thiophene-2-yl)ethylamine to **50** (Chart 7.3) [55]. The structure of this cage system is given in chapter 7.5.1. In Chart 7.4 we show as a further example [3]catenate **51** [56] with a defined distance between the two metal centers due to the stiff butadiyne units. The oxidative coupling reaction of terminal alkynes is important for the formation of such species [56, 57]. The cavitand **52**, in which two chiral [1.1.1]orthocyclophane units are bridged by three hexa-2,4-diyne units, is also based on the homocoupling of two terminal alkynes to a 1,3-butadiyne unit [58]. In Chart 7.4, for the sake of clarity only the meso-isomer is shown. A simple molecular channel model system, **53**, was constructed by spanning two 4,13-diaza-18-crown-6 ethers with two hexa-2,4-diyne units in rigid opposition [59]. The key step for the synthesis of **53** was a Hay-type coupling reaction of N,N'-bis(propargyl)-4,13-diaza-18-crown-6 ether.

7.4
Reactivity

7.4.1
Base-induced Prototropic Rearrangements

The availability of **3(2)**–**6(2)** from a Glaser-type coupling of 1,5-hexadiyne (**1(2)**) was used by Sondheimer and Wolovsky [26] to generate the fully conjugated cyclic

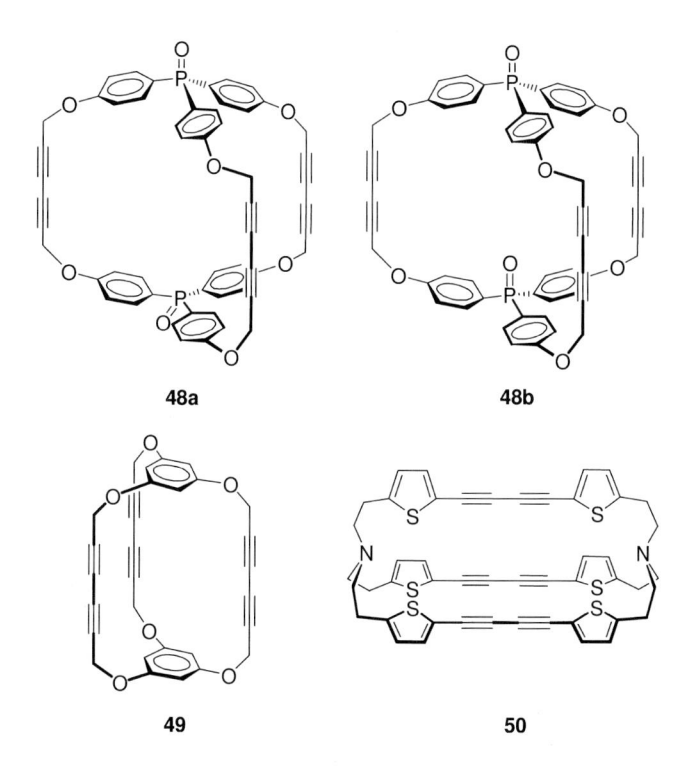

Chart 7.3. Macrobicycles **48**, **49** and **50** with three butadiyne units.

systems **54–57** by treating **3(2)–6(2)** with *t*-BuOK in *t*-BuOH (Scheme 7.12). The conjugated systems were characterized by their elemental analyses and their UV spectra.

When **2(2)** was treated in the same way biphenylene (**58**) and the bisdehydro[12]annulenes **59** and **60** were generated (Scheme 7.13). Evidence for the structures of **59** and **60** was obtained by ^{1}H NMR-, IR- and UV-spectroscopy [27, 60]. It is interesting to note that the treatment of **3(2)** with a strong base provides apart from **54** (50%) a small amount (7%) of triphenylene [61]. The transannular reaction to triphenylene was rationalized [61] by assuming a nonplanar isomer of **54**.

Further starting points for fully unsaturated cyclic systems proved to be the dimers and trimers of 1,5,9-decatriyne, the macrocycles **61** and **62** (see Scheme 7.14) which were obtained by Glaser coupling of the decatriyne [62].

A benzene solution of the cyclic dimer **61** when refluxed with a saturated solution of *t*-BuOK in *t*-BuOH gave a red solution of the bisdehydro[20]annulene **63** [62]. The cyclic trimer **62** was rearranged similarly to tridehydro[30]annulene **64**. Both species show characteristic UV–vis spectra [62]. The bisdehydro[20]annulene was partially hydrogenated in benzene solution over a Lindlar catalyst (quinoline-poisoned palladium calcium carbonate catalyst). This procedure led to the [20]annulene **65**. Similarly, the reduction of tridehydro[30]annulene **64** with Lindlar catalyst yielded the [30]annulene **66** [62].

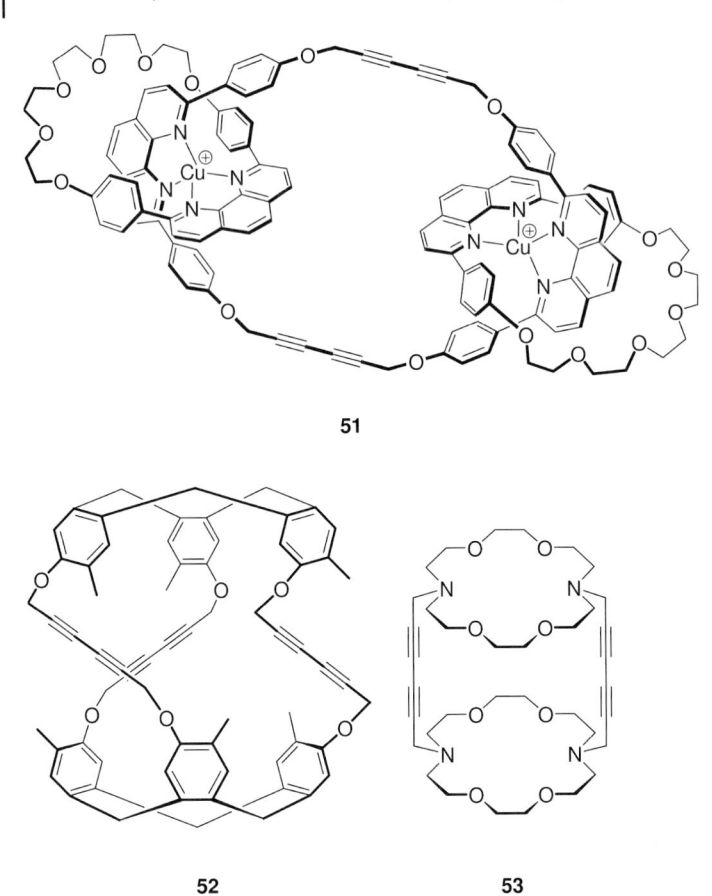

Chart 7.4. [3]Catenate **51**, Cavitands **52** and **53** with butadiyne units.

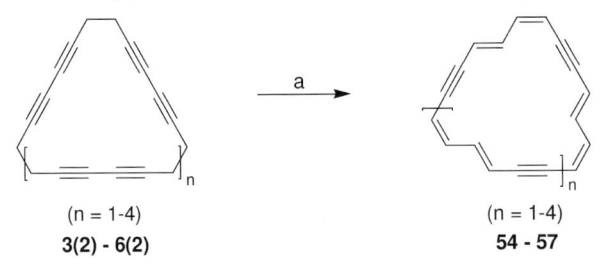

Scheme 7.12. Base-induced generation of fully conjugated cyclic systems **54–57**. (a) t-BuOK/t-BuOH (1–3%).

Treating tetrayne **67** with a saturated solution of t-BuOK in t-BuOH at 40 °C for 1 min yielded an isomeric mixture of bisdehydro[16]annulenes from which the isomer **69** could be separated in ca. 5% yield. The tetracyclic product **68** was isolated in 15% yield and identified (Scheme 7.15). It was formed in 50% yield when **67** was treated with boiling ethanolic KOH solution [63].

7.4 Reactivity | 307

2(2) → **58** (7.4%) + **59** (1.5%) + **60** (0.6%)

Scheme 7.13. Base-induced generation of **58**, **59**, and **60** starting from **2(2)**. (a) t-BuOK/t-BuOH.

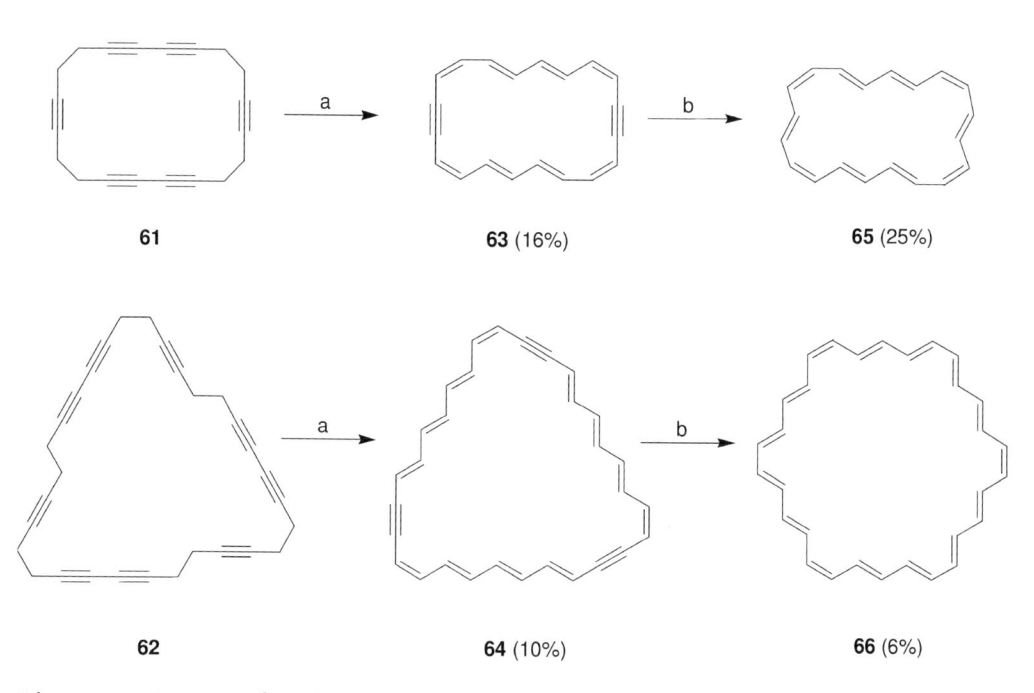

61 →(a) **63** (16%) →(b) **65** (25%)

62 →(a) **64** (10%) →(b) **66** (6%)

Scheme 7.14. Generation of annulenes **65** and **66** by base-induced rearrangement and subsequent hydrogenation of **61** and **62**. (a) t-BuOK/t-BuOH, (b) H_2, Lindlar catalyst.

67 →(a) **68** + **69** + isomers

Scheme 7.15. Base-induced generation of **68** and **69** starting from **67**. (a) t-BuOK/t-BuOH.

7.4.2
Elimination Reactions

Instead of prototropic rearrangements base-induced eliminations were also used to generate fully conjugated cyclic π-systems. As an example we mention the elimination from 3,5,10,12-cyclotetradecatetrayne-1,8-diol dimethanesulfonate (**70**). Treatment of **70** with methanolic KOH solution yielded 1,5,9-tridehydro[14]annulene (**71**) as side product and the bicyclic species **72** as main product (15–20%) (Scheme 7.16a) [64]. The structure of **72** was assigned based upon the hydrogenation products and the UV–vis spectrum. When the reaction was carried out under more vigorous conditions only the bicyclic systems **73**–**75** could be isolated (Scheme 7.16b). For the reaction mechanism it is assumed that **72** is formed first. The transformation of **72** to the benzazulene **73** may proceed by hydride ion transfer from methoxide to yield **76** (Scheme 7.17). From this intermediate **73** and **75** might be generated as shown in Scheme 7.17. The methoxy-5,6-benzazulene **74** might arise from **72** by nucleophilic addition of a methoxide ion, followed by ring closure and protonation. Related to the reactions summarized in Scheme 7.16 is a reaction sequence shown in the following section.

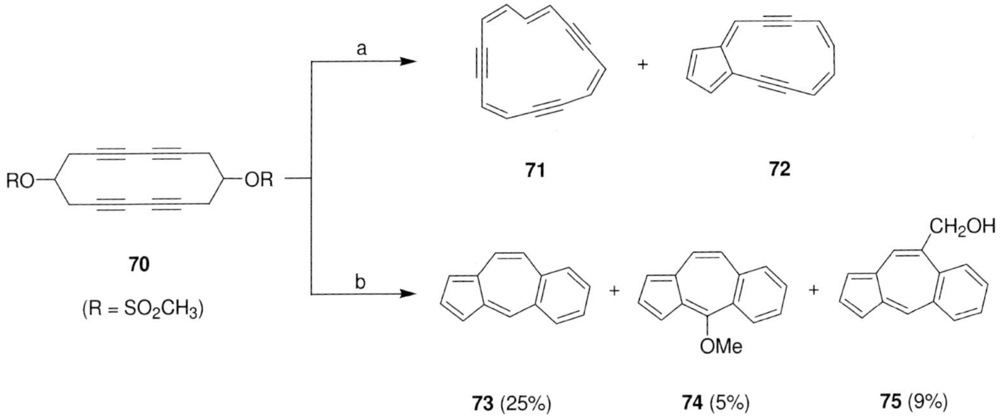

Scheme 7.16. Different modes of base-induced eliminations of **70**. (a) 2% KOH in 97% aq. MeOH, 10–20 °C, (b) 7% KOH in 95% aq. MeOH/DMSO, reflux.

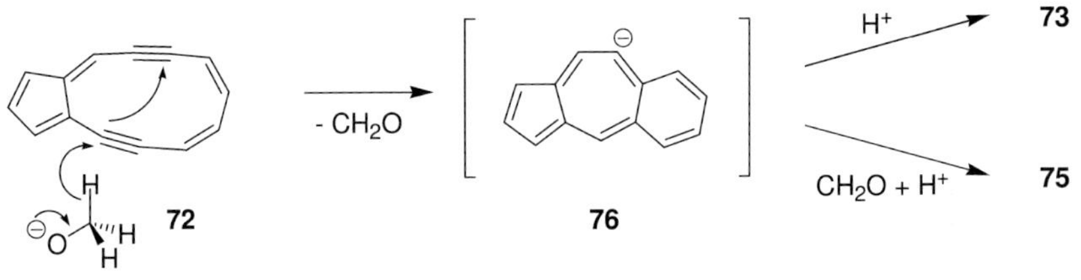

Scheme 7.17. Rationalization for the generation of **73** and **75** by hydride transfer.

7.4.3
Reductions of Cyclic 1,3-Butadiynes

A ring closure which also produces a bicyclic system was reported by Myers and Goldberg [65]. In connection with the synthesis of the bicyclic core of the chromoprotein antitumor agents kedarcidin and neocarzinostatin, a transannular reductive cyclization of the cyclic tetrayne **77a** (Scheme 7.18) was investigated. Treatment of solutions of **77a** in THF at 0 °C with $NaAlH_2(OCH_2CH_2OCH_3)_2$ afforded a mixture of the bicyclic product **78** and two monocycles (**79**, **80**). Rigorously dried solutions of **77a** yielded only **79** (43%) and **80** (8%). In the presence of protic solvents (MeOH) **78** was also generated. Deuterium incorporation experiments using $[D_4]$-methanol showed that all three products were formed by hydride addition to the proximal carbon of the acetylene followed by deuteration as shown in Scheme 7.18.

Hydrogenations of **2(n)–6(n)** yielded the fully hydrogenated ring systems [25]. The same holds for the macrocyclic systems **67** and **81–83** (Chart 7.5) [66].

More interesting are reactions of compounds in which the two 1,3-butadiyne units are oriented in close proximity. Although the systems **84** and **89** do not

Scheme 7.18. Reduction of **77a**. (a) $NaAlH_2(OCH_2CH_2OCH_3)_2$, $CH_3OH(D)$, THF, 0 °C (30%).

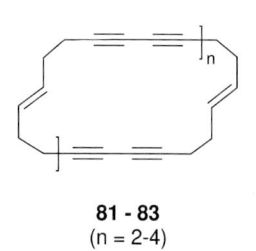

81 - 83
(n = 2-4)

Chart 7.5. Macrocyclic systems **81–83**.

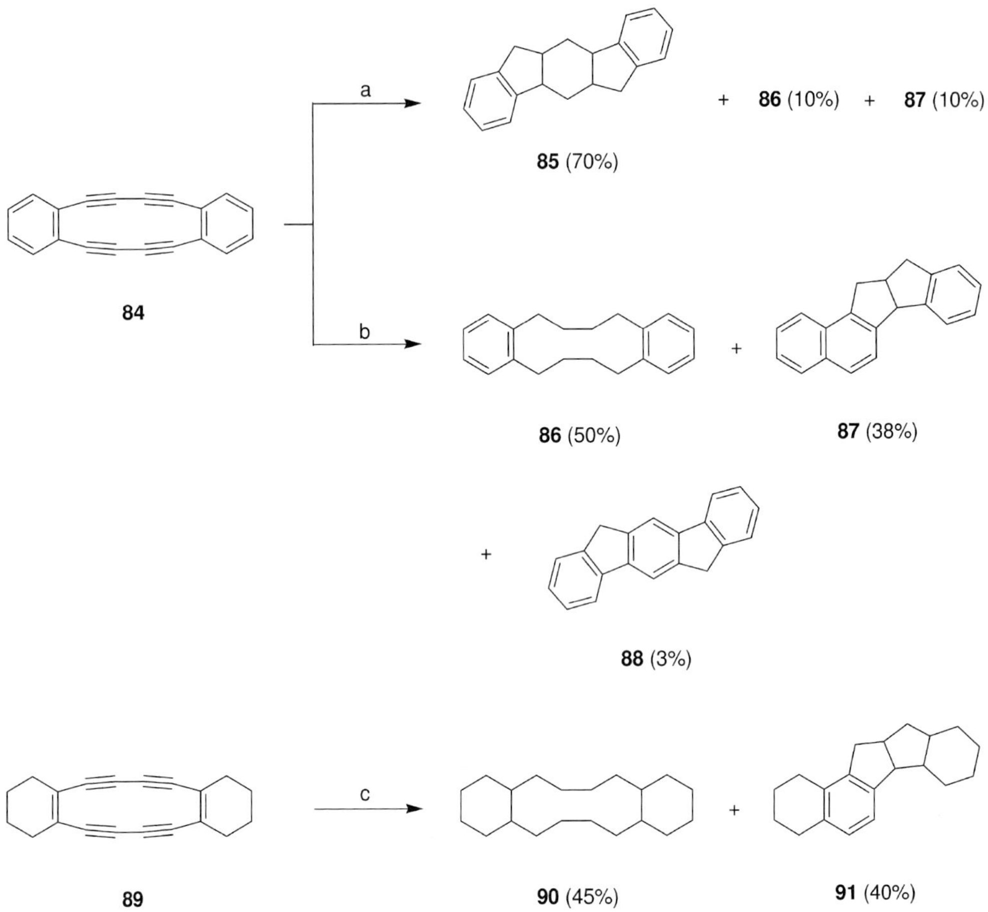

Scheme 7.19. Reduction of **84** and **89** under different conditions. (a) Na, NH$_3$, (b) Pd/C, H$_2$, (c) Pt, H$_2$.

belong directly to our topic we list their hydrogenation and reduction products (Scheme 7.19) because they nicely demonstrate that 1,3-butadiynes in close proximity tend to undergo transannular reactions [67, 68].

The hydrogenation of **84** was carried out over Pd/charcoal to yield **86** as main product in which both 1,3-butadiyne units were converted into tetramethylene moieties. The second major product was assigned to **87** [67]. It is interesting to note that **88** was the only detectable product of the Pd-induced dehydrogenation of **86** at 320 °C. The Na/NH$_3$ reduction of **84** yielded as main product **85** and as minor products **86** and **87** [67]. Hydrogenation of **89** in EtOAc and AcOH over a Pt catalyst afforded about equal amounts of **90** and **91** [68].

7.4.4
Acid-induced Transannular Reactions

The examples given in Scheme 7.19 demonstrate that cyclic systems in which 1,3-butadiynes are situated in close proximity are prone to transannular reactions. These observations were corroborated by the reactions of **10** and **27** with concentrated HCl.

The reaction of **10(3)c** with concentrated HCl in the presence of EtOH (1:1) yielded one product whose properties revealed that two equivalents of HCl had been added. Based on the ^{13}C NMR data it was evident that the sp carbons of **10(3)c** were changed to sp^2 carbons. The NMR data also revealed that the original C_s symmetry of **10(3)c** was retained. The fact that the product could be oxidized by DDQ to a new species which gave single crystals was crucial for the structure determination. X-ray investigations revealed structure **93** for the oxidation product (Scheme 7.20) [32]. Thus, the primary product was the tricyclic system **92**.

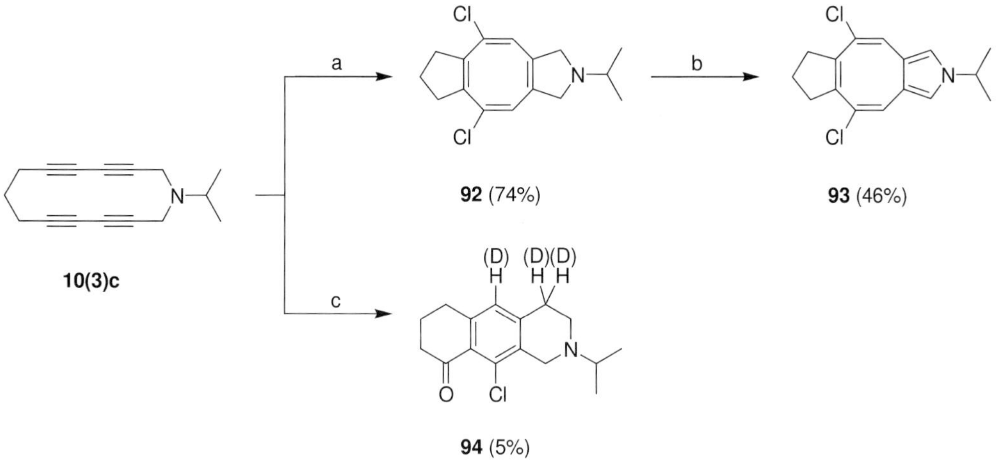

Scheme 7.20. Acid-induced transannular reactions of **10(3)c**.
(a) HCl (conc.)/EtOH (1:1), 80 °C, (b) DDQ/CHCl$_3$, (c) HCl (conc.) (DCl), 80 °C.

A comparison with the products obtained from two 1,3-butadiyne units incorporated in a 12-membered scaffold (Scheme 7.19) shows that if more flexibility is available, as in the 14-membered ring of **10(3)c**, a 5–8–5 ring closure is possible.

The preferred ring closure between the sp termini of the 1,3-butadiyne units to yield an eight-membered ring is also in accord with the shorter transannular distances of the terminal sp centers (310 pm, see Table 7.1) as compared to the central sp centers (340 pm, see Table 7.1) in **27c** or **10(3)**. A possible mechanism for the HCl addition to **10(3)c** is presented in Scheme 7.21. One assumes that in solution

Table 7.1. Comparison of the most relevant X-ray structural data of **2(3)**, **2(4)**, **17** and **27c**.

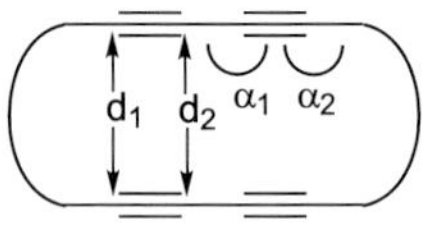

β_1-β_n : angles at saturated tether atoms

	2(3)[b]	**2(4)**	**17**[b]	**27c**[b]
d_1 [pm]	309.8(2)	417.2(2)[a]	269.6(3)	308.1(2)
d_2 [pm]	339.0(2)	425.1(2)[a]	323.6(4)	336.5(2)
C≡C [pm][a]	119.4(2)	118.8(3)	120.0(4)	119.7(2)
C_{sp}–C_{sp} [pm]	138.5(3)	138.5(3)[a]	138.7(3)	138.2(2)
α_1 [°][a]	176.7(3)	174.0(4)	166.6(4)	173.9(1)
α_2 [°][a]	172.9(3)	177.1(4)	166.3(4)	173.0(1)
β_1–β_n [°]	114.2(3)–115.9(3)	110.2(3)–113.7(4)	107.7(3)–108.2(3)	111.9(2)–115.9(2)

a Mean value.
b Molecule shows center of inversion.

the nitrogen atom is first protonated. (For the sake of simplicity we omit this protonation in Schemes 7.21, 7.22 and 7.26). Based on quantum chemical calculations [69] it seems reasonable that the protonation at C4 is accompanied by an immediate bond formation between C13 and C3 to yield **95**. The resulting bicyclic 1,2,3-butatrienyl cation is further stabilized by ring closure from C6 to C10 to yield the bicyclic vinyl cation **96**. This species adds a chloride anion to yield **97**, a valence isomer of **98**. Further HCl addition finally leads to **92**. When DCl was used instead of HCl [D_2]-**92** was obtained deuterated in the α-positions to the chlorine substituents. This result supports the mechanism proposed in Scheme 7.21.

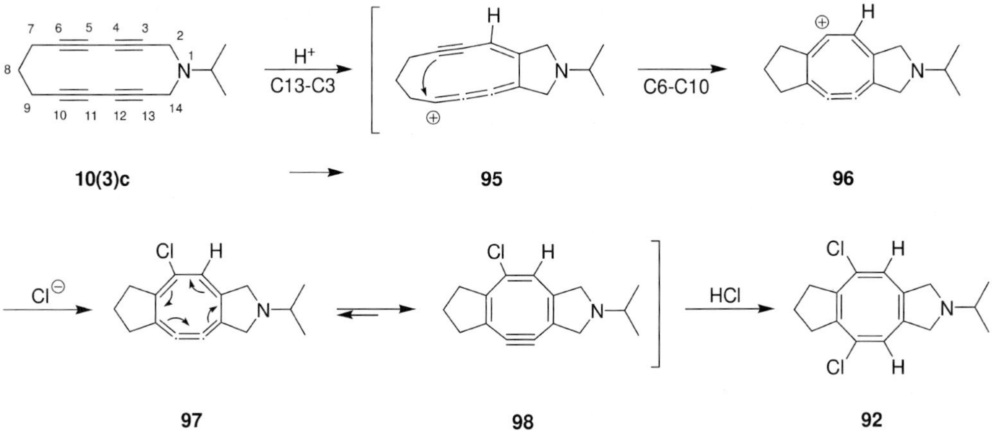

Scheme 7.21. Rationalization of the acid-induced generation of **92**.

Another mode of ring closure was found when **10(3)c** was treated with concentrated HCl without the presence of EtOH; however, the product **94** (Scheme 7.20) was isolated only in low yields. An experiment with DCl in D_2O reveals the hydrogens indicated in **94** are replaced by deuterium. The structural assignment of **94** was confirmed by X-ray analysis of a single crystal. To rationalize the occurrence of a 6–6–6 ring closure modus the mechanism shown in Scheme 7.22 was proposed. The occurrence of **94** was explained by a protonation of **10(3)c** at C3 accompanied by a ring closure between C4 and C13 giving rise to the bicyclic cation **99** incorporated in the larger ring. This species is trapped by the chloride anion yielding **100**. Further protonation accompanied by transannular ring closure yields the vinyl cation **101** which adds water to afford **102**. The latter rearranges to **94**. Another possibility to explain the isolation of **94** is given in the literature [32]; however, quantum chemical calculations [69] favor the path shown in Scheme 7.22.

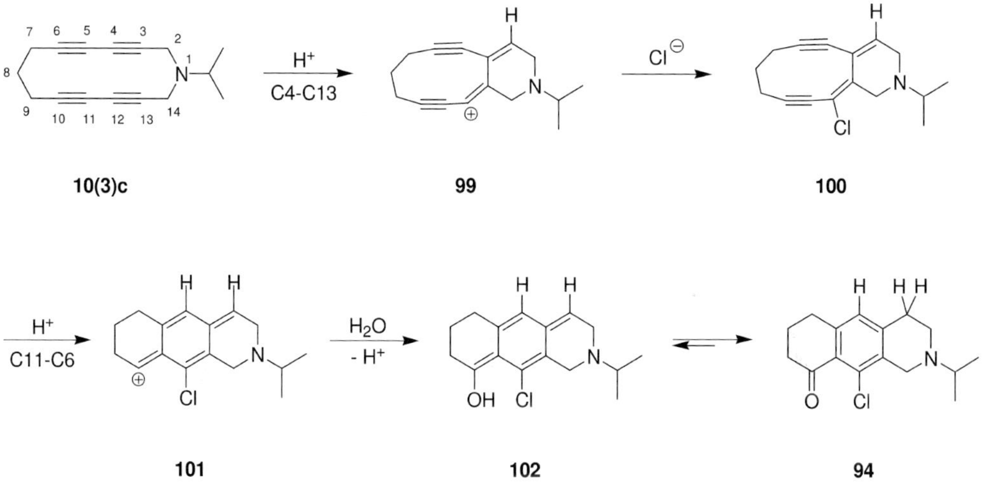

Scheme 7.22. Rationalization for HCl-induced generation of **94**.

For **27c** an analogous ring closure to **10(3)c** was observed [70]. Two products could be detected, the major one proved to be **103** and the minor one **104** (Scheme 7.23). The structural assignment is based on an X-ray investigation of single crystals of the oxidation product **105** which was isolated during work-up of the reaction mixture in the presence of air. This procedure yielded **105** and small amounts of **106** and **107** (Scheme 7.23). All products could be identified by their spectroscopic properties [70]. Analogous to **93** the structure of **105** revealed a central eight-membered ring in the boat conformation. In the case of the HCl addition to **27c** both possible regioisomers **103** and **104** could be detected.

In **10(4)c** and **10(5)c** the alkane tether of the rings is elongated as compared to **10(3)c**. This leads to a nonparallel orientation of the 1,3-butadiyne units with distances between the terminal sp centers of 310 pm and about 400 to 430 pm [33],

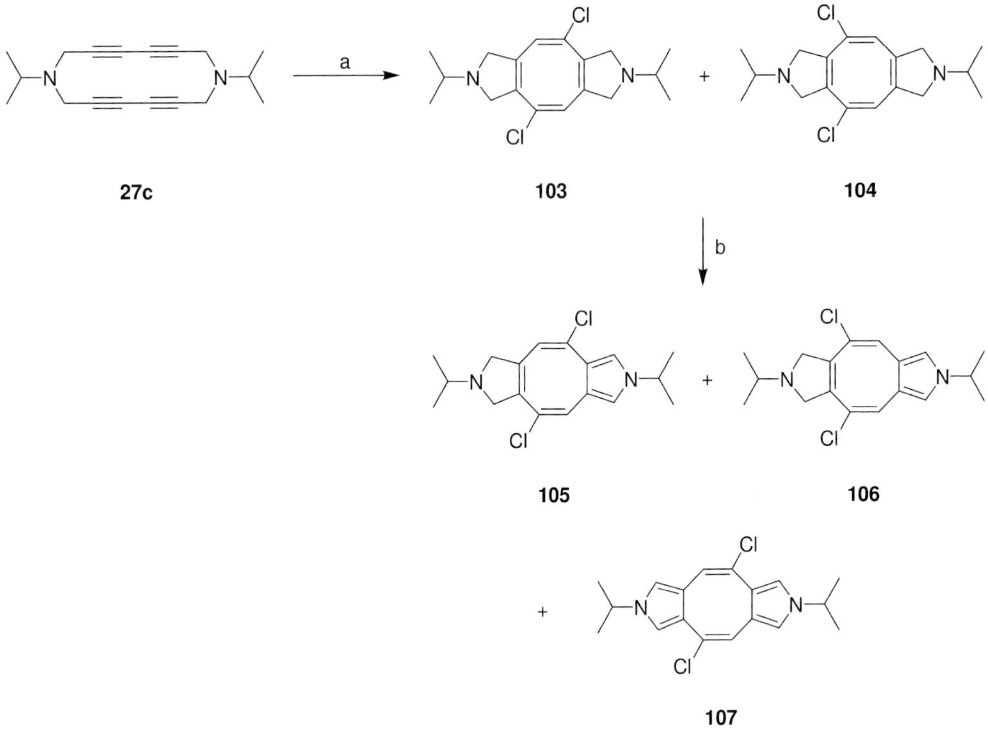

Scheme 7.23. HCl-induced transannular reactions of **27c**. (a) HCl (conc.) (63%), (b) air (19%).

respectively. Treatment of **10(5)c** with concentrated HCl in the presence of EtOH (2:1) yielded a product to which a tricyclic structure with one 2,5-dihydropyrrole ring and an aromatic ring substituted with an alkyne unit attached to a pentamethylene chain was assigned (**108**) [33]. When **108** was treated with concentrated HCl/EtOH in the ratio of 2:1 the substances **109** and **110** were isolated. Treatment of **108** with DCl/EtOD afforded the corresponding deuterated products [D_2]-(**109**) and [D_3]-(**110**) (see Scheme 7.24) [33]. The hydrochlorination reaction carried out with **10(4)c** led to **111** (Scheme 7.25).

For the reaction mechanism one assumes a protonation at C4 with concomitant ring closure between C15 and C3 to yield the bicyclic system **112**. This system is further stabilized by a ring closure between C6 and C14 to yield the phenyl cation **113** which is finally trapped by a chloride anion to afford **108** (Scheme 7.26).

These studies show that if the two 1,3-butadiyne units are in close proximity on one side, only three of the four alkyne units are involved in the transannular ring closure.

The oxidative coupling (Hay protocol) of di(1,1'-dimethylpropyn-2-yl)amine (**114**) did not yield the anticipated cycle with three 1,3-butadiyne units, **115**, but the tris(2,2,5,5-tetramethylpyrrolo)-[a,e,i]-3,7,11-tridehydro[12]annulene (**116**) as depicted in Scheme 7.27 [71]. The UV–vis spectrum of **116** is almost identical to that of **60** [27]. It is assumed by the authors that in **115** the termini of the

Scheme 7.24. HCl-induced transannular reactions of **10(5)c**. (a) HCl + H₂O (D₂O)/EtOH (1:2), (b) HCl + H₂O (D₂O)/EtOH (2:1).

Scheme 7.25. HCl-induced transannular reaction of **10(4)c**. (a) HCl (conc.)/EtOH (2:1) (20%).

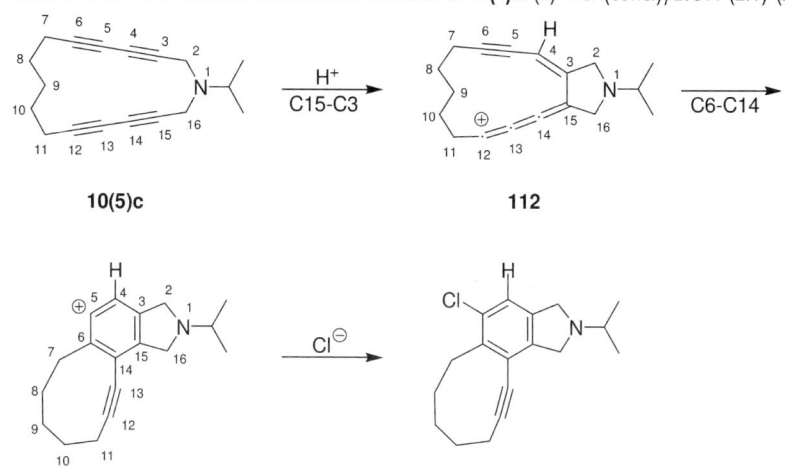

Scheme 7.26. Rationalization for the generation of **108**.

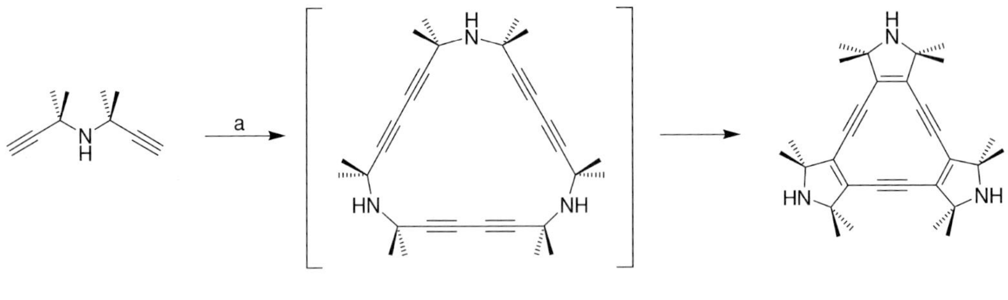

Scheme 7.27. Oxidative coupling of **114**, valence isomerization of **115** to generate **116**. (a) CuCl, pyridine, O_2 (7%).

1,3-butadiyne units are forced close together due to the methyl groups so that valence isomerization occurs during the reaction.

7.5
Spectroscopic and Structural Investigations

7.5.1
Structural Investigations

Among the various X-ray structures of molecules bearing 1,3-butadiyne moieties, in this section we would like to focus on two aspects: firstly, highly strained cyclotetraynes attract attention. Besides their intrinsic highly energetic triple bond system a bending of the butadiyne moiety results in a further increase in energy. Interesting structural and spectroscopic properties are elucidated and discussed. Secondly, molecules bearing butadiyne moieties as rigid spacers often show cavities on the molecular level that are able to host guest molecules. Going to a supramolecular level, several of these molecules, being of appropriate geometry, can interconnect their cavities by stacking on top of each other. As a result, in the solid state channels are built that can be filled with a plethora of different guest molecules.

The first structural investigations on cyclic tetraynes were carried out in 1957 by Sondheimer [23b]. These studies revealed for **2(3)** the presence of a center of symmetry that implies a chair conformation in the solid state, as is known for cyclohexane and 1,6-cyclodecadiyne. Later, more accurate results on the solid state geometry of **2(3)** and **2(4)** were obtained [72]. In contrast to **2(3)**, the larger homologue **2(4)** adopts a twisted conformation. Structural data has also been obtained for the even more strained cyclotetraynes **17**, **19** and **77b** (R = TIPS) that are tethered by only two saturated carbon atoms [37, 65]. In Fig. 7.1 a comparison between the X-ray structures of **2(3)** and **17** showing the top and a side view is displayed. The most relevant parameters and their definition for **2(3)**, **2(4)** and **17** are given in

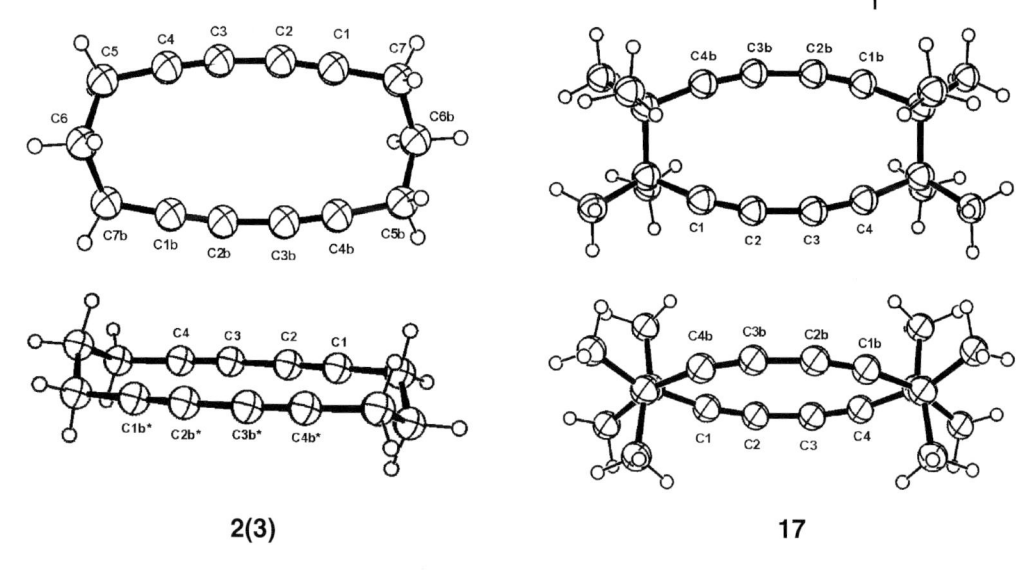

Figure 7.1. ORTEP plots of the molecular structures of **2(3)** and **17** (50% ellipsoid probability), top and side view.

Table 7.1 (see chapter 7.4.4). Due to large standard deviations we abstain from a discussion of **77b** [65c]. The average bond lengths were found to be 119–120 pm for the triple bonds and 139 pm for the sp–sp single bonds. All the 1,3-butadiyne moieties deviate considerably from linearity, giving rise to an ellipsoidal shape of the molecule. In **17** the bending is even much stronger than in **2(3)** and **2(4)**. The CCC bond angles at the sp^3 centers in **2(3)** vary between 114° and 116°, while in the less-strained 16-membered ring **2(4)** the CCC bond angles at the saturated centers are found between 110° and 114°. In contrast to this widening, the CCC bond angles at the sp^3 centers in **17** adopt values of about 108°. This angle is even smaller than the usual tetrahedral angle of 109.5° and may be the result of the two geminal Me groups forcing the butadiyne moieties to shorter distances. However, the molecule tries to decrease the strongly repulsive interaction between the triple bonds. As is shown in Fig. 7.1 (side view), a further bending of the butadiyne units (up and down) occurs forming an essentially planar arch. This 12-membered tetrayne can be formally viewed as a somewhat flattened chair cyclohexane in which two opposite C_3 bridges have each been replaced by a bent butadiyne unit. As a result, the Me groups at the saturated ring carbon atoms occupy axial and equatorial positions. A potential ring interconversion should exchange these groups. Such a readily occurring conformational change was the reason why the X-ray analysis of **17** had to be performed at −160 °C to prevent disorder [37].

Structural investigations were also carried out on the nitrogen-containing analogs **10(n)** and **27b–e** [32, 33, 39]. The X-ray structures of the cyclotetraynes **27b–e** with two nitrogen atoms in each tether reveal a crystallographic center of inversion. Thus, all molecules adopt a chair conformation in the solid state like the all-

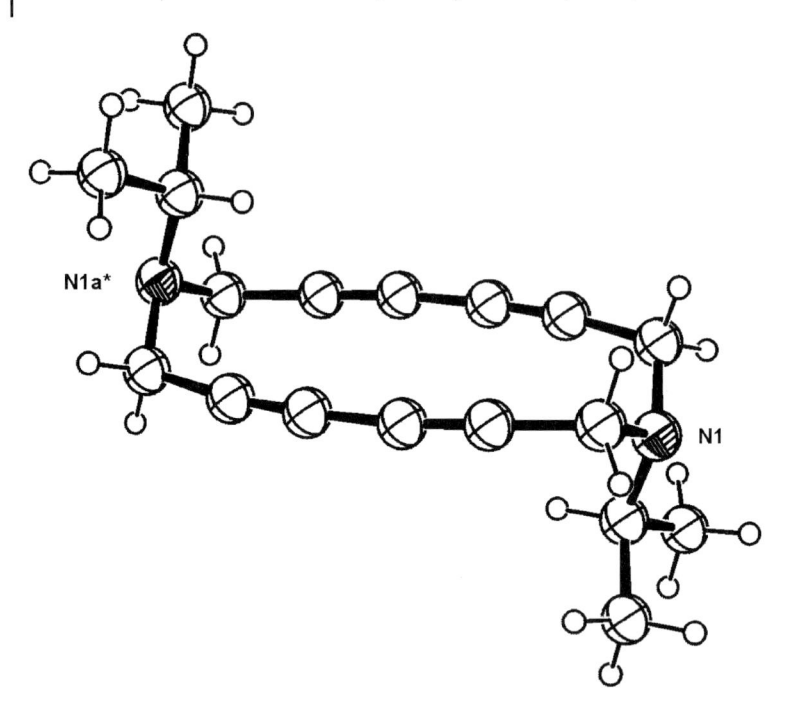

Figure 7.2. ORTEP plot of the molecular structure of **27c** (50% ellipsoid probability).

carbon congener **2(3)**. The relevant structural features are very similar to the values observed for **2(3)** (Table 7.1). The most remarkable geometrical feature is the di-axial substitution of the cyclic system in **27b–e** that is in sharp contrast to the commonly preferred equatorial conformation (Fig. 7.2).

This unexpected behavior was also observed in the case of various 1,6-disubstituted diazacyclodeca-3,8-diynes [73]. The *axial* orientation of the substituents at the nitrogen atom(s) in **10** or **27** cause a *gauche* arrangement between the lone pair at the nitrogen atom and the CH_2 groups. This implies an *anti* arrangement with the $\sigma(CC)$ bond (Fig. 7.3, left).

In the case of an *equatorial* orientation of the substituent on the nitrogen atom(s) in **10** or **27** the adjacent *axial* $\sigma(CH)$ bonds and the nitrogen lone pair(s) would show an *anti* arrangement (Fig. 7.3, right) whereas the lone pair(s) could be *gauche* to the endocyclic $\sigma(CC)$ bond(s). Both *anti* interactions, the $\sigma(CC)$-n and the $\sigma(CH)$-n interaction, are antibonding. Because the $\sigma(CH)$-n interaction is stronger than the $\sigma(CC)$-n interaction [74] the lone pairs prefer the *equatorial* position. It should be mentioned that in N-substituted piperidines and piperazines the antibonding hyperconjugative effects are overruled by strong *synaxial* 1,3-interactions between NR and *axial* CH bonds forcing the substituent R into the *equatorial* position. Further effects which favor the *axial* conformation are back-strain effects which are shown in Fig. 7.4 for **10(3)a** (R = CH_3). For the *equatorial* orientation of

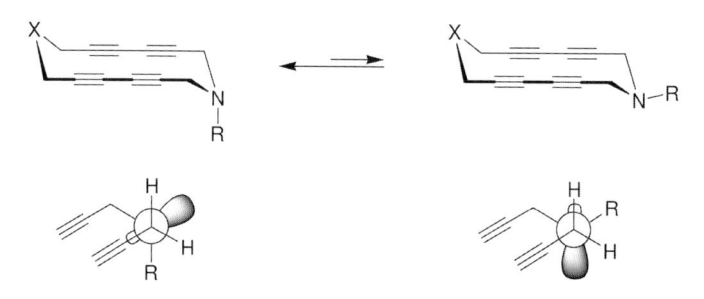

Figure 7.3. Nitrogen inversion at **10(3)** or **27** and the corresponding vicinal interactions of the lone pairs for axial (left) and equatorial (right) orientation.

the methyl group there are five repulsive *synaxial* 1,3-interactions, whereas for an *axial* orientation only three such interactions occur.

As discussed above, the 1,3-butadiyne unit often serves as a rigid spacer to keep a distinct distance between two functional moieties. Together with adjacent CH_2 groups that are collinear with the diacetylene moiety, distances of 660–680 pm are established. Heteroatoms and adjacent aromatic cycles can increase these values tremendously. Figure 7.5 displays two examples: in **49** two electron-rich aromatic cycles are held parallel by three butadiyne units. In the resulting cavity an electron-deficient acetonitrile molecule is hosted [50]. In **50** the three diacetylene moieties keep two tripodal thiophene-containing amines at a distinct distance [55]. Both structures show a slight torsion along the molecular C_3 axis as the majority of compounds with a three butadiyne framework do.

Whereas the above presented examples such as **49** and **50** show a compact structure with a cavity inside, large monocyclic systems such as tetrachalcogenacyclotetraynes and hexachalcogenacyclohexaynes **29(n)** and **33(n)–35(n)** [40] adopt a more open structure with a large free accessible space. This opens up the possibility of supramolecular organization.

In contrast to cyclic tetrachalcogenadiynes [75], the conformations of cyclic tetrachalcogenatetraynes [76] are not mainly determined by electronic effects. In the smaller counterparts (cyclic tetrathia- and tetraselenadiynes) where only one triple bond links the chalcogen centers the repulsive interaction between the lone pairs of the chalcogens is the structure-determining interaction [75, 76]. This repulsive interaction is remarkably reduced by the longer distance spanned by the two acetylene units. The favored perpendicular conformation is circumvented by steric or

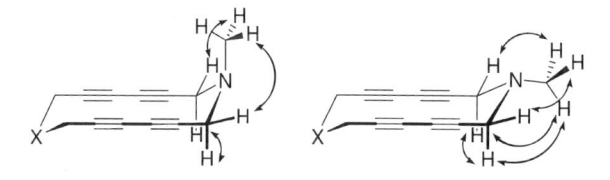

Figure 7.4. Back-strain effects in axial and equatorial conformers of **10(3)** and **27**, respectively.

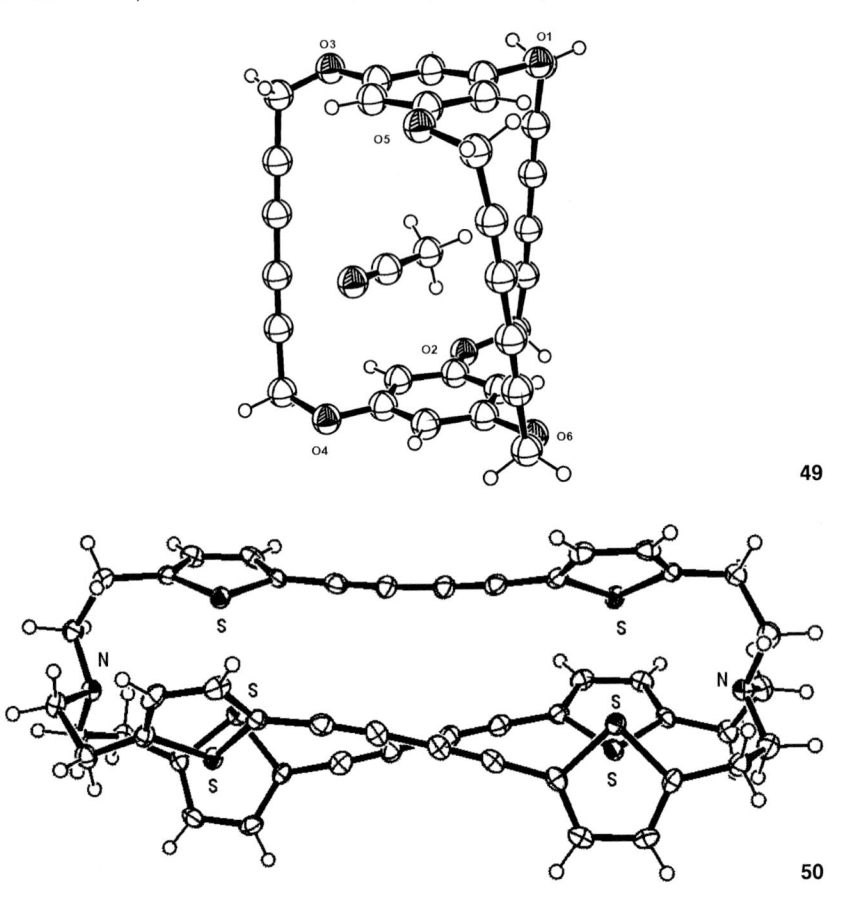

Figure 7.5. ORTEP plot of the molecular structures of **49** (with included acetonitrile) and **50** (50% ellipsoid probability).

packing effects. Recent quantum chemical calculations using density functional theory (B3LYP) confirm this view [77]. They show that the rotational barriers of the corresponding butadiyne systems amount only to about 50% of the monoacetylenic ones. In the case of a sulfur-containing model system (MeSC$_4$SMe) a value of 10.6 kJ mol^{-1} was found for the rotational barrier whereas for the corresponding selenium congener (MeSeC$_4$SeMe) this value decreases to 8.0 kJ mol^{-1}. This behavior can readily be understood: the overlap between the π^* orbital of the butadiyne moiety and the lone pairs of sulfur is, due to their comparable size, more efficient than the overlap with the corresponding lone pairs of selenium. This effect seems to be the dominant one although the lone pairs of selenium are slightly higher in energy [77].

A systematic discussion of the molecular structures of tetraynes and hexaynes is not meaningful due to the considerably large number of conformations available at room temperature. The structures are determined by the three rigid donor substituted butadiyne units. The three bridging alkane chains try to adopt energetically favored zigzag conformations. As a result, some kind of elastic cycle is obtained containing stiff and rather flexible units. Therefore we focus our discussion on their self-assembly properties and their packing in the solid state [78].

In their smaller counterparts, the cyclic tetrathia- and tetraselenadiynes, it was observed that chalcogen–chalcogen contacts are responsible for the occurrence of columnar or even tubular stacks in the solid state [76]. Whereas the cyclic diynes are too small to include guest molecules in the hollow core the corresponding tetraynes offer an enlarged cavity. In the case of **29(5)** the four chalcogen atoms are located at the four edges of a rather flat rectangle. In the solid state, the cycles stack in an AAA fashion in such a way that the chalcogen centers of one cycle keep in close contact with chalcogen centers of neighboring stacks. As a result, a channel-like structure arises [40, 78]. By recrystallizing **29(5)** from hexane we found no solvent molecules in these cavities (modification α). In contrast, by using toluene we observed the inclusion of one aromatic molecule per two cyclotetrayne molecules in a regular manner (modification β) as shown in Fig. 7.6 (side and top view). However, the channel diameter is still too small to house the aromatic ring, only the methyl group of the guest is included. The arene unit itself is situated between two molecules of the stack. Calculations of the solvent-accessible volume carried out with the PLATON program [79] revealed that 13.1% (α) and 16.1% (β), respectively, of the crystal volume may be occupied by solvent. Similar experiments with the corresponding selenium congener **33(5)** afforded an analogous solid-state structure including hexane as guest which could only be detected as disordered electron density.

In contrast to only two butadiyne moieties the three butadiyne units of hexachalcogenahexaynes provide a larger cavity giving rise to larger channels [40]. For the hexaselenacyclohexayne **35(4)**, two different kinds of supramolecular organizations

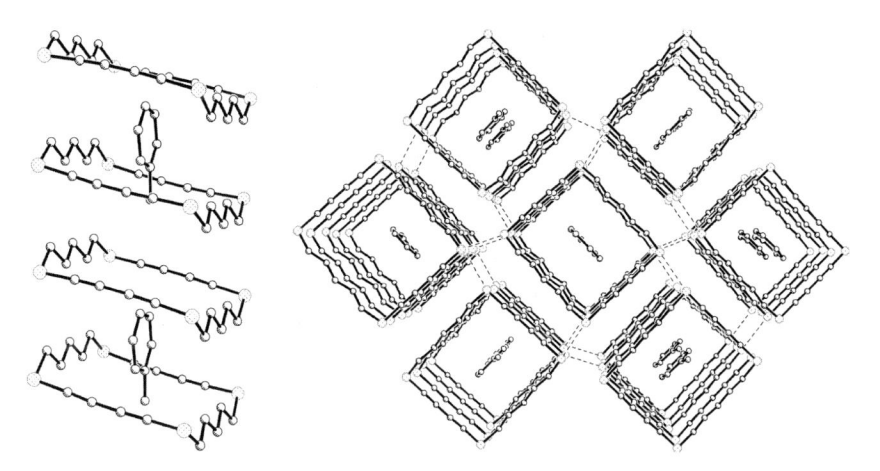

Figure 7.6. Columnar structure of **29(5)** in the solid state with included toluene guest molecules: (left) side view, (right) top view. H atoms are omitted for the sake of clarity.

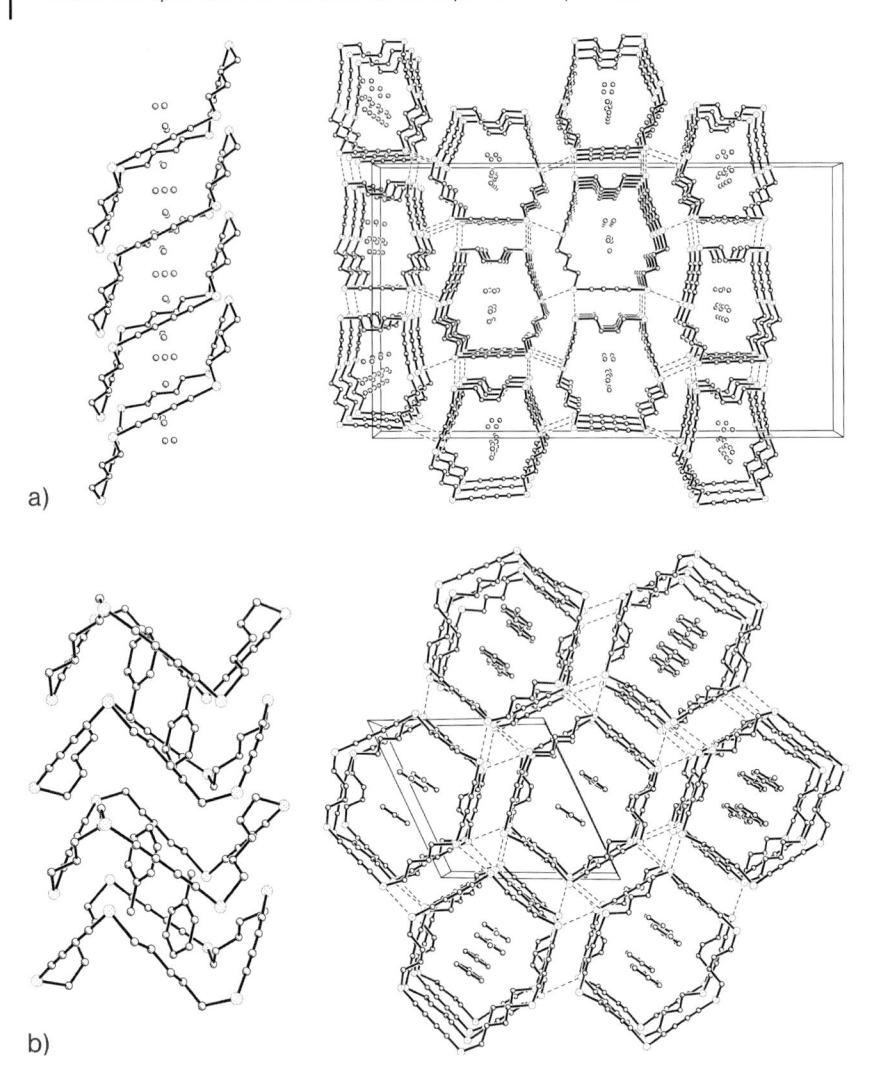

Figure 7.7. Two totally different modifications of **35(4)**: (a) with included disordered hexane molecules (side and top view), (b) with included toluene molecules (side and top view). Short Se···Se contacts are indicated as well as the unit cell dimensions. H atoms are omitted for the sake of clarity.

were elucidated. In the first case disordered hexane molecules are incorporated, and AAA stacking of the molecules is observed (modification α, Figure 7.7(a)). In the second case showing an ABAB stacking toluene guest molecules are accomodated at well-defined positions (modification β). Here, in contrast to the above-mentioned cyclotetrayne one molecule of aromatic guest per macrocycle is included (Figure 7.7(b)). Calculations of the solvent-accessible volume [79] revealed values of 19.2% for α and 24.3% for β, respectively.

The two modifications show extremely close intermolecular Se···Se distances between two neighboring stacks of molecules (α: 360 and 378 pm; β: 366 and 368 pm). These distances are much shorter than the sum of the van der Waals radii (400 pm) [80]. Examining the directionality of these interactions shows that the 4p lone pairs at the Se centers directly coincide with the directions of the Se–C σ^* orbitals of the neighboring cycles. Besides isotropic dispersion forces between the electron-rich chalcogen centers, this interaction is the thermodynamic basis of the observed structures. Quantum chemical calculations at the MP2 level concerning the interaction energies between Me$_2$X and MeXC$_2$H units (X = S, Se) in the most favored arrangement revealed values of 10 kJ mol^{-1} (S) and 15 kJ mol^{-1} (Se) [81]. The corresponding interchalcogen distances are 363 pm (S) and 363 pm (Se), shorter than the sum of the van der Waals radii, as anticipated from experiment.

In the case of the 33-membered hexaselenacyclohexayne **35(5)** a systematic study was carried out in order to examine the solid-state behavior dependent on the kind of included guest molecule [40]. The results of this study are shown in Table 7.2. In total, 12 different guest molecules were included, aliphatic as well as aromatic, and electron-rich as well as electron-poor. All structures have very similar primitive unit cells, some of them leading to a higher symmetric R centered lattice. Each structure is described by a greek letter (from α to μ). In the cases where the trigonal

Table 7.2. Systematic study of the solid state behavior of **35(5)** depending on the included guest molecules.

Structure	Guest molecule	Crystal system	Volume [%][a]	Angle ϕ [°][b]
35(5)α	furan	trigonal	18.9	37.8
35(5)β	hexane	trigonal	19.2	37.6
35(5)γ	thiophene	trigonal	20.0	36.6
35(5)δ	benzene	trigonal	20.3	36.1
35(5)ε	aniline	trigonal	20.4	36.0
35(5)ζ	chlorobenzene	trigonal	20.7	35.9
35(5)η	nitrobenzene	triclinic	20.8	35.5
35(5)θ	toluene	triclinic	20.8	35.3
35(5)ι	anisole	triclinic	21.0	35.1
35(5)κ	p-xylene	triclinic	23.3	34.1
35(5)λ	4-bromoanisole	triclinic	23.5	33.6
35(5)μ	mesitylene	triclinic	25.6	30.6

[a] Solvent-accessible volume of the unit cell, calculated with PLATON [79].
[b] Angle ϕ between the medium plane of the macrocycle and the SeC$_4$Se unit(s).

space group $R\bar{3}$ was observed, the guest molecules could not be localized at well-defined positions. In the other cases when the guests were found on well-defined positions the trigonal symmetry of the crystal was not present and a triclinic space group resulted ($P\bar{1}$). The fact that many solvent molecules are incorporated in an ordered fashion can be ascribed to weak C–H $\cdots \pi$ interactions [82] between the alkane chains of the macrocycle and the π system of the aromatic guest. We also list in Table 7.2 the solvent-accessible volume calculated with the PLATON program [79] and the angle ϕ between the medium plane of the macrocycle and the SeC$_4$Se unit(s). Table 7.2 unambiguously illustrates that there is a straightforward, almost linear correlation between the solvent-accessible volume of the crystal and the angle ϕ. The larger the guest, the smaller is ϕ, and the more the tubes are stretched resulting in larger cavities. Thus, neither a potential bending of the butadiynyl fragments nor a contribution of the saturated alkyl linkers is responsible for the increased cavity size. The most important factor is the variation of the torsional angle at the donor substituted butadiyne unit serving as a hinge [40].

Figure 7.8 compares the shapes of the solid-state structures of **35(5)** containing toluene and mesitylene. In the former the solvent-accessible volume [79] is 20.8% of the volume of the unit cell, whereas in the latter this value amounts to 25.6%. The more voluminous guest mesitylene with the two further methyl groups affords an increase in the channel diameter. In crystallographic parameters, this means a decrease of the axis in the stacking direction (short axis) by 2.3%, whereas the two longer axes of the unit cell increase by 3.2%. As a result, the solvent-accessible volume increases by 23% compared with the toluene-containing structure [40]. In contrast, if the channels are occupied by a sea of highly disordered hexane molecules which are totally flexible, the value for the solvent-accessible volume decreases to 19.2%. The sulfur congener **34(5)** showed quasi-isomorphous structures when similar experiments with hexane and toluene were carried out. The flexibility of these "elastic" cycles [40] – a term we recently introduced in contrast to shape-persistent macrocycles – is due to a low force constant when changing the torsional angle in R–X–C≡C-C≡C–X–R units.

7.5.2
NMR Spectroscopic Investigations

For analytical purposes the ^{13}C NMR data of the 1,3-butadiyne units are important. Unfortunately for most cyclic hydrocarbons such as **3–6**, **61** and **62** the ^{13}C NMR data are unknown because they were prepared prior to the availability of ^{13}C NMR spectra for unlabeled species. Fortunately the data which are available for **2(3)** [72], **2(4)** [72], **10(3)**–**10(5)** [32, 33], **15** [35], **17** [36], **19** [37], **22–24** [38], **27** [39], **29** [40], **34** [40] and **35** [40] reveal that the chemical shift of the sp carbons varies only between 60 and 80 ppm if no heteroatoms are bound to the sp centers. In a study of acyclic polyalkynes, capped by XCH$_3$ (X = S, Se, Te) [77, 83], it was found that the C(sp) centers directly bound to the chalcogen center are deshielded. In the case of 2,7-ditelluraocta-3,5-diyne the resonance for the terminal sp carbons is found at 41 ppm (CDCl$_3$), whereas the signal for the central carbons is found at

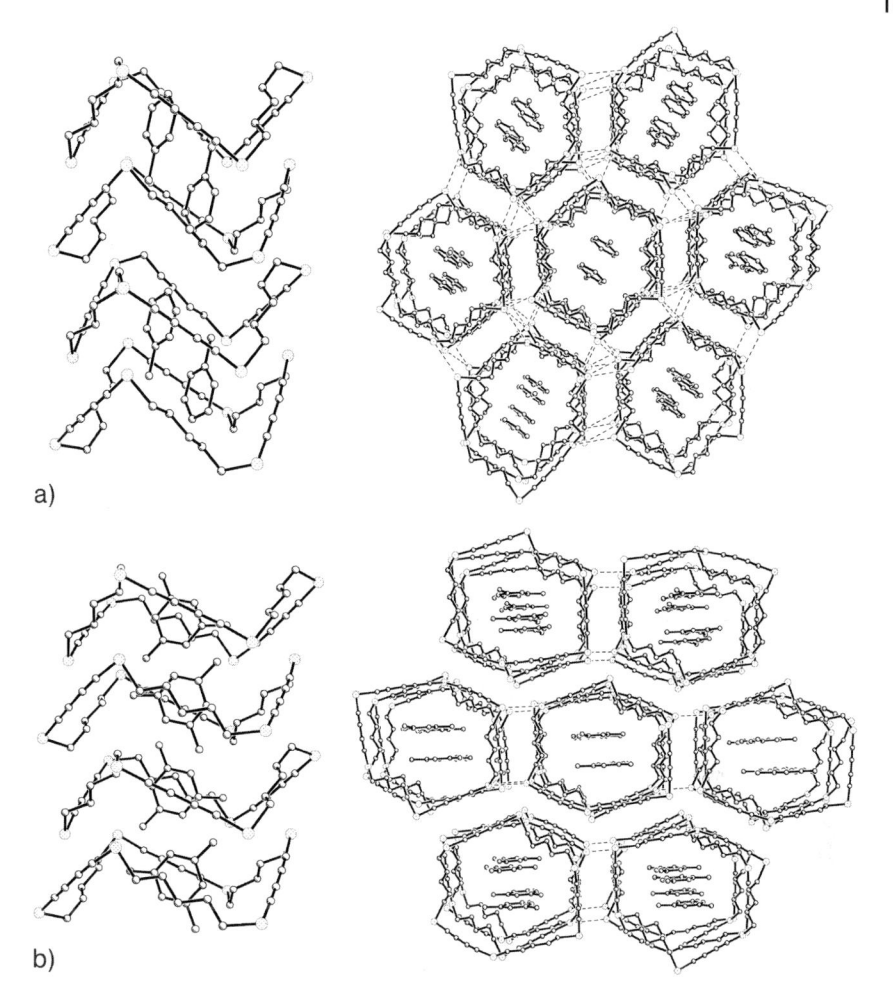

Figure 7.8. Two different modifications of **35(5)**: (a) with included toluene molecules (side and top view), (b) with included mesitylene molecules (side and top view). In the case of the larger guest, the cycles are more flattened giving rise to a larger cavity. Short Se···Se contacts are indicated. H atoms are omitted for the sake of clarity.

96 ppm. The corresponding values for the selenium congener are 65 and 85 ppm and for the sulfur congener 75 and 80 ppm. The high field shift of the terminal ^{13}C signal in the case of tellurium substitution was ascribed to the so-called "heavy atom effect" [84] and can also be understood in terms of increased electron density. Solvents which decrease this electron density are able to change these shifts significantly, as also encountered in the case of iodoalkynes [85, 86].

Dynamic NMR studies were carried out on **10(3)a** (R = C_2H_5) and **27a–27d** [70] to investigate the inversion process between chair and boat conformation of the 14-membered rings (see Fig. 7.3). The coalescence temperature for the signals of the diastereotopic hydrogens of the ring CH_2 groups for **10(3)b** (R = C_2H_5) was reported to be $T_c = 210$ K [32]. From this result a $\Delta G^{\ddagger}$ value of 44 kJ mol^{-1} for the chair–boat interconversion was estimated. For **27a–27d** the $\Delta G^{\ddagger}$ values for this interconversion were in the range 42–46 kJ mol^{-1} [70]. These results are very similar to those reported for the 1,6-diazacyclodeca-3,8-diyne-derivatives (46 kJ mol^{-1}) [73].

For the boat–chair equilibrium of **27a–27c** values of ΔG were in the range 0.4–1.7 kJ mol^{-1} in favor of the chair conformation [70]. This result is in line with quantum chemical calculations on the parent system using the DFT(6-311G*) method [69].

7.5.3
Photoelectron and Electron Absorption Spectra of Cyclic 1,3-Diynes

The He(I) photoelectron (PE) spectra of 1,3-butadiyne and simple substituted products have been studied extensively by Heilbronner et al. [87]. The PE spectrum of the parent system shows two ionization events from the π orbitals (e_g, e_u) at 10.17 eV (e_g) and 12.62 eV (e_u). Due to the thermal lability of polyynes only two PE spectra of cyclic tetraynes have so far been measured: octamethyl-1,3,7,9-dodecatetrayne (**17**) [88] and 1,3,9,11-hexadecatetrayne (**2(4)**) [89]. The latter is shown in Fig. 7.9. Both spectra exhibit two peaks close together in the low energy region. They can be assigned to four ionization events from the "in plane" and "out of plane" π MOs of two parallel or nearly parallel 1,3-butadiyne units corresponding to four ionizations from b_{2g}, a_u, b_{3g} and b_{1u} orbitals assuming D_{2h} symmetry for both species.

The longest wavelength band of the electronic absorption spectrum of 1,3-butadiyne was assigned to the vibrational progressions of two forbidden electronic transitions $^1\Sigma_g^- \leftarrow {}^1\Sigma_g^+$ and $^1\Delta_u \leftarrow {}^1\Sigma_g^+$ which overlap each other [90]. The peaks appear at 243.4, 231.5, 220.7 and 211.1 nm in the gas phase spectrum [90]. The

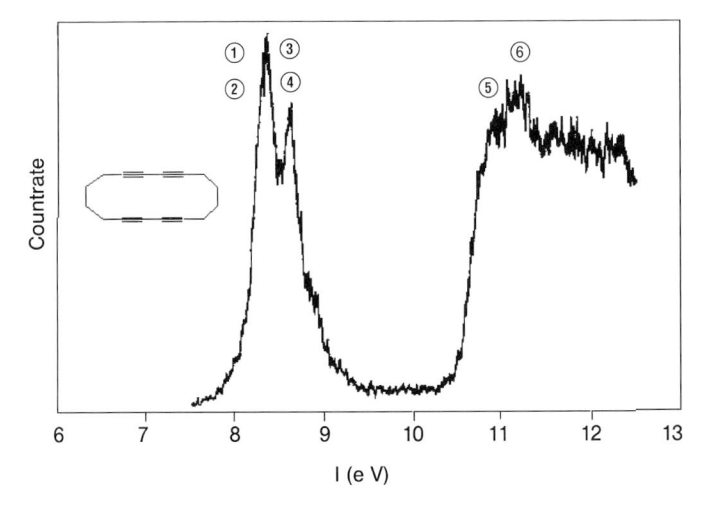

Figure 7.9. He(I) photoelectron spectrum of **2(4)**.

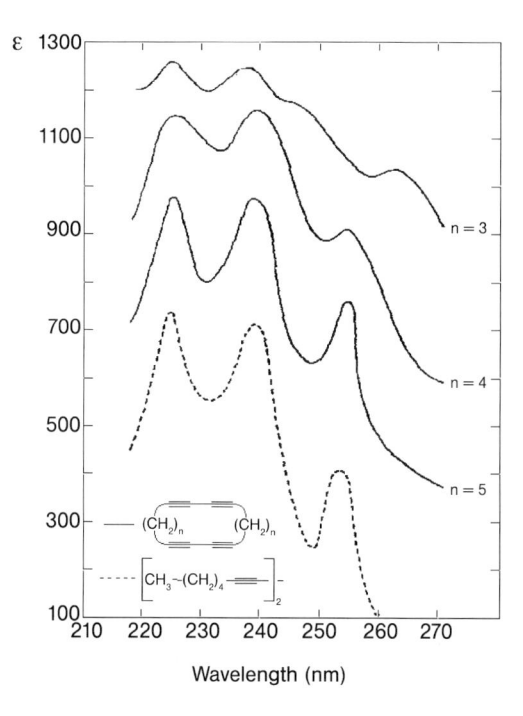

Figure 7.10. UV absorption spectra of **2(3)** to **2(5)** in isooctane. The ε-values of dodeca-6,8-diyne have been doubled, those of **2(5)** have been increased by 300, those of **2(4)** by 550 and that of **2(3)** by 650.

first three vibrational peaks are also found with commercial spectrometers in solutions of pentane. Sondheimer et al. [23b] correlated the long wavelength band of **2(3)** to **2(5)** with that of tetradeca-6,8-diyne. This comparison is shown in Fig. 7.10. It is found that the first vibrational band is considerably shifted to lower energy. This change is reminiscent to that found in the absorption spectra of (n.n)paracyclophanes [91].

7.6 Concluding Remarks and Outlook

The first carbon-rich cycles with two and more 1,3-butadiyne units were prepared as precursors for cyclic conjugated annulenes. After these goals were successfully achieved the 1,3-butadiyne moieties were used as rigid building blocks in cyclic and bicyclic systems, helping to generate large cavities able to host guest molecules. Cyclic systems with two 1,3-butadiyne units also serve as ideal models to study the reaction modes of 1,3-butadiyne units in close proximity. Reductive and acid-induced cyclization experiments were undertaken, radical-, photochemical- and metal-induced cyclization reactions are still waiting to be investigated.

(X = S, Se, Te)

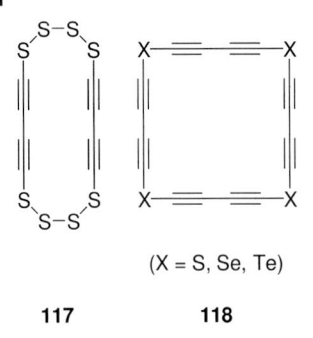

117   118

Chart 7.6. Hitherto unsynthesized binary carbon–chalcogen compounds **117** and **118**.

Chalcogen centers adjacent to stiff butadiyne units proved to favor the formation of tubular structures in the solid state that are able to host a large variety of guest molecules. A further replacement of alkyl tethers in tetrachalcogenacyclotetraynes by chalcogen atoms or even butadiyne moieties would lead to highly interesting binary carbon–chalcogen compounds such as C_8S_8 (**117**) and the pericyclyne $C_{16}X_4$ (**118**), respectively (Chart 7.6). Apart from amazing structures one expects that these molecules could be an ideal source for reactive C_4 species. In this field more activity is anticipated.

7.7
Experimental: Selected Procedures

7.7.1
General Procedure for the Preparation of N,N′-Dialkyl-1,8-diazacyclotetradeca-3,5,10,12-tetraynes (27a–e)

To a suspension of K_2CO_3 (40 g) in dry THF (1L) were added simultaneously within 8–10 h a solution of 1,6-dibromohexa-2,4-diyne (**25**) (118 mmol) in THF (500 mL) and the alkylamine (**26**) (118 mmol) in THF (500 mL) at 40–50 °C. After the addition was complete the mixture was stirred for 2 d. The reaction was carried out under argon and with the exclusion of light. For work up the solution was filtered and concentrated on the rotary evaporator at 30 °C to 100–200 mL, then absorbed on 100 g Alox III and chromatographed on silica gel using cyclohexane/EtOAc mixtures.

7.7.2
General Procedure for the Cadiot–Chodkiewicz Coupling During the Syntheses of Azacyclotetraynes 10(n)

CuI (4 mmol) was dissolved in pyrrolidine (50 mL) under magnetic stirring. At 0 °C the diyne **1(n)** (20 mmol) was added to the solution and the bromo-alkyne **7**

(40 mmol) was added dropwise through a syringe into the reaction mixture over a period of 2 h. After stirring for 30 min at 0 °C, the resulting solution was poured into a stirred mixture of ice (200 g) and Et$_2$O (100 mL). Concentrated HCl was added slowly until the green color of the resulting mixture turned slightly red. The ether layer was separated quickly, and the acidic aqueous solution was extracted with Et$_2$O (2 × 50 mL). The combined organic layers were washed twice with a saturated aqueous solution of NaHCO$_3$ and brine, and dried over anhydrous Na$_2$SO$_4$. After filtration the solvent was evaporated, and the crude product was purified by column chromatography on Alox III using mixtures of petroleum ether/Et$_2$O as eluant.

7.7.3
General Procedure for the Preparation of the Tetrachalcogenacyclotetraynes 29(n) and 33(n) and Hexachalcogenacyclohexaynes 34(n) and 35(n)

To a solution of bis(trimethyl)silylbutadiyne (**30**) (13 mmol) in THF (100 mL) under argon was added MeLi/LiBr in Et$_2$O (1.5 M) (30 mmol) dropwise at −78 °C over a period of 20 min. The solution was stirred for 3 h at −78 °C. To anhydrous THF (500 mL), the resulting suspension of dilithium butadiynide in THF and a solution of the corresponding dithiocyanatoalkane **31(n)** and diselenocyanatoalkane **32(n)** (14 mmol), respectively, in THF (100 mL) were added dropwise simultaneously at −40 °C over a period of 4 h. After complete addition, the reaction mixture was allowed to warm up to room temperature overnight. The solvent was removed by rotary evaporation at 30 °C. Polymers and salts were removed by flash filtration (SiO$_2$ with 3%(v/v) NEt$_3$, toluene as eluant). The solvent was removed by rotary evaporation. Column chromatography of the crude product (SiO$_2$, hexane/Et$_2$O or hexane/CH$_2$Cl$_2$ as eluant) afforded the cyclic tetraynes and in most cases the cyclic hexaynes.

7.7.4
General Procedures for the Preparation of Hexadiyne-bridged Macrobicycles

7.7.4.1 Alternative A
To a solution of trispropargylic ether (1.23 mmol) in O$_2$-free pyridine (1230 mL) is added anhydrous CuCl (123 mmol) and anhydrous CuCl$_2$ (15 mmol). The mixture is stirred for 48 h at 0 °C. After evaporation of the solvent the crude product is purified by chromatography and crystallization.

7.7.4.2 Alternative B
A suspension of Cu(OAc)$_2$ · 2H$_2$O (62 mmol) in CH$_3$CN (500 mL) was heated to 60 °C. Then a solution of the corresponding trispropargylic ether (6.2 mmol) in CH$_3$CN was added with stirring and the mixture was kept at 60 °C for 8 h. It was subsequently poured into water (1 L), the precipitate was filtered and washed with water (500 mL). The solid was dried *in vacuo* and purified by column chromatography.

Acknowledgments

We thank all our coworkers and colleagues who contributed to the results reported in this chapter. Furthermore, we thank the Deutsche Forschungsgemeinschaft, the Fonds der Chemischen Industrie and the BASF Aktiengesellschaft, Ludwigshafen, for financial support. D. B. W. is grateful to the Studienstiftung des deutschen Volkes for a graduate fellowship and to the Alexander von Humboldt Foundation for a Feodor Lynen Research Fellowship.

Abbreviations

Ac	acetyl
B3LYP	Becke's three parameter exchange functional combined with the correlation functional of Lee, Yang and Parr
DDQ	2,3-dichloro-5,6-dicyano-p-benzoquinone
DMF	dimethylformamide
DMSO	dimethylsulfoxide
Et	ethyl
IR	infrared
Me	methyl
MO	molecular orbital
NMR	nuclear magnetic resonance
PE	photoelectron
rt	room temperature
t-Bu	$tert$-butyl
THP	tetrahydropyranyl
THF	tetrahydrofuran
TIPS	triisopropylsilyl
TMEDA	N,N,N',N',-tetramethylethylenediamine
UV	ultraviolet

References

1 (a) F. SONDHEIMER, I. C. CALDER, J. A. ELIX, Y. GAONI, P. J. GARRATT, K. GROHMANN, G. DIMAIO, J. MAYER, M. V. SARGENT, R. WOLOVSKY, *Special Publication No 21*, The Chemical Society, London, **1967**; (b) F. SONDHEIMER, *Pure Appl. Chem.* **1963**, *7*, 363–388; (c) R. M. McQUILKIN, P. J. GARRATT, F. SONDHEIMER, *J. Am. Chem. Soc.* **1970**, *92*, 6682–6683.

2 M. NAKAGAWA in *Topics in Nonbenzenoid Aromatic Chemistry* (T. NOZOE, R. BRESLOW, K. HAFNER, S. ITO, I. MURATA (eds.)) Hirokawa Publishing Co., Tokyo, **1973**, Vol. *1*, pp. 191–219.

3 (a) H. W. KROTO, J. R. HEATH, S. C. O'BRIEN, R. F. CURL, R. E. SMALLEY, *Nature* **1985**, *318*, 162–163; (b) W. KRÄTSCHMER, L. D. LAMB, K. FOSTIROPOULOS, D. R. HUFFMAN, *Nature* **1990**, *347*, 354–358.

4 F. DIEDERICH, Y. RUBIN, *Angew. Chem.* **1992**, *104*, 1123–1146; *Angew. Chem. Int. Ed. Engl.* **1992**, *31*, 1101–1123.

5 (a) A. DE MEIJERE (ed.), Carbon Rich Compounds I, *Top. Curr. Chem.* **1998**, *196*; (b) A. DE MEIJERE (ed.), Carbon Rich Compounds II, Macrocyclic Oligoacetylenes and Other Linearly Conjugated Systems, *Top. Curr. Chem.* **1999**, *201*.

6 *Metal-catalyzed Cross-coupling Reactions*, F. Diederich, A. de Meijere (eds.), Wiley-VCH, Weinheim, 2nd edn., **2004**.

7 (a) L. Brandsma, *Preparative Acetylene Chemistry*, L. Brandsma, Elsevier, Amsterdam, 2nd edn., **1988**; (b) L. Brandsma, *Synthesis of Acetylenes, Allenes and Cumulenes*, Elsevier, Amsterdam, **1981**; (c) L. Brandsma, *Best Synthetic Methods: Synthesis of Acetylenes, Allenes and Cumulenes: Methods and Techniques*, Elsevier, Amsterdam, **2004**.

8 C. S. Jones, M. J. O'Connor, M. M. Haley, in *Acetylene Chemistry – Chemistry, Biology, and Material Science*, F. Diederich, P. J. Stang, R. R. Tykwinski (eds.), Wiley-VCH, Weinheim, **2005**, pp. 303–385.

9 A. de Meijere, S. I. Kozhushkov, in Carbon Rich Compounds, *Top. Curr. Chem.* **1999**, *201*, 1–42.

10 (a) D. Venkataraman, S. Lee, J. Zhang, J. S. Moore, *Nature* **1994**, *371*, 591–593; (b) Y. Tobe, N. Utsumi, K. Kawabata, A. Nagano, K. Adachi, S. Araki, M. Sonoda, K. Hirose, K. Naemura, *J. Am. Chem. Soc.* **2002**, *124*, 5350–5364.

11 (a) O. Henze, D. Lentz, A. D. Schlüter, *Chem. Eur. J.* **2000**, *6*, 2362–2367; (b) S. Höger, D. L. Morrison, V. Enkelmann, *J. Am. Chem. Soc.* **2002**, *124*, 6734–6736.

12 R. Gleiter, D. Kratz, W. Schäfer, V. Schehlmann, *J. Am. Chem. Soc.* **1991**, *113*, 9258–9264.

13 (a) R. Gleiter, J. Ritter, *Angew. Chem.* **1994**, *106*, 2550–2552; *Angew. Chem. Int. Ed. Engl.* **1994**, *33*, 2470–2472; (b) W. R. Roth, T. Wasser, R. Gleiter, H. Weigl, *Liebigs Ann.-Recl.* **1997**, 1329–1331; (c) G. Haberhauer, R. Gleiter, *J. Am. Chem. Soc.* **1999**, *121*, 4664–4668.

14 J. Ritter, R. Gleiter, *Liebigs Ann.-Recl.* **1997**, 1179–1188.

15 (a) G. Wegner, *Z. Naturforsch. b* **1969**, *24*, 824–832; (b) G. Wegner, *Makromol. Chem.* **1971**, *145*, 85–94; (c) G. Wegner, *Makromol. Chem.* **1972**, *154*, 35–48.

16 V. Enkelmann, *Adv. Polym. Sci.* **1984**, *63*, 91–136. For recent results of diynes with stacks for which ϕ is close to 90° see Ref. 17.

17 G. W. Coates, A. R. Dunn, L. M. Henling, D. A. Dougherty, R. H. Grubbs, *Angew. Chem.* **1997**, *109*, 290–293; *Angew. Chem. Int. Ed. Engl.* **1997**, *36*, 248–251.

18 (a) C. Glaser, *Ber. Dtsch. Chem. Ges.* **1869**, *2*, 422–424; (b) C. Glaser, *Ann. Chem. Pharm.* **1870**, *154*, 137–171.

19 G. Eglinton, A. R. Galbraith, *Chem. Ind.* **1956**, 737–738.

20 A. S. Hay, *J. Org. Chem.* **1962**, *27*, 3320–3321.

21 (a) A. Krebs in *Chemistry of Acetylenes*, H. G. Viehe (ed.), Marcel Dekker, New York **1969**, 987–1062; (b) G. Eglinton, W. McCrae, *Adv. Org. Chem.* **1963**, *4*, 225–328; (c) K. Sonogashira, in *Comprehensive Organic Chemistry*, B. M. Trost, I. Fleming (eds.), Pergamon Press, Oxford, **1991**, Vol. 3, pp. 551–561.

22 P. Siemsen, R. C. Livingston, F. Diederich, *Angew. Chem.* **2000**, *112*, 2740–2767; *Angew. Chem. Int. Ed.* **2000**, *39*, 2632–2657.

23 (a) F. Sondheimer, Y. Amiel, R. Wolovsky, *J. Am. Chem. Soc.* **1957**, *79*, 4247–4248; (b) F. Sondheimer, Y. Amiel, R. Wolovsky, *J. Am. Chem. Soc.* **1957**, *79*, 6263–6267.

24 F. Sondheimer, Y. Amiel, *J. Am. Chem. Soc.* **1957**, *79*, 5817–5820.

25 F. Sondheimer, Y. Amiel, R. Wolovsky, *J. Am. Chem. Soc.* **1959**, *81*, 4600–4606.

26 F. Sondheimer, R. Wolovsky, *J. Am. Chem. Soc.* **1962**, *84*, 260–269.

27 (a) R. Wolovsky, F. Sondheimer, *J. Am. Chem. Soc.* **1962**, *84*, 2844–2845; (b) R. Wolovsky, F. Sondheimer, *J. Am. Chem. Soc.* **1965**, *87*, 5720–5727.

28 M. Nakagawa, in *The Chemistry of Triple-bonded Functional Groups*, P. Patai (ed.), Wiley, New York, **1978**, Vol. I, pp. 635–712.

29 S. Misumi, T. Kaneda, in *The Chemistry of Triple-bonded Functional Groups*, P. Patai (ed.), Wiley, New York, **1978**, Vol. I, pp. 713–737.

30 (a) W. Chodkiewicz, P. Cadiot, C. R. Hebed, *Seances Acad. Sci.* **1955**, *241*, 1055–1057; (b) W. Chodkiewicz, *Ann. Chim.* **1957**, *2*, 819–869.

31 P. Cadiot, W. Chodkiewicz, in *Chemistry of Acetylenes*, H. G. Viehe (ed.), Marcel Dekker, New York, 1969, pp. 597–647.

32 E. M. Schmidt, R. Gleiter, F. Rominger, *Chem. Eur. J.* 2003, 9, 1814–1822.

33 E. M. Schmidt, R. Gleiter, F. Rominger, *Eur. J. Org. Chem.* 2004, 2818–2825.

34 G. Fuhrmann, T. Debaerdemaeker, P. Bäuerle, *Chem. Commun.* 2003, 948–949.

35 D. Kuch, Diplomarbeit, Universität Heidelberg, 2004.

36 L. T. Scott, G. J. DeCicco, J. L. Hyun, G. Reinhardt, *J. Am. Chem. Soc.* 1985, 107, 6546–6555.

37 (a) M. Kaftory, I. Agmon, M. Ladika, P. J. Stang, *J. Am. Chem. Soc.* 1987, 109, 782–787; (b) K. N. Houk, L. T. Scott, N. G. Rondan, D. C. Spellmeyer, G. Reinhardt, J. L. Hyun, G. J. DeCicco, R. Weiss, M. H. M. Chen, L. S. Bass, J. Clardy, F. S. Jørgensen, T. A. Eaton, V. Sarkozi, C. M. Petit, L. Ng, K. D. Jordan, *J. Am. Chem. Soc.* 1985, 107, 6556–6562.

38 K. Negishi, M. Unno, H. Matsumoto, *Chem. Lett.* 2004, 33, 430–431.

39 K. Hövermann, J. Ritter, F. Rominger, B. Nuber, R. Gleiter, *Eur. J. Org. Chem.* 2000, 2291–2294.

40 D. B. Werz, R. Gleiter, F. Rominger, *J. Org. Chem.* 2004, 69, 2945–2952.

41 (a) S. Claude, J.-M. Lehn, F. Schmidt, J.-P. Vigneron, *J. Chem. Soc., Chem. Commun.* 1991, 1182–1185; (b) P. Čudić, M. Žinić, V. Tomišić, V. Simeon, J. P. Vigneron, J.-M. Lehn, *J. Chem. Soc., Chem. Commun.* 1995, 1073–1075.

42 A. Lorente, M. Fernández-Saiz, J.-M. Lehn, J. P. Vigneron, *Tetrahedron Lett.* 1995, 36, 8279–8282.

43 (a) K. M. Neder, H. W. Whitlock, Jr., *J. Am. Chem. Soc.* 1990, 112, 9412–9414; (b) S. P. Miller, H. W. Whitlock, Jr., *J. Am. Chem. Soc.* 1984, 106, 1492–1494.

44 B. J. Whitlock, H. W. Whitlock, Jr., *Tetrahedron Lett.* 1988, 29, 6047–6050.

45 R. Bürli, A. Vasella, *Angew. Chem.* 1997, 109, 1945–1946; *Angew. Chem. Int. Ed. Engl.* 1997, 36, 1852–1853.

46 M. Srinivasan, S. Sankararaman, H. Hopf, I. Dix, P. G. Jones, *J. Org. Chem.* 2001, 66, 4299–4303.

47 R. M. Moriarty, D. Pavlović, *J. Org. Chem.* 2004, 69, 5501–5504.

48 A. de Meijere, F. Jaekel, A. Simon, H. Borrmann, J. Köhler, D. Johnels, L. T. Scott, *J. Am. Chem. Soc.* 1991, 113, 3935–3941.

49 D. O'Krongly, S. R. Denmeade, M. Y. Chiang, R. Breslow, *J. Am. Chem. Soc.* 1985, 107, 5544–5545.

50 (a) F. Vögtle, R. Berscheid, W. Schnick, *J. Chem. Soc., Chem. Commun.* 1991, 414–416; (b) R. Berscheid, M. Nieger, F. Vögtle, *J. Chem. Soc., Chem. Commun.* 1991, 1364–1366.

51 R. Berscheid, F. Vögtle, *Synthesis* 1992, 58–62.

52 (a) R. Berscheid, M. Nieger, F. Vögtle, *Chem. Ber.* 1992, 125, 1687–1695; (b) R. Berscheid, M. Nieger, F. Vögtle, *Chem. Ber.* 1992, 125, 2539–2552.

53 F. Vögtle, I. Michel, R. Berscheid, M. Nieger, K. Rissanen, S. Kotila, K. Airola, N. Armaroli, M. Maestri, V. Balzani, *Liebigs Ann.* 1996, 1697–1704.

54 (a) B. P. Friedrichsen, H. W. Whitlock, *J. Am. Chem. Soc.* 1989, 111, 9132–9134; (b) B. P. Friedrichsen, D. R. Powell, H. W. Whitlock, *J. Am. Chem. Soc.* 1990, 112, 8931–8941.

55 M. Baier, Dissertation, Universität Heidelberg, 2005.

56 (a) C. O. Dietrich-Buchecker, J. Guilhelm, A. K. Khemiss, J.-P. Kintzinger, C. Pascard, J.-P. Sauvage, *Angew. Chem.* 1987, 99, 711–714; *Angew. Chem. Int. Ed. Engl.* 1987, 26, 661–663; (b) C. O. Dietrich-Buchecker, J.-P. Sauvage, *Chem. Rev.* 1987, 87, 795–810.

57 C. O. Dietrich-Buchecker, A. Khemiss, J.-P. Sauvage, *J. Chem. Soc., Chem. Commun.* 1986, 1376–1378.

58 D. J. Cram, M. E. Tanner, S. J. Keipert, C. B. Knobler, *J. Am. Chem. Soc.* 1991, 113, 8909–8916.

59 A. Nakano, Y. Li, P. Geoffroy, M. Kim, J. L. Atwood, S. Bott, H. Zhang, L. Echegoyen, G. W. Gokel, *Tetrahedron Lett.* 1989, 30, 5099–5102.

60 F. Sondheimer, R. Wolovsky, P. J. Garratt, I. C. Calder, *J. Am. Chem. Soc.* **1966**, *88*, 2610.

61 Y. Amiel, F. Sondheimer, *Chem. Ind.* **1960**, 1162.

62 F. Sondheimer, Y. Gaoni, *J. Am. Chem. Soc.* **1962**, *84*, 3520–3526.

63 F. Sondheimer, Y. Gaoni, *J. Am. Chem. Soc.* **1961**, *83*, 4863–4864.

64 (a) J. Mayer, F. Sondheimer, *J. Am. Chem. Soc.* **1966**, *88*, 602–603; (b) J. Mayer, F. Sondheimer, *J. Am. Chem. Soc.* **1966**, *88*, 603–604.

65 (a) A. G. Myers, S. D. Goldberg, *Tetrahedron Lett.* **1998**, *39*, 9633–9636; (b) A. G. Myers, S. D. Goldberg, *Angew. Chem.* **2000**, *112*, 2844–2847; *Angew. Chem. Int. Ed.* **2000**, *39*, 2732–2735. (c) A. G. Myers, personal communication, **2004**.

66 F. Sondheimer, R. Wolovsky, D. A. Ben-Efraim, *J. Am. Chem. Soc.* **1961**, *83*, 1686–1691.

67 O. M. Behr, G. Eglinton, A. R. Galbraith, R. A. Raphael, *J. Chem. Soc.* **1960**, 3614–3625.

68 G. M. Pilling, F. Sondheimer, *J. Am. Chem. Soc.* **1971**, *93*, 1970–1977.

69 B. Esser, Diplomarbeit, Universität Heidelberg, **2004**.

70 R. Gleiter, K. Hövermann, F. Rominger, to be published; (b) K. Hövermann, Dissertation, Universität Heidelberg **1999**.

71 J. H. van Roosmalen, E. Jones, H. J. Kevelam, *Tetrahedron Lett.* **1972**, *13*, 1865–1868.

72 R. Gleiter, R. Merger, J. Chavez, T. Oeser, H. Irngartinger, H. Pritzkow, B. Nuber, *Eur. J. Org. Chem.* **1999**, 2841–2843.

73 J. Ritter, R. Gleiter, H. Irngartinger, T. Oeser, *J. Am. Chem. Soc.* **1997**, *119*, 10599–10607.

74 R. Hoffmann, L. Radom, J. A. Pople, P. v. R. Schleyer, W. J. Hehre, L. Salem, *J. Am. Chem. Soc.* **1972**, *94*, 6221–6223.

75 C. Benisch, S. Bethke, R. Gleiter, T. Oeser, H. Pritzkow, F. Rominger, *Eur. J. Org. Chem.* **2000**, 2479–2488.

76 D. B. Werz, R. Gleiter, F. Rominger, *J. Org. Chem.* **2002**, *67*, 4290–4297.

77 D. B. Werz, R. Gleiter, *J. Org. Chem.* **2003**, *68*, 9400–9405.

78 (a) D. B. Werz, R. Gleiter, F. Rominger, *J. Am. Chem. Soc.* **2002**, *124*, 10638–10639; (b) R. Gleiter, D. B. Werz, B. J. Rausch, *Chem. Eur. J.* **2003**, *9*, 2676–2683; (c) R. Gleiter, D. B. Werz, *Chem. Lett.* **2005**, *34*, 126–131.

79 A. L. Spek, A. M. C. T. PLATON; Utrecht University: Utrecht, The Netherlands, **1998**.

80 L. Pauling, *The Nature of the Chemical Bond and the Structure of Molecules and Crystals*, Cornell University Press: Ithaca, NY, 3rd edn., **1973**.

81 C. Bleiholder, Diplomarbeit, Universität Heidelberg, **2004**.

82 (a) M. Nishio, M. Hirota, Y. Umezawa, *The CH/π Interaction*, Wiley-VCH: New York, **1998**; (b) Y. Umezawa, S. Tsuboyama, K. Honda, J. Uzawa, M. Nishio, *Bull. Chem. Soc. Jpn.* **1998**, *71*, 1207–1213.

83 D. B. Werz, R. Gleiter, F. Rominger, *Organometallics* **2003**, *22*, 843–849.

84 M. Kaupp, O. L. Malkina, V. G. Malkin, P. Pyykkö, *Chem. Eur. J.* **1998**, *4*, 118–126.

85 K. Gao, N. S. Goroff, *J. Am. Chem. Soc.* **2000**, *122*, 9320–9321.

86 J. Hlavatý, L. Kavan, M. Štícha, *J. Chem. Soc., Perkin Trans. 1*, **2002**, 705–706.

87 (a) G. Bieri, F. Burger, E. Heilbronner, J. P. Maier, *Helv. Chim. Acta* **1977**, *60*, 2213–2233; (b) F. Brogli, E. Heilbronner, V. Hornung, E. Kloster-Jensen, *Helv. Chim. Acta* **1973**, *56*, 2171–2178; (c) C. Baker, D. W. Turner, *Chem. Commun.* **1967**, 797–799.

88 C. Santiago, K. N. Houk, G. J. DeCicco, L. T. Scott, *J. Am. Chem. Soc.* **1978**, *100*, 692–696.

89 J. Chávez, Dissertation, Universität Heidelberg, **1995**.

90 (a) E. Kloster-Jensen, *Angew. Chem.* **1972**, *84*, 483–485; *Angew. Chem. Int. Ed. Engl.* **1972**, *11*, 438–439; (b) E. Kloster-Jensen, H.-J. Haink, H. Christen, *Helv. Chim. Acta* **1974**, *57*, 1731–1744.

91 D. J. Cram, N. L. Allinger, H. Steinberg, *J. Am. Chem. Soc.* **1954**, *76*, 6132–6141.

8
Carbon-rich Compounds: Computational Considerations

Peter R. Schreiner

8.1
Introduction

Computational methods have become an indispensable tool for chemists, and they are ideally suited for highly regular structures consisting of only two "light" elements such as carbon and hydrogen. As this is the situation found in carbon-rich materials, theory should be particularly important when studying these. However, this is not yet the case – much is left to be done in terms of applying computational methods to understand better and, in particular, to predict the properties of the ever growing class of new carbon-based materials such as polyaromatic hydrocarbons (PAHs), graphenes, fullerenes, nanotubes, diamondoids etc. This is not to say that there is not a wealth of computational papers on these structures already, but there are only few papers that are conceptual and therefore of *predictive* quality.

In the present chapter we attempt to present the current status of computational chemistry as applied to carbon-rich structures. Rather than offering an extensive introduction to the arsenal of computational chemistry, we only introduce the very basic concepts of the underlying theory and make some general remarks regarding carbon-rich compounds. There are excellent books on computational methodology that give details on the most commonly used methods and their underlying theory [1, 2]. Since most carbon-rich materials are too large in molecular weight (carbon sheets, graphenes, diamondoids, etc.) to be computed explicitly, we will elaborate on several key structural elements (alkyne units, polyynes, polyadamantanes) and extrapolate to the larger systems whenever possible.

Many low molecular weight benchmark structures such as adamantane (**1**), diamantane (**2**) as well as the next several higher diamondoids (**3–6**), cubane (**7**), allene (**8**), propyne (**9**), benzene (**10**), naphthalene (**11**), and anthracene (**12**) (Scheme 8.1) are not all carbon-rich per definition but they serve as excellent models to test the underlying theory and to extrapolate to the much larger carbon-rich, often "infinite" structures. A further distinction can be made according to hybridization, whereby conjugated (sp and sp^2) structures are best put in one category and non-

Carbon-Rich Compounds. Edited by Michael M. Haley and Rik R. Tykwinski
Copyright © 2006 WILEY-VCH Verlag GmbH & Co. KGaA, Weinheim
ISBN: 3-527-31224-2

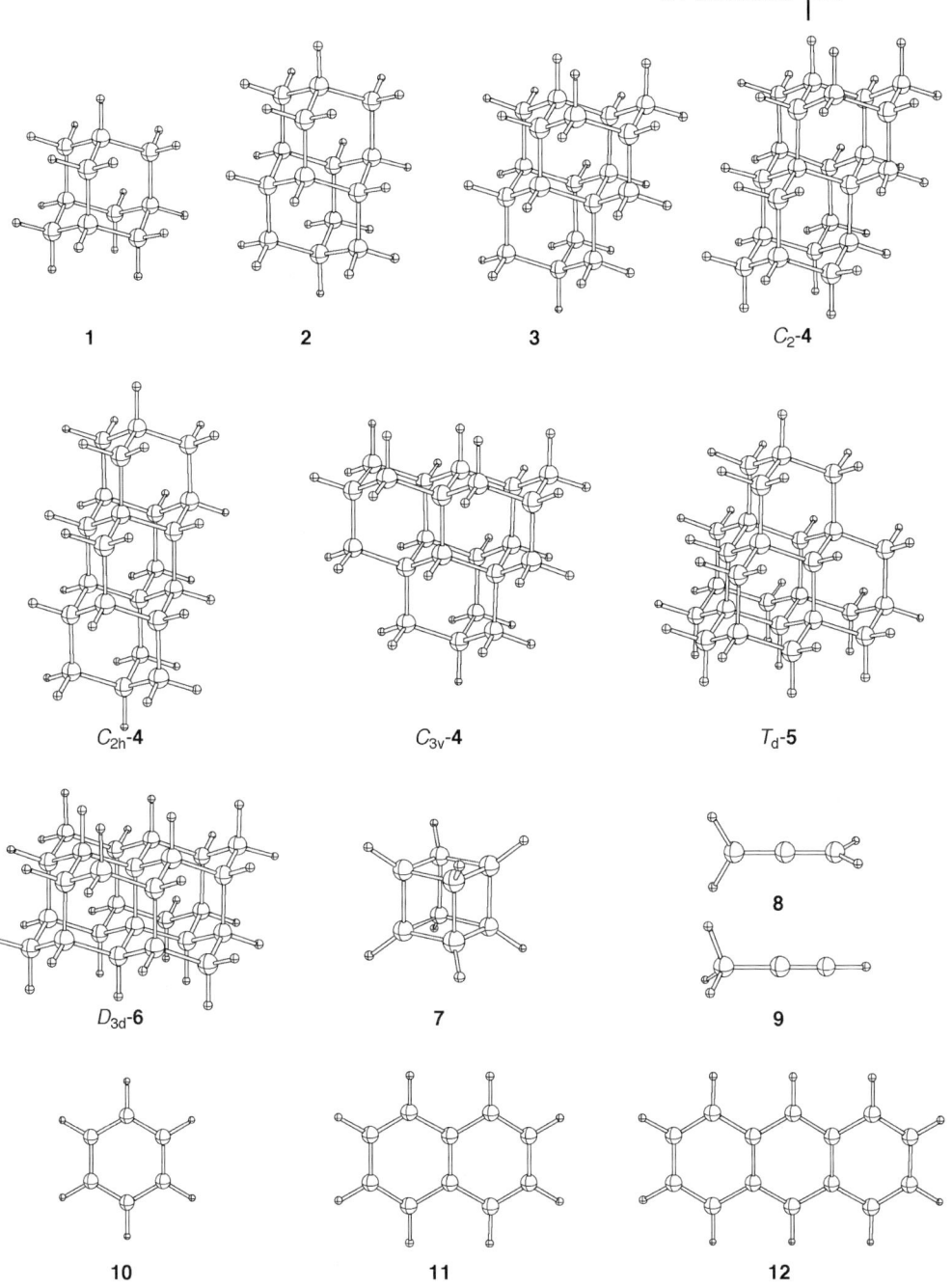

Scheme 8.1. Model compounds.

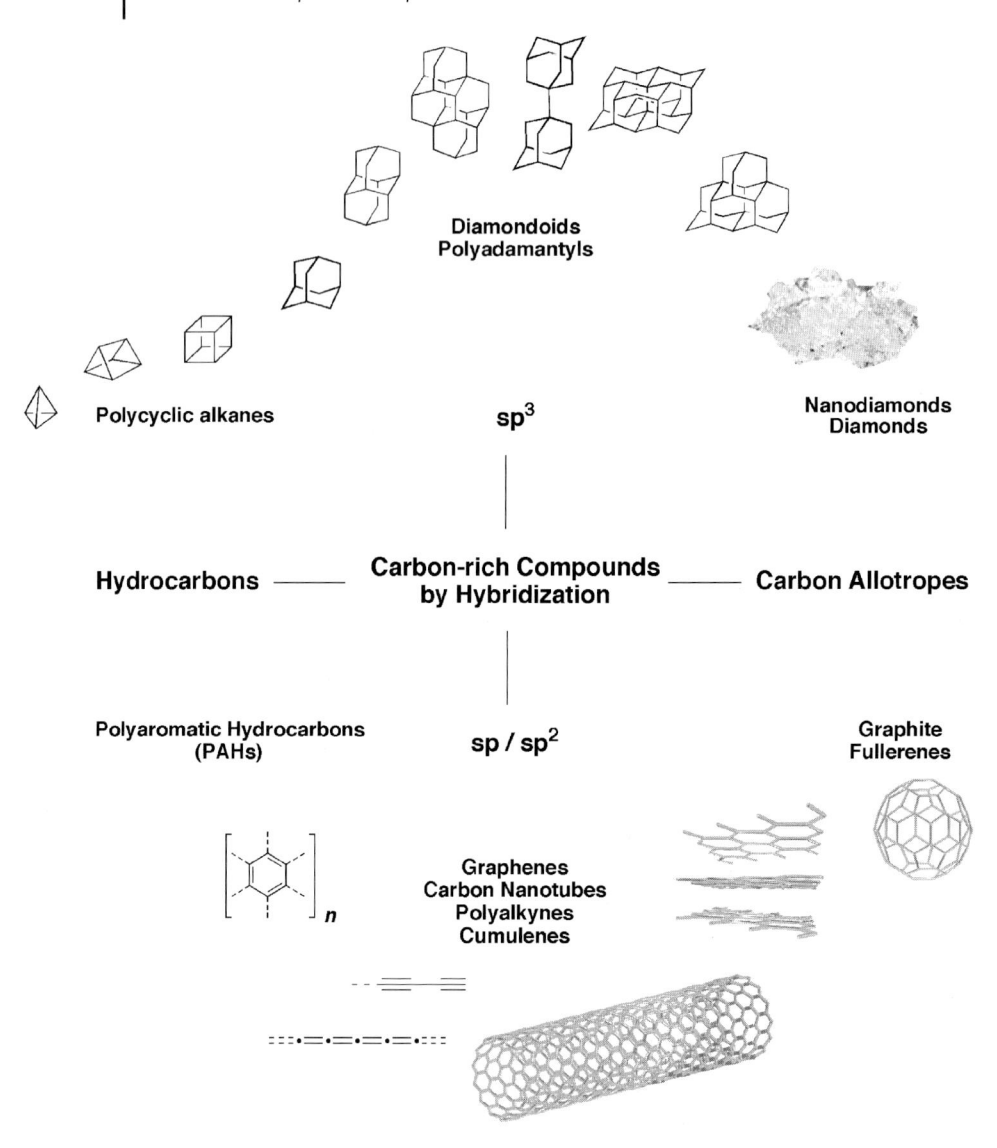

Scheme 8.2. Classification of hydrocarbons through hybridization.

conjugated (sp³) structures in another (Scheme 8.2). This classification immediately reveals something about the intrinsic reactivity patterns of these families of similarly hybridized structures.

At the same time a severe problem arises in relation to infinite systems such as graphite and diamond. These can also be modeled but with methods that are typically different from those for isolated molecular structures. At this point in time, infinite systems are still the domain of physicists so that we will concentrate in this chapter on finite systems of well-defined size and shape. However, this separation is artificial and is expected to become smaller with ever increasing computational power and the interest of chemists in materials properties. In going from finite to infinite systems, an obvious question arises: where are the transitions (and what are they?) for, e.g., PAHs to graphenes and graphite on the one hand and from adamantane to diamond on the other? What measurable and computable criteria can be used to define these transitions? Are there properties that only occur at the transition? Many of these questions are still unanswered, mostly because many molecular structures, albeit often considered very large from the chemist's viewpoint, are still rather small (0.1–10 nm) so that they do not necessarily exhibit predictable macromolecular (on a scale larger than 100 nm) properties; isolated molecular properties do not directly relate to macromolecular properties. The borders of these transitions were successfully explored by studying ever-larger PAHs as outlined in, for instance, Chapter 3. Surprisingly little (just shy of saying nearly nothing!) is known about the property transitions between adamantane, the core structure of all diamond-like molecules, lower (up to tetramantane **4**) and higher diamondoids (starting at pentamantane **5**), and macroscopic (mostly industrial) nanodiamonds that have enormous practical relevance [3]. The synthetic inaccessibility of higher diamondoids is the main reason why the chemistry of these fascinating cage hydrocarbons is virtually undeveloped. As higher diamondoids are now available in sizeable quantities from crude oil, the quest for understanding their structures and properties arises [4].

Unstrained hydrocarbons with (formally) sp^3-hybridized carbons are characterized by highly stable, tetrahedral building blocks found in alkanes, and, in the present context, in diamondoids and consequently diamond itself. These systems become increasingly more stable with increasing size; at the same time, low molecular weight sp^3 carbon compounds are considerably more stable than those composed of sp and sp^2 carbons (Fig. 8.1). That is, while macroscopic diamond is thermodynamically less stable than graphite, the same is not true for much lighter diamondoids such as those depicted in Schemes 8.1 and 8.2. The crossover point between diamondoid and graphite-like structures occurs at a number of approximately 10^5 carbon atoms and a diamond diameter of about 5 nm. Hence, in terms of thermodynamic stability up to this number of carbon atoms, and, as a consequence for (mechanical) applications in nanotechnology, diamondoids are much more important than graphenes, or even fullerenes or nanotubes. It is surprising that diamondoid technologies are just emerging but this is entirely due to our limited abilities to synthesize and modify larger diamondoid structures.

Issues of molecular dynamics approaches will not be covered in the present chapter. Defects often lead to properties that are considerably different from homogeneous structures. While this is an important aspect for doping regular structures with defects to tune the properties of the respective materials (e.g., defect diamond as a semiconductor), we will not treat these types of structures in the present chapter.

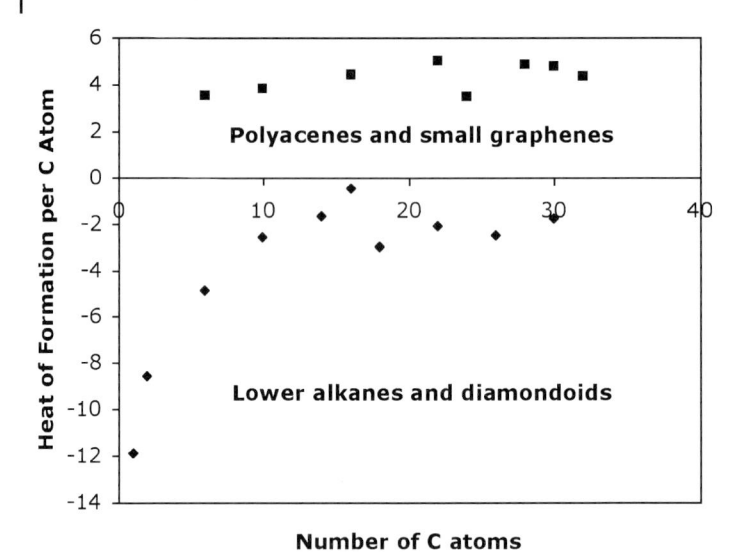

Figure 8.1. Cross-over in thermodynamic stabilities between diamondoids and graphenes [5]. Heat of formation in kcal mol^{-1}.

8.2
Basic Aspects

Applying computational techniques to chemical problems first requires a careful choice of the theoretical method. Basic knowledge of the capabilities and, more importantly, the drawbacks of the various methods is an absolute necessity. No practising chemist, however, can be expected to be well-versed in the language and fine details of the computational theory; we take the approach of briefly reminding the reader of the underlying concepts of a particular method and the often less well documented limitations. Our survey is neither comprehensive nor complete; we discuss the most common methods only. A second important aspect is which properties of carbon-rich molecules and materials can be computed accurately. As it is clear that molecular properties do not necessarily translate directly into macroscopic properties, we will only discuss molecular properties for which theoretical methods are well established.

The following sections summarize the most common theoretical methods with respect to the above criteria. As sometimes computational costs prevent high-level computations on large size systems such as carbon-rich compounds, we recommend either to carefully validate the applicability of "low"-level computation or to reduce the chemical problem to simple systems in order to obtain meaningful results.

8.2.1
Molecular Mechanics (MM)

Molecular mechanics (MM) [6] force fields (MMFF), originating in the 1930s [7], all use, in principle, empirical formulas to approximate the interatomic interactions in an average fashion with a variable set of corrective terms to account mostly for electronic interactions that are typically not included. Specific force fields combining molecular mechanics and electronic structure theory approaches (e.g., based on Hückel valence theory) have also been developed [8]. The empirical corrections are usually derived from comparisons with experimental data or high-level computations. As a consequence of the simplicity of the model, these computations are very fast and allow large "real" systems to be computed. Owing to the enormous effort that goes into the parametrization, virtually all force fields are tailor-made to particular classes of compounds (hydrocarbons, carbohydrates, peptides, etc.). The first force fields have explicitly been developed for determining the heats of formation of hydrocarbons and these programs have proven their value. Nowadays $\Delta H_f°$ values as well as the corresponding structures can be computed very accurately with *ab initio* methods so that MMFF approaches are mostly used for very large structures (and, in particular, their large numbers of isomers) for the evaluation of their macroscopic properties. Every force field is as good as its parametrization set and vice versa. That is, it is always possible to develop a new and highly accurate force field for a particular set of compounds but this force field should then only be applied to compounds that are comparable in their structures and properties. In this context, MMFF computations are attractive to get fast, out of the box molecular structures and heats of formation that are typically in excellent agreement with higher level computations at very much reduced computational costs; therefore, highly accurate force fields have also been developed for carbon-rich structure such as fullerenes, nanotubes, and graphenes [8, 9]. The computed structures can also be characterized as minima (transition structures are typically not part of the parametrization step and should therefore not be computed with common force fields) by computing the infrared and Raman vibrational frequencies through second derivatives of the energy with respect to all coordinates. These harmonic vibrational frequencies, as well as the computed heats of formation, can be used to judge the quality of the respective force field and to aid in the experimental structure elucidation. The mechanical properties such as elasticity can also be evaluated on the basis of molecular mechanics simulations of axial and torsional deformations of single-walled carbon nanotubes, for instance, to predict shear moduli and the mechanical consequences of bending and buckling [10].

8.2.2
Wavefunction Theory

The description of electron motion and electronic states, which is at the heart of all of chemistry, is included in wavefunction theory, also referred to as self-consistent-field (SCF) or, in honor of its originators, Hartree-Fock (HF) theory [11]. This also includes density functional theory *if* one uses densities derived from SCF densities, which is common but not a precondition [2]; therefore we treat density functional theory separately. Many wavefunction based approaches date back to the time when supercomputer desktops were not available and scientists had to reduce the computational effort by approximating the underlying equations with data from experiment. This approach and its applications to carbon-rich structures is outlined in the following section on semiempirical methods. Ways to include explicit electron correlation are introduced next, followed by a section on density functional theory. The applications in the second half of this chapter should then further illustrate the advantages and shortcomings of the various methods as applied to carbon-rich structures.

8.2.3
Semiempirical Methods

In contrast to molecular mechanics force fields, modern semiempirical methods are classified as an SCF electron-structure theory (wavefunction-based) method [12]. Older (pre-HF) semiempirical approaches such as extended Hückel theory, which can be classified as a one-electron effective Hamiltonian approach, involve drastic approximations but rely on the researcher's intuition and ability to extrapolate from simple computations to meaningful chemistry. This method is not used much these days but still plays a role in determining the band structures of organic polymers, most of which are carbon-rich by definition [13].

In line with the chemists' valence electron models of chemical bonding, only these electrons are considered in the computational procedures. To avoid the unphysical computation of every electron interaction with *every* other electron, multicenter integrals are neglected and empirical corrections from experiments added. The integrals are either determined directly from experimental data or are computed from the corresponding analytical formulas or suitable parametric expressions. These simplifications speed up the computations considerably and still include most electronic effects, in particular electron correlation that is introduced not explicitly but through the "back door" by means of fitting the parameters to experimental data. The interpretation of semiempirical computations is straightforward as all typically relevant properties can be computed using the underlying physical principles (eigenvalues of operators). There are many flavors of semiempirical theory and they continue to be developed because many very large structures can simply not be computed with *ab initio* or DFT methods [14]. The common abbreviations stand either for the place of their development (such as the popular Austin Model 1, AM1) [15] or their level of approximation (e.g., MNDO, modified neglect of differential diatomic overlap) [16]. As semiempirical MO methods are parametrized to reproduce experimental reference data or high-level theoretical predictions, the reference properties are best selected such that they are representative for the intended applications. The quality of semiempirical results strongly depends on the effort put into the parametrization [17].

Table 8.1. Mean absolute errors for molecules from the G2 neutral test set (148 molecules included; due to missing parameters for MNDO, AM1, and PM3 the data refer to 146, 142, and 144 molecules, respectively) [18].

Method	$\Delta\Delta H_f^0$ (kcal mol^{-1})
LDA (SVWN)	91.2
BP86	20.2
MNDO	9.3
AM1	7.8
BLYP	7.1
PM3	7.0
B3LYP	3.1
G2(MP2)	2.0
G2	1.6

In its widely used form, the MNDO approximation and its variants AM1 and PM3 are suitable for the computation of molecules consisting of first row elements, such as hydrocarbons (Table 8.1). The computational effort (often referred to by computational chemists as "cost") scales approximately with N^2 (N = number of heavy atoms). These methods serve as efficient tools for searching large conformational spaces for, e.g., subsequent higher-level computations; however, errors in the computations are less systematic than in *ab initio* methods. This is particularly evident when an error cannot be related to a physically measurable quantity (expectation value).

A key advantage of semiempirical methods is that they give heats of formation directly; for hydrocarbons, the errors are of the order of 5–6 kcal mol^{-1} for AM1 and MNDO. Small hydrocarbon cycles are typically computed to be too stable, and sterically overcrowded structures are predicted too high in energy. This is a consequence of the fact that semiempirical methods do not describe weak interactions, e.g., those arising from London dispersion forces, well; thus, they would not be suitable to describe, for instance, the intermolecular interactions of all-carbon sheets or onion-like single-wall carbon nanotubes. Semiempirical methods have been used very recently for studying the multitude of isomers of all-carbon structures such as C_{20} (*vide infra*) because exhaustive isomer searching would currently be impractical at reasonably high-level *ab initio* and even DFT levels of theory [19].

Nucleus-independent chemical shifts, introduced by Schleyer et al. as a measure for aromaticity [20], can also be computed at the GIAO-MNDO level of theory, making this useful tool also available to much larger aromatic structures such as graphenes and fullerenes [21]. The *ab initio* reference data are generally reproduced well, but four-membered rings are still problematic. The aromatic properties of Hückel-type hydrocarbons, Möbius aromatics, and three-dimensional cage compounds are well described and can therefore also be assessed for much larger structures.

8.2.4
Hartree-Fock Theory

Self-consistent field computations considering all electrons in such a way that a single electron experiences the other electrons as an average field are referred to as Hartree-Fock (HF) theory. HF theory minimizes the interactions of all electrons with respect to the atomic coordinates, and adheres strictly to the variational theorem, that is, that any arbitrary (wave) function used in the Schrödinger equation will give an energy (E_{HF}) that is *higher* than the "true" energy (E) of the real system. Hence, the energies computed with HF theory present an upper bound of the true energy.

The energy difference between the true energy of a system and the E_{HF} is the electron correlation energy (E_{corr}) that arises from the spontaneous repulsion of the electrons that goes beyond the average field assumption. The correlation energy is therefore the seemingly simple difference $E = E_{HF} + E_{corr}$ but all so-called post-HF methods attempt to recover this small but important energy term (*vide infra*). Part of the idea of semiempirical theories was to parametrize the semiempirical equations in such a way that most of the correlation energy can be recovered very simply. Naturally this leads to a diversion of theory (and theoreticians!) whereby one side attempts to solve the Schrödinger equation as rigorously as possible (keeping in mind that it can only be solved for the simplest of systems) while the other tries to devise efficient and practical ways to probe the theory in determining molecular properties.

The omission of electron correlation has serious consequences. As the spontaneous repulsion of electrons occupying the same space at the same time is only averaged in HF theory, geometries at this level usually have bond lengths that are too short. Highly delocalized systems such as carbocations, polyaromatics, and strained rings are generally not described well. Bond-breaking/bond-making processes are also described poorly because the effects of multideterminantal descriptions are not included in the basic HF formalism; dissociation energies have larger errors.

Despite these deficiencies, HF theory has come a long way. First, virtually all post-HF (i.e., correlation) methods rely on a HF reference wavefunction that is amended with elaborate schemes to evaluate the correlation energy. Many density functional theory (DFT) approaches also determine the HF electron density first before it is used as the operand in the DFT treatment. Moreover, systematic errors such as constant amounts of electron correlation energy that is left out in, for instance, relative energy comparisons of isomers, often cancel and HF energies are a good estimate of the relative energies of structurally related species. The use of balanced but hypothetical isodesmic (the various bonds on the left- and right-hand side of the equation are rather similar) equations makes use of this advantageous error cancellation. This approach is particularly useful for hydrocarbons and carbon-rich structures as the bonding motifs are often very similar and repetitive; the errors in the bond energies, etc. are assumed to be transferable and therefore are likely to cancel when isodesmic equations are used.

Problems arise when open-shell species are considered. These are often formulated as intermediates or transition structures that are either true radicals, biradicals and higher unpaired electron states or have large open-shell character. Such species are usually invoked in high-temperature thermal reactions of hydrocarbons or carbon-rich structure precursors. Examples are the mechanisms of soot or fullerene/nanotube formation. Some polyaromatic hydrocarbons (e.g., the polyacenes, the higher homologues of **10–12**) [22] have very low-lying open-shell singlet states that can, in some instances, even be the ground state [22, 23]; hence, homologs higher than hexacene have not been fully characterized. Pentacene is a semiconductor that can be used for field-effect transistors owing to its ability to carry high charge [24]. Although the electronic properties of large oligoacenes have been extensively examined computationally (and often at the HF level of theory in older studies), the electronic properties, especially the band gaps and HOMO–LUMO energy separations are still highly controversial [22, 25]. Large polyacenes are predicted either to behave as one-dimensional organic conductors with a zero band gap or to possess triplet ground states [26]. In such instances it is often very difficult to converge (minimize) the HF wavefunction because of (near) degeneracies of HOMO and LUMO: either MO could be doubly occupied leading to energetically very similar solutions of the energy eigenvalue equation. As a consequence, the energy (and often also the geometry) of such species is very sensitive to the energies and shapes (coefficients) of all remaining orbitals.

Most diamondoid C–H bond activation methods rely on radical abstractions, so that it is important to consider the performance of HF methods when dealing with open-shell species when, for instance, hydrogen abstraction mechanisms are studied [27]. When computing open-shell species one has to invoke *unrestricted* (UHF) Hartree-Fock theory as opposed to the normally employed *restricted* (RHF) formalism in which all electrons are paired and the RHF wavefunctions are eigenfunctions of the spin operator S^2. This is no longer the case when a UHF wavefunction is employed, which, however, includes important effects of spin polarization by computing the SCF wavefunction separately for each spin, leading to different molecular orbitals for the α and β electrons. As UHF wavefunctions are not eigenfunctions of S^2, they contain some degree of "spin contamination" from higher spin states that are included in the energy computation and therefore artificially lower the energy. A good measure of this spin contamination is the expectation value of S^2, namely $\langle S^2 \rangle$ which must have eigenvalues of $S(S+1)$ of the sum of all spin magnetic moments S of all unpaired electrons; hence, a radical (doublet) gives $\langle S^2 \rangle = 0.75$, a biradical (triplet) 2.00 etc. Many *ab initio* programs report the $\langle S^2 \rangle$ when UHF is invoked and one should check if the deviation from the expected value is larger than 5–10% (depending on the case). Strongly contaminated wavefunctions will not give proper energies and are meaningless.

8.2.5
Symmetry

As many typical carbon-rich structures are highly symmetric or are built up from repetitive symmetric subunits (Scheme 8.1), symmetry should also be employed in the computations, not only to satisfy the experimental observations but also, more importantly, to speed up the computations by reducing the dimensionality of the coordinate space to be optimized with respect to the total energy. Consider cubane (**7**) that formally has a 42-dimensional potential energy surface that would be rather difficult to rationalize. In reality we know that there are only two degrees of freedom (assuming a perfect cube) that could either be the C–H and the C–C distances or, in an equivalent representation, the distance of the carbon and hydrogen atoms from the center of the cube. Without going into mathematical details it is obvious that the proper consideration of symmetry very much reduces the computational complexity and should therefore always be employed. It often makes searches for transition structures much easier as the flexibility of a particular structure is reduced. A stationary point should in any event be characterized by computing the second derivative, i.e., the Hessian matrix in its full dimensionality, to make sure that it is a minimum (no imaginary frequencies) or a transition structure (one imaginary frequency only).

One should be cautious, however, of some difficult and rare (for closed-shell systems) cases when the molecular symmetry and the wavefunction symmetry are *not* the same. A trivial example would be the two possible closed-shell singlet states of methylene ($:CH_2$) that are both C_{2v} symmetric. The state with the doubly occupied methylene $3a_1$ orbital is the lowest lying singlet state of methylene, labeled $\tilde{a}\ ^1A_1$, with the open-shell $\tilde{b}\ ^1B_1$ and the closed–shell $\tilde{c}\ ^1A_1$ states being higher lying singlet states [28, 29]. If such a situation occurs it is imperative to use symmetry to assign the computed states properly by their symmetry labels (they are all *A* in C_1). Further computations are needed to make sure that all states of interest are found (*vide infra*) and properly assigned.

8.2.6
Basis Sets

All wavefunction based methods use mathematical functions to describe the hydrogen-type orbitals; each molecular orbital is constructed as a linear combination of the basis functions with coefficients that minimize the iterative HF procedure. This is usually done either with Slater-type-orbitals (STOs, semiempirical and some DFT methods) or products of primitive Gaussian-type-orbitals (GTOs, virtually all remaining approaches) that approximate the STO behavior of the hydrogen orbitals. STOs are normally used in computationally inexpensive semiempirical methods where the integral computation is not the computational bottleneck. Owing to their quadratic exponent GTOs are easier to integrate but they are not good in the chemically irrelevant cusp region. Many GTOs are needed to properly describe the regions in space far away from the atomic centers because STOs have long delocalization "tails." This is particularly important for highly delocalized structures such as PAHs and weak interactions (e.g., the interactions between graphene sheets). Hence, one has to attempt to use as large a basis as computationally feasible; however, this may lead to severe HF wavefunction convergence problems. As the orbital space is saturated with basis functions, there are many solutions (orbital occupations) that have similar energies and it is often difficult to converge to the desired lowest-energy solution or to any solution. This is particularly true for highly delocalized and symmetric structures. One possible way out of this dilemma is to optimize the wavefunction with a minimal basis set (e.g., STO-3G) which leads to well separated MOs. This solution can then be taken as a better (than the typical Hückel or semiempirical) initial guess for the orbital coefficients that comprise the wavefunction for the next higher basis set. Diffuse ("+") functions, which are often necessary for the computation of carbanions, also often lead to convergence problems because they have a large radial extent that can span several atoms [30]. A similar strategy therefore suggests converging the wavefunction without the use of diffuse functions first and using this as a good guess to add these additional, often important, functions in a second step.

The choice of basis set can be crucial and is particularly important for post-HF methods described briefly in the following section. HF itself and DFT are far less basis set dependent but this should not be taken as an excuse not to check a few basis sets for those levels as well. The popular Pople-type basis sets [31] (e.g., 6-31G*) have proven their value for HF and DFT and are well behaved for structures involving carbon and hydrogen. The correlation-consistent basis sets (cc-pVXZ) [32] that are also sometimes used in conjunction with DFT computations were explicitly designed to recover as much correlation energy as possible for highly correlated methods and are therefore not necessarily guaranteed to perform well with DFT.

8.2.7
Electron Correlation Methods

Although HF theory is quite remarkable in describing molecular structures and energies, it is clear that the simplified assumption of one electron moving in the average field of all the others is not sufficient to take into account important electronic effects that do make all the difference in chemistry, such as delocalization, polarization, aromaticity etc. HF theory recovers approximately 99.5% of the total energy of a system which, for instance, amounts in benzene to about 14.5×10^3 kcal mol^{-1} {−230.70967 au at HF/6-31+G(d)}; half a percent then still amounts to about 700 kcal mol^{-1}. In view of these values the 20–30 kcal mol^{-1} aromaticity computed for most prototypical aromatic molecules emphasizes the problem and the absolute necessity to recover as much as possible to be able to make fair energy comparisons. Of course, *relative* energy comparisons alleviate many of these inadequacies *if* the computational errors are transferable between the different species that are compared. This is often the case for energy differences between isomers (see, however, the discussion for propyne and allene below), but certainly not for chemical reactions that involve bond-breaking and bond-making processes. As noted above, isodesmic equations have proven their value in systematically canceling the errors but better computational models are needed for accurate numbers. There are many elaborate ways to include electron correlation into chemical computations. This is not the place to elaborate on them but a few comments with respect to their applicability to carbon-rich systems should be made.

The simplest way of including electron correlation is *Møller-Plesset* (MP) *perturbation theory* (also referred to as MBPT = many-body perturbation theory) that introduces electron correlation as a small perturbation to the HF Hamiltonian [33]. The energy and wavefunction are expressed in infinite additive terms (k) that correspond to the order of the MP perturbation; the sum of the zero and first order energy terms represent the unperturbed HF energy. This series is often truncated with the second term (MP2, $k = 2$) with the serious consequence that the variational theorem is *not* fulfilled (the computed energy is *not* an upper bound of the true energy of a given system). Secondly, including higher levels of MP theory (MP3, MP4, etc.) leads to a nonconvergent series at infinite order that does not converge to the true energy. One of the dangers of MP theory is that the underlying assumption is that the perturbation is small; hence, although MP theory lends itself to describing highly delocalized systems such as PAHs, it still suffers from the inadequacies of HF in describing these.

Despite serious and obvious limitations, the MP approach works remarkably well and MP wavefunctions recover a large fraction of the electron correlation energy. For instance, the total energy of benzene using second-order MP theory {simply MP2/6-31+G(d)} is −231.47203 au or about 480 kcal mol^{-1}, which is a significant improvement over HF. With respect to carbon-rich compounds that can be quite large, MP2 is a sensible level of theory because it is well behaved in that one is well aware of its deficiencies. Although MP2 scales reasonably well – compared to more elaborate schemes to recover electron correlation – with about N^5, this is still a very serious limitation. Structures are described very accurately and offer a significant improvement over HF because, for instance, bond lengths are longer at MP2 and hence closer to experiment. Highly delocalized systems are also described very well but convergence problems may arise when using large basis sets, which are, however, a requirement for accurate MP computations.

Configuration interaction (CI) is the next higher and theoretically logical development to include electron correlation. The way to do this in CI theory is to generate selected electron excitations in terms of configuration state functions (CSFs) from the HF reference wavefunction; most typical are single and double excitations leading to the acronym CISD. Including all excitations, which is only practical for very small (currently about 8–12 electron systems, depending on the basis set) [34] molecules, is then referred to as full configuration interaction (FCI). While CISD gives excellent results for small systems, it is generally not a suitable method for increasingly larger structures because of the enormous computational effort (CISD scales with about N^6) and, more severely, because it is not size-extensive due to its truncation at the SD excitation level. That is, the amount of electron correlation energy recovered does not scale proportionally with molecular size. CI methods are therefore excellent benchmark methods for small molecules but they are impractical for large carbon-rich compounds. Configuration interaction treatments can also be easily extended to multi-determinantal problems but these require even more care and an exorbitant computational effort that makes their application to carbon-rich structures currently impossible [35, 36]. It needs to be pointed out, however, that multireference CI approaches are ideally suited to determine, for instance, electronic spectra, as they can deal with excited states in a straightforward fashion [37].

Coupled cluster theory (CC) [38], often referred to as the "gold standard" of computational quantum chemistry, includes electron correlation through expanding the HF reference wavefunction through a cluster operator that is weighted by its cluster amplitudes [39]. Similar to other electron correlation methods the infinite series expansion is truncated at a certain level with single and double substitutions (CCSD) being the most common. The effect of triple excitations is often estimated with perturbation theory and this is expressed by placing "T" in parentheses as in CCSD(T). This level of theory has proven to be extremely accurate, provided that large well-balanced basis sets (preferably those designed for high levels of electron correlation such as the cc-pVXZ series of basis sets) [32] are used. Although coupled cluster wavefunctions are nonvariational, the effect of a finite basis set is typically much larger than the potential error introduced by not having a variational method. Coupled cluster methods are size extensive so that they are, in principle, also applicable, at least to determine the energies of pre-optimized structures, to ever larger systems. This is only hampered by the exorbitant computational effort as CCSD(T), for instance, scales with N^7; so-called local or even linear scaling methods are being developed and show great promise for the application of this mathematically elegant and accurate approach to large systems [40]. A huge advantage of coupled cluster methods is, in contrast to MP and CI treatments, the insensitivity to the quality of the reference wavefunction. That is, even structures with pronounced multireference character (*vide infra*) can often be described with the formally single-determinant coupled cluster approach. Sometimes problems arise from unrealistically large singles amplitudes when near-degeneracies of orbitals lead to state problems and CCSD(T) fails to give accurate answers; these problems are usually diagnosed by inspection of the so-called T_1 diagnostic terms that should be below 0.02 [41]. One way out of this dilemma is to use a different set of orbitals, for instance, Brückner-type orbitals {the approach is then often referred to as BD and BD(T)} where the singles amplitudes vanish by rotation and linear combination of the HF orbitals. Applications will be shown below.

As a guide, the following row exemplifies the improvements in quantum chemical accuracy by inclusion of increasingly larger fractions of the electron correlation energy. This also correlates with the computational effort. However, as, e.g., MP2 gives quite good structures, energies can be determined at the MP4 level, as is commonly done.

HF < MP2 < CISD < MP4 < CCSD(T)

8.2.8
Density Functional Theory

Although, in principle, an exact quantum mechanical method (if the true functional were known), we treat density functional theory (DFT) separately from traditional *ab initio* methods because the arbitrariness in choosing the functional combinations makes the logical and systematic improvement of DFT currently rather difficult [2]. This is also not the place to elaborate on the various functionals; there is some discussion of this in the applications section. We also note that DFT *is* largely the domain of materials physics for which this method had been applied long before chemists began using it for molecules. It has taken so long for chemists to pick up DFT because physicists typically use plane waves and pseudopotentials for large structures and these are simply not very accurate for single molecules (an example for graphenes is given in Ref. [42]).

Something should be said about the use of B3LYP that is currently the most commonly applied DFT method for finite systems. B3LYP is a hybrid method that corrects the combination of the HF and Slater exchange functionals with the Becke88 gradient correction to exchange. The weights on the exchange and correlation functionals (VWN and LYP) are determined by minimizing the root mean square error over the G1 molecular set [43, 44]; B3LYP has proven to be robust and generally applicable to a large variety of chemical problems. Nevertheless, as emphasized in the next section and in the applications below, it is not always the best choice for a given problem, in particular, when dealing with carbon-rich structures. Also, while classical ab initio methods systematically overestimate barrier heights, B3LYP typically (and not necessarily systematically) underestimates them [45]. Some of these problems are associated with the open-shell character (as in, e.g., polyacenes, *vide supra*) of larger carbon as well as hydrocarbon structures for which the self-interaction part in the B3LYP functional is energetically overemphasized; while chemical structures are still acceptable, the energies can err significantly. Indeed, for ever larger to infinite systems, local DFT approaches with different functionals are often preferred [46].

8.2.9
Validation

Computational methods are typically employed to rationalize experimental findings such as molecular structures and spectroscopic data (*vide infra*), which also serve as benchmarks for assessing the quality of the methods used. At the same time, it is clear that the true strength of computational chemistry is the *prediction* of new, hitherto unknown structures and reactions. In an ideal situation, this would greatly reduce experimental efforts and consequently time and monetary investments; however, this is not yet done routinely, especially when the predictions cannot readily be probed. The key problem is that although many computational methods are well tested on a limited set of structures, new and unusual bonding arrangements, as often found in carbon-rich materials (bent bonds,

```
   ┌─────────────┐
   │ Prediction  │
 ┌─┴─────────────┴─┐
 │  Interpretation │
┌┴─────────────────┴┐
│                   │
│   **Validation**  │
│                   │
└───────────────────┘
```

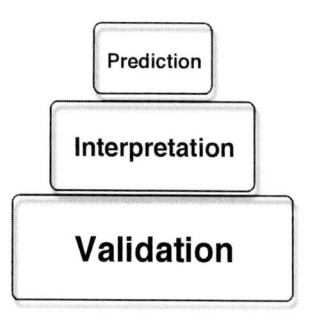

Figure 8.2. Approaching computational chemistry applications systematically.

electron-richness, open-shell character etc.), often bear surprises and thus expose sometimes the weaknesses of particular methods. It is therefore imperative to keep in mind the progression in applying computational methods as depicted in Fig. 8.2. This implies that most of the work when dealing with new structural patterns goes into validating the computational approach. Unfortunately, this is often overlooked or even ignored and the results are sometimes meaningless. The appropriateness of the validation should then be used to interpret the experimental data, ideally from an angle for which the experimentalists have not had a handle on. For instance, chemical computations give good estimates of properties such as electronic charges that are neither experimentally observable nor are they the expectation value of a quantum mechanical operator. That is, charges are only a wonderful and very useful construction to rationalize structures and interactions but they are not "real". Nevertheless, the polarization of a molecule is something that can be computed rather reliably and this helps the interpretation significantly. Even the popular and exhaustively used molecular orbital picture and the interactions that arise from combining orbitals are only mathematical but nevertheless immensely useful constructions to aid the computational solution of the Schrödinger equation. Keeping all of this in mind, it is clear why predictions are equally difficult and dangerous, and only possible if the computational approach has been properly validated and the interpretation of the computations tested against *high quality* experimental data.

With respect to ever-larger carbon-rich structures (consider the natural transitions from polyacenes over graphenes to graphite and polyadamantanes over diamondoids to nanodiamonds), this poses additional problems because quantum effects give way to macroscopic properties. Computations silently and incorrectly sometimes imply that this is a straightforward transition by, for instance, converting atomic units as applied to single molecules to an entire mole of the same (e.g., "kcal mol^{-1}"). All of thermodynamics relies on being able to make this transition and we should not forget that this is a huge leap of faith (in partition functions and statistical mechanics!). The prediction of macroscopic properties such as flexibility, hardness, mechanical strength, packing etc. is much more difficult as it is not entirely clear yet how the atomistic and molecular picture translates into macroscopic structures and properties. Much work remains to be done in this direction, but the especially well-structured and often highly symmetrical carbon-rich materials may pave the way by considering that there *is* a smooth transition for example between graphenes and diamondoids. A perfect macroscopic diamond is a single molecule that can, in principle, be computed with traditional methods and comparisons can be made.

8.3
Applications

8.3.1
Alkynes, Cumulenes, and the sp–sp² Transition

In the spirit of validating a theoretical approach before using it for the interpretation and even the prediction of molecular properties, it is instructive to compare the results for a seemingly simple pair of molecules: allene and propyne. The presence and comparison of single, double, and triple bonds in two very different bonding arrangements already indicates that the energies and geometries of these structures should be sensitive to electron correlation and basis set effects. This can be readily understood with the high electron density (and therefore the requirement for a method to account for proper electron repulsion) of the triple bond on one side and the "normal" electron distribution of a single bond on the other; the two double bonds of allene are exactly in between. These effects, which are present at all "low-cost" levels of theory (DFT and MP2), are amplified when going to larger polyynes or cumulenes, as was pointed out in several recent studies [47–49]. The relative energy difference between allene (**8**) and propyne (**9**) is not only severely in error at all commonly used DFT levels (B3LYP, BLYP, BP86), but these methods also give the wrong sign, i.e., the incorrect energetic ordering of **8** and **9** (Table 8.2)[1]. Insight into this and related problems involving DFT can thus be gained through the systematic study of allene and propyne as well as their higher analogs at the highest level of theory possible.

It has previously been speculated that higher cumulenes may become nonlinear as the length of the carbon chain increases [50]. These studies were carried out at the RHF/4-31G level of theory for geometry optimizations and single-point energies computed through MP4(SDQ)/6-31G. The authors computed very low bending frequencies that were even negative (indicative of linear transition structures connecting two nonlinear minima) in many cases. A higher-level follow-up study explicitly examined the CCC bending frequencies at the SCF, MP2, and CISD levels with moderately sized basis sets (6-31G, DZ, and TZ) [51]. This revealed that the optimizations of the higher cumulenes are very much basis set and method dependent with CISD/6-31G, CISD/TZ, MP2/6-31G, and MP2/TZ yielding sizeable imaginary CCC bending frequencies.

---

1) The error in relative energies of allene and propyne by DFT methods first has been noted by J. A. Pople at the Symposium on Theoretical Organic Chemistry, 25th Central Regional Meeting of the American Chemical Society, Pittsburgh, PA, Oct 5, 1993.

Table 8.2. Comparison of the energy difference between allene (**8**) and propyne (**9**, more stable experimentally by 1.4 kcal mol^{-1}) at various levels of theory [48]. Values in parentheses include ZPVE.

Level of theory	ΔE (8–9) (kcal mol^{-1})
HF/6-31G(d)	−2.0 (−1.6)
HF/6-311+G(d,p)	−2.3 (−1.8)
BLYP/6-31G(d)	4.1 (4.1)
BLYP/6-311+G(d,p)	2.7 (3.1)
B3LYP/6-31G(d)	2.8 (3.0)
B3LYP/6-311+G(d,p)	1.6 (2.0)
BP86/6-31G(d)	4.1 (4.2)
BP86/6-311+G(d,p)	3.1 (3.4)
KMLYP/6-31G(d)	(1.2)[a]
KMLYP/6-311+G(d,p)	(0.5)[a]
MP2/cc-pVDZ	−4.6 (−4.4)
MP2/cc-pVTZ	−4.5 (−4.3)
CCSD(T)/cc-pVDZ	−0.5 (−0.3)
CCSD(T)/cc-pVTZ	−1.0 (−0.7)
CCSD(T)/cc-pVQZ//MP2/cc-pVTZ	−1.4
experimental	−1.4

[a] Enthalpies at 298 K, from Ref. [49].

Although HF theory seems to be doing an excellent job, this may be considered a "Pauling point", where fortuitous error cancellation leads to an apparently reasonable result. This is already obvious from the changes in energies when using a larger basis set, as normally the results are expected to improve systematically with an increase in basis set size. As shown below, these hidden errors surface when going to larger cumulenes and polyalkynes. Furthermore, the computed HF/6-31G(d) C–C bond lengths (Scheme 8.3) differ considerably from experiment and high-level computations: the multiple bonds are much too short due to incomplete inclusion of (repulsive) electron correlation effects.

It is clear that at all the examined DFT levels of theory perform poorly, as the relative energetics of the two isomers are reversed. MP2 seriously overestimates the stability of **9** compared to **8**; even the very large basis set MP2/cc-pVQZ//MP2/cc-pVTZ computation gives an energy separation of −4.8 kcal mol^{-1}. This is *not* a geometrical problem because the critical multiple bond lengths at the various levels do not differ much (Scheme 8.3) but is instead due to the incomplete treatment of dynamical electron correlation.

It is comforting to see that the best method available for computing energies does get the energy separation between **8** and **9** right {CCSD(T)/cc-pVQZ//MP2/cc-pVTZ}, both in magnitude and in sign. When the optimization level is in-

8

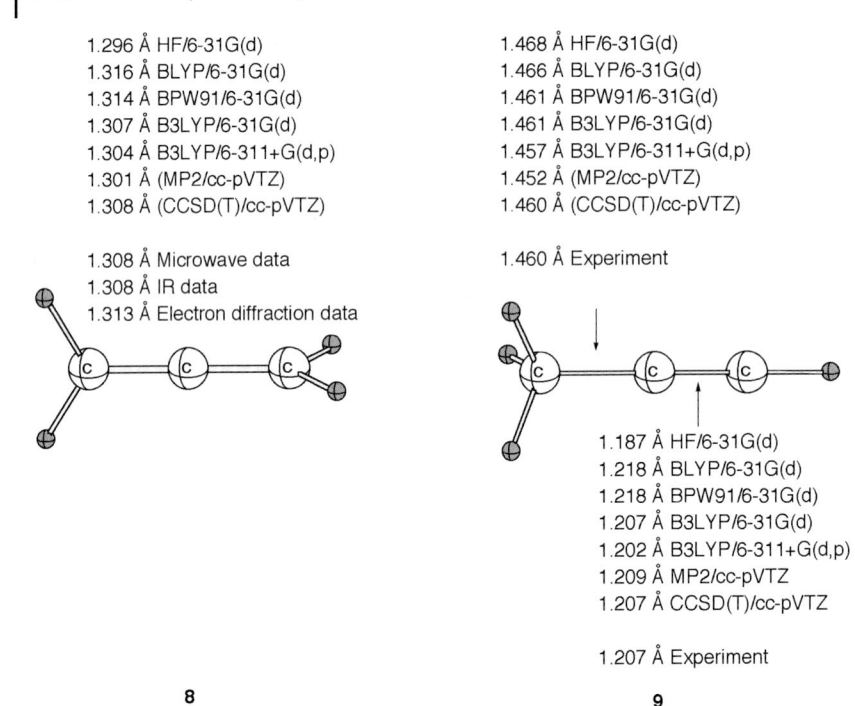

1.296 Å HF/6-31G(d)
1.316 Å BLYP/6-31G(d)
1.314 Å BPW91/6-31G(d)
1.307 Å B3LYP/6-31G(d)
1.304 Å B3LYP/6-311+G(d,p)
1.301 Å (MP2/cc-pVTZ)
1.308 Å (CCSD(T)/cc-pVTZ)

1.308 Å Microwave data
1.308 Å IR data
1.313 Å Electron diffraction data

9

1.468 Å HF/6-31G(d)
1.466 Å BLYP/6-31G(d)
1.461 Å BPW91/6-31G(d)
1.461 Å B3LYP/6-31G(d)
1.457 Å B3LYP/6-311+G(d,p)
1.452 Å (MP2/cc-pVTZ)
1.460 Å (CCSD(T)/cc-pVTZ)

1.460 Å Experiment

1.187 Å HF/6-31G(d)
1.218 Å BLYP/6-31G(d)
1.218 Å BPW91/6-31G(d)
1.207 Å B3LYP/6-31G(d)
1.202 Å B3LYP/6-311+G(d,p)
1.209 Å MP2/cc-pVTZ
1.207 Å CCSD(T)/cc-pVTZ

1.207 Å Experiment

Scheme 8.3. Comparison of the C–C bond lengths of allene (**8**) and propyne (**9**) at different levels of theory vs. experiment.

creased to CCSD(T)/cc-pVQZ//CCSD(T)/cc-pVTZ, which has only minor effects on the structure (Scheme 8.3), the energy separation comes to within 0.02 kcal mol^{-1} of the experiment, which is less than the experimental uncertainty. DFT optimized geometries generally are also quite good and this is important when considering much larger carbon-rich systems that incorporate multiple bonds. As all common DFT levels reverse the relative energies of allene and propyne, we proposed that there is a systematic deficiency in the current DFT formulation of functionals incorporating variable HF exchange contributions [48]. We examined how the DFT relative energy difference change with respect to altering the Becke and Hartree-Fock exchange contributions: BXLYP (X describes the amount of HF exchange included) single-point energy computations were performed at the B3LYP/6-311+G(d,p) geometries while varying the contribution of the Becke DFT and HF exchange term. These computations show a smooth decline in the error in the isomerization energy difference as the percentage of HF exchange is increased; the optimal amount of Becke exchange to include in BXLYP computations is 13.4%. This value was obtained via a polynomial fit of the data and subsequent interpolation to the experimental ΔE of -1.4 kcal mol^{-1}. This contrasts the current implementation of B3LYP that includes 72% Becke exchange. To understand these incorrect DFT relative energy orderings, an energy decomposition analysis was performed in which the total DFT energy was examined in terms of one-electron, total coulomb, HF exchange, DFT exchange, and DFT correlation contributions. The results of this analysis show erroneous DFT exchange behavior as the Becke exchange term is decreased from 90% to 60% of the total exchange term. The lower end of this scale, where the HF exchange contribution begins to dominate, shows a smooth decline in the DFT exchange included in the total energy. This is also the region that most accurately reproduces ΔE; hence, the errors observed with the C_3H_4 isomers are associated with the DFT exchange term and, by minimizing the contribution to the appropriate level, experimental and high level theoretical results may be more accurately reproduced. One explanation for this finding is the delocalization of the DFT exchange hole over symmetric molecules such as allene [52] that leads to an overestimation of the exchange self-interaction terms. Gradient corrected functionals, which localize exchange holes, typically overestimate the exchange energy where the true Kohn-Sham exchange hole would be delocalized; hence, cumulenes display this overstabilization problem, whereas more localized polyynes are more accurately described by gradient corrected functionals.

Another probably much more robust alternative is the development of DFT approaches that explicitly address the self-interaction problem. KMLYP (KM stands for the developers Kung and Musgrave) is such a hybrid DFT formulation that attempts to reduce self-interaction errors [53]; it utilizes the HF and Slater exchange functionals to compute the exchange energy and the VWN and LYP functionals for the correlation energy. In contrast to many of the other DFT methods, it does not include the Becke gradient correction to exchange [44]. The weights of the functional combinations were chosen to reproduce the ground-state energy of atomic hydrogen, to minimize the self-interaction energy, and the electron affinity of atomic oxygen, to include exchange and correlation effects in systems with significant electron–electron interactions that are not present in atomic hydrogen, except as a self-interaction error. Hence, this is a generally applicable technique that was not optimized for cumulenes or polyynes. KMLYP has been shown to be successful in determining molecular structures, activation barriers, and thermochemical properties [53, 54].

As B3LYP consists of mainly gradient corrected Slater exchange with a small contribution from HF exchange and BXLYP consists mainly of HF exchange with a small contribution from gradient corrected Slater exchange, KMLYP balances these two components by using approximately equal mixtures of HF and Slater exchange. Hence it is expected to do reasonably well for determining the energy difference between **8** and **9** [49]. Indeed, KMLYP represents an improvement over other DFT formulations for this particular problem although it still reverses the energetic ordering of these two isomers (Table 8.2). Systematic comparisons, also with respect to basis set effects, are depicted in Fig. 8.3. Although the overall effect of basis set size is rather small and should not be overinterpreted, it is remarkable that this is one of the few cases where the choice of basis set is *more* critical for the DFT methods than for the explicitly correlated methods.

What happens when larger $C_{2n+1}H_4$ cumulene/polyyne energy differences are considered? Are the errors magnified or do they cancel? Unfortunately, the first is

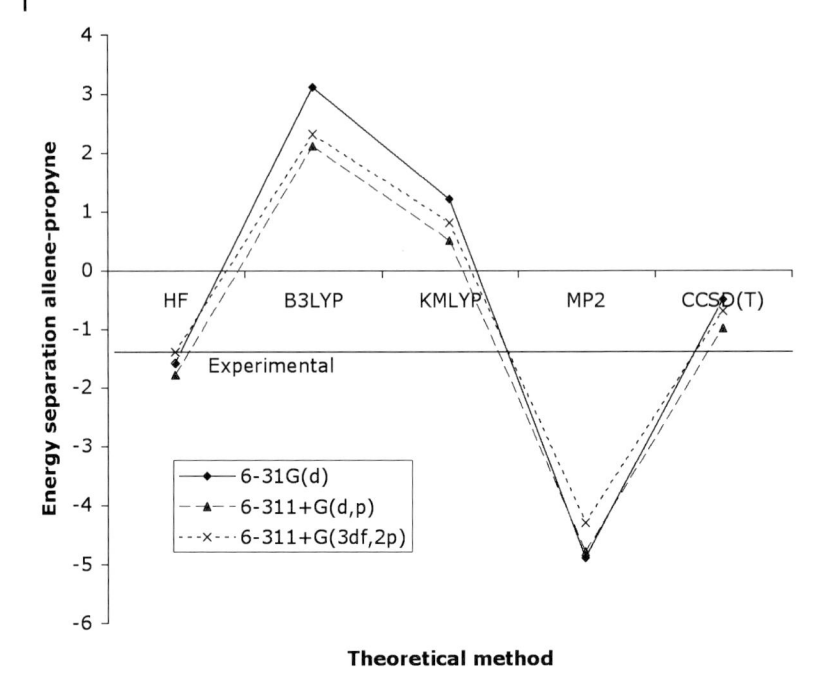

Figure 8.3. Isomerization energy difference (in kcal mol^{-1}) of allene and propyne at different levels of theory using three different basis sets.

true: the errors add up and practically get out of hand with HF and the common DFT functionals (Fig. 8.4); even the newly formulated BXLYP performs poorly for longer chains. The notable exception is the KMLYP approach, apparently because of fortuitous error cancellation in the HF and Slater exchange parts. Although this is also not "the right answer for the right reason", it implies that the KMLYP method is particularly useful for describing large systems that involve subtle competition between localization and delocalization.

The above analysis finds its direct application when determining large carbon-rich structures that are comparable in energy. Earlier systematic computational comparisons of the potentially doubly aromatic cyclic isomers of C_{10}, C_{14}, and C_{18} (**13–16**, Scheme 8.4) at the HF, MP2, and various DFT levels (LSDA, BLYP, B3LYP) with 6-31G(d) and some other basis sets already showed a clear trend towards electron-localized polyyne structures when going from C_{10} to C_{18} [47, 55]. The MP2 approach is invalid as C_{18} shows an RHF $\rightarrow$ UHF expansion instability making the perturbation treatment invalid. All "uncorrected" (for the error discussed above) DFT methods find that a cumulenic $D_{(n/2)h}$ structure is the energy minimum. When correcting for the allene–propyne error, however, the flattened circular polyyne of C_{9h} symmetry (**15**) is approximately 20 kcal mol^{-1} more stable than the cumulenic isomers. Structures **13**, **14**, and **16** are not minima, but are transi-

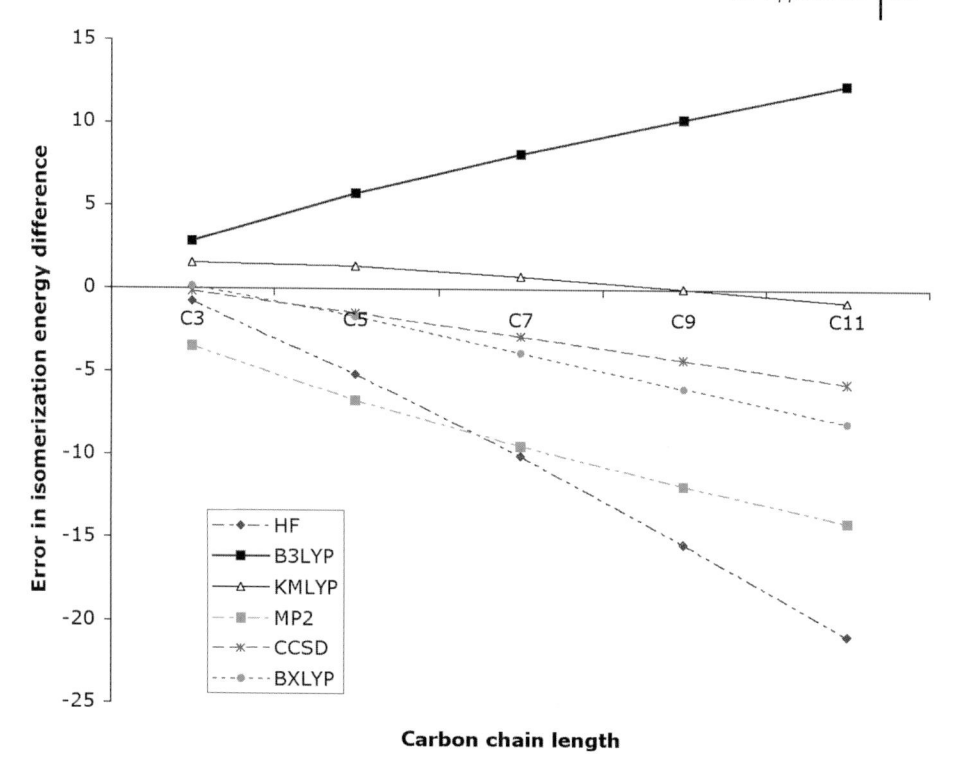

Figure 8.4. Isomerization energy difference (in kcal mol^{-1}) of allene and propyne relative to CCSD(T)cc-pVTZ//MP2/cc-pVTZ benchmark results (reproducing the experimental energy separation of **8** and **9** of 1.4 kcal mol^{-1}) at different levels of theory. For HF, B3LYP, BXLYP, and KMLYP a 6-311+G(3df,2p) basis set was utilized; MP2 and CCSD energies were determined with a cc-pVTZ basis.

tion states for various isomerization processes of **15**; these low-lying transition structures (e.g., **13** is just a few kcal mol^{-1} above **15**) equalize all carbons in **15**. The following trends in cyclocarbons were established: all $(4n+2)$ π-electron clusters favor "flattened" bond-angle alternating structures, while the cumulenic $D_{(n/2)h}$ structures are lowest in energy for small all-carbon cycles. As the value of

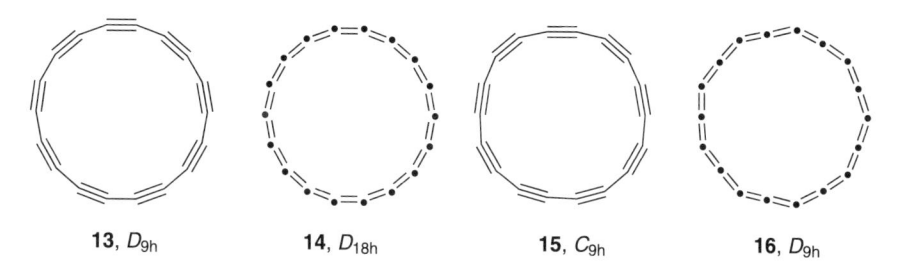

13, D_{9h} **14**, D_{18h} **15**, C_{9h} **16**, D_{9h}

Scheme 8.4. C$_{18}$ polyyne vs. allenic structures.

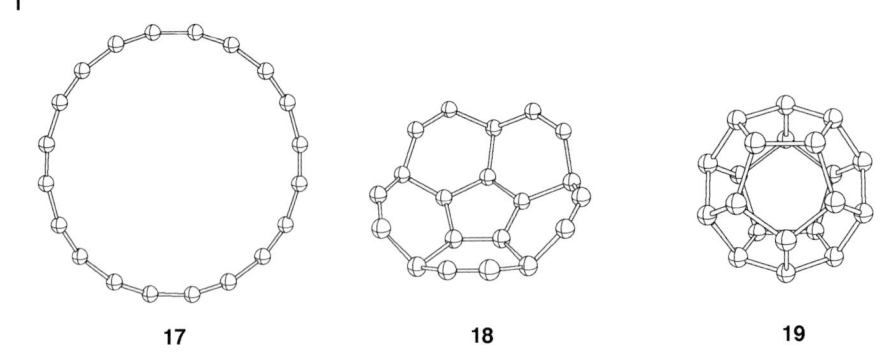

Scheme 8.5. Three selected ring (**17**), bowl (**18**), and cage (**19**) C_{20} isomers.

n increases, the energy of the polyynic $C_{(n/2)h}$ structures fall below those of the respective cumulenes; thus C_{18} is a cyclic C_{9h} symmetric polyyne (**15**) [47].

A second example concerns the long-standing question of the relative energies of the cyclic, bowl, and cage isomers **17–19** of C_{20} (Scheme 8.5, Table 8.3) which include carbon atoms in all degrees of hybridization [56–58]. The viability of dodecahedral **19** as the smallest regular fullerene consisting of five-membered rings was only recently shown by gas-phase experiments and identification by anion photoelectron spectroscopy [59]. The experimental assignments were further confirmed by comparison of the computed vibronic fine structure [60] and systematic computations of the free energies, electron affinities, and vibrational progressions of a number of isomers [61]. Depending on the sophistication of the theoretical methods employed, ring **17** and bowl **18** structures have been computed to be

Table 8.3. Relative energies of ring (**17**), bowl (**18**), and cage (**19**) C_{20} isomers at different levels of theory in kcal mol^{-1} [57].

Method	Geometry	17	18	19
PM3[19, 67]	PM3	0.0	70.3	176.0
AM1	AM1	0.0	137.9	295.1
B3LYP/6-31G[67]	B3LYP/6-31G	0.0	30.0	93.4
B3LYP/6-311G(d)[67]	B3LYP/6-311G(d)	0.0	9.2	43.8
MP2/TZVd	MP2/TZVd	48.0	0.0	0.7
MP2/TZV2d	MP2/TZV2d	47.5	15.2	0.0
MP2/TZV2d1f	MP2/TZV2d1f	48.4	12.5	0.0
MP2/TZV2d1f	HF/6-31G*	60.2	0.0	15.9
MR-MP2/TZV2d1f	MP2/TZV2d1f	58.3	0.0	4.4
MR-MP2/TZV2d1f	HF/6-31G*	69.2	0.0	29.3
QMC[66]	HF/6-31G*	23.1	0.0	48.2

more stable than the cage **19**; some early (lower-level) investigations even favored bicyclic rings and linear chains [58]. To assist the further characterization of the C_{20} isomers, the low-energy electron-scattering resonant structures [62], the optical absorption [63], IR, Raman, and NMR chemical shifts also have been computed [64]. Cage **19** and parts of it have also been computationally examined as nanotube building blocks [65]. The relative energetics of **17–19** were also examined in all-electron fixed-node quantum Monte Carlo computations, which agree well with valence-only quantum Monte Carlo and MP2 results with large basis sets; these studies place **18** at the lowest energy with **17** (+23.1 kcal mol^{-1}) and **19** (+48.2 kcal mol^{-1}) being much higher [66].

As noted before, the isomerization energy difference for **17–19** is challenging because single, double, double allenic, and triple bonds have to be treated equally well. The differentiation between a cyclic cumulene and an alternating triple-bonded structure for **17** is equally challenging as for **13–16**. Furthermore, **19** is particularly difficult for all theoretical methods because of structures with different symmetries are all energetically close-lying and interchange between different levels of theory. Additional problems arise from the high strain around the formal double bonds in **19** that display varying degrees of open-shell character; hence, it will be important to include contributions from static and dynamic correlation effects [57].

Not surprisingly, the two semiempirical approaches PM3 and AM1 fail miserably in determining realistic relative energies of **17–19**. Both methods incorrectly (compared to high level computations) place the cyclic polyyne (**17**) much lower than both the bowl (**18**) and the cage (**19**) structures. Despite these deficiencies it is discomforting to see that studies are being carried out at the semiempirical level dealing with many more C_{20} isomers that are equally unlikely to be represented well by this approach [19]. The sensitivity of the isomer energies to electron correlation effects is obvious from HF computations with medium-sized basis sets which places the ring 20–30 and the bowl 60–90 kcal mol^{-1} above the cyclic polyyne. Density functional theory methods display similar deficiencies as different formulations of DFT place either the ring (B3LYP), the bowl (B3PW91) or the cage (LDA) as the lowest isomer [66]. At the explicitly correlated levels of theory (MP2, CCSD, and QMC) the geometries were not optimized at the same level but were rather taken from lower quality, mostly HF/6-31G(d) optimizations. Furthermore, only modest size basis sets were employed for computing the energies at the MP2 and coupled cluster levels [68]. However, it has been shown recently that optimized geometries at the correlated levels are particularly important for higher level energy points and a consistent placement of the isomer energies [57]. Notwithstanding the notion that the three isomers represent borderline cases for the application of single-reference perturbation theory (MP2), the structures were optimized at the MP2 level using a large valence triple-ζ AO basis set with additional polarization functions (TZV2df). The cage isomer **19** turned out to be the most sensitive with respect to geometry optimization; hence, a D_{2h} structure ensues in contrast to the previously determined more localized HF/6-31G(d) structure that only displays C_2 symmetry with large bond alternations. The poor quality of the HF structures is also apparent when these are used for MP2/TZV2d1f single point energies computations: this interchanges the energies of **18** and **19**, indicating that the geometries are far from being converged.

Multireference second-order perturbation theory (MR-MP2) was then utilized for a much more elaborate treatment of this tricky problem; this approach also explicitly includes important higher excitations [69]. The multireference treatment was indicated by the three isomers showing relatively large amplitudes (ca. 0.1) in the first-order MP2 wavefunctions. Note that triple, quadruple and higher excitations are explicitly considered in MR-MP2. At the MR-MP2/TZV2df level the bowl is the most stable structure but the energy difference to the cage is rather small (4.4 kcal mol^{-1}); this level places the cyclic polyyne much higher in energy (58.3 kcal mol^{-1}). The change in the energetic ordering compared to the single-reference MP2 approach can be understood in terms of MP2 overestimating the correlation energies for systems with (partially) open-shell character that is largest in the cage. The larger sensitivity of the results for the cage with respect to variations in the correlation treatment is also in line with the stronger basis set effects seen for the cage relative to the bowl. This is further supported by MR-MP2/TZV2df computations on the HF/6-31G* geometries: the bowl is significantly more favored over the other two structures with the cage increasing dramatically in relative energy. Hence, within the error limits of the highest theoretical level available today, the bowl (**18**) and cage (**19**) are approximately isoenergetic. It is discouraging to see that DFT fails badly for this and related problems but this might just as well point to systematic improvements in the current DFT formulations.

8.3.2
Cyclization Reactions of Polyunsaturated Systems

A logical extension of the first example is the validation and application of computational methods to other systems containing single, double, and triple bonds. All of these are present in the structures of enediynes, enyne-allenes and related systems; these are, per definition, carbon-rich as the parent reactions involve highly unsaturated moieties. The ring closure reactions of these and related polyunsaturated systems (Scheme 8.6), which are known as Bergman (cycloaromatization) [70], Myers-Saito [71], and Schmittel-cyclizations [72], are of particular significance both for medicinal and organic materials chemistry [73]. Such, or structurally related moieties, lead to DNA-cleaving biradicals, i.e., they are potent anti-tumor drugs [74]. The parent reactions are now well understood both experimentally and theoretically [75–79]. Although many modes of cyclization can be envisaged, only three principally different ones are known experimentally (Scheme 8.6). As we have demonstrated in a very recent paper [80], other entirely unexplored cyclization modes of enediyne and enyne-allene substrates are possible, leading to five- and seven-membered rings [81, 82]. These findings bear important implications for materials science and nanoelectronics applications as organic materials (as noted in several other chapters of this book, five-membered rings introduce curvature in PAHs) may be fabricated in a more flexible and predictable way [83]. Also,

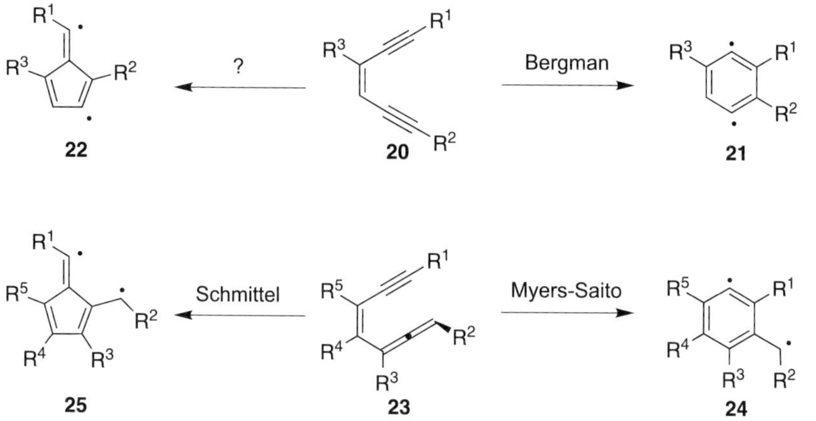

Scheme 8.6. Cyclization reactions of enediynes and enyne-allenes.

carbon-rich materials are often built up using structures with varying degrees of polyalkynes moieties or substituents; these structures are typically treated thermally (see Chapters 3, 4 and 12) to give the desired materials.

As noted above, true predictions are difficult and require that the computational results are validated against accurate experimental numbers and that the interpretation of the collective findings is self-consistent. Reactions or structures which are similar enough to allow a relatively safe interpretation and extrapolation, i.e., "families of reactions" are helpful in this respect. Such a "family" emerges when the above transformations are viewed within the larger framework of Cope-type reactions [84–86]. Within this line of thinking, Bergman, Myers-Saito, and many related reactions are classified in the broader scheme of these pericyclic transformations. As a consequence, one may ask why there are subtle but distinct differences in these types of reactions. For instance, while it is generally accepted that the Cope rearrangement of **26** does not involve an intermediate, it is the *p*-didehydrobenzene biradical **21** that is able to abstract hydrogens from simple (e.g., cyclohexadiene) or complex (e.g., DNA) H-donors in the cyclization of **20** (Fig. 8.5). When these reactions are viewed as related, one can predict the involvement of biradical intermediates.

For a systematic study of many related systems, a practical but nevertheless reliable approach had to be identified. It is clear that the reactions of the type depicted in Fig. 8.5 inherently have some diradical and hence multi-determinantal character. Therefore, semiempirical, HF and MP2 methods are unsuitable approaches because a single-reference will not be able to describe all the species involved equally well. DFT may be an option but it is also single-reference; however, as it is not the wavefunction that is the operand but the electron density, DFT may still do a reasonable job in describing these reactions. The simple take on this is that the density may be "well-behaved" even if the wavefunction is not. For instance, DFT has been proven to work exceptionally well for carbenes although singlet carbenes are

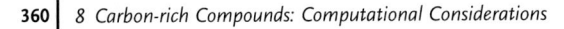

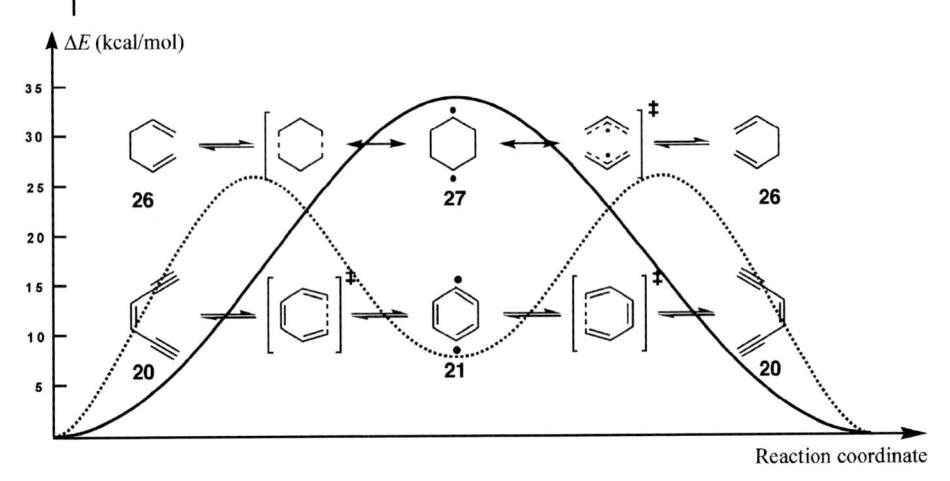

Figure 8.5. The Cope rearrangement paths for 1,5-hexadiene **26** and enediyne **20**.

also multideterminantal in character [29]. The reasoning behind this is that for wavefunction based methods it makes all the difference whether the two electrons in the two orbitals are placed in different orbitals or in one only [87]. For the density, however, which integrates over all orbitals and all electrons, it makes very little difference and this state can be described well within the DFT framework. Spin-separated biradicals are more difficult as the pairing of the electrons in one orbital only gives ylide-like structures that are very different from their biradical cousins [77]. Hence, the validation of DFT methods (which flavor to use?) must be carefully done and be gauged against accurate experimental data and theoretical methods.

Of course, a proper treatment would utilize multireference methods, but these are very time consuming and limited to model structures only [36, 88]. Furthermore, dynamic electron correlation is also very important in these reactions so that, for instance, simple CASSCF (complete active space SCF) computations are meaningless, as they mostly include static correlation. Perturbative corrections to CASSCF wavefunctions such as CASPT2 (CAS perturbation theory of second order) are necessary; however, multireference perturbation methods do not improve significantly the singlet–triplet energy separations of many biradicals as compared to CASSCF results [89].

We tested a large array of various functional combinations against available experimental data that contained heats of formation, activation barriers, and reaction energies (Fig. 8.6) [80]. While the best results are obtained with combinations including the correlation functional of Lee, Yang, and Parr (LYP), the choice of the exchange functional has a smaller effect on the quality of the DFT results. Since it was shown earlier that the solutions for *pure* DFT functionals are somewhat less prone to symmetry breaking [90], MPWLYP, G96LYP and BLYP are the best candidates for a compromise between chemical accuracy and computational demand. The inferior performance of pure vs. hybrid functionals in the UBS scheme has

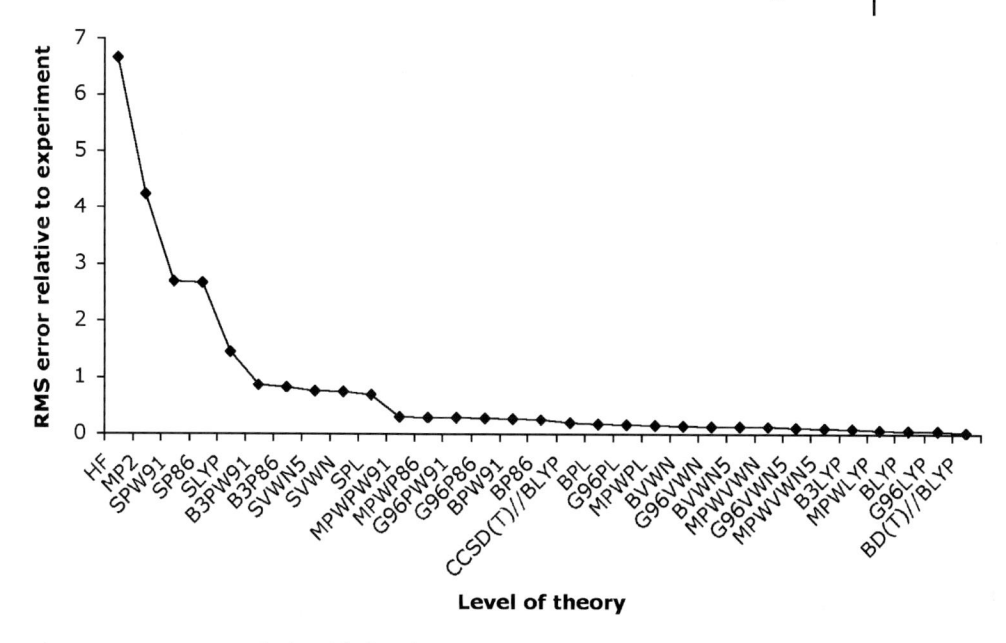

Figure 8.6. RMS error (in kcal mol^{-1}) for a large series of DFT energy single points on BLYP/6-31G(d) optimized geometries for several reference reactions and heats of formation; for details see Ref. [80]. Comparison with high-level CCSD(T)/cc-pVDZ and BD(T)/cc-pVDZ data.

been attributed to the self-interaction error, which mimics static correlation in DFT theory, being greater in pure than in hybrid functionals [78, 91]. We have had excellent experience in selecting BLYP as the method of choice in these and related problems [76, 77, 80–82, 84, 85, 92, 93] and this finding was later independently shown to be true by the best agreement of BLYP infrared intensities [79, 94] and energies [95] with experiment. It was also shown that the BLYP description of the potential surface is much closer to benchmark CCSD(T) and multi-reference CI computations [96].

The basis set dependence was also checked by systematically enlarging the basis from DZ to TZ type valence descriptions (Fig. 8.7). As often found for DFT, the behavior on increasing the basis set becomes erratic, in particular for pure functionals (BLYP and G96LYP were tested), where cc-pVTZ worsens the cc-pVDZ results (the same is found for the Pople-type basis sets 6-311G** vs. 6-31G*). B3LYP behaves more consistently, but does not necessarily give improved results with a better basis set (compare B3LYP/6-311G** vs. B3LYP/6-31++G**).

As demonstrated in Fig. 8.6, the quality of the results is further improved by single-point energy computations on the BLYP/6-31G* structures utilizing the Brückner-doubles coupled cluster approach including triple excitations perturbatively {BCCD(T), also called BD(T)} based on BS-UHF reference wavefunctions. In contrast, coupled-cluster computations with unmodified UHF molecular orbi-

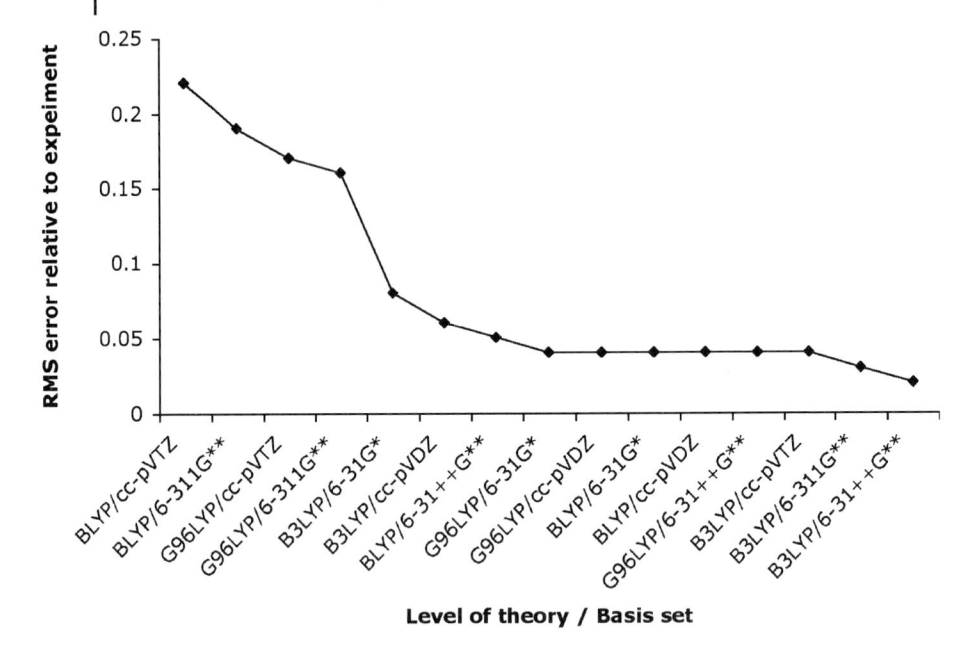

Figure 8.7. RMS error (in kcal mol^{-1}) for the BLYP, B3LYP, and G96LYP functional: energy single points on BLYP/6-31G(d) optimized geometries for several reference reactions and heats of formation; for details see Ref. [80].

tals {CCSD(T)} give poor results although they often deal well with multireference problems when the effect of triple excitations is included; this is due to the fact that in some open-shell cases CCSD(T) overestimates the single cluster-amplitudes [97]. The use of Brückner orbitals, in which the singles contribution is zero due to orbital rotation, leads to a significant improvement [98]. The BD(T) approach is clearly superior since the reference determinant is progressively rotated leading to an optimum mixing, generating a reference determinant with the proper symmetry to describe the biradical. UCCSD(T) or closed shell BD(T) methodologies are also indicated when calculating properties beyond energies and geometries, since RHF based CCSD(T) calculations can suffer from orbital instabilities leading to spurious vibrational frequencies [99].

The use of the UBS DFT methodology gives reasonable results for the cyclizations of polyunsaturated systems. Whereas geometry optimizations with medium-sized basis sets are acceptable, the energies must be computed with basis sets of triple-ζ quality or better. The use of single-point coupled-cluster energy computations using Brückner theory BD(T) on the DFT geometries provides even better and more reliable results.

With this validated methodology at hand, our research group has systematically computed new reaction modes of polyunsaturated carbon-rich systems. For instance, we evaluated alternatives to the experimentally known cyclization modes of enediynes and enyne-allenes, not giving the expected aromatic biradical but also five- (e.g., **22**, Scheme 8.6) and, for the enyne-allenes, seven-membered rings [80, 82, 85, 100]. Heterosubstituted systems were also systematically examined and led to the prediction of a large number of new cyclization reactions that await their experimental realization [81, 93, 101].

We have also recently shown that Bergman (Scheme 8.7) and related reactions (not shown) [84, 85] of polyunsaturated hydrocarbons, with a 1,3,5-hexatriene skeleton form a branch inside a larger "Cope" reaction family characterized by a common 1,5-hexadiene structural unit. The examination of this whole *family of reactions* allowed us to derive a very simple rule for involvement of transient biradicals in Cope-like reactions of hydrocarbons: *A non-concerted reaction takes place when biradical intermediates are stabilized either by allyl or aromatic resonance* [84, 85].

These conclusions bear important implications for thermal reactions of unsaturated hydrocarbons and carbon-rich structures. For instance, irregularities in carbon nanotubes such as Stone-Wales defects play a significant role in determining the properties of these fascinating structures [102].

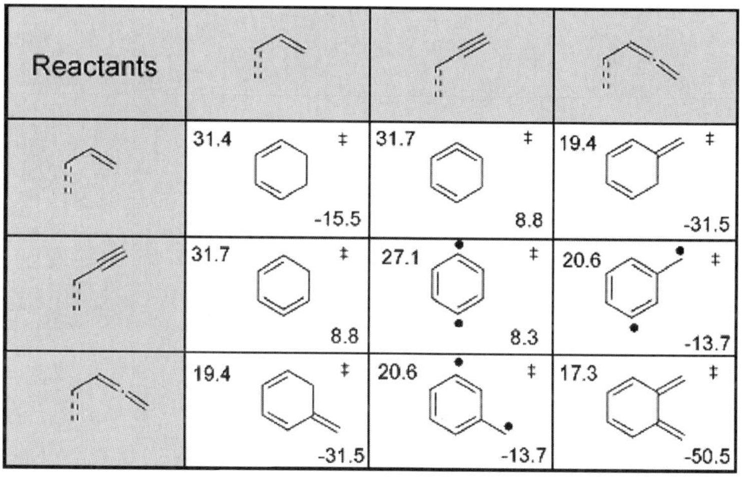

Reactants						
	31.4	-15.5	31.7	8.8	19.4	-31.5
	31.7	8.8	27.1	8.3	20.6	-13.7
	19.4	-31.5	20.6	-13.7	17.3	-50.5

Scheme 8.7. Part of the Cope reaction family for the 1,3,5-hexatriene reaction family: left-top corners: activation enthalpies; right-bottom corners reaction enthalpies (ΔH_0) at BD(T)/cc-pVDZ//UBLYP/6-31G(d).

8.3.3
Expanded (Super)molecules

A particularly creative application of predictive computations is the concept of building known molecules with single or double carbon–carbon bonds from larger building blocks to give so-called expanded molecules or supermolecules (*vide infra*) of the same gross topology and symmetry but larger size. For instance, a single

C–C bond can be replaced by an alkyne C≡C unit to retain the local symmetry; by analogy, a C=C double can be substituted by a larger =C=C= unit to retain the double bond character and the orientation of the terminating groups. A C≡C triple bond would have to be replaced by one ≡C–C≡ unit. Greater expansions can be achieved by using multiples of these single replacement units. As the resulting structures for the most part retain the symmetry and the original gross topology of the original molecules, these structures have been termed "carbomers" or carbo^n-molecules [103]. Straightforward examples of this concept are the carbomers of methane (carbo-methane **28**) and benzene (carbo-benzene **30**) that is the parent of a new class of carbo-aromatics (Scheme 8.8). This general concept applies from linear poly(acetylenes) to poly(triacetylenes) [104] and from dendralenes to expanded dendralenes [105]. Many of these structures have been examined using DFT methods and the results have to be taken with some caution due to the inability of DFT to properly distinguish between acetylenic and allenic moieties (*vide supra*). For instance, as the formal carbo-cyclohexatriene is an isomer of carbo-benzene **30**, this point needs to be elaborated at decisive levels of theory in the near future.

With this concept in mind, the [N]pericyclynes (**32–35**), several of which have been prepared, can also be viewed as carbomers of the cycloalkanes (i.e., carbo-cycloalkanes) [106, 107]. Apart from the obvious, there are, however, distinct differences between pericyclynes and cycloalkanes. The most striking feature about the pericyclynes, which is absent in the cycloalkanes, is their ability to display two forms (in- and out-of-plane) of homoconjugation. Whether this formal homoconjugation in [5]pericyclyne (**34**) results in homoaromaticity (10 π electrons both in and perpendicular to the plane of the molecule) is vividly debated in the literature. While decamethyl [5]pericyclyne, which displays an envelope shape in its crystal structure, was estimated to be stabilized by an appreciable 6 kcal mol^{-1} [108], later DFT computations [109], which, however, favor a planar form, suggest that, based on geometric and magnetic criteria of aromaticity, it is not homoaromatic [110]. The analysis of the electron localization function (ELF) [111] at the B3PW91/6-31G** level suggests low but recognizable cyclic electron delocalization, that is, however, mostly due to σ-delocalization so that the pericyclynes are indeed not considered to be homoaromatic [107].

Other species that are potentially aromatic such as the annulenes and radialenes have also been discussed, based on DFT computations and several experimentally prepared structures [112, 113]. The key conclusion of these studies is that aromaticity is an important consideration for these types of structures but the considerable strain often found in these molecules can override some of the aromatic stabilization. Owing to their cyclic delocalization virtually all carbomers are excellent electron acceptors, as evident from their high electron affinities [113].

Superb experimental skills led to the construction of the first substituted *three-dimensional* alkane carbomer, C$_{56}$ carbo-cubane (**38**, the experimental structure carries eight methoxy groups for every hydrogen), containing linear *dialkyne* (–C≡C–C≡C–, buta-1,3-diynediyl) fragments as the first member of the expanded

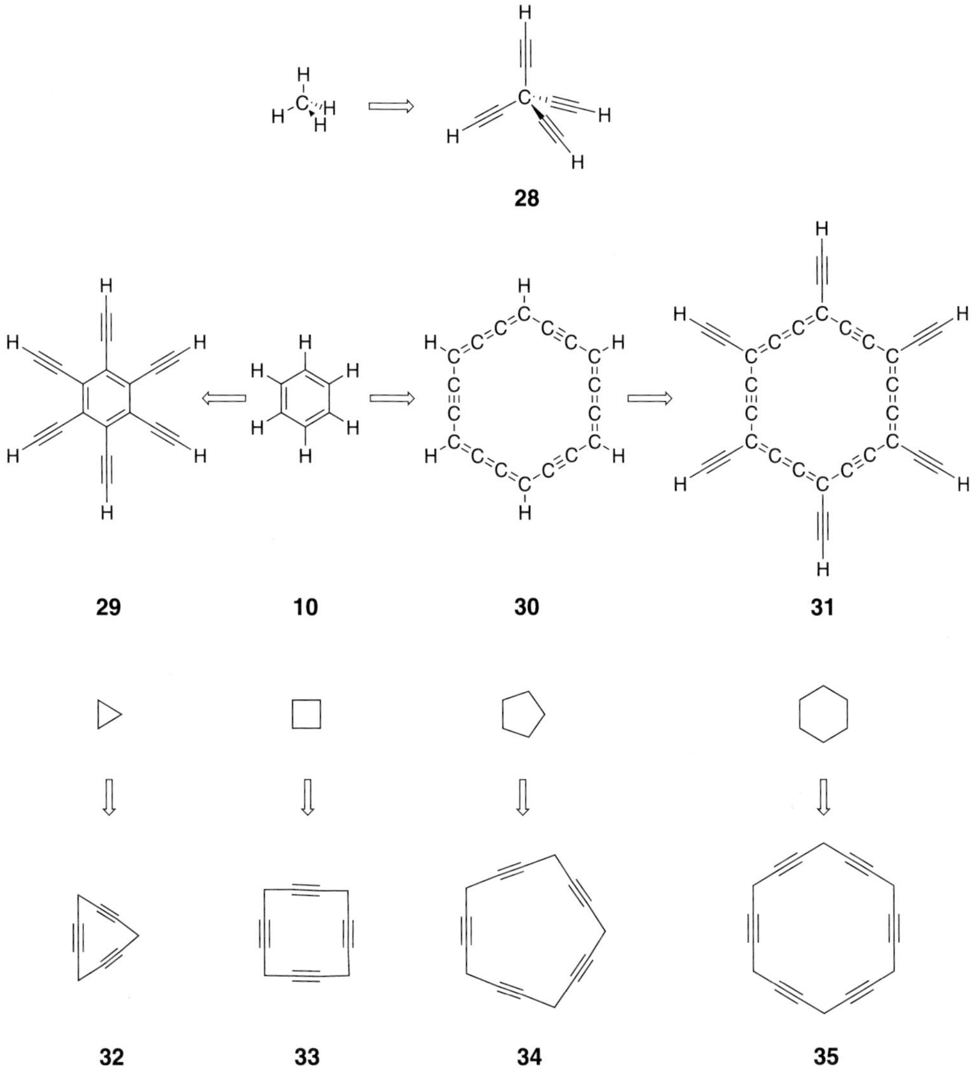

Scheme 8.8. Carbomers of methane and benzene and related isomers.

polyhedranes (Scheme 8.9) [114]. Highly strained **38** emphasizes that it is, in principle, possible to build hydrocarbons representing several of the platonic solids utilizing multiple alkyne units [115]; these types of molecules are expected to display interesting properties for material applications. High-resolution Fourier-transform ion cyclotron mass spectrometric studies suggest the formation of fullerene ions C_{56}^* that characteristically lose C_2 units and undergo ion–molecule coalescence reactions to form higher fullerene ions [114]. Hence, a second feature of carbo-

34   **35**   **36**

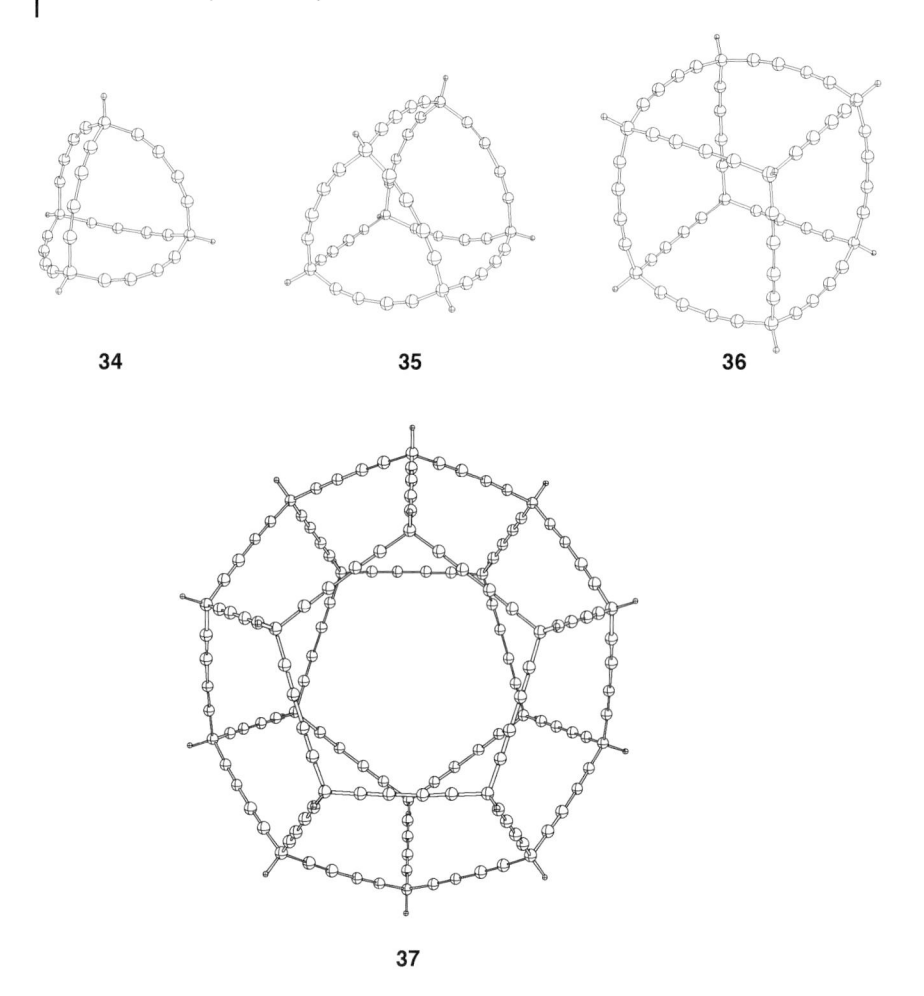

37

Scheme 8.9. Four of the five platonic solid-type polyhedrane carbomers.

alkanes is the possibility of making other carbon-rich structures through thermal rearrangement reactions. Finally, the larger inner cavity of these expanded molecules might be host to a variety of small molecules or atoms [116].

In marked contrast to highly strained but kinetically very stable cubane (**7**), **38**-(OMe)$_8$ is explosive and kinetically extremely labile. The theoretically determined (HF/6-31G*) heats of formation of **36**–**38** through isodesmic/homodesmotic equations [117] as well as their strain energies, which were determined by computing the energy difference to hypothetically unstrained reference molecules (computed via group additivities), suggest that the expanded molecules generally follow the strain order as their parent structures. Owing to the greater flexibility of the diyne unit, the expanded structures are, however, considerably *less* strained. Thus, the strain energies for **36–38** are 111.8, 113.1, and 107.8 kcal mol^{-1}, respectively. As the strain energy of **7** is 155–161 kcal mol^{-1} [118] the reason for the instability of **38**-(OMe)$_8$ must lie in its substitution pattern that allows stabilization of otherwise kinetically higher lying transition structures or intermediates along the decomposition pathways.

The geometries at HF/6-31G* and the B3LYP/6-31G* level are rather consistent and do not vary much from 2,4-hexadiyne as a reference structure. Although the X-ray structure of **38**-(OMe)$_8$ is not available due to its instability, the structural data of some smaller expanded molecules such as carbo-cyclobutane compare favorably with their computed structures at these levels of theory; molecular mechanics geometries, determined with an amended MM3 force field, also agree well with these estimates [115]. The advantage of having this modified force field description available is the possibility of optimizing the geometry of expanded dodecahedrane **39**, which already contains 100 carbon atoms, or possibly an even larger carbon-rich structure with the hexadiyne structural element. As hexadiyne expanded carbo-cyclopentane and carbo-cyclohexane are virtually unstrained utilizing this MM3 force field, the strain in **39** is only 10.9 kcal mol^{-1}. Even larger expanded structures like the C$_{300}$H$_{60}$ icosahedrane have also been examined and gave very low strain energies per diyne unit (<1 kcal mol^{-1}).

8.3.4
Diamondoids and Related sp^3-Structures

As noted above, the materials chemistry of sp^3 carbons mostly concentrates on fascinating and often highly strained hydrocarbons on the small scale and virtually unstrained diamond on the large scale. The spectrum in between with diamondoids (cage carbons with frameworks resembling the diamond lattice) and polyadamantyls (connected adamantane or larger diamondoid cages, cf. Scheme 8.2 top) has largely been left unexplored. This is surprising in view of the fact that diamondoids are, as noted in the introduction (Fig. 8.1), considerably *more* stable than their graphene or PAH counterparts. They are also much less reactive and display many highly desirable materials properties, e.g., polymers with high rigidity [119], and hybrid materials [120]. All-carbon "onions" were shown to rearrange into diamondoid structures at high temperature; these structures are stable upon re-cooling to room temperature [121].

Part of the reason why diamondoid structures have not been fully explored for molecular and materials chemistry is our limited ability to make larger diamondoids by chemical synthesis, although thermodynamics tells us that it should not be that difficult. Schleyer's milestone achievement of a simple preparation of adamantane via the "stabilomeric" rearrangement of tetrahydrodicyclopentadiene catalyzed by Lewis acids [122] triggered huge efforts to make higher diamondoids through thermodynamically driven isomerizations. Another successful example is the preparation of diamantane from norbornene [123] and norbornadiene dimers [124]. Unfortunately, the stabilomer approach is not efficient for the synthesis of higher diamondoids because these possess multiple energetically close low-lying minima so that mixtures of isomers form. To date only triamantane [125] and C_{2h}-tetramantane have been prepared synthetically [126].

As Nature offers diamondoids in large quantities from crude oil [4, 127], one ought to explore their chemistry especially in view of their potential applications in nanoelectronic devices [128]. The first challenge is to understand systematically the reactivity patterns of diamondoids, especially with respect to their selective peripheral C–H bond functionalization. This difficulty is emphasized when one considers that even triamantane (**3**) reacts with typical electrophiles (e.g., Br_2) with very low selectivity [129]. What alternatives are there – will ionic, radical, and radical ionic C–H activation reactions eventually lead to higher C–H bond selectivities? These questions can, in part, be answered by computational methods when considering the very different stabilities of the cations, radicals, and radical cations of the respective diamondoids in the first step. These purely thermodynamic stabilities very often translate nicely into selectivities, at least for cationic structures. As this is often not the case for radicals, transition structures also have to be considered which makes the prediction of selectivities far more elaborate [130].

We recently assessed the reactivity of diamantane (**2**) both computationally and experimentally and extrapolated our findings to the functionalizations of higher diamondoids [27]. The very few available heats of formation for some of the lower diamondoids were taken as a validation for the B3LYP/6-31G(d) optimizations and homodesmotic energy evaluations. The ΔH_{f298} computed for **2** is -34.1 kcal mol^{-1}, in excellent agreement with the experimental value of -34.87 ± 0.58 kcal mol^{-1} [131]; for triamantane **3** we find a ΔH_{f298} of -43.2 ± 1.8 kcal mol^{-1}. For higher diamondoids the absolute value of ΔH_{f298} continuously increases and follows the decrease in the strain energies (note that although diamondoids are comprised of almost perfectly tetrahedral geometries around their carbon atoms, they are, as parent adamantane, not strain-free [132]).

8.3.4.1 Diamondoidyl Cations

Higher diamondoids contain several nonequivalent tertiary C–H bonds making selective C–H substitutions rather difficult. It was shown previously that the positional selectivities for the C–H substitutions in cage compounds under ionic conditions correlate reasonably well with the stabilities of the respective carbocations [133]. As expected, the relative stabilities (Scheme 8.10) of diamondoidyl cations vs. the 1-adamantyl cation (1-Ad$^+$) increase continuously with cage size owing to better dispersion of charge [27]. The increasing stabilities of tertiary cations enhance the reactivities of the higher diamondoids relative to adamantane under ionic conditions: Triamantane reacts rapidly with bromine at 0 °C [129], whereas the bromination of adamantane requires longer reaction times or substantial heating [134].

8.3.4.2 Diamondoidyl Radicals

As noted above, the correlation of the relative stabilities of hydrocarbon radicals with the selectivities in C–H substitution reactions is poor as hydrogen atom abstractions are mostly entropy controlled [130, 135]. Hence, the analogous radical

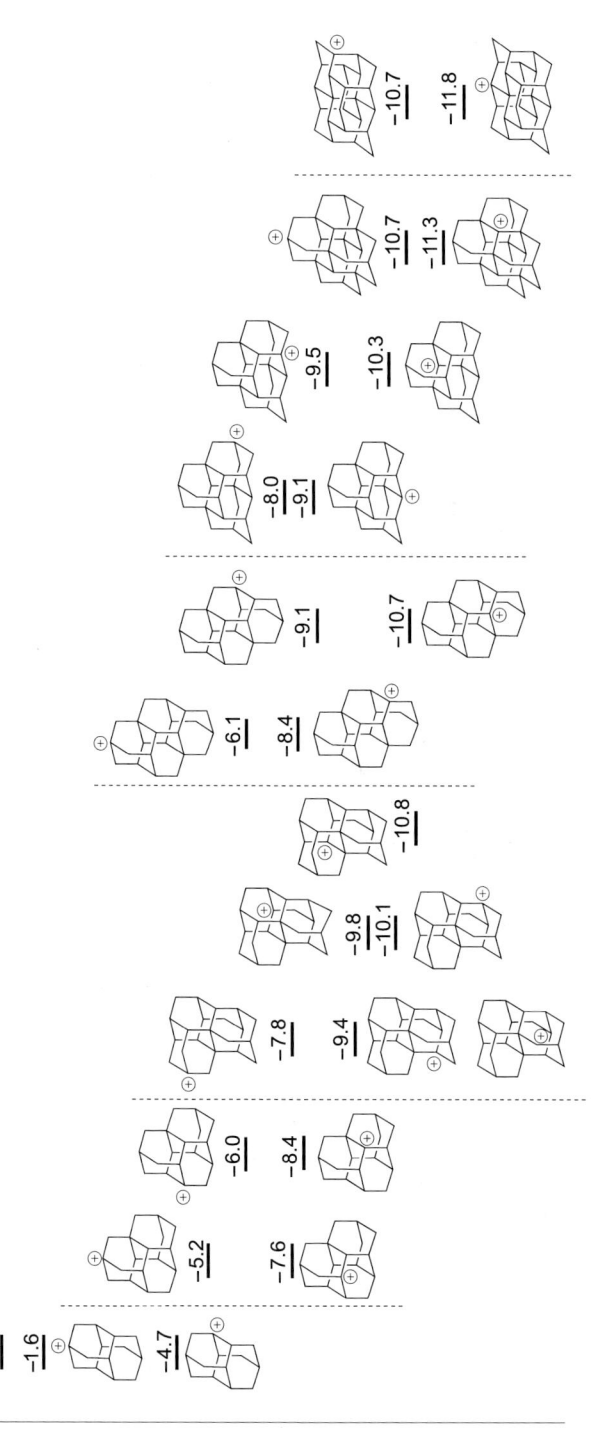

Scheme 8.10. Relative energies of diamondoidyl cations in kcal mol^{-1} according to the equation 1-Ad$^+$ + diamondoid → AdH + *tert*-diamondoidyl$^+$ at B3LYP/6-31G(d).

stabilities of the structures depicted in Scheme 8.10 vary within only 1 kcal mol^{-1} making free radical substitutions in diamondoids depend much more on the structure of the abstracting radical than on the C–H bond strengths.

8.3.4.3 Diamondoid Radical Cations

Despite common suspicions about the selectivities of reactions involving hydrocarbon radical cations, these reactions are often the most selective relative to cationic and radical transformations [130, 136, 137]. Single-electron transfer (SET) oxidations of hydrocarbons to radical cations require strong oxidants as the oxidation potentials of saturated hydrocarbons are rather high. As the HOMOs of cage hydrocarbons are typically highly delocalized, their radical cations also display pronounced spin/charge delocalizations over the entire cage. This lowers the ionization potentials (IPs) of cage hydrocarbons substantially: while the computed IP of diamantane **2** is 209.2 kcal mol^{-1} (experimental: 206.0 kcal mol^{-1}) [138] it is only 182.3 kcal mol^{-1} for hexamantane **6**. The radical cations usually undergo proton loss in solution from the position with the most elongated tertiary C–H bond. With the structures depicted in Scheme 8.11, one can therefore predict the C–H bond that will be broken from the optimized geometry of the respective diamondoid radical cation. This subsequent deprotonation usually produces the respective hydrocarbon radicals much more selectively than through direct radical H-abstractions [130, 137, 139]. Although the higher diamondoids are only slightly strained, they display remarkably low adiabatic IPs (177–184 kcal mol^{-1}) hence, SET oxidations with conventional single-electron chemical oxidants are predicted to be practically valuable.

8.3.4.4 Diamondoid Supermolecules

A virtually undiscovered class of molecules is that of the polyadamantyl supermolecules, which can also be viewed under the heading of expanded (super)molecules (*vide supra*), in this case relating the T_d core structure of adamantane (**1**) to methane. Diadamantyl (**40**, also referred to as 1,1'-biadamantane or diadamantane, not to be confused with diamantane **2**) then is the expanded form of ethane, triadamantyl (**41**) that of propane and so on (Scheme 8.11 and Fig. 8.2) [140]. The adamantyl building blocks thus replace the atoms of the simple alkanes. This is an attractive molecular as well as modular *Aufbau*-principle for carbon-rich sp^3-materials because of the aforementioned extraordinary stability of the lower diamondoids (the adamantyls can be readily replaced by larger diamondoidyl modules) and the relationship of these expanded structures (perhaps to be called superalkanes) to the diamond lattice itself (**44**) [140, 141] (Scheme 8.12). Although some of these structures are known (**38**) or have been prepared very recently (**41** [141], **42** [140], **43** [141]) by straightforward Wurtz-Fittig coupling reactions (solubility is usually a problem of the larger polyadamantyls), their chemistry is practically unexplored. To the best of our knowledge, there are no computations regarding these species although it would be interesting to examine whether the analogy provided here really holds. For instance, will **42** display *anti* and *gauche* conformers as found for its parent n-butane? Will **43** be more stable thermodynamically than its isomer

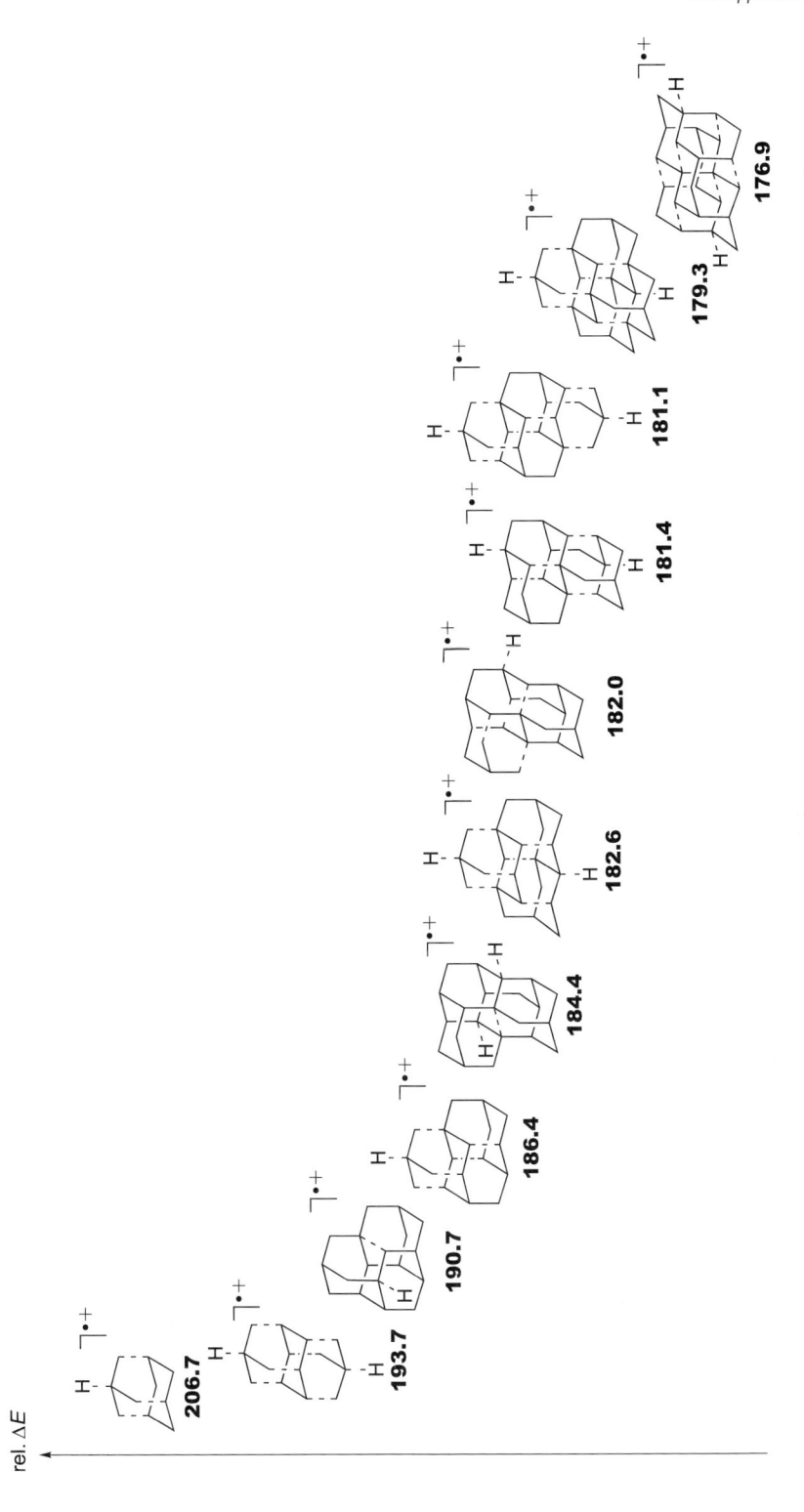

Scheme 8.11. The B3LYP/6-31G(d) energies of the relaxed diamondoid radical cations relative to the respective neutral hydrocarbons (ΔE, kcal mol^{-1}).

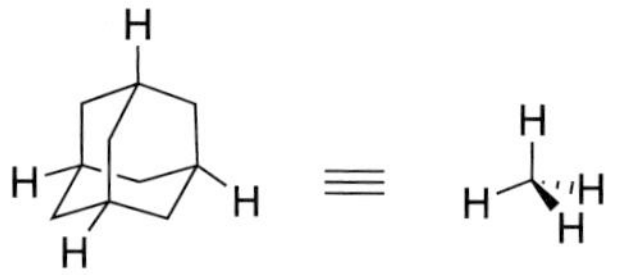

1

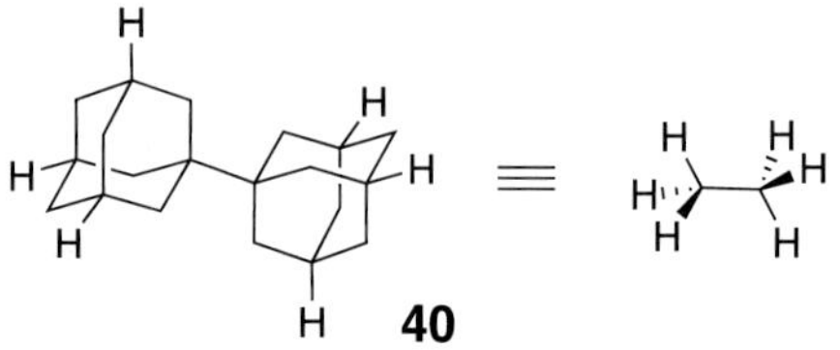

40

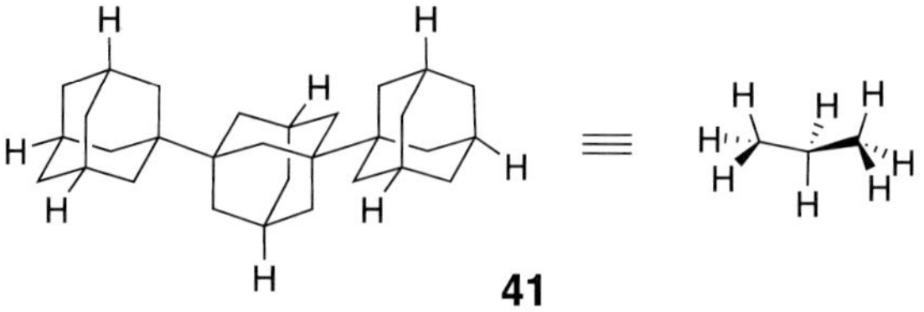

41

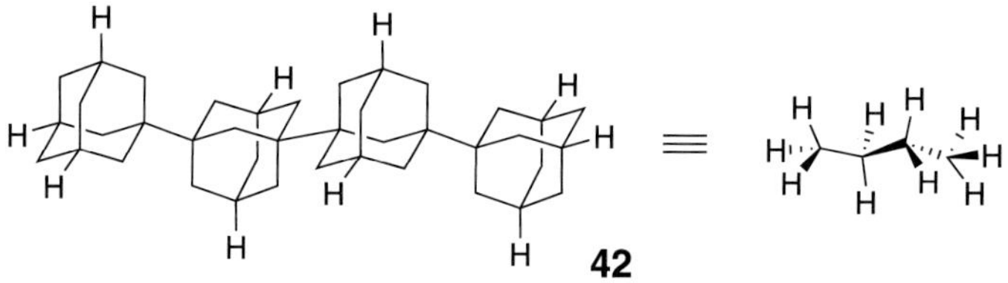

42

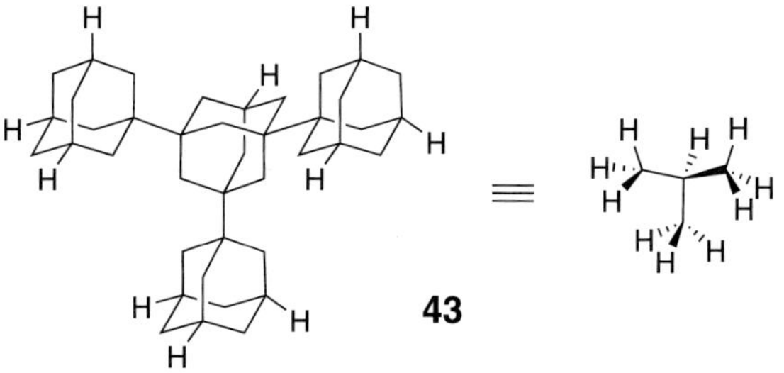

43

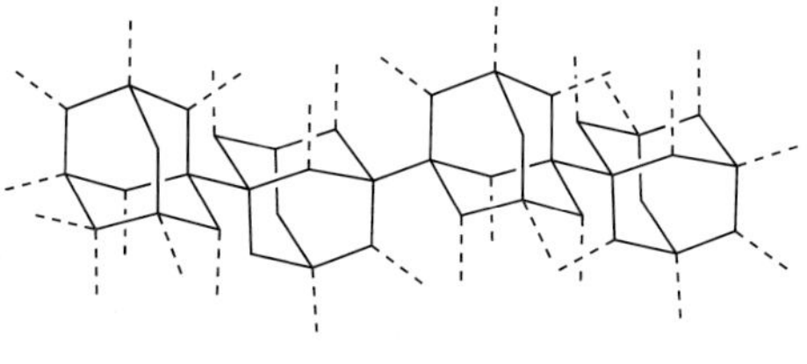

44

Scheme 8.12. Analogy between polyadamantyls and alkanes; polyadamantyls as part of the diamond lattice.

42 as isobutane is more stable than n-butane? Will a cyclohexane analog show the same ring inversion behavior and preference for large equatorial substitutents? Many of these fascinating aspects can be studied with computational methods at sufficiently high levels so that it is only a question of time when researchers begin to explore fully the chemistry of polyadamantantes as wonderful sp^3-carbon-rich building blocks.

8.4
Conclusions and Outlook

Computational methods are ideally suited to examine and, after proper validation, predict the structures and reactivities of carbon-rich materials. They also aid in identifying newly prepared structures by matching experimental and computed spectra. The single molecule picture often used in gas-phase computations is quite adequate but methods need to be developed to deal with much larger assemblies. While unsaturated sp- and sp^2-structures have received much attention, studies regarding larger sp^3-carbon (diamondoid) systems are still scarce. Efforts should be pointed in this direction.

Density functional methods are very valuable when dealing with carbon-rich structures but they are not error free. A good strategy builds on a proper validation, DFT optimized structures, and energies determined at a higher level. With the continuing development of linear scaling high-level *ab initio* methods, one expects also a considerable impact on carbon-rich structures from these well behaved and easily, as well as rationally, improvable approaches.

Acknowledgments

I thank Volker Lutz for initial literature surveys and discussions and Andrey Fokin for the data in Schemes 8.10 and 8.11. This work was supported by the National Science Foundation (CHE-0209857).

Abbreviations

ΔH_{f298}	Heat of formation at 298 K
AM1	Austin model 1 (a semiempirical theory)
AO	Atomic orbital
B3LYP	Becke's three-parameter hybrid exchange functional with the Lee-Yang-Parr correlation functional
B3P86	Becke's three-parameter hybrid exchange functional with gradient corrections of Perdew coupled with Perdew's 1981 local correlation functional

B3PW91		Becke's three-parameter hybrid exchange functional with Perdew and Wang's 1991 gradient corrected correlation functional
BD		Brückner doubles coupled cluster expansion including single and double excitations with Brückner-type orbitals
BD(T)		BD with perturbatively included triple excitations
BLYP		Becke exchange functional and Lee-Yang-Parr correlation functional
BP86		Becke exchange functional with gradient corrections of Perdew coupled with Perdew's 1981 local correlation functional
BPL		Becke exchange functional and Perdew's local correlation functional
BVWN		Becke exchange functional and Vosko, Wolk, Nusair correlation functional
BXLYP		Variable (x) amount of HF exchange included in the BLYP functional
CAS		Complete active space
CC		Coupled cluster theory
cc-pVXZ		correlation consistent polarized valence x-zeta basis sets (X = double/O, triple/T, quadrople/Q, etc)
CCSD		Coupled cluster theory including single and double excitations
CCSD(T)		CCSD including perturbatively determined triple excitations
CI		Configuration interaction
CISD		Configuration interaction including single and double excitations
CSF		Configuration state function
DFT		Density functional theory
DZ		double zeta basis set
E_{corr}		Correlation energy
E_{hf}		HF energy
ELF		Electron localization function
FCI		Full configuration interaction
FF		Force field
G96LYP		Gill's 1996 exchange functional and Lee-Yang-Parr correlation functional
G96P86		Gill's 1996 exchange functional with gradient corrections of Perdew coupled with Perdew's 1981 local correlation functional
G96PL		Gill's 1996 exchange functional and Perdew's local correlation functional
G96VWN		Gill's 1996 exchange functional and Vosko, Wolk, Nusair correlation functional
G96VWN5		Gill's 1996 exchange functional and Vosko, Wolk, Nusair correlation functional from 1980 which first the Coperley-Alder solution of the uniform electron gas
GIAO		Gauge invariant atomic orbitals
Gn		Gaussian-n set of reference molecules
GTO		Gaussian-type orbital
HF		Hartree-Fock
HOMO		Highest occupied molecular orbital
IP		Ionization potential

KMLYP	Kang-Musgrave exchange functional and Lee-Yang-Parr correlation functional
LDA	Local density appoximation
LSDA	Local spin density approximation (= Slater exchange functional and VWN correlation = SVWN)
LUMO	Lowest unoccupied molecular orbital
LYP	Lee-Yang-Parr correlation functional
MBPT	Many-body perturbation theory
MM	Molecular mechanics
MNDO	Modified neglect of differential diatomic overlap
MO	Molecular orbital
MPn	nth-order Møller-Plesset perturbation theory
MPWLYP	Barone and Adamo's Becke-style one parameter functional modified with Perdew-Wang 1991 exchange and the Lee-Yang-Parr correlation functional
MPWP86	Barone and Adamo's Becke-style one parameter functional modified with Perdew-Wang 1991 exchange coupled with Perdew's 1981 local correlation functional
MPWPL	Barone and Adamo's Becke-style one parameter functional modified with Perdew-Wang 1991 exchange and Perdew's local correlation functional
MPWPW91	Barone and Adamo's Becke-style one parameter functional modified with Perdew-Wang 1991 exchange with Perdew and Wang's 1991 gradient corrected correlation functional
MPWVWN	Barone and Adamo's Becke-style one parameter functional modified with Perdew-Wang 1991 exchange and Vosko, Wolk, Nusair correlation functional
MR-	Multireference
NICS	Nucleus independent chemical shieldings
PAH	Polyaromatic hydrocarbon
PM3	Parametrized model 3
QMC	Quantum Monte Carlo method
RHF	Restricted HF
RMS	Root mean-square
S	Spin operator
SCF	Self-consistent field (analogous to HF theory)
SET	Single-electron transfer
SLYP	Slater exchange and the Lee-Yang-Parr correlation functional
SP86	Slater exchange with Perdew's 1981 local correlation functional
SPL	Slater exchange and Perdew's local correlation functional
SPW91	Slater exchange with Perdew and Wang's 1991 gradient corrected correlation functional
STO	Slater-type orbital
SVWN	Slater exchange with Vosko, Wilk, Nusair correlation functional
TZ	triple zeta basis set

UBS	Unrestricted broken-spin
UHF	Unrestricted HF
VWN	Vosko, Wilk, Nusair correlation functional
ZPVE	Zero-point vibrational energy

References

1 C. J. Cramer, *Introduction to Computational Chemistry*, Wiley-Interscience, Hoboken, **2004**; F. Jensen, *Introduction to Computational Chemistry*, John Wiley & Sons, Chichester, **1999**; P. v. R. Schleyer, N. L. Allinger, T. Clark, J. Gasteiger, P. A. Kollman, H. F. Schaefer III, P. R. Schreiner, *The Encyclopedia of Computational Chemistry*, John Wiley & Sons, Chichester, **1998**.

2 W. Koch, M. C. Holthausen, *A Chemist's Guide to Density Functional Theory*, 2nd edn., Wiley-VCH, Weinheim, **2001**.

3 V. V. Danilenko, *Phys. Solid State* **2004**, *46*, 595–599.

4 J. E. P. Dahl, J. M. Moldowan, T. M. Peakman, J. C. Clardy, E. Lobkovsky, M. M. Olmstead, P. W. May, T. J. Davis, J. W. Steeds, K. E. Peters, A. Pepper, A. Ekuan, R. M. K. Carlson, *Angew. Chem. Int. Ed.* **2003**, *42*, 2040–2044; J. E. Dahl, S. G. Liu, R. M. K. Carlson, *Science* **2003**, *299*, 96–99; H. Hopf, *Angew. Chem. Int. Ed.* **2003**, *42*, 2000–2002.

5 P. Badziag, W. S. Verwoerd, W. P. Ellis, N. R. Greiner, *Nature* **1990**, *343*, 244–245.

6 U. Burkert, N. L. Allinger, *Molecular Mechanics*, American Chemical Society, Washington, D.C., **1982**.

7 C. F. Kettering, L. W. Shutts, D. H. Andrews, *Phys. Rev.* **1930**, *36*, 531–543; J. B. Hendrickson, *J. Am. Chem. Soc.* **1961**, *83*, 4537–4547.

8 C. Mapelli, C. Castiglioni, G. Zerbi, K. Mullen, *Phys. Rev. B* **1999**, *60*, 12710–12725; K. Ohno, R. Takahashi, *Chem. Phys. Lett.* **2002**, *356*, 409–422.

9 B. Borstnik, B. Lukman, *Fullerene Sci. Technol.* **1994**, *2*, 357–366.

10 A. Sears, R. C. Batra, *Phys. Rev. B* **2004**, *69*, 235406–235406–235410.

11 P. M. W. Gill, in *The Encyclopedia of Computational Chemistry*, Vol. 1, P. v. R. Schleyer, N. L. Allinger, T. Clark, J. Gasteiger, P. A. Kollman, H. F. Schaefer III, P. R. Schreiner (eds.), John Wiley & Sons, Chichester, **1998**, pp. 678–689.

12 M. J. S. Dewar, *The PMO Theory of Organic Chemistry*, Plenum, New York, **1975**; J. J. P. Stewart, in *Reviews in Computational Chemistry*, Vol. 1, K. B. Lipkowitz, D. B. Boyd (eds.), Wiley-VCH, New York, **1990**, p. 45.

13 H. Genin, R. Hoffmann, *Macromolecules* **1998**, *31*, 444–455.

14 T. Clark, *J. Mol. Struct. (THEOCHEM)* **2000**, *530*, 1–10.

15 M. J. S. Dewar, E. G. Zoebisch, E. F. Healy, J. J. P. Stewart, *J. Am. Chem. Soc.* **1985**, *107*, 3902–3909.

16 M. J. S. Dewar, W. Thiel, *J. Am. Chem. Soc.* **1977**, *99*, 4899–4907.

17 W. Thiel, in *Thermochemistry from Semiempirical Molecular Orbital Theory*, ACS Symposium Series, Vol. 677, K. K. Irikura, D. J. Frurip (eds.), American Chemical Society, Washington, DC, **1998**, pp. 142–161.

18 L. A. Curtiss, K. Raghavachari, G. W. Trucks, J. A. Pople, *J. Chem. Phys.* **1991**, *94*, 7221–7230; L. A. Curtiss, K. Raghavachari, J. A. Pople, *J. Chem. Phys.* **1993**, *98*, 1293–1298; L. A. Curtiss, K. Raghavachari, J. A. Pople, *J. Chem. Phys.* **1995**, *103*, 4192–4200; L. A. Curtiss, K. Raghavachari, P. C. Redfern, J. A. Pople, *J. Chem. Phys.* **1997**, *106*, 1063–1079.

19 J. I. Chavez, M. M. Carrillo, K. A. Beran, *J. Comput. Chem.* **2004**, *25*, 322–327.

20 P. v. R. Schleyer, C. Maerker, A. Dransfeld, H. Jiao, N. J. R. v. E. Hommes, *J. Am. Chem. Soc.* **1996**, *118*, 6317–6318.

21 S. Patchkovskii, W. Thiel, *J. Mol. Model.* **2000**, *6*, 67–75.

22 M. Bendikov, H. M. Duong, K. Starkey, K. N. Houk, E. A. Carter, F. Wudl, *J. Am. Chem. Soc.* **2004**, *126*, 7416–7417; M. Bendikov, H. M. Duong, K. Starkey, K. N. Houk, E. A. Carter, F. Wudl, *J. Am. Chem. Soc.* **2004**, *126*, 10493–10493 (correction).

23 W. T. Borden, E. R. Davidson, *J. Am. Chem. Soc.* **1977**, *99*, 4587–4594; T. Bally, W. T. Borden, in *Reviews in Computational Chemistry*, Vol. 13 (Eds.: K. B. Lipkowitz, D. B. Boyd), Wiley-VCH, Weinheim, **1999**, pp. 1–97; M. Filatov, S. Shaik, *J. Phys. Chem.* **1999**, *103*, 8885–8889.

24 H. Meng, M. Bendikov, G. Mitchell, R. Helgeson, F. Wudl, Z. Bao, T. Siegrist, C. Kloc, C. H. Chen, *Adv. Mater.* **2003**, *15*, 1090–1093.

25 P. v. R. Schleyer, M. Manoharan, H. Jiao, F. Stahl, *Org. Lett.* **2001**, *3*, 3643–3646; K. B. Wiberg, *J. Org. Chem.* **1997**, *62*, 5720–5727.

26 K. Hannewald, V. M. Stojanovic, J. M. T. Schellekens, P. A. Bobbert, *Synth. Met.* **2003**, *137*, 891–892.

27 A. A. Fokin, B. A. Tkachenko, P. A. Gunchenko, D. V. Gusev, P. R. Schreiner, *Chem. Eur. J.* **2005**, *11*, 7091–7101.

28 H. F. Bettinger, P. v. R. Schleyer, P. R. Schreiner, H. F. Schaefer III, in *Modern Electronic Structure Theory and Applications in Organic Chemistry*, E. R. Davidson (ed.), World Scientific, River Edge, NJ, **1997**, pp. 89–171.

29 H. F. Bettinger, P. R. Schreiner, P. v. R. Schleyer, H. F. Schaefer, in *The Encyclopedia of Computational Chemistry*, P. v. R. Schleyer, N. L. Allinger, T. Clark, J. Gasteiger, P. A. Kollman, H. F. Schaefer, P. R. Schreiner (eds.), John Wiley & Sons, Chichester, **1998**, pp. 183–196.

30 T. Clark, J. Chandrasekhar, G. W. Spitznagel, P. v. R. Schleyer, *J. Comput. Chem.* **1983**, *4*, 294–301.

31 J. S. Binkley, J. A. Pople, W. J. Hehre, *J. Am. Chem. Soc.* **1980**, *102*, 939–947; P. C. Hariharan, J. A. Pople, *Theor. Chim. Acta* **1973**, *28*, 213–218; W. J. Hehre, R. F. Stewart, J. A. Pople, *J. Chem. Phys.* **1969**, *51*, 2657–2664.

32 T. H. Dunning Jr., K. A. Peterson, D. E. Woon, in *The Encyclopedia of Computational Chemistry*, Vol. 1, P. v. R. Schleyer, N. L. Allinger, T. Clark, J. Gasteiger, P. A. Kollman, H. F. Schaefer, P. R. Schreiner (eds.), John Wiley & Sons, Chichester, **1998**, pp. 88–115.

33 D. Cremer, in *The Encyclopedia of Computational Chemistry*, Vol. 3, P. v. R. Schleyer, N. L. Allinger, T. Clark, J. Gasteiger, P. A. Kollman, H. F. Schaefer III, P. R. Schreiner (eds.), John Wiley & Sons, Chichester, **1998**, pp. 1706–1735; C. Møller, M. S. Plesset, *Phys. Rev.* **1934**, *46*, 618–622.

34 C. D. Sherrill, T. J. Van Huis, Y. Yamaguchi, H. F. Schaefer, *J. Mol. Struct. (Theochem)* **1997**, *400*, 139–156.

35 M. Hanrath, B. Engels, *Chem. Phys.* **1997**, *225*, 197–202.

36 B. Engels, M. Hanrath, C. Lennartz, *Comput. Chem.* **2001**, *25*, 15–38.

37 H. Lischka, R. Shepard, F. B. Brown, I. Shavitt, *Int. J. Quant. Chem.* **1981**, *515*, 91–100; H. J. Werner, P. J. Knowles, *J. Chem. Phys.* **1991**, *94*, 1264–1270.

38 J. Cizek, *J. Chem. Phys.* **1966**, *45*, 4256–4266; G. D. Purvis, R. J. Bartlett, *J. Chem. Phys.* **1981**, *75*, 1284–1292; M. Urban, J. Noga, S. J. Cole, R. J. Bartlett, *J. Chem. Phys.* **1985**, *83*, 4041–4046; R. J. Bartlett, J. Noga, *Chem. Phys. Lett.* **1988**, *150*, 29–36; R. J. Bartlett, J. D. Watts, S. A. Kucharski, J. Noga, *Chem. Phys. Lett.* **1990**, *165*, 513–522.

39 J. Gauss, in *The Encyclopedia of Computational Chemistry*, P. v. R. Schleyer, N. L. Allinger, T. Clark, J. Gasteiger, P. A. Kollman, H. F.

Schaefer III, P. R. Schreiner (eds.), John Wiley & Sons, Chichester, **1998**, pp. 615–636.

40 J. L. Ozment, A. M. Schmiedekamp, L. A. Schultz-Merkel, R. H. Smith, C. J. Michejda, *J. Am. Chem. Soc.* **1991**, *113*, 397–405; K. R. Bates, A. D. Daniels, G. E. Scuseria, *J. Chem. Phys.* **1998**, *109*, 3308–3312; N. Flocke, R. J. Bartlett, *J. Chem. Phys.* **2004**, *121*, 10935–10944; T. Korona, K. Pfluger, H. J. Werner, *Phys. Chem. Chem. Phys.* **2004**, *6*, 2059–2065; M. Schutz, F. R. Manby, *Phys. Chem. Chem. Phys.* **2003**, *5*, 3349–3358; H. J. Werner, F. R. Manby, P. J. Knowles, *J. Chem. Phys.* **2003**, *118*, 8149–8160; M. Schutz, *Phys. Chem. Chem. Phys.* **2002**, *4*, 3941–3947; W. Z. Liang, Y. H. Shao, C. Ochsenfeld, A. T. Bell, M. Head-Gordon, *Chem. Phys. Lett.* **2002**, *358*, 43–50; M. Schutz, *J. Chem. Phys.* **2002**, *116*, 8772–8785; S. Li, J. Ma, Y. Jiang, *J. Comput. Chem.* **2002**, *23*, 237–244; D. A. Mazziotti, *J. Chem. Phys.* **2001**, *115*, 8305–8311; M. Schutz, H. J. Werner, *J. Chem. Phys.* **2001**, *114*, 661–681; M. Schutz, *J. Chem. Phys.* **2000**, *113*, 9986–10001; K. N. Kudin, G. E. Scuseria, *Phys. Rev. B* **2000**, *61*, 16440–16453; G. E. Scuseria, P. Y. Ayala, *J. Chem. Phys.* **1999**, *111*, 8330–8343.

41 T. J. Lee, A. P. Rendell, P. R. Taylor, *J. Phys. Chem.* **1990**, *94*, 5463–5468; T. J. Lee, J. E. Rice, G. E. Scuseria, H. F. Schaefer, *Theoret. Chim. Acta* **1989**, *75*, 81–98.

42 E. J. Duplock, M. Scheffler, P. J. D. Lindan, *Phys. Rev. Lett.* **2004**, *92*, 225502-225502-225510.

43 A. D. Becke, *J. Chem. Phys.* **1993**, *98*, 1372–1377; A. D. Becke, *J. Chem. Phys.* **1993**, *98*, 5648–5652.

44 A. D. Becke, *Phys. Rev. A* **1988**, *38*, 3098–3100.

45 B. J. Lynch, D. G. Truhlar, *J. Phys. Chem.* **2001**, *105*, 2936–2941.

46 B. I. Dunlap, J. C. Boettger, *J. Phys. B: At. Mol. Phys.* **1996**, *29*, 4907–4913.

47 D. A. Plattner, K. N. Houk, *J. Am. Chem. Soc.* **1995**, *117*, 4405–4406.

48 H. L. Woodcock, H. F. Schaefer, P. R. Schreiner, *J. Phys. Chem. A* **2002**, *106*, 11923–11931.

49 A. E. Pomerantz, J. H. Han, C. B. Musgrave, *J. Phys. Chem. A* **2004**, *108*, 4030–4035.

50 C. Liang, L. C. Allen, *J. Am. Chem. Soc.* **1991**, *113*, 1873–1878.

51 C. Liang, Y. Xie, H. F. Schaefer, K. S. Sim, H. S. Kim, *J. Am. Chem. Soc.* **1991**, *113*, 2452–2459.

52 O. V. Gritsenko, B. Ensing, P. R. T. Schipper, E. J. J. Baerends, *J. Phys. Chem. A* **2000**, *104*, 8558–8565.

53 J. K. Kang, C. B. Musgrave, *J. Chem. Phys.* **2001**, *115*, 11040–11051.

54 J. K. Kang, C. B. Musgrave, *J. Chem. Phys.* **2002**, *116*, 9907–9913; J. K. Kang, *J. Appl. Phys.* **2002**, *91*, 34083414; J. P. Senosiain, C. B. Musgrave, D. M. Golden, *J. Phys. Chem. A* **2001**, *105*, 1669–1675.

55 F. Diederich, Y. Rubin, C. B. Knobler, R. L. Whetten, K. E. Schriver, K. N. Houk, Y. Li, *Science* **1989**, *245*, 1088–1090.

56 P. R. Taylor, E. Bylaska, J. H. Weare, R. Kawai, *Chem. Phys. Lett.* **1995**, *235*, 558–563; M. C. Domene, P. W. Fowler, D. Mitchell, G. Seifert, F. Zerbetto, *J. Phys. Chem. A* **1997**, *101*, 8339–8344; Y. L. An, S. L. Qian, *J. Mol. Struct. (THEOCHEM)* **1998**, *432*, 115–120.

57 S. Grimme, C. Muck-Lichtenfeld, *ChemPhysChem* **2002**, *3*, 207–209.

58 Z. Chen, T. Heine, H. Jiao, A. Hirsch, W. Thiel, P. v. R. Schleyer, *Chem. Eur. J.* **2004**, *10*, 963–970.

59 H. Prinzbach, A. Weller, P. Landenberger, F. Wahl, J. Worth, L. T. Scott, M. Gelmont, D. Olevano, B. von Issendorff, *Nature* **2000**, *407*, 60–63.

60 M. Saito, Y. Miyamoto, *Phys. Rev. B* **2002**, *65*, 165434/165431–165434/165410; M. Saito, Y. Miyamoto, *Phys. Rev. Lett.* **2001**, *87*, 035503/035501–035503/035504.

61 J. Lu, S. Re, Y. Choe, S. Nagase, Y. Zhou, R. Han, L. Peng, X. Zhang, X. Zhao, *Phys. Rev. B* **2003**, *67*, 125415-125415-125417.

62 F. A. Gianturco, G. Y. Kashenock,

R. R. Lucchese, N. Sanna, *J. Chem. Phys.* **2002**, *116*, 2811–2824.

63 A. Castro, M. A. L. Marques, J. A. Alonso, G. F. Bertsch, K. Yabana, A. Rubio, *J. Chem. Phys.* **2002**, *116*, 1930–1933.

64 A. H. Romero, D. Sebastiani, R. Ramirez, M. Kiwi, *Chem. Phys. Lett.* **2002**, *366*, 134.

65 Z. Slanina, F. Uhlik, *Fullerenes Nanotub. Carb. Nanostruct.* **2002**, *10*, 207–215; Z. Slanina, X. Zhao, F. Uhlik, *Mater. Sci. Eng., B* **2002**, *96*, 164–168.

66 S. Sokolova, A. Luchow, J. B. Anderson, *Chem. Phys. Lett.* **2000**, *323*, 229–233.

67 C. Allison, K. A. Beran, *J. Mol. Struct. (THEOCHEM)* **2004**, *680*, 59–63.

68 K. Raghavachari, D. L. Strout, G. K. Odom, G. E. Scuseria, J. A. Pople, B. G. Johnson, P. M. W. Gill, *Chem. Phys. Lett.* **1993**, *214*, 357–361.

69 S. Grimme, M. Waletzke, *Phys. Chem. Chem. Phys.* **2000**, *2*, 2075–2081; R. B. Murphy, R. P. Messmer, *Chem. Phys. Lett.* **1991**, *183*, 443–448.

70 R. R. Jones, R. G. Bergman, *J. Am. Chem. Soc.* **1972**, *94*, 660–661; R. G. Bergman, *Acc. Chem. Res.* **1973**, *6*, 25–31.

71 A. G. Myers, E. Y. Kuo, N. S. Finney, *J. Am. Chem. Soc.* **1989**, *111*, 8057–8059; R. Nagata, H. Yamanaka, E. Okazaki, I. Saito, *Tetrahedron Lett.* **1989**, *30*, 4995–4998.

72 T. Gillmann, T. Hülsen, W. Massa, S. Wocadlo, *Synlett* **1995**, 1257–1259; M. Schmittel, M. Strittmatter, S. Kiau, *Tetrahedron Lett.* **1995**, *36*, 4975–4978.

73 K. C. Nicolaou, P. Maligres, J. Shin, E. de Leon, D. Rideout, *J. Am. Chem. Soc.* **1990**, *112*, 7825–7826; K. C. Nicolaou, W.-M. Dai, *Angew. Chem.* **1991**, *103*, 1453–1481; M. E. Maier, *Synlett* **1995**, 13–26; A. L. Smith, K. C. Nicolaou, *J. Med. Chem.* **1996**, *39*, 2103–2117.

74 A. G. Myers, P. S. Dragovich, *J. Am. Chem. Soc.* **1992**, *114*, 5859–5860; M. Schmittel, M. Maywald, M. Strittmatter, *Synlett* **1997**, 165–166;

M. F. Semmelhack, Y. Gu, D. M. Ho, *Tetrahedron Lett.* **1997**, *38*, 5583–5586.

75 A. C. Scheiner, H. F. Schaefer III, B. Liu, *J. Am. Chem. Soc.* **1989**, *111*, 3118–3124; E. Kraka, D. Cremer, *J. Am. Chem. Soc.* **1994**, *116*, 4929–4936; R. Lindh, B. J. Persson, *J. Am. Chem. Soc.* **1994**, *116*, 4963–4969; P. G. Wenthold, R. R. Squires, *J. Am. Chem. Soc.* **1994**, *116*, 6401–6412; R. Lindh, T. J. Lee, A. Bernhardsson, B. J. Persson, G. Karlström, *J. Am. Chem. Soc.* **1995**, *117*, 7186–7194; W.-C. Chen, N.-y. Chang, C.-h. Yu, *J. Phys. Chem. A* **1998**, *102*, 2584–2593; C. J. Cramer, *J. Am. Chem. Soc.* **1998**, *120*, 6261–6269; E. Kraka, D. Cremer, *J. Am. Chem. Soc.* **2000**, *122*, 8245–8264; F. Stahl, D. Moran, P. v. R. Schleyer, M. Prall, P. R. Schreiner, *J. Org. Chem.* **2002**, *67*, 1453–1461; N. Koga, K. Morokuma, *J. Am. Chem. Soc.* **1991**, *113*, 1907–1911; M. Winkler, H. H. Wenk, W. Sander, in *Reactive Intermediates*, R. A. Moss, M. S. Platz, M. Jones Jr. (eds.), Wiley-Interscience, Hoboken, New Jersey, **2004**, pp. 741–794.

76 P. R. Schreiner, *J. Am. Chem. Soc.* **1998**, *120*, 4184–4190.

77 J. M. Galbraith, P. R. Schreiner, N. Harris, W. Wei, A. Wittkopp, S. Shaik, *Chem. Eur. J.* **2000**, *6*, 1446–1454.

78 J. Gräfenstein, A. M. Hjerpe, E. Kraka, D. Cremer, *J. Phys. Chem. A* **2000**, *104*, 1748–1761.

79 H. H. Wenk, M. Winkler, W. Sander, *Angew. Chem. Int. Ed.* **2003**, *42*, 502–528.

80 M. Prall, A. Wittkopp, P. R. Schreiner, *J. Phys. Chem. A* **2001**, *105*, 9265–9274.

81 B. König, W. Pitsch, M. Klein, R. Vasold, M. Prall, P. R. Schreiner, *J. Org. Chem.* **2001**, *66*, 1742–1746.

82 P. R. Schreiner, M. Prall, *J. Am. Chem. Soc.* **1999**, *121*, 8615–8627.

83 D. M. Bowles, G. J. Palmer, C. A. Landis, J. L. Scott, J. E. Anthony, *Tetrahedron* **2001**, *57*, 3753–3760.

84 A. Navarro-Vazquez, M. Prall, P. R. Schreiner, *Org. Lett.* **2004**, *6*, 2981–2984.

85 P. R. Schreiner, A. Navarro-Vázquez, M. Prall, *Acc. Chem. Res.* **2005**, *38*, 29–37.
86 K. N. Houk, Y. Li, J. D. Evanseck, *Angew. Chem. Int. Ed. Engl.* **1992**, *31*, 682–708; H. Hopf, *Classics in Hydrocarbon Chemistry*, Wiley-VCH, Weinheim, **2000**.
87 P. R. Schreiner, W. L. Karney, P. v. R. Schleyer, W. T. Borden, T. P. Hamilton, H. F. Schaefer, *J. Org. Chem.* **1996**, *61*, 7030–7039.
88 B. Engels, M. Hanrath, *J. Am. Chem. Soc.* **1998**, *120*, 6356–6361.
89 S. Koseki, Y. Fujimura, M. Hirama, *J. Phys. Chem. A* **1999**, *103*, 7672–7675.
90 J. D. Goddard, X. Chen, G. Orlova, *J. Phys. Chem. A* **1999**, *103*, 4078–4084.
91 V. Polo, E. Kraka, D. Cremer, *Theor. Chem. Acc.* **2002**, *107*, 291–303.
92 S. P. de Visser, M. Filatov, P. R. Schreiner, S. Shaik, *Eur. J. Org. Chem.* **2003**, 4199–4204; M. Prall, A. Krüger, P. R. Schreiner, H. Hopf, *Chem. Eur. J.* **2001**, *7*, 4386–4394; P. R. Schreiner, *Chem. Commun.* **1998**, 483–484; M. Prall, P. R. Schreiner, in *High-Performance Computing*, Springer, Heidelberg, **2000**, pp. 194–211.
93 S. P. Kawatkar, P. R. Schreiner, *Org. Lett.* **2002**, *4*, 3643–3646; M. Prall, A. Wittkopp, A. A. Fokin, P. R. Schreiner, *J. Comput. Chem.* **2001**, *22*, 1605–1614.
94 W. Sander, M. Exner, M. Winkler, A. Balster, A. Hjerpe, E. Kraka, D. Cremer, *J. Am. Chem. Soc.* **2002**, *124*, 13072–13079.
95 J. J. Nash, K. E. Nizzi, A. Adeuya, M. J. Yurkovich, C. J. Cramer, H. I. Kenttaemaa, *J. Am. Chem. Soc.* **2005**, *127*, 5760–5761.
96 M. Winkler, W. Sander, *J. Phys. Chem. A* **2001**, *105*, 10422–10432.
97 T. D. Crawford, E. Kraka, J. F. Stanton, D. Cremer, *J. Chem. Phys.* **2001**, *114*, 10638–10650.
98 T. H. Dunning Jr., *J. Chem. Phys.* **1989**, *90*, 1007–1023.
99 T. D. Crawford, T. J. Lee, N. C. Handy, H. F. Schaefer III, *J. Chem. Phys.* **1997**, *107*, 9980–9984; T. D. Crawford, J. F. Stanton, W. D. Allen, H. F. Schaefer, *J. Chem. Phys.* **1997**, *107*, 10626–10632.
100 P. R. Schreiner, M. Prall, V. Lutz, *Angew. Chem. Int. Ed.* **2003**, *42*, 5757–5760.
101 B. H. Bui, P. R. Schreiner, *Org. Lett.* **2003**, *5*, 4871–4874.
102 M. Yoon, S. Han, G. Kim, S. B. Lee, S. Berber, E. Osawa, J. Ihm, M. Terrones, F. Banhart, J.-C. Charlier, N. Grobert, H. Terrones, P. M. Ajayan, D. Tomanek, *Phys. Rev. Lett.* **2004**, *92*, 015504–015504–015504; H. T. Yang, L. Yang, J. Chen, J. Dong, *Phys. Lett. A* **2004**, *325*, 287–293; Z. Slanina, F. Uhlik, L. Adamowicz, S. L. Lee, *Int. J. Quantum Chem.* **2004**, *99*, 634–639; Y. Miyamoto, A. Rubio, S. Berber, M. Yoon, D. Tomanek, *Phys. Rev. B* **2004**, *69*, 121413–121413–121414; R. W. Alder, J. N. Harvey, *J. Am. Chem. Soc.* **2004**, *126*, 2490–2494; L. G. Zhou, S. Q. Shi, *Appl. Phys. Lett.* **2003**, *83*, 1222–1224; Y. F. Zhao, Y. Lin, B. I. Yakobson, *Phys. Rev. B* **2003**, *68*, 233403–233403–233404; H. F. Bettinger, B. I. Yakobson, G. E. Scuseria, *J. Am. Chem. Soc.* **2003**, *125*, 5572–5580; G. G. Samsonidze, B. I. Yakobson, *Comput. Mater. Sci.* **2002**, *23*, 62–72; P. I. Dosa, A. Schleifenbaum, K. P. C. Vollhardt, *Org. Lett.* **2001**, *3*, 1017–1020; D. Babic, N. Trinajstic, *J. Mol. Graph.* **2001**, *19*, 210–215; Z. Slanina, X. Zhao, F. Uhlik, M. Ozawa, E. Osawa, *J. Organomet. Chem.* **2000**, *599*, 57–61; H. Ueno, S. Osawa, E. Osawa, K. Takeuchi, *Fullerene Sci. Technol.* **1998**, *6*, 319–338; E. Osawa, Z. Slanina, K. Honda, *Mol. Cryst. Liq. Cryst. Sci. Technol. Sect. C-Mol. Mater.* **1998**, *10*, 1–8; T. Y. Astakhova, G. A. Vinogradov, *Fullerene Sci. Technol.* **1998**, *6*, 1037–1055; P. A. Marcos, M. J. Lopez, A. Rubio, J. A. Alonso, *Chem. Phys. Lett.* **1997**, *273*, 367–370; M. Terrones, H. Terrones, *Fullerene Sci. Technol.* **1996**, *4*, 517–533; E. Osawa, K. Honda, *Fullerene Sci. Technol.* **1996**, *4*, 939–961; D.

Mitchell, P. W. Fowler, F. Zerbetto, *J. Phys. B-At. Mol. Opt. Phys.* **1996**, *29*, 4895–4906; P. W. Fowler, D. E. Manolopoulos, G. Orlandi, F. Zerbetto, *J. Chem. Soc., Faraday Trans.* **1995**, *91*, 1421–1423; D. Babic, S. Bassoli, M. Casartelli, F. Cataldo, A. Graovac, O. Ori, B. York, *Mol. Simul.* **1995**, *14*, 395–401; S. J. Austin, P. W. Fowler, D. E. Manolopoulos, F. Zerbetto, *Chem. Phys. Lett.* **1995**, *235*, 146–151.

103 R. Chauvin, *Tetrahedron Lett.* **1995**, *36*, 397–400; R. Chauvin, *Tetrahedron Lett.* **1995**, *36*, 401–404.

104 M. Schreiber, J. Anthony, F. Diederich, M. E. Spahr, R. Nesper, M. Hubrich, F. Bommeli, L. DeGiorgi, C. Bosshard, P. Gunter, M. Colussi, U. W. Suter, C. Boudon, J.-P. Gisselbrecht, M. Gross, *Adv. Mater.* **1994**, *6*, 786–790.

105 Y. Zhao, R. McDonald, R. R. Tykwinski, *J. Org. Chem.* **2002**, *67*, 2805–2812; E. Burri, F. Diederich, M. Bronsted-Nielsen, *Helv. Chim. Acta* **2002**, *85*, 2169–2182.

106 L. T. Scott, G. J. DeCicco, J. L. Hyun, G. Reinhardt, *J. Am. Chem. Soc.* **1985**, *107*, 6546–6555; L. T. Scott, G. J. DeCicco, J. L. Hyun, G. Reinhardt, *J. Am. Chem. Soc.* **1983**, *105*, 7760–7761; P. J. Stang, F. Diederich, *Modern Acetylene Chemistry*, VCH, Weinheim, **1996**; L. Maurette, C. Godard, S. Frau, C. Lepetit, M. Soleilhavoup, R. Chauvin, *Chem. Eur. J.* **2001**, *7*, 1165–1170; M. Brake, V. Enkelmann, U. H. F. Bunz, *J. Org. Chem.* **1996**, *61*, 1190; M. J. S. Dewar, M. K. Holloway, *J. Chem. Soc. Chem. Commun.* **1984**, 1188–1193; K. N. Houk, L. T. Scott, N. G. Rondan, D. C. Spellmeyer, J. L. Hyun, G. Reinhardt, G. J. DeCicco, R. Weiss, H. M. O. Marle, L. S. Bass, J. Clardy, F. S. Jorgensen, T. A. Eaton, V. Sarkosi, C. M. Petit, L. Nq, K. D. Jordan, *J. Am. Chem. Soc.* **1985**, *107*, 6556–6562.

107 C. Lepetit, B. Silvi, R. Chauvin, *J. Phys. Chem. A* **2003**, *107*, 464–473.

108 L. T. Scott, M. J. Cooney, D. W. Rogers, K. Dejroongruang, *J. Am. Chem. Soc.* **1988**, *110*, 7244; L. T. Scott, M. J. Cooney, in *Modern Acetylene Chemistry*, P. J. Stang, F. Diederich (eds.), **1995**, pp. 321–351; L. J. Schaad, B. A. Hess Jr., L. T. Scott, *J. Phys. Org. Chem.* **1993**, *6*, 316–321.

109 R. V. Williams, H. A. Kurtz, *Adv. Phys. Org. Chem.* **1994**, *29*, 273–331.

110 H. Jiao, N. J. R. v. E. Hommes, P. von Ragué Schleyer, A. de Meijere, *J. Org. Chem.* **1996**, *61*, 2826–2828.

111 A. D. Becke, K. E. Edgecombe, *J. Chem. Phys.* **1990**, *92*, 5397–5403.

112 C. Godard, C. Lepetit, R. Chauvin, *Chem. Commun.* **2000**, 1833–1834; C. Lepetit, C. Godard, R. Chauvin, *New J. Chem.* **2001**, *25*, 572–580.

113 C. Lepetit, M. B. Nielsen, F. Diederich, R. Chauvin, *Chem. Eur. J.* **2003**, *9*, 5056–5066.

114 P. Manini, W. Amrein, V. Gramlich, F. Diederich, *Angew. Chem. Int. Ed.* **2002**, *41*, 4339–4342.

115 P. D. Jarowski, F. Diederich, K. N. Houk, *J. Org. Chem.* **2005**, *70*, 1671–1678.

116 D. J. Cram, J. M. Cram, *Container Molecules and Their Guests*, The Royal Society of Chemistry, Cambridge, **1994**.

117 W. J. Hehre, L. Radom, P. v. R. Schleyer, J. A. Pople, *Ab Initio Molecular Orbital Theory*, Wiley Interscience, New York, **1986**.

118 A. A. Fokin, B. A. Tkachenko, P. A. Gunchenko, P. R. Schreiner, *Angew. Chem. Int. Ed.* **2005**, *44*, 146–149.

119 L. J. Mathias, J. J. Jensen, V. T. Reichert, C. M. Lewis, G. L. Tullos, in *Step-Growth Polymers for High-Performance Materials*, Vol. 624, J. L. Hedrick, J. W. Labodie (eds.), American Chemical Society, Washington, **1996**, pp. 197–207.

120 D. Ranganathan, V. Haridas, K. P. Madhusudanan, R. Roy, R. Nagaraj, G. B. John, M. B. Sukhaswami, *Angew. Chem. Int. Ed. Engl.* **1996**, *35*, 1105–1107; D. Ranganathan, V. Haridas, R. Nagaraj, I. L. Karle, *J. Org. Chem* **2000**, *65*, 4415–4422.

121 M. Zaiser, F. Banhart, *Phys. Rev.*

Lett. **1997**, *79*, 3680–3683; F. Banhart, P. M. Ajayan, *Nature* **1996**, *382*, 433–435.

122 P. v. R. Schleyer, *J. Am. Chem. Soc.* **1957**, *79*, 3292.

123 C. Cupas, P. v. R. Schleyer, D. J. Trecker, *J. Am. Chem. Soc.* **1965**, *87*, 917–918.

124 T. M. Gund, P. v. R. Schleyer, G. D. Unruh, G. J. Gleicher, *J. Org. Chem.* **1974**, *39*, 2995–3003.

125 V. Z. Williams, P. v. R. Schleyer, G. J. Gleicher, L. B. Rodewald, *J. Am. Chem. Soc.* **1966**, *88*, 3862–3863.

126 W. Burns, T. R. B. Mitchell, M. A. McKervey, J. J. Rooney, G. Ferguson, P. Roberts, *J. Chem. Soc., Chem. Commun.* **1976**, 893–895.

127 R. Lin, Z. A. Wilk, *Fuel* **1995**, *74*, 1512–1521.

128 G. C. McIntosh, M. Yoon, S. Berber, D. Tomanek, *Phys. Rev. B* **2004**, *70*, 045401–045401–045408.

129 F. Hollowood, A. Karim, M. A. McKervey, P. McSweeney, *J. Chem. Soc., Chem. Commun.* **1978**, 306–308.

130 A. A. Fokin, P. R. Schreiner, *Chem. Rev.* **2002**, *102*, 1551–1593.

131 T. Clark, T. M. Knox, M. A. McKervey, H. Mackle, J. J. Rooney, *J. Am. Chem. Soc.* **1979**, *101*, 2404–2410.

132 P. v. R. Schleyer, J. E. Williams, K. R. Blanchard Jr., *J. Am. Chem. Soc.* **1970**, *92*, 2377–2386.

133 E. Osawa, E. M. Engler, S. A. Goleski, Y. Inamoto, G. J. Kent, M. Kuesch, P. v. R. Schleyer, *J. Org. Chem.* **1980**, *45*, 984–991.

134 H. Stetter, C. Wulff, *Chem. Ber.* **1960**, *93*, 1366–1371.

135 M. Finn, R. Friedline, N. K. Suleman, C. J. Wohl, J. M. Tanko, *J. Am. Chem. Soc.* **2004**, *126*, 7578–7584.

136 P. R. Schreiner, A. A. Fokin, P. v. R. Schleyer, H. F. Schaefer III, in *Conceptual Trends in Quantum Chemistry*, E. Kryachko (ed.), Kluwer, Dordrecht, **2003**, pp. 369–386; P. R. Schreiner, A. A. Fokin, *Chem. Rec.* **2004**, *3*, 247–257; P. R. Schreiner, A. Wittkopp, P. A. Gunchenko, A. I. Yaroshinsky, S. A. Peleshanko, A. A. Fokin, *Chem. Eur. J.* **2001**, *7*, 2739–2744.

137 A. A. Fokin, B. A. Tkachenko, T. E. Shubina, P. A. Gunchenko, D. V. Gusev, J. K. Vohs, G. H. Robinson, A. G. Yurchenko, P. R. Schreiner, *Eur. J. Org. Chem.* **2002**, 3844–3849; A. A. Fokin, T. E. Shubina, P. A. Gunchenko, S. D. Isaev, A. G. Yurchenko, P. R. Schreiner, *J. Am. Chem. Soc.* **2002**, *124*, 10718–10727; A. A. Fokin, P. R. Schreiner, P. A. Gunchenko, S. A. Peleshanko, T. E. Shubina, S. D. Isaev, P. V. Tarasenko, N. I. Kulik, H.-M. Schiebel, A. G. Yurchenko, *J. Am. Chem. Soc.* **2000**, *122*, 7317–7326; A. A. Fokin, P. A. Gunchenko, S. A. Peleshanko, P. v. R. Schleyer, P. R. Schreiner, *Eur. J. Org. Chem.* **1999**, 855–860.

138 M. J. S. Dewar, S. D. Worley, *J. Chem. Phys.* **1969**, *50*, 654–667.

139 A. A. Fokin, P. R. Schreiner, P. v. R. Schleyer, P. A. Gunchenko, *J. Org. Chem.* **1998**, *63*, 6494–6502.

140 T. Ishizone, H. Tajima, S. Matsuoka, S. Nakahama, *Tetrahedron Lett.* **2001**, *42*, 8645–8647.

141 P. R. Schreiner, A. A. Fokin, P. A. Gunchenko, **2005**, manuscript in preparation.

9
Fullerene Reactivity – Fullerene Cations and Open-Cage Fullerenes

Toshikazu Kitagawa, Yasujiro Murata, and Koichi Komatsu

9.1
Introduction

Among the carbon-rich compounds, which are attracting current interest for their great potential as functional materials, the spherical carbon clusters, fullerenes, and their organic derivatives can be considered the most typical.

During the past 15 years since C_{60} became available in macroscopic quantities in 1990 [1], a wide variety of its derivatives have been synthesized as part of the explosive development of the study of its chemistry [2]. Various organic reactions have been reported, most of which are cycloadditions, nucleophilic additions, and radical additions. Fullerenes, as represented by C_{60}, are now commonly accepted to behave as electron-deficient olefins, hence there have been numerous studies on their anions. This has led to a situation where the other equally important species, the fullerene cations, have been left unexplored for nearly a decade in spite of their significance in both fundamental and application studies. Clearly, a systematic study of this class of species is needed.

Another challenging and yet uncultivated subject in fullerene chemistry is that related to the organic synthesis of endohedral fullerenes. The development of the study on this class of compounds has been much retarded compared with the wealth of studies on the transformation of fullerenes on their outer surface. These particular fullerenes seem to be waiting for the approach of organic chemists.

In this chapter, these two most intriguing subjects in the current study of fullerene chemistry are discussed including the most up to date results of the authors' own research.

9.2
Functionalized Fullerene Cations

In the history of the chemistry of fullerene cations, the first preparation of a radical cation of C_{76}, a non-functionalized fullerene cation, was achieved by Reed et al. [3]. C_{76}, which has an oxidation potential of 0.81 V (vs. Fc/Fc^{+}) [4] can be oxidized by

Carbon-Rich Compounds. Edited by Michael M. Haley and Rik R. Tykwinski
Copyright © 2006 WILEY-VCH Verlag GmbH & Co. KGaA, Weinheim
ISBN: 3-527-31224-2

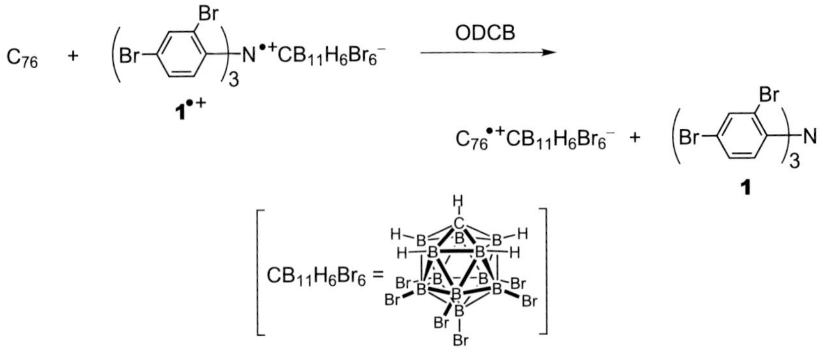

Scheme 9.1. Isolation of C_{76} radical cation salt.

the radical cation of tris(2,4-dibromophenyl)amine $1^{•+}$. By using a carborane anion, $CB_{11}H_6Br_6^-$, one of the least nucleophilic anions known to date, as the counteranion, the salt $C_{76}^{•+}CB_{11}H_6Br_6^-$ was isolated as a dark brown solid (Scheme 9.1).

C_{60} has a higher oxidation potential (1.26 V vs. Fc/Fc^+ [5]) but can be oxidized by the hexabromo-N-phenylcarbazole radical cation [6], which is a stronger oxidizing reagent. Although $C_{60}^{•+}CB_{11}H_6X_6^-$ (X = Cl or Br) is stable in solution, the isolation of a pure crystalline salt was difficult, due to the possible bimolecular disproportionation of $C_{60}^{•+}$ into C_{60}^{2+} and C_{60} at high concentrations.

9.2.1
Derivatized C_{60} and C_{70} Cations

Derivatized fullerene cations, such as RC_{60}^+ and RC_{70}^+, are formally produced by adding a cationic species R^+ to an sp^2 hybridized carbon of a fullerene core. Cations with additional attached groups, such as $R_nC_{60}^+$ (n: odd number) are also conceivable. Unlike radical cations, these cations are carbenium ions (trivalent carbocations) and have no unpaired electrons. Besides the addition of R^+ to C_{60} (method (a)), four other methods ((b)–(e)), shown in Scheme 9.2, are currently known for the generation of RC_{60}^+. As described in the following sections, these approaches have been successfully employed to produce the derivatized C_{60} and C_{70} cations, either as long-lived species or as reactive intermediates.

9.2.1.1 Generation of Derivatized C_{60} and C_{70} Cations as Long-lived Species
Monoalkylated (RC_{60}^+ and RC_{70}^+), monoprotonated (HC_{60}^+), and pentaarylated ($Ar_5C_{60}^+$) fullerene cations have thus far been prepared as stable solutions or isolated as solids using the methods shown in Scheme 9.2.

Monoalkylated C_{60} Cation (RC_{60}^+). Derivatized C_{60} cations with an alkyl group attached, RC_{60}^+, can be generated from the corresponding fullerenols 1,4-RC_{60}OH (**2a–d**) in strong acid (Scheme 9.3). Ionization of alcohols and alkyl halides in a superacid, such as magic acid (FSO_3H–SbF_5), is a well established method, devel-

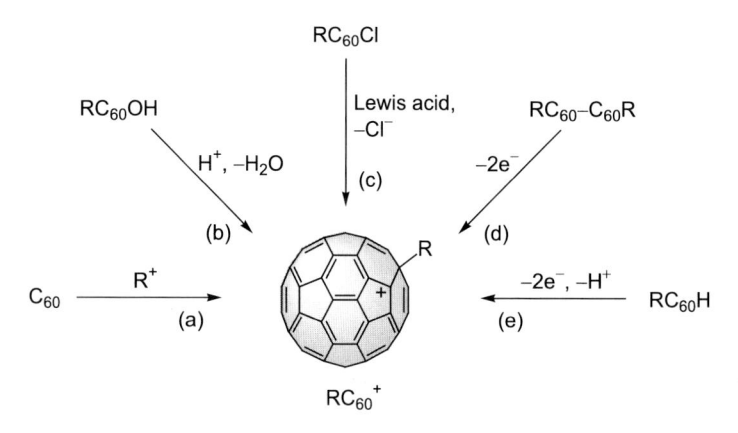

Scheme 9.2. Methods for the generation of RC_{60}^+.

Scheme 9.3. Generation of alkylated C_{60} cations by the ionization of fullerenol precursors.

a: R = CH_2Cl, b: R = $CHCl_2$,
c: R = CCl_3, d: R = CCl_2CH_2Cl

oped as the result of extensive research by Olah in the 1960s, for the generation of carbocations as long-lived species. The use of this method for preparing C_{60} cations, however, may suffer from the undesirable oxidation of the fullerene core, since superacids are typically strong oxidants [7, 8]. For example, when $RC_{60}OH$ was dissolved in FSO_3H-SbF_5 (1:1) a paramagnetic dark green solution was formed, which is most likely due to the generation of $C_{60}^{·+}$ via further oxidation of the initially formed RC_{60}^+. On the other hand, triflic acid (CF_3SO_3H), although a weaker acid, is suitable for the ionization of fullerenols because of its nonoxidizing nature. Thus, solutions of **2a–d**, which were synthesized by the method described in Section 9.2.1.2, in triflic acid give stable reddish purple solutions of cations **3a–d** [9], which show characteristic absorption maxima at around 500, 790, and 1200 nm. Quenching these solutions with methanol led to the regioselective bonding of a methoxy group to form 1,4-$RC_{60}OMe$.

The ^{13}C NMR spectra of **3a–d** show signals corresponding to a cationic center at δ 180.4, 175.6, 171.8, and 174.9, respectively. In each spectrum 30 sp^2 carbon signals are observed at δ 136–154. The intensity of the signal for the cationic center is half that of the majority of other sp^2 carbons, indicating that the molecule has C_s symmetry and that the positive carbon is adjacent to the unique sp^3 carbon on the C_{60} cage. The relatively high-field shift of these signals, compared with aryl-substituted carbenium ions (for example, the triphenylmethyl cation and 9-phenylfluorenyl cations show resonances at δ 211.6 [10] and 224.2 [11], respectively), suggests appreciable charge delocalization. In addition to the resonance delocalization over the C_{60} cage, the intramolecular coordination of the lone pair electrons of the chlorine atom in the R group to the cationic center is suggested. The ionization of **2b** was complete also in FSO_3H and H_2SO_4. The cationic center showed essentially the same chemical shifts in three acids with different acidities (acidity function H_0; FSO_3H, -15.1; CF_3SO_3H, -14.1; H_2SO_4, -12 [12]), showing that **3b** is free from contact ion pairing. The complete ionization of **2b** in H_2SO_4 means that **3b** is such a stable cation that no superacid is required for its generation.

Derivatized C_{60} cations can also be generated from the corresponding dimer or hydrofullerene by treatment with sufficiently strong oxidants (Scheme 9.2 (d) and (e)). This was demonstrated by the oxidation of dimer **4** and dihydrofullerene derivative **5** containing $CH_2P(O)(OEt)_2$ group(s) by sulfuric acid (Scheme 9.4) [13]. ^{13}C NMR spectra in H_2SO_4–CF_3SO_3H show the formation of a cation with C_s symmetry, which exhibits a cationic center at δ 174.67. The observation of ^{13}C–^{31}P coupling ($J = 6.6$ Hz) and a ^{31}P NMR signal at 7 ppm more down-field than the precursor **5** suggests the coordination of a lone pair of electrons of an oxygen atom to the cationic center to form a five-membered cyclic structure. Furthermore, based on a comparison of the observed ^{13}C chemical shifts with the GIAO-calculated values, it is likely that the phosphoryl oxygen is protonated to give a dication **6**. The formation of this cation from dimer **4** can be explained by a homolytic cleavage of the central C–C bond, the dissociation free energy of which is 13 kcal mol^{-1} (27 °C) [14], followed by a single-electron oxidation. In the reaction of **5**, initial single-electron oxidation occurs to form a radical cation **5$^{·+}$**, which is then deprotonated and further oxidized.

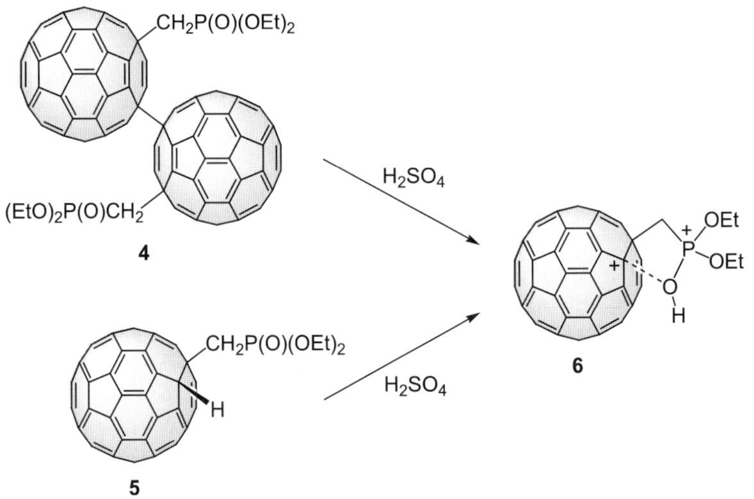

Scheme 9.4. Generation of a derivatized C_{60} cation in its protonated form by the oxidation of a dimer or a dihydrofullerene derivative in sulfuric acid.

9.2 Functionalized Fullerene Cations

Monoalkylated C_{70} Cation (RC_{70}^+). The first alkylated C_{70} cation $CHCl_2-C_{70}^+$ (**7**) was synthesized very recently by the ionization of a C_{70} derivative **8** containing a dichloromethyl and a hydroxy group at C-7 and C-23, respectively (Scheme 9.5) [15]. A solution of this alcohol in CF_3SO_3H at room temperature shows a reddish brown color, NMR analysis indicated complete conversion to a C_1-symmetric single isomer of cation $CHCl_2-C_{70}^+$. Among the five different carbons (A–E) on the C_{70} cage, the position of the $CHCl_2$ group in **7** was determined to be D on the basis of the symmetry requirement (C_1) and the calculated energies. Three low-field signals were observed at δ 198.16, 177.49, and 175.81, which were assigned to C-8, C-6, and C-21, respectively, based on DFT GIAO calculations. Other sp^2 carbons appeared at δ 128–156. Mulliken charges were shown to be mainly distributed on carbons close to the unique sp^3 carbon in the cage, as well as on the carbon atoms in region E, where the pyramidalization of the C–C bonds is relatively small.

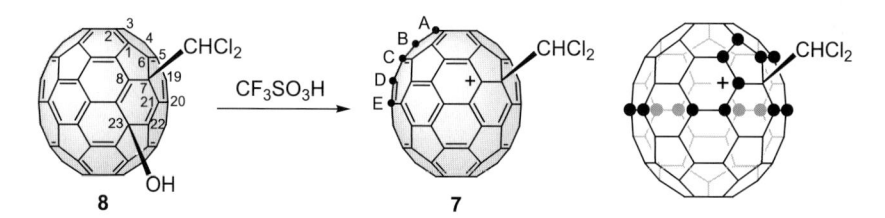

Scheme 9.5. Generation of an alkylated C_{70} cation. The plus sign designates the most downfield shifted carbon. Dots represent carbons possessing a Mulliken charge greater than 0.025 (B3LYP/6-31G*).

Protonated C_{60} (HC_{60}^+). Protonation of C_{60} produces the cation HC_{60}^+, which is the simplest form of functionalized C_{60} cation. By using a nonoxidizing superacid $H(CB_{11}H_6Cl_6)$, whose conjugate base is an exceptionally inert and weakly nucleophilic carborane anion, $CB_{11}H_6Cl_6^-$, Reed et al. prepared HC_{60}^+ as a stable cation both in solution and in the solid state (Scheme 9.6) [6]. In ODCB solution, HC_{60}^+ shows only a single ^{13}C NMR signal at 0.08 ppm more downfield relative to C_{60}. This suggests fluxional behavior due to rapid 1,2 shifts of a proton over the C_{60} cage. Proton migration is frozen in the solid state, as indicated by the observation of distinct signals of a positive sp^2 carbon and an sp^3 carbon at δ 182 and 56, respectively. Solid state $^{13}C-^{13}C$ 2D NMR measurements confirmed the direct connection of these carbon atoms [16], with the cationic center located at C-2 as in the case of chloroalkylated C_{60} cations **3a–d**.

Pentaarylated C_{60} Cations ($Ar_5C_{60}^+$). Fullerene cations with five attached aryl groups were generated by Birkett et al. by dechlorinating $Ar_5C_{60}Cl$ (**9a**: Ar = Ph, **9b**: Ar = 4-FC_6H_4) with $AlCl_3$ (Scheme 9.7) [17]. The ^1H and ^{13}C NMR spectra of the resulting purple-red solutions in $CS_2-CD_2Cl_2$ or $CDCl_3$ indicated the formation of a C_s-symmetric cation **10**, which can be explained by the initial formation

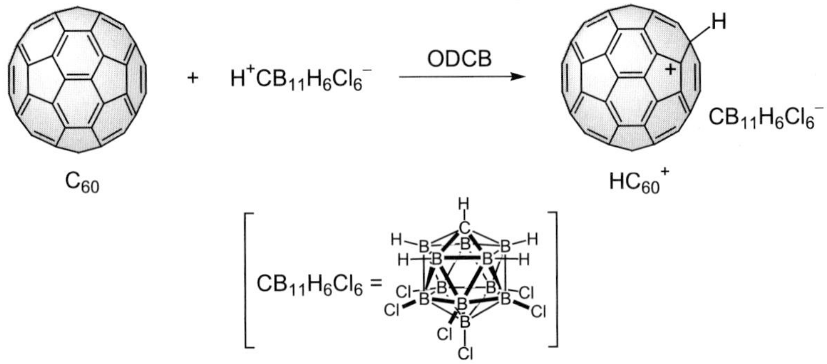

Scheme 9.6. Preparation of the hexachlorocarborane anion salt of HC_{60}^+.

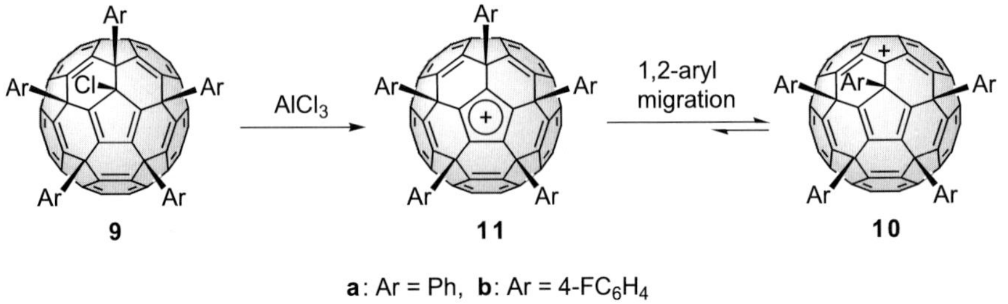

Scheme 9.7. Generation of $Ar_5C_{60}^+$.

of a cyclopentadienyl cation **11**, followed by a 1,2-aryl shift to avoid the unfavorable formation of an antiaromatic 4π electron system. Proton NOE experiments showed equal saturation transfer to all five aromatic rings of **10**, suggesting the exchange of the aryl groups due to an equilibrium between this cation and a small amount of **11**. In the ^{13}C NMR spectra, the cationic centers of **10a** and **10b** were observed at δ 173.74 and 171.67, respectively. These values are relatively high-field for a carbocationic center and likely reflect the contribution of this equilibrium.

NMR studies of ^3He-encapsulating compounds showed a reduced ^3He shielding in $^3He@10b$ compared to $^3He@9b$ [18], which implies that the π-system of the cation **10b** is less aromatic relative to its precursor **9b**. This can be explained by reduced delocalization in the six-membered rings (shaded rings in Scheme 9.8) adjacent to the cationic center as a result of the diminished resonance contribution of the antiaromatic cyclopentadienyl cation structure **10b′**.

9.2.1.2 Reactions via RC_{60}^+ and RC_{70}^+ Intermediates

The involvement of a fullerene cation as an intermediate has been proposed in some reactions. In the examples shown below, the cationic intermediates are generated either by an electrophilic addition to C_{60} (Scheme 9.2(a)), or by the ionization of a C_{60}–Cl bond (Scheme 9.2(c)).

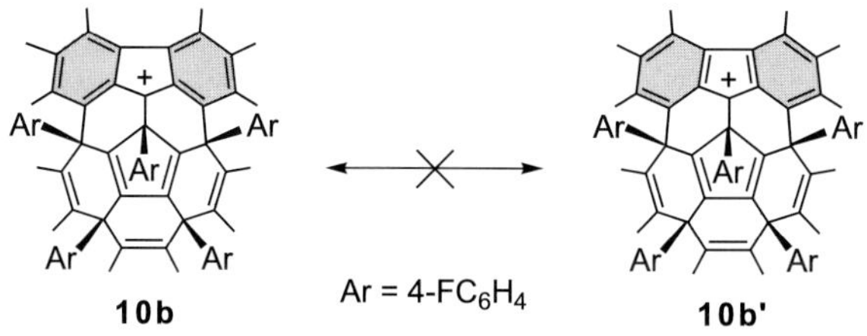

Scheme 9.8. Adverse resonance of $Ar_5C_{60}^+$.

Addition of Electrophiles to C_{60} and C_{70}. Although the formation of stable carbanions by a nucleophilic addition to C_{60} and C_{70} is well known, addition of electrophiles, which leads to derivatized C_{60} cations, is quite rare even though this type of addition commonly occurs in alkenes. Earlier reports on the reaction of C_{60} with $NO_2^+RCO_2^-$ (R = Ph, 4-BrC_6H_4) proposed the electrophilic addition of nitronium ions, which, after aqueous work-up, eventually gave polyhydroxylated C_{60} (Scheme 9.9(a)) [19]. When C_{60} is treated with $AlCl_3$ in benzene, a mixture of polyhydrophenylated C_{60} is obtained (Scheme 9.9(b)) [7, 20]. In this reaction, the initial protonation of C_{60} and subsequent phenylation of the intermediate cation HC_{60}^+ were proposed. The formation of multiadducts in both reactions clearly demonstrates the difficulties confronted in controlling the reaction.

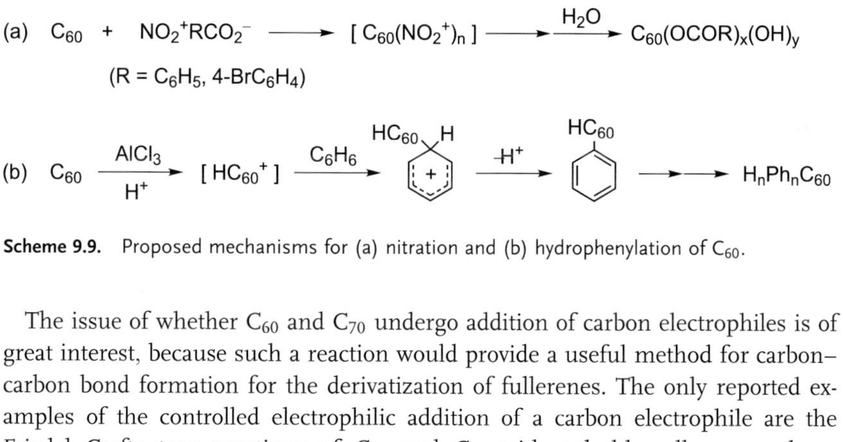

Scheme 9.9. Proposed mechanisms for (a) nitration and (b) hydrophenylation of C_{60}.

The issue of whether C_{60} and C_{70} undergo addition of carbon electrophiles is of great interest, because such a reaction would provide a useful method for carbon–carbon bond formation for the derivatization of fullerenes. The only reported examples of the controlled electrophilic addition of a carbon electrophile are the Friedel–Crafts type reactions of C_{60} and C_{70} with polychloroalkanes such as CH_2Cl_2, $CHCl_3$, CCl_4, and $Cl_2CHCHCl_2$, in the presence of a large excess of $AlCl_3$ (Scheme 9.10) [9, 21]. These chloroalkanes serve as the source of the electrophile as well as the solvent for the reaction. The electrophilic attack of a carbocationic complex, $R^+ \cdot AlCl_4^-$, forms an intermediate fullerene cation, RC_{60}^+ (**3a–d**) or $CHCl_2$-C_{70}^+ (**7**), which is then trapped by a chloride ion. When $Cl_2CHCHCl_2$ is used as the solvent, the initially formed cation $^+CHClCHCl_2$ (or its complex with $AlCl_4^-$) rearranges to $^+CCl_2CH_2Cl$, which is more stable due to the p(π) donation of unshared 3p electrons of the two chlorine atoms [22, 23].

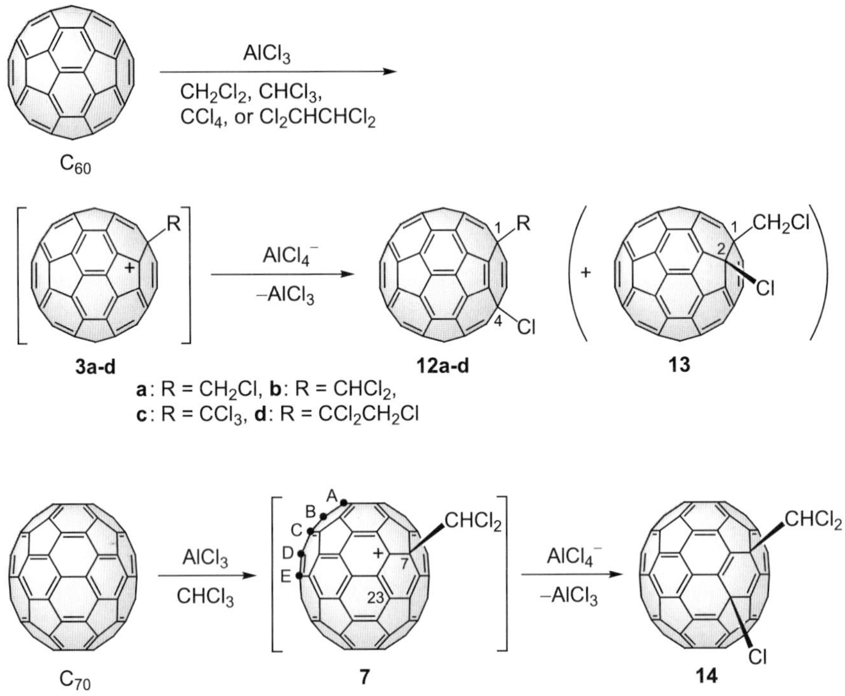

Scheme 9.10. Addition of polychloroalkanes to C_{60} and C_{70} under Friedel–Crafts conditions.

These addition reactions are highly regioselective. The reactions of C_{60} gave only the 1,4-adduct (**12b–d**), with an exception that a 1:1 mixture of 1,2- and 1,4-adducts (**12a** and **13**) was formed when CH_2Cl_2 was used as solvent [9, 21]. This is consistent with theoretical predictions. That is, among the possible isomers of $C_{60}H_2$ the 1,2-isomer is the most stable and the 1,4-isomer is the second most stable [24], whereas the order of calculated energies is reversed when the attached hydrogens are replaced with large groups [25]. The preferred formation of 1,4- and 1,2-isomers also occurs when nucleophiles are added to C_{60} [26]. On the other hand, C_{70} shows different addition patterns toward nucleophiles and electrophiles. Most nucleophilic additions [26f, 27, 28] and cycloadditions [27, 29] to C_{70} favor highly pyramidalized carbons A and B, whereas Cl_2CH^+ adds preferentially at carbon D, which can be attributed to the large HOMO coefficient of this carbon [15]. Trapping of the resulting cation **7** occurs exclusively at C-23, due to the large LUMO coefficient of this carbon as well as to the high stability of the resulting adduct **14** [15]. Adduct **14** is a derivative of 7,23-$C_{70}H_2$, which, on the basis of *ab initio* calculations, is the most stable derivative among the $C_{70}H_2$ isomers with nonadjacent hydrogen atoms [24a–c, 30].

Ionic Dissociation of $RC_{60}Cl$ and $RC_{70}Cl$. Trapping of an intermediate fullerene cation, generated by the heterolytic cleavage of RC_{60}–Cl or RC_{70}–Cl bonds under

9.2 Functionalized Fullerene Cations

various conditions, is a useful method for transformation to new fullerene derivatives. The heterolysis occurs either under solvolytic conditions in highly polar solvents or in the presence of an added Lewis acid.

A fact that demonstrates facile ionization of fullerene–Cl bonds is the S_N1 solvolysis reaction of **12a–d** and **14** in CF_3CH_2OH–anisole to form solvent substitution products **15–18** (Scheme 9.11) [15, 21], in which ionization occurs with the assistance of the highly polar CF_3CH_2OH. An alternative S_N2'-type mechanism, which involves a direct nucleophilic attack by a solvent molecule and the concomitant elimination of Cl^- (Scheme 9.11), is considered to be unimportant, on the basis of the poor nucleophilicity of the solvent and the absence of any reaction in pure anisole.

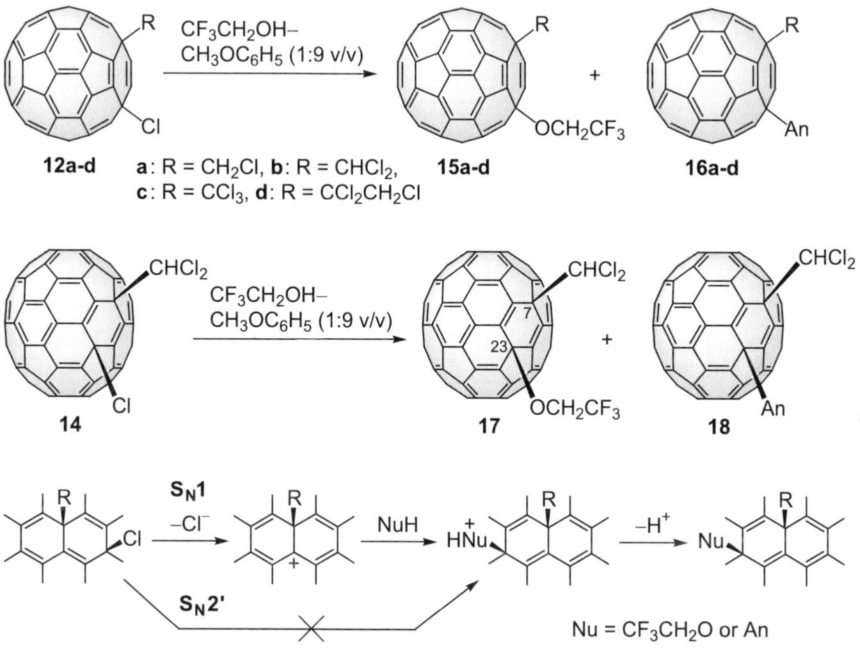

Scheme 9.11. Solvolysis of **12** and **14** and the proposed mechanism. An = $4\text{-}CH_3OC_6H_4$.

Similarly, **12a–d** and **14** are hydrolyzed cleanly to the corresponding fullerenols, **2a–d** and **8** (Scheme 9.12), when their solutions in CS_2 or benzene are passed through a column of silica gel [9, 15]. As previously described, these fullerenols are excellent precursors for the study of monoalkylated fullerene cations under stable-ion conditions.

Treatment of chlorinated fullerenes, such as **12a–d**, **14**, and **19**, with $AlCl_3$ or $FeCl_3$ in the presence of an aromatic compound gave the substitution products, as a result of the abstraction of a chloride ion and the subsequent electrophilic attack of the intermediate cation to the aromatic ring (Scheme 9.13) [21, 31]. The chloride

$$1,4\text{-}RC_{60}Cl\ (\mathbf{12a\text{-}d}) \xrightarrow[CS_2 \text{ or } C_6H_6]{H_2O,\ SiO_2} 1,4\text{-}RC_{60}OH\ (\mathbf{2a\text{-}d})$$

$$7,23\text{-}CHCl_2\text{-}C_{70}Cl\ (\mathbf{14}) \xrightarrow[CS_2]{H_2O,\ SiO_2} 7,23\text{-}CHCl_2\text{-}C_{70}OH\ (\mathbf{8})$$

Scheme 9.12. Silica gel assisted hydrolysis of **12** and **14**.

ion can also be abstracted from $RC_{60}Cl$ by Ag^+ [21]. Unlike the ionization using $AlCl_3$, these reactions can be carried out in the presence of compounds containing a hydroxy group. Thus, **15** was obtained by the treatment of **12** with $AgBF_4$ in the presence of CF_3CH_2OH.

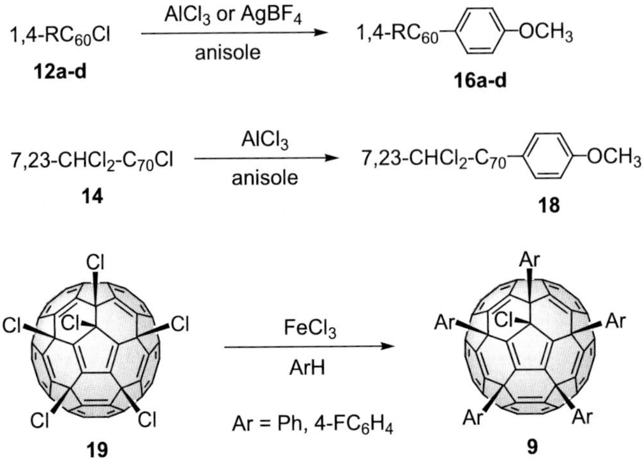

Scheme 9.13. Electrophilic substitution of aromatics by derivatized fullerene cations under Friedel–Crafts conditions.

9.2.1.3 Thermodynamic Stabilities of $RC_{60}{}^+$ and $RC_{70}{}^+$

The unexpected stabilities of the alkylated C_{60} and C_{70} cations (**3a–d**, **7**, $HC_{60}{}^+$, and **10**) allowed preparation of these cations either in a strong acid or in an organic solvent, which enabled them to be analyzed by NMR and UV–vis–NIR spectroscopy. A quantitative evaluation of the thermodynamic stabilities of **3a–d** and **7**, by two different methods, has been carried out.

The first approach to evaluating their stability involves measurements of the rates of solvolysis, as shown in Scheme 9.11 [15, 21]. Typically, the transition state of the rate-determining step of S_N1 reactions is assumed to closely resemble the intermediate ion pair, on the basis of the Hammond postulate [32]. Thus, the free energy of activation of this reaction, $\Delta G^{\ddagger}$, reflects the relative thermodynamic stabilities of the intermediate carbocations. The $\Delta G^{\ddag}$s for the solvolysis of **12a–d** and **14** in CF_3CH_2OH–anisole (1:9 v/v) were found to be close to each other, ranging from 23.8 to 25.0 kcal mol^{-1} at 25 °C. No significant difference was observed between **12b** and **14** ($\Delta G^{\ddag} = 24.0$ and 24.1 kcal mol^{-1}, respectively), which have the same addend but different cages. These $\Delta G^{\ddag}$s are only slightly lower than that observed for t-BuCl under the same conditions (26.9 kcal mol^{-1}), indicating that the monoalkylated C_{60} and C_{70} cations (**3a–d** and **7**) have approximately the same stability as that of the *tert*-butyl cation.

The second measure of cation stability is pK_R^+, which is defined as the negative logarithm of the equilibrium constant for hydrolysis (Scheme 9.14) and is correlated to the free energy difference between the cation and the corresponding alcohol. The generation of cations **3a–d** from fullerenols is complete in H_2SO_4, but the degree of ionization decreases when a weaker acid, such as p-toluenesulfonic acid, is added. By spectrometrically measuring equilibrium constants in mixtures of various ratios of H_2SO_4–TsOH, the pK_R^+ of **3a–d** were determined to be -15.7 to -15.8 [25]. These values are comparable to -16.4 [33a], -15.5 [33b], and -14.7 [33c] and -16.27 [34], observed for the *tert*-butyl cation and the tris(4-nitrophenyl)-methyl cation, respectively, in agreement with the result of the solvolysis rate measurements.

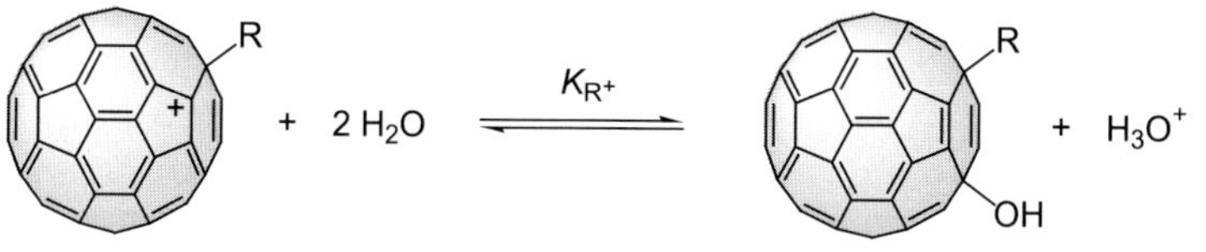

$$K_{R^+} = [RC_{60}OH][H_3O^+]/[RC_{60}^+]$$

$$pK_{R^+} = -\log K_{R^+}$$

Scheme 9.14. Reversible hydrolysis of alkylated C_{60} cations.

9.2.2
Aza[60]fullerene Cation $C_{59}N^+$

9.2.2.1 Generation of $C_{59}N^+$

Azafullerenes are the nitrogen analogs of fullerenes, in which one or more of the cage carbon atoms is substituted with nitrogen atom(s). The simplest members of this family, $C_{59}N$ and $C_{69}N$, are the only examples of heterofullerenes that have been prepared in bulk quantities and fully characterized. Owing to the presence of a nitrogen atom, the neutral forms of these fullerenes have a dangling bond, and consequently they exist in the form of radicals ($C_{59}N^{\bullet}$ or $C_{69}N^{\bullet}$) or their dimers. Two independent methods have been developed for the synthesis of $(C_{59}N)_2$ by Wudl [35, 36] and Hirsch [37]. Wudl proposed the involvement of cat-

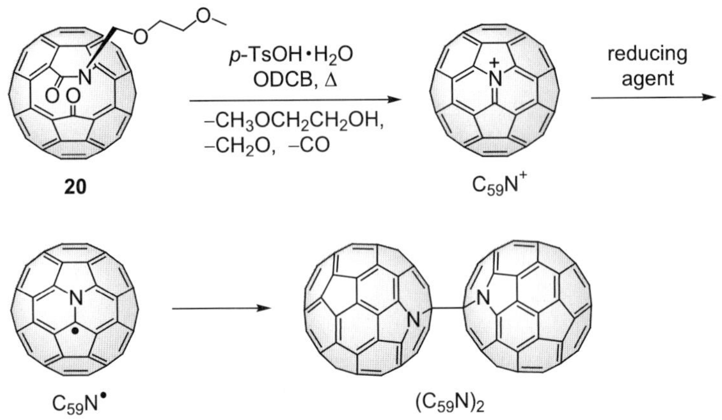

Scheme 9.15. Synthesis of aza[60]fullerene dimer $(C_{59}N)_2$.

ion $C_{59}N^+$ as an intermediate, which is then reduced to radical $C_{59}N^\bullet$ by water or 2-methoxyethanol, in their preparation of $(C_{59}N)_2$ starting from ketolactam **20** [35] (Scheme 9.15).

The dimer $(C_{59}N)_2$ is a good precursor for the monomeric cation $C_{59}N^+$. Isolation of the Ag^+ bis-carborane complex salt of $C_{59}N^+$ was achieved by Reed et al. As illustrated in Scheme 9.16, $(C_{59}N)_2$ is oxidized by the radical cation of hexabromo-N-phenylcarbazole to give analytically pure, dark green crystals, which are reasonably stable in air [38]. This oxidation is reversible, as evidenced by the slow reduction to $(C_{59}N)_2$ by zinc amalgam pellets. The salt was fully characterized by both spectroscopic and X-ray analyses. The ^{13}C NMR spectrum of the salt in $Cl_2CDCDCl_2$ showed 31 peaks, as expected from C_s symmetry. The most downfield shifted signal appeared at δ 153.8, which was assigned to the carbon atom adjacent to the nitrogen on the basis of a comparison with the DFT-calculated spectrum. This chemical shift indicates a greater delocalization of positive charge in $C_{59}N^+$, compared to HC_{60}^+ (δ 182.0) [6, 16] and RC_{60}^+ (δ 172–180) [9], due to the resonance contribution of an iminium ion structure (Scheme 9.17).

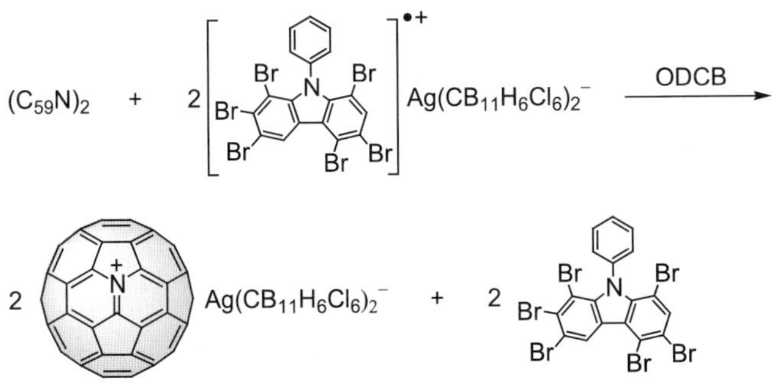

Scheme 9.16. Isolation of a $C_{59}N^+$ salt. See Scheme 9.6 for the structure of $CB_{11}H_6Cl_6^-$.

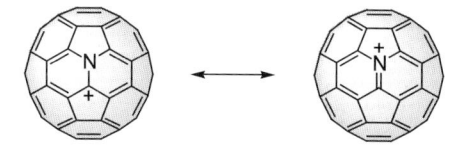

Scheme 9.17. Resonance structures of $C_{59}N^+$.

The cation $C_{59}N^+$ can also be conveniently generated as a reactive intermediate by heating or irradiating the dimer in the presence of p-toluenesulfonic acid in air (Scheme 9.18). Due to the low bond dissociation energy (18 kcal mol^{-1}) [39] the central carbon–carbon bond of the dimer is susceptible to both thermal and photochemical homolysis. Oxidation of the resulting radical $C_{59}N^\cdot$ occurs in the presence of O_2 (or added chloranil) to give cation $C_{59}N^+$. The presence of p-toluenesulfonic acid is essential, because it probably traps the reduced form of oxygen [40]. The cation acts as an electrophilic reagent that is useful for further transformation. Thus, generating $C_{59}N^+$ in the presence of a nucleophile (NuH) gives a derivatized aza[60]fullerene as the final product. Examples for such a transformation are described in the next section. The formation of $C_{59}N^+$ is a clean process, as evidenced by the nearly quantitative formation of the product $RC_{59}N-Nu$, when efficient nucleophiles are employed.

Scheme 9.18. Generation of $C_{59}N^+$ and its reaction with a nucleophile (NuH).

9.2.2.2 Reactions via $C_{59}N^+$

Nucleophilic trapping of $C_{59}N^+$ is one of the two major methods that are currently known for the synthesis of aza[60]fullerene derivatives, and this approach has been more extensively investigated when compared with the other method, which involves the trapping of radical $C_{59}N^\cdot$. The cation $C_{59}N^+$ has the highest reactivity at the most positive carbon, which is adjacent to the nitrogen atom across a 6–6 bond. Thus, the chemical nature of $C_{59}N^+$ is determined by its carbenium ion characteristics and resembles that of $RC_{60}{}^+$.

Treatment of $(C_{59}N)_2$ with an electron-rich aromatic compound in the presence of an excess of p-toluenesulfonic acid at high temperature (~150 °C) under air provides an arylated aza[60]fullerene **21** (Scheme 9.19) [40–46]. The intermediate cation $C_{59}N^+$, which is formed *in situ* from $(C_{59}N)_2$, undergoes electrophilic aromatic substitution. ODCB is inert to electrophilic attack by $C_{59}N^+$ and thus can be used as the solvent. Only para-substitution products **21a–d** are obtained from monosub-

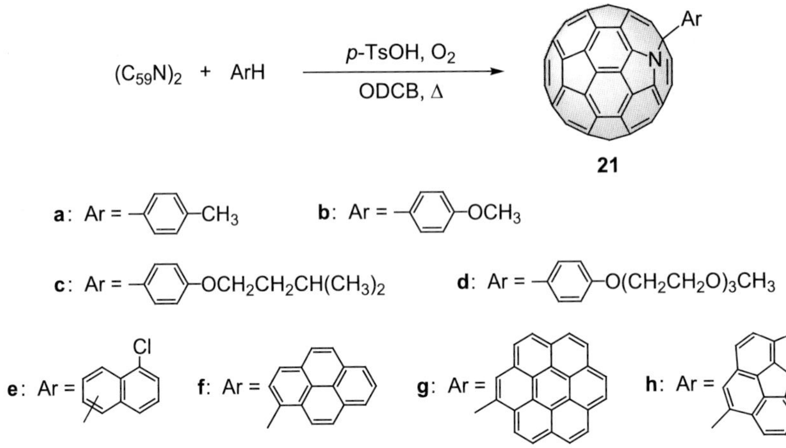

396 | *9 Fullerene Reactivity – Fullerene Cations and Open-Cage Fullerenes*

$(C_{59}N)_2$ + ArH $\xrightarrow[\text{ODCB, }\Delta]{p\text{-TsOH, }O_2}$ **21**

a: Ar = –C₆H₄–CH₃   **b**: Ar = –C₆H₄–OCH₃

c: Ar = –C₆H₄–OCH₂CH₂CH(CH₃)₂   **d**: Ar = –C₆H₄–O(CH₂CH₂O)₃CH₃

e: Ar = chloronaphthyl   **f**: Ar = pyrenyl   **g**: Ar = coronenyl   **h**: Ar = corannulenyl

Scheme 9.19. Electrophilic substitution of aromatics with $C_{59}N^+$.

stituted benzenes [41–43], whereas a mixture of isomers **21e** is formed from 1-chloronaphthalene [41]. The electrophilic attack of $C_{59}N^+$ on condensed aromatics such as pyrene, coronene, and corannulene leads to dyads **21f–h** [44]. Fluorescence studies indicate that photoexcitation of the arene moiety of these dyads causes highly efficient energy transfer from the arene singlet excited state to the singlet excited state of $C_{59}N$, due to the strong coupling of the two π-systems that are directly attached.

Reactions with toluene-d_5 and anisole-d_5 to form ring-deuterated derivatives of **22a** and **22b** (Scheme 9.20) were found to show primary deuterium isotope effects (k_H/k_D) of 1.18 ± 0.04 and 1.52 ± 0.05, respectively, demonstrating that proton elimination from an arenium ion intermediate **23** is a rate-determining step [47].

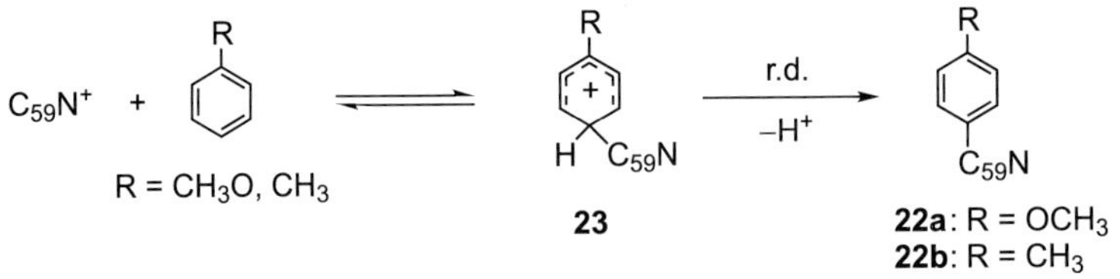

$C_{59}N^+$ + ArR ⇌ **23** $\xrightarrow[-H^+]{\text{r.d.}}$ **22a**: R = OCH₃; **22b**: R = CH₃

R = CH₃O, CH₃

Scheme 9.20. Mechanism of the electrophilic aromatic substitution with $C_{59}N^+$.

An ether, such as **24** can be produced, when the intermediate cation $C_{59}N^+$ is trapped by an alcohol (Scheme 9.21) [42]. This product contains a carbon atom attached to an oxygen and a nitrogen atom, providing a potential precursor of open-cage fullerenes [42].

The formation of Mannich bases **25** results from the reaction of $(C_{59}N)_2$ with enolizable ketones, aldehydes, and carboxylic esters (**26–29**) (Scheme 9.22) [42, 48, 49]. The reaction is reasonably interpreted in terms of the electrophilic addition of $C_{59}N^+$ intermediate to the enol form of the carbonyl compound. Dicarbonyl

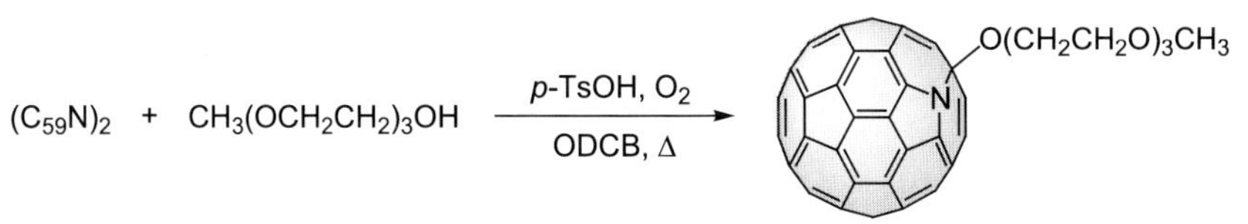

$(C_{59}N)_2$ + $CH_3(OCH_2CH_2)_3OH$ →[p-TsOH, O_2 / ODCB, Δ] **24** (fullerene with $O(CH_2CH_2O)_3CH_3$ substituent)

Scheme 9.21. Synthesis of an alkoxyhydroaza[60]fullerene.

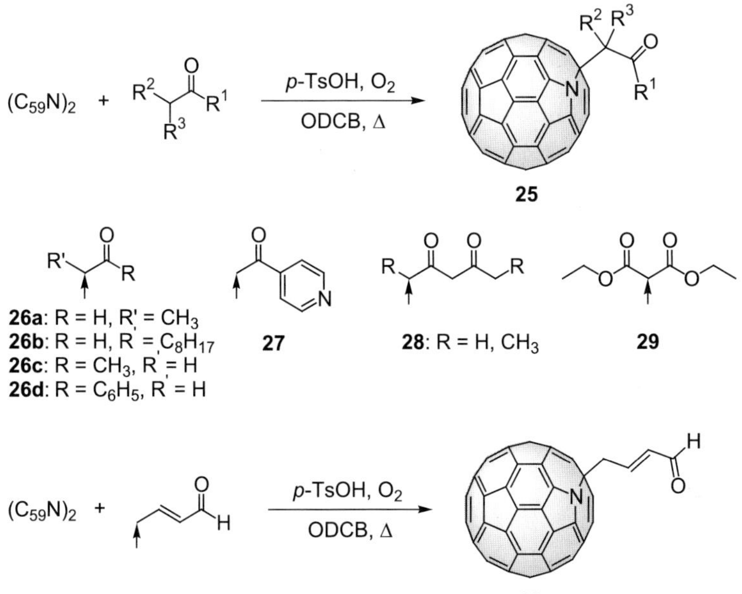

$(C_{59}N)_2$ + R^2-C(=O)-CHR^1R^3 →[p-TsOH, O_2 / ODCB, Δ] **25**

26a: R = H, R' = CH_3
26b: R = H, R' = C_8H_{17}
26c: R = CH_3, R' = H
26d: R = C_6H_5, R' = H

27 (4-acetylpyridine)

28: R = H, CH_3 (1,3-diketone)

29 (malonate diester)

$(C_{59}N)_2$ + crotonaldehyde →[p-TsOH, O_2 / ODCB, Δ] **30**

Scheme 9.22. Reactions of $C_{59}N^+$ with enolizable carbonyl compounds. Arrows indicate the position of the reaction with $C_{59}N^+$.

compounds **28** and **29** react with $C_{59}N^+$ in a similar manner. Diketone **28** reacts at the terminal rather than the central carbon atom. Although the latter carbon atom has more acidic protons, attack at the former carbon is more favorable due to the thermodynamic stability of the produced cationic intermediate. Similarly, crotonaldehyde reacts at the vinylogous position with $C_{59}N^+$ to form adduct **30** [42].

Using similar additions, the first fullerene–heterofullerene dyad **31** was synthesized by linking the C_{60} and $C_{59}N$ cores through a malonic acid ester containing an acetyl function (Scheme 9.23) [50]. The distinct difference in the reactivities of C_{60} and $C_{59}N$ enabled the two units to be connected by consecutive Bingel cyclopropanation [51] and the electrophilic reaction of $C_{59}N^+$ with the acetyl group.

The formation of different products, **32** and **33**, depending on the reaction conditions, has recently been reported in the reaction of $C_{59}N^+$ and benzyltrimethylsilane **34** (Scheme 9.24) [52]. The thermal reaction of $(C_{59}N)_2$ with **34** in the presence of p-toluenesulfonic acid under air afforded the electrophilic aromatic

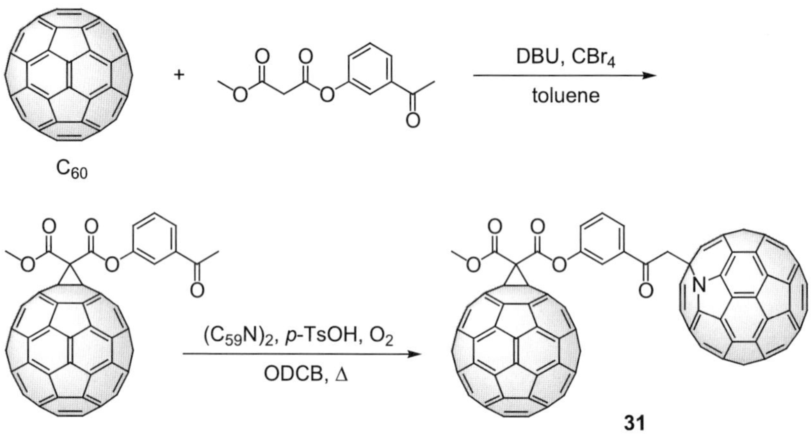

Scheme 9.23. Synthesis of a C_{60}–C_{59}N dyad.

substitution product **32**. On the other hand, an N-oxide with a benzyl group attached at the adjacent carbon atom, **33**, was formed by irradiation (>300 nm) at 0 °C. In this reaction, a photo-induced electron transfer occurs from **34** to $C_{59}N^+$, followed by desilylation and radical coupling to form adduct **35**. As **35** is formed, it is oxidized to **33** in the presence of O_2 by self-initiated photooxygenation.

9.2.3
Electrophilic Functionalization of Carbon Nanotubes

The Friedel–Crafts type addition of chloroform, which was observed for C_{60} [9, 21] and C_{70} [15], also occurs on the sidewall of single-walled carbon nanotubes (SWNTs) (Scheme 9.25) [53, 54]. In this transformation SWNTs were first ground with $AlCl_3$ to produce thinner bundles of nanotubes. The reaction of this mixture with $CHCl_3$ under reflux gave functionalized nanotubes **36**, with $CHCl_2$ and Cl groups covalently attached. This material, although difficult to analyze spectroscopically owing to its labile nature and poor solubility, was converted to the hydroxy derivative **37**, which could be characterized by the observation of the infrared absorption of the OH stretching vibration. The side-chain transformation of **37** by treatment with propionyl chloride afforded carbon nanotubes with an ester function, **38**, which could be characterized not only by its solid state IR spectrum but also, owing to its improved solubility, by solution NMR analysis.

9.3
Open-Cage Fullerenes to Endohedral Fullerenes

Endohedral fullerenes are the closed-cage carbon molecules encapsulating a guest species such as metal ions, metal carbides, metal nitrides, noble gas atoms, and

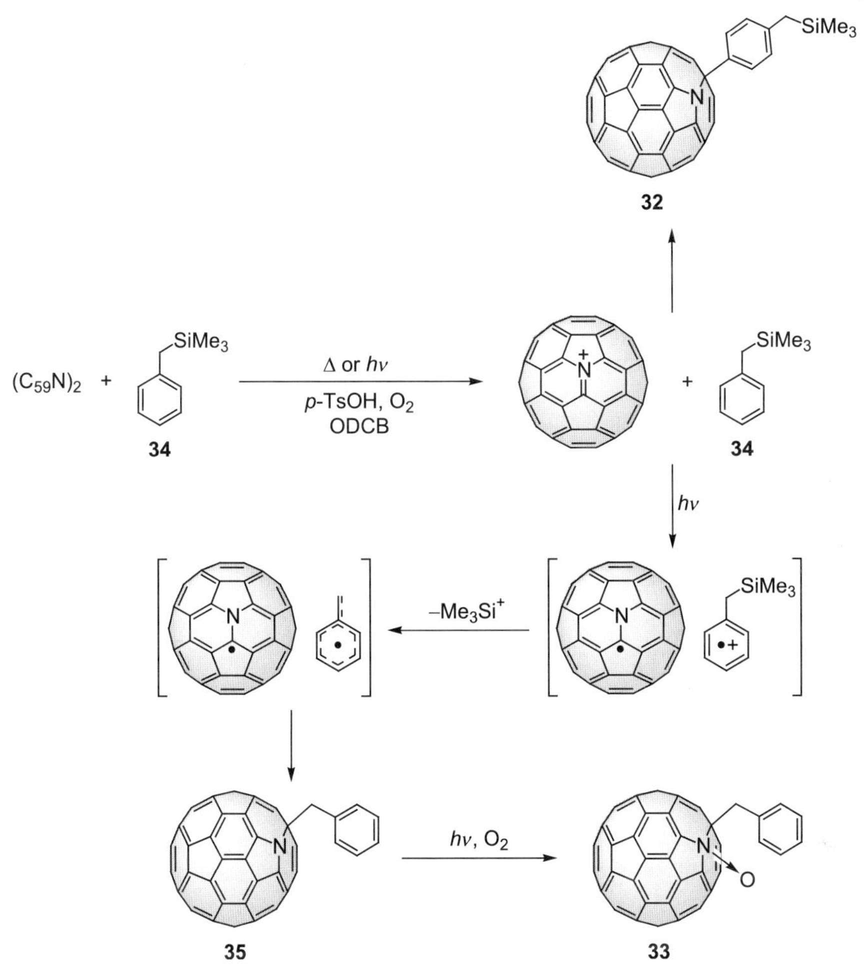

Scheme 9.24. Electrophilic aromatic substitution and photo-induced electron transfer between $C_{59}N^+$ and benzyltrimethylsilane.

nitrogen and phosphorus atoms [55]. They are of great importance not only for scientific curiosity but for their potential use in various fields such as molecular electronics [56], medical science (as an MRI contrast agent) [57], and NMR analysis [58]. However, development of their application has been hampered by difficulty in their production. So far, their production has relied only on methods such as hardly controllable co-vaporization of carbon and metal atoms, high-pressure/high-temperature treatment with gases, or ion-beam implantation, all of which yield only milligram quantities of pure product after laborious isolation procedures [55].

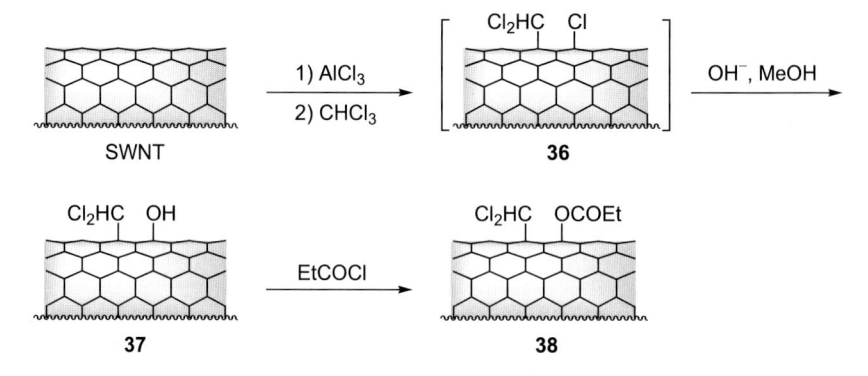

Scheme 9.25. Sidewall functionalization of single-walled carbon nanotubes by electrophilic addition.

One promising method to overcome such limitation would be a "molecular surgery" of fullerene, which involves opening an orifice on the fullerene cage, insertion of a guest through the orifice, and closure of the orifice by the use of techniques of organic synthesis [59, 60]. The first open-cage fullerene **20** was synthesized by Wudl and co-workers in 1995 (Fig. 9.1) [36]. However, the orifice constructed by an 11-membered ring was too small even for a ^{3}He atom to pass through upon heating to 200 °C [60]. In 1996, Rubin and co-workers achieved another open-cage fullerene **39** by a sequential intramolecular reaction pathway, and transformed it into a cobalt complex **40** [61]. The structure of **40** was determined by X-ray crystallography, the first for a derivative of an open-cage fullerene. Insertion of the cobalt ion into the inside of the fullerene cage was attempted, but it was unsuccessful even by activation with heating to 400 °C [60] or under pressure up to 40 kbar [62].

Using C_{60} derivatives having an orifice with a ketolactam or "bisfulleroid" structure as starting materials, several reactions to enlarge the orifice were recently developed. Since reviews of open-cage fullerenes based on C_{60} and C_{70} [63, 64] reported before 1998 have already been published [46, 60], this article mainly focuses on recent developments in the synthesis of derivatives of open-cage C_{60} with an orifice larger than an 11-membered ring and on the insertion of small molecules into them. A recent achievement of the closure of the orifice with retention of molecular hydrogen inside the C_{60} cage will also be discussed.

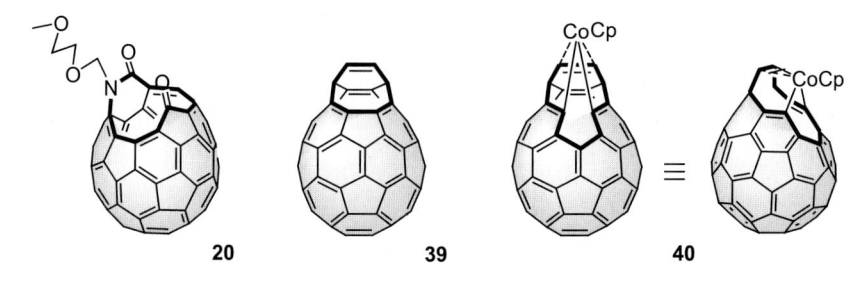

Figure 9.1. Examples of open-cage C_{60} derivatives.

9.3.1
Synthesis of Open-Cage Fullerene Derivatives

9.3.1.1 An Open-Cage Fullerene with a 14-Membered-Ring Orifice

In 1999, Rubin and co-workers reported open-cage C_{60} derivative **41** having a 14-membered-ring orifice [65]. Based on the concept of a retro [2+2+2] he had proposed (Scheme 9.26) [60, 66], he attempted the reaction of fullerene C_{60} with diazidobutadiene **42** in order to obtain **43** (Scheme 9.27). It was expected that threefold addition of the reactive moieties in **42** to three "6–6" double bonds in one hexagon of C_{60} would result in the formation of a fully saturated and strained planar cyclohexane ring on C_{60} (**44**) [67], which would undergo a facile [2+2+2] ring-opening reaction to yield **43** (Scheme 9.1). Diazidobutadiene **42** actually reacted with the three 6–6 bonds on the same six-membered ring of C_{60} in ODCB at 55 °C. However, the final product isolated in 12% was **41** instead of **43**. After extrusion of two nitrogen molecules from **44** followed by insertion of the nitrogen radicals into the nearest 5–6 bonds, the resulting electron-rich butadiene unit on C_{60} cage reacted with oxygen along with dehydrogenation to afford open-cage fullerene derivative **41**. The orifice of **41** is a 14-membered ring with a rather elliptic shape, which was the largest one at that time [68].

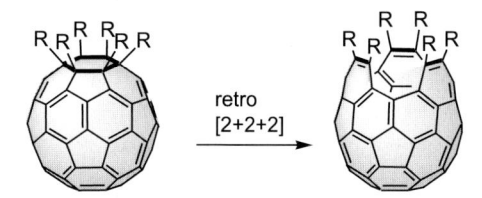

Scheme 9.26. Conceptual representation of a possible way to open-cage C_{60}.

9.3.1.2 Open-Cage Fullerenes with a 12-Membered-Ring Orifice

After the first report by Rubin [61] on the elegant synthesis of open-cage fullerene ("bisfulleroid") **39** having an eight-membered-ring orifice, several synthetic routes were developed for a variety of its derivatives **45–49** (Fig. 9.2) [69–74]. These are nickel-promoted ene-diyne cycloaddition on C_{60} [70], [4+2] cycloaddition of palladacyclopentadiene with C_{60} [71], and thermal reaction of C_{60} with phthalazine [72] and a 1,2,3-triazine derivative [73]. The reaction of C_{60} with a 1,2,4-triazine derivative gave a nitrogen-containing derivative **49**, whose structure was clearly determined by X-ray crystallography showing that the eight-membered-ring orifice has a tub-form with the C1–C2–C3–C4 and C3–C4–C5–C6 dihedral angles being $-39.0(4)$ and $38.8(4)°$, respectively (Fig. 9.3) [74].

The eight-membered-ring orifice in some of the derivatives **46–49** was enlarged by selective C=C bond cleavage with singlet oxygen generated photochemically. Murata and co-workers reported that compound **46** is readily oxidized via **50** to afford diketone **51** having a 12-membered-ring orifice (Scheme 9.28) [75]. Komatsu

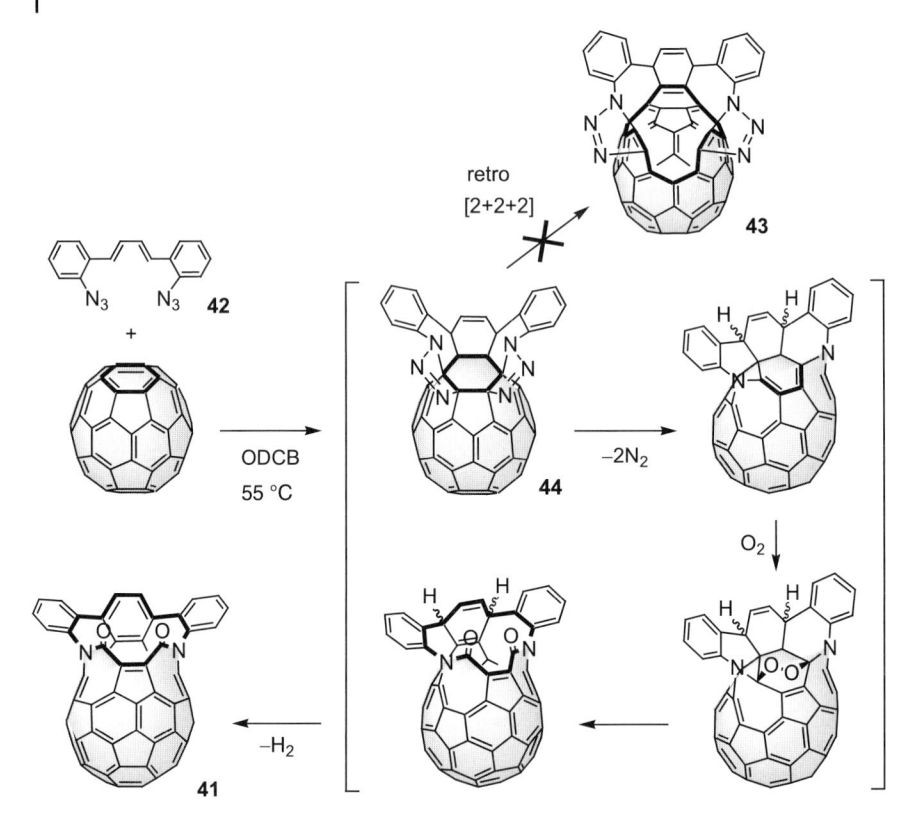

Scheme 9.27. Possible reaction pathway leading to **41**.

and co-workers independently found that compound **47** is also oxidized to give enol **52** via keto–enol tautomerization (Fig. 9.4) [76]. Triketone **53** [73] and diketones **54** and **55** [74] were obtained in the same way. In these reactions, singlet oxygen generated by irradiation of visible light adds to one of the C=C bonds on the rim of the eight-membered-ring orifice, which is distorted and has the relatively large values for HOMO coefficients, to give dioxetane intermediate such as **50**. This reaction is in contrast to the photooxygenation of pristine C_{60}, which affords fullerene epoxide $C_{60}O$ [77]. However the size of the orifice is still too small for a small atom or molecule to pass through.

9.3.1.3 Enlargement of the Orifice by Insertion of a Sulfur Atom

In 2003, further enlargement of the orifice of diketone **54** was achieved by Komatsu and co-workers [74]. Cyclic voltammetry of diketone **54** exhibited at least four reversible reduction waves in a similar way to that of C_{60} itself, in spite of such a severely torn out π-system. Furthermore, the first reduction potential was even less negative than C_{60} by 0.2 V, indicating that **54** is a stronger π-acceptor than C_{60}. The coefficients of the LUMO of **54** are relatively localized on the conju-

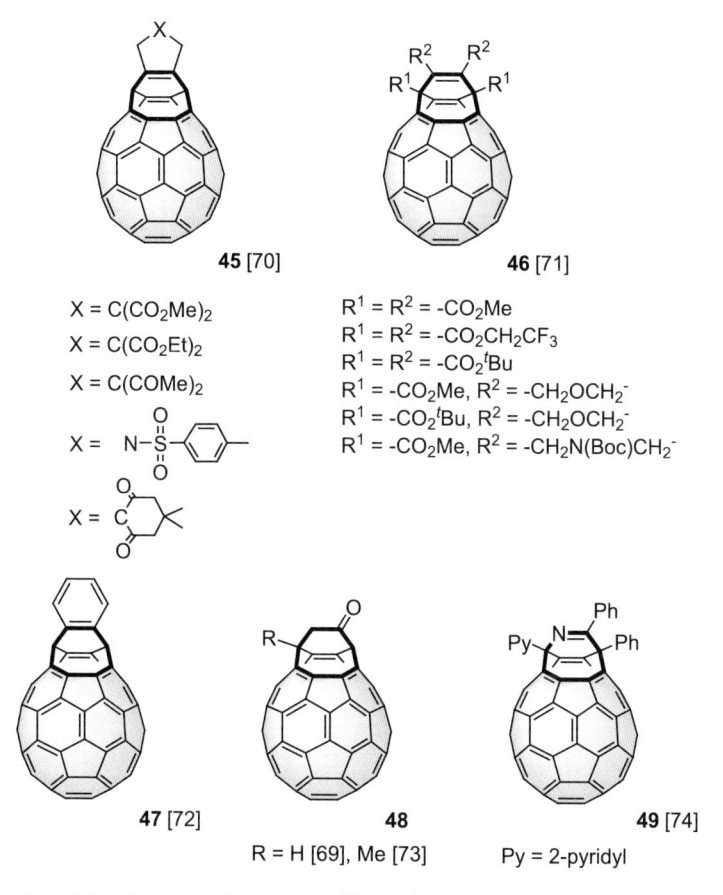

9.3 Open-Cage Fullerenes to Endohedral Fullerenes

45 [70]

X = C(CO$_2$Me)$_2$
X = C(CO$_2$Et)$_2$
X = C(COMe)$_2$

X = N–S(=O)$_2$–C$_6$H$_4$–CH$_3$

X = (dimedone-like diketone)

46 [71]

R^1 = R^2 = -CO$_2$Me
R^1 = R^2 = -CO$_2$CH$_2$CF$_3$
R^1 = R^2 = -CO$_2^t$Bu
R^1 = -CO$_2$Me, R^2 = -CH$_2$OCH$_2^-$
R^1 = -CO$_2^t$Bu, R^2 = -CH$_2$OCH$_2^-$
R^1 = -CO$_2$Me, R^2 = -CH$_2$N(Boc)CH$_2^-$

47 [72]

48
R = H [69], Me [73]

49 [74]
Py = 2-pyridyl

Figure 9.2. Examples of derivatives of **39** having an eight-membered-ring orifice.

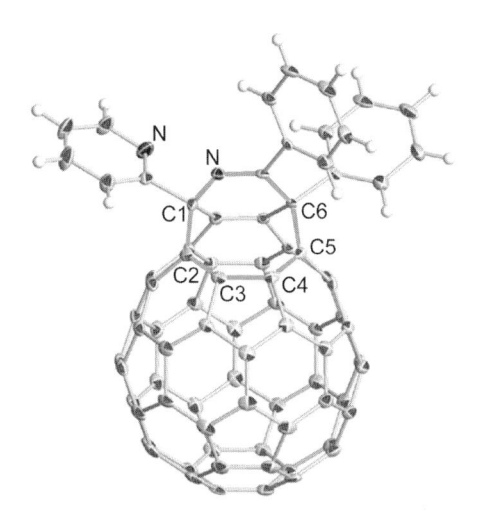

Figure 9.3. X-ray structure of **49**.

Scheme 9.28. Photooxygenative ring enlargement of **46**.

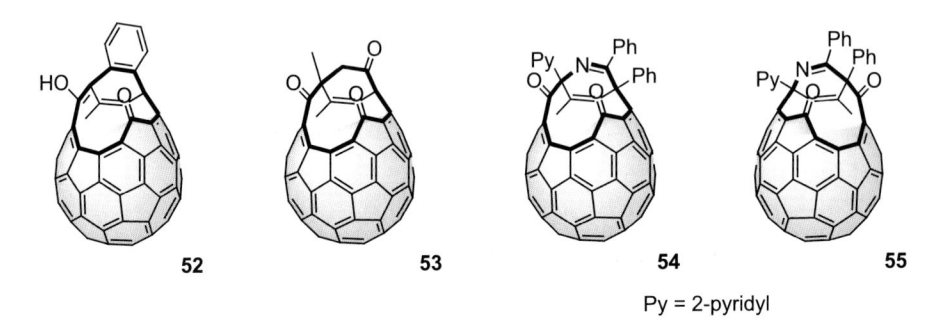

$R^1 = R^2 = -CO_2Me$ 95%
$R^1 = R^2 = -CO_2CH_2CF_3$ 87%
$R^1 = R^2 = -CO_2{}^tBu$
$R^1 = -CO_2Me, R^2 = -CH_2OCH_2-$
$R^1 = -CO_2{}^tBu, R^2 = -CH_2OCH_2-$
$R^1 = -CO_2Me, R^2 = -CH_2N(Boc)CH_2-$

Py = 2-pyridyl

Figure 9.4. Open-cage fullerene derivatives having a 12-membered-ring orifice.

gated butadiene moiety on the rim of the 12-membered-ring orifice, suggesting that this part of the molecule should be activated by a π-electron donor. Thus, in the presence of tetrakisdimethylaminoethylene (TDAE) in ODCB at 180 °C, a reaction of **54** with elemental sulfur proceeded smoothly resulting in sulfur atom insertion to the butadiene unit to give **56** having a 13-membered-ring orifice in 77% yield (Scheme 9.29). The structure of **56** was unambiguously determined by X-ray crystallography (Fig. 9.5).

The versatility of this procedure was subsequently demonstrated by application to ketolactam **20** to afford **57** in 72% yield under similar reaction conditions by Orfanopoulos and co-workers (Scheme 9.30) [78].

9.3.1.4 Enlargement of the Orifice by Reaction with Aromatic Hydrazine or Diamine

In 2003, Iwamatsu and co-workers reported another reaction to make the orifice of diketone **51** larger than that of **56**. The reaction of **51** with aromatic hydrazines gave **58** having a 16-membered-ring orifice in 50–89% yields (Scheme 9.31) [79].

Scheme 9.29. Insertion of a sulfur atom to enlarge the 12-membered-ring orifice.

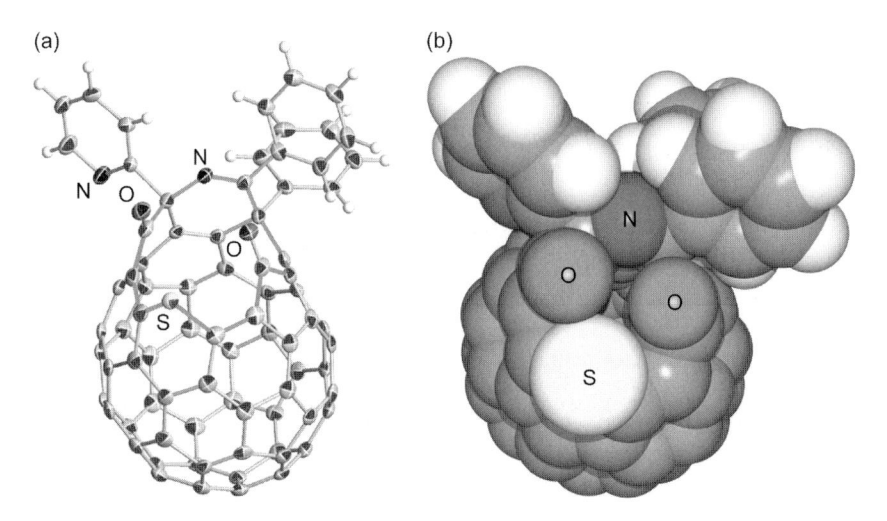

Figure 9.5. X-Ray Structure of **56**; (a) Side-view and (b) Space-filling model.

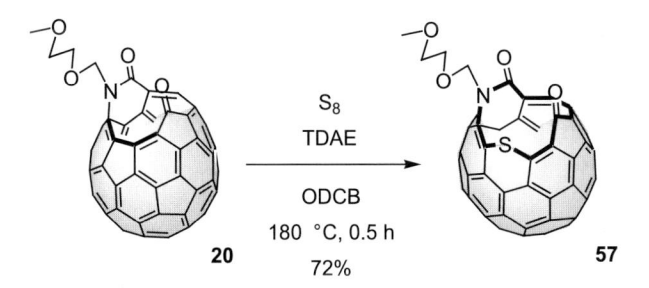

Scheme 9.30. Insertion of a sulfur atom into ketolactam **20**.

The structure of **58** was assigned by careful NMR measurements including ^{1}H, ^{13}C, and ^{15}N NMR, DEPT, HMBC, HMQC, and INADEQUATE techniques using the sample enriched with ^{13}C or ^{15}N. One of the double bonds on the rim of the orifice of **51** was assumed to be selectively cleaved to form a hydrazone with a

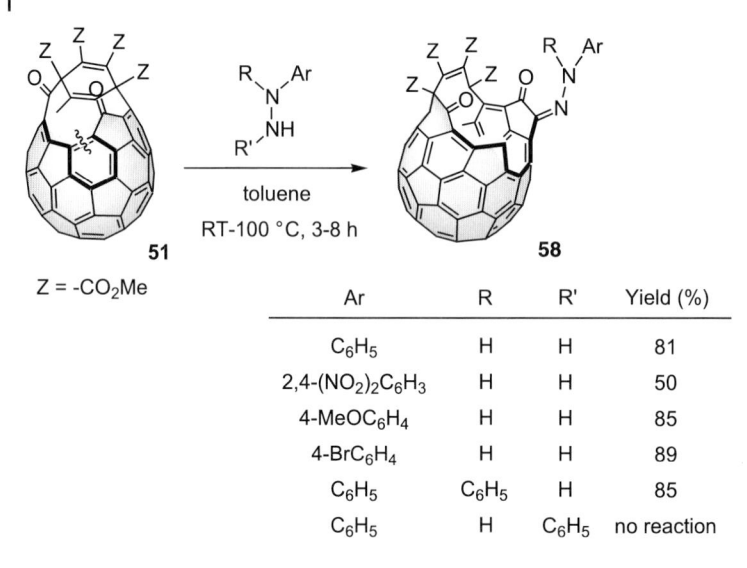

Scheme 9.31. Reaction of open-cage C$_{60}$ derivative **51** with aromatic hydrazines.

Ar	R	R'	Yield (%)
C$_6$H$_5$	H	H	81
2,4-(NO$_2$)$_2$C$_6$H$_3$	H	H	50
4-MeOC$_6$H$_4$	H	H	85
4-BrC$_6$H$_4$	H	H	89
C$_6$H$_5$	C$_6$H$_5$	H	85
C$_6$H$_5$	H	C$_6$H$_5$	no reaction

Z = -CO$_2$Me

methylene unit by migration of two hydrogen atoms from the hydrazine to the fullerene cage. For this reaction, the NH$_2$ group in the hydrazine is essential since the reaction with 1,2-diphenyl hydrazine without terminal NH$_2$ group resulted in complete recovery of **51**. However, the mechanism of this reaction is not yet clear.

The similar reaction was applied to ketolactam **20** by Iwamatsu and co-workers in 2003 [80] and tested by Orfanopoulos and co-workers in 2004 [81] (Scheme 9.32) to give a product with a 15-membered-ring orifice, whose structure was claimed to be **59** by the former and **60** by the latter. The difference in the structure between **59** and **60** arose from which of the two C=C double bonds, (a) or (b), was cleaved. Iwamatsu's assignment assuming the cleavage of C=C bond (a) was made in a similar manner as described for **58**, based on the fact that an α,β-unsaturated carbonyl moiety exists in both **20** and **51**. On the other hand, Orfanopoulos reported that the reaction should have occurred on the C=C bond (b) to afford **60** based on the NOE and the C–H coupling analysis of the product. The answer to this question has to wait until single crystal X-ray analysis of the product is made.

In 2004, Iwamatsu and co-workers reported that even larger orifices can be formed by the regioselective multiple cage scissions of **51** (Scheme 9.33) [82]. The reaction of **51** with 4,5-dimethyl-1,2-phenylenediamine at room temperature in toluene gave **61**, with a 16-membered-ring orifice, in 85% yield. The same reaction at 60 °C gave another open-cage C$_{60}$ derivative **62**, with a 20-membered-ring orifice, in 45% yield. The transformation from **61** to **62**, in 52% yield, was also promoted by excess of the diamine or pyridine added as a base. The structural assignments of **61** and **62** were carefully made by spectral analyses using MS, UV–vis, IR, ^{1}H and ^{13}C NMR, with various NMR techniques such as COSY, HMQC, and HMBC. Al-

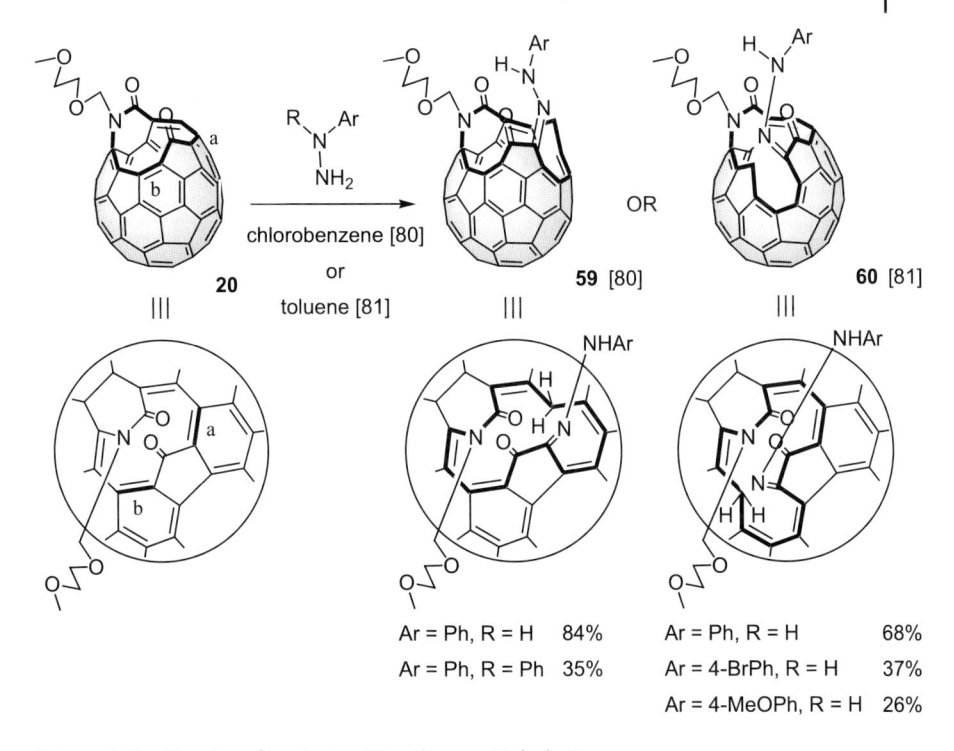

9.3 Open-Cage Fullerenes to Endohedral Fullerenes

Ar = Ph, R = H   84%
Ar = Ph, R = Ph   35%

Ar = Ph, R = H   68%
Ar = 4-BrPh, R = H   37%
Ar = 4-MeOPh, R = H   26%

Scheme 9.32. Reaction of ketolactam **20** with aromatic hydrazines.

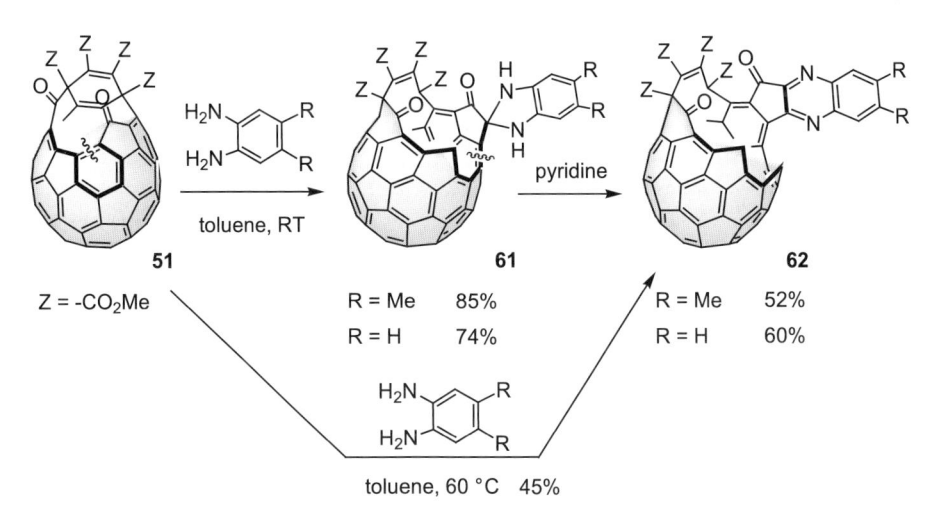

Z = -CO$_2$Me

R = Me   85%
R = H   74%

R = Me   52%
R = H   60%

toluene, 60 °C   45%

Scheme 9.33. Regioselective multiple cage scissions of **51**.

though the mechanism for the reaction is not yet clear, the size of the orifice is the largest among open-cage fullerene derivatives reported so far.

The reaction with o-phenylenediamine was applied to ketolactam **20** in the presence of pyridine at 80 °C in chlorobenzene to afford **64** with a 19-membered-ring orifice in 37% yield in addition to **63** which is the initial cage scission product formed in 7% yield (Scheme 9.34) [83]. The theoretical calculations at the B3LYP/6-31G* level of theory showed that the size of the orifice of **64** is slightly smaller than that of **62**.

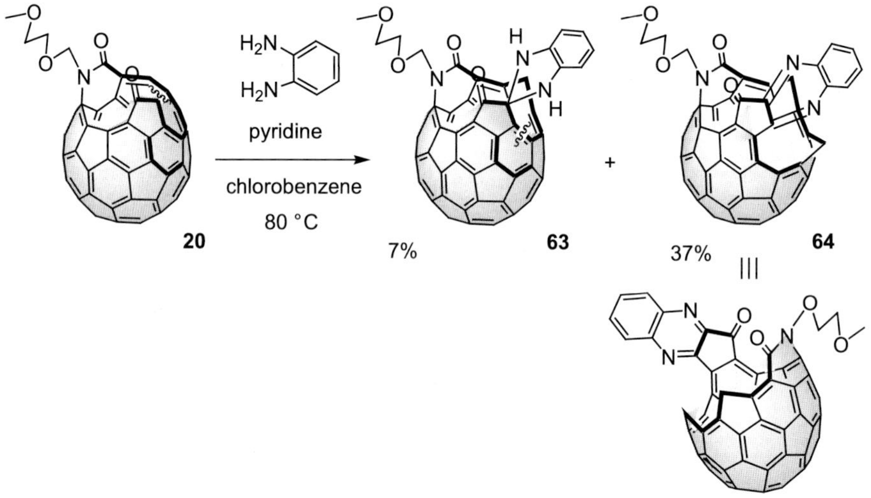

Scheme 9.34. Regioselective multiple cage scissions of **20**.

9.3.2
Insertion of a Small Guest into Open-Cage Fullerene Derivatives

9.3.2.1 Insertion of a Helium Atom and a Hydrogen Molecule

For pristine C_{60}, insertion of a helium atom (^{3}He) required drastic conditions such as 650 °C and 3000 atm to give ^{3}He@C_{60} in an incorporation rate of 0.1% [58, 84]. The incorporation was improved to 1% when C_{60} was pre-treated with KCN; the reason for this is as yet unknown [85]. In contrast, theoretical calculations showed that insertion of a helium atom or a hydrogen molecule into open-cage fullerene **65** would take place under milder conditions such as at 124 or 397 °C for helium or hydrogen, respectively (Scheme 9.35) [59]. Actually, such insertion into **65**, which was obtained in a similar manner as **41**, was achieved for the first time in 2001 by Rubin and coworkers [59]. The treatment of **65** as a crystalline powder with ^{3}He (288–305 °C, ca. 475 atm, 7.5 h) gave the ^{3}He-encapsulating molecule, ^{3}He@**65**, in 1.5% yield. The NMR chemical shift of the ^{3}He incorporated in **65** appeared at $\Delta\delta = -10.10$ ppm relative to the free dissolved ^{3}He gas. The ^{3}He atom slowly escaped when ^{3}He@**65** was heated above 80 °C in a solution of

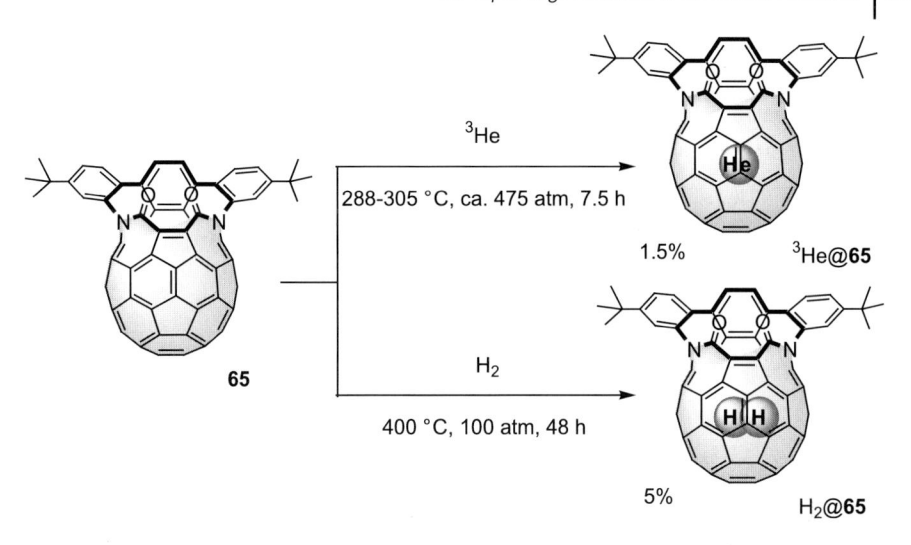

9.3 Open-Cage Fullerenes to Endohedral Fullerenes

Scheme 9.35. Insertion of a helium atom and a hydrogen molecule into **65**.

ODCB-d_4. The rate constant followed first order kinetics and was determined at 80, 100, 120, and 130 °C, with the resulting Arrhenius plot giving the activation energy of 24.6 ± 0.8 kcal mol^{-1}, which is in good agreement with the calculated value (Table 9.1). In a similar manner, molecular hydrogen was introduced (400 °C, ca. 100 atm, 48 h) into **65** in 5%, with about 30% of **65** decomposing under these conditions. The chemical shift of the hydrogen inside **65** was $\delta = -5.43$ from TMS, i.e., $\Delta\delta = -9.96$ ppm relative to free hydrogen.

The 100% encapsulation of molecular hydrogen was achieved into **56** by Komatsu and co-workers in 2003 [86]. Theoretical calculations at the B3LYP/6-31G**//B3LYP/3-21G level of theory showed that the energy required for He, Ne, H$_2$, and Ar to be inserted into **56** is lower than that for **65** (Table 9.1), indicating that the effective size of the orifice of **56** is larger than that of **65**. Although calcu-

Table 9.1. Predicted activation barriers for insertion and cage escape, and energies of encapsulation for the neutral guests He, Ne, H$_2$, N$_2$, and Ar gases inside the open-cage fullerenes **56** and **65** (B3LYP/6-31G**//B3LYP/3-21G, kcal mol^{-1}).

Guest	Barrier to insertion		Energy of encapsulation		Barrier to escape	
	56 [86]	65 [59]	56 [86]	65 [59]	56 [86]	65 [59]
He	18.9	24.5	+0.3	+0.2	18.6	24.3
Ne	26.2	40.6	−1.0	−1.1	27.2	41.7
H$_2$	30.1	41.4	+1.4	+1.4	28.7	40.0
Ar	97.7	136.3	+6.1	+6.1	91.6	130.2

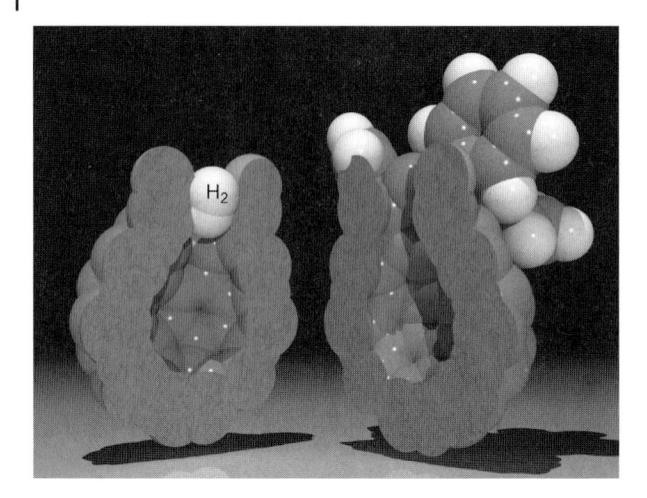

Figure 9.6. Cutout view of space-filling model of the B3LYP/3-21G transition structure for the insertion of molecular hydrogen into **56**.

lations indicated that the neck of the open-cage fullerene **56** was as narrow as shown in the transition-state structure (Fig. 9.6), the treatment of a crystalline powder of **56** with hydrogen gas (200 °C, 800 atm, 8 h) afforded the hydrogen-incorporated molecule H$_2$@**56** in 100% yield without any decomposition of **56** (Scheme 9.36). The NMR signal of the encapsulated hydrogen appeared at $\delta = -7.25$, i.e., $\Delta\delta = -11.79$ ppm relative to free hydrogen, as a sharp singlet. The solid-state ^{1}H NMR study revealed that the motional anisotropy of the encapsulated hydrogen inside **56** is quite small [87]. Furthermore, the hydrogen inside **56** was directly observed by single crystal synchrotron X-ray analysis, by the use of

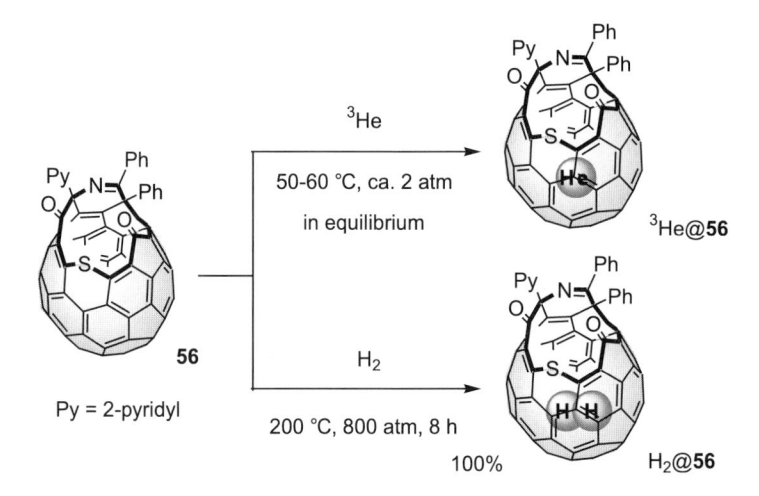

Scheme 9.36. Insertion of a helium atom and a hydrogen molecule into **56**.

the maximum entropy method, to be located at the center of the fullerene cage [88]. Although H$_2$@**56** was stable at room temperature, escape of the hydrogen slowly took place at temperatures above 160 °C. The activation energy for the escape of the hydrogen was determined as 34.3 ± 0.7 kcal mol^{-1} from the Arrhenius plot of the rate constants measured at 160, 170, 180, and 190 °C in ODCB-d_4. In the MALDI-TOF MS spectrum, the molecular ion peak for H$_2$@**56** was clearly observed. In addition, when higher laser power was used the peak for H$_2$@C$_{60}$ was observed, indicating that closure of the orifice of H$_2$@**56** takes place in the gas-phase by the energy of the irradiating laser.

Insertion of a helium atom into **56** was found to take place under conditions near room temperature and 1 atm, reaching an incorporation rate of 0.1% (Scheme 9.36) [89]. In the ^{3}He NMR spectra, a signal of the encapsulated ^{3}He in **56** appeared at $\delta = -11.86$ relative to dissolved ^{3}He gas. The escape rates of the encapsulated ^{3}He were determined at 30, 40, 50, and 60 °C, giving an activation energy of 22.8 ± 0.7 kcal mol^{-1}. This value was lower than that for ^{3}He@**65**, confirming that the orifice of **56** is larger than that of **65**. It is to be noted that the equilibrium constant for the insertion/escape of ^{3}He was experimentally determined for the first time. These values are $(2.7 - 3.4) \times 10^{-4}$ atm^{-1} at 50–60 °C, which is rather comparable to the value calculated by MP2 for C$_{60}$ itself $(6.4 \times 10^{-4}$ atm$^{-1})$ assuming that ^{3}He is in the insertion/escape equilibrium with pristine C$_{60}$ [90].

9.3.2.2 Insertion of a Water Molecule

In 2004, Iwamatsu and co-workers reported that a water molecule can enter into **62** (R = Me) which has a 20-membered-ring orifice (Scheme 9.37) [82]. This incorporation of water took place in a solution of 1,1,2,2-tetrachloroethane-d_2 or CDCl$_3$ at ambient temperature under normal pressure. A singlet NMR signal attributed to the encapsulated H$_2$O appeared at $\delta = -11.4$, and disappeared on treatment with D$_2$O. From the integrated value of the signal, the rate of encapsulation of H$_2$O was shown to be 75%, which decreased reversibly upon heating the solution. Although elemental analysis of the isolated solid indicated the structure to be the monohydrate, H$_2$O@**62** (R = Me), the molecular ion peak corresponding to H$_2$O@**62** (R = Me) was not detected in the MS spectrum. Thus, the molecule of H$_2$O can get in and out in solution but escapes from **62** in the gas phase.

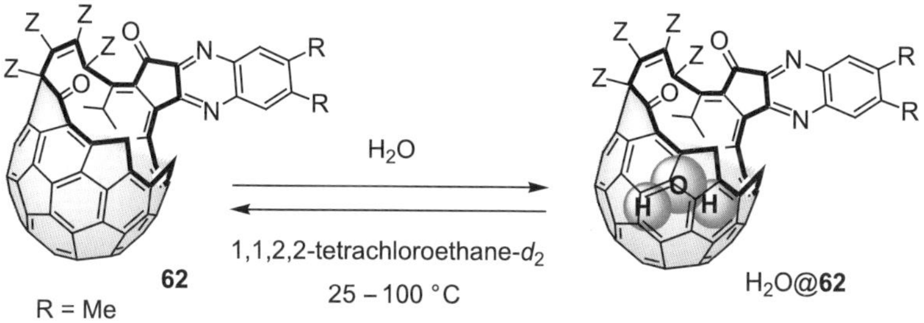

Scheme 9.37. Insertion of a water molecule into **62**.

Encapsulation of a molecule of H_2O into **64** also took place. Reflecting the smaller size of the orifice of **64** than that of **62**, the encapsulation rate in **64** at room temperature was less than 10% in contrast to 75% for **62**. However, the yield of $H_2O@64$ reached 85% under reflux in a mixture of toluene and water (Scheme 9.38). The NMR chemical shift of the water proton appeared at $\delta = -10.0$ in the ^1H NMR spectrum and the intensity of the signal decreased upon treatment with D_2O. In the ESI MS spectra, the molecular ion peaks for $H_2O@64$ and $D_2O@64$ were clearly observed with the signal intensity being consistent with the encapsulation ratio estimated by ^1H NMR [83].

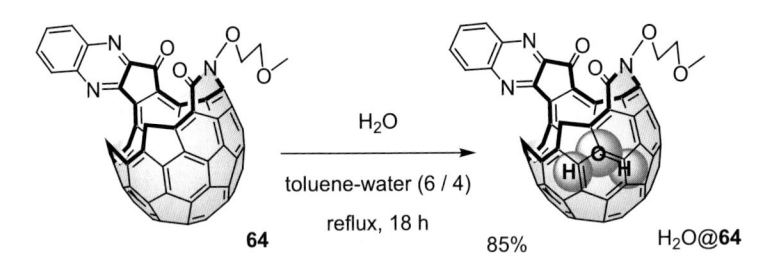

Scheme 9.38. Insertion of a water molecule into **64**.

When the methoxyethoxymethyl (MEM) group in **64** was removed, the resulting compound **66** showed a different encapsulation ratio for H_2O, which was only 35% in comparison to 85% in **64** (Scheme 9.39). The "amphiphilic" character of the MEM group was considered to be more important than the steric hindrance or the size of the orifice [83].

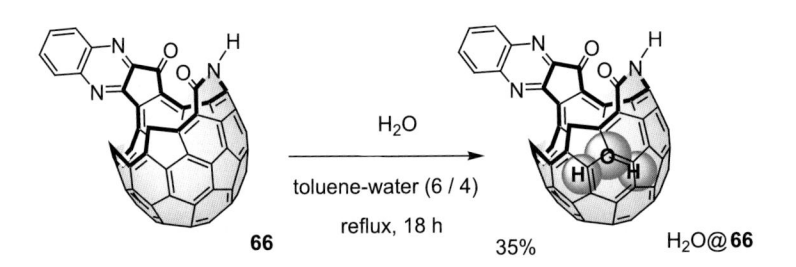

Scheme 9.39. Insertion of a water molecule into **66**.

9.3.3
Synthesis of an Endohedral Fullerene by Closure of the Orifice

If the orifice of an open-cage fullerene encapsulating a guest species can be completely closed while retaining the guest inside, the "molecular surgery" approach toward the synthesis of endohedral fullerenes is actualized. In 2005, Komatsu and co-workers reported such closure of the orifice of $H_2@56$ by a four-step sequence of

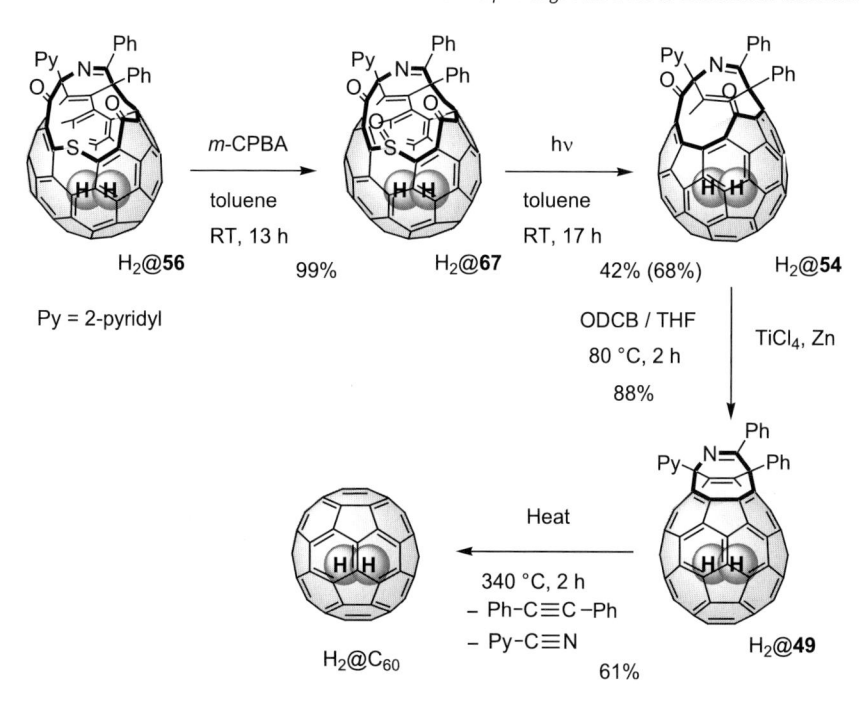

Scheme 9.40. Closure of the orifice of H$_2$@**56** yielding H$_2$@C$_{60}$.

organic reactions (Scheme 9.40) [91]. First, the sulfur atom in H$_2$@**56** was oxidized by m-chloroperbenzoic acid (m-CPBA) to give sulfoxide H$_2$@**67** in 99% yield in order to make the size-reduction of the 13-membered-ring orifice of H$_2$@**56** easy. Then the sulfinyl group in H$_2$@**67** was removed by irradiation with visible light to afford H$_2$@**54** with the orifice reduced to a 12-membered ring in 42% yield (68% yield based on consumed H$_2$@**67**), retaining 100% of encapsulated hydrogen as examined by ^{1}H NMR. Next, the two carbonyl groups in H$_2$@**54** were coupled by Ti(0) according to McMurry's method [92] to give H$_2$@**49** with an eight-membered-ring orifice in 88% yield. Finally, thermal reaction of H$_2$@**49** at 340 °C for 2 h afforded C$_{60}$ encapsulating hydrogen, H$_2$@C$_{60}$, in 61% yield.

The resulting H$_2$@C$_{60}$ was stable upon heating at 500 °C for 10 min with no release of the encapsulated hydrogen molecule. The ^{1}H NMR signal for the encapsulated hydrogen of H$_2$@C$_{60}$ in ODCB-d_4 was observed at $\delta = -1.44$, which is shifted 5.98 ppm upfield relative to the signal of dissolved free hydrogen. In the ^{13}C NMR spectrum, a very small downfield shift (0.078 ppm) was observed for H$_2$@C$_{60}$ ($\delta = 142.844$) compared with empty C$_{60}$ ($\delta = 142.766$). Almost no difference was observed in the UV–vis and IR spectra and in the electrochemical behavior between H$_2$@C$_{60}$ and empty C$_{60}$, indicating that the interaction between the encapsulated hydrogen and the π-system of the outer cage is quite small. In order to examine the chemical reactivity of H$_2$@C$_{60}$, the dimerization of H$_2$@C$_{60}$ was conducted under solid-state mechanochemical conditions [93] to give dimer **68**,

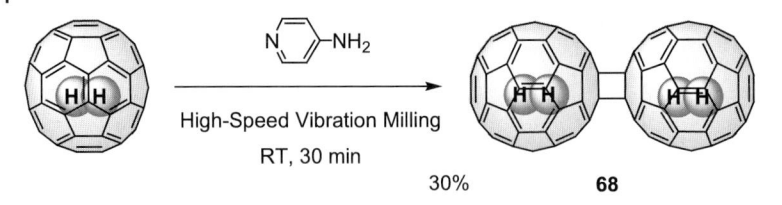

Scheme 9.41. Mechanochemical dimerization of $H_2@C_{60}$.

$(H_2@C_{60})_2$, in 30% yield, which was the same as that for the empty C_{60} (Scheme 9.41). The ^{1}H NMR signal for the encapsulated H_2 in **68** appeared at $\delta = -4.04$. Thus, the encapsulated hydrogen molecule in $H_2@C_{60}$ can be used as an NMR probe toward the chemical modification of the fullerene cage like ^{3}He in the ^{3}He NMR [58].

9.4
Summary and Conclusion

In contrast to the high stability of fullerene anions [8e, 9b], cationic species of fullerenes have been considered to be far more difficult to generate, based on the well-known resistance of C_{60} toward oxidation. Besides the radical cation of fullerenes, two types of nonradical, cationic fullerenes have been prepared to date. One is derivatized C_{60} and C_{70} cations, which have one or more attached groups on the cage, and the other is the aza[60]fullerene cation $C_{59}N^+$. Both types of cations could be generated as long-lived species, and some of them have been isolated as pure salts. Although $C_{59}N^+$ can be stabilized by a significant resonance contribution of the iminium ion structure, no such stabilization is available for alkylated C_{60} and C_{70} cations. However, precise evaluation of the thermodynamic stabilities has demonstrated that even such cations have stabilities similar to tertiary alkyl cation such as the *tert*-butyl cation. Derivatized fullerene cations and the azafullerene cation are useful as reaction intermediates, as is demonstrated by its formation as a reaction intermediate and trapping with a nucleophile, which provides an efficient method for the synthesis of exohedral fullerene derivatives.

Several open-cage fullerene derivatives have been synthesized by selective cage scission reactions along with insertion of a sulfur atom on a rim of the orifice to enlarge it. Small guest species such as a helium atom and hydrogen and water molecules have been encapsulated into them under various different conditions. Then, closure of the orifice of a particular open-cage fullerene encapsulating molecular hydrogen was achieved to give $H_2@C_{60}$ via the molecular surgery approach.

Open-cage fullerene derivatives are useful not only for the synthesis of endohedral fullerenes as described in this chapter but also for the synthesis of heterofullerenes such as $C_{59}NH$ [94] and $(C_{59}N)_2$ [35]. Thus, there are possibilities that novel heterofullerenes can be obtained from open-cage fullerenes if reactions are developed to insert heteroatoms on the rim of the orifice and to close the orifice with the heteroatoms kept on the fullerene cage. Furthermore, an open-cage fullerene has been used to modify a metal surface such as Au(111) by the formation of self-assembled monolayers [95].

Insertion of a metal ion into an open-cage fullerene still remains a challenge in this field of chemistry. By theoretical calculations, the lithium ion is sufficiently small to be inserted into open-cage fullerene **41** [96]. However, there is a possibility that lithium ion coordinates to the carbonyl groups in open-cage fullerenes such as **41** and **56**. Thus, further developments in transformations of functional groups in the orifice of open-cage fullerenes are required. In connection with the developments in the total synthesis of fullerenes [97] and their partial structure [98], a variety of open-cage fullerenes will become available for such purposes in future. The development in this field is highly expected.

9.5
Experimental: Selected Procedures

9.5.1
Synthesis of 1-Chloro-4-dichloromethyl-1,4-dihydro[60]fullerene (12b)

To a solution of C_{60} (0.30 g, 0.42 mmol) in dry $CHCl_3$ (300 mL) was added $AlCl_3$ (5.0 g, 38 mmol). The mixture was stirred for 2 h and quenched by the addition of 120 mL of cold water. The product was extracted with CS_2, and the organic layer was washed with 5% $NaHCO_3$ and water and dried ($MgSO_4$). Evaporation of the solvent gave a dark solid, which was purified by MPLC (SiO_2, quick elution with CS_2 at $-20\ ^\circ$C) to give 0.24 g (68%) of **12b** as a dark brown solid.

9.5.2
Synthesis of 1-Dichloromethyl-4-hydroxy-1,4-dihydro[60]fullerene (2b)

Silica gel (230–400 mesh, 5 g) was suspended in a CS_2 solution (30 mL) of crude chloride **12b**, obtained from 0.30 g (0.42 mmol) of C_{60}. The CS_2 was evaporated under vacuum, and the residual silica gel, on which the sample was adsorbed, was charged at the inlet end of a chromatography column (60 mm id × 500 mm), filled with 230–400 mesh silica gel. The column was eluted with benzene at a rate of 20 mL min^{-1}. The movement of the sample in the column was visually monitored from the outside of the glass column. Fractions containing the major component were collected and evaporated to give fullerenol **2b** (0.18 g, 53% based on C_{60}) as a dark solid.

9.5.3
Generation of Dichloromethylated [60]Fullerene Cation (3b)

Fullerenol **2b** (20 mg) was dissolved in 1 mL of CF_3SO_3H. The NMR spectrum of the resulting reddish purple solution of **3b** was recorded at room temperature using cyclohexane-d_{12}, sealed in a glass capillary, as a deuterium lock and an external reference.

9.5.4
Synthesis of an Open-Cage Fullerene with an Eight-membered-ring Orifice (49)

A mixture of C_{60} (50 mg, 0.069 mmol) and 5,6-diphenyl-3-(2-pyridyl)-1,2,4-triazine (21 mg, 0.068 mmol) in ODCB (4 mL) was refluxed at 180 °C for 17 h under an argon atmosphere. The resulting dark purple solution was directly subjected to flash column chromatography over silica gel. Elution with CS_2 gave unreacted C_{60} (20 mg, 41%) while the following elution with CS_2-ethyl acetate (20:1) gave open-cage fullerene derivative **49** (35 mg, 0.035 mmol, 50% based on C_{60}) as a brown powder.

9.5.5
Synthesis of an Open-Cage Fullerene with a 12-Membered-ring Orifice (54)

A solution of compound **49** (66 mg, 0.066 mmol) in CCl_4 (65 mL) in a Pyrex flask was irradiated with a high-pressure mercury lamp (500 W) from the distance of 20 cm for 6 h under air. The resulting brown solution was evaporated and the residual black solid was dissolved in ODCB (3 mL), which was subjected to preparative HPLC using a Cosmosil 5PBB column (10 mm id × 250 mm) eluted with ODCB (flow rate, 2 mL min^{-1}) to afford open-cage fullerene derivatives **54** (40 mg, 0.038 mmol, 60%) and **55** (21 mg, 0.020 mmol, 31%) after nine recycles, both as brown powders.

9.5.6
Synthesis of an Open-Cage Fullerene with a 13-Membered-ring Orifice (56)

To a heated and stirred solution of compound **54** (32 mg, 0.031 mmol) and elemental sulfur (8 mg, 0.031 mmol as S_8) in ODCB (15 mL) was added tetrakis(dimethylamino)ethylene (7.1 μL, 0.031 mmol) at 180 °C. The solution was refluxed at 180 °C for 30 min. Then the resulting dark red-brown solution was concentrated to ca. 3 mL, which was added to pentane (30 mL) with vigorous stirring to give a brown precipitate. The precipitate was collected by centrifuge and dissolved in ODCB (2 mL). The resulting solution was subjected to flash chromatography on silica gel eluted with toluene–ethyl acetate (30:1) to give the open-cage fullerene derivative **56** (25 mg, 0.023 mmol, 77%) as a brown powder.

Abbreviations

An	4-methoxyphenyl
m-CPBA	m-chloroperbenzoic acid
DBU	1,8-diazabicyclo[5.4.0]undec-7-ene

DFT density functional theory
Fc ferrocene
GIAO gauge-independent atomic orbital
MEM methoxyethoxymethyl
ODCB o-dichlorobenzene
Py 2-pyridyl
SWNT single-walled carbon nanotube
TDAE tetrakisdimethylaminoethylene

References

1 W. Krätschmer, L. D. Lamb, K. Fostiropoulos, D. R. Huffman, Nature **1990**, 347, 354–358.
2 (a) A. Hirsch, M. Brettreich, Fullerenes: Chemistry and Reactions, Wiley-VCH, Weinheim, **2005**; (b) Topics in Current Chemistry, 199. Fullerenes and Related Structures, A. Hirsch (ed.), Springer, Berlin, **1999**; (c) The Chemistry of Fullerenes, R. Taylor (ed.), World Scientific, Singapore, **1995**; (d) A. Hirsch, The Chemistry of the Fullerenes, Thieme, Stuttgart, **1994**; (e) Fullerenes: Chemistry, Physics, and Technology, K. M. Kadish, R. S. Ruoff (eds.), John Wiley & Sons, New York, **2000**.
3 R. D. Bolskar, R. S. Mathur, C. A. Reed, J. Am. Chem. Soc. **1996**, 118, 13093–13094.
4 Y. Yang, F. Arias, L. Echegoyen, L. P. F. Chibante, S. Flanagan, A. Robertson, L. J. Wilson, J. Am. Chem. Soc. **1995**, 117, 7801–7804.
5 Q. Xie, F. Arias, L. Echegoyen, J. Am. Chem. Soc. **1993**, 115, 9818–9819.
6 C. A. Reed, K.-C. Kim, R. D. Bolskar, L. J. Mueller, Science **2000**, 289, 101–104.
7 G. A. Olah, I. Bucsi, R. Aniszfeld, G. K. S. Prakash, Carbon **1992**, 30, 1203–1211.
8 (a) G. P. Miller, C. S. Hsu, H. Thomann, L. Y. Chiang, M. Bernardo, Mater. Res. Soc. Symp. Proc. **1992**, 247, 293–300; (b) F. Cataldo, Spectrochim. Acta, Part A **1995**, 51, 405–414; (c) H. Thomann, M. Bernardo, G. P. Miller, J. Am. Chem. Soc. **1992**, 114, 6593–6594; (d) S. G. Kukolich, D. R. Huffman, Chem. Phys. Lett. **1991**, 182, 263–265; (e) C. A. Reed, R. D. Bolskar, Chem. Rev. **2000**, 100, 1075–1119.
9 (a) T. Kitagawa, H. Sakamoto, K. Takeuchi, J. Am. Chem. Soc. **1999**, 121, 4298–4299; (b) T. Kitagawa, K. Takeuchi, Bull. Chem. Soc. Jpn. **2001**, 74, 785–800.
10 B. Abarca, G. Asensio, R. Ballesteros, T. Varea, J. Org. Chem. **1991**, 56, 3224–3229.
11 G. A. Olah, G. K. S. Prakash, G. Liang, P. W. Westerman, K. Kunde, J. Chandrasekhar, P. v. R. Schleyer, J. Am. Chem. Soc. **1980**, 102, 4485–4492.
12 G. A. Olah, G. K. S. Prakash, J. Sommer, Superacids, John Wiley & Sons, New York, **1985**.
13 Y. Murata, F. Cheng, T. Kitagawa, K. Komatsu, J. Am. Chem. Soc. **2004**, 126, 8874–8875.
14 F. Cheng, Y. Murata, K. Komatsu, Org. Lett. **2002**, 4, 2541–2544.
15 T. Kitagawa, Y. Lee, N. Masaoka, K. Komatsu, Angew. Chem. **2005**, 117, 1422–1425; Angew. Chem. Int. Ed. **2005**, 44, 1398–1401.
16 L. J. Mueller, D. W. Elliott, K.-C. Kim, C. A. Reed, P. D. W. Boyd, J. Am. Chem. Soc. **2002**, 124, 9360–9361.
17 A. G. Avent, P. R. Birkett, H. W. Kroto, R. Taylor, D. R. M. Walton, Chem. Commun. **1998**, 2153–2154.
18 P. R. Birkett, M. Bühl, A. Khong, M. Saunders, R. Taylor, J. Chem. Soc., Perkin Trans. 2 **1999**, 2037–2039.
19 (a) L. Y. Chiang, R. B. Upasani, J. W. Swirczewski, J. Am. Chem. Soc.

1992, *114*, 10154–10157; (b) Ref. 2c, Ch. 5.

20 (a) G. A. Olah, I. Bucsi, D. S. Ha, R. Aniszfeld, C. S. Lee, G. K. S. Prakash, *Fullerene Sci. Tech.* **1997**, *5*, 389–405; (b) G. A. Olah, I. Bucsi, C. Lambert, R. Aniszfeld, N. J. Trivedi, D. K. Sensharma, G. K. S. Prakash, *J. Am. Chem. Soc.* **1991**, *113*, 9387–9388.

21 T. Kitagawa, Y. Lee, M. Hanamura, H. Sakamoto, H. Konno, K. Takeuchi, K. Komatsu, *Chem. Commun.* **2002**, 3062–3063.

22 G. A. Olah, G. Rasul, L. Heiliger, G. K. S. Prakash, *J. Am. Chem. Soc.* **1996**, *118*, 3580–3583.

23 G. Frenking, S. Fau, C. M. Marchand, H. Grützmacher, *J. Am. Chem. Soc.* **1997**, *119*, 6648–6655.

24 (a) C. C. Henderson, C. M. Rohlfing, P. A. Cahill, *Chem. Phys. Lett.* **1993**, *213*, 383–388; (b) *Electronic Structure Calculations on Fullerenes and Their Derivatives*, J. Cioslowski (ed.), Oxford University Press, New York, **1995**, Ch. 9; (c) Ref. 2c, Ch. 4; (d) C. C. Henderson, P. A. Cahill, *Chem. Phys. Lett.* **1992**, *198*, 570–576; (e) N. Matsuzawa, D. A. Dixon, T. Fukunaga, *J. Phys. Chem.* **1992**, *96*, 7594–7604.

25 T. Kitagawa, Y. Lee, M. Hanamura, K. Takeuchi, K. Komatsu, unpublished results.

26 (a) T. Kitagawa, T. Tanaka, Y. Takata, K. Takeuchi, K. Komatsu, *J. Org. Chem.* **1995**, *60*, 1490–1491; (b) T. Kitagawa, T. Tanaka, Y. Takata, K. Takeuchi, K. Komatsu, *Tetrahedron* **1997**, *53*, 9965–9976; (c) T. Kitagawa, T. Tanaka, H. Murakita, K. Takeuchi, *J. Org. Chem.* **1999**, *64*, 2–3; (d) T. Kitagawa, T. Tanaka, H. Murakita, A. Nishikawa, K. Takeuchi, *Tetrahedron* **2001**, *57*, 3537–3547; (e) P. J. Fagan, P. J. Krusic, D. H. Evans, S. A. Lerke, E. Johnston, *J. Am. Chem. Soc.* **1992**, *114*, 9697–9699; (f) A. Hirsch, T. Grösser, A. Skiebe, A. Soi, *Chem. Ber.* **1993**, *126*, 1061–1067.

27 C. Thilgen, A. Herrmann, F. Diederich, *Angew. Chem.* **1997**, *109*, 2362–2374; *Angew. Chem. Int. Ed. Engl.* **1997**, *36*, 2268–2280 and references cited therein.

28 (a) M. Sawamura, H. Iikura, A. Hirai, E. Nakamura, *J. Am. Chem. Soc.* **1998**, *120*, 8285–8286; (b) Z. Wang, M. S. Meier, *J. Org. Chem.* **2003**, *68*, 3043–3048; (c) Z. Wang, M. S. Meier, *J. Org. Chem.* **2004**, *69*, 2178–2180.

29 (a) C. Bellavia-Lund, F. Wudl, *J. Am. Chem. Soc.* **1997**, *119*, 943–946; (b) M. S. Meier, G.-W. Wang, R. C. Haddon, C. P. Brock, M. A. Lloyd, J. P. Selegue, *J. Am. Chem. Soc.* **1998**, *120*, 2337–2342.

30 (a) H. R. Karfunkel, A. Hirsch, *Angew. Chem.* **1992**, *104*, 1529–1531; *Angew. Chem. Int. Ed. Engl.* **1992**, *31*, 1468–1470; (b) C. C. Henderson, C. M. Rohlfing, K. T. Gillen, P. A. Cahill, *Science* **1994**, *264*, 397–399.

31 (a) R. Taylor, *Synlett* **2000**, 776–793; (b) P. R. Birkett, A. G. Avent, A. D. Darwish, I. Hahn, H. W. Kroto, G. J. Langley, J. O'Loughlin, R. Taylor, D. R. M. Walton, *J. Chem. Soc., Perkin Trans. 2* **1997**, 1121–1125; (c) A. G. Avent, P. R. Birkett, J. D. Crane, A. D. Darwish, G. J. Langley, H. W. Kroto, R. Taylor, D. R. M. Walton, *J. Chem. Soc., Chem. Commun.* **1994**, 1463–1464; (d) Ref. 2c, Ch. 8.

32 M. B. Smith, J. March, *Advanced Organic Chemistry*, John Wiley & Sons, New York, **2001**, pp. 432–433.

33 (a) M. M. Toteva, J. P. Richard, *J. Am. Chem. Soc.* **1996**, *118*, 11434–11445; (b) D. D. M. Wayner, D. J. McPhee, D. Griller, *J. Am. Chem. Soc.* **1988**, *110*, 132–137; (c) E. M. Arnett, T. C. Hofelich, *J. Am. Chem. Soc.* **1983**, *105*, 2889–2895.

34 N. C. Deno, J. J. Jaruzelski, A. Schriesheim, *J. Am. Chem. Soc.* **1955**, *77*, 3044–3051.

35 J. C. Hummelen, B. Knight, J. Pavlovich, R. González, F. Wudl, *Science* **1995**, *269*, 1554–1556.

36 J. C. Hummelen, M. Prato, F. Wudl, *J. Am. Chem. Soc.* **1995**, *117*, 7003–7004.

37 B. Nuber, A. Hirsch, *Chem. Commun.* **1996**, 1421–1422.

38 K.-C. Kim, F. Hauke, A. Hirsch,

P. D. W. Boyd, E. Carter, R. S. Armstrong, P. A. Lay, C. A. Reed, *J. Am. Chem. Soc.* **2003**, *125*, 4024–4025.

39 C. Bellavia-Lund, R. González, J. C. Hummelen, R. G. Hicks, A. Sastre, F. Wudl, *J. Am. Chem. Soc.* **1997**, *119*, 2946–2947.

40 U. Reuther, A. Hirsch, *Carbon* **2000**, *38*, 1539–1549.

41 B. Nuber, A. Hirsch, *Chem. Commun.* **1998**, 405–406.

42 F. Hauke, A. Hirsch, *Tetrahedron* **2001**, *57*, 3697–3708.

43 U. Reuther, A. Hirsch, *Chem. Commun.* **1998**, 1401–1402.

44 F. Hauke, S. Atalick, D. M. Guldi, J. Mack, L. T. Scott, A. Hirsch, *Chem. Commun.* **2004**, 766–767.

45 A. Hirsch, B. Nuber, *Acc. Chem. Res.* **1999**, *32*, 795–804.

46 J. C. Hummelen, C. Bellavia-Lund, F. Wudl, *Top. Curr. Chem.* **1999**, *199*, 93–134.

47 G. C. Vougioukalakis, N. Chronakis, M. Orfanopoulos, *Org. Lett.* **2003**, *5*, 4603–4606.

48 F. Hauke, A. Hirsch, *Chem. Commun.* **1999**, 2199–2200.

49 F. Hauke, A. Swartz, D. M. Guldi, A. Hirsch, *J. Mater. Chem.* **2002**, *12*, 2088–2094.

50 F. Hauke, M. Á. Herranz, L. Echegoyen, D. Guldi, A. Hirsch, S. Atalick, *Chem. Commun.* **2004**, 600–601.

51 C. Bingel, *Chem. Ber.* **1993**, *126*, 1957–1959.

52 G. C. Vougioukalakis, M. Orfanopoulos, *J. Am. Chem. Soc.* **2004**, *126*, 15956–15957.

53 N. Tagmatarchis, V. Georgakilas, M. Prato, H. Shinohara, *Chem. Commun.* **2002**, 2010–2011.

54 N. Tagmatarchis, V. Georgakilas, D. Tasis, M. Prato, H. Shinohara, *AIP Conf. Proceed.* **2003**, *685*, 287–290.

55 (a) *Endofullerenes: A New Family of Carbon Clusters*, T. Akasaka, S. Nagase (eds.), Kluwer Academic Publishers, Dordrecht, **2002**; (b) H. Shinohara, *Fullerenes: Chemistry, Physics and Technology*, K. M. Kadish, R. S. Ruoff (eds.), Wiley-VCH, Weinheim, **2000**, 357–393; (c) H. Shinohara, *Rep. Prog. Phys.* **2000**, *63*, 843–892; (d) S. Liu, S. Sun, *J. Organomet. Chem.* **2000**, *599*, 74–86; (e) S. Nagase, K. Kobayashi, T. Akasaka, *Bull. Chem. Soc. Jpn.* **1996**, *69*, 2131–2142.

56 S. Kobayashi, S. Mori, S. Iida, H. Ando, T. Takenobu, Y. Taguchi, A. Fujiwara, A. Taninaka, H. Shinohara, Y. Iwasa, *J. Am. Chem. Soc.* **2003**, *125*, 8116–8117.

57 (a) H. Kato, Y. Kanazawa, M. Okumura, A. Taninaka, T. Yokawa, H. Shinohara, *J. Am. Chem. Soc.* **2003**, *125*, 4391–4397; (b) É. Tóth, R. D. Bolskar, A. Borel, G. González, L. Helm, A. E. Merbach, B. Sitharaman, L. J. Wilson, *J. Am. Chem. Soc.* **2005**, *127*, 799–805.

58 (a) M. Saunders, R. J. Cross, H. A. Jiménez-Vázquez, R. Shimshi, A. Khong, *Science* **1996**, *271*, 1693–1697; (b) M. Saunders, H. A. Jiménez-Vázquez, R. J. Cross, S. Mroczkowski, D. I. Freedberg, F. A. L. Anet, *Nature* **1994**, *367*, 256–258.

59 Y. Rubin, T. Jarrosson, G.-W. Wang, M. D. Bartberger, K. N. Houk, G. Schick, M. Saunders, R. J. Cross, *Angew. Chem.* **2001**, *113*, 1591–1594; *Angew. Chem. Int. Ed.* **2001**, *40*, 1543–1546.

60 Y. Rubin, *Top. Curr. Chem.* **1999**, *199*, 67–91.

61 M. J. Arce, A. L. Viado, Y.-Z. An, S. I. Khan, Y. Rubin, *J. Am. Chem. Soc.* **1996**, *118*, 3775–3776.

62 C. M. Edwards, I. S. Butler, W. Qian, Y. Rubin, *J. Mol. Struct.* **1998**, *442*, 169–174.

63 P. R. Birkett, A. G. Avent, A. D. Darwish, H. W. Kroto, R. Taylor, D. R. M. Walton, *J. Chem. Soc., Chem. Commun.* **1995**, 1869–1870.

64 K. Hasharoni, C. Bellavia-Lund, M. Keshavarz-K., G. Srdanov, F. Wudl, *J. Am. Chem. Soc.* **1997**, *119*, 11128–11129.

65 G. Schick, T. Jarrosson, Y. Rubin, *Angew. Chem.* **1999**, *111*, 2508–2512; *Angew. Chem. Int. Ed.* **1999**, *38*, 2360–2363.

66 Y. Rubin, *Chem. Eur. J.* **1997**, *3*, 1009–1016.

67 Y. Rubin, P. S. Ganapathi, A. Franz, Y.-Z. An, W. Qian, R. Neier, *Chem. Eur. J.* **1999**, *5*, 3162–3184.

68 J.-F. Nierengarten, *Angew. Chem.* **2001**, *113*, 3061–3062; *Angew. Chem. Int. Ed.* **2001**, *40*, 2973–2974.

69 W. Qian, M. D. Bartberger, S. J. Pastor, K. N. Houk, C. L. Wilkins, Y. Rubin, *J. Am. Chem. Soc.* **2000**, *122*, 8333–8334.

70 T.-Y. Hsiao, K. C. Santhosh, K.-F. Liou, C.-H. Cheng, *J. Am. Chem. Soc.* **1998**, *120*, 12232–12236.

71 (a) H. Inoue, H. Yamaguchi, T. Suzuki, T. Akasaka, S. Murata, *Synlett* **2000**, 1178–1180; (b) S. Iwamatsu, P. S. Vijayalakshmi, M. Hamajima, C. H. Suresh, N. Koga, T. Suzuki, S. Murata, *Org. Lett.* **2002**, *4*, 1217–1220.

72 Y. Murata, N. Kato, K. Komatsu, *J. Org. Chem.* **2001**, *66*, 7235–7239.

73 Y. Murata, M. Murata, K. Komatsu, *J. Org. Chem.* **2001**, *66*, 8187–8191.

74 Y. Murata, M. Murata, K. Komatsu, *Chem. Eur. J.* **2003**, *9*, 1600–1609.

75 H. Inoue, H. Yamaguchi, S. Iwamatsu, T. Uozaki, T. Suzuki, T. Akasaka, S. Nagase, S. Murata, *Tetrahedron Lett.* **2001**, *42*, 895–897.

76 Y. Murata, K. Komatsu, *Chem. Lett.* **2001**, *30*, 896–897.

77 K. M. Creegan, J. L. Robbins, W. K. Robbins, J. M. Millar, R. D. Sherwood, P. J. Tindall, D. M. Cox, A. B. Smith III, J. P. McCauley Jr., D. R. Jones, R. T. Gallagher, *J. Am. Chem. Soc.* **1992**, *114*, 1103–1105.

78 G. C. Vougioukalakis, K. Prassides, M. Orfanopoulos, *Org. Lett.* **2004**, *6*, 1245–1247.

79 S. Iwamatsu, F. Ono, S. Murata, *Chem. Commun.* **2003**, 1268–1269.

80 S. Iwamatsu, F. Ono, S. Murata, *Chem. Lett.* **2003**, *32*, 614–615.

81 G. C. Vougioukalakis, K. Prassides, J. M. Campanera, M. I. Heggie, M. Orfanopoulos, *J. Org. Chem.* **2004**, *69*, 4524–4526.

82 S. Iwamatsu, T. Uozaki, K. Kobayashi, S. Re, S. Nagase, S. Murata, *J. Am. Chem. Soc.* **2004**, *126*, 2668–2669.

83 S. Iwamatsu, S. Murata, *Tetrahedron Lett.* **2004**, *45*, 6391–6394.

84 M. Saunders, H. A. Jiménez-Vázquez, R. J. Cross, S. Mroczkowski, M. L. Gross, D. E. Giblin, R. J. Poreda, *J. Am. Chem. Soc.* **1994**, *116*, 2193–2194.

85 R. J. Cross, A. Khong, M. Saunders, *J. Org. Chem.* **2003**, *68*, 8281–8283.

86 Y. Murata, M. Murata, K. Komatsu, *J. Am. Chem. Soc.* **2003**, *125*, 7152–7153.

87 M. Carravetta, Y. Murata, M. Murata, I. Heinmaa, R. Stern, A. Tontcheva, A. Samoson, Y. Rubin, K. Komatsu, M. H. Levitt, *J. Am. Chem. Soc.* **2004**, *126*, 4092–4093.

88 H. Sawa, Y. Wakabayashi, Y. Murata, M. Murata, K. Komatsu, *Angew. Chem.* **2005**, *117*, 2017–2019; *Angew. Chem. Int. Ed.* **2005**, *44*, 1981–1983.

89 C. M. Stanisky, R. J. Cross, M. Saunders, M. Murata, Y. Murata, K. Komatsu, *J. Am. Chem. Soc.* **2005**, *127*, 299–302.

90 S. Patchkovskii, W. Thiel, *J. Chem. Phys.* **1997**, *106*, 1796–1799.

91 K. Komatsu, M. Murata, Y. Murata, *Science* **2005**, *307*, 238–240.

92 J. E. McMurry, *Chem. Rev.* **1989**, *89*, 1513–1524.

93 G.-W. Wang, K. Komatsu, Y. Murata, M. Shiro, *Nature* **1997**, *387*, 583–586.

94 M. Keshavarz-K, R. González, R. G. Hicks, G. Srdanov, V. I. Srdanov, T. G. Collins, J. C. Hummelen, C. Bellavia-Lund, J. Pavlovich, F. Wudl, K. Holczer, *Nature* **1996**, *383*, 147–150.

95 S. Yoshimoto, E. Tsutsumi, Y. Honda, Y. Murata, M. Murata, K. Komatsu, O. Ito, K. Itaya, *Angew. Chem.* **2004**, *116*, 3106–3109; *Angew. Chem. Int. Ed.* **2004**, *43*, 3044–3047.

96 S. Irle, Y. Rubin, K. Morokuma, *J. Phys. Chem. A* **2002**, *106*, 680–688.

97 L. T. Scott, M. M. Boorum, B. J. McMahon, S. Hagen, J. Mack, J. Blank, H. Wegner, A. Meijere, *Science* **2002**, *295*, 1500–1503.

98 H. Sakurai, T. Daiko, T. Hirao, *Science* **2003**, *301*, 1878.

10
Polyynes

Vivian W. W. Yam and C. H. Tao

10.1
Introduction

The chemistry of carbon has long been one of the most fascinating fields in the history of chemistry and has attracted enormous attention, not only from chemists but also from scientists in different fields [1–7]. The discovery of the new allotropes, the fullerenes [8], which was honored by the award of the Nobel Prize in Chemistry in 1996 [1–3], has further boosted research into carbon chemistry.

Along with the new chemistry of the allotropes, the synthesis and study of polymeric carbon–carbon triple bonds, referred to as carbyne or chaoite, remain a challenging task (Scheme 10.1) [9–15]. Instead, shorter analogs of end-capped carbyne and their corresponding metal complexes have been synthesized by various groups and have been reviewed recently [16–20]. The rigidity and structural linearity render the C≡C moiety a promising construction unit for carbon-rich compounds [21]. For instance, carbon-rich phenylenes of different topologies can be prepared by cobalt-catalyzed cycloisomerization of branched polyynyl molecules [22–27].

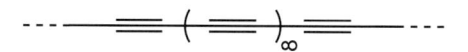

Scheme 10.1. The polymeric carbon–carbon triple bond allotrope "carbyne" or "chaoite".

In addition, there are well-established coupling reaction protocols for terminal alkynes [28]. For instance, terminal alkynes can be readily coupled with aryl halides in high yields using palladium/Cu-catalyzed Sonogashira or Stille coupling conditions; they can also be self-coupled via copper-catalyzed Hay or Eglinton coupling reactions. All these advantages have made the C≡C unit appropriate for the construction of carbon-rich compounds of different dimensions. Despite the extensive use and wide applicability of various coupling methods, there has still been a growing interest and demand for improved methods. Many concentrate on the design of appropriate catalyst systems [29–31], while considerable efforts have been

Carbon-Rich Compounds. Edited by Michael M. Haley and Rik R. Tykwinski
Copyright © 2006 WILEY-VCH Verlag GmbH & Co. KGaA, Weinheim
ISBN: 3-527-31224-2

made to avoid the use of a copper co-catalyst [32]. Recently, metal-free coupling reactions between aryl halides and terminal alkynes have also been reported [33, 34]. These new synthetic protocols are targeted to prevent or reduce homocoupling of terminal alkynes or to shorten the reaction times, and it is obvious that such advances are important in the synthesis and study of polyynyl materials.

10.1.1
Organic Polyynes

Since the first attempts to synthesize long sp-carbon chains by Bohlmann and Jones in the 1950s [35–38], the chemical and electronic properties of this class of compounds have been extensively studied. A series of end-capped oligoynes, in which the carbon chains are kept at a distance by bulky and spherical dendrimer endgroups with a chain length of up to 20 carbon atoms, has been synthesized by Hirsch and coworkers [39]. Apart from dendrimer end-capped oligoynes, which mainly help to stabilize the system, many different functional end-groups have been incorporated into the two ends of sp carbon chains by chemists in order to study the electronic properties and explore possible applications of this class of compounds (Scheme 10.2) [10, 40–43]. The electronic structures of organic

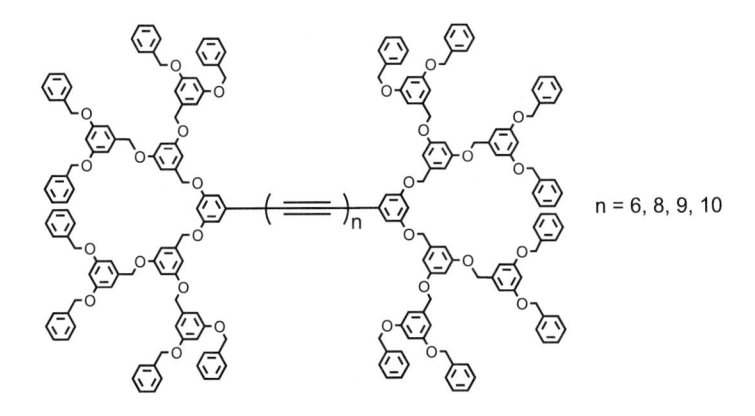

Scheme 10.2. Examples of oligoynes with different endgroups.

polyynyl chains with various end-groups have been studied by Bohlmann [35, 36], Walton [44, 45], Hirsch [12, 13, 39], and Tykwinski [46, 47]. The electronic absorption spectra of the polyynes generally show highly structured absorption bands, with the lowest energy absorption bands showing a bathochromic shift with increasing number of C≡C units. The absorption wavelength for the hypothetical carbyne has been estimated by Hirsch and coworkers to be around 569 nm. Instead of the commonly found even number of C≡C bonds in this class of compounds, Diederich and coworkers have been interested in the synthesis of long oligoynes incorporating an odd number of C≡C bonds [48].

Recently, the exploration of oligo(phenylene ethynylene)s as wires and devices has drawn much attention due to their relatively high stability compared to oligoynes. Sulfur-containing alligator clips have recently been introduced into oligo(phenylene ethynylene)s by Tour et al. [49, 50], and Kushmerick and Shashidhar [51] to facilitate electronic measurements on self-assembled monolayers (SAM) of these "nano-wires" using scanning tunnelling microscopy (STM) or the crossed-wire junction technique (Scheme 10.3).

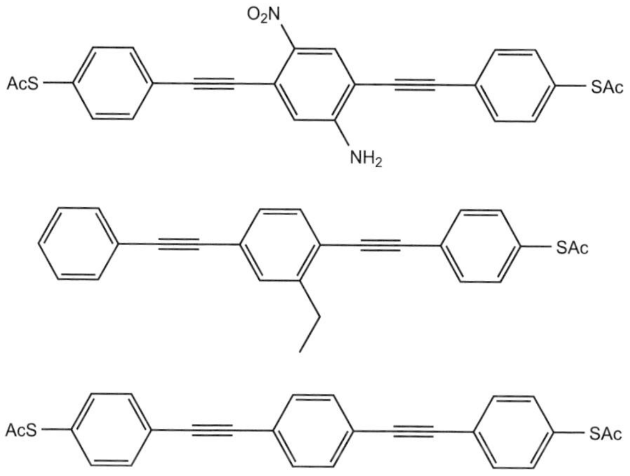

Scheme 10.3. Examples of oligo(phenylene ethynylene)s with thioacetate as alligator clips.

A change in the linkage from the para-position to the meta- or ortho-positions leads to the zig-zag systems of Grubbs and Kratz [52], or the helical folding polymers of Moore and coworkers (Scheme 10.4) [53]. It has been proposed that the formation of the helical structures of the latter is driven by solvophobic interactions that are sensitive to chain length, solvent quality, and temperature.

In addition to the one-dimensional wire-like compounds, a considerable number of two-dimensional and three-dimensional architectures have also been built, using the chemistry of the alkynyl units in an elegant manner, by the groups of Bunz [54], Diederich [55–57], de Meijere [58], Moore [59], Tykwinski [46, 47], Haley [60], Vollhardt [22–27] and others [61, 62]. In particular, a series of perylene-terminated phenylene ethynylene dendrimers with energy gradients was

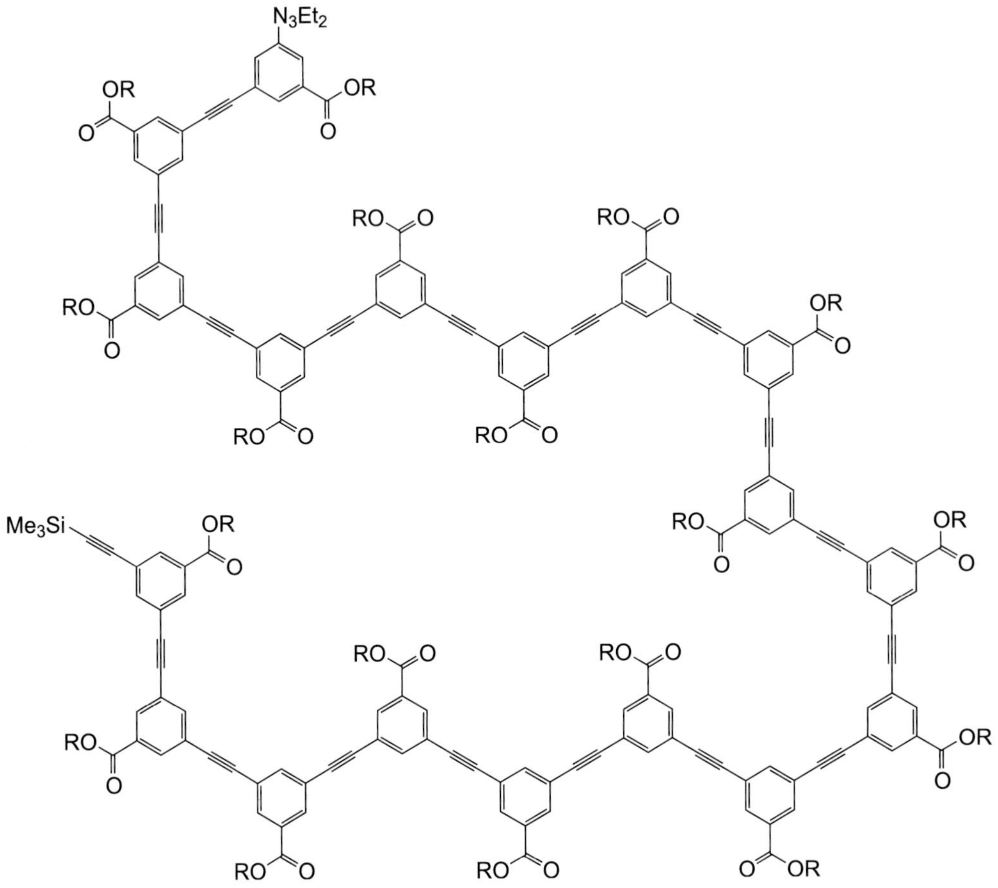

Scheme 10.4. Helical folding polymers by Moore and coworkers.

successfully synthesized and isolated by Moore and coworkers (Scheme 10.5) [63]; the luminescence and energy transfer properties have been evaluated. The light harvesting ability of these compounds increases with increasing generation, however, the efficiency of the energy transfer decreases with increasing generation. The rate constant for energy transfer for one of the perylene-terminated dendrimer was calculated to be 1.9×10^{11} s^{-1} and it was proposed that such ultrafast energy transfer could be attributed to the energy gradient from the periphery to the perylene focal point.

10.1.2
Organometallic Polyynyl Complexes

The alkynyl group is also known for its ability to interact with transition metal centers through pπ–dπ overlap and appears to be a promising candidate for the

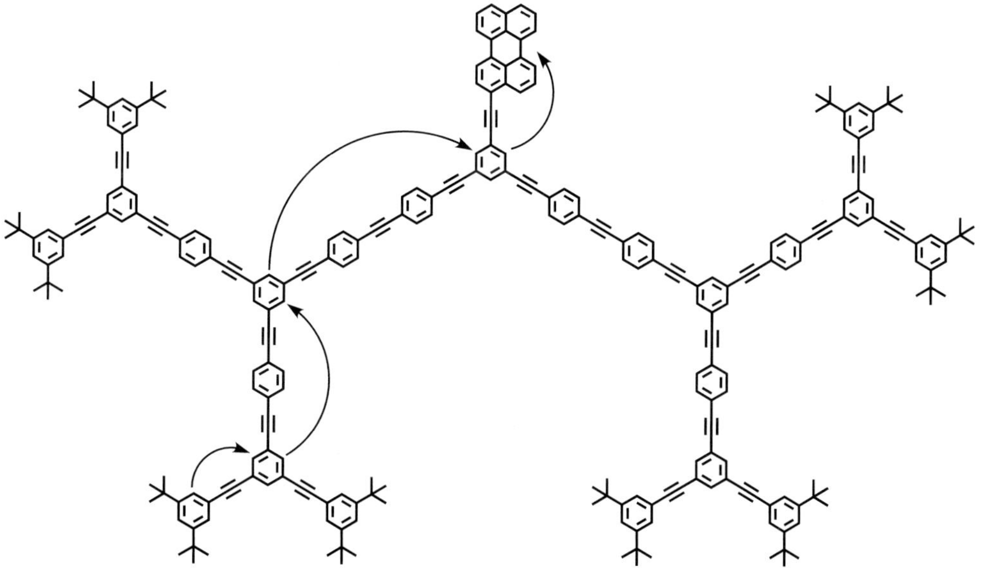

Scheme 10.5. A perylene-terminated phenylene ethynylene dendrimers with energy gradient. Arrows indicate direction of energy transfer.

construction of carbon-rich metal-containing materials while its inherent rigidity would be beneficial in the preparation of luminescent materials. The emission origins of these metal alkynyl complexes may include intraligand (IL) states, ligand-field (LF) states, metal-to-ligand charge transfer (MLCT) states, ligand-to-ligand charge transfer (LLCT) states and metal-metal-to-ligand charge transfer (MMLCT) states [64]. Recently, the interplay of the relative energy of these plausible excited states in organometallic complexes has aroused much attention. It is envisioned that the construction of carbon-rich metal alkynyl complexes may find potential applications in the field of OLEDs and sensor and molecular devices [65]. In this chapter, we will focus on the luminescence aspect of alkynyl complexes with different metal centers, the structure–property relationship between the polyynyl backbones and general synthetic protocols towards metal alkynyls.

10.2
Synthetic Routes Towards Transition Metal Alkynyl Complexes

Several general synthetic routes towards transition metal alkynyls have been developed in the past few decades. Some of the most commonly employed methods are discussed below while selected experimental procedures are summarized at the end of this chapter.

10.2.1
Copper-catalyzed Dehydrohalogenation

One of the most commonly employed and convenient methods for the synthesis of transition metal alkynyls is the copper-catalyzed dehydrohalogenation reaction between the metal halides and terminal alkynes, which involves the use of alkylamine as the base together with a trace amount of copper(I) halide as catalyst (Scheme 10.6). In fact, this route was developed by Hagihara and coworkers for the synthesis of the first soluble transition metal alkynyl polymers with molecular weights (M_w) of more than 10^6 [66]. The mechanism, which has been studied by Yamamoto et al. [67–69], involves the transmetallation of the organic alkyne from the d^{10} metal catalyst to the metal halide with the elimination of hydrogen halide which was then reacted with the alkylamine. This synthetic protocol has been widely employed in the synthesis of d^8 and d^{10} transition metal alkynyls of palladium(II), platinum(II), gold(I) and mercury(II). The advantages of this method lie in its ease of manipulation and the absence of highly reactive and potentially dangerous organometallic reagents.

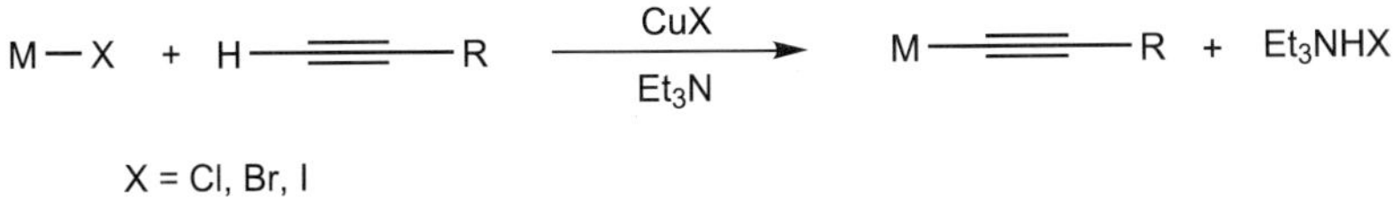

$$M-X + H-\!\!\!\equiv\!\!\!-R \xrightarrow[\text{Et}_3\text{N}]{\text{CuX}} M-\!\!\!\equiv\!\!\!-R + \text{Et}_3\text{NHX}$$

$$X = \text{Cl, Br, I}$$

Scheme 10.6. Preparation of metal alkynyls via copper-catalyzed dehydrohalogenation.

10.2.2
Metal Halides with Alkynylating Agents

Another common synthetic method for transition metal alkynyls is the reaction between the metal halide and alkynylating agents, including lithium, sodium, magnesium and copper(I) alkynyl reagents (Scheme 10.7). The use of such alkynylating reagents is particularly important for metal halides, especially those that are inert to the copper-catalyzed dehydrohalogenation reaction. However, these compounds are highly moisture- and air-sensitive. The preparation of metal alkynyls via this synthetic route requires strictly anaerobic and anhydrous conditions. Moreover, the strongly basic nature of these alkynylating reagents may limit the applicability and lower the yield of the reaction. In view of the above mentioned limitations, the more stable and mild trialkylstannylalkynes appear to be a suit-

$$M-X + M'-\!\!\!\equiv\!\!\!-R \longrightarrow M-\!\!\!\equiv\!\!\!-R + M'X$$

$$M' = \text{Li, Na, MgX, Cu, SnR}_3$$
$$X = \text{Cl, Br, I}$$

Scheme 10.7. Preparation of metal alkynyls via metal halides with alkynylating agents.

able alkynylating reagent for a wider spectrum of metals such as rhodium(I) and cobalt(III) which are not compatible with the more reactive alkynylating reagents.

10.2.3
Metal-free Methods with Terminal Alkynes

In place of the commonly used metal halides as the metal precursors, transition metal complexes with labile leaving groups have also been used for the preparations of metal alkynyls (Scheme 10.8). The presence of good leaving groups, for example, N_2, MeCN or H_2O ligands, would provide potential sites for ligand substitutions. These kinds of reactions are usually carried out in the presence of sodium hydroxide or trialkylamine which act as a base to deprotonate the terminal alkynes.

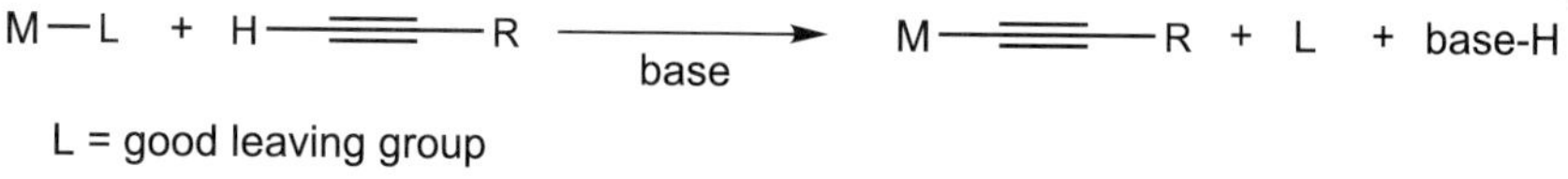

M—L + H≡≡≡R $\xrightarrow{\text{base}}$ M≡≡≡R + L + base-H

L = good leaving group

Scheme 10.8. Preparation of metal alkynyls via metal-free methods with terminal alkynes.

This synthetic protocol, although not widely applicable to various transition metals, provides a convenient route for the preparation of metal alkynyl complexes without the use of Cu(I) catalyst or strongly basic alkynylating reagents (Scheme 10.8). This metal-free method not only opens up the possibility of using a wider range of alkynyl ligands in the preparation of metal alkynyls but also eliminates possible π-coordination of the catalyst to the alkynyl units.

10.3
Rhenium(I) Alkynyls

In 1968, Bruce and coworkers reported the first preparation of rhenium(I) alkynyl complexes in low yield by treating rhenium carbonyl bromide with organolithium reagents [70]. Later, Gladysz and coworkers reported the synthesis of an extensive series of wire-like rhenium(I) alkynyl complexes containing two [(η^5-C_5Me_5)-Re(NO)(PPh$_3$)] moieties bridged by polyynyl C_4, C_8, C_{12} and C_{20} ligands using homocoupling reactions (Scheme 10.9) [71–79]. In particular, the C_4-bridged dirhenium complex, [(η^5-C_5Me_5)-Re(NO)(PPh$_3$)(C≡C–C≡C)(PPh$_3$)(NO)Re(η^5-C_5Me_5)], has been isolated in three oxidation states and the magnetic, optical and electrochemical properties investigated. EPR spectra of the dirhenium complexes show undecets with $A_{iso, Re}$ values half of those of related monorhenium radical cations, indicating spin delocalization over two rhenium centers [79]. The limiting HOMO–LUMO energy gap of each series of R–(C≡C)$_n$–R could be

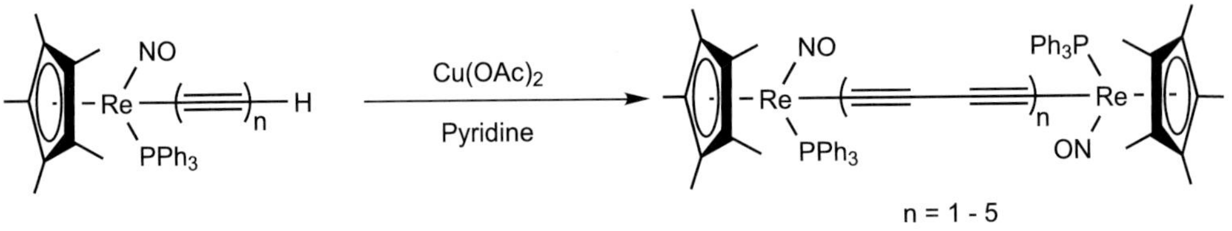
NO
Re
H
n
PPh3
Cu(OAc)2
Pyridine
NO
Re
PPh3
Ph3P
Re
ON
n
n = 1 - 5

Scheme 10.9. A series of wire-like complexes with [(η^5-C_5Me_5)–Re(NO)(PPh_3)] endgroups by Gladysz and coworkers.

estimated by plots of energies versus $1/n$, and a value of 565 nm was reported for this class of rhenium(I) polyynyl complexes, comparable to that reported by Hirsch and coworkers for their dendrimer end-capped polyynes.

In 1995, Yam and coworkers reported the first series of luminescent rhenium(I) diimine alkynyl complexes [$Re(CO)_3$($^tBu_2$bpy)(C≡C–R)] (R = alkyl or aryl) (Scheme 10.10) [80, 81] that contain the rhenium(I) tricarbonyl diimine system first developed by Wrighton et al. [82]. The electronic and photophysical properties have also been studied. In the early experiments, this class of complexes was synthesized using lithium alkynylating agents generated *in situ* from ^{n}BuLi and the corresponding terminal alkynes under an inert atmosphere. Later, a more general and improved approach to the synthesis of this class of compounds was developed, which is applicable for the synthesis of rhenium(I) wire-like species with a terminal acetylenic proton and with a wider variety of diimine ligands. Thus an extensive series of rhenium(I) alkynyls, [$Re(CO)_3$(N^N)$(C{\equiv}C)_n$–R″] (N^N = bpy, $^tBu_2$bpy, Me_2bpy, $(CF_3)_2$bpy, phen; $n = 1$ or 2) [83, 84], was prepared (Scheme 10.11).

Extension of this work to the triynyl system was accomplished by reacting the butadiynyl complexes with an acetylenic proton, [$Re(CO)_3$(N^N)(C≡C–C≡CH)], with an excess of bromo- or iodoalkynes in the presence of a catalytic amount of

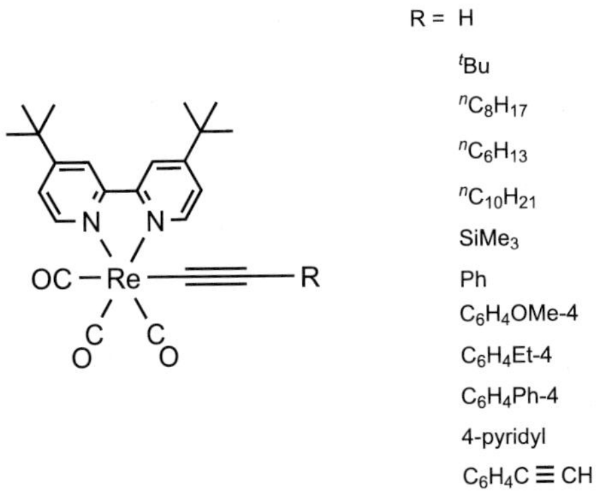
N
N
OC
Re
R
C
C
O
O
R = H
^{t}Bu
^{n}C8H17
^{n}C6H13
^{n}C10H21
SiMe3
Ph
C6H4OMe-4
C6H4Et-4
C6H4Ph-4
4-pyridyl
C6H4C≡CH

Scheme 10.10. A series of luminescent rhenium(I) diimine alkynyl complexes [$Re(CO)_3$($^tBu_2$bpy)(C≡C–R)] by Yam and coworkers.

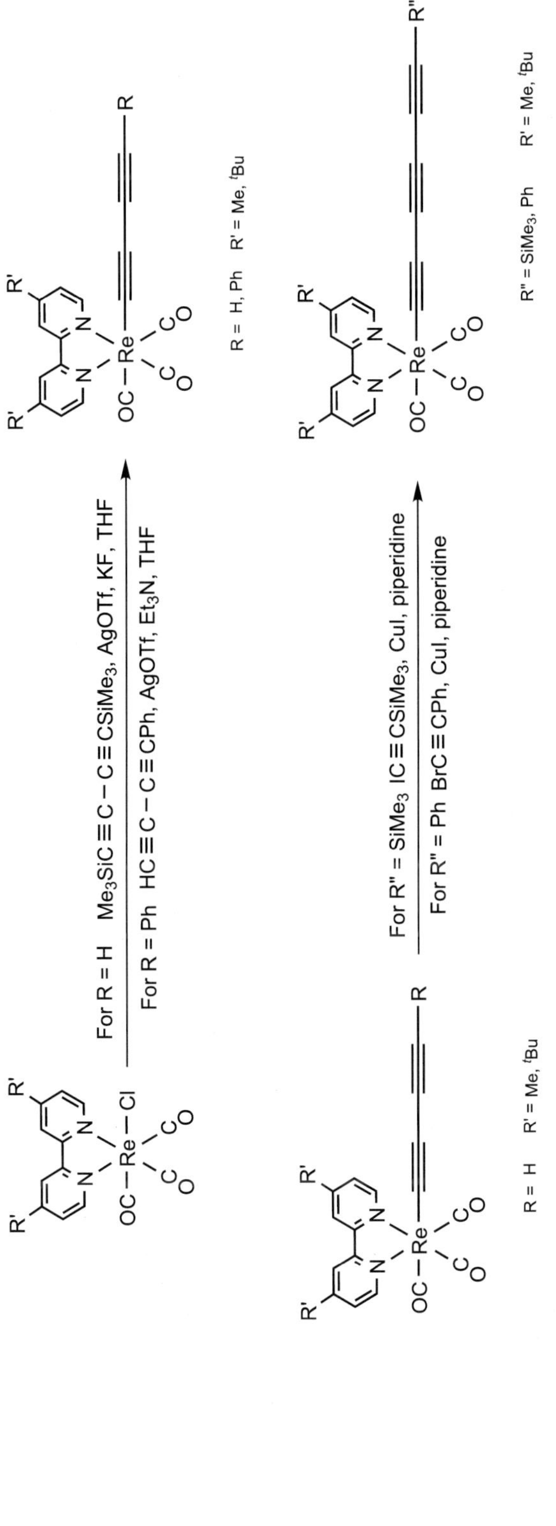

Scheme 10.11. Synthesis of a series of luminescent rhenium(I) diimine polyynyl complexes.

copper(I) catalyst (Scheme 10.11) [85]. In addition, the dinuclear C_4 and C_8 complexes of rhenium(I) tricarbonyl diimine system have also been prepared via the copper-catalyzed oxidative homo-coupling reaction of the monoynyl and diynyl complexes, respectively (Scheme 10.12) [86, 87]. The solubility of the dinuclear complexes decreases with increase in the number of C≡C units, as is commonly observed in organometallic polyynyl complexes, and the tetraynyl complex [Re(CO)$_3$(bpy)–C≡C–C≡C–C≡C–C≡C–Re(CO)$_3$(bpy)] is found to be insoluble in common organic solvents due to the absence of solubilizing ^{t}Bu groups on the bpy ligands [88].

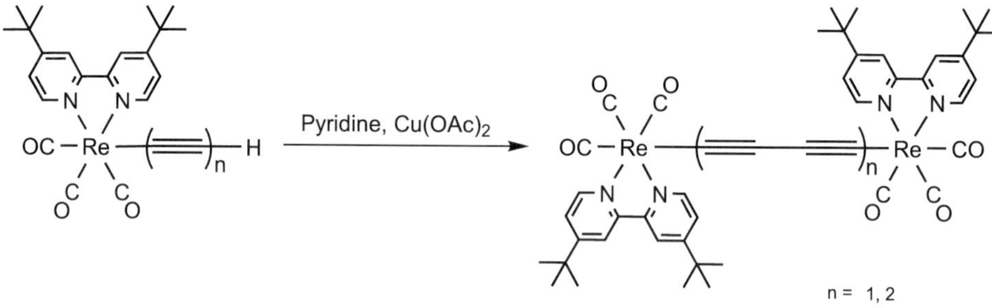

Scheme 10.12. Synthesis of a dinuclear luminescent rhenium(I) diimine wire-like complexes.

Upon photo-excitation at $\lambda > 350$ nm, all the rhenium(I) alkynyls exhibit intense orange-red luminescence in the solid state and in solution at room temperature (Table 10.1). The emission energies of the rhenium alkynyls are generally lower than those of the chlororhenium(I) precursors, [Re(CO)$_3$(N^N)Cl], probably due to the presence of the strong π-donating alkynyl group. The luminescence energies for this class of compounds have been found to be dependent on the nature of the substituents, both the diimine and the alkynyl ligands. For instance, higher energy MLCT absorption bands as well as ^{3}MLCT emission bands are observed when the electron-donating ability of the substituents on the diimine ligands increases. The same trend in the electronic absorption and emission energy is also observed when the electron-donating ability of substituents on the alkynyl ligand decreases. Such a trend in both absorption and emission energies would indicate a HOMO of mainly rhenium(I) character with some contribution of alkynyl character and a LUMO of mainly diimine character and the emissions have been assigned as originating from MLCT [dπ(Re) $\rightarrow \pi^*$(N^N)] triplet states, mixed with some ^{3}LLCT [π(C≡C) $\rightarrow \pi^*$(N^N)] character.

Another interesting feature of this class of luminophores is that a blue shift in absorption and emission energy is noted on going from the monoynyl complexes to the diynyl complexes and to the triynyl complexes, which is in contrast to the "trivial" red-shift observed upon increasing the π-conjugation in organic systems. Investigations of the rhenium(I) alkynyls by molecular orbital calculations [85] have shown that the HOMO is dominated by the antibonding character of the

Table 10.1. Electronic absorption and photophysical data of selected Re(I) diimine alkynyl complexes.

Complex	Absorption λ/nm (ε/dm^3 mol^{-1} cm^{-1})	Medium (T/K)	Emission λ_{em}/nm (τ_0/μs)	Ref.
[Re(^{t}Bu$_2$bpy)(CO)$_3$Cl]	292 sh (22 090), 314 sh (8 270), 384 (3 905)[a]	Solid (298) Solid (77) THF (298)	547 (0.90) 510 624 (0.06)	80
[Re(bpy)(CO)$_3$–C≡C–Ph]	286 (37 250), 298 sh (30 160), 326 sh (10 430), 420 (3 240)[b]	Solid (298) Solid (77) CH$_2$Cl$_2$ (298)	585 (0.19) 570 654 (<0.1)	89
[Re(bpy)(CO)$_3$–C≡C–C$_6$H$_4$–Cl-4]	284 (39 880), 298 sh (35 380), 324 sh (13 330), 420 (2 710)[b]	Solid (298) Solid (77) CH$_2$Cl$_2$ (298)	580 (<0.1) 565 650 (<0.1)	64
[Re(bpy)(CO)$_3$–C≡C–C$_6$H$_4$–C$_8$H$_{17}$-4]	278 (36 120), 298 sh (29 200), 328 sh (9 570), 424 (2 430)[b]	Solid (298) Solid (77) THF (298)	597 (0.27) 588 668 (<0.1)	64
[Re(^{t}Bu$_2$bpy)(CO)$_3$–C≡C–Ph]	325 sh (11 550), 419 (3 485)[a]	Solid (298) Solid (77) THF (298)	600 (0.24) 585 688 (0.20)	80
[Re(^{t}Bu$_2$bpy)(CO)$_3$–C≡C–C≡C–H]	248 (16 410), 284 (17 130), 404 (3 470)[a]	Solid (298) Solid (77) THF (298)	565 (<0.1) 580 620 (<0.1)	84
[Re(^{t}Bu$_2$bpy)(CO)$_3$–C≡C–C≡C–Ph]	248 (52 260), 298 (48 570), 340 (14 610), 416 (3 220)[a]	Solid (298) Solid (77) THF (298)	570 (<0.1) 570 (1.90, 0.32)[c] 625 (<0.1)	84
[Re(^{t}Bu$_2$bpy)(CO)$_3$–C≡C–C≡C–C≡C–SiMe$_3$]	250 (3 940), 296 (3 290), 350 sh (4 370), 412 (2 160)[a]	Solid (298) Solid (77) THF (298)	585 (<0.05) 575 596 (0.1)	85
[Re(^{t}Bu$_2$bpy)(CO)$_3$–C≡C–C≡C–C≡C–Ph]	260 (60 740), 282 (70 290), 298 sh (63 100), 352 (13 370), 378 (13 570), 426 (4 540)[a]	Solid (298) Solid (77) THF (298)	568 (<0.1) 588 (1.20, 0.11)[c] 632 (<0.1)	85

[a] Measured in THF at 298 K.
[b] Measured in CH$_2$Cl$_2$ at 298 K.
[c] Biexponential decay.

Re–(C≡C)$_n$Ph ($n = 2, 3$) moiety resulting from the overlap between dπ(Re) and [π(C≡C)$_n$Ph] orbitals, while the LUMO is mainly π^*(diimine). This is in accordance with the experimental findings and supports the assignment of the lowest-energy excited state of predominantly MLCT [dπ(Re) → π^*(N^N)] character. The unusual blue-shift upon increasing the π-conjugation in the alkynyl chain is probably a consequence of decreased overlap integral between the dπ(Re) and π(C≡C–C≡C–C≡C–Ph) relative to that of dπ(Re) and the π(C≡C–C≡C–Ph) and π(C≡C–Ph) orbitals respectively which leads to a larger HOMO–LUMO gap. With rational modifications on the alkynyl and diimine ligands, the emission color of these rhenium(I) alkynyl complexes could be readily tuned.

10.4
Platinum(II) Alkynyls

10.4.1
Platinum(II) Phosphine Systems

The construction of carbon-rich materials using platinum(II) phosphine systems is probably one of the most active areas of research in the past decade. The *trans*-square planar geometry adopted by the platinum(II) center provides a versatile motif for the construction of one-dimensional linear structures, while the corresponding *cis*-geometry could serve as connectors at an angle.

Since the successful synthesis of the first soluble alkynyl polymer of platinum in 1977 by Hagihara and coworkers (Scheme 10.13) [66], subsequent reports on platinum- and palladium-containing polyynyl polymers by both Hagihara and Lewis are good examples of the linearity of the *trans*-orientation [66, 90, 91].

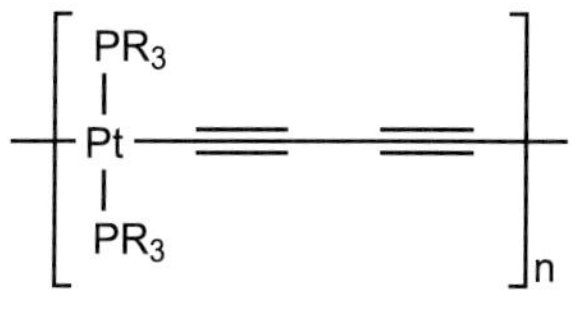

Scheme 10.13. The first soluble platinum-containing alkynyl polymer by Hagihara and coworkers.

A series of "insulated" molecular wires reported by Gladysz and coworkers with oligoynyl C≡C units surrounded by long bridging phosphine ligands in double-helical or lateral fashions are of particular interest (Scheme 10.14) [92–94]. More recently, a series of tetraplatinum bis(polyynediyl) complexes with laterally arrayed sp carbon chains was successfully isolated by the same group [95]. The "arrayed" complexes are found to have electronic absorption spectra similar to those of the diplatinum counterparts, indicating only minor or even no chain–chain electronic interactions. The electronic structure and chain-length effects in this class of mo-

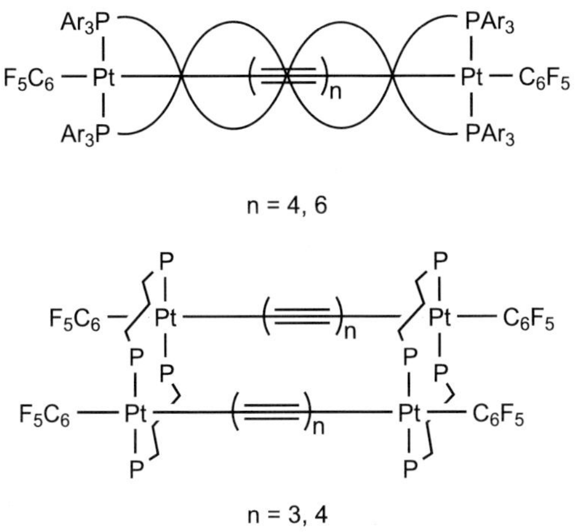

Scheme 10.14. Examples of "insulated" and "arrayed" polyynyl complexes with platinum-containing endgroups by Gladysz and coworkers.

lecular wires have been recently studied by DFT investigation, and the low-energy absorptions are assigned as IL($\pi \rightarrow \pi^*$) transitions of the polyynyl chain [96].

The auxiliary phosphine ligands also play an important role in controlling the physical properties of platinum(II) alkynyl complexes. Monomeric complexes of *trans*-[Pt(PR$_3$)$_2$(C≡CR')$_2$] usually exist in the crystalline state at room temperature. Cooper et al. recently communicated the synthesis and isolation of platinum alkynyls in the liquid state at room temperature using trioctylphosphine as the auxiliary ligands [97]. It was proposed that these liquids may possess enhanced nonlinear optical properties.

Apart from one-dimensional rigid-rod metal complexes, a number of molecular architectures have been reported by Youngs [98–100] and Bruce [101] using the *cis*-platinum(II) phosphine system that is locked by chelating phosphines (Scheme 10.15). Recently, organometallic assemblies with binaphthyl alkynyl moieties have been reported by Lin and coworkers, in which several chiral platinum-containing molecular polygons with binaphthyl alkynyl moieties have been synthesized (Scheme 10.16). Similar to the earlier reports by Pu and coworkers on the related organic architectures containing binaphthyl moieties [102], in which intriguing catalytic properties were observed, one of the platinum-containing molecular triangles has been shown to catalyze enantioselective diethylzinc additions to aromatic aldehydes [103]. The electronic absorption spectra are dominated by the metal-perturbed IL($\pi \rightarrow \pi^*$) transitions of the binaphthyl alkynyl units [104–106]. Energy transfer from the phenylene or phenyleneethynylene dendrons to the binaphthyl locus has also been reported [107–110]. All these are good examples of using the platinum(II) phosphine systems for the construction of carbon-rich molecular materials and their applications in various fields. In addition, the relatively high

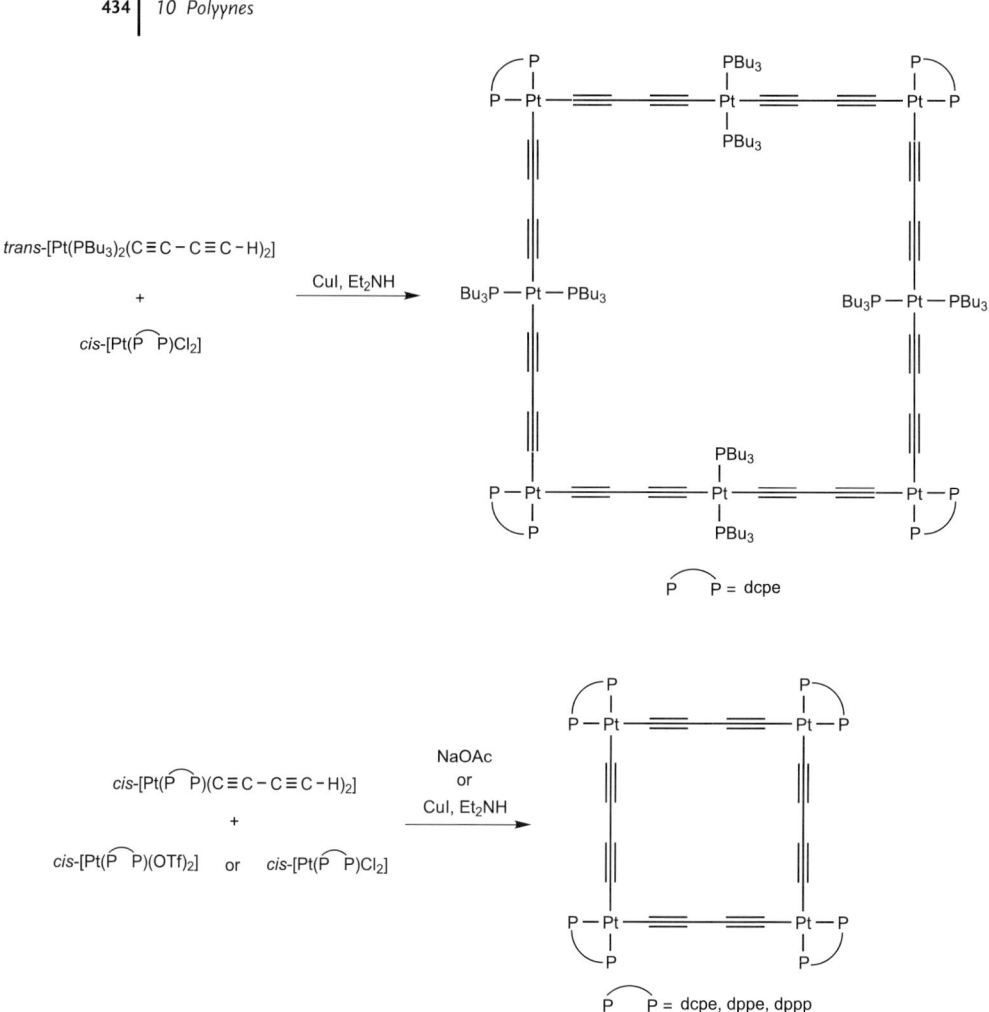

Scheme 10.15. Synthesis of platinum-containing molecular squares with butadiynyl ligands.

solubility and stability of the platinum(II) phosphine systems have made them appropriate entities to accompany the C≡C group.

Owing to the advances in the synthesis of organoplatinum(II) alkynyl compounds as well as the intriguing physical and photophysical properties exhibited by these materials, the evaluation of their structure–property relationship is therefore attractive and feasible. The spectroscopic properties of mononuclear nickel(II), palladium(II) and platinum(II) alkynyl complexes of the type trans-$[M(C\equiv CR)_2L_2]$ (R = alkyl, aryl; L = phosphine, stibine) have been widely investigated [111–114]. The electronic absorption spectra of this class of complexes have been reported by Masai et al. [115], while the emission properties of a series of closely related platinum(II) complexes were reported by Demas and coworkers [116]. The lowest

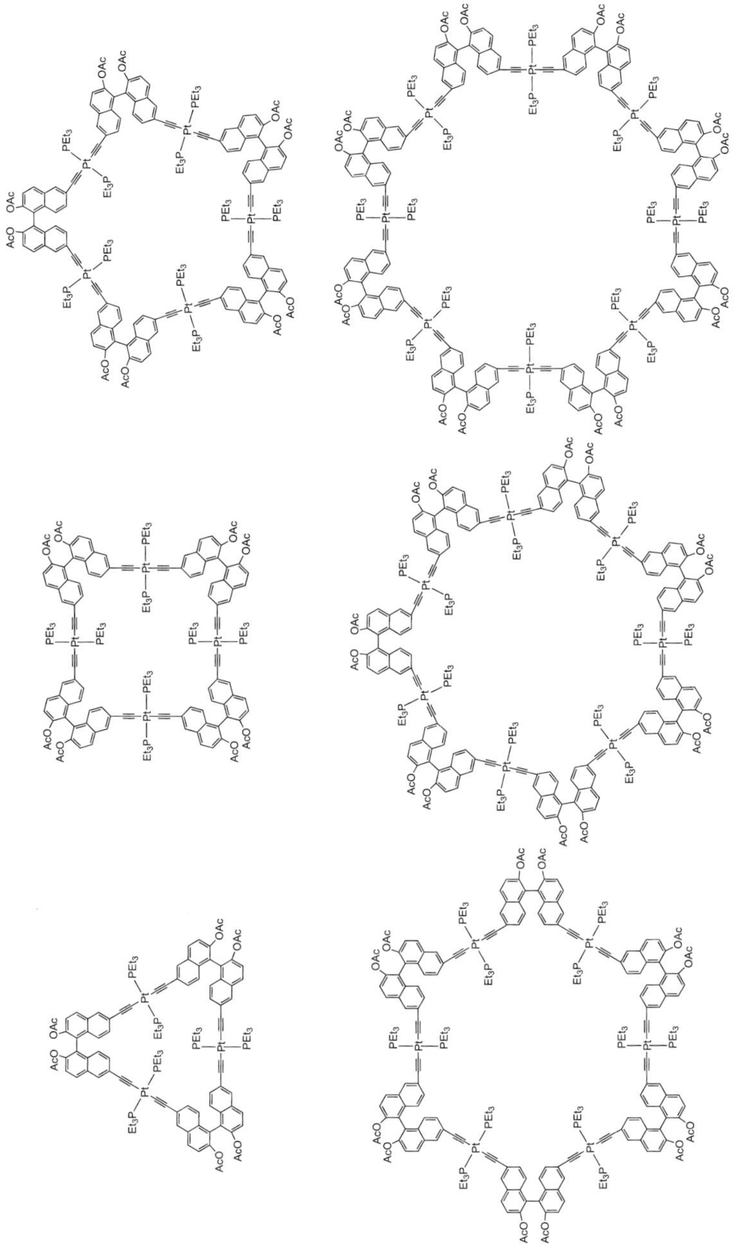

Scheme 10.16. Chiral platinum-containing molecular polygons with potential enantioselective catalytic activities.

energy excited states in these square-planar monomeric compounds are assigned to be of ^{3}MLCT character.

Other than the monomeric systems mentioned above that mainly focused on the fundamental spectroscopic properties of this class of compounds, polymeric systems of platinum- and palladium-containing alkynyl conjugated polymers and oligomers have also attracted much interest because of their long-lived ^{3}IL and ^{3}MLCT emissions [117–120]. This characteristic leads to the development of materials that may be suitable for the construction of high-efficiency energy harvesting systems and organic light emitting diodes [65]. After the successful isolation of this class of polymers in the late 1970s by Hagihara and coworkers, incorporation of Group 8, 9 and 10 metals into the one dimensional rigid-rod backbone has been achieved by Lewis [90, 91], Raithby [117–120], Marder [121], Gladysz [71–74, 92, 93] and others [16].

In particular, the luminescence properties of some of the platinum and palladium-containing polymers have been studied by Lewis, Raithby and Friend (Scheme 10.17) [89, 90, 117–119]. They have also investigated platinum and palladium-containing polymers with different heterocyclic organic spacers and studied their photophysical properties. In general, the emissions of this class of polymers have been reported to originate from the ^{3}IL states of the organic backbones. This work was recently extended by Wong and coworkers [120]. Very recently, a congenerous series of monodisperse oligomers prepared via an iterative-

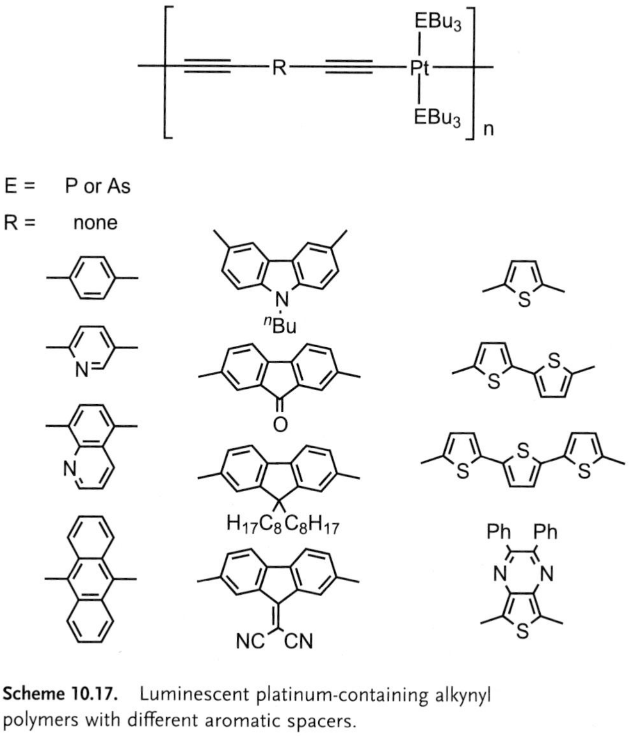

Scheme 10.17. Luminescent platinum-containing alkynyl polymers with different aromatic spacers.

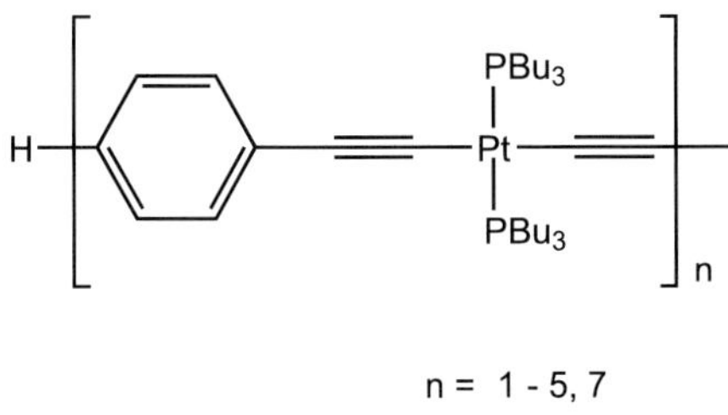

Scheme 10.18. Monodisperse luminescent platinum alkynyl oligomers.

convergent approach has been reported by Schanze and coworkers (Scheme 10.18) [122]. These oligomers were found to be luminescent in fluid solution at room temperature with lifetimes in the microsecond range. Red shifts were observed in the electronic absorption and emission spectra with increasing number of repeating units, and the conjugation length of this class of compounds was found to be around six repeat units. The triplet exciton has been shown to be much less delocalized than the singlet exciton in these π-conjugated polymers, and it was estimated to be confined to two or at most three repeat units in a typical π-conjugated system [123].

In view of the growing attention to materials with unit dispersity and their applications in the various fields, recent interest has been directed towards the extension of the platinum(II) alkynyl systems towards carbon-rich multinuclear platinum- and palladium-containing branched molecules and organometallic dendrimers. Their relatively high solubilities and processabilities compared with their rod-like counterparts are of particular interest. Early reports include the works of Takahashi [124] and Stang [125] on the construction of metallodendrimers using the respective triethynylmesitylene and triethynylbenzene as the organic framework. Both divergent [124, 126] and convergent [127] syntheses of metallodendrimers of this class using a series of Cu-catalyzed dehydrohalogenation reactions have been reported. Recently the synthesis of the sixth-generation platinum-containing organometallic dendrimers of up to 189 platinum atoms, via a divergent approach, has been reported [126]. The radius of the sixth-generation dendrimer was found to be 56.8 ± 0.2 Å by small angle neutron scattering (SANS). This result is in accord with a 6 nm radius calculated by molecular modeling and is a verification of the rigid structure imposed by the alkynyl backbone.

The electronic absorption, luminescence and energy transfer properties of monodisperse branched organometallic complexes and dendrimers have been relatively unexplored, which is in contrast to that of the organometallic linear polymers described earlier [90, 91, 117–120, 122]. In addressing this relatively unexplored area, a series of related branched organometallic chloroplatinum(II) and chloropalladium(II) complexes was synthesized by the Cu(I)-catalyzed dehydrohalogenation reaction of *trans*-[Pt(PEt$_3$)$_2$Cl$_2$] and the corresponding branched alkynes and their photophysical properties studied [128]. Despite their non-emissive behavior at room temperature, these chloroplatinum(II) complexes are versatile

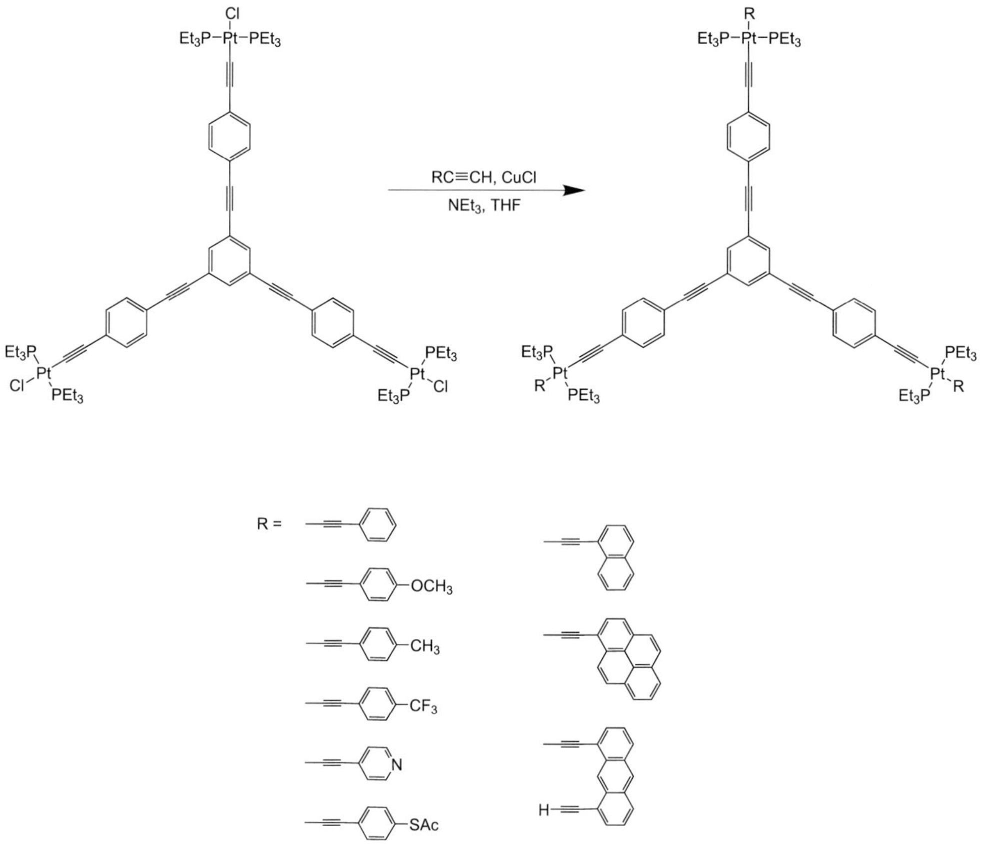

Scheme 10.19. Synthesis of multinuclear carbon-rich platinum(II) complexes of branched alkynyl.

precursors for the preparation of luminescent multinuclear assemblies as well as monodisperse luminescent organometallic dendrimers (Scheme 10.19) [129]. Subsequent substitutions of the remaining chloro ligands with various aromatic alkynyls gave a series of luminescent branched carbon-rich platinum(II) alkynyl complexes. Slight red shifts in the absorption and emission energies were observed on going from the chloropalladium(II) complex to the chloroplatinum(II) analog, which is indicative of the involvement of MLCT[$d\pi$(Pt) → π^*(C≡CR)] character in the excited states. On the contrary, the lowest energy absorption bands of the branched alkynyl complexes with aryl-alkynyl ligands, which appear at ca. 364–368 nm (Table 10.2), were found to be insensitive to the electron-withdrawing or -donating nature of the substituents on the peripheral ligands. Thus, these absorption bands were best described as an admixture of IL[$\pi → \pi^*$(C≡CR)] and MLCT [$d\pi$(Pt) → π^*(C≡CR)] transitions with predominantly IL character.

Upon photo-excitation, the branched organometallic complexes were found to emit in the green to red region at room temperature with exceptionally long life-

Table 10.2. Electronic absorption and photophysical data of selected Pd(II) and Pt(II) complexes of branched alkynyls.

Complex	Absorption[a] λ/nm (ε/dm^3 mol^{-1} cm^{-1})	Medium (T/K)	Emission λ_{em}/nm (τ_o/μs)	Ref.
[1,3,5-{Cl(PEt$_3$)$_2$PdC≡CC$_6$H$_4$C≡C}$_3$C$_6$H$_3$]	274 (74 555), 332 (153 525), 352 (168 720)	CH$_2$Cl$_2$ (298) solid (298) solid (77) glass (77)[b]	–[c] –[c] 527 (340) 517 (1610)	128
[1,3,5-{Cl(PEt$_3$)$_2$PtC≡CC$_6$H$_4$C≡C}$_3$C$_6$H$_3$]	290 sh (45 190), 342 sh (168 300), 362 (194 170)	CH$_2$Cl$_2$ (298) solid (298) solid (77) glass (77)[b]	–[c] –[c] 537 (81.0) 528 (430)	128
[1,3,5-{PhC≡C(PEt$_3$)$_2$PtC≡CC$_6$H$_4$C≡C}$_3$C$_6$H$_3$]	268 (78 850), 298 (72 470), 338 sh (141 600), 366 (228 900)	CH$_2$Cl$_2$ (298) solid (298) solid (77) glass (77)[b]	532 (49.7) 533 (0.2) 538 (74.5) 533 (359)	129
[1,3,5-{MeC$_6$H$_4$C≡C(PEt$_3$)$_2$PtC≡CC$_6$H$_4$C≡C}$_3$C$_6$H$_3$]	268 (79 610), 298 (71 610), 338 sh (134 590), 364 (213 650)	CH$_2$Cl$_2$ (298) solid (298) solid (77) glass (77)[b]	532 (41.4) 530 (0.15) 538 (115) 528 (409)	129
[1,3,5-{MeOC$_6$H$_4$C≡C(PEt$_3$)$_2$PtC≡CC$_6$H$_4$C≡C}$_3$C$_6$H$_3$]	268 (88 870), 298 (78 380), 338 sh (140 530), 368 (235 480)	CH$_2$Cl$_2$ (298) solid (298) solid (77) glass (77)[b]	535 (45.6) 532 (7.1) 530 (127) 530 (394)	129
[1,3,5-{F$_3$CC$_6$H$_4$C≡C(PEt$_3$)$_2$PtC≡CC$_6$H$_4$C≡C}$_3$C$_6$H$_3$]	274 (55 640), 298 (59 180), 338 sh (121 900), 366 (195 090)	CH$_2$Cl$_2$ (298) solid (298) solid (77) glass (77)[b]	532 (60.8) 533 (0.14) 537 (30.9) 532 (381)	129
[1,3,5-{NpC≡C(PEt$_3$)$_2$PtC≡CC$_6$H$_4$C≡C}$_3$C$_6$H$_3$]	308 (78 380), 370 (244 950)	CH$_2$Cl$_2$ (298) solid (298) solid (77) glass (77)[b]	546 (60.8) –[c] 548 (13.4) 543 (438)	129
[1,3,5-{PyrC≡C(PEt$_3$)$_2$PtC≡CC$_6$H$_4$C≡C}$_3$C$_6$H$_3$]	280 (92 765), 292 (129 200), 352 sh (138 440), 372 sh (196 190), 388 (218 465), 400 (197 640)	CH$_2$Cl$_2$ (298) solid (298) solid (77) glass (77)[b]	660 (45.2) –[c] –[c] 655 (244)	129
[1,3,5-{AnC≡C(PEt$_3$)$_2$PtC≡CC$_6$H$_4$C≡C}$_3$C$_6$H$_3$]	256 (251 010), 270 sh (181 495), 278 sh (154 015), 302 sh (74 025), 342 sh (146 190), 364 (174 590), 404 (60 145), 428 (53 460)	CH$_2$Cl$_2$ (298) solid (298) solid (77) glass (77)[b]	744 (37.2) –[c] –[c] 735 (96.5)	129

[a] Measured in CH$_2$Cl$_2$ at 298 K.
[b] Measured in EtOH–MeOH (4:1, v/v) glass.
[c] Nonemissive.

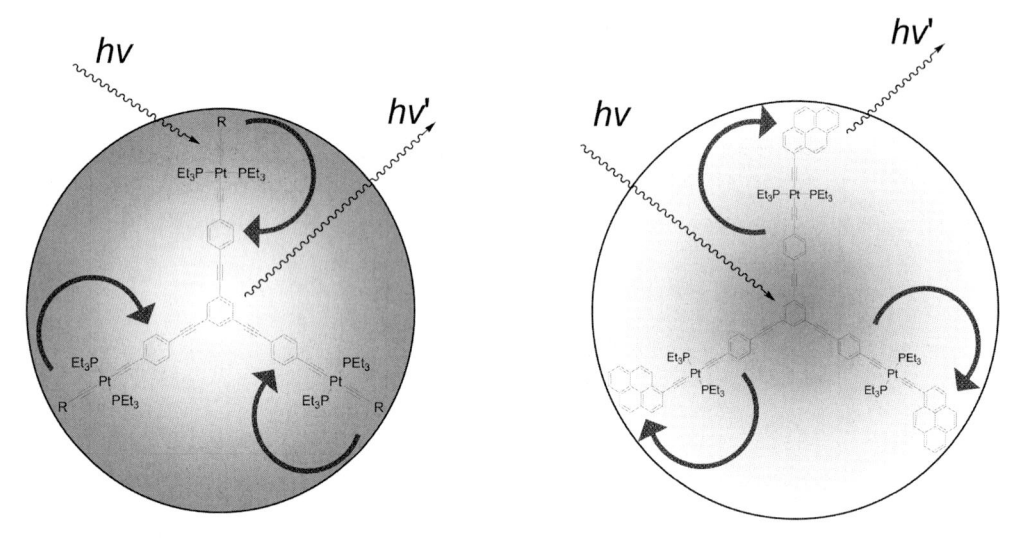

Figure 10.1. Energy transfer in multinuclear carbon-rich platinum(II) complexes of branched alkynyls. Reproduced with permission from [132].

times that are indicative of their triplet parentage. Similar to the electronic absorption, the emission maxima of the branched platinum(II) alkynyl complexes were independent of the nature of the electron-withdrawing or -donating substituents on the peripheral aryl-alkynyl ligands. Instead, incorporation of polyaromatic alkynyl ligands on the periphery led to a change in the emission origin of these branched complexes, in which the emissive states originally localized in the central $[C_6H_3(C{\equiv}CC_6H_4C{\equiv}C)_3]$ moiety are shifted to the lower-lying states of the polyaromatic alkynyl ligands on the periphery. Thus, by systematic variation of the alkynyl ligands on the periphery, the emission energies as well as the direction of energy transfer of these branched complexes could be readily tuned (Fig. 10.1). The lowest-lying emissive states were assigned as derived from predominantly ^{3}IL states either of the central $[C_6H_3(C{\equiv}CC_6H_4C{\equiv}C)_3]$ moiety or the peripheral polyaromatic alkynyl ligands, mixed with some ^{3}MLCT$[d\pi(Pt) \rightarrow \pi^*(C{\equiv}CR)]$ character.

10.4.2
Platinum(II) Polypyridine Systems

Besides the platinum and palladium diphosphine systems, platinum polypyridine complexes have also attracted attention in recent years [130–162]. Their changes in spectroscopic and luminescence properties brought about by Pt$\cdots$Pt and π–π interactions are of particular interest [138–142]. For instance, the square planar platinum(II) diimine complex [Pt(bpy)Cl$_2$] exhibits polymorphic behavior and rich photophysical properties [142–144]. These, together with the recent growing development of platinum alkynyl systems [98, 122, 125, 145–147], mainly due to the unique properties of the C≡C group, including versatile reactivity, peculiar nonlinear optical, liquid crystal and photophysical properties [98, 122, 125, 145, 147], have led to the synthesis of highly luminescent platinum polypyridyl alkynyl complexes. In 1994, Che and coworkers reported the synthesis of a luminescent platinum diimine alkynyl complex, [Pt(phen)(C≡CC$_6$H$_5$)$_2$] [130], and the emissions were assigned as derived from dπ(Pt) → π*(diimine) MLCT states [163]. This new class of highly luminescent compounds has attracted enormous attention. Similar classes of complexes have also been subsequently investigated by Eisenberg [132, 133], Schanze [134] and Raithby [136, 137]. Insight into the spectroscopic origins of these complexes was obtained via systematic variations of the electron-donating or -withdrawing substituents on either the diimine or alkynyl ligands by various researchers. In general, red shifts in both the absorption and emission bands were observed for complexes with increasing electron-withdrawing diimine ligands. On the contrary, the electronic absorption and emission bands shift to higher energy with increasing electron-withdrawing ability of the alkynyl ligands. These findings lend support for the MLCT assignment for this system. Recent reports on platinum(II) diimine complexes include the incorporation of pyrenyl and naphthyl acetylides into platinum(II) diimine complexes, which showed that solvents play an important role in controlling the lowest energy excited state in these multichromophoric platinum(II) complexes [148, 149].

Apart from the platinum(II) diimine system, analogous platinum(II) systems with planar tridentate ligands are also hot topics in current chemical research. Their useful physical and biological properties, such as interaction with DNA and proteins, were first reported by Lippard and coworkers [150, 151]. Subsequent reports by Che and coworkers described the luminescence properties of a series of platinum(II) complexes with terpyridine ligands, [Pt(trpy)X]$^{n+}$ (X = Cl, Br, I, N$_3$, SCN, n = 1; X = NH$_3$, n = 2) [152]. McMillin and coworkers recently reported the photophysical properties of a series of platinum(II) terpyridyl complexes with different counter ions [153]. These classes of platinum(II) complexes have also been found to exhibit rich polymorphic properties and have been studied by Gray and coworkers [154].

The platinum(II) terpyridine systems are found to exhibit rich luminescence attributed to dπ(Pt) → π*(N^N^N) ^{3}MLCT origin in the solid state [155]. Quenching of their ^{3}MLCT luminescence by low-lying ligand-field states (LF) or ligand-to-ligand charge transfer (LLCT) states has also been reported [155], and the interplay of their relative energies has been a topic of recent interest and may lead to applications in different fields.

The strong σ-donating ability of the alkynyl unit that is capable of raising the energy of the LF excited states, together with the interest in metal alkynyl systems, has led to the synthesis of the first series of luminescent alkynyl complexes of platinum(II) terpyridine, [Pt(trpy)C≡C–R]$^+$, which was reported by Yam and coworkers [156]. The complexes were prepared in reasonable yields by reacting the platinum(II) terpyridyl complex, [Pt(trpy)(NCMe)](OTf)$_2$, with various aryl-alkynes in the presence of sodium hydroxide or triethylamine in methanol (Scheme 10.20). The use of this synthetic route was analogous to that for the preparation of

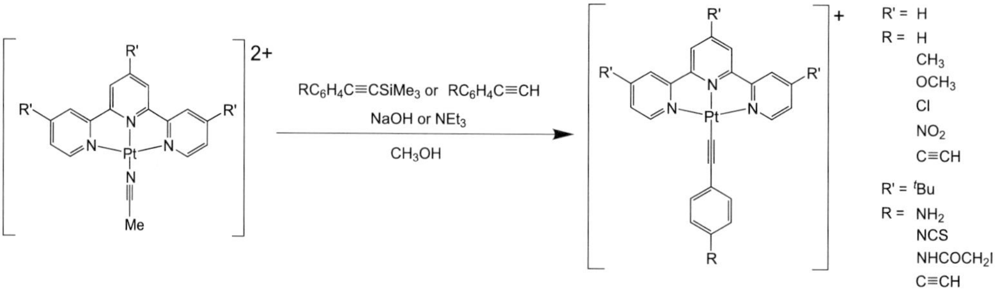

Scheme 10.20. Synthesis of mononuclear platinum(II) terpyridyl alkynyl complexes.

rhenium(I) alkynyl complexes, employing precursor complexes with labile acetonitrile ligand that could be readily replaced by other ligands.

Low-energy absorption bands in the range 456–480 nm were noted for this class of complexes with extinction coefficients on the order of 10^3 dm^3 mol^{-1} cm^{-1} (Table 10.3). The absorption energy was dependent on the nature of the alkynyl ligand. Lower energy absorption was observed with the more electron-rich alkynyl groups, and this energy trend was in line with an assignment of $[d\pi(Pt) \rightarrow \pi^*(trpy)]$ MLCT transitions, mixed with a LLCT$[\pi(C\equiv C) \rightarrow \pi^*(trpy)]$ character [156].

Photo-excitation into the low-energy MLCT/LLCT envelope of these complexes gave intense phosphorescence, and their triplet parentage was confirmed by their large Stokes shifts as well as microsecond range lifetimes. All the alkynyl complexes emit at lower energies with respect to their chloro analog, [Pt(trpy)Cl]$^+$ [156], attributed to the relatively stronger electron-donating effect of the alkynyl ligand. The emission energies were found to follow a trend similar to that observed in electronic absorption studies, suggesting their ^{3}MLCT$[d\pi(Pt) \rightarrow \pi^*(trpy)]$ origin (Table 10.3), with some mixing of a ^{3}LLCT$[\pi(C\equiv C) \rightarrow \pi^*(trpy)]$ character. Platinum(II) terpyridyl alkynyl complexes with a pendant crown ether have also been synthesized. The spectral changes upon addition of guest metal ions were followed by both electronic absorption and emission studies [156].

The versatility and the generalization of the synthetic route of this class of platinum(II) terpyridine complexes have enabled the synthesis of molecular wire-like [(^{t}Bu$_3$trpy)Pt(C≡C)$_n$Pt(^{t}Bu$_3$trpy)]$^{2+}$ system with up to four C≡C units (Scheme 10.21) [158]. Solubilizing *tert*-butyl groups introduced on the terpyridine ligand not only enable purification by column chromatography but also eliminate the influence of Pt···Pt and π–π interactions on their luminescence behavior. It is noteworthy that a blue-shift in the low-energy absorption band was observed with increasing number of C≡C units (Fig. 10.2), which is in contrast to the commonly observed red-shift upon increasing conjugation. This atypical phenomenon has also been observed in the case of the rhenium(I) diimine alkynyl system [85].

The monoynyl complex was found to emit at a longer wavelength compared to that of the diynyl complex (Table 10.3), which is similar to the trend observed in

Table 10.3. Electronic absorption and photophysical data of selected Pt(II) terpyridyl alkynyl complexes.

Complex	Absorption[a] λ/nm (ε/dm^3 mol^{-1} cm^{-1})	Medium (T/K)	Emission λ_{em}/nm (τ_o/μs)	Ref.
[Pt(trpy)C≡C–Ph]PF$_6$	272 (33 410), 286 (23 860), 312 (12 450), 328 (12 280), 342 (14 440), 432 (4 430)	CH$_3$CN (298) solid (298) solid (77) glass (77)[b]	630 (0.5) 800 (<0.1) 830 (0.7) 530 (13.3)	156
[Pt(trpy)C≡C–C$_6$H$_4$–CH$_3$-4]PF$_6$	272 (36 150), 312 (12 380), 330 (12 390), 344 (13 870), 412 sh (3 580), 456 (4 380)	CH$_3$CN (298) solid (298) solid (77) glass (77)[b]	665 (0.1) 720 (<0.1) 740 (1.0) 545 (13.8)	156
[Pt(trpy)C≡C–C$_6$H$_4$–OCH$_3$-4]PF$_6$	272 (39 310), 308 (13 920), 330 (13 500), 344 (13 140), 416 sh (3 500), 474 (4 700)	CH$_3$CN (298) solid (298) solid (77) glass (77)[b]	–[c] –[c] –[c] 580 (11.8)	156
[Pt(trpy)C≡C–C$_6$H$_4$–Cl-4]PF$_6$	272 (36 550), 286 (27 830), 312 (12 530), 328 (12 490), 342 (14 500), 432 (4 790)	CH$_3$CN (298) solid (298) solid (77) glass (77)[b]	620 (0.5) 710 (<0.1) 790 (0.8) 530 (15.0)	156
[Pt(trpy)C≡C–C$_6$H$_4$–NO$_2$-4]PF$_6$	272 (33 500), 286 (23 780), 310 (24 500), 326 (12 360), 340 sh (14 640), 417 (9 740)	CH$_3$CN (298) solid (298) solid (77) glass (77)[b]	560 (<0.1) 720 (<0.1) 790 (0.7) 510 (240)	156
[(^{t}Bu$_3$trpy)Pt–C≡C–Pt(^{t}Bu$_3$trpy)](OTf)$_2$	314 (29 820), 328 (29 180), 342 (23 155), 484 (10 175), 522 (7 615)	CH$_3$CN (298) solid (298) solid (77) glass (77)[b]	625 615 (0.2) 604 (2.6) 572 (8.1)	158
[(^{t}Bu$_3$trpy)Pt–C≡C–C≡C–Pt(^{t}Bu$_3$trpy)](OTf)$_2$	310 (32 210), 322 (32 475), 338 (38 110), 436 (13 510), 466 (16 670)	CH$_3$CN (298) solid (298) solid (77) glass (77)[b]	625 606 (0.2) 604 (1.1) 550 (6.5)	158
[(^{t}Bu$_3$trpy)Pt–C≡C–C≡C–C≡C–Pt(^{t}Bu$_3$trpy)](PF$_6$)$_2$	308 (45 705), 324 (38 590), 338 (44 980), 428 (21 690), 452 (20 480)	CH$_3$CN (298) solid (298) solid (77) glass (77)[b]	589 605 (0.2) 607 (1.0) 582 (1.1)	158

[a] Measured in acetonitrile at 298 K.
[b] Measured in butyronitrile glass (concentration 5 × 10^{-6} mol dm^{-3}).
[c] Nonemissive.

Intramolecular Pt-Pt Distance = 5.16 Å

Intramolecular Pt-Pt Distance = 7.71 Å

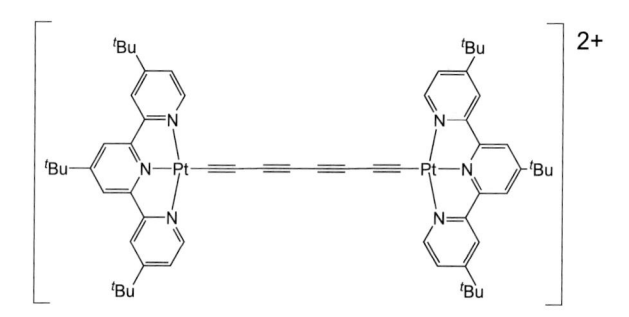

Intramolecular Pt-Pt Distance = 12.83 Å

Scheme 10.21. Molecular wire-like platinum(II) terpyridyl complexes of different lengths.

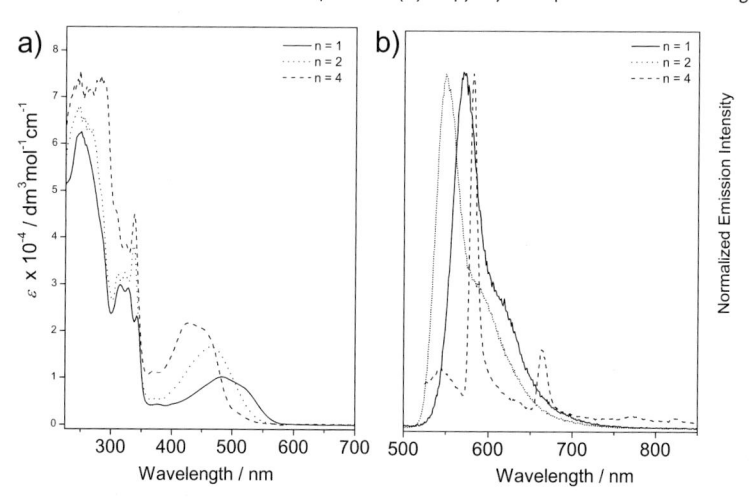

Figure 10.2. (a) Electronic absorption and emission spectra of [(^{t}Bu$_3$trpy)Pt(C≡C)$_n$Pt(^{t}Bu$_3$trpy)]$^{2+}$ in acetonitrile at room temperature. (b) Emission spectra of [(^{t}Bu$_3$trpy)Pt(C≡C)$_n$Pt(^{t}Bu$_3$trpy)]$^{2+}$ in butyronitrile at 77 K. Reproduced with permission from [158].

electronic absorption studies. At 77 K, vibronic-structured emission bands were observed for the mono- and diynyl complexes with vibrational progressional spacings of ca. 1250 cm^{-1}, typical of the $v(C \doteq C)$ and $v(C \doteq N)$ stretching modes of the terpyridyl ligand, and the emission was assigned as originating from excited states of predominantly ^{3}MLCT[$d\pi$(Pt) → π^*(trpy)] character, mixed with a certain degree of ^{3}LLCT[π(C≡C) → π^*(trpy)]/^{3}IL[π → π^*(C≡C)] character. Different vibrational progressional spacings of ca. 2100 cm^{-1} were observed at low temperature for the triynyl complex, corresponding to the v(C≡C) stretching modes of the alkynyl units. This was suggestive of an emissive state of a largely ^{3}IL character. It was proposed that upon increasing the π conjugation of the alkynyl chain, the origin of the emission changed in character, with an increasing contribution from the ^{3}IL[π → π^*(C≡C)] excited state, leading to an emission that was predominantly ^{3}IL in character.

Yam and coworkers have recently reported a mononuclear diynyl platinum(II) complex [Pt(trpy)(C≡C–C≡C–H)]$^+$ [159] which was found to exist in two crystal forms, dark green and red. The dark green form exists as a linear chain with the platinum atoms equally spaced with short intermolecular Pt···Pt distances of 3.388 Å, while the platinum atoms in the red form show a zigzag arrangement to give dimeric structures with alternating Pt···Pt distances of 3.394 and 3.648 Å (Fig. 10.3). Drastic color changes and emission wavelength and intensity changes were observed in the solutions of this complex by changing the solvent composition. Upon increasing the diethyl ether content in acetone or acetonitrile solution of the complex, the color of the solution changed dramatically from yellow to green to blue. A growth of a new band at 615 nm with a concomitant drop in intensity of the band at 416 nm was observed in the electronic absorption spectra with increasing diethyl ether content (Fig. 10.4). Such color changes were ascribed to the formation of aggregates upon increasing the nonsolvent content in solution. The new absorption band at 615 nm was assigned as an MMLCT transition of the aggregated species. The emission intensity enhancement of a new emission band at 785 nm was found to be closely related to the absorbance changes upon increasing the non solvent content of the mixture (Fig. 10.5), thus the phosphorescence at 785 nm was assigned as originating from MMLCT excited states. Such drastic solvent-induced sensitive changes in color and on–off switching of emission via aggregation of metal–metal bonded and π-stacked species have been suggested to provide a basis for the design of versatile reporters and probes of environmental changes, and they may find potential applications in materials science and sensor technology. Recent work by Yam and coworkers has also demonstrated the versatility of the self-assembly of this system in reporting microenvironment changes in a polymer system [160].

A related class of tridentate cyclometalated platinum(II) alkynyl complexes, [Pt(C^N^N)(C≡C–R)], was recently reported by Che and coworkers (Scheme 10.22) [161, 162]. The complexes were prepared by the reaction between the organic alkynes and the corresponding cyclometalated chloroplatinum(II) precursors [Pt(C^N^N)Cl] in the presence of copper(I) catalyst and trialkylamine.

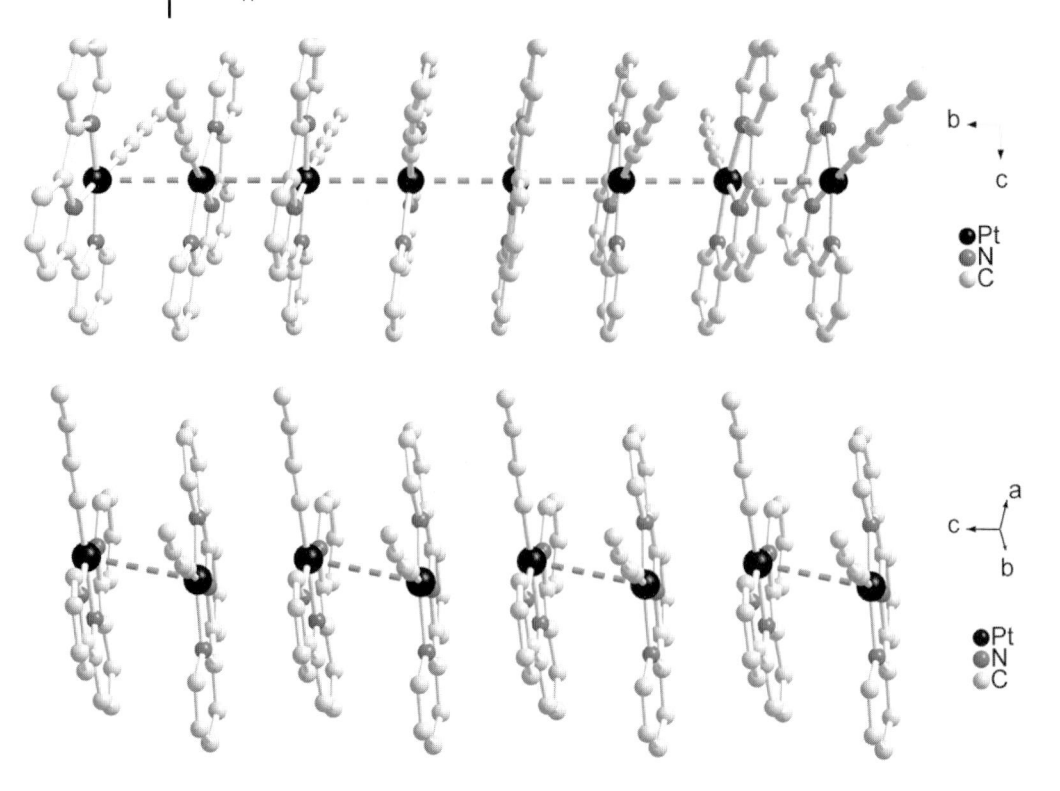

Figure 10.3. Perspective view of complex cations of [Pt(trpy)(C≡C–C≡C–H)]$^+$ in dark green form (top) exhibiting extended linear chain packing and in red form (bottom) showing zigzag arrangement. Hydrogen atoms are omitted for clarity. Reproduced with permission from [159].

It is noteworthy that a blue-shift in the lowest energy absorption bands of these platinum(II) complexes has also been observed with an increase in the number of C≡C units in the oligoynyl ligand, similar to the findings first reported by Yam and coworkers in the rhenium(I) diimine [85] and platinum(II) terpyridine alkynyl systems [158]. These complexes also display luminescence both in the solid state and in fluid solutions at room temperature. Similar to the [Pt(diimine)(C≡C–R)$_2$] [130, 132, 133] and [Pt(trpy)(C≡C–R)]$^+$ congeners [156], the emission energies of this class of compounds are dependent on the nature of the chelating ligand and the alkynyl ligand. Lowest energy emissive states of ^{3}MLCT[d(Pt) → π*(C^N^N)], ^{3}IL(C^N^N) or ^{3}IL(C≡C) were assigned. By rational design and modifications of the cyclometalating ligand and alkynyl ligand, the relative energies of the excited states could be varied, which in turn facilitates tuning of emission energies of these complexes. This class of complexes has been found to serve as potential dopants in OLED applications in view of the wide spectral coverage of emissions and high thermal stabilities [162].

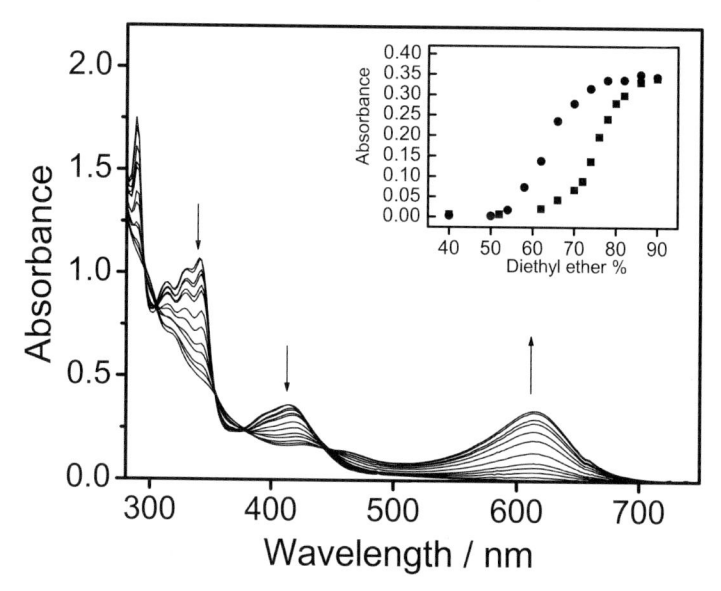

Figure 10.4. UV–vis absorption changes of [Pt(trpy)(C≡C–C≡C–H)]PF$_6$ (concentration = 1.47×10^{-4} M) in acetonitrile with increasing diethyl ether content. Inset: Plot of absorbance vs. diethyl ether composition in acetonitrile at 615 nm (■) and in acetone at 610 nm (●). Reproduced with permission from [64].

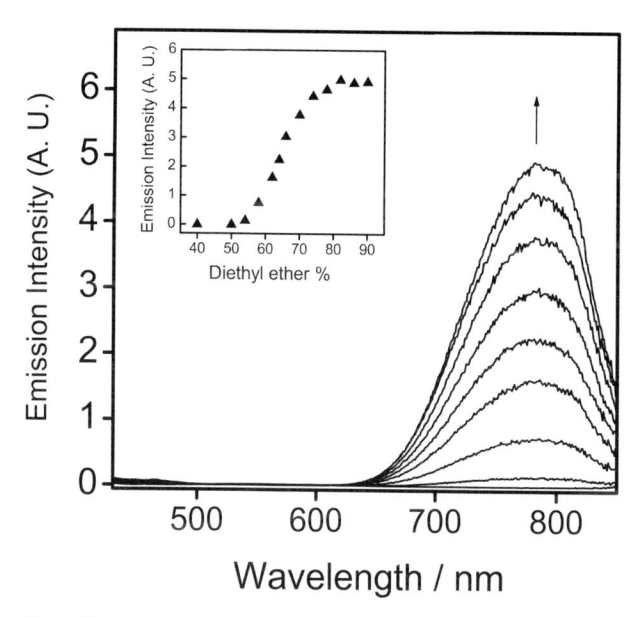

Figure 10.5. Emission enhancement of [Pt(trpy)(C≡C–C≡C–H)]PF$_6$ (concentration = 1.47×10^{-4} M) in acetone upon increasing diethyl ether content. Inset: Plot of corrected emission intensity as a function of diethyl ether composition (▲). Reproduced with permission from [64].

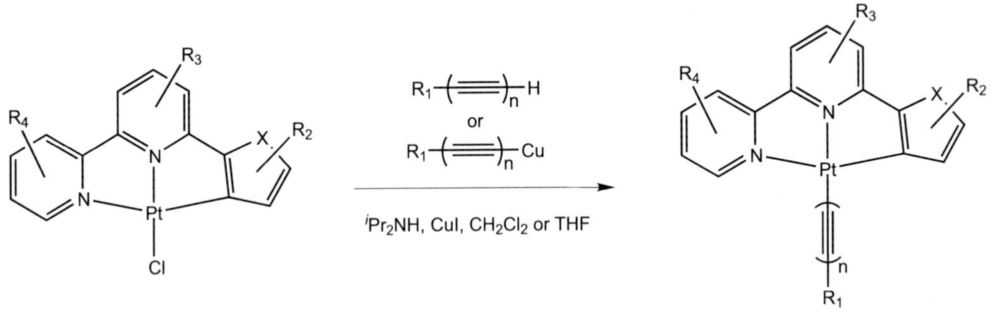

R_1 = ^{t}Bu	R_2 = H	R_3 = H	R_4 = H	X = HC=CH	n = 1 - 4
SiMe$_3$	CH$_3$	Ph	^{t}Bu	S	
Ph	F	^{t}Bu		O	
C$_6$H$_4$-CH$_3$-4	CF$_3$	C$_6$H$_4$-CH$_3$-4			
C$_6$H$_4$-OCH$_3$-4		C$_6$H$_4$-OCH$_3$-4			
C$_6$H$_4$-Cl-4		C$_6$H$_4$-Cl-4			
C$_6$H$_4$-F-4		C$_6$H$_4$-NO$_2$-4			
C$_6$H$_4$-NO$_2$-4		CO$_2$Et			
C$_6$F$_5$					
pyrenyl-1					

Scheme 10.22. Synthesis of luminescent cyclometalated platinum(II) alkynyl complexes.

10.5
Copper(I) and Silver(I) Alkynyls

Alkynyl complexes of copper(I) and silver(I) are usually characterized by their high nuclearity, and unusual and numerous bonding modes (Scheme 10.23).

Among various coordination modes displayed by copper(I) and silver(I) alkynyl complexes, the μ_3-η^1 bonding mode would probably be one of the most commonly observed coordination motifs. The syntheses of *triangulo* copper(I) alkynyl complexes, [Cu$_3$(μ-dppm)$_3$(μ_3-η^1-C≡C–Ph)]$^{2+}$ and [Cu$_3$(μ-dppm)$_3$(μ_3-η^1-C≡C–Ph)$_2$]$^+$, were first reported by Gimeno and coworkers, with X-ray crystal structures solved confirming the μ_3-η^1-coordination mode [164, 165]. The luminescence and photophysical properties of these clusters have been extensively investigated by Yam and coworkers. A series of luminescent mono-capped trinuclear alkynyl complexes [Cu$_3$(μ-dppm)$_3$(μ_3-η^1-C≡C–R)]$^{2+}$ along with the bi-capped analogues [Cu$_3$(μ-dppm)$_3$(μ_3-η^1-C≡C–R)$_2$]$^+$ were successfully synthesized from the reaction of the dinuclear [Cu$_2$(μ-dppm)$_2$(MeCN)$_2$]$^{2+}$ precursor and various alkynes in the presence of ^{n}BuLi, sodium hydroxide or triethylamine in the correct stoichiometric ratio (Scheme 10.24) [20, 166–171]. Other than the *triangulo* Cu$_3$ alkynyl complexes, corresponding analogues of *triangulo* Ag$_3$ complexes have been successfully synthesized and isolated using similar synthetic methodology [169].

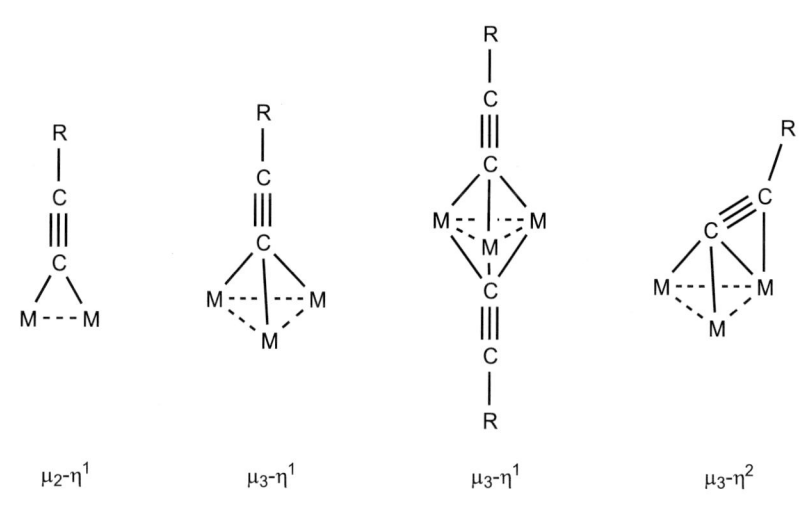

Scheme 10.23. Examples of bonding modes exhibited by copper(I) and silver(I) alkynyl complexes.

$\mu_2\text{-}\eta^1$ $\mu_3\text{-}\eta^1$ $\mu_3\text{-}\eta^1$ $\mu_3\text{-}\eta^2$

By the systematic variations of the bidendate phosphine and alkynyl ligands in these d^{10} complexes and subsequent investigations of their electronic absorption and emission properties (Table 10.4), insight into the excited state origin of this system could be obtained. The electronic absorption spectra of these trinuclear complexes are dominated by the IL($\pi \to \pi^*$) transitions of the phosphine and alkynyl ligands. These complexes are found to be strong emitters at room temperature upon photo-excitation. In both mono-capped and bi-capped systems, lower emission energies are noted whenever the alkynyl ligand is more electron-rich. The trend in the emission energies implies a lowest-energy emissive state that is of predominantly ^{3}LMCT [(C≡C–R) $\to$ M$_3$] character. However, mixing to a certain extent of metal-centered [nd^9(n + 1)s^1] character is also likely, provided that there are substantial M–M interactions within the cluster. In addition, the lower emission energy in the mono-capped complexes compared to their corresponding bi-capped analogs would further support the proposed assignment. It is likely that there is mixing of some ^{3}IL[$\pi \to \pi^*$(C≡CR)] character in their lowest-lying emissive state for the complexes with electron-withdrawing substituents on the alkynyl ligands.

Comparison of the photophysical properties of the Cu(I) and Ag(I) clusters lends additional evidence to support the spectroscopic assignments of the excited states in these multinuclear complexes. The Ag(I) alkynyl complexes, both in the mono-capped and bi-capped fashion, generally show a blue-shift of emission wavelength compared to their copper(I) analogs (Table 10.4) and this is again in favor of the predominantly ^{3}LMCT[(C≡CR) $\to$ M$_3$] assignment of their emissive states. However, mixing of the ^{3}IL[$\pi \to \pi^*$(C≡CR)] is also possible, based on the highly structured emission bands with exceptionally long lifetimes, as well as the higher-lying silver-centered acceptor orbitals in the *triangulo* Ag$_3$ complexes.

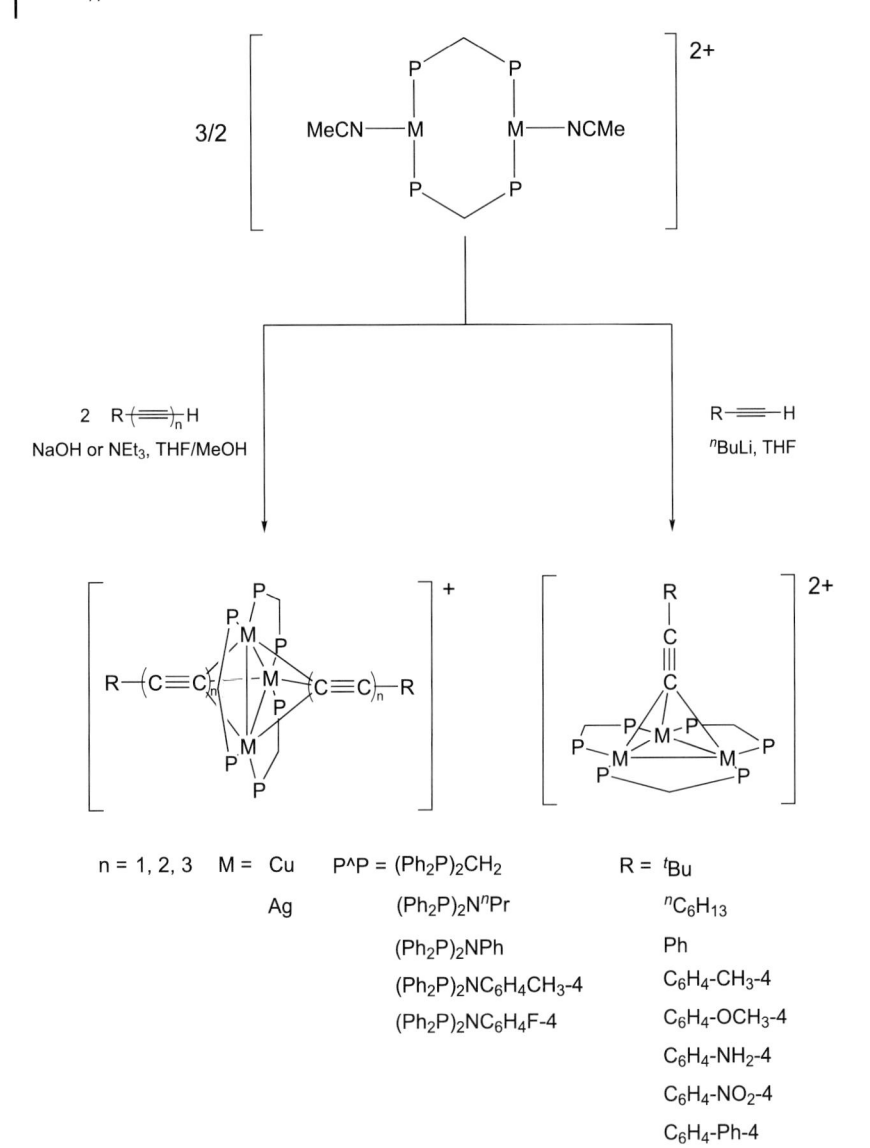

Scheme 10.24. Synthesis of *triangulo-* Cu(I) or Ag(I) alkynyl complexes.

Other than the μ_3-η^1 bonding mode, a novel tetranuclear complex [Cu$_4$(μ-dppm)$_4$(μ_4-η^1,η^2-C≡C–)]$^{2+}$ was synthesized by reaction of the precursor complex [Cu$_2$(μ-dppm)$_2$(MeCN)$_2$]$^{2+}$ with trimethylsilylacetylene in the presence of ^{n}BuLi in THF under an inert atmosphere of nitrogen [172]. This complex was recently found to contain an oscillating C≡C unit within the Cu$_4$ rectangle (Scheme 10.25) [173]. Flipping of the dppm ligands and oscillation of the C≡C unit within the Cu$_4$ array were revealed using variable temperature ^{1}H and ^{31}P NMR techniques with

10.5 Copper(I) and Silver(I) Alkynyls

Table 10.4. Photophysical data of selected Cu(I) and Ag(I) alkynyl complexes.

Complex	Medium (T/K)	Emission λ_{em}/nm (τ_o/μs)	Ref.
[Cu$_3$(μ-dppm)$_3$ (μ$_3$-η^1-C≡C–Ph)](BF$_4$)$_2$	solid (298) solid (77) acetone (298)	500 (21) 492, 530 sh 499 (15)	167
[Cu$_3$(μ-dppm)$_3$(μ$_3$-η^1-C≡C–C$_6$H$_4$–OCH$_3$-4)](BF$_4$)$_2$	solid (298) solid (77) acetone (298)	471, 523 sh, 585 (2.9) 495, 538 sh, 600 sh (2.9) 483 (5.5)	171
[Cu$_3$(μ-dppm)$_3$(μ$_3$-η^1-C≡C–C$_6$H$_4$–NH$_2$-4)](BF$_4$)$_2$	solid (298) solid (77) acetone (298)	418 (3.5) 508 504, 564 sh (4.8)	171
[Cu$_3$(μ-dppm)$_3$(μ$_3$-η^1-C≡C–^{t}Bu)](PF$_6$)$_2$	solid (298) solid (77) acetone (298)	627 (14) 450, 570 sh, 692 640 (2.6)	171
[Cu$_3$(μ-dppm)$_3$(μ$_3$-η^1-C≡C–^{n}C$_6$H$_{13}$)](BF$_4$)$_2$	solid (298) solid (77) acetone (298)	601 (24.4) 540 sh, 640 434 (0.46), 650 (1.54)	171
[Cu$_3$(μ-dppm)$_3$(μ$_3$-η^1-C≡C–Ph)$_2$]PF$_6$	solid (298) solid (77) acetone (298)	493 (14) 485, 525 sh 495 (5.9)	166
[Cu$_3$(μ-dppm)$_3$(μ$_3$-η^1-C≡C–C$_6$H$_4$–OCH$_3$-4)$_2$]PF$_6$	solid (298) solid (77) acetone (298)	450, 482 (63.8) 450, 478, 490 sh, 525 481 (11.5)	171
[Cu$_3$(μ-dppm)$_3$(μ$_3$-η^1-C≡C–^{t}Bu)$_2$]PF$_6$	solid (298) solid (77) acetone (298)	450 (0.44), 540 (1.7) 450, 530 sh 444 (0.24), 580 sh (16)	166
[Cu$_3$(μ-dppm)$_3$(μ$_3$-η^1-C≡C–C≡C–Ph)$_2$]PF$_6$	solid (298) solid (77) acetone (298)	540, 601, 685 sh (<0.1) 547, 615 540, 596 sh (19.2)	175
[Cu$_3$(μ-dppm)$_3$(μ$_3$-η^1-C≡C–C≡C–C$_6$H$_4$–CH$_3$-4)$_2$]PF$_6$	solid (298) solid (77) acetone (298)	580, 655 (<0.1) 530, 581, 655 531, 586 sh (19.4)	175
[Cu$_3$(μ-dppm)$_3$(μ$_3$-η^1-C≡C–C≡C–C$_6$H$_4$–OCH$_3$-4)$_2$]PF$_6$	solid (298) solid (77) acetone (298)	592, 676 (<0.1) 520, 580 523, 583 sh (21.5)	175
[Ag$_3$(μ-dppm)$_3$ (μ$_3$-η^1-C≡C–Ph)](BF$_4$)$_2$	solid (298) solid (77)	430, 449, 467, 488 sh (93.0) 428, 462, 469, 495, 515, 536 sh	169
[Ag$_3$(μ-dppm)$_3$(μ$_3$-η^1-C≡C–C$_6$H$_4$–OCH$_3$-4)](BF$_4$)$_2$	solid (298) solid (77)	464 (5.2) 443, 464, 482, 508 sh	169
[Cu$_3$(μ-dppm)$_3$(μ$_3$-η^1-C≡C–C$_6$H$_4$–C≡C-4) Cu$_3$(μ-dppm)$_3$](BF$_4$)$_4$	solid (298) solid (77)	583 (222) 582	174
[Ag$_3$(μ-dppm)$_3$(μ$_3$-η^1-C≡C–C$_6$H$_4$–C≡C-4) Ag$_3$(μ-dppm)$_3$](BF$_4$)$_4$	solid (298) solid (77)	513 (351) 515	174

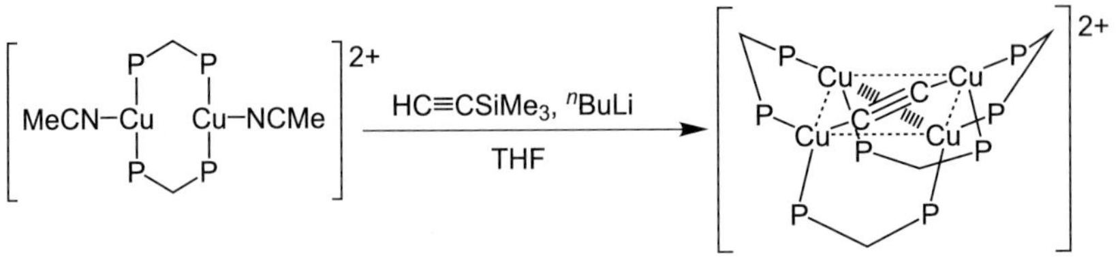

Scheme 10.25. Synthesis of a tetranuclear complex [Cu$_4$(μ-dppm)$_4$(μ_4-η^1,η^2-C≡C-)]$^{2+}$.

experimentally determined $\Delta G^{\ddagger}$ values of ca. 47 kJ mol^{-1} and 38 kJ mol^{-1}, respectively, which are comparable to those estimated by DFT calculations.

In addition, attempts have also been made to synthesize luminescent alkynyl complexes of Cu(I) and Ag(I) with higher nuclearity. This was achieved by employing 1,4-diethynylbenzene as the rigid bridging ligand to give the corresponding dumb-bell shaped hexanuclear Cu(I) and Ag(I) alkynyl complexes (Scheme 10.26) [174]. The Cu(I) and Ag(I) complexes, similar to their trinuclear analogs, are strong emitters in the yellow-orange and greenish-yellow region (Table 10.4). Again, the blue-shift of absorption and emission energies on going from the Cu$_6$ cluster to the Ag$_6$ analog is in line with the assignment of a mixed d → s/LMCT[(C≡CR) → M$_3$] origin, with mixing of a ^{3}IL[π → π^*(C≡CR)] character.

Recently, extension of the monoynyl backbone to diynyl [175] and triynyl [176] complexes of this class has also been successfully accomplished. The structured emission bands with vibrational progressional spacings of ~1800–2200 cm^{-1}, which are typical of the ν(C≡C) stretches in the ground state, are suggestive of the

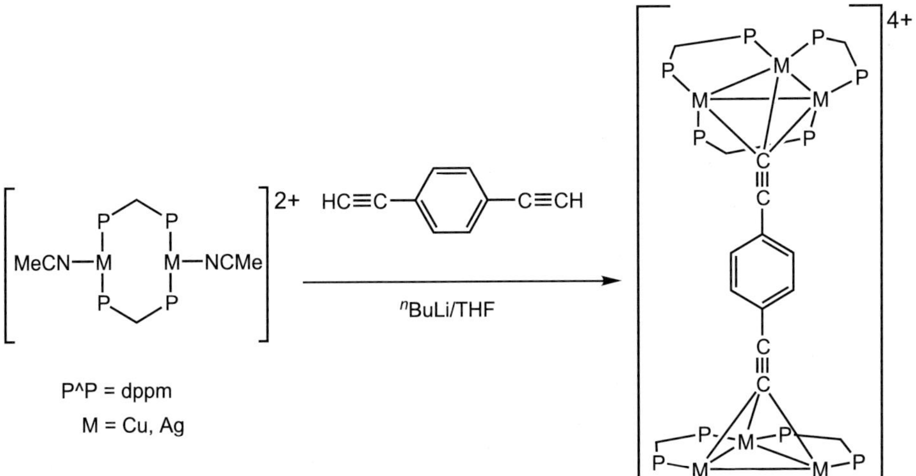

Scheme 10.26. Synthesis of hexanuclear alkynyl complexes of Cu(I) and Ag(I).

increased ^{3}IL[$\pi \to \pi^*$(C≡CR)] character in the lowest energy excited states relative to their monoynyl counterparts. Better insight into the electronic structure, in particular to explain the spectroscopic, structural, and electrochemical properties of the complexes, has been provided through molecular orbital calculations and quantum chemical calculations of the DFT type on some of these complexes. In general, four nearly degenerate MOs were located in the HOMO region, while two nearly degenerate MOs were found in the LUMO region. In the case of the diynyl model complex, two of the four HOMOs were predominantly localized on the C$_4$ carbon chains with slight contribution from the Cu(I) centers and the other two are mainly localized on the *triangulo*-Cu$_3$ array. The LUMOs were formed by π-type carbon orbitals with contributing copper orbitals. The DFT results are again supportive of a fairly mixed excited state consisting of ^{3}LMCT, ^{3}MC[nd^9(n + 1)s^1] and ^{3}IL[$\pi \to \pi^*$(C≡C–C≡CR)] character, with a growing contribution of the ^{3}IL character for the complexes with a more extended π-conjugation in the alkynyl ligand.

10.6
Gold(I) Alkynyls

Despite being in the same group as Cu and Ag, the Au complexes of alkynyl tend to be quite different from them in terms of the molecular structures and bonding mode. Common coordination geometries of Au(I) include two-coordination in a linear fashion and three-coordination in a trigonal planar arrangement. Key features in gold complexes include their short Au···Au contacts due to relativistic and correlation effects [177–179] and the resulting wide range of molecular structures displayed by gold complexes [180, 181]. These characteristics have rendered the chemistry of gold a topical field of research.

Two-coordinate Au(I) alkynyl complexes with linear geometry could be readily found in the literature with most of them focused on systems stabilized with phosphine ligands. This class of compounds is usually prepared by reacting the organic alkynes with the chlorogold(I) complex [R$_3$PAuCl] in the presence of a base in methanolic solutions. Another synthetic methodology, which has commonly been employed in the synthesis of gold alkynyls, is the depolymerization of gold polymer [{Au(C≡C)}$_n$] with electron-rich ligands such as phosphines.

The first report on the luminescence properties of Au(I) alkynyl complexes was described by Che and coworkers, in which the luminescence behavior of the dinuclear Au(I) alkynyl complex [Au$_2$(μ-dppe)$_2$(C≡C–Ph)$_2$] was reported (Scheme 10.27) [182]. Short intermolecular Au···Au contacts of 3.153(2) Å were observed in the X-ray crystal structure. The 550 nm emission of the solid sample at room temperature was described as originating from a $(d_{\delta^*})^1(p_\sigma)^1$ triplet excited state arising from the interactions between the two Au atoms. Other Au(I) alkynyl complexes with various bridging phosphine ligands have also been prepared by depolymerization reaction of [{Au(C≡C–Ph)}$_n$] [183] and their low-energy emission was proposed to be emanating from the $^3[(d_{\delta^*})^1(p_\sigma)^1]$ excited states.

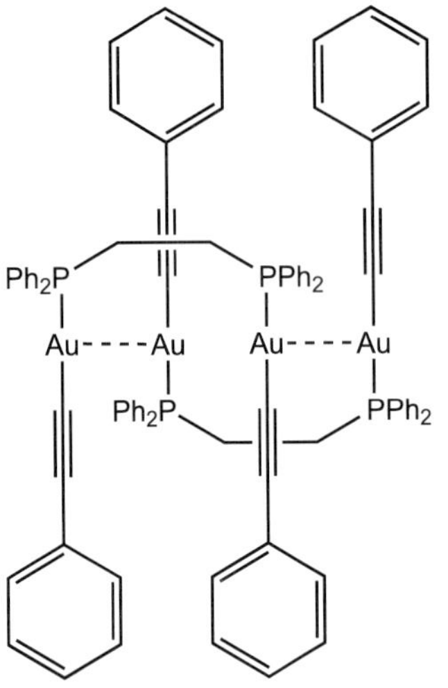

Scheme 10.27. Dinuclear Au(I) alkynyl complex [Au$_2$(μ-dppe)$_2$(C≡C–Ph)$_2$] with short intermolecular Au···Au contacts.

The photophysical properties of this class of compounds have since then been extensively investigated by the groups of Che [183–186], Yam [19, 20], Mingos [187], Puddephatt [87, 188–190] and others [101] with emissive states ranging from ^{3}IL to ^{3}MLCT and admixtures of these manifolds. Inter- and intramolecular Au···Au interactions also play a pivotal role in the photophysical characteristics of Au(I) alkynyl complexes. A red shift in the solid-state emission energies relative to that in solution is commonly observed for complexes with such interactions [87, 183, 188, 189]. In addition, the nonlinear optical (NLO) properties of a number of di- and trinuclear Au(I) alkynyl complexes with di- and tri-ethynylbenzene cores have been investigated by Humphrey and coworkers [191–194]. These complexes may serve as potential NLO materials with reasonable first and second hyperpolarizabilities. Introduction of nitro group into these systems was suggested to enhance the optical nonlinearity.

Au(I) alkynyl systems, with the linear geometry adopted by the Au(I) centers and the rich photophysical properties, appear to be ideal candidates for the construction of luminescent molecular wires. A number of dinuclear Au(I) alkynyl rod-like complexes bridged by an ethynyl or butadiynyl moiety have been synthesized and their photophysical properties reported by the groups of Mingos and Yam [195]. More recently, the synthesis of a related series of dinuclear Au(I) polyynyl complexes was reported by Che and coworkers and their photophysical properties were also investigated (Scheme 10.28) [184, 185]. The importance of using tricyclohexylphosphine as the ancillary ligand in these gold(I) polyynyl complexes has been highlighted by the authors. By using such bulky phosphine ligands in gold(I) alkynyl complex formation, one could help to restrict oligomerization via Au(I)···Au(I) interactions and, consequently, minimize their influence on the

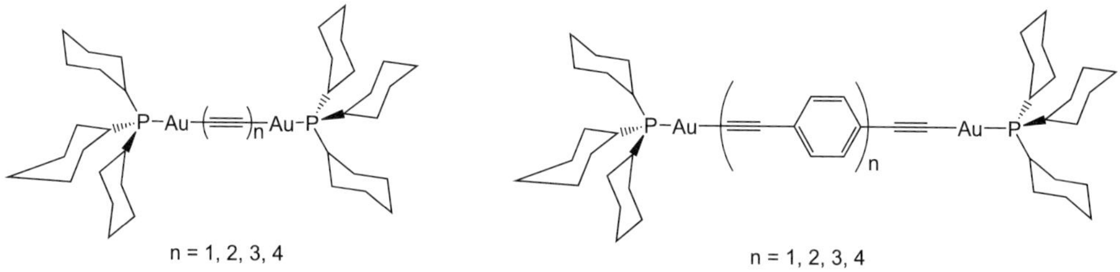

Scheme 10.28. Dinuclear Au(I) polyynyl wire-like complexes with tricyclohexylphosphines.

spectroscopic properties of the Au(I) alkynyl complexes. In addition, the spectroscopic interfering IL($\pi \rightarrow \pi^*$) states, which would be present in triarylphosphines, could be eliminated and the luminescence properties could be unequivocally assigned as emanating from the bridging polyynyl chain. The complexes were found to be emissive at room temperature in the blue region. Spectroscopic studies showed well-resolved structured electronic absorption and emission bands in these complexes with vibrational progressional spacings of about 2000 cm^{-1}, corresponding to the ν(C≡C) stretch. As expected, these emissions were assigned as derived from excited states with essentially ^{3}IL($\pi \rightarrow \pi^*$) character of the polyynyl unit. It was suggested that the enhanced accessibility to the spin-forbidden triplet state through the incorporation of the heavy gold atoms would provide the impetus for future study. Besides, a congenerous mononuclear Au(I) alkynyl complex, [(Cy)$_3$PAu(C≡CC$_6$H$_4$–NO$_2$)], has also been reported to exist as two polymorphs with distinct photophysical properties that are related to the orientation of the molecular dipoles [186].

With the recent developments and growing interest in metal-based chemosensors [196–198], some work toward the utilization of the luminescent Au(I) system in chemosensing has been described [199]. A series of gold(I) calix[4]crown-5 alkynyl complexes has been reported by Yam et al. [200, 201]. These calix[4]crown-5 alkynyl complexes were found to bind K$^+$ ions preferentially over Na$^+$ ions with UV–vis absorption spectral changes. It is noteworthy that the crown size of these alkynyl complexes could be fine-tuned by a subtle change in the steric demands on the triarylphosphine ligands, providing a firm basis for the further development of these classes of selective metalloreceptors for various metal ions.

Other than alkynyl complexes stabilized by phosphine ligands, mononuclear homoleptic dialkynylaurate(I) complexes, [RC≡C–Au–C≡CR]$^-$, have also been reported in the literature [101, 202–207]. Compared to the Au(I) alkynyl complexes with phosphine as the ancillary ligand, the luminescence properties of this class of anionic Au(I) alkynyl complexes are relatively unexplored. Highly structured absorption and emission bands were observed for such complexes with vibrational progressional spacings assignable to ν(C≡C) stretch and the emission origin has been assigned as derived from ^{3}IL[$\pi \rightarrow \pi^*$(C≡C)] excited states [204, 207]. In addition, a novel catenane structure was reported by Mingos et al., where the alkynyl ligands adopt both η^1 and η^2 bonding modes (Scheme 10.29) [208].

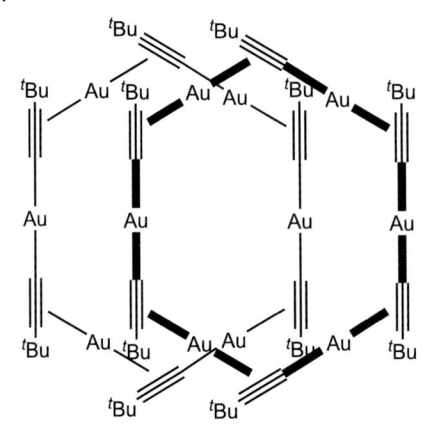

Scheme 10.29. Schematic diagram of the [2]-catanane [{Au(C≡C^{t}Bu)$_6$}$_2$].

Gold(I) alkynyl complexes with a novel planar η^2,η^2-coordination mode have recently been reported by Yam and coworkers using diethynylcalix[4]crown-6 in a 1,3-alternate conformation as the building block [209]. Treatment of the diethynylcalix[4]crown-6 with Au(tht)Cl in the presence of triethylamine as base yielded a series of novel tetranuclear Au(I) alkynyl complexes with planar η^2,η^2-bonding mode. The four gold(I) centers were arranged in a rhomboidal array and capped by the diethynylcalix[4]crown-6 ligands on the two ends. Two Au(I) atoms are σ-bonded to the alkynyl units while the other two Au(I) atoms are each π-coordinated to two alkynyl units in a η^2,η^2-sandwich fashion (Fig. 10.6). Short Au···Au contacts between adjacent Au(I) atoms were observed in the X-ray crystal structure, indicating the presence of significant Au···Au interactions.

These tetranuclear complexes are strong emitters in the orange region at room temperature with luminescence quantum yields of up to 0.22. Red-shifts in emis-

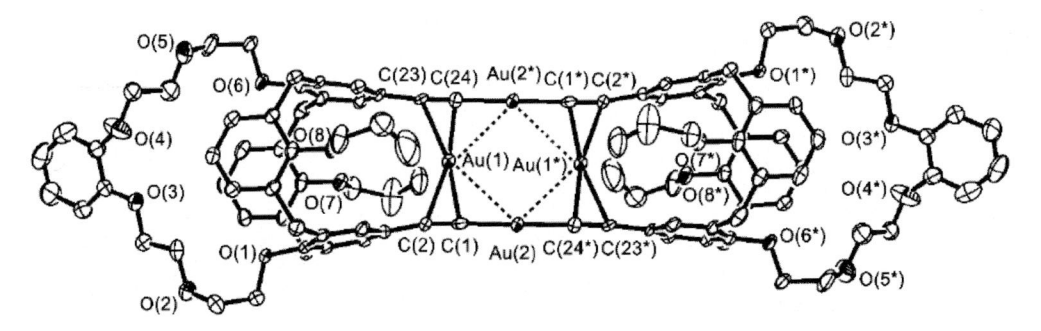

Figure 10.6. Perspective view of a Au(I) alkynyl complex in a novel planar η^2,η^2-coordination mode using diethynylcalix[4]crown-6 as the building block. Thermal ellipsoids are shown at the 30% probability level. Reproduced with permission from [209].

sion bands were noted in the solid state (592–611 nm) relative to that recorded in solution (587–588 nm), which may be attributed to the presence of intermolecular Au···Au interactions in the solid state. In view of the large Stokes shifts and microsecond range lifetimes, the emissions in these Au(I) supramolecular assemblies have been tentatively assigned as derived from states of metal-cluster-centered (ds/dp) character that are modified by Au···Au interactions, mixed with metal-perturbed intraligand $\pi \rightarrow \pi^*$(C≡C) states.

10.7
Mixed-metal Alkynyls

The study of electronic communication through conjugated organic spacers has attracted much attention in the past decade [210, 211]. Apart from mixed-valence systems, mixed-metal compounds bridged by rigid conjugated spacers appear to be an ideal platform to study the electron-transfer process. In particular, redox-active metal complexes have attracted enormous attention due to their ease in probing the electronic communication with various electrochemical techniques. The electron and energy transfer properties of a series of Ru(II)–Os(II) mixed-metal complexes with rigid-rod polyphenylene bridges have been extensively investigated by Balzani and De Cola (Scheme 10.30) [212] representing classical examples of mixed-metal complexes for such studies.

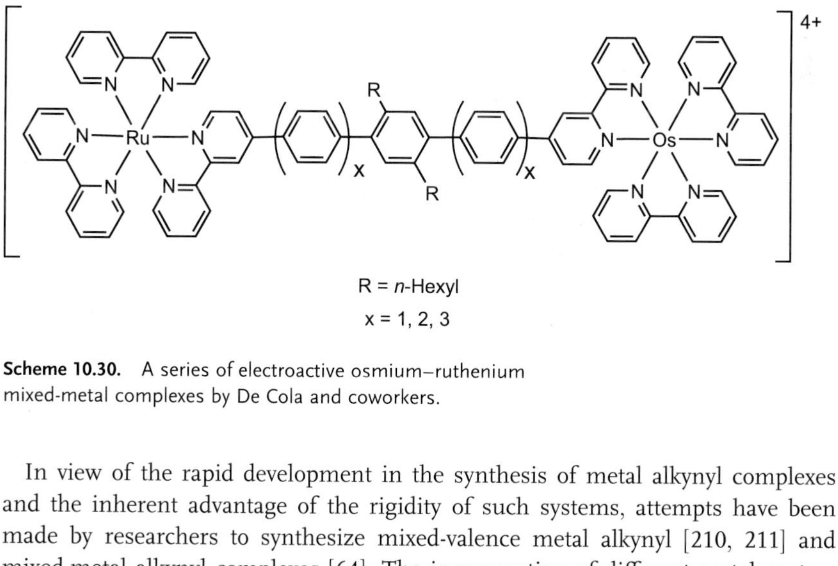

R = n-Hexyl
x = 1, 2, 3

Scheme 10.30. A series of electroactive osmium–ruthenium mixed-metal complexes by De Cola and coworkers.

In view of the rapid development in the synthesis of metal alkynyl complexes and the inherent advantage of the rigidity of such systems, attempts have been made by researchers to synthesize mixed-valence metal alkynyl [210, 211] and mixed-metal alkynyl complexes [64]. The incorporation of different metal centers to the organic spacer, which is usually achieved by the metalloligand approach, may provide a good handle to investigate the electron transfer and communication across the molecule. Isolation of mixed-valence homometallic species have been accomplished by Gladysz, Lapinte and Bruce using their electroactive signature end-groups of [(η^5-C$_5$Me$_5$)Re(NO)(PPh$_3$)], [(η^5-C$_5$Me$_5$)Fe(dppe)] and [(η^5-C$_5$H$_5$)Ru(PPh$_3$)$_2$], respectively. In a recent collaboration between Gladysz and Lapinte, the rhenium–iron mixed-metal alkynyl complex with a butadiynyl bridge, [(η^5-C$_5$Me$_5$)Re(NO)(PPh$_3$)(C≡C–C≡C)(η^5-C$_5$Me$_5$)Fe(dppe)], was synthesized using the metalloligand approach [213]. Both electronic and magnetic interactions were found between remote unpaired spins on the rhenium and iron endgroups in the oxidized species.

On the other hand, with the recent development of various luminescent metal alkynyl systems, attempts have been made to construct luminescent mixed-metal alkynyl assemblies. Perturbation of the emission characteristics by the second metal center would also yield interesting properties as well as providing further insights into the origin of their spectroscopic properties.

The robustness of the rhenium(I) diimine alkynyl systems and rich photophysical behavior have rendered them suitable as metalloligands for the synthesis of mixed-metal complexes. It is well-known that organometallic alkynes exhibit rich coordination chemistry with Cu(I), Ag(I) and Au(I) [214–218], however, photophysical properties of these π-coordinated compounds are rare. Recent work by Yam and coworkers has shown that luminescent mixed-metal alkynyl complexes could be synthesized by the metalloligand approach using the rhenium(I) diimine alkynyl complexes as the η^2-ligand. Reaction of the rhenium(I) diimine alkynyl complex [Re(bpy)(CO)$_3$C≡CPh] with [M(MeCN)$_4$]PF$_6$ in THF at room temperature in an inert atmosphere afforded mixed-metal Re(I)–Cu(I) or –Ag(I) alkynyl complexes (Scheme 10.31) [89]. Their photophysical properties have also been studied. These luminescent mixed-metal complexes were found to emit from their ^{3}MLCT[dπ(Re) → π*(N^N)] manifolds with emission bands blue-shifted relative to their mononuclear precursors (Table 10.5). This has been attributed to the stabilization of the dπ(Re) orbital as a consequence of the weaker π-donating ability of the alkynyl unit upon coordination to the d^{10} metal centers.

Exploration of the construction of mixed-metal luminescent materials using rhenium(I) diimine alkynyl complexes in a terminal mode has also been of interest

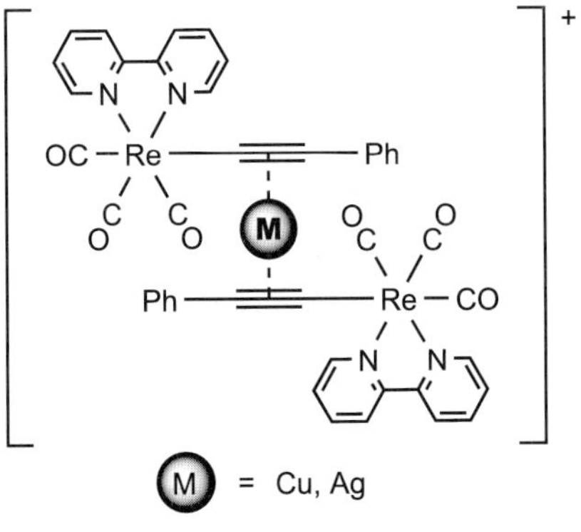

Scheme 10.31. Schematic diagram of Re(I)–Cu(I) and Re(I)–Ag(I) mixed-metal complexes.

Table 10.5. Electronic absorption and photophysical data of Re(I)–Cu(I) and –Ag(I) mixed metal alkynyl complexes. Data adapted from [89].

Complex	Absorption[a] λ/nm (ε/dm^3 mol^{-1} cm^{-1})	Medium (T/K)	Emission λ_{em}/nm (τ_o/μs)
[Re(bpy)(CO)$_3$–C≡C–Ph]	286 (37 250), 298 sh (30 160), 326 sh (10 430), 420 (3 240)	solid (298) solid (77) CH$_2$Cl$_2$ (298)	585 (0.19) 570 654 (<0.1)
[{η^2-Re(CO)$_3$(bpy)(C≡CPh)}$_2$Cu]$^+$	246 (28 160), 288 (29 000), 396 (6 050)	solid (298) solid (77) CH$_2$Cl$_2$ (298)	550 (0.35) 540 590 (0.18)
[{η^2-Re(CO)$_3$(bpy)(C≡CPh)}$_2$Ag]$^+$	250 (26 890), 292 (28 240), 396 (5 930)	solid (298) solid (77) CH$_2$Cl$_2$ (298)	535 (<0.1) 533 600 (0.16)

[a] In CH$_2$Cl$_2$ at 298 K.

and has been facilitated by the isolation of the rhenium(I) alkynyl complexes with terminal acetylenic proton [Re(N^N)(CO)$_3$C≡C–R–C≡C–H]. As an extension of the *triangulo* Cu(I) and Ag(I) system, a series of Re(I)–Cu(I) and –Ag(I) alkynyl complexes with *triangulo* d^{10} metal cores end-capped by rhenium(I) alkynyls was successfully synthesized. The metalloligand approach was used, reacting the d^{10} precursor complexes [M(P^P)(MeCN)$_2$]$^{2+}$ with the metalloligands [Re(N^N)(CO)$_3$C≡C–R–C≡C–H] in the presence of an excess of potassium hydroxide as the base (Scheme 10.32) [219, 220].

In addition, a series of heterodecanuclear assemblies of Re(I)–Ag(I) alkynyls has been isolated from the reaction between [Ag$_2$(μ-dppm)$_2$(MeCN)$_2$]$^{2+}$ and the metalloligand [Re(N^N)(CO)$_3$C≡C–C≡C–H] in the presence of potassium hydroxide (Scheme 10.33) [221]. Each silver atom is linked to two bridging dppm ligands while the rhenium(I) butadiynyl units are found to be coordinated to three silver atoms in a μ_3,η^1-bridging mode. Subtle changes in the substituents on the metalloligands could lead to the formation of entirely different multinuclear architectures. The successful synthesis and isolation of these multinuclear mixed-metal assemblies has shown that the metalloligand approach is a versatile and promising strategy towards the construction of multinuclear luminescent materials. These Re(I)–Cu(I) and –Ag(I) multinuclear complexes were found to be emissive at room temperature upon photo-excitation. Similar to their homometallic precursors, the emission energies of these carbon-rich hybrids are found to be readily tunable by systematic variations on the bridging phosphine ligands, the identity of the d^{10} metal centers, and the nature of the polyynyl and diimine ligands on the Re(I) centers. The emission energy trends are compatible with the predominantly ^{3}MLCT[dπ(Re) → π^*(N^N)] assignment of the lowest energy emissive states

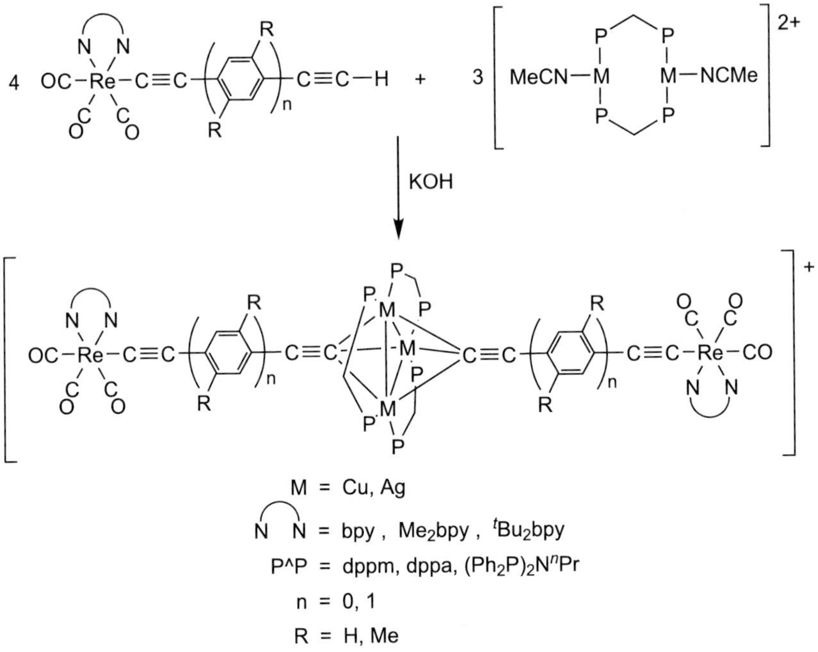

Scheme 10.32. Synthesis of Re(I)–Cu(I) and Re(I)–Ag(I) mixed metal complexes via the metalloligand approach.

with mixing of some ^{3}LLCT[π(C≡C) → π^*(N^N)] character that has been perturbed by the M$_3$ or M$_6$ unit.

Apart from the rhenium(I) mixed-metal complexes with coinage metals, the design and synthesis of mixed-metal species of metal centers with various properties have also attracted much attention. Lapinte and coworkers reported the synthesis and characterization of a series of di- and tri-nuclear iron-containing organometallic complexes with electroactive [(C$_5$Me$_5$)Fe(dppe)] moieties linked by alkynyl bridging ligands (Scheme 10.34) [18]. These Fe(II) alkynyl complexes could be isolated with multiple oxidation states and were found to possess intriguing properties, such as efficient long-distance electron transfer and strong magnetic exchange interaction between remote iron termini.

An electro-switchable luminescent mixed-metal alkynyl complex, [(CO)$_3$(bpy)Re(C≡CC$_6$H$_4$C≡C)Fe(C$_5$Me$_5$)(dppe)], was recently reported by Lapinte and Yam (Scheme 10.35) [222]. Unlike the Re(I)–Cu(I) and –Ag(I) mixed-metal complexes, the Re(I)–Fe(II) complex was found to be nonemissive both in the solid state and in fluid solution (Table 10.6). This has been ascribed to the presence of intramolecular reductive electron transfer and energy transfer quenching pathways for the ^{3}MLCT [dπ(Re) → π^*(bpy)] emissive state via the low-lying MLCT and LF excited states of the Fe(II) moiety. On the contrary, the oxidized form of the mixed-metal complex, [(CO)$_3$(bpy)Re(C≡CC$_6$H$_4$C≡C)Fe(C$_5$Me$_5$)(dppe)]PF$_6$, was found to emit at ca. 590 nm which was assigned as a ^{3}MLCT [dπ(Re) → π^*(bpy)]

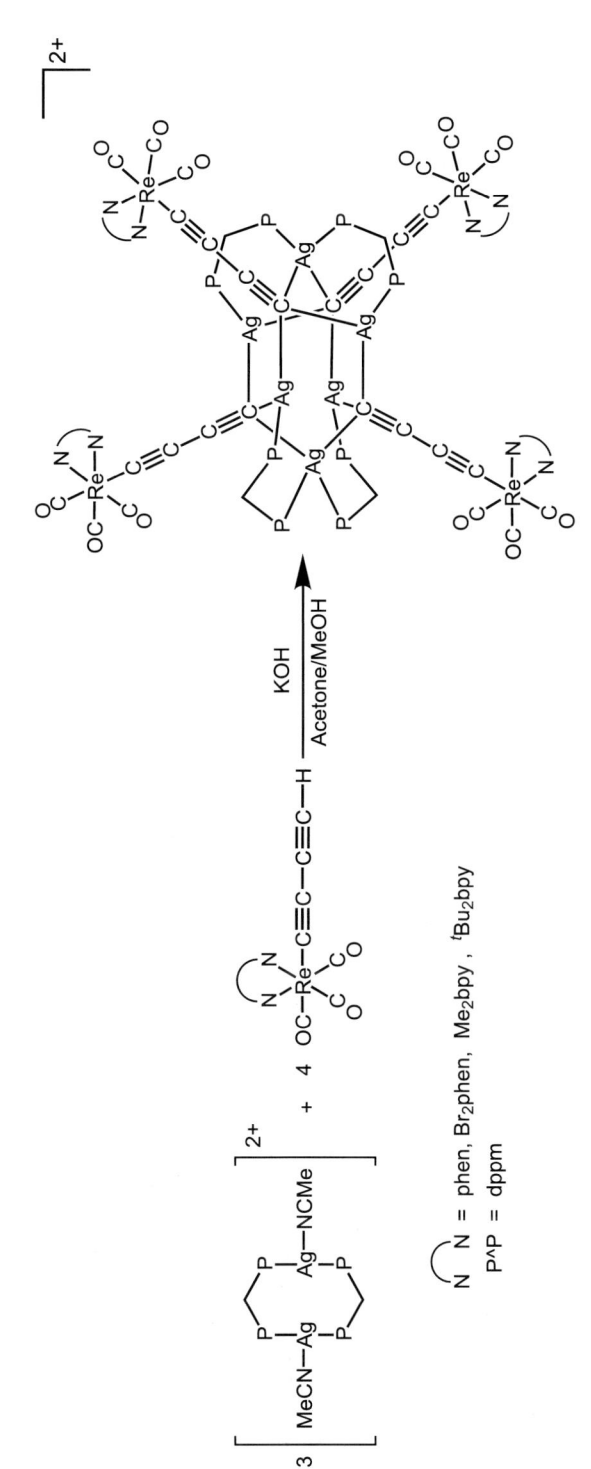

Scheme 10.33. Synthesis of novel heterodecanuclear Re(I)–Ag(I) alkynyl complexes.

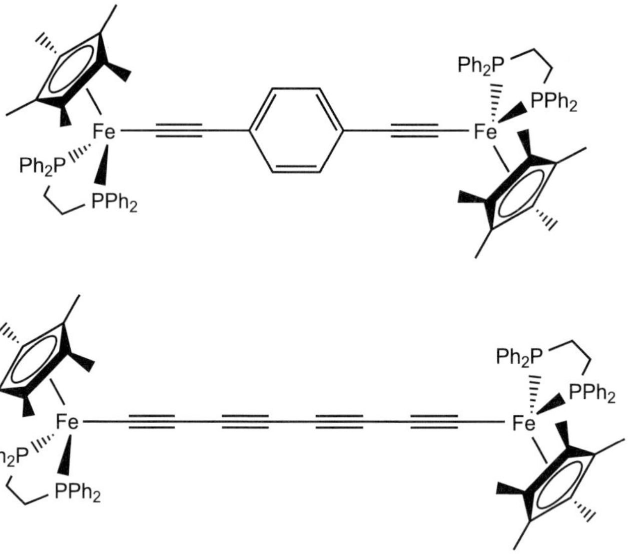

Scheme 10.34. Examples of iron-containing alkynyl complexes with electroactive [(C$_5$Me$_5$)Fe(dppe)] endgroups by Lapinte and coworkers.

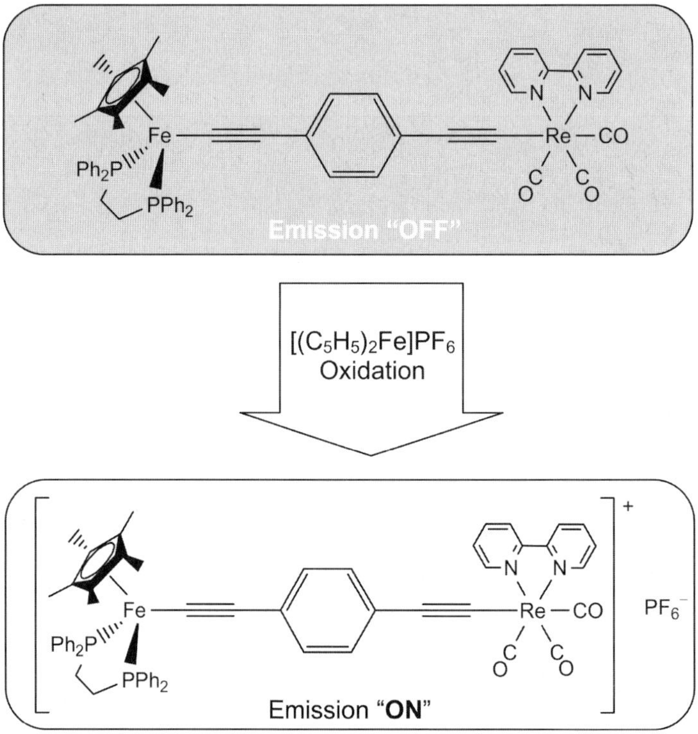

Scheme 10.35. A rhenium–iron mixed-metal complex with electro-switchable photoluminescence activity.

Table 10.6. Electronic absorption and emission data of Re–Fe mixed-metal alkynyl complexes in dichloromethane solution. Data adapted from [222].

Complex	Absorption[a] λ/nm (ε/dm^3 mol^{-1} cm^{-1})	Emission λ_{em}/nm
[Re(bpy)(CO)$_3$–C≡C–C$_6$H$_4$–C≡C–Fe(C$_5$Me$_5$)(dppe)]	290 (39 900), 390 (22 000), 420 sh (20 100)	nonemissive
[Re(bpy)(CO)$_3$–C≡C–C$_6$H$_4$–C≡C–Fe(C$_5$Me$_5$)(dppe)]PF$_6$	292 (49 000), 340 sh (25 300), 432 (9 100), 474 sh (7 500), 830 (7 600), 1389 (22), 1773 (210)	590
[Re(bpy)(CO)$_3$–C≡C–C$_6$H$_4$–CH=C=Fe(C$_5$Me$_5$)(dppe)]PF$_6$	300 (41 100), 348 (43 600), 358 (46 000), 420 (3 805)	605
[Re(bpy)(CO)$_3$–C≡C–C$_6$H$_4$–C≡CH]	298 (49 740), 328 sh (28 640), 420 (2 860)	650
[Fe(C$_5$Me$_5$)(dppe)C≡C–C$_6$H$_5$]	245 (34 000), 348 (12 000)	not measured
[Fe(C$_5$Me$_5$)(dppe)C≡C–C$_6$H$_5$]PF$_6$	241 (34 000), 267 (37 000), 575 (2 200), 663 (2 900)	not measured

[a] Measured in CH$_2$Cl$_2$ at 298 K.

emission. The regeneration of ^{3}MLCT emission in the Re(I)–Fe(III) mixed-metal complex has been rationalized by the fact that the Fe(III) center, being electron deficient, would no longer be able to quench the ^{3}MLCT state by reductive electron-transfer. In addition, low-lying MLCT and LF states were removed upon oxidation of Fe(II) to Fe(III). This has further been supported by the higher emission energy observed for [(CO)$_3$(bpy)Re(C≡CC$_6$H$_4$C≡C)Fe(C$_5$Me$_5$)(dppe)]PF$_6$ (590 nm) relative to the related rhenium(I) precursor complex [(CO)$_3$(bpy)Re(C≡CC$_6$H$_4$C≡CH)] (650 nm), which is in line with the poorer σ- and π-donating ability of the Fe(III) moiety for a ^{3}MLCT [dπ(Re) → π*(bpy)] emission. Alternatively, one can visualize this as the stabilization of the dπ(Fe) orbital in the Fe(III) state, which would lead to the absence of low-lying LF and MLCT states that would quench the triplet MLCT emission. These assignments have been further supported by DFT calculations by Halet [222]. It is envisioned that heterometallic systems with redox-active metal centers will provide a promising and versatile strategy for the design of electroswitchable molecular materials.

Apart from one-dimensional linear structures, mixed-metal branched complexes of rhenium(I)–palladium(II) have also been reported by Yam and coworkers recently (Scheme 10.36) [223]. The trinuclear mixed-metal complexes were prepared by Cu-catalyzed dehydrohalogenation reaction of the Re(I) metalloligand with an excess of trans-[Pd(PEt$_3$)$_2$Cl$_2$]. The formation of undesirable homo-coupled by-products was minimized by the use of strictly anaerobic conditions using Schlenk techniques and the controlled dropwise addition of the metalloligand to a solution of trans-[Pd(PEt$_3$)$_2$Cl$_2$]. The electronic absorption, luminescence and electrochemical properties of these mixed-metal alkynyls have been investigated (Table 10.7). The intense broad absorption bands of the Re(I) precursors at ca. 410–430 nm have been assigned as the MLCT[dπ(Re) → π*(diimine)] transition mixed with

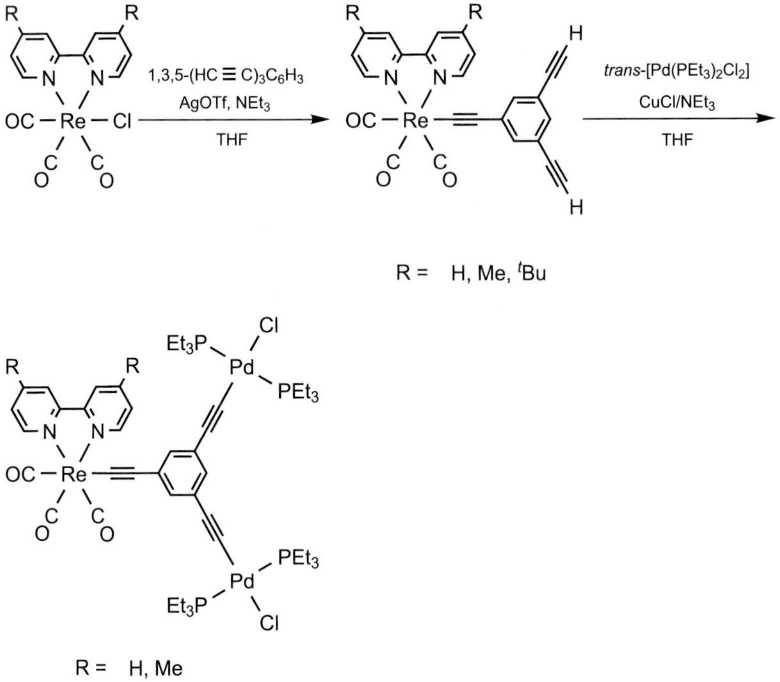

Scheme 10.36. Mixed-metal branched complexes of rhenium(I)–palladium(II) – potential building blocks for heterometallic metallodendrimers.

some alkynyl-to-diimine LLCT[π(C≡C) → π^*(diimine)] character. Red-shifts in the MLCT bands were reported on going from the Re(I) precursors to their corresponding mixed-metal complexes, which have been rationalized by the increased electron-donating ability of the 1,3,5-triethynylbenzene unit upon incorporation of the electron-rich [Pd(PEt$_3$)$_2$Cl] moieties that consequently rendered the dπ(Re) orbital higher-lying in energy. DFT calculations on these complexes supported the mixed MLCT/LLCT assignment of their lowest energy transitions. These mixed-metal complexes were found to be strongly emissive in the orange region, with lowest energy emissive states assigned to be of predominantly ^{3}MLCT[dπ → π^*(diimine)] character. Similar to the branched platinum(II) alkynyl complexes discussed in the preceding section, the chloro groups in these mixed-metal complexes could also be readily replaced by other ligands which rendered them potential building blocks for heterometallic metallodendrimers.

10.8
Conclusions

In this chapter, the synthesis and luminescence properties of selected transition metal alkynyls have been described. The alkynyl group appears to be a promising

Table 10.7. Electronic absorption and photophysical data of Re(I)–Pd(II) mixed-metal alkynyl complexes. Data adapted from [223].

Complex	Absorption[a] λ/nm (ε/dm^3 mol^{-1} cm^{-1})	Medium (T/K)	Emission λ_{em}/nm (τ_o/μs)
[1,3-(HC≡C)$_2$-5-{(bpy)(CO)$_3$ReC≡C}C$_6$H$_3$]	252 (30 410), 294 (32 010), 332 sh (12 740), 418 (2 590)	THF (298) solid (298) solid (77) glass (77)[b]	636 (<0.1) 568 (0.28) 564 (1.25, 0.22)[c] 568 (3.27, 0.70)[c]
[1,3-(HC≡C)$_2$-5-{(Me$_2$bpy)(CO)$_3$ReC≡C}C$_6$H$_3$]	244 (52 540), 294 (41 180), 334 sh (12 940), 408 (3 020)	THF (298) solid (298) solid (77) glass (77)[b]	625 (<0.1) 573 (0.20) 559 (0.38) 555 (3.36, 1.00)[c]
[1,3-(HC≡C)$_2$-5-{(^{t}Bu$_2$bpy)(CO)$_3$ReC≡C}C$_6$H$_3$]	240 (51 640), 294 (39 680), 334 sh (12 100), 408 (2 840)	THF (298) solid (298) solid (77) glass (77)[b]	625 (<0.1) 570 (0.26) 573 (0.36, 0.21)[c] 538 (2.56, 0.70)[c]
[1,3-{Cl(PEt$_3$)$_2$PdC≡C}$_2$-5-{(bpy)(CO)$_3$ReC≡C}C$_6$H$_3$]	268 (88 870), 298 (78 380), 338 sh (140 530), 368 (235 480)	THF (298) solid (298) solid (77) glass (77)[b]	639 (<0.1) 598 (0.26) 594 (0.62) 565 (3.13, 0.78)[c]
[1,3-{Cl(PEt$_3$)$_2$PdC≡C}$_2$-5-{(Me$_2$bpy)(CO)$_3$ReC≡C}C$_6$H$_3$]	236 (57 080), 284 (103 890), 336 sh (16 240), 416 (2 250)	THF (298) solid (298) solid (77) glass (77)[b]	628 (<0.1) 600 (<0.1) 589 (0.73) 546 (4.39, 0.73)[c]

[a] Measured in THF at 298 K.
[b] Measured in EtOH–MeOH (4:1, v/v) glass.
[c] Biexponential decay.

and versatile building block for the construction of luminescent carbon-rich materials. Through the incorporation of the structurally rigid and strongly electron-donating alkynyl units into transition metal complex systems, nonradiative deactivation pathways via low-lying d–d LF states could be suppressed, which accounts for the rich and interesting photophysical properties of these transition metal alkynyls. By changing the substituents on the alkynyl and auxiliary ligands, the electronic and steric properties of these complexes could be modified and the nature of the lowest energy excited state could also be probed and tuned. The understanding of the fundamental aspects of these electronic absorption and luminescence properties will provide the groundwork for the rational design of new carbon-rich luminescent materials for the development of molecular devices, sensors, OLEDs and others.

10.9
Experimental: Selected Procedures

10.9.1
Synthesis of [(^{t}Bu$_2$bpy)(CO)$_3$ReC≡C–C≡C–C≡CPh]

A mixture of [(^{t}Bu$_2$bpy)(CO)$_3$ReC≡C–C≡C–H] (100 mg, 0.17 mmol), BrC≡CPh (0.17 mmol), piperidine (0.67 mL, 6.80 mmol), and CuI (0.0051 mmol) in THF (40 mL) was stirred at room temperature in an inert atmosphere of nitrogen for 15 min, after which the solvent was removed under vacuum. The product was then purified by column chromatography on silica gel using dichloromethane–petroleum ether (1:1 v/v) as eluant. Subsequent recrystallization from dichloromethane–n-hexane gave orange crystals of [(^{t}Bu$_2$bpy)(CO)$_3$ReC≡C–C≡C–C≡CPh]. Yield: 41 mg, 35%.

10.9.2
Synthesis of [1,3,5-{PhC≡CPt(PEt$_3$)$_2$C≡CC$_6$H$_4$C≡C}$_3$C$_6$H$_3$]

[1,3,5-{ClPt(PEt$_3$)$_2$C≡CC$_6$H$_4$C≡C}$_3$C$_6$H$_3$] (142 mg, 0.077 mmol) and phenylacetylene (26 mg, 0.252 mmol) were dissolved in a mixture of THF (20 mL) and diethylamine (10 mL) in an inert atmosphere of nitrogen. To this reaction mixture was added CuCl (5 mg) as a catalyst. The pale yellow mixture was then stirred overnight at room temperature, and the solvent was removed under reduced pressure. The greenish-yellow gummy residue was then redissolved in dichloromethane and washed successively with brine, deionized water and dried over anhydrous sodium sulfate. This was filtered, and the solvent was removed under reduced pressure. The yellow residue was chromatographed on basic aluminum oxide (50–200 microns) using dichloromethane as the eluant. Subsequent recrystallization of the crude product with dichloromethane–n-hexane afforded [1,3,5-{PhC≡CPt(PEt$_3$)$_2$C≡CC$_6$H$_4$C≡C}$_3$C$_6$H$_3$] as a pale yellow powder. Yield: 68 mg, 43%.

10.9.3
Synthesis of [Pt(trpy)C≡C–Ph]PF$_6$

To a stirred solution of phenylacetylene (44 mg, 0.43 mmol) in methanol was added sodium hydroxide (23 mg, 0.58 mmol). The resultant solution was stirred at room temperature for 30 min. [Pt(trpy)(MeCN)](OTf)$_2$ (300 mg, 0.39 mmol) was added to the reaction mixture, which turned to a deep red solution immediately and was then stirred for 12 h at room temperature. The mixture was filtered, and a saturated solution of ammonium hexafluorophosphate in methanol was added. The product was isolated, washed with methanol, and dried. Subsequent recrystallization by diffusion of diethyl ether vapor into an acetonitrile solution of the product gave [Pt(trpy)C≡C–Ph]PF$_6$ as dark purple crystals. Yield: 174 mg, 60%.

10.9.4
Synthesis of [Cu$_3$(μ-dppm)$_3$(μ_3-η^1-C≡C-^{t}Bu)$_2$]PF$_6$

A solution of the complex [Cu$_2$(μ-dppm)$_2$(MeCN)$_2$](PF$_6$)$_2$ (127 mg, 0.10 mmol) in THF/MeOH (v/v 3:1; 12 mL) was treated with 3,3-dimethylbut-1-yne (16.2 µL, 0.13 mmol) and an excess of KOH and stirred at room temperature for 24 h. After evaporation to dryness, the resulting solid was extracted with THF (3 × 8 mL), and the solution was filtered and reduced in volume. Subsequent diffusion of diethyl ether into the concentrated solution gave [Cu$_3$(μ-dppm)$_3$(μ_3-η^1-C≡C-^{t}Bu)$_2$]PF$_6$ as air-stable colorless crystals. Yield: 67 mg, 40%.

10.9.5
Synthesis of [{η^2-Re(CO)$_3$(bpy)(C≡CPh)}$_2$Ag]PF$_6$

[Re(bpy)(CO)$_3$–C≡C–Ph] (100 mg, 0.19 mmol) and [Ag(MeCN)$_4$]PF$_6$ (39 mg, 0.085 mmol) were stirred in THF (30 mL) at room temperature in an inert atmosphere of nitrogen for 30 min. The yellow solution was filtered and the filtrate was reduced in volume under reduced pressure. Subsequent recrystallization from dichloromethane–n-hexane gave yellow crystals of [{η^2-Re(CO)$_3$(bpy)(C≡CPh)}$_2$Ag]PF$_6$. Yield: 56 mg, 45%.

10.9.6
Synthesis of [Ag$_6$(μ-dppm)$_4${μ_3-C≡C–C≡C–Re(Me$_2$bpy)(CO)$_3$}$_4$](PF$_6$)$_2$

This was prepared by the reaction of [Ag$_2$(μ-dppm)$_2$(MeCN)$_2$](PF$_6$)$_2$ (120 mg, 0.09 mmol) with [Re(CO)$_3$(Me$_2$bpy)(C≡C–C≡C–H)] (60 mg, 0.12 mmol) in the presence of excess KOH in acetone/MeOH (15 mL/5 mL). The mixture was stirred at room temperature for 24 h. After evaporation to dryness, the resulting solid residue was extracted with acetone and the solution was filtered and concentrated. Layering n-hexane onto an acetone solution of the complex gave [Ag$_6$(μ-dppm)$_4${μ_3-C≡C–C≡C–Re(Me$_2$bpy)(CO)$_3$}$_4$](PF$_6$)$_2$ as orange crystals. Yield: 103 mg, 78%.

10.9.7
Synthesis of [1,3-{Cl(PEt$_3$)$_2$PdC≡C}$_2$-5-{(bpy)(CO)$_3$ReC≡C}C$_6$H$_3$]

To a suspension of trans-[Pd(PEt$_3$)$_2$Cl$_2$] (211 mg, 0.51 mmol), CuCl (0.50 mg, 0.0051 mmol), and Et$_3$N (52 mg, 0.51 mmol) in THF (20 mL) was added dropwise a solution of [1,3-(HC≡C)$_2$-5-{(bpy)(CO)$_3$ReC≡C}C$_6$H$_3$] (98 mg, 0.17 mmol) in THF (20 mL). The reaction mixture was allowed to stir at room temperature in an inert atmosphere of nitrogen for 2 h. The orange suspension was filtered, and the orange filtrate was reduced in volume under reduced pressure. The residue was then purified by column chromatography on silica gel using dichloromethane as eluant. After removal of the first band, which contained unreacted [1,3-(HC≡C)$_2$-5-{(bpy)(CO)$_3$ReC≡C}C$_6$H$_3$], ethyl acetate was used to elute out the second band, which contained [1,3-{Cl(PEt$_3$)$_2$PdC≡C}$_2$-5-{(bpy)(CO)$_3$ReC≡C}C$_6$H$_3$] as the desired product. Subsequent recrystallization from vapor diffusion of diethyl ether into a dichloromethane solution of [1,3-{Cl(PEt$_3$)$_2$PdC≡C}$_2$-5-{(bpy)(CO)$_3$ReC≡C}C$_6$H$_3$] gave orange crystals. Yield: 66 mg, 29%.

Acknowledgments

V.W.W.Y. acknowledges support from the University Development Fund of the University of Hong Kong, the University of Hong Kong Foundation for Education Development and Research Limited, and the Research Grants Council of the Hong Kong Special Administrative Region, China.

Abbreviations and Acronyms

Ac	acetyl
Ar	aromatic group
bpy	2,2-bipyridine
Br$_2$phen	5,6-dibromo-1,10-phenanthroline
^{t}Bu$_2$bpy	4,4'-di-*tert*-butyl-2,2'-bipyridine
^{t}Bu$_3$trpy	4,4',4''-tri-*tert*-butyl-2,2':6',2''-terpyridine
(CF$_3$)$_2$bpy	4,4'-bis-trifluoromethyl-2,2'-bipyridine
Cy	cyclohexyl
dcpe	bis(dicyclohexylphosphino)ethane
DFT	density functional theory
dppa	N,N-bis(diphenylphosphino)amine
dppe	bis(diphenylphosphino)ethane
dppm	bis(diphenylphosphino)methane
dppp	bis(diphenylphosphino)propane
HOMO	highest occupied molecular orbital
IL	intraligand
LF	ligand-field
LLCT	ligand-to-ligand charge transfer
LMCT	ligand-to-metal charge transfer
LUMO	lowest unoccupied molecular orbital
Me$_2$bpy	4,4'-dimethyl-2,2'-bipyridine
MLCT	metal-to-ligand charge transfer
MMLCT	metal-metal-to-ligand charge transfer
MO	molecular orbital
NLO	nonlinear optical
OLED	organic light emitting device
OTf	trifluoromethyl sulfonate
phen	1,10-phenanthroline
THF	tetrahydrofuran
tht	tetrahydrothiophene
trpy	2,2':6',2''-terpyridine

References

1 R. F. Curl, *Angew. Chem.* **1997**, *109*, 1636–1647; *Angew. Chem. Int. Ed. Engl.* **1997**, *36*, 1566–1576.
2 H. W. Kroto, *Angew. Chem.* **1997**, *109*, 1648–1664; *Angew. Chem. Int. Ed. Engl.* **1997**, *36*, 1578–1593.
3 R. E. Smalley, *Angew. Chem.* **1997**, *109*, 1666–1673; *Angew. Chem. Int. Ed. Engl.* **1997**, *36*, 1594–1601.
4 F. Diederich, Y. Rubin, *Angew. Chem.* **1992**, *104*, 1123; *Angew. Chem. Int. Ed. Engl.* **1992**, *31*, 1101–1123.
5 F. Diederich, *Nature* **1994**, *369*, 199–207.
6 R. Faust, *Angew. Chem.* **1998**, *110*, 2985–2988; *Angew. Chem. Int. Ed.* **1998**, *37*, 2825–2828.
7 U. H. F. Bunz, Y. Rubin, Y. Tobe, *Chem. Soc. Rev.* **1999**, *28*, 107–119.
8 H. W. Kroto, J. R. Heath, S. C. O'Brian, R. F. Curl, R. E. Smalley, *Nature* **1985**, *318*, 162–163.
9 J. M. Hunter, J. L. Fye, E. J. Roskamp, M. F. Jarrold, *J. Phys. Chem.* **1994**, *98*, 1810–1818.
10 R. J. Lagow, J. J. Kampa, H.-C. Wei, S. L. Battle, J. W. Genge, D. A. Laude, C. J. Harper, R. Bau, R. C. Stevens, J. F. Haw, E. Munson, *Science* **1995**, *267*, 362–367.
11 K.-H. Homman, *Angew. Chem.* **1998**, *110*, 2572–2590; *Angew. Chem. Int. Ed.* **1998**, *37*, 2434–2451.
12 T. Grösser, A. Hirsch, *Angew. Chem.* **1993**, *105*, 1390; *Angew. Chem. Int. Ed. Engl.* **1993**, *32*, 1340–1342.
13 G. Schermann, T. Grösser, F. Hampel, A. Hirsch, *Chem. Eur. J.* **1997**, *3*, 1105–1112.
14 P. P. K. Smith, P. R. Buseck, *Science* **1982**, *216*, 984–986.
15 K. Akagi, M. Nishiguchi, H. Shirakawa, Y. Furukawa, I. Harada, *Synth. Met.* **1987**, *17*, 557–562.
16 N. J. Long, C. K. Williams, *Angew. Chem.* **2003**, *115*, 2690–2722; *Angew. Chem. Int. Ed.* **2003**, *42*, 2586–2617.
17 S. Szafert, J. A. Gladysz, *Chem. Rev.* **2003**, *103*, 4175–4205.
18 F. Paul, C. Lapinte, *Coord. Chem. Rev.* **1998**, *178–180*, 431–509.
19 V. W. W. Yam, *Acc. Chem. Res.* **2002**, *35*, 555–563.
20 V. W. W. Yam, K. K. W. Lo, K. M. C. Wong, *J. Organomet. Chem.* **1999**, *578*, 3–30.
21 *Carbon Rich Compounds II*, A. de Meijere (ed.), Springer, Berlin, **1999**.
22 R. Boese, A. J. Matzger, D. L. Mohler, K. P. C. Vollhardt, *Angew. Chem.* **1995**, *107*, 1630–1633; *Angew. Chem. Int. Ed. Engl.* **1995**, *34*, 1478–1481.
23 C. Eickmeier, H. Junga, A. J. Matzger, F. Scherhag, M. Shim, K. P. C. Vollhardt, *Angew. Chem.* **1997**, *109*, 2194–2199; *Angew. Chem. Int. Ed. Engl.* **1997**, *36*, 2103–2108.
24 C. Eickmeier, D. Holmes, H. Junga, A. J. Matzger, F. Scherhag, M. Shim, K. P. C. Vollhardt, *Angew. Chem.* **1999**, *111*, 856–860; *Angew. Chem. Int. Ed. Engl.* **1999**, *38*, 800–804.
25 S. Han, D. R. Anderson, A. D. Bond, H. V. Chu, R. L. Disch, D. Holmes, J. M. Schulman, S. J. Teat, K. P. C. Vollhardt, G. D. Whitener, *Angew. Chem.* **2002**, *114*, 3361–3364; *Angew. Chem. Int. Ed.* **2002**, *41*, 3227–3230.
26 D. T.-Y. Bong, L. Gentric, D. Holmes, A. J. Matzger, F. Scherhag, K. P. C. Vollhardt, *Chem. Commun.* **2002**, 278–279.
27 P. I. Dosa, G. D. Whitener, K. P. C. Vollhardt, A. D. Bond, S. J. Teat, *Org. Lett.* **2002**, *4*, 2075–2078.
28 P. Siemsen, R. C. Livingston, F. Diederich, *Angew. Chem.* **2000**, *112*, 2740–2767; *Angew. Chem. Int. Ed.* **2000**, *39*, 2632–2657.
29 A. Köllhofer, T. Pullmann, H. Plenio, *Angew. Chem.* **2003**, *115*, 1086–1088; *Angew. Chem. Int. Ed.* **2003**, *42*, 1056–1058.
30 A. Elangovan, Y.-H. Wang, T.-I. Ho, *Org. Lett.* **2003**, *5*, 1841–1844.
31 Z. Novák, A. Szabó, J. Répási, A. Kotschy, *J. Org. Chem.* **2003**, *68*, 3327–3329.
32 B. Liang, M. Dai, J. Chen, Z. Yang, *J. Org. Chem.* **2005**, *70*, 391–393.

33 N. E. Leadbeater, M. Marco, B. J. Tominack, *Org. Lett.* **2003**, *5*, 3919–3922.
34 P. Appukkuttan, W. Dehaen, E. Van der Eycken, *Eur. J. Org. Chem.* **2003**, 4713–4716.
35 F. Bohlmann, *Angew. Chem.* **1955**, *67*, 389–394.
36 F. Bohlmann, P. Herbst, H. Gleinig, *Chem. Ber.* **1961**, *94*, 948–957.
37 E. R. H. Jones, J. M. Thompson, M. C. Whiting, *J. Chem. Soc.* **1957**, 2012–2017.
38 E. R. H. Jones, H. H. Lee, M. C. Whiting, *J. Chem. Soc.* **1960**, 3483–3489.
39 T. Gibtner, F. Hampel, J.-P. Gisselbrecht, A. Hirsch, *Chem. Eur. J.* **2002**, *8*, 408–432.
40 L.-H. Lee, V. Lynch, R. J. Lagow, *J. Chem. Soc., Perkin Trans. 1* **2000**, 2805–2809.
41 T. Müller, J. Hulliger, W. Seichter, E. Weber, T. Weber, M. Wübbenhorst, *Chem. Eur. J.* **2000**, *6*, 54–61.
42 S. Eisler, A. D. Slepkov, E. Elliott, T. Luu, R. McDonald, F. A. Hegmann, R. R. Tykwinski, *J. Am. Chem. Soc.* **2005**, *127*, 2666–2667.
43 Y. Zhao, R. McDonald, R. R. Tykwinski, *J. Org. Chem.* **2002**, *67*, 2805–2812.
44 R. Eastmond, T. R. Johnson, D. R. M. Walton, *Tetrahedron* **1972**, *28*, 4601–4616.
45 T. R. Johnson, D. R. M. Walton, *Tetrahedron* **1972**, *28*, 5221–5236.
46 S. Eisler, N. Chahal, R. McDonald, R. R. Tykwinski, *Chem. Eur. J.* **2003**, *9*, 2542–2550.
47 A. L. K. Shi Shun, E. T. Chernick, S. Eisler, R. R. Tykwinski, *J. Org. Chem.* **2003**, *68*, 1339–1347.
48 Y. Rubin, S. S. Lin, C. B. Knobler, J. Anthony, A. M. Boldi, F. Diederich, *J. Am. Chem. Soc.* **1991**, *113*, 6943–6949.
49 J. M. Tour, *Acc. Chem. Res.* **2000**, *33*, 791–804 and references therein.
50 J. M. Tour, A. M. Rawlett, M. Kozaki, Y. Yao, R. C. Jagessar, S. M. Dirk, D. W. Price, M. A. Reed, C.-W. Zhou, J. Chen, W. Wang, I. Campbell, *Chem. Eur. J.* **2001**, *7*, 5118–5134.
51 J. G. Kushmerick, D. B. Holt, S. K. Pollack, M. A. Ratner, J. C. Yang, T. L. Schull, J. Naciri, M. H. Moore, R. Shashidhar, *J. Am. Chem. Soc.* **2002**, *124*, 10654–10655.
52 R. H. Grubbs, D. Kratz, *Chem. Ber.* **1993**, *126*, 149–157.
53 J. C. Nelson, J. G. Saven, J. S. Moore, P. G. Wolynes, *Science* **1997**, *277*, 1793–1796.
54 U. H. F. Bunz, *J. Organomet. Chem.* **2003**, *683*, 269–287.
55 F. Diederich, Y. Rubin, C. B. Knobler, R. L. Whetten, K. E. Schriver, K. N. Houk, Y. Li, *Science* **1989**, *245*, 1088–1090.
56 F. Diederich, *Chem. Commun.* **2001**, 219–227 and references therein.
57 J.-P. Gisselbrecht, N. N. P. Moonen, C. Boudon, M. B. Nielsen, F. Diederich, M. Gross, *Eur. J. Org. Chem.* **2004**, 2959–2972 and references therein.
58 R. Haag, A. De Meijere in *Modern Arene Chemistry*, D. Astruc (ed.), Wiley-VCH, Weinheim, **2002**, pp. 32–52.
59 D. Zhao, J. S. Moore, *Chem. Commun.* **2003**, 807–818 and references therein.
60 J. A. Marsden, G. J. Palmer, M. M. Haley, *Eur. J. Org. Chem.* **2003**, 2355–2369 and references therein.
61 L. T. Scott, M. J. Cooney, *Modern Acetylene Chemistry*, P. J. Stang, F. Diederich (eds.), VCH, Weinheim, **1995**, 321–351 and references therein.
62 H. Hopf, *Angew. Chem.* **2003**, *115*, 2928–2931; *Angew. Chem. Int. Ed.* **2003**, *42*, 2822–2825 and references therein.
63 C. Devadoss, P. Bharathi, J. S. Moore, *J. Am. Chem. Soc.* **1996**, *118*, 9635–9644.
64 V. W. W. Yam, *J. Organomet. Chem.* **2004**, *689*, 1393–1401.
65 J. S. Wilson, A. S. Dhoot, A. J. A. B. Seeley, M. S. Khan, A. Köhler, R. H. Friend, *Nature* **2001**, *413*, 828–831.
66 K. Sonogashira, S. Takahashi, N. Hagihara, *Macromolecules* **1977**, *10*, 879–880.

67 K. Osakada, T. Yamamoto, *Coord. Chem. Rev.* **2000**, *198*, 379–399.
68 K. Osakada, M. Hamada, T. Yamamoto, *Organometallics* **2000**, *19*, 458–468.
69 K. Osakada, R. Sakata, T. Yamamoto, *Organometallics* **1997**, *16*, 5354–5364.
70 M. I. Bruce, D. A. Harbourne, F. Waugh, F. G. A. Stone, *J. Chem. Soc. A* **1968**, 356–359.
71 W. Weng, T. Bartik, M. T. Johnson, A. M. Arif, J. A. Gladysz, *Organometallics* **1995**, *14*, 889–897.
72 T. Bartik, B. Bartik, M. Brady, R. Dembinski, J. A. Gladysz, *Angew. Chem.* **1996**, *108*, 467–469; *Angew. Chem. Int. Ed. Engl.* **1996**, *35*, 414–417.
73 M. Brady, W. Weng, Y. Zhou, J. W. Seyler, A. J. Amoroso, A. M. Arif, M. Böhme, G. Frenking, J. A. Gladysz, *J. Am. Chem. Soc.* **1997**, *119*, 775–788.
74 W. E. Meyer, A. J. Amoroso, C. R. Horn, M. Jaeger, J. A. Gladysz, *Organometallics* **2001**, *20*, 1115–1127.
75 R. Dembinski, T. Lis, S. Szafert, C. L. Mayne, T. Bartik, J. A. Gladysz, *J. Organomet. Chem.* **1999**, *578*, 229–246.
76 R. Dembinski, S. Szafert, P. Haquette, T. Lis, J. A. Gladysz, *Organometallics* **1999**, *18*, 5438–5440.
77 T. Bartik, W. Weng, J. A. Ramsden, S. Szafert, S. B. Falloon, A. M. Arif, J. A. Gladysz, *J. Am. Chem. Soc.* **1998**, *120*, 11071–11081.
78 S. B. Falloon, S. Szafert, A. M. Arif, J. A. Gladysz, *Chem. Eur. J.* **1998**, *4*, 1033–1042.
79 R. Dembinski, T. Bartik, B. Bartik, M. Jaeger, J. A. Gladysz, *J. Am. Chem. Soc.* **2000**, *122*, 810–822.
80 V. W. W. Yam, V. C. Y. Lau, K. K. Cheung, *Organometallics* **1995**, *14*, 2749–2753.
81 V. W. W. Yam, V. C. Y. Lau, K. K. Cheung, *Organometallics* **1996**, *15*, 1740–1744.
82 M. Wrighton, D. L. Morse, *J. Am. Chem. Soc.* **1974**, *96*, 998–1003.
83 V. W. W. Yam, K. M. C. Wong, S. H. F. Chong, V. C. Y. Lau, S. C. F. Lam, L. Zhang, K. K. Cheung, *J. Organomet. Chem.* **2003**, *670*, 205–220.
84 V. W. W. Yam, S. H. F. Chong, K. K. Cheung, *Chem. Commun.* **1998**, 2121–2122.
85 V. W. W. Yam, S. H. F. Chong, C. C. Ko, K. K. Cheung, *Organometallics* **2000**, *19*, 5092–5097.
86 V. W. W. Yam, *Chem. Commun.* **2001**, 789–796.
87 W. J. Hunks, M.-A. MacDonald, M. C. Jennings, R. J. Puddephatt, *Organometallics* **2000**, *19*, 5063–5070.
88 S. H. F. Chong, Ph. D. Thesis, The University of Hong Kong **2001**.
89 V. W. W. Yam, S. H. F. Chong, K. M. C. Wong, K. K. Cheung, *Chem. Commun.* **1999**, 1013–1014.
90 S. J. Davies, B. F. G. Johnson, M. S. Khan, J. Lewis, *J. Chem. Soc., Chem. Commun.* **1991**, 187–188.
91 B. F. G. Johnson, A. K. Kakkar, M. S. Khan, J. Lewis, A. E. Dray, R. H. Friend, F. Wittmann, *J. Mater. Chem.* **1991**, *1*, 485–486.
92 J. Stahl, J. C. Bohling, E. B. Bauer, T. B. Peters, W. Mohr, J. M. Martin-Alvarez, F. Hampel, J. A. Gladysz, *Angew. Chem.* **2002**, *114*, 1951–1957; *Angew. Chem. Int. Ed.* **2002**, *41*, 1871–1876.
93 W. Mohr, J. Stahl, F. Hampel, J. A. Gladysz, *Chem. Eur. J.* **2003**, *9*, 3324–3340.
94 G. R. Owen, J. Stahl, F. Hampel, J. A. Gladysz, *Organometallics* **2004**, *23*, 5889–5892.
95 G. R. Owen, F. Hampel, J. A. Gladysz, *Organometallics* **2004**, *23*, 5893–5895.
96 F. Zhuravlev, J. A. Gladysz, *Chem. Eur. J.* **2004**, *10*, 6510–6522.
97 T. M. Cooper, B. C. Hall, A. R. Burke, J. E. Rogers, D. G. McLean, J. E. Slagle, P. A. Fleitz, *Chem. Mater.* **2004**, *16*, 3215–3217.
98 S. M. AlQaisi, K. J. Galat, M. Chai, D. G. Ray, III, P. L. Rinaldi, C. A. Tessier, W. J. Youngs, *J. Am. Chem. Soc.* **1998**, *120*, 12149–12150.
99 D. Zhang, D. B. McConville, C. A. Tessier, W. J. Youngs, *Organometallics* **1997**, *16*, 824–825.
100 J. D. Bradshaw, L. Guo, C. A.

Tessier, W. J. Youngs, *Organometallics* **1996**, *15*, 2582–2584.
101 M. I. Bruce, B. C. Hall, B. W. Skelton, M. E. Smith, A. H. White, *J. Chem. Soc., Dalton Trans.* **2002**, 995–1001.
102 L. Pu, *Chem. Rev.* **2004**, *104*, 1687–1716 and references therein.
103 S. J. Lee, A. Hu, W. Lin, *J. Am. Chem. Soc.* **2002**, *124*, 12948–12949.
104 S. J. Lee, W. Lin, *J. Am. Chem. Soc.* **2002**, *124*, 4554–4555.
105 H. Jiang, W. Lin, *J. Am. Chem. Soc.* **2003**, *125*, 8084–8085.
106 S. J. Lee, C. R. Luman, F. N. Castellano, W. Lin, *Chem. Commun.* **2003**, 2124–2125.
107 L.-Z. Gong, Q.-S. Hu, L. Pu, *J. Org. Chem.* **2001**, *66*, 2358–2367.
108 Q.-S. Hu, V. Pugh, M. Sabat, L. Pu, *J. Org. Chem.* **1999**, *64*, 7528–7536.
109 V. Pugh, Q. S. Hu, L. Pu, *Angew. Chem.* **2000**, *112*, 3784–3787; *Angew. Chem. Int. Ed.* **2000**, *39*, 3638–3641.
110 V. J. Pugh, Q.-S. Hu, X. Zuo, F. D. Lewis, L. Pu, *J. Org. Chem.* **2001**, *66*, 6136–6140.
111 J. Chatt, B. L. Shaw, *J. Chem. Soc.* **1959**, 4020–4033.
112 U. Belluco, *Organometallic and Coordination Chemistry of Platinum*, Academic Press, London and New York, **1974**.
113 H. D. Empsall, B. L. Shaw, A. J. Stringer, *J. Organomet. Chem.* **1975**, *94*, 131–138.
114 J. H. Nelson, A. W. Verstuyft, J. D. Kelly, H. B. Jonassen, *Inorg. Chem.* **1974**, *13*, 27–33.
115 H. Masai, K. Sonogashira, N. Hagihara, *Bull. Chem. Soc. Jpn.* **1971**, *44*, 2226–2230.
116 L. Sacksteder, E. Baralt, B. A. DeGraff, C. M. Lukehart, J. N. Demas, *Inorg. Chem.* **1991**, *30*, 2468–2476.
117 M. S. Khan, A. K. Kakkar, N. J. Long, J. Lewis, P. R. Raithby, P. Nguyen, T. B. Marder, F. Wittmann, R. H. Friend, *J. Mater. Chem.* **1994**, *4*, 1227–1232.
118 D. Beljonne, H. F. Wittmann, A. Köhler, S. Graham, M. Younus, J. Lewis, P. R. Raithby, M. S. Khan, R. H. Friend, J. L. Brédas, *J. Chem. Phys.* **1996**, *105*, 3868–3877.
119 N. Chawdhury, A. Köhler, R. H. Friend, M. Younus, N. J. Long, P. R. Raithby, J. Lewis, *Macromolecules* **1998**, *31*, 722–727.
120 N. Chawdhury, A. Köhler, R. H. Friend, W.-Y. Wong, J. Lewis, M. Younus, P. R. Raithby, T. C. Corcoran, M. R. A. Al-Mandhary, M. S. Khan, *J. Chem. Phys.* **1999**, *110*, 4963–4970.
121 T. B. Marder, G. Lesley, Z. Yuan, H. B. Fyfe, P. Chow, G. Stringer, I. R. Jobe, N. J. Taylor, I. D. Williams, S. K. Kurtz, *Materials for Nonlinear Optics: Chemical Perspectives*, G. D. Stucky, S. R. Marder, J. Sohn (eds.), ACS Symposium Series 455, American Chemical Society, Washington D. C., **1991**, p. 605.
122 Y. Liu, S. Jiang, K. Glusac, D. H. Powell, D. F. Anderson, K. S. Schanze, *J. Am. Chem. Soc.* **2002**, *124*, 12412–12413.
123 K. Haskins-Glusac, M. R. Pinto, C. Tan, K. S. Schanze, *J. Am. Chem. Soc.* **2004**, *126*, 14964–14971.
124 N. Ohshiro, F. Takei, K. Onitsuka, S. Takahashi, *J. Organomet. Chem.* **1998**, *569*, 195–202.
125 S. Leininger, P. J. Stang, S. Huang, *Organometallics* **1998**, *17*, 3981–3987.
126 K. Onitsuka, A. Shimizu, S. Takahashi, *Chem. Commun.* **2003**, 280–281.
127 K. Onitsuka, M. Fujimoto, N. Ohshiro, S. Takahashi, *Angew. Chem.* **1999**, *111*, 737–739; *Angew. Chem. Int. Ed.* **1999**, *38*, 689–692.
128 V. W. W. Yam, C. H. Tao, L. Zhang, K. M. C. Wong, K. K. Cheung, *Organometallics* **2001**, *20*, 453–459.
129 C. H. Tao, N. Zhu, V. W. W. Yam, *Chem. Eur. J.* **2005**, *11*, 1647–1657.
130 C.-W. Chan, L.-K. Cheng, C.-M. Che, *Coord. Chem. Rev.* **1994**, *132*, 87–97.
131 M. Hissler, W. B. Connick, D. K. Geiger, J. E. McGarrah, D. Lipa, R. J. Lachicotte, R. Eisenberg, *Inorg. Chem.* **2000**, *39*, 447–457.
132 T. J. Wadas, R. J. Lachicotte, R. Eisenberg, *Inorg. Chem.* **2003**, *42*, 3772–3778.

133 J. E. McGarrah, R. Eisenberg, *Inorg. Chem.* **2003**, *42*, 4355–4365.
134 C. E. Whittle, J. A. Weinstein, M. W. George, K. S. Schanze, *Inorg. Chem.* **2001**, *40*, 4053–4062.
135 Y. Kang, J. Lee, D. Song, S. Wang, *J. Chem. Soc., Dalton Trans.* **2003**, 3493–3499.
136 C. J. Adams, S. L. James, X. Liu, P. R. Raithby, L. J. Yellowlees, *J. Chem. Soc., Dalton Trans.* **2000**, 63–67.
137 S. L. James, M. Younus, P. R. Raithby, J. Lewis, *J. Organomet. Chem.* **1997**, *543*, 233–235.
138 V. H. Houlding, V. M. Miskowski, *Coord. Chem. Rev.* **1991**, *111*, 145–152.
139 V. M. Miskowski, V. H. Houlding, *Inorg. Chem.* **1989**, *28*, 1529–1533.
140 V. M. Miskowski, V. H. Houlding, *Inorg. Chem.* **1991**, *30*, 4446–4452.
141 J. Biedermann, G. Gliemann, U. Klement, K.-J. Range, M. Zabel, *Inorg. Chem.* **1990**, *29*, 1884–1888.
142 R. S. Osborn, D. Rogers, *J. Chem. Soc., Dalton Trans.* **1974**, 1002–1004.
143 R. H. Herber, M. Croft, M. J. Coyer, B. Bilash, A. Sahiner, *Inorg. Chem.* **1994**, *33*, 2422–2426.
144 W. B. Connick, L. M. Henling, R. E. Marsh, H. B. Gray, *Inorg. Chem.* **1996**, *35*, 6261–6265.
145 H. Lang, A. del Villar, *J. Organomet. Chem.* **2003**, *670*, 45–55.
146 N. Ohshiro, F. Takei, K. Onitsuka, S. Takahashi, *Chem. Lett.* **1996**, 871–872.
147 K. Onitsuka, S. Yamamoto, S. Takahashi, *Angew. Chem.* **1999**, *111*, 129–131; *Angew. Chem. Int. Ed.* **1999**, *38*, 174–176.
148 I. E. Pomestchenko, C. R. Luman, M. Hissler, R. Ziessel, F. N. Castellano, *Inorg. Chem.* **2003**, *42*, 1394–1396.
149 I. E. Pomestchenko, F. N. Castellano, *J. Phys. Chem. A* **2004**, *108*, 3485–3492.
150 S. J. Lippard, *Acc. Chem. Res.* **1978**, *11*, 211–217.
151 M. Howe-Grant, S. J. Lippard, *Biochemistry* **1979**, *18*, 5762–5769.
152 H.-K. Yip, L.-K. Cheng, K.-K. Cheung, C.-M. Che, *J. Chem. Soc., Dalton Trans.* **1993**, 2933–2938.
153 R. Büchner, C. T. Cunningham, J. S. Field, R. J. Haines, D. R. McMillin, G. C. Summerton, *J. Chem. Soc., Dalton Trans.* **1999**, 711–717.
154 J. A. Bailey, M. G. Hill, R. E. Marsh, V. M. Miskowski, W. P. Schaefer, H. B. Gray, *Inorg. Chem.* **1995**, *34*, 4591–4599.
155 D. R. McMillin, J. J. Moore, *Coord. Chem. Rev.* **2002**, *229*, 113–121.
156 V. W. W. Yam, R. P. L. Tang, K. M. C. Wong, K. K. Cheung, *Organometallics* **2001**, *20*, 4476–4482.
157 T. K. Aldridge, E. M. Stacy, D. R. McMillin, *Inorg. Chem.* **1994**, *33*, 722–727.
158 V. W. W. Yam, K. M. C. Wong, N. Zhu, *Angew. Chem.* **2003**, *115*, 1438–1441; *Angew. Chem. Int. Ed.* **2003**, *42*, 1400–1403.
159 V. W. W. Yam, K. M. C. Wong, N. Zhu, *J. Am. Chem. Soc.* **2002**, *124*, 6506–6507.
160 C. Yu, K. M. C. Wong, K. H. Y. Chan, V. W. W. Yam, *Angew. Chem.* **2005**, *117*, 801–804; *Angew. Chem. Int. Ed.* **2005**, *44*, 791–794.
161 W. Lu, B.-X. Mi, M. C. W. Chan, Z. Hui, N. Zhu, S.-T. Lee, C.-M. Che, *Chem. Commun.* **2002**, 206–207.
162 W. Lu, B.-X. Mi, M. C. W. Chan, Z. Hui, C.-M. Che, N. Zhu, S.-T. Lee, *J. Am. Chem. Soc.* **2004**, *126*, 4958–4971.
163 S. C. Chan, M. C. W. Chan, Y. Wang, C.-M. Che, K.-K. Cheung, N. Zhu, *Chem. Eur. J.* **2001**, *7*, 4180–4190.
164 M. P. Gamasa, J. Gimeno, E. Lastra, A. Aguirre, S. García-Granda, *J. Organomet. Chem.* **1989**, *378*, C11–C14.
165 J. Díez, M. P. Gamasa, J. Gimeno, A. Aguirre, S. García-Granda, *Organometallics* **1991**, *10*, 380–382.
166 V. W. W. Yam, W. K. Lee, T. F. Lai, *Organometallics* **1993**, *12*, 2383–2387.
167 V. W. W. Yam, W. K. Lee, K. K. Cheung, B. Crystall, D. Phillips, *J. Chem. Soc., Dalton Trans.* **1996**, 3283–3287.
168 V. W. W. Yam, W. K. M. Fung, M. T. Wong, *Organometallics* **1997**, *16*, 1772–1778.

169 V. W. W. Yam, W. K. M. Fung, K. K. Cheung, *Organometallics* **1997**, *16*, 2032–2037.
170 V. W. W. Yam, W. K. M. Fung, K. K. Cheung, *Organometallics* **1998**, *17*, 3293–3298.
171 V. W. W. Yam, W. K. M. Fung, K. K. Cheung, *J. Cluster Sci.* **1999**, *10*, 37–69.
172 V. W. W. Yam, W. K. M. Fung, K. K. Cheung, *Angew. Chem.* **1996**, *108*, 1213–1215; *Angew. Chem. Int. Ed. Engl.* **1996**, *35*, 1100–1102.
173 W. Y. Lo, C. H. Lam, W. K. M. Fung, H. Z. Sun, V. W. W. Yam, D. Balcells, F. Maseras, O. Eisenstein, *Chem. Commun.* **2003**, 1260–1261.
174 V. W. W. Yam, W. K. M. Fung, K. K. Cheung, *Chem. Commun.* **1997**, 963–964.
175 W. Y. Lo, C. H. Lam, V. W. W. Yam, N. Zhu, K. K. Cheung, S. Fathallah, S. Messaoudi, B. Le Guennic, S. Kahlal, J. F. Halet, *J. Am. Chem. Soc.* **2004**, *126*, 7300–7310.
176 W. Y. Lo, Ph. D. Thesis, The University of Hong Kong **2004**.
177 P. Pyykkö, *Angew. Chem.* **2002**, *114*, 3723–3728; *Angew. Chem. Int. Ed.* **2002**, *41*, 3573–3578.
178 P. Pyykkö, *Angew. Chem.* **2004**, *116*, 4512–4557; *Angew. Chem. Int. Ed.* **2004**, *43*, 4412–4456.
179 H. Schmidbaur, *Gold Bull.* **2000**, *33*, 3–10.
180 R. J. Puddephatt, *Comprehensive Coordination Chemistry*, Vol. 5, Sir G. Wilkinson, R. D. Gillard, J. A. McCleverty (eds.), Pergamon Press, Oxford, 1987, pp. 861–923.
181 A. Grohmann, H. Schmidbaur, *Comprehensive Organometallic Chemistry II*, Vol. 3, E. W. Abel, F. G. A. Stone and G. Wilkinson (eds.), Pergamon, Oxford, 1995, pp. 1–56.
182 D. Li, X. Hong, C.-M. Che, W.-C. Lo, S.-M. Peng, *J. Chem. Soc., Dalton Trans.* **1993**, 2929–2932.
183 C.-M. Che, H.-K. Yip, W.-C. Lo, S.-M. Peng, *Polyhedron* **1994**, *13*, 887–890.
184 C.-M. Che, H.-Y. Chao, V. M. Miskowski, Y. Li, K.-K. Cheung, *J. Am. Chem. Soc.* **2001**, *123*, 4985–4991.
185 W. Lu, H.-F. Xiang, N. Zhu, C.-M. Che, *Organometallics* **2002**, *21*, 2343–2346.
186 W. Lu, N. Zhu, C.-M. Che, *J. Am. Chem. Soc.* **2003**, *125*, 16081–16088.
187 D. M. P. Mingos, R. Vilar, D. Rais, *J. Organomet. Chem.* **2002**, *641*, 126–133.
188 M. J. Irwin, J. J. Vittal, R. J. Puddephatt, *Organometallics* **1997**, *16*, 3541–3547.
189 M.-A. MacDonald, R. J. Puddephatt, G. P. A. Yap, *Organometallics* **2000**, *19*, 2194–2199.
190 W. J. Hunks, J. Lapierre, H. A. Jenkins, R. J. Puddephatt, *J. Chem. Soc., Dalton Trans.* **2002**, 2885–2889.
191 I. R. Whittall, M. G. Humphrey, S. Houbrechts, J. Maes, A. Persoons, S. Schmid, D. C. R. Hockless, *J. Organomet. Chem.* **1997**, *544*, 277–283.
192 S. K. Hurst, N. T. Lucas, M. G. Humphrey, T. Isoshima, K. Wostyn, I. Asselberghs, K. Clays, A. Persoons, M. Samoc, B. Luther-Davies, *Inorg. Chim. Acta* **2003**, *350*, 62–76.
193 S. K. Hurst, M. P. Cifuentes, A. M. McDonagh, M. G. Humphrey, M. Samoc, B. Luther-Davies, I. Asselberghs, A. Persoons, *J. Organomet. Chem.* **2002**, *642*, 259–267.
194 C. E. Powell, M. G. Humphrey, *Coord. Chem. Rev.* **2004**, *248*, 725–756.
195 T. E. Müller, S. W.-K. Choi, D. M. P. Mingos, D. Murphy, D. J. Williams, V. W. W. Yam, *J. Organomet. Chem.* **1994**, *484*, 209–224.
196 A. W. Czarnik, *Fluorescent Chemosensors for Ion and Molecular Recognition*, American Chemical Society, Washington DC, **1993**.
197 L. Fabbrizzi, A. Poggi, *Chem. Soc. Rev.* **1995**, *24*, 197–202.
198 A. P. de Silva, H. Q. N. Gunaratne, T. Gunnlaugsson, A. J. M. Huxley, C. P. McCoy, J. T. Rademacher, T. E. Rice, *Chem. Rev.* **1997**, *97*, 1515–1566.
199 V. W. W. Yam, C. K. Li, C. L. Chan, *Angew. Chem.* **1998**, *110*, 3041–3044; *Angew. Chem. Int. Ed.* **1998**, *37*, 2857–2859.
200 V. W. W. Yam, K. L. Cheung, L. H. Yuan, K. M. C. Wong, K. K. Cheung, *Chem. Commun.* **2000**, 1513–1514.

201 V. W. W. Yam, S. K. Yip, L. H. Yuan, K. L. Cheung, N. Zhu, K. K. Cheung, *Organometallics* **2003**, *22*, 2630–2637.
202 J. Vicente, M.-T. Chicote, M. D. Abrisqueta, P. G. Jones, *Organometallics* **1997**, *16*, 5628–5636.
203 J. Vicente, M.-T. Chicote, M. M. Alvarez-Falcón, M. A. Fox, D. Bautista, *Organometallics* **2003**, *22*, 4792–4797.
204 M. Ferrer, L. Rodríguez, O. Rossell, F. Pina, J. C. Lima, M. F. Bardia, X. Solans, *J. Organomet. Chem.* **2003**, *678*, 82–89.
205 J. Vicente, M. T. Chicote, M. D. Abrisqueta, M. M. Alvarez-Falcón, *J. Organomet. Chem.* **2002**, *663*, 40–45.
206 M. Shiotsuka, Y. Yamamoto, S. Okuno, M. Kitou, K. Nozaki, S. Onaka, *Chem. Commun.* **2002**, 590–591.
207 Y. Yamamoto, M. Shiotsuka, S. Okuno, S. Onaka, *Chem. Lett.* **2004**, *33*, 210–211.
208 D. M. P. Mingos, J. Yau, S. Menzer, D. J. Williams, *Angew. Chem.* **1995**, *107*, 2045–2047; *Angew. Chem. Int. Ed. Engl.* **1995**, *34*, 1894–1895.
209 S. K. Yip, E. C. C. Cheng, L. H. Yuan, N. Zhu, V. W. W. Yam, *Angew. Chem.* **2004**, *116*, 5062–5065; *Angew. Chem. Int. Ed.* **2004**, *43*, 4954–4957.
210 V. Balzani, A. Juris, M. Venturi, S. Campagna, S. Serroni, *Chem. Rev.* **1996**, *96*, 759–833 and references therein.
211 L. De Cola, P. Belser, *Coord. Chem. Rev.* **1998**, *177*, 301–346 and references therein.
212 B. Schlicke, P. Belser, L. De Cola, E. Sabbioni, V. Balzani, *J. Am. Chem. Soc.* **1999**, *121*, 4207–4214.
213 F. Paul, W. E. Meyer, L. Toupet, H. Jiao, J. A. Gladysz, C. Lapinte, *J. Am. Chem. Soc.* **2000**, *122*, 9405–9414.
214 O. M. Abu Salah, M. I. Bruce, *J. Chem Soc., Dalton Trans.* **1974**, 2302–2304.
215 H. Lang, K. Köhler, S. Blau, *Coord. Chem. Rev.* **1995**, *143*, 113–168.
216 Y. Zhu, O. Clot, M. O. Wolf, G. P. A. Yap, *J. Am. Chem. Soc.* **1998**, *120*, 1812–1821.
217 C. Müller, J. A. Whiteford, P. J. Stang, *J. Am. Chem. Soc.* **1998**, *120*, 9827–9837.
218 S. Mihan, K. Sünkel, W. Beck, *Chem. Eur. J.* **1999**, *5*, 745–753.
219 V. W. W. Yam, W. Y. Lo, C. H. Lam, W. K. M. Fung, K. M. C. Wong, V. C. Y. Lau, N. Zhu, *Coord. Chem. Rev.* **2003**, *245*, 39–47.
220 V. W. W. Yam, W. K. M. Fung, K. M. C. Wong, V. C. Y. Lau, K. K. Cheung, *Chem. Commun.* **1998**, 777–778.
221 V. W. W. Yam, W. Y. Lo, N. Zhu, *Chem. Commun.* **2003**, 2446–2447.
222 K. M. C. Wong, S. C. F. Lam, C. C. Ko, N. Zhu, V. W. W. Yam, S. Roué, C. Lapinte, S. Fathallah, K. Costuas, S. Kahlal, J. F. Halet, *Inorg. Chem.* **2003**, *42*, 7086–7097.
223 S. H. F. Chong, S. C. F. Lam, V. W. W. Yam, N. Zhu, K. K. Cheung, S. Fathallah, K. Costuas, J. F. Halet, *Organometallics* **2004**, *23*, 4924–4933.

11
Defined-length Carbon-rich Conjugated Oligomers

Herbert Meier

11.1
Introduction

Because of their interesting optical, electrical and optoelectronic properties, conjugated oligomers represent target compounds for many applications in materials science; moreover, they are model compounds for the corresponding polymers [1]. Carbon-rich extended π electron systems can consist of building blocks such as olefinic double bonds, triple bonds, benzene rings or higher condensed aromatic ring systems. Accordingly the *repeat units* of conjugated oligomers are single or composite units of such building blocks. Scheme 11.1 summarizes a few typical examples.

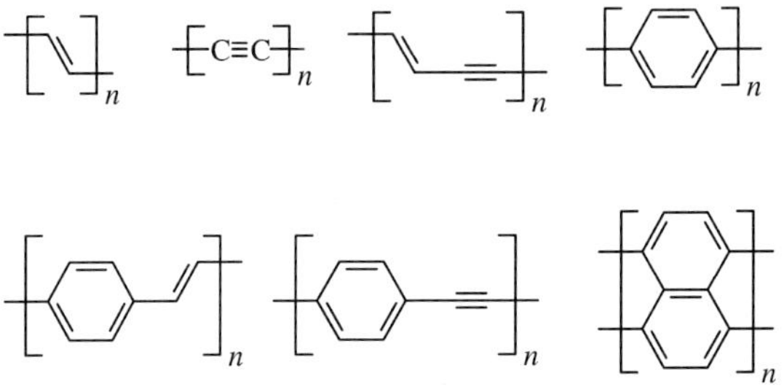

Scheme 11.1. Selected repeat units for conjugated oligomers.

The exactly defined *length* of such oligomers is due to the monodisperse character of the compounds having an exactly defined number of repeat units n. Moreover, the distance L between the terminal positions of the conjugated system can be described as a linear function of n, provided that conformational effects do not considerably affect the extent of the oligomeric molecules. Scheme 11.2 illustrates $L(n)$ for the series of oligo(1,4-phenyleneethynylene)s (OPE) **1** and oligo(1,4-phenylenevinylene)s (OPV) **2**.

Carbon-Rich Compounds. Edited by Michael M. Haley and Rik R. Tykwinski
Copyright © 2006 WILEY-VCH Verlag GmbH & Co. KGaA, Weinheim
ISBN: 3-527-31224-2

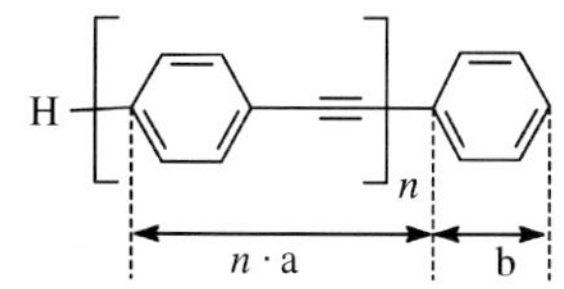

1

$$L = n \cdot a + b \qquad a = 0.68,\; b = 0.28$$

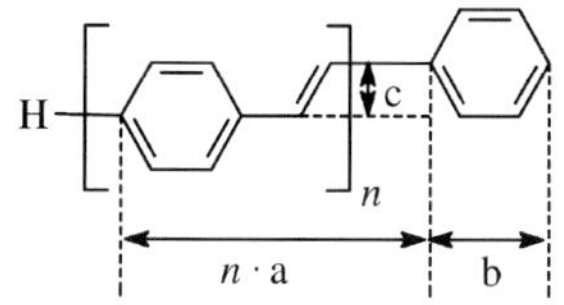

2

$$na + b \leq L \leq \sqrt{n^2(a^2 + c^2) + b^2 + 2nab}$$

$$L \approx n \cdot a + b$$

$a = 0.66$
$b = 0.28$
$c = 0.11$

Scheme 11.2. Length *L* (nm) of OPE chains **1** and OPV chains **2** as a linear function of the number of repeat units, *n*. The parameters *a*, *b* and *c* correspond to average values obtained from crystal structure data, which are registered for related compounds in the CCDC.

The benzene rings in **1** and **2** can show torsions along the chain, but that does not affect *L*. However, the situation is somewhat more complex for **2** than for **1**, because different olefinic *conformers* can differ to some extent in their length *L*. Two (neighboring) olefinic double bonds of **2** can have a *cisoid* or *transoid* orientation. Thus, the number *N* of conformers increases strongly with the increasing number *n* of repeat units (Fig. 11.1). The pentadecamer **2** ($n = 15$), the highest known monodisperse OPV [2], has 8256 such conformers. Because of symmetry criteria, the number *N* is lower than $2^{n-1} = 2^{14} = 16384$. According to the relation $L(n)$ in Scheme 11.2, all 8256 conformers have almost the same length *L*, namely $L = 10.2 \pm 0.1$ nm. Therefore a linear equation $L(n)$ is also reasonable for OPVs **2**.

Since the energy for deformation of bond angles is relatively low, one can expect in higher oligomers some deviation from the ideal geometry obtained for the repeat unit in small molecules. A striking example is represented in Scheme 11.3 by the oligoynes **3** (OY). Despite the expected 180° angle, even a cyclic system, namely cyclooctadecanonayne (**4**) could be generated [3, 4]. The bond angle in **4** is reduced to about 160° which causes a strain energy of $18 \times 4 = 72$ kcal mol^{-1}. The structure with alternating bond lengths is more favorable than the corresponding cumulene structure with equal bond lengths [3, 4].

The length *L* in **1** and **2** characterizes not only the size of these rod-like nanopar-

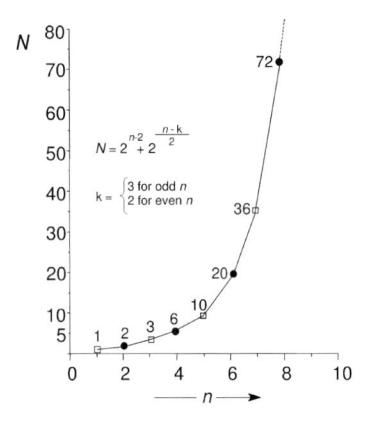

Figure 11.1. Number N of conformers of a symmetric oligomer chain with two possible orientations for each additional repeat unit n.

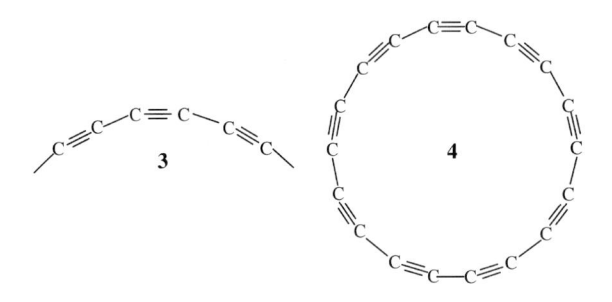

$$N = 2^{n-2} + 2^{\frac{n-k}{2}}$$

$$k = \begin{cases} 3 \text{ for odd } n \\ 2 \text{ for even } n \end{cases}$$

Scheme 11.3. Bend in oligoynes **3** which ultimately enables the formation of the cyclic structure **4**.

ticles, but it also represents the decisive parameter for oligomeric chromophores in nonlinear optics (NLO), particularly in the function for the second hyperpolarizability $\gamma = f(L)$.

The oligo(1,4-phenylene)s (OP) comply also with a linear relation $L(n)$; however, as soon as cross-conjugated compounds with *meta* linked benzene rings are considered, L becomes meaningless, because a variety of conformers of different shape exists.

Apart from the linear arrangement of repeat units, cyclic, star-shaped and dendritic arrangements can be realized. Scheme 11.4 shows some examples based on stilbenoid building blocks.

The rigidity of the substructures leads in all of these oligomers to molecules with a well-defined size: length of the rods (**2**), diameter of the discs (**5, 6** and first and second generation of **7**) and diameter and height of the cylindric shape of the higher dendrimers (third, fourth and fifth generation of **7**). Reports on areno-condensed annulenes **5** [5], star-shaped stilbenoid compounds **6** [6–11] and stilbenoid dendrimers **7** [12–16] are given elsewhere; this chapter is

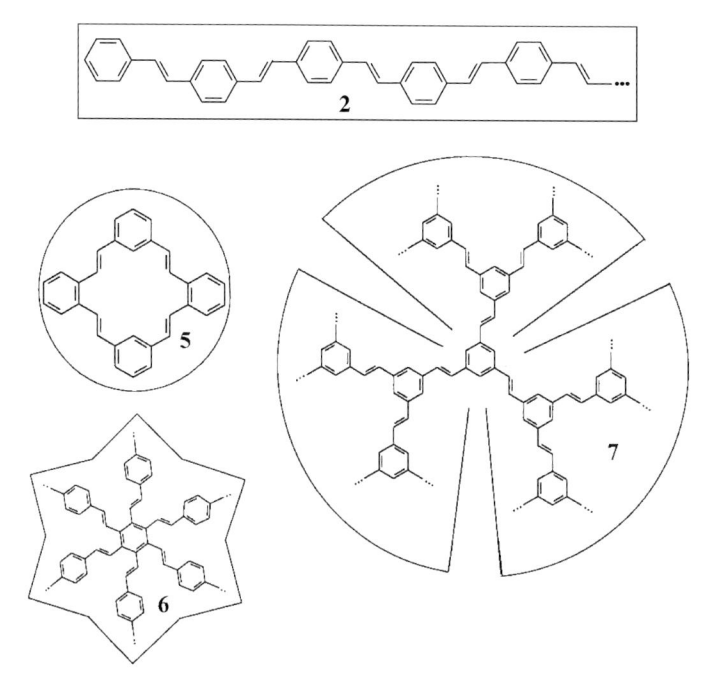

Scheme 11.4. Oligomers consisting of (E)-stilbene units: rigid rods **2**, discs **5**, stars **6**, and dendrimers **7**.

focused on linear systems, namely on oligo(1,4-phenylenevinylene)s **2** and oligo-(1,4-phenyleneethynylene)s **1**.

11.2
Oligo(1,4-phenylenevinylene)s OPV

The unsubstituted OPVs [1a] become very soon ($n > 3$) hardly soluble in organic solvents; therefore solubilizing groups like alkyl or alkoxy groups have to be attached. The 2,5-dipropoxy substituted OPVs **2a–j**, shown in Scheme 11.5 [17–25], have been studied in detail.

2	a	b	c	d	e	f	g	h	i	j
n	1	2	3	4	5	6	7	8	11	15

Scheme 11.5. Oligo(2,5-dipropoxy-1,4-phenylenevinylene)s.

Propoxy groups enhance sufficiently the solubility and processability of these materials. Much longer and/or branched alkoxy groups would be even better for this purpose; however, they impair somewhat the tunneling of charges from chain to chain [26]. Another useful strategy consists of the attachment of one long and one short sidechain to the benzene ring or in an alternating sequence of substituted and unsubstituted benzene rings.

11.2.1
Synthesis

The construction of OPV chains is based on the repetitive formation of carbon–carbon (CC) double or single bonds. Table 11.1 gives a survey of the reactions which are most often used for this purpose.

The Knoevenagel condensation with elimination of water requires activated methylene groups like CH_2–CN. Such an activation is not necessary for the Siegrist reaction in which aniline is eliminated. Due to the size of the latter leaving group, the antiperiplanar mechanism guarantees an extremely high stereoselectivity. The kinetically controlled Siegrist reaction leads to trans/cis ratios up to 1000:1 [27], whereas the thermodynamic trans/cis equilibrium of stilbenoid compounds is

Table 11.1. Selection of commonly used CC coupling reactions for the generation of OPVs.

Formation of CC double bonds

Ar–C=X	+	Y=C–Ar	Reaction
–C=O		H_2C–	Knoevenagel
–C=NPh		H_2C–	Siegrist
–C=O		Ph_3P=C–	Wittig
–C=O		$(RO)_2$P(=O)–C–	Horner
–C=O		O=C–	McMurry

Formation of CC single bonds

Ar–X	+	Y–C=C–Ar	Reaction
–Hal		H–	Heck
–Hal		R_3Sn–	Stille
–B(OH)$_2$		Hal–	Suzuki

in the range 95:5. The Wittig-Horner reactions promise high yields, but a lower stereoselectivity, particularly the Wittig variant. The same statement is valid for the McMurry method which is often used to double the length of the chain. The Pd-catalyzed Heck, Stille and Suzuki couplings are also not free of structural defects so that careful purification processes become inevitable. In many cases a combination of several methods is favorable.

In principle, the coupling techniques can be achieved with AB or with AA and BB components (Scheme 11.6).

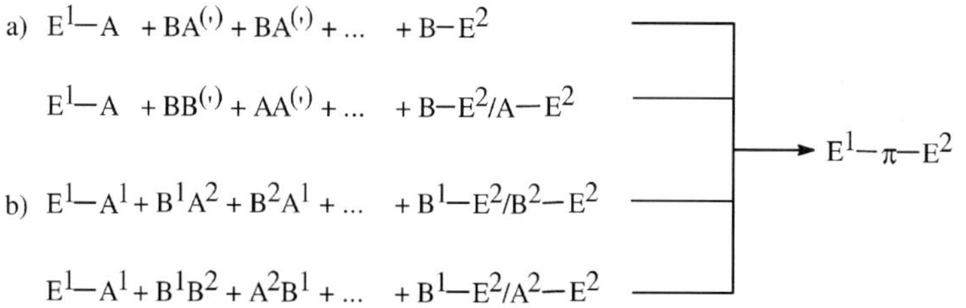

a) $E^1-A + BA^{(')} + BA^{(')} + ... + B-E^2$

$E^1-A + BB^{(')} + AA^{(')} + ... + B-E^2/A-E^2$

$\longrightarrow E^1-\pi-E^2$

b) $E^1-A^1 + B^1A^2 + B^2A^1 + ... + B^1-E^2/B^2-E^2$

$E^1-A^1 + B^1B^2 + A^2B^1 + ... + B^1-E^2/A^2-E^2$

Scheme 11.6. Chain construction $E^1-\pi-E^2$ by repetitive CC coupling steps with AB or AA and BB monomers: (a) selective coupling by applying a protection/deprotection technique: reaction of A of E^1-A with B of BA', deprotection of A', reaction with BA', etc. or reaction of A of E^1-A with B of BB', deprotection of B', reaction with A of AA', etc.; (b) selective coupling by applying an orthogonal strategy: reaction of A^1 of E^1-A^1 with B^1 of B^1A^2, reaction of A^2 with B^2 of B^2A^1, etc. or reaction of A^1 of E^1-A^1 with B^1B^2, reaction of B^2 with A^2 of A^2B^1, etc.

In order to perform selective coupling reactions, a *protection/deprotection technique* A → A' → A and/or B → B' → B has to be applied. An alternative strategy, the so-called *orthogonal method* uses different functions A^1, A^2 and B^1, B^2 which permit selective reactions $A^1 + B^1$ and $A^2 + B^2$ but not $A^1 + B^2$ and $A^2 + B^1$. E^1 and E^2 are the desired endgroups; they are normally introduced in the first and the very last reaction steps; however, a *bidirectional growth strategy*, which starts in the center of the chain can also be applied.

The pentadecamer OPV **2j** (Scheme 11.7) serves here as an example for the first strategy (Scheme 11.6a) [2, 24]. The Schiff base **8** represents the start segment and the methyl component **9** the extension reagent for the construction of the chain. The Siegrist reaction of **8** and **9** in a strongly alkaline medium and the subsequent spontaneous deprotection of the aldehyde function yields **10**. Transformation of **10** to the Schiff base **11** and a second Siegrist reaction followed by deprotection leads to **12** and by repetition of the same procedure to **13** and **14**. Finally **14** is dimerized by a McMurry reaction. Up to the formation of **13** the yields are high; then they decrease when the chains become very long. The red precipitate **2j** obtained from THF has a pure trans configuration. The ^{1}H NMR spectrum of the mother liquor shows an additional isomer with a cis configured central double bond. The trans configured double bonds generated by the Siegrist reactions are not affected by the conditions used in the McMurry reaction.

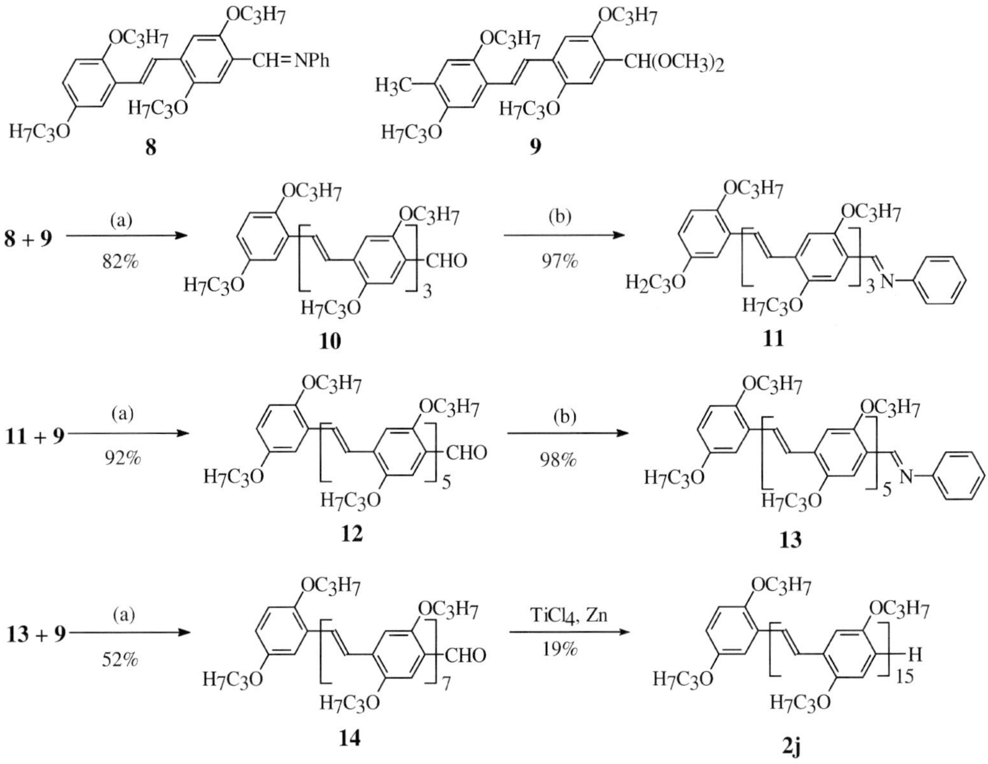

Scheme 11.7. Synthesis of the pentadecamer OPV **2j**: (a) 1. KO*t*-Bu, DMF, 2. H⁺; (b) C₆H₅–NH₂.

The *trans* selectivity is a very important point for the generation of OPVs. If **2j** were synthesized by a stepwise formation of 15 double bonds and a selectivity of 95% trans and 5% cis for each step, which corresponds roughly to Horner reactions in the stilbenoid series, the amount of all-trans **2j** would be

$$\binom{15}{15} \cdot \left(\frac{95}{100}\right)^{15} \cdot \left(1 - \frac{95}{100}\right)^{0} \cdot 100\% = 46\%$$

The remaining 54% of **2j** would contain at least one cis configuration. These figures demonstrate how important synthetic strategies and purification processes are for the generation of constitutionally and configurationally pure products and their application in materials science.

A nice example of an orthogonal approach was published by Yu et al. [28]. A sequence of Heck and Horner reactions furnishes an OPV series with alternating substituted and unsubstituted benzene rings. A start segment **15** and two bifunctional extension reagents **16** and **17** were used as shown in Scheme 11.8. The yields of the five steps are good; some disadvantages of the method are due to the

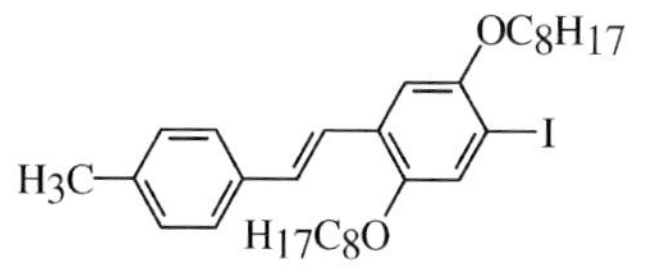

11.2 Oligo(1,4-phenylenevinylene)s OPV

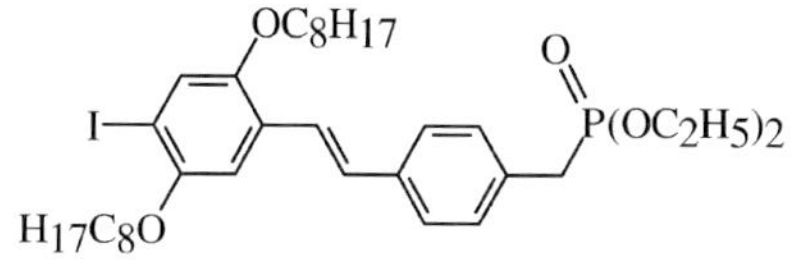

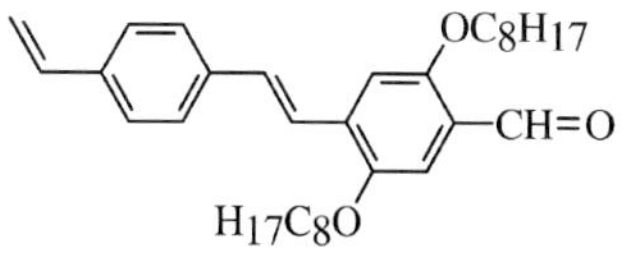

[Structures 15, 16, 17 shown with OC$_8$H$_{17}$ and H$_{17}$C$_8$O substituents]

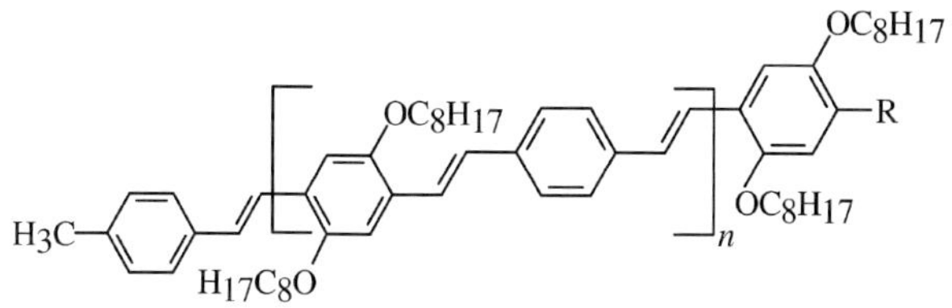

15: H$_3$C–C$_6$H$_4$–CH=CH–C$_6$H$_2$(OC$_8$H$_{17}$)(OC$_8$H$_{17}$)–I

16: I–C$_6$H$_2$(OC$_8$H$_{17}$)(OC$_8$H$_{17}$)–CH=CH–C$_6$H$_4$–CH$_2$–P(=O)(OC$_2$H$_5$)$_2$

17: CH$_2$=CH–C$_6$H$_4$–CH=CH–C$_6$H$_2$(OC$_8$H$_{17}$)(OC$_8$H$_{17}$)–CH=O

[Structure 18a–e: oligomer with methyl endgroup, multiple OC$_8$H$_{17}$-substituted phenylenevinylene units, and R endgroup]

18	a	b	c	d	e
n	1	2	3	4	5
R	CHO	I	CHO	I	CHO

$$15 \xrightarrow[\text{(a)}]{17} 18a \xrightarrow[\text{(b)}]{16} 18b \xrightarrow[\text{(a)}]{17} 18c \xrightarrow[\text{(b)}]{16} 18d \xrightarrow[\text{(a)}]{17} 18e$$

Scheme 11.8. Preparation of the OPV series **18a–e** by an orthogonal approach: (a) Pd(OAc)$_2$, P(o-tolyl)$_3$, NBu$_3$, DMF; (b) NaH, DME.

formation of regioisomers in the Heck reactions and small amounts of cis isomers in the Heck as well as in the Horner reactions. However, careful purification can solve this problem. Another critical feature of the obtained series **18a–18e** is manifested by different endgroups; thus for absorption, fluorescence and NLO studies, the series with odd numbers n should be regarded separately from the series with even numbers n.

In principle, all synthetic procedures that have been developed for PPVs can be applied for the preparation of OPVs. However, it is often difficult to stop the reactions at an early stage and moreover it is difficult to separate accurately the oligodisperse reaction products. Nevertheless, the most common PPV preparations by 1,2-elimination reactions in prepolymers will be mentioned here:

- Wessling-Zimmermann route [29]
- Gilch route [30]
- Vanderzande route [31]

A common feature of these three methods is the polymerization of 1,4-xylylene intermediates to 1,4-phenyleneethylenes which are then subjected to thermal 1,2-elimination reactions of SR_2/HCl [29], HCl [30, 32] or RSOH [31]. The Gilch route proved to be very valuable for technical processes.

Different precursor routes applying metathesis reactions were developed by Grubbs et al. [33] and Bazan et al. [34].

Thorn-Csányi and coworkers succeeded in the acyclic diene metathesis (AD-MET) of 1,4-divinylbenzene **19**. The primary obtained oligomers **20** ($n = 2$–6) have reactive vinyl endgroups (Scheme 11.9). An end-capping cross metathesis with 3-hexene furnished the oligomers **21** ($n = 2$–6) which bear terminal 1-butenyl groups. A Schrock-type Mo-alkylidene complex served as catalyst. The almost completely trans configured oligomers could be separated by repeated column chromatography [35, 36].

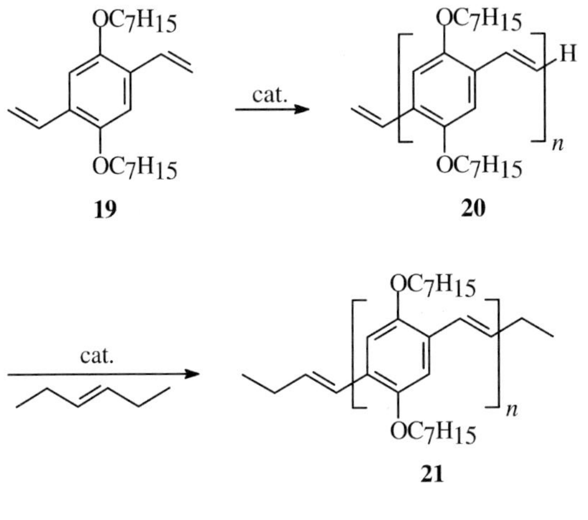

Scheme 11.9. Acyclic diene metathesis leading to the oligomers **20** and end capping to **21** by cross metathesis (cat: Mo-alkylidene complex of the Schrock-type).

An elegant cation–anion coupling technique (Scheme 11.10) was found by Klärner, Müllen et al. [37]. Repetitive coupling of the type **22** + **23** and termination with the benzylcyanide anion **26** led to the series **27a–c** ($n = 1$–3) which exhibits increasing dipole moment as a function of length.

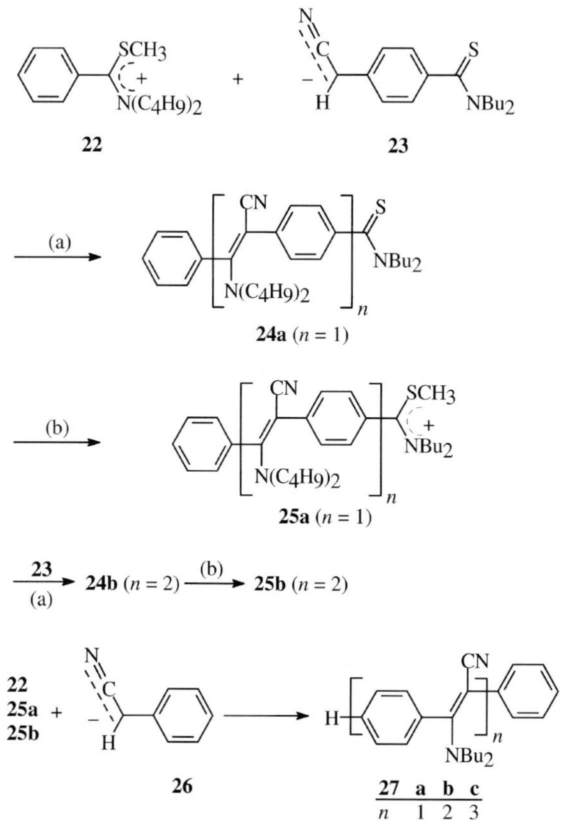

Scheme 11.10. Cation–anion coupling technique for the generation of the OPVs **27a–c** having olefinic push–pull segments: (a) CH$_2$ group deprotonated by DMF, (b) (H$_3$C)$_3$O$^+$BF$_4^-$.

The simple one-pot method for the successive construction of oligomers leads to E/Z mixtures with a relatively low isomerization barrier [24, 37].

Finally the application of the Ramberg-Bäcklund reaction for the preparation of OPVs will be mentioned [38].

After the primary publications on the unsubstituted OPV series by Drehfahl, Hörhold et al. [39, 40] more than 1000 papers appeared on OPVs and PPVs. Within the scope of this chapter, only a selection can be given. As mentioned above, solubilizing groups are indispensable; however, they can provoke higher torsional angles along the chain. Therefore various attempts have been made to confine the substitution to the terminal benzene rings of the OPV chain (Scheme 11.11). Particularly, series with t-butyl groups in the 3- and 5-position [41, 42] have found many applications.

An opposite strategy makes use of the selective attachment of sidechains at the center of the OPV chain. Scheme 11.12 shows a bidirectional synthetic approach to

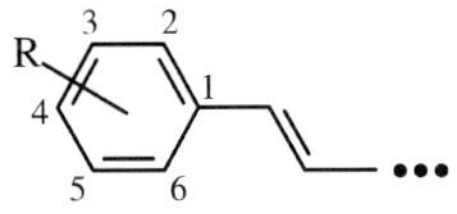

11 Defined-length Carbon-rich Conjugated Oligomers

3,5-di-*t*-butyl	[38, 41, 42]
3,5-diisopropyl	[43]
4-CH$_3$	[39]
4-alkoxy	[44]
3,4,5-trialkoxy	[45]
4-methylthio	[46]
4-*t*-butylthio	[47, 48]
4-acetylthio	[46]
2-cyano	[49]
4-diethoxyphosphoryl	[50]
2,5-dipropoxy/4'-formyl	[51]
4-methyl-2,5-dioctyloxy/4'-formyl	[52, 53]
4-methyl-2,5-dioctyloxy/4'-diethoxymethylsilyl	[52]

Scheme 11.11. OPVs with exclusive substitution at the terminal benzene rings.

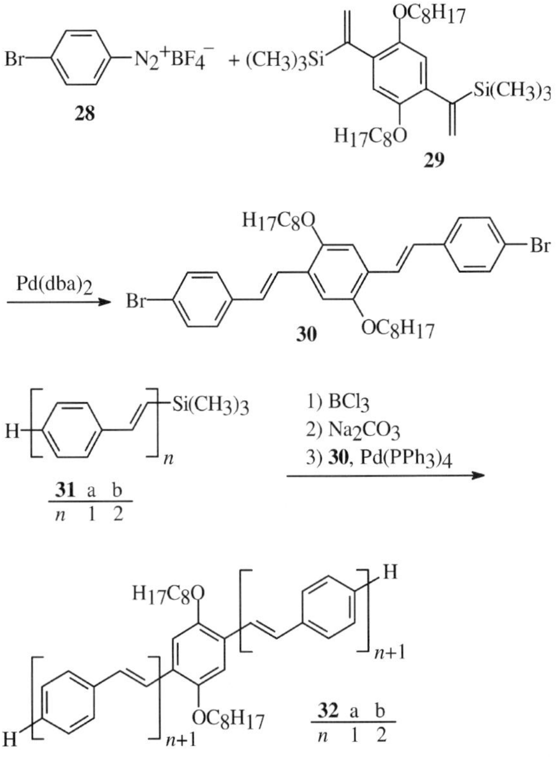

Scheme 11.12. Preparation of the OPVs **32a,b** by cross coupling reactions.

such systems [54]. The diazonium salt **28** reacts twice chemo-, regio- and stereo-selectively with the silyl component **29**. The product **30** can then be subjected to a further cross coupling with **31a,b** (Suzuki-Miyaura type). The desired OPVs **32a,b** are obtained in good yields by the bidirectional growth method. The compound **32a** and the corresponding dioctyl compound were originally synthesized by Wittig reactions [55].

OPVs with 2,5-dioctyloxy substitution on a central ring as well as the analogous 2-(2-ethylhexyloxy)-5-methoxy systems [56–59] were used in many physicochemical studies. Moreover, OPVs with a selective substitution at the center and at the terminal rings have to be quoted here [46, 48, 54, 60–65].

Nevertheless, the majority of OPVs has a regular structure with one of the repeat units shown in Table 11.2.

A variety of different functional groups, chromophores and electrophores, particularly electron-withdrawing substituents, which enhance the electron affinity, have been attached to the olefinic or aromatic part of OPV scaffolds:

CN [37, 63, 64, 66, 87–90], NO_2 [64, 66], SO_2R [64, 90, 91], 1,3,4-oxadiazoles [64, 92–94], thiophenes and oligothiophenes [95, 96], 1,3,5-triazines [97], NR_2 [37, 61, 66, 98–100], silyl or alkoxysilyl [53, 101–104], phenyl [105, 106], phanes [107, 108], ferrocene [109, 110], porphyrene [111], fullerene [71, 112], SO_3H [108, 113–115].

The introduction of SO_3H groups provides water/alkali soluble OPVs. The property of OPVs as molecular wires can be used when chromophores/electrophores

Table 11.2. OPVs with regular repeat units.

R	References	R	References
C_7H_{15}	66, 67	C_6H_{13}	78–80
OC_3H_7	17–25	C_9H_{19}	81
OC_6H_{13}	68	OC_6H_{13}	82
OC_7H_{15}	36, 69, 70	OC_8H_{17}	28, 83–86
OC_8H_{17}	71		
$OC_{10}H_{21}$	72		
$H_3C-CH(C_2H_5)-CH_2-O-$	73–76		
$H_3C-CH(C_2H_5)-CH_2-O-$ / $H_3C-CH(C_2H_5)-O-$	77		

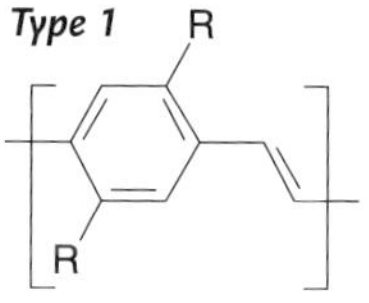

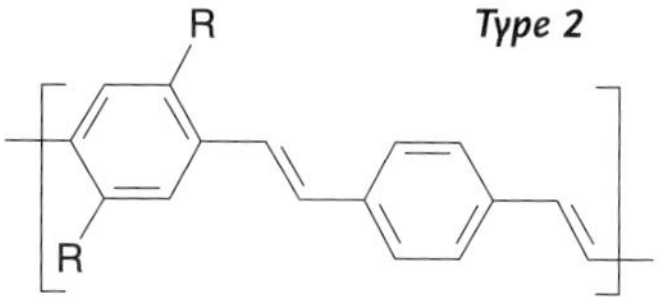

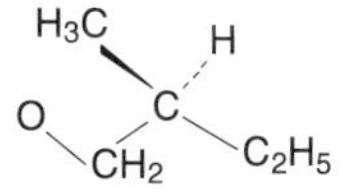

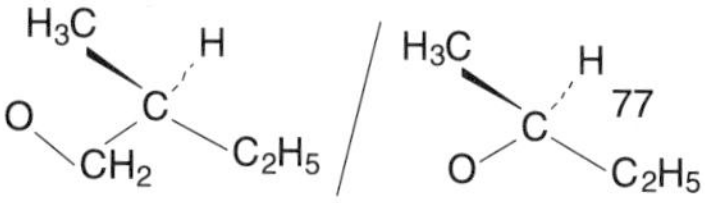

are attached to the chain ends. Energy and charge transfer can compete or complement each other. Recently an interesting study was published on a molecular OPV wire which connects a fullerene C_{60} with an extended tetrathiafulvalene [116]. Exciting the isolated C_{60} chromophore or the OPV leads quantitatively to C_{60} (S_1) which then powers an exothermic electron transfer to yield charge-separated radical ion pairs $C_{60}^{-\cdot}$–OPV–e_xTTF$^{+\cdot}$.

11.2.2
Absorption and Fluorescence

A special aspect of structure–property relationships in a series of conjugated oligomers is due to the fact that a systematic change of certain properties $P(n)$ can be expected for increasing numbers n of repeat units. Either P or its first derivative P' converge to a *limiting value*:

$$P(n) \rightarrow P_\infty (n \rightarrow \infty)$$

or

$$\frac{dP(n)}{dn} \rightarrow P'_\infty (n \rightarrow \infty)$$

Figure 11.2 shows the UV/Vis spectra of some selected OPVs of the series **2a–j** and Fig. 11.3 demonstrates the approach of the absorption and fluorescence maxima of

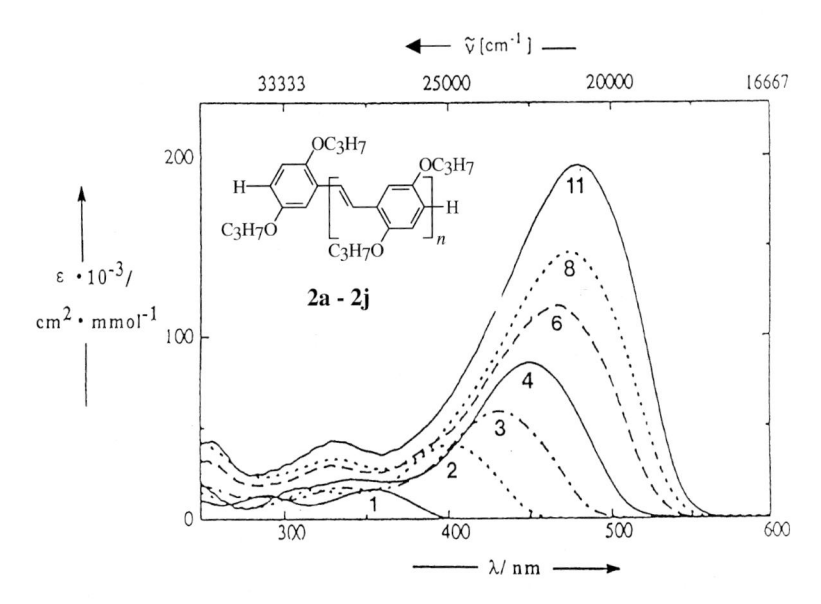

Figure 11.2. UV/Vis spectra of some selected OPVs ($n = 1$–4, 6, 8, 11) of the series **2a–j** in CHCl$_3$ [17]. (See Scheme 11.5.)

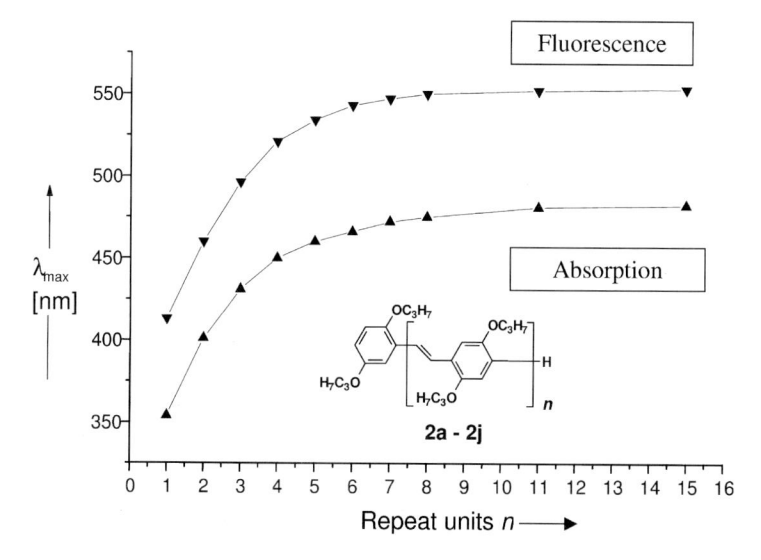

Figure 11.3. Plot of the absorption and fluorescence maxima of **2a–j** in CHCl$_3$ versus the numbers n of repeat units.

2a–j to limiting values, which are in agreement with the values of the corresponding polymer.

On the basis of 4 or 5 members of the series, a reliable prediction of the λ_∞ values is feasible. Exponential functions as natural growth functions proved to be much better for this purpose than the conventional hyperbolic approximations for the transition energy $E = E(S_1) - E(S_0)$ [2, 17, 18, 74]. The correlation of E with n^{-1} fits quite well for OPVs with low n, but it fails for the *saturation domain* and for the extrapolation to the polymer ($n \to \infty$) [2].

The following functions for the transitions $S_0 \leftrightarrows S_1$ can be recommended:

$$E(n) = E_\infty + (E_1 - E_\infty)e^{-a(n-1)}$$
$$\lambda(n) = \lambda_\infty - (\lambda_\infty - \lambda_1)e^{-b(n-1)}$$

whereby E_1 and λ_1 are the values for $n = 1$ and E_∞ and λ_∞ the limiting values:

$$\lim_{n \to \infty} E(n) = E_\infty$$

$$\lim_{n \to \infty} \lambda(n) = \lambda_\infty$$

The total effect of conjugation, represented by

$$\Delta E = E_1 - E_\infty$$
$$\Delta \lambda = \lambda_\infty - \lambda_1$$

answers the question, how much is an absorption or a fluorescence shifted between the first member of the series ($n = 1$) and an ideal, infinite long conjugated chain?

The effective conjugation length n_{ECL}, defined by

$$|\lambda_\infty - \lambda(n_{ECL})| < 1 \text{ nm},$$

can be calculated by the equation

$$n_{ECL} = b^{-1} \ln \Delta\lambda + 1$$

n_{ECL} characterizes the rate of the convergence. The corresponding absorption or fluorescence maxima $\lambda(n_{ECL})$ represent the minimum HOMO–LUMO gap (bandgap) of the series. Apart from the importance of this figure for applications in materials science, it provides a quality criterion for the preparation of the corresponding polymers. Polymer syntheses are always subject to structural defects. If λ_{max} of the polymer is lower than $\lambda(n_{ECL})$, the number of defects will be so high that the majority of defect-free regions of the polymer chain has n values smaller than n_{ECL}.

The following results were obtained for the series **2a–j** in $CHCl_3$ [2, 18]:

Absorption $\quad \lim_{n\to\infty} \lambda_{max} = \lambda_\infty = 481 \pm 4$ nm

$\qquad\qquad\quad \lim_{n\to\infty} E = E_\infty = 2.58 \pm 0.02$ eV

Fluorescence $\quad \lim_{n\to\infty} \lambda_{max} = \lambda_\infty = 552 \pm 4$ nm

$\qquad\qquad\quad \lim_{n\to\infty} E = E_\infty = 2.24 \pm 0.02$ eV

The effective conjugation length amounts to

$$n_{ECL} = 11$$

Other OPV series give similar results in comparison to **2a–j** [18, 74]. Nevertheless, one has to be aware of the fact that sidechains have an electronic and steric impact on the absorbance $A(\lambda)$. Endgroups, which are involved in the chromophore also change $A(\lambda)$.

In order to exclude aggregation effects, the absorbance A should be measured at low concentrations in a good solvent. A simple method to detect small effects consists of measurements with a constant product of concentration c and path length d:

$$A\varepsilon^{-1} = cd = 10^{-1}c \cdot 10d = 10^{-2}c \cdot 10^2 d$$

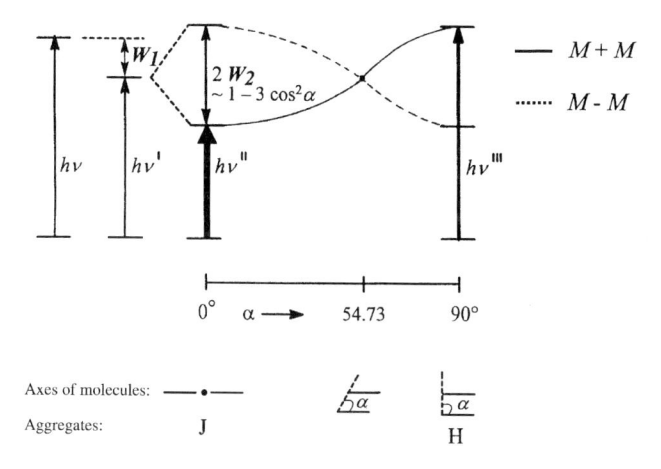

Figure 11.4. Change in the transition energies $E(S_0 \rightarrow S_1)$ for aggregates of OPV molecules in which the transition moments M are located in the longitudinal axes of the parallel oriented molecules [1z].

A systematic change in the obtained curves $A(\lambda)$ indicates an aggregation/dissociation phenomenon.

Particularly deceptive are aggregates whose absorption is similar to the monomer absorption. Figure 11.4 shows the changes in the transition energies $E(S_0 \rightarrow S_1)$ for aggregates. Van der Waals interaction W_1 leads to an energy level that is subjected to a Davidov splitting. The transition moments M lie in the longitudinal axis of the OPV molecules so that an allowed $(M + M)$ and a forbidden $(M - M)$ transition results for (dimeric) aggregates [1z].

The transition energy E for the allowed excitation depends on the orientation of the molecules in the aggregate; E is lowest for pure J aggregates ($\alpha = 0°$) and highest for pure H aggregates ($\alpha = 90°$). The function W_2 in Fig. 11.4 reveals that an angle of 54.73° (magic angle) causes an E value for the aggregate which resembles the value of the monomer chain. Apart from the electronic effect of aggregation, a steric effect may be relevant, when the molecules have a restricted torsional angle in the aggregates and consequently show a red-shifted absorption.

Various calculations (force field, semiempirical quantum mechanical, *ab initio*) were made to determine the geometry of the groundstate S_0 and the relaxed excited states S_1 and T_1 of OPVs and in particular to obtain the energy $E(n)$ of the electron excitation $S_0 \rightarrow S_1$. A recently published time-dependent DFT study [117] shows good aggreement between the calculated $E(n)$ and the measured values for low numbers n of repeat units. In contrast to the empirical *e*-function, the theoretical approach to the convergence region at higher n is still a problem.

The majority of OPVs shows strong fluorescence. Figure 11.5 depicts the fluorescence quantum yields ϕ_F of **2a–i** in CH_2Cl_2. The average fluorescence lifetime τ_0 of such OPVs decreases with increasing n from 1–2 ns to about 0.5 ns [74]. According to the Strickler-Berg relationship, an enhancement of the radiative decay

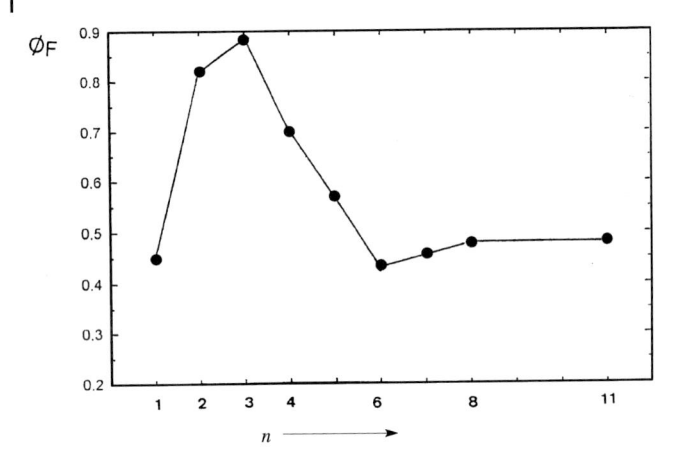

Figure 11.5. Fluorescence quantum yields ϕ_F obtained for the OPVs **2a–i** ($n = 1-8, 11$) in CH_2Cl_2 at room temperature.

with increasing n can be expected, but a much stronger increase in nonradiative decay occurs for higher n. The fluorescence behavior in neat films can be quite different [118].

Laser scanning confocal microscopy was used to characterize the emission of polycrystalline OPV films. Apart from the normal emission profile, a low energy component due to an excimer emission was detected [119].

11.2.3
Applications in Materials Science

OPVs are suitable for various applications in materials science: Organic light emitting diodes (OLED), field-effect transistors (FET), semiconductors (doped), photoconductors, solar cells, photovoltaic devices, optical brighteners, laser dyes, nonlinear optics (NLO), optical switching, imaging techniques, photoresists and liquid crystals [1a–e, 1j–o, 1r, 1v, 27, 120]. Among these applications, two fields will be selected here, namely NLO and electroluminescence studies.

NLO materials with large molecular hyperpolarizabilities would be particularly useful for the performance of optical signal processing, transferring and switching techniques. In particular an acceleration of data transfer rates seems to be highly promising.

Figure 11.6 gives a short introduction to *nonlinear optics* of OPVs [27]. A light wave polarizes the molecules in a mode which corresponds to the periodicity of the E vector of the lightwave; however, the function $P(t)$ is not so symmetrical as $E(t)$. If for example E^1 is a donor and E^2 an acceptor group, the electrons are more easily shifted in the direction of E^2 than in the opposite direction. A Fourier transformation of the periodic function $P(t)$ leads not only to the original frequency ω, but also to the double and triple frequency. Hence, nonlinear optics provide a

11.2 Oligo(1,4-phenylenevinylene)s OPV

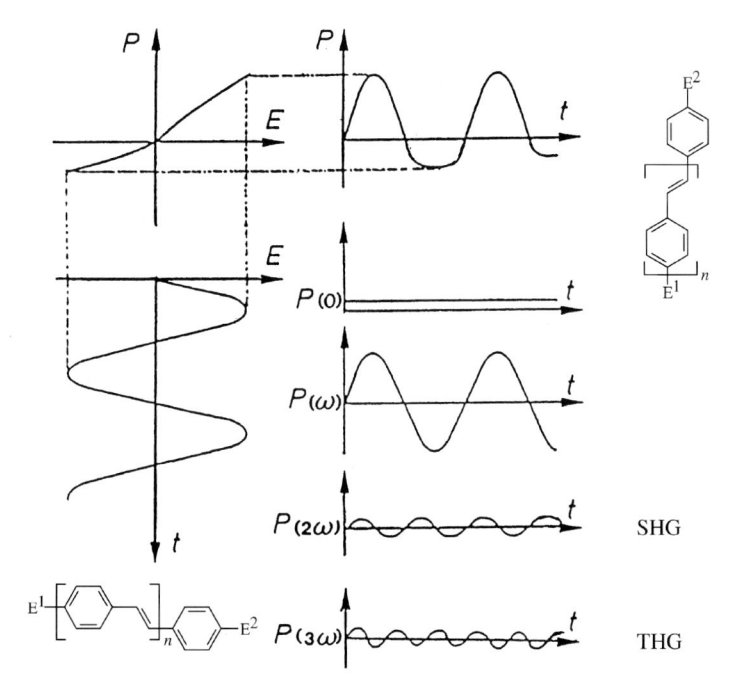

Figure 11.6. Explanation of nonlinear optics of OPVs (SHG: Second harmonic generation; THG: Third harmonic generation) [27].

method for frequency doubling (SHG) and tripling (THG). The equation $P(E)$ for the bulk system

$$P = \varepsilon_0(\chi^{(1)} E + \chi^{(2)} EE + \chi^{(3)} EEE + \cdots)$$

has an analogy for the induced dipole moment in the molecules:

$$\mu_{ind} = \alpha E + \beta EE + \gamma EEE + \cdots$$

where $\chi^{(n)}$ are the susceptibilities of nth order, α is the linear polarizability, β and γ the first and second hyperpolarizability.

Centrosymmetric molecules (crystals) must obey the relation

$$P(-E) = -P(E)$$

Therefore they do not show a frequency doubling. A series of symmetrical end-functionalized OPVs has been measured [45, 121–123]. In all these cases $\chi^{(3)}$ and γ increase with increasing length L of the oligomer. The general scaling law with $\lambda^x{}_{max}$, originally proposed by Agrawal, Flytzanis et al. [124], can be wrong, because there are series where λ_{max} decreases with increasing L (or n) [125]. A plot of

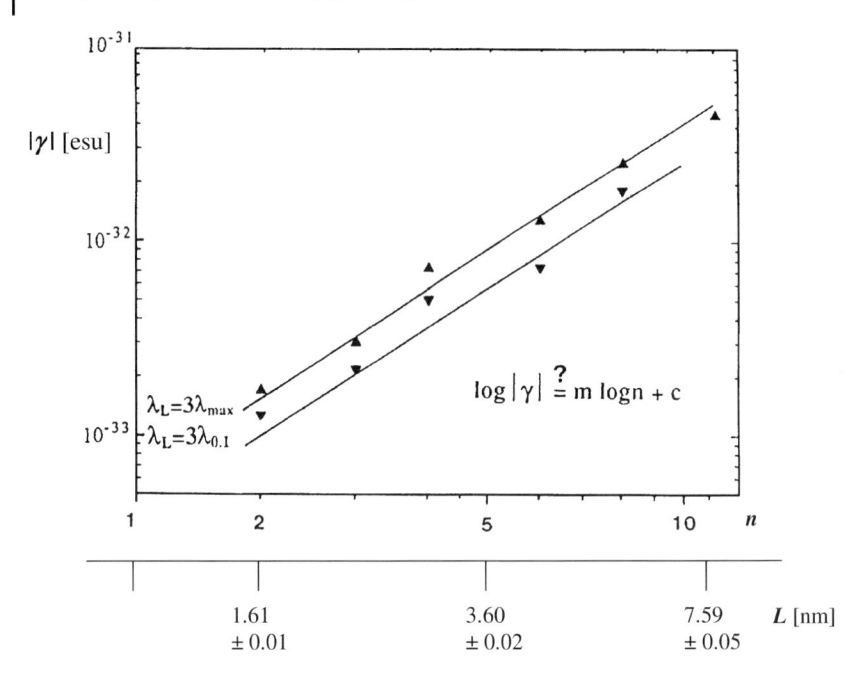

Figure 11.7. THG measurement of selected numbers of the OPVs **2a–j** (in polystyrene matrix). The laser wavelength was tuned according to the absorption maxima (▲) or according to the long-wavelength edge (▼) of the absorption [126, 127].

$|\gamma|$ versus n or L, obtained for selected members of the series **2a–j**, is depicted in Fig. 11.7 [126, 127]. In contrast to $\lambda_{max}(n)$, which approaches the limiting value $\lambda_\infty = 481$ nm, γ should increase steadily with increasing n and only the slope (first derivative) should converge. The slope of the double logarithmic plot of Fig. 11.7 seems to be constant; however, neither the monomer **2a** ($n = 1$) nor the polymer **2p** (with an average n of 30) have γ values which fit on the linear plot. The linear correlation is confined to certain ranges of n and the convergence range seems to have very high numbers of repeat units. Obviously, the effective conjugation length n_{ECL} of **2** is much higher for nonlinear optics than for linear optics. Whereas S_0 and S_1 are responsible for absorption and fluorescence, THG has to be explained by an *essential state model*, which contains more states (S_0, S_1, S_2,...). Another feature of the plot in Fig. 11.7 is that the γ values for a certain compound are somewhat higher, when a laser wavelength λ_L is used, which is equal to $3\lambda_{max}$ (resonance case), than for $\lambda_L = 3\lambda_{0.1}$. The wavelength $\lambda_{0.1}$ corresponds to the value $\varepsilon = 0.1\varepsilon_{max}$ at the long-wavelength end and represents the $0 \to 0$ transition which is normally not resolved at room temperature.

Small-molecule based *organic light emitting diodes* (OLEDs) have been on the market for several years. A good LED has to fulfil several preconditions: Injection efficiencies and charge mobilities for electrons and holes should be balanced, turn-on voltage and current should be low, the probability for recombination of positive and negative charges should be high in the emitter zone and should lead to a high luminescence efficiency, the emission should produce a bright color at the desired wavelength and the materials of the device should be stable towards oxygen, water, light, etc. OPVs fulfil several but not all of these requirements, as numerous studies have revealed [19, 20, 22, 44, 81, 87, 128–130].

Conjugated polymers are *a priori* insulators and become semiconductors by oxidative or reductive doping. The theory of their electrical conductivity is based on solitons, polarons or bipolarons. OPVs, in particular those with electron-donating substituents, have high-lying HOMOs. Therefore the injection of positive charges (holes) works quite well, provided that the work function of the anode matches the energy of the HOMO. The same should be valid for the electron injection into the LUMO which occurs at the cathode. However, OPVs are essentially hole-transporting materials. Electron-withdrawing substituents like CN groups improve the electron transport so that the hole transport in OPVs is better balanced by the opposite electron flow. Scheme 11.13 illustrates OPV polarons and bipolarons on a molecular basis. Although the charges are delocalized, P^+ and P^{2+} have a confined extension [131, 132].

Scheme 11.13. Quinoid substructures in OPV molecules representing polarons P^+ and bipolarons P^{2+}.

Figure 11.8 shows the operating mode of electroluminescence in a single-layer OLED. According to solid state physics, polarons and bipolarons have partly occupied mid-gap levels. The electron–hole capture (recombination of P^+ and P^-) furnishes excitons, whose lengths in the OPV chain are confined by the exciton–phonon coupling. The singlet excitons can be deactivated by the emission of light.

The quantum yield η_{EL} of the electroluminescence, that means the number of photons generated per injected hole or electron, is determined by three factors:

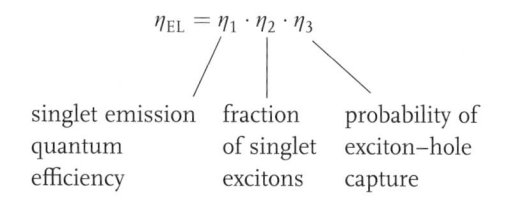

$$\eta_{EL} = \eta_1 \cdot \eta_2 \cdot \eta_3$$

- η_1: singlet emission quantum efficiency
- η_2: fraction of singlet excitons
- η_3: probability of exciton–hole capture

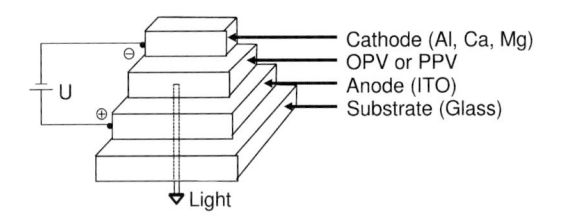

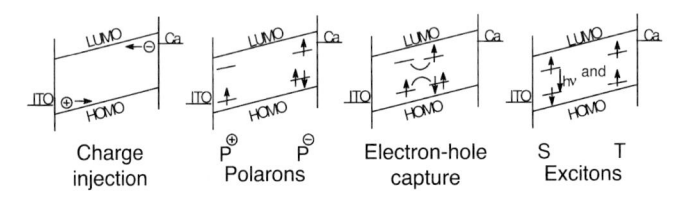

Figure 11.8. Schematic set-up of a single-layer LED and operating mode of the electroluminescence.

Originally it was thought that η_2 had a maximum of 25% according to one singlet and three equivalent nonradiative triplet states. However, it has turned out recently that η_2 can be much higher [133, 134]; a maximum value of 42% was established for PPV [133]. The formation cross section of singlet and triplet excitons depends on the optical gap [133]. The overall quantum efficiencies η_{EL} are very much lower and depend strongly on the special material and the device. A modern OLED consists of a transparent and a reflecting electrode, a layer for injection and transport of electrons, a layer for injection and transport of holes and an emitter layer in between. Moreover, each layer can be a composite material. Several deactivation processes of the S_1 states can compete with the fluorescence, namely internal conversion IC (multiphonon emission) $S_1 \rightarrow S_0$, intersystem crossing ISC ($S_1 \rightarrow T_1$) and quenching processes by interaction of an exciton with another exciton, a polaron or a bipolaron. The IC and ISC tendencies should increase to limiting values for increasing n. A systematic study of the OPV series **2b–g** [19, 20, 22] was performed in a single-layer LED. Figure 11.9(a) shows the band scheme obtained by cyclic voltammetry measurements. To visualize the charge injection at the electrodes, the workfunctions of the anode (ITO: indium tin oxide) and the cathode (Al) are displayed as well. The injection barriers to both electrodes are lowered with increasing n (decreasing electrochemical bandgap). The majority of charge carriers in OPVs are holes. Figure 11.9(b) depicts the injected current versus applied field characteristics.

The turn-on current shifts to lower electric fields with increasing n – according to the facilitated injection and transport of holes (oxidation demonstrated by the arrows in Fig. 11.9(a)). Figure 11.9(c) shows the relative intensities of the electroluminescence as a function of the current density. The highest value $\eta = 10^{-4}$ was measured for **2c** ($n = 3$). The relative electroluminescence intensities (Fig. 11.9(c)) at a current density of 1 mA cm^{-2} show a dependence on n which

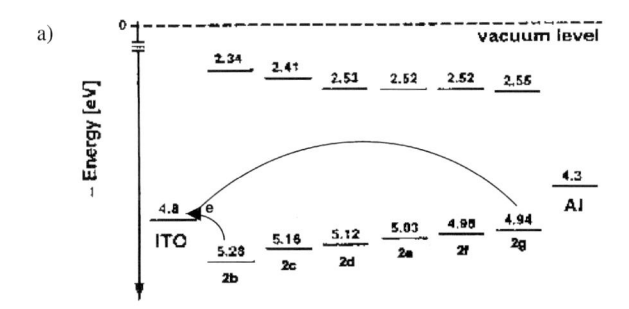

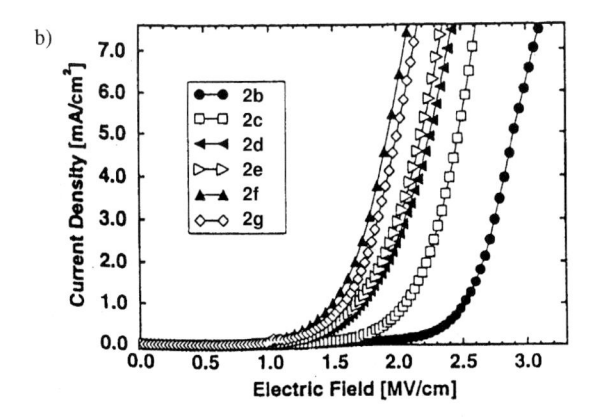

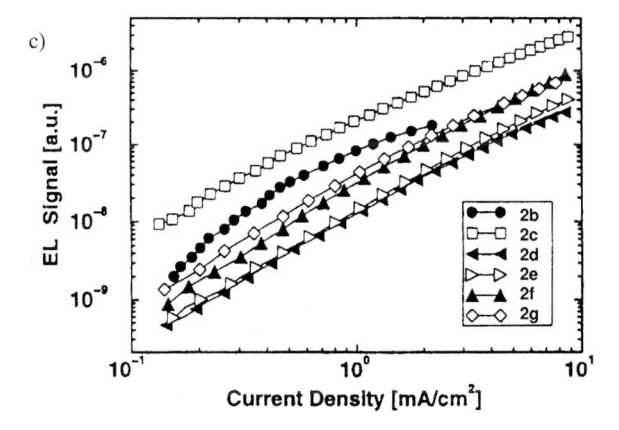

Figure 11.9. (a) Band scheme of the OPVs **2b–g** obtained by cyclic voltammetry; (b) current density versus electric field of single-layer LEDs with a 30 wt.% oligomer concentration in polystyrene; (c) relative electroluminescence intensities as a function of the current density. (Preparation of the OPV layer as described above, thickness 140–160 nm) [22].

resembles $\phi_F(n)$ (Fig. 11.5). After the maximum is reached for $n = 3$, a strong decrease is observed, before the values increase again and approach a limiting value for high n, which is much lower than the maximum for $n = 3$. A maximum luminance of approximately 100 cd m^{-2} was reached at a current of 50 mA cm^{-2}. On the whole, the electroluminescence intensities do not show a uniform behavior in the series **2b–g**. Apart from the radiationless processes competing with the fluorescence, different interchain interactions may have a decisive influence. The measurements shown in Fig. 9(b) and (c) were made in homogeneous amorphous films obtained by spin-coating the oligomers in a polystyrene matrix [22]. Evaporated films [19, 20] are difficult to obtain from the higher oligomers and exhibit an unfavorable crystallization tendency for the lower oligomers.

Finally it will be mentioned that the electroluminescence EL of OPV systems can be red-shifted in comparison to the photoluminescence PL (Fig. 11.10). An explanation is provided by the emission of J aggregates or excimers; another reason may be the recombination of P^+ and P^- in low-lying energy sites close to the cathode, because the number and mobility of P^+ is much higher than the number and mobility of P^- for OPV series.

The thermal stability of OPVs is relatively high; however, irradiation with $\lambda = 254$ nm leads, in the absence of oxygen, to photopolymerization and photocrosslinking, a radical process, which is typical for stilbenoid compounds [27]. Other stilbene photoreactions like trans ⇌ cis isomerizations or [π^6a]cyclizations cannot be observed or play a marginal role. In the presence of oxygen, small amounts of OPV in the triplet state T_1 live long enough and have sufficient energy to produce singlet oxygen

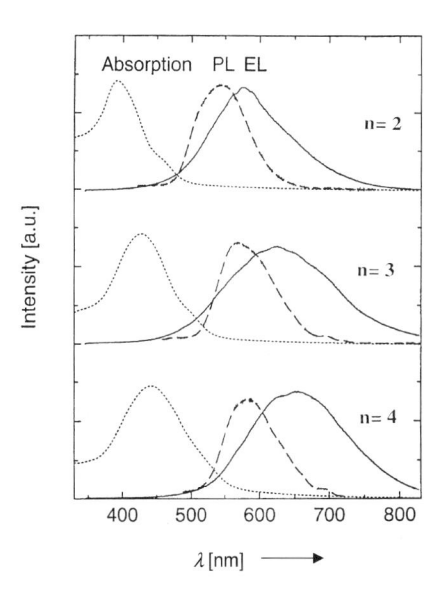

Figure 11.10. Absorption, fluorescence (PL) and electroluminescence (EL) of thin films of the OPVs **2b** ($n = 2$), **2c** ($n = 3$) and **2d** ($n = 4$) [19].

$$\text{OPV}(S_0) \xrightarrow{h\nu} \text{OPV}(S_1) \rightsquigarrow \text{OPV}(T_1)$$

$$\text{OPV}(T_1) + {}^3O_2 \rightarrow \text{OPV}(S_0) + {}^1O_2({}^1\Delta_g)$$

which then attacks the olefinic bonds and cleaves the OPV chain [135] (^{1}H NMR spectroscopic measurements reveal the primary formation of substituted benzaldehydes). Electron-withdrawing groups at the olefinic double bonds do not prevent the energy transfer (photosensitization), but they decrease the reactivity of the olefinic double bonds.

11.3
Oligo(1,4-phenyleneethynylene)s OPE

Oligo(1,4-phenyleneethynylene)s (OPE) **1** without solubilizing sidechains or endgroups very quickly ($n > 3$) become hardly soluble in organic solvents. Meier and coworkers have studied in detail the 2,5-dipropoxy substituted OPEs **1a–i** (Scheme 11.14) which up to now represent the longest known OPE series [136, 137]. Due to the identical substitution, this series can be easily compared with the OPV series **2a–j**.

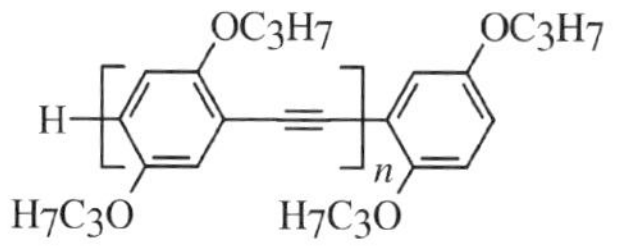

1	a	b	c	d	e	f	g	h	i
n	1	2	3	4	5	6	7	8	10

Scheme 11.14. Oligo(2,5-dipropoxy-1,4-phenyleneethynylene)s **1a–i**.

11.3.1
Synthesis

In contrast to the variety of synthetic procedures, which proved to be important for the preparation of OPVs, the Sonogashira-Hagihara reaction represents by far the dominating process for the generation of OPEs. The mechanism of the Pd-catalyzed CC coupling between an iodoarene R^1–I and an ethynylarene R^2–C≡CH is outlined in Scheme 11.15.

A simple example for this reaction is shown in Scheme 11.16, starting from 1,4-dipropoxybenzene (**33**) [136, 137]. The mono- and diiodo compounds **34** and **35** can be easily prepared. The coupling reaction of **35** with trimethylsilylacetylene yields **36**, which is deprotected to **37**. The twofold Sonogashira-Hagihara reaction of **37** and **34** then furnishes the target compound **1b** in an excellent yield [24].

Diyne **37** and diiodobenzene **35** can react in a polycondensation according to the AA + BB process (Scheme 11.6). The same is true for the AB system 1-ethynyl-4-

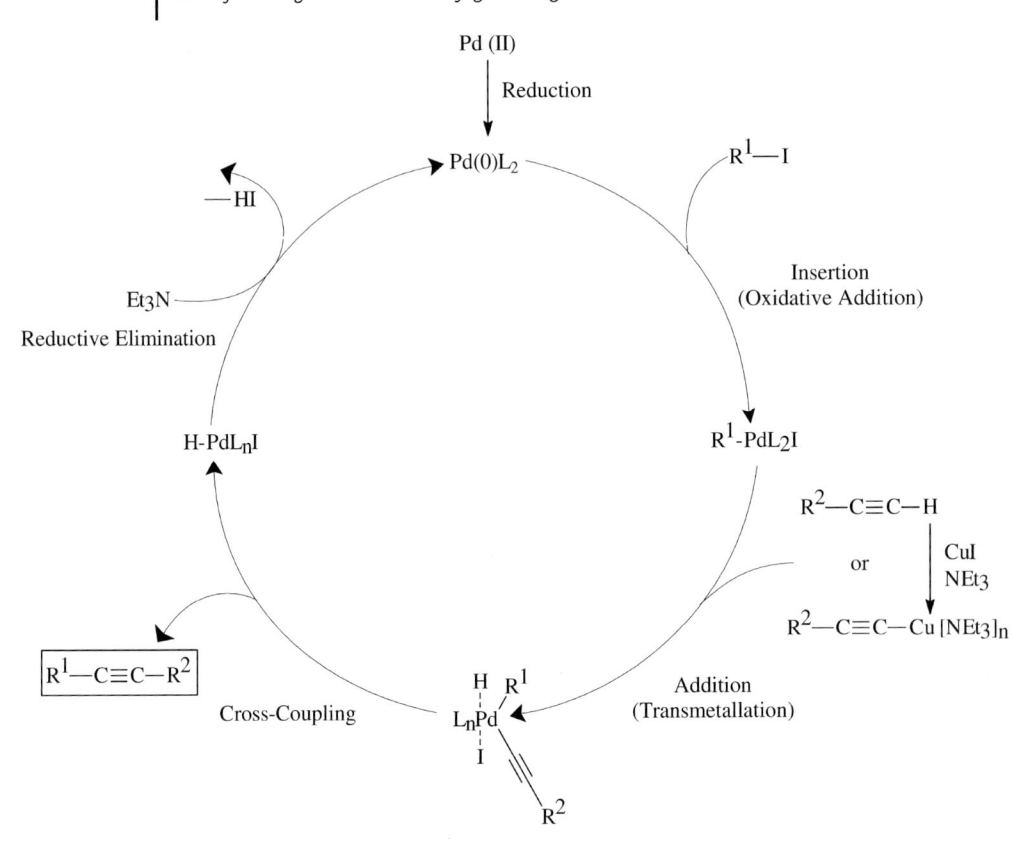

Scheme 11.15. Catalytic circle of the Sonogashira-Hagihara reaction (Instead of iodoarenes R^1I, bromo compounds or electrophiles with other leaving groups can be used; CuI as cocatalyst can be omitted when iodobenzene derivatives are used.)

iodo-2,5-dipropoxybenzene. The MALDI-TOF mass spectrum reveals, in the latter case, an oligomer mixture with 3–15 repeat units [138]. There are no hints of a cyclic product which would have a belt structure. Cyclooligomers consisting of 1,4-phenyleneethynylene units ($n = 6, 7, 8, 9$) were obtained by Oda et al. from the corresponding cycloalkenes by bromination/dehydrobromination reactions [139, 140].

An enrichment of monodisperse OPEs from oligodisperse mixtures by GPC is laborious; therefore Meier and coworkers used selective reactions with different reactive sites and different protecting groups for the preparation of the higher OPEs **1c–i**. The iodine in compound **38** (Scheme 11.17) is much more reactive than the bromine; it couples even without CuI cocatalyst. Thus, Br is a "dormant" substituent which can be used in a later reaction step under more severe conditions. The trimethylsilyl group in **39** can be cleaved by the action of bases like

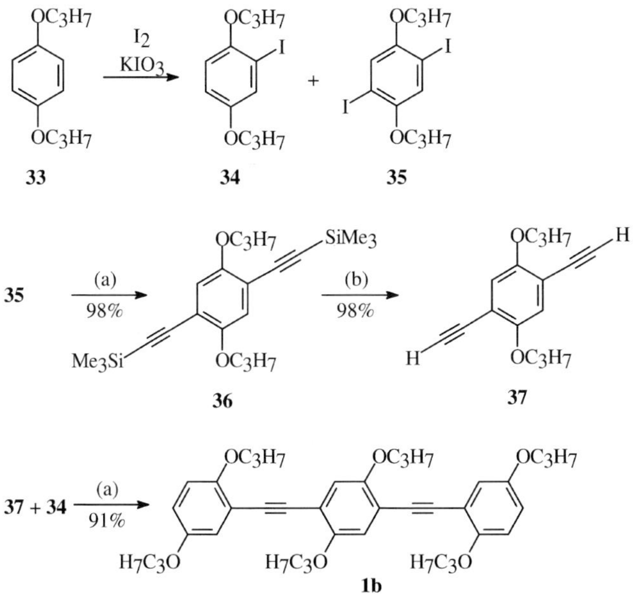

Scheme 11.16. Preparation of OPE **1b** (n = 2): (a) HC≡C–SiMe₃, Pd(PPh₃)₂Cl₂, CuI, piperidine; (b) NaOH, CH₃OH, THF.

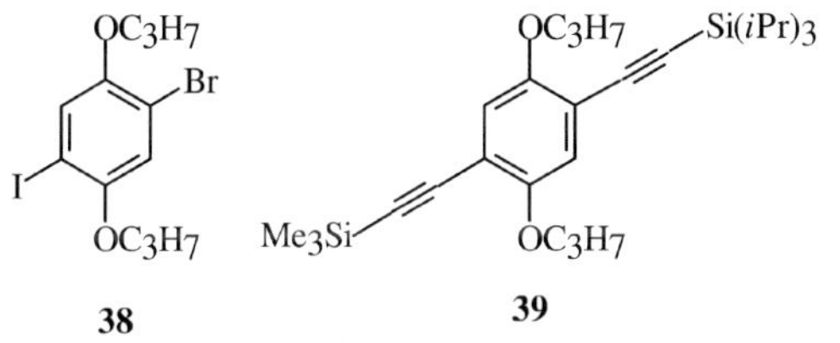

Scheme 11.17. Building blocks for selective CC coupling reactions.

K_2CO_3 or NaOH which leave the triisopropylsilylethynyl group intact. A deprotection of the C≡CSi(iPr)₃ group can be achieved by Bu₄NF in a later reaction step. Accordingly the functional groups I/Br and C≡CSiMe₃/C≡CSi(iPr)₃ provide in **38**, **39** and higher intermediates different reactive sites which can be used for chemoselective coupling reactions. Thus, the nine members of the OPE series **1a–i** could be obtained by relatively few selective reaction steps [136, 137].

The unsubstituted short OPEs (n ≤ 3) were originally prepared by Drefahl and Plötner [141] by bromination/dehydrobromination of the corresponding OPVs. Misumi applied the twofold ethynylation of p-benzoquinone and transformed the resulting cyclohexa-2,5-diene-1,4-diol ring to a benzene ring by reduction with SnCl₂ [142]. Both methods were later seldom used, because of the versatility of the Sonogashira-Hagihara reaction. Tour and coworkers [143, 144] and Godt and

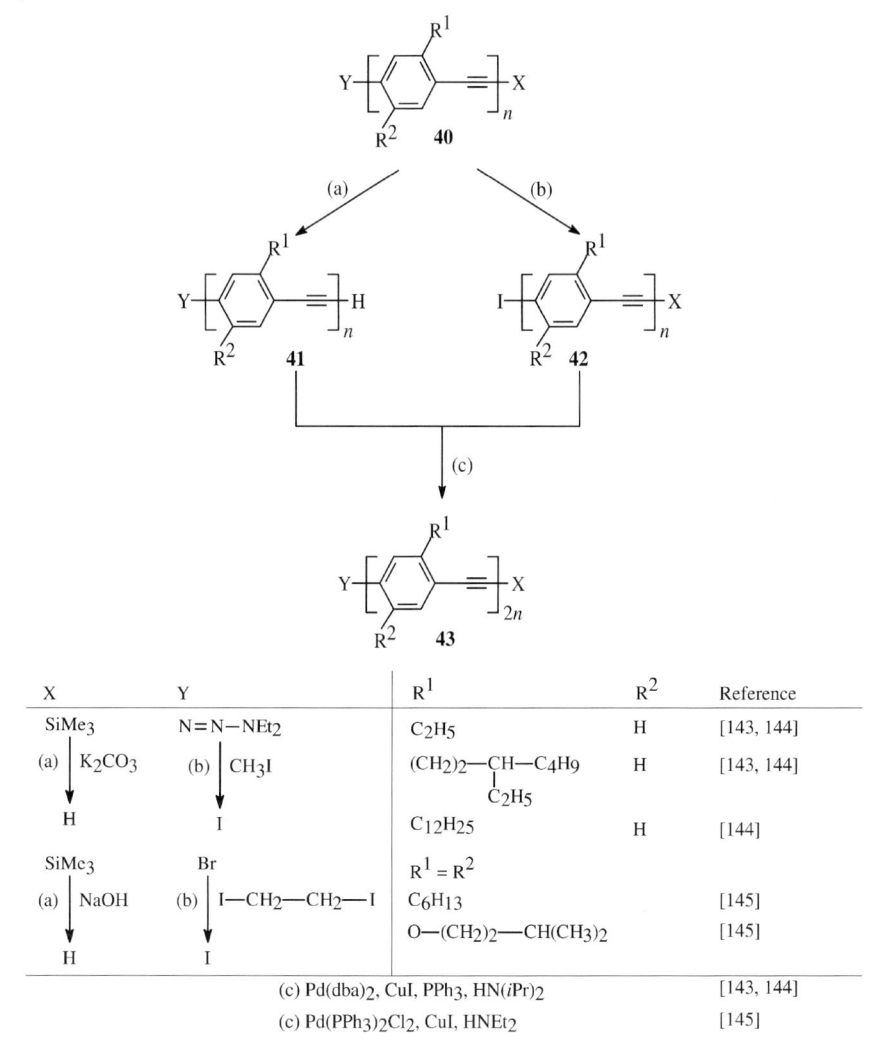

Scheme 11.18. Iterative divergent–convergent strategies for the preparation of the OPEs **43**.

X	Y	R^1	R^2	Reference
SiMe₃	N=N—NEt₂	C₂H₅	H	[143, 144]

(a) ↓ K₂CO₃
H
	(b) ↓ CH₃I
	I
		(CH₂)₂—CH—C₄H₉
		         |
		        C₂H₅
			H	[143, 144]

		C₁₂H₂₅	H	[144]
SiMe₃	Br	$R^1 = R^2$		
(a) ↓ NaOH H	(b) ↓ I—CH₂—CH₂—I I	C₆H₁₃ O—(CH₂)₂—CH(CH₃)₂		[145] [145]
	(c) Pd(dba)₂, CuI, PPh₃, HN(iPr)₂			[143, 144]
	(c) Pd(PPh₃)₂Cl₂, CuI, HNEt₂			[145]

coworkers [145] elaborated iterative divergent/convergent strategies which permit a doubling of the length of the OPE chain (Scheme 11.18). The obtained oligomers have n values of 1, 2, 4, 8 and, for the better soluble systems, even 16. In principle, the compounds **40** ($n = 1, 2, 4, 8$) are masked AB systems which can be activated on one side to yield **41** and **42**, respectively. Subsequent cross-coupling furnishes **43**. The hexadecamer **43** ($2n = 16$, X: SiMe₃, Y: N₃Et₂) has – including the endgroups – a calculated length of 12.8 nm [143, 144]; it represents a rigid rod nanoparticle.

The replacement of Br by I on the benzene ring is not trivial. Under severer conditions, the bromo compounds can give the coupling reaction as well; but in this

11.3 Oligo(1,4-phenyleneethynylene)s OPE

case somewhat higher quantities of diynes are formed as by-products. They result from an oxidative CC coupling (Hay coupling) triggered by Pd^{2+} or Cu^{2+} or from an adventitious presence of atmospheric oxygen. Finally, a modified procedure was suggested in which the protection of the ethynyl function was performed on one side with triisopropylsilyl groups (TIPS) and on the other side with polar hydroxymethyl groups. This orthogonal strategy is based on a deprotection with F^- at the TIPS side and on an oxidative deprotection with MnO_2/KOH at the CH_2OH side [146]. As long as the obtained OPEs **43** contain (protected) ethynyl groups, they can enter end-capping processes with iodobenzene derivatives [143–147].

In principle, all methods used for the generation of poly(1,4-phenyleneethynylene)s (PPE) can be also used for the OPE synthesis; however, it is often difficult to stop the polycondensation at an early stage and it is always difficult to separate uniform oligomers from oligodisperse mixtures.

Acyclic diyne metathesis (ADIMET) is the method of choice for the preparation of PPEs. After the first successful attempts with a tungsten carbyne complex [148], Bunz found that molybdenum carbyne complexes of the Schrock-type are highly suitable for this purpose **44 → 45, 46** (Scheme 11.19) [149]. Alkoxy substituents facilitate the generation of oligomers with a low molecular mass [149].

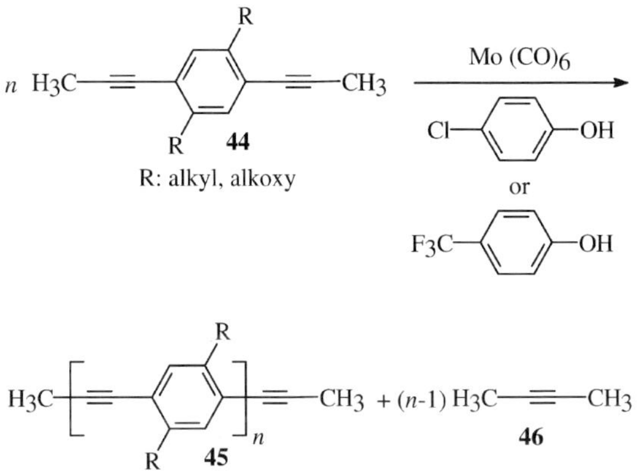

Scheme 11.19. Acyclic diyne metathesis for the formation of OPEs and PPEs.

The application of an AA-BB polycondensation of the Suzuki-type (Scheme 11.20) permits the generation of PPE chains with alternating substituted and unsubstituted benzene rings (**47 + 48 → 49**) [150].

However, as already mentioned above, the great majority of OPEs were prepared by Sonogashira-Hagihara reactions. Several useful modifications appeared in the literature, for example the application of phase transfer conditions [151], which suppress the Hay coupling. In order to obtain libraries of compounds, combinatorial protocols have been developed, where the CC coupling occurs in solution or on a solid support (modified Merrifield resin) [152, 153].

According to the synthetic procedures (Schemes 11.18–11.20), many OPEs have terminal ethynyl or trialkylsilylethynyl groups and/or I/Br substituents. Addition-

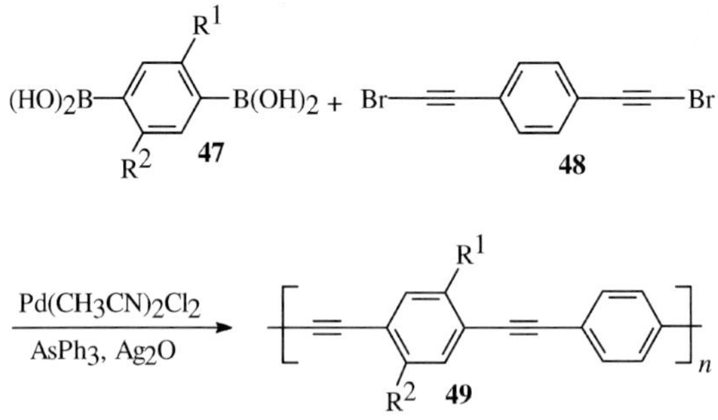

Scheme 11.20. PPEs obtained by Suzuki-type AA–BB polycondensation reactions.

ally, many compounds with other functional groups, chromophores or electrophores attached to the OPE chain (direct or on a linker) were investigated, for example: NR_2 [154, 155], OH [156–163], SR/SH [164–179], COOR [147, 156–158, 160, 180–184], additional C≡CR groups [185–187], 2-naphthyl [181, 188], steroid [189], fullerene [154, 155], heterocycles [190–198], ferrocene and other metal complexes [175, 176, 179, 199–204].

Despite the large number of studies on OPEs, few systematically extended series with $n > 3$ exist (see Table 11.3).

The SH/SR functional groups serve for the self-assembly of the OPE chains on gold surfaces (electrodes). The chromophores/electrophores in the terminal chain positions permit the investigation of energy and/or charge transfer processes. The mixed-metal complex **50** studied by Harriman et al. [202] represents a nice example for an intramolecular triplet-energy transfer by a Dexter-type electron exchange (Scheme 11.21). The localized triplet state of the Ru complex transfers its energy through bond to the Os complex. The triplet energy of the spacer ($n = 1$–5) is obviously too high to disturb the superexchange mechanism. The transfer rate k decreases exponentially – at least in the range up to a Ru–Os distance of $r = 5.2$ nm ($n = 5$).

$$k = Ae^{-\beta r}$$

The attenuation factor β is remarkably small, namely 0.11 ± 0.01 A^{-1}. An additional description of such systems is provided in Chapter 2.

11.3.2
Absorption and Fluorescence

The convergence behavior of the absorption and the fluorescence observed for the OPE series **1a–i** is similar to the results of the OPV series **2a–j**. Figure 11.11 shows the approach of λ_{max} to λ_∞ for increasing n. The exponential fit leads to the following data [137]:

11.3 Oligo(1,4-phenyleneethynylene)s OPE

Table 11.3. Series with at least four members of OPEs with regular repeat units.

$$E^1 \mathrm{-\!\!\!-\!\!\!-} \left[\underset{5\ R_i}{\overset{2}{\bigcirc}} \mathrm{-\!\!\equiv\!\!-} \right]_n E^2$$

End groups —E^1	—E^2	Solubilizing sidechains R in the repeat units	Number n of repeat units	References
Br—	—SiMe₃	2,5–C₆H₁₃	1, 2, 4, 8	145
Br—	—SiMe₃	2,5–O–(CH₂)₂–CH(CH₃)₂	1, 2, 4, 8	145
H—	(phenyl with OC₃H₇ and H₇C₃O)	2,5–OC₃H₇	1–8, 10	136, 137
Me₃Si–≡–	—SiMe₃	2,5–OC₆H₁₃	1, 3, 5, 7, 9	165
AcS–(phenyl)–≡–	–(phenyl)–SAc	2,5–OC₆H₁₃	1, 3, 5, 7	165
H–≡–	—Si(iPr)₃	—	1, 2, 3, 4	205
Me₃Si–≡–	—Si(iPr)₃	—	1, 2, 3, 4	205
Et₂N–N=N—	—SiMe₃	2–C₁₂H₂₅	1, 2, 4, 8, 16	144
Et₂N–N=N—	—H	2–C₁₂H₂₅	1, 2, 4, 8	144
Et₂N–N=N—	—SiMe₃	2–(CH₂)₂–CH(C₂H₅)–C₄H₉	1, 2, 4, 8, 16	143, 144
Et₂N–N=N—	—H	2–(CH₂)₂–CH(C₂H₅)–C₄H₉	1, 2, 4, 8	143, 144
Et₂N–N=N—	—SiMe₃	2–C₂H₅	1, 2, 4, 8	143, 144
Et₂N–N=N—	—H	2–C₂H₅	1, 2, 4	143, 144
Me₃Si–≡–(phenyl)–≡–Me₃Si (3,5-disubstituted)	SiMe₃–≡–(phenyl)–≡–SiMe₃ (3,5-disubstituted)	—	1, 2, 3, 4	185
Fe(cyclopentadienyl)	—SCH₃	—	1, 2, 3, 4	179

Absorption $\quad \lim_{n\to\infty} \lambda_{\max} = \lambda_\infty = 438 \pm 3$ nm

$\lim_{n\to\infty} E = E_\infty = 2.83 \pm 0.02$ eV

$\Delta E = 0.837$ eV

$n_{ECL} = 10$

Fluorescence $\quad \lim_{n\to\infty} \lambda_{\max} = \lambda_\infty = 475 \pm 3$ nm

$\lim_{n\to\infty} E = E_\infty = 2.61 \pm 0.02$ eV

$\Delta E = 0.732$ eV

$n_{ECL} = 10$

Scheme 11.21. Mixed-metal complexes with variable OPE spacer length for he triplet-energy transfer Ru → Os.

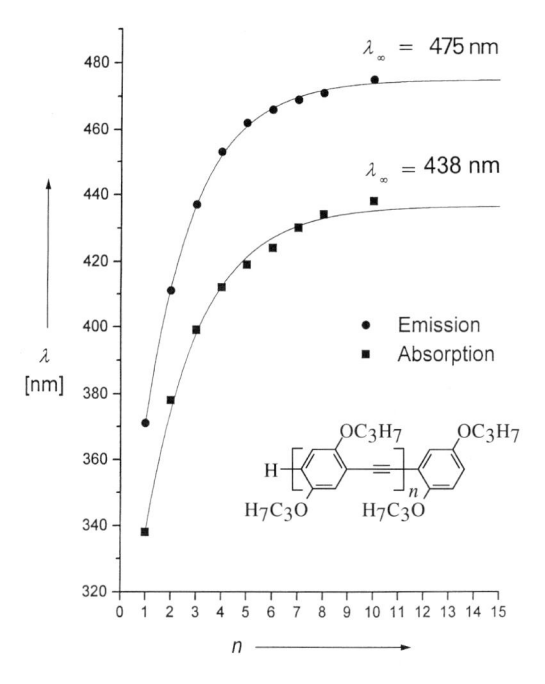

Figure 11.11. Plot of the absorption and fluorescence maxima of the OPEs **1a–i** versus the numbers n of repeat units [137].

The values for $\Delta E = E_1 - E_\infty$ found for the OPEs **1a–i** are somewhat smaller than those found for the related OPV series **2a–j** (see Section 11.2.2). The effect of conjugation is a little bit higher in the olefinic series, because the bond lengths and the corresponding resonance integrals differ more in the acetylenic series. The effective conjugation length $n_{ECL} = 10$ is almost the same as in the OPV series **2a–j** ($n_{ECL} = 11$).

The OPEs exhibit an aggregation tendency in chloroform that is much more relevant for the fluorescence than for the absorption. The emission of excited aggregates/excimers leads to additional bands at longer wavelengths, the intensities of which decrease with decreasing concentration (Fig. 11.12).

Other OPE series give similar results [18]. The series with 3-ethylheptyl substituents in the repeat unit and diethyltriazenyl and trimethylsilylethynyl endgroups (Table 11.3) for example gave the following values for the absorption maxima in CH_2Cl_2: $\lambda_\infty = 375$ nm, $\Delta E = 0.41$ eV, $n_{ECL} = 5$.

The fast rate of convergence, expressed in the low value for the effective conjugation length is not necessarily just a consequence of the smaller ΔE value, it can also be due to higher torsional angles along the chain, which are caused by long and/or branched sidechains. Moreover, the triazenyl endgroups have considerable influence on the energy of the electron transitions $S_0 \rightarrow S_1$; the corresponding OPEs with iodine endgroups exhibit blue-shifted absorption bands, the same λ_∞ value, but much longer effective conjugation lengths [144].

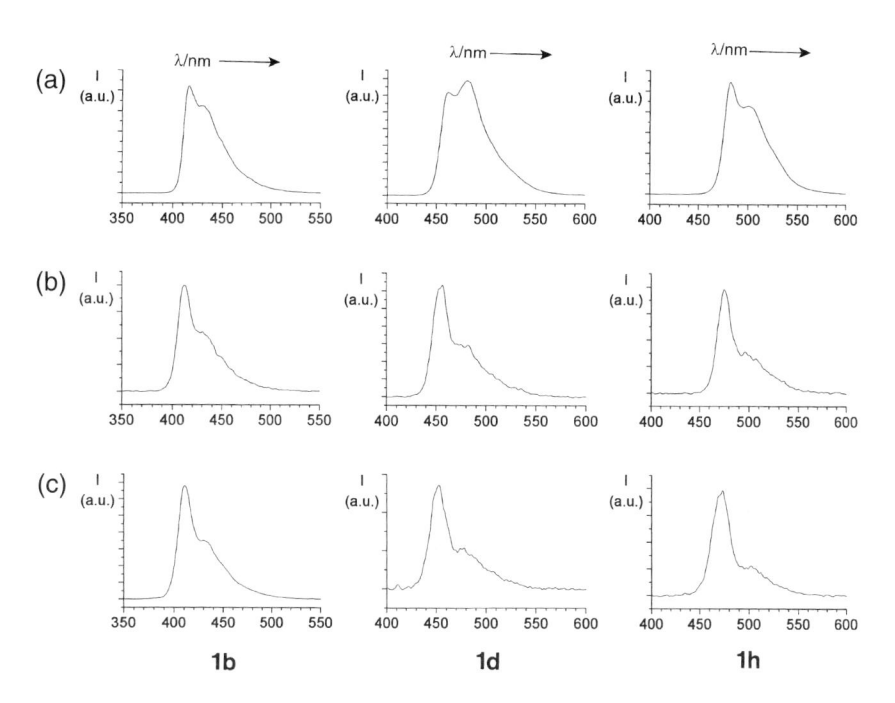

Figure 11.12. Fluorescence spectra of **1b** ($n = 2$), **1d** ($n = 4$) and **1h** ($n = 8$) in $CHCl_3$; concentration: (a) 10^{-4} M, (b) 10^{-5} M, (c) 10^{-6} M [137].

11.3.3
Applications in Materials Science

OPEs have found similar applications as OPVs in various fields of materials science. (See Section 11.2.3.) The emphasis within this chapter will again be on nonlinear optics (NLO) [163, 206, 207] and on electroluminescence [136, 156, 157, 160, 169, 188, 207].

We measured the third harmonic generation (THG) of **1a–e** using polystyrene matrices and variable laser wavelengths [136]. In order to understand the influence of the chain length on nonlinear optical properties, it is better to discuss the second hyperpolarizability γ, which is a molecular parameter, than the macroscopic susceptibility $\chi^{(3)}$, which is a bulk quantity. For the determination of $|\gamma_{res}|$, the imaginary part of the total third-order susceptibility $|\chi_{res}^{(3)}|$ of the films (thickness 50 and 212 nm) was taken. The result is depicted in Fig. 11.13 [136] in a double logarithmic plot of $|\gamma_{res}|$ versus the length L of the OPE chains; L is a linear function of the number n of repeat units. (See Section 11.1.)

A comparison of the OPE series **1a–e** with the corresponding OPV series **2** reveals that the γ values of the series **1** are higher; **1d** ($n = 4$), for example, has a $|\gamma_{res}|$ value of 3.93×10^{-32} esu, whereas the corresponding compound **2d** ($n = 4$) has a $|\gamma_{res}|$ of 7.34×10^{-33} esu. In contrast to Fig. 11.7, the slope of the curve depicted in Fig. 11.13 decreases with increasing length L. However, according to essential state models the limiting value for the slope should be reached for much higher oligomers **1**.

The electroluminescence of OPEs, observed in LEDs, shows blue-shifted emissions in comparison to OPVs with the same length and the same substitution. Apart from the difference in the HOMO–LUMO gap (band gap) and the electron correlation, the absolute values of the HOMO and LUMO energies are different, so that the work functions of certain electrodes result in a somewhat more difficult hole injection and a less difficult electron injection for the OPEs in comparison to the OPVs (Fig. 11.14) [152, 208, 209]. Nevertheless, hole transport seems to be the

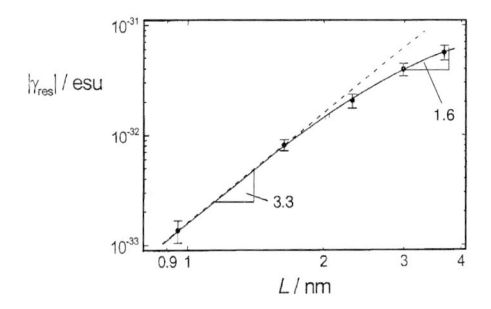

Figure 11.13. Double logarithmic plot of the hyperpolarizabilities $|\gamma_{res}|$ of the OPEs **1a–e** versus chain length L. (The dotted line corresponds to the linear equation $\log|\gamma| = 3.3 \log L + c$) [208].

11.4 OPVs and OPEs with Terminal Donor–Acceptor Substitution

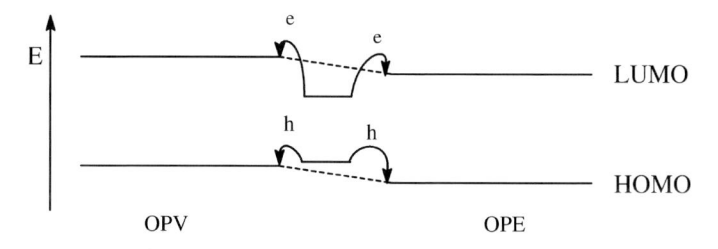

Figure 11.14. Hole and electron injection on OPVs and OPEs.

predominant conduction mechanism in 2,5-dialkoxy OPVs and OPEs. A careful study of an OPE series concerning the turn-on current and the emissive intensities of such OLEDs is still lacking.

11.4
OPVs and OPEs with Terminal Donor–Acceptor Substitution

Push-pull substituted conjugated systems represent a special class of oligomers [1z]. The *extension of the conjugation* by increasing numbers n of repeat units is accompanied by the decreasing effect of the *intramolecular charge transfer* (ICT). In terms of the valence bond theory (VB), one has to distinguish between symmetrical cyanines **51** with degenerate resonance structures and oligomers with an electroneutral and a zwitterionic resonance structure like **52** and **53** (Scheme 11.22). Involved *p*-quinoid structures render the intramolecular charge transfer more and more unfavorable, the longer the conjugated chain.

Meier and coworkers have studied in detail the OPVs and OPEs with terminal donor–acceptor substitution (DAOPVs and DAOPEs) shown in Scheme 11.23 [210–213].

Scheme 11.22. Conjugated oligomers with terminal donor–acceptor substitution: charged cyanines (symmetrical polymethine dyes) **51** below the so-called cyanine limit; DAOPVs **52**/DAOPEs **53** having electroneutral and zwitterionic resonance structures.

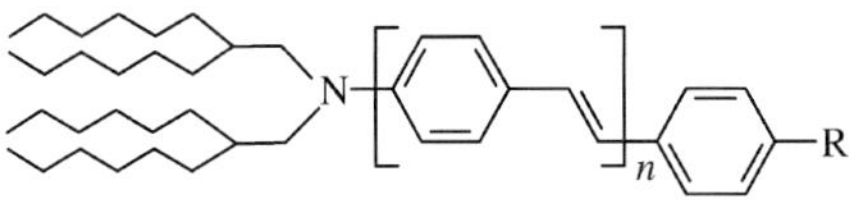

52	a	b	c	d	e
R	H	CN	CHO	NO_2	$CH=C(CN)_2$

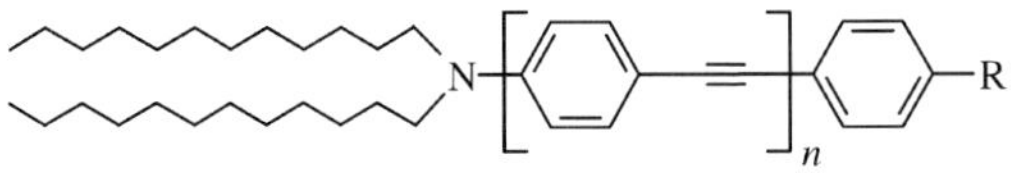

53	a	b	c	d	e
R	H	CN	CHO	NO_2	$CH=C(CN)_2$

Scheme 11.23. DAOPVs **52b–e** and DAOPEs **53b–e** and for comparison the simply donor-substituted systems **52a** and **53a** (n varies from 1 to 4 and for some series from 1 to 5).

11.4.1
Synthesis

Dialkylamino groups with long (branched) alkyl substituents are not only strong donors, they also guarantee a good solubility, so that sidechains in the repeat units can be omitted – at least up to $n = 5$. The synthetic strategy outlined in Table 11.4 is based on the stepwise extension of a fundamental oligomer series ($n = 1, 2, 3, 4,\ldots$) which starts on the donor side and finishes with an end-capping reaction, in which various acceptor groups (A = CN, CHO, NO_2, CH = $C(CN)_2$) are introduced [210–213].

A construction set consisting of 4-aminobenzaldehyde **54** and the phosphonates **55** and **56** was needed for the DAOPV series. The Wittig-Horner reaction of aldehyde **54** and phosphonate **55** followed by a deprotection yielded aldehyde **52c** ($n = 1$) which was extended by the next Wittig-Horner reaction to **52c** ($n = 2$), etc. The end-capping reactions with the phosphonates **56a,b,d,e** furnished then the target series **52a,b,d** with the corresponding acceptor group A. Thus the aldehyde series **52c** ($n = 1-4$) serves as the fundamental series for the coupled and convergent synthesis of the DAOPVs [24].

The five DAOPE series were generated in an analogous way. 4-Aminophenylethyne **57** reacted in a Sonogashira-Hagihara process with the iodine component **58**. Deprotection of the silylated alkyne yielded **60** ($n = 1$) and repetition of the procedure **60** ($n = 2$), etc. By this method, the fundamental OPE series **60** ($n = 1-4$) was obtained. End-capping with the iodine compounds **59a–d** furnished the target series **53a–e** with different acceptor groups. The best method to prepare the series **52e** ($n = 1-4$) and **53e** ($n = 1-4$) is the condensation reaction of malonodinitrile with the aldehydes **52c** and **53c**, respectively [213].

Synthetic strategies for the preparation of polymers according to the AB or AA + BB type of polycondensation reactions (Scheme 11.6), combined with end-capping processes, can also be applied here, provided that a sufficient separation of the generated oligomers is feasible. Müllen and coworkers [214] published a

11.4 OPVs and OPEs with Terminal Donor–Acceptor Substitution

Table 11.4. Preparation of DAOPVs and DAOPEs.

		OPVs			OPEs	
Initial construction set of compounds	54	R₂N—⌬—CHO		57	R₂N—⌬—≡—H	
	55	HC(OR)₂—⌬—PO(OR)₂		58	I—⌬—≡—SiMe₃	
	56	A—⌬—PO(OR)₂		59	I—⌬—A	
Fundamental series ($n = 1, 2, 3, 4, \ldots$)	52c	R₂N—[⌬—CH=CH]ₙ—⌬—CHO Wittig-Horner reaction Deprotection 87 – 44%		60	R₂N—[⌬—≡]ₙ—⌬—≡—H Sonogashira-Hagihara reaction Deprotection 91 – 40%	
Capping to target series $n = 1$–4, (5)	52a,b,d,e	R₂N—[⌬—CH=CH]ₙ—⌬—A Wittig-Horner 94 – 24%		53a–e	R₂N—[⌬—≡]ₙ—⌬—A Sonogashira-Hagihara 98 – 39%	

series of OPEs with terminal dimethylamino/nitro substitution. The solubility was achieved by hexyl chains on the nonterminal benzene rings and the separation was performed by HPCL.

11.4.2
Absorption and Fluorescence

The UV/Vis absorption of conjugated chromophores is strongly influenced by *push–pull effects*. Table 11.5 summarizes the colors and the absorption maxima of the *trans*-stilbenes **52a–f** ($n = 1$). Increasing acceptor strength causes increasing

Table 11.5. Color and long-wavelength absorption maxima in CHCl₃ of the *trans*-stilbenes **52a–f** ($n = 1$) with bis(2-hexyloctyl)amino groups as electron donors and various electron acceptor groups R [213].

Compound	R	Color of the crystals	λ_{max} [nm]
52a	H	colorless	366
52b	CN	yellow	401
52c	CHO	orange	423
52d	NO₂	red	461
52e	CH=C(CN)₂	dark red	525
52f	C(CN)=C(CN)₂	blue	670

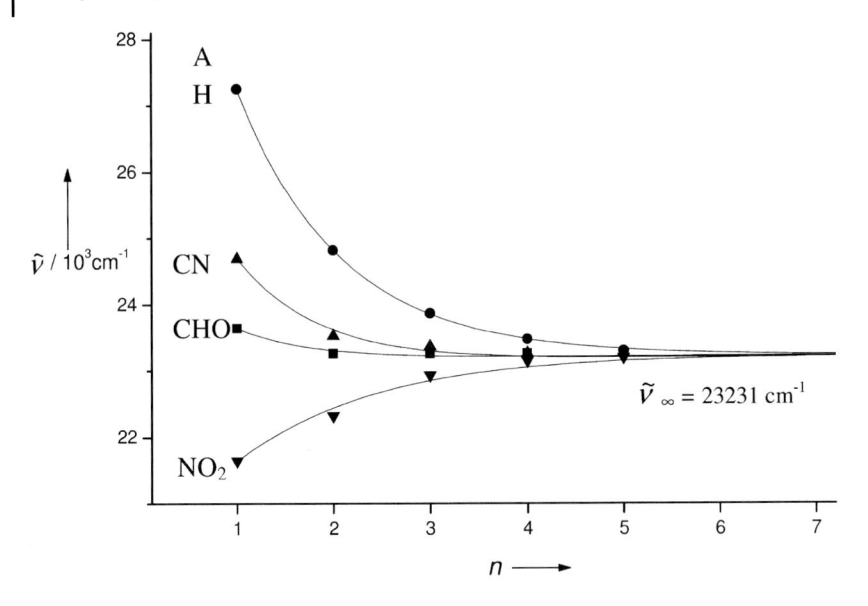

Figure 11.15. Long-wavelength absorption maxima of the series **52a–d** [$n = 1–4, (5)$] in CHCl$_3$ [211].

intramolecular charge transfer (ICT) and consequently an increasing bathochromic shift of the absorption and fluorescence [213]. The major reason for this effect is due to the decrease of the electron interaction (electron correlation), which is not included in single-electron wavefunctions; the change in the HOMO LUMO gap plays a minor role [1z].

Now, the crucial question is, how is the electron transition $S_0 \rightarrow S_1$ of such a push–pull system changed, when the chromophores in **52** are extended ($n = 1, 2, 3, \ldots$)? Figure 11.15 reveals the expected red-shift in the simply donor-substituted series **52a** ($n = 1–5$). This effect is preserved for **52b** with CN groups as weak acceptors. In the formyl series **52c**, the absorption maxima are almost independent of the size of the conjugated chromophore and the nitro series **52d** shows a distinct hypsochromic effect. The latter effect is even more pronounced for the dicyanovinyl series [210, 211, 213]. How can this unexpected behavior be explained? We found that the blue-shift in the DAOPV series is restricted to measurements in normal organic solvents like CHCl$_3$. As soon as a protonation occurs, the hypsochromic effect is reversed to a bathochromic effect (Fig. 11.16).

Protonation of the amino group in **52** obviously provokes the disappearance of the ICT. The convergence behavior of the transition energy $E_{DA}(n)$ can be split into two terms. The first term $E_D(n)$ comprises the *extension of the conjugation* with increasing numbers n of repeat units. It is the single term for the series **52a** and always causes a *bathochromic shift*. The second term, $\Delta E_{DA}(n)$, takes into account the *effect of the ICT* on the transition energies in the push–pull series **52b–d**. The decrease in the ICT with increasing n causes a *hypsochromic shift*. The two

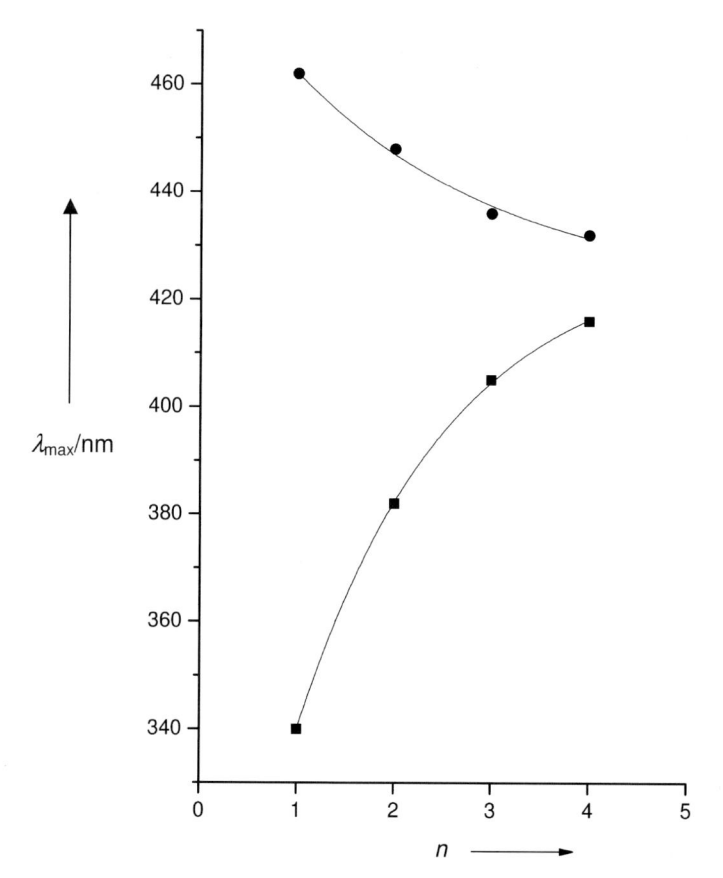

Figure 11.16. Reversal of the hypsochromic effect by protonation: absorption maxima of the series **52d** in CHCl$_3$ (—●—) and in CHCl$_3$/CF$_3$COOH (10:1) —■— [211].

effects are thus opposite to each other and the question arises, which effect predominates? Figure 11.17 contains the answer. Relatively weak acceptors such as CN lead to an overall bathochromic shift. Strong acceptors such as NO$_2$ provoke an overall hypsochromic effect, because the reduced effect of the ICT cannot be compensated by the "conjugation effect". In the formyl series **52c** ($n = 1-4$), the two effects annihilate each other almost completely [210, 211].

The algorithm discussed in Section 11.2.2 is now represented by the following equations:

$$E_{DA}(n) = E_D(n) - \Delta E_{DA}(n)$$
$$E_{DA}(n) = E_\infty + [E_D(1) - E_\infty]e^{-a(n-1)} - [E_D(1) - E_{DA}(1)]e^{-\Delta a(n-1)}$$

The common value E_∞ amounts to 2.88 eV and the effective conjugation length n_{ECL} is very small, namely 6, 4, 2 and 5 for **52a–d** [211].

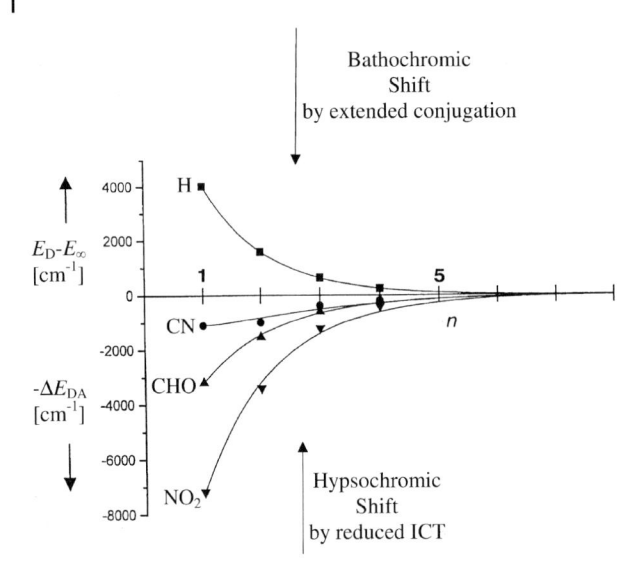

Figure 11.17. Bathochromic and hypsochromic contributions to the transition energies $E(S_0 \rightarrow S_1)$ of **52a–d** [211].

The absorption measurements of the DAOPE series **53a–d** gave very similar results [212, 1z]. The two opposite influences of conjugation extension and ICT decrease are illustrated in Fig. 11.18.

The NO$_2$ and the CHO series **53d** and **53c** exhibit an overall hypsochromic effect

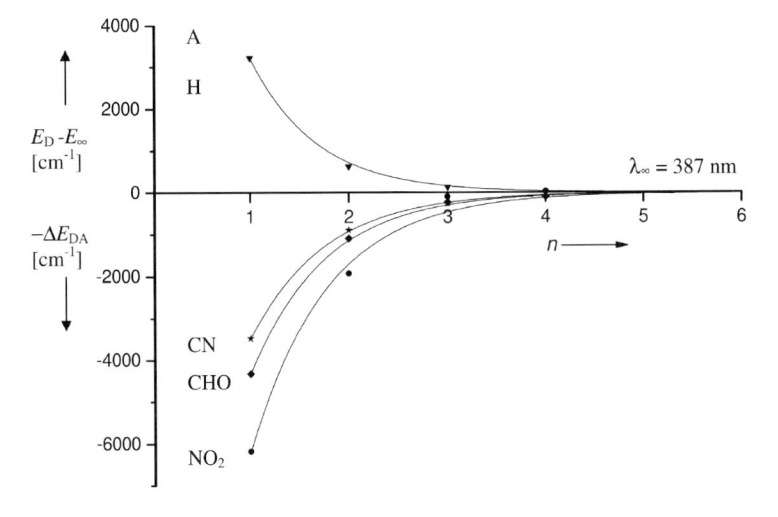

Figure 11.18. Bathochromic effect $[E_D - E_\infty]$ due to the extension of the conjugation and opposite hypsochromic effect $[-\Delta E_{DA}]$ caused by the decrease in the ICT upon increasing the number n of repeat units in the DAOPE series **53a–d** [212].

11.4 OPVs and OPEs with Terminal Donor–Acceptor Substitution

for increasing n. The CN series is almost balanced, that means there is a small dependence of the λ_{max} values on the size of the chromophore. On the whole, the OPE chain is somewhat more prone to a blue-shift than the OPV chain, when the D–A distance grows.

The AM1/INDO-S calculations [210–212] reveal that the HOMOs of **52** and **53** are localized on the donor side and the LUMOs on the acceptor side. The electron excitation HOMO → LUMO is therefore connected with a strong intramolecular charge transfer (ICT). However, the fraction of the HOMO → LUMO transition in the long-wavelength band decreases fast with increasing length of the chain. The overlap densities of HOMO and LUMO and the energy differences to the neighboring orbitals N-HOMO and N-LUMO (HOMO−1 and LUMO+1) decrease with increasing numbers n of repeat units. Thus, the transitions HOMO → N-LUMO, N-HOMO → LUMO and N-HOMO → N-LUMO become important, but these transitions have a small ICT or no ICT at all. Figure 11.19 shows the rapidly de-

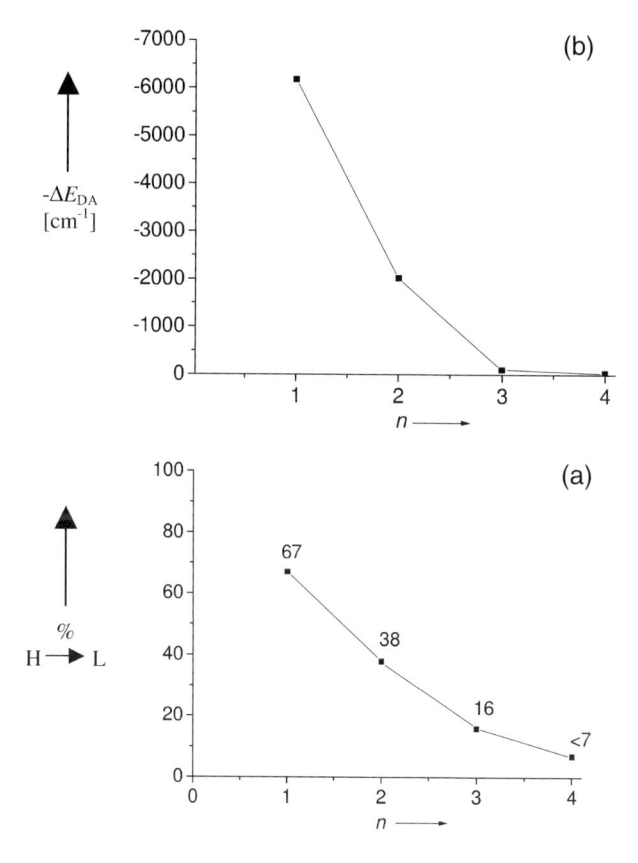

Figure 11.19. (a) Decrease in the fraction of the HOMO → LUMO transition in $S_0 \rightarrow S_1$ of **53d** and (b) corresponding decrease in the ICT correction term ΔE_{DA} on the long-wavelength absorption of **53d** ($n = 1–4$) [212].

creasing percentage of HOMO → LUMO transition in the series **53d** and the corresponding influence of the ICT term ΔE_{DA} on the long-wavelength absorption $S_0 \to S_1$ [212].

Analogous results were obtained for the DAOPV series **52** [210, 211]. Summarizing one can say: The stronger the acceptor in **52** and **53** the stronger the ICT and the more red-shifted the absorption. With increasing length, the ICT decreases and consequently the ICT correction term $\Delta E_{DA}(n)$ for the transition energy decreases. Thus, the bathochromic shift, that results from the extension of the chromophore can be diminished, annihilated or even reversed to a hypsochromic effect.

In principle, the same observations can be made in the fluorescence spectra. The NO_2 series **52d** ($n = 1–5$) for example shows a pronounced hypsochromic shift with increasing n. The effect is with $\Delta \lambda_{max} = 65$ nm unequivocal, but the solvation has a much bigger influence on the fluorescence than on the absorption. The Stokes shift varies dramatically within the series; it amounts in dioxane to about 5800 cm^{-1} for **52d** ($n = 2$), but only to about 4200 cm^{-1} for **52d** ($n = 4$) [138]. Therefore an exact correlation of the fluorescence maxima with n is difficult.

The nonlinear optics of the push–pull series **52** and **53** are under investigation. However, it is already clear that the β and γ values increase with increasing length L of the chains, even when λ_{max} decreases [215]. (See also Sections 11.2.3 and 11.3.3.)

11.5
Summary and Outlook

Conjugated oligomers, which consist of CC double or triple bonds and/or 1,4-phenylene units, form *rigid rods* with a well defined *length L*. The extent of the chromophores/electrophores can be calculated by linear functions $L(n)$, where n is the number of repeat units. Scheme 11.24 gives a survey of such oligomers with $k, l, m = 0, 1$. Many more conjugated series can be conceived with $k, l, m > 1$

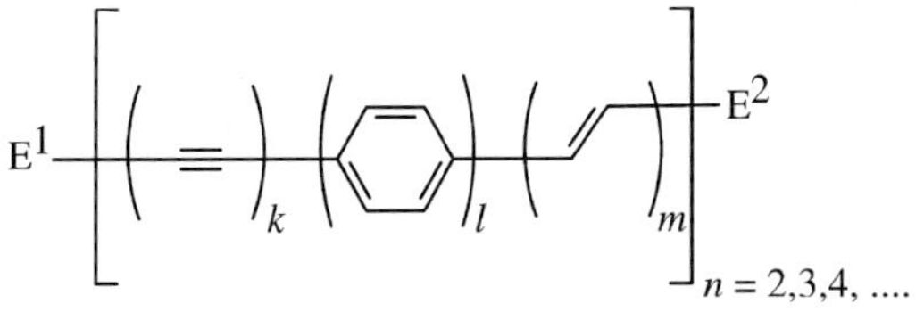

k	l	m	
0	0	1	Oligoenes (OE)
0	1	0	Oligophenylenes (OP)
1	0	0	Oligoynes (OY)
0	1	1	Oligo(phenylenvinylene)s (OPV)
1	1	0	Oligo(phenyleneethynylene)s (OPE)
1	0	1	Oligo(enyne)s (OEY)
1	1	1	Oligo(phenylenevinyleneethynylene)s (OPVE)

Scheme 11.24. Conjugated oligomers which form rigid rods.

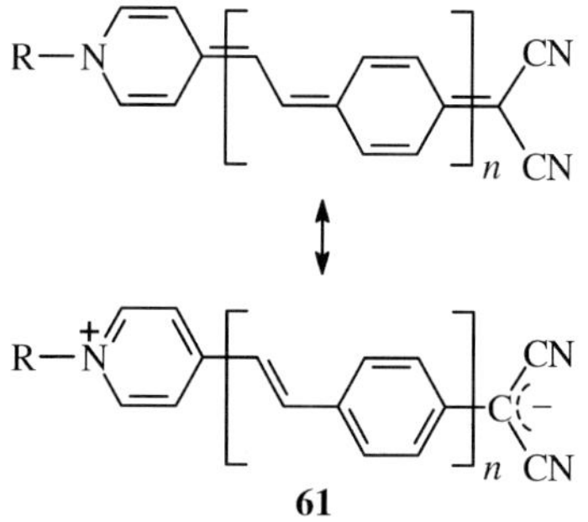

Scheme 11.25. Push–pull substituted OPVs whose zwitterionic resonance structure is compatible with aromatic subunits in the conjugated chain.

and with other arylene subunits like napththylene, anthrylene, phenanthrylene, pyrenylene, perylenylene, fluorenylene, etc.; moreover, nonbenzenoid aromatics such as azulene or cyclobutadiene complexes can replace the benzene ring in Scheme 11.24.

This chapter is focused on OPVs and OPEs. Monodisperse oligomers in these series are known up to a size of more than 10 nm. The *preparative approach* to these *nanoparticles* starts normally with the aromatic ring systems, which are then connected by olefinic or acetylenic bridges. A variety of convergent or divergent methods is available for this purpose – in particular for the OPV chains; the OPE chains were predominantly synthesized by applying the Sonogashira-Hagihara reaction.

Sidechains on the benzene rings, as for example alkyl, alkoxy or dialkylamino groups, enhance the solubility – an effect which is important for the synthesis and processing of the materials.

A special aspect of conjugated oligomers consists of the expectance that certain properties P approach a limiting value P_∞ for increasing numbers n of repeat units. The *saturation range* is then characterized by the *effective conjugation length* n_{ECL}. Exponential functions proved to be excellent fit functions for this behavior. Linear optics (absorption, fluorescence) provide good examples for such a *convergence*, whereas in nonlinear optics (NLO) another type of convergence is also possible, namely the *approach of the slope* dP/dn to a *limiting value* P'_∞.

Substituents, even single functional groups in the terminal positions of the conjugated chain, can have a strong influence on many properties. In particular, this was demonstrated by *push–pull systems* with a donor (D) on one end and an acceptor (A) on the other end of the chain: DAOPVs and DAOPEs. A striking result was found for the UV/Vis absorption characteristics of such compounds which can exhibit a *bathochromic* or a *hypsochromic* effect for increasing numbers n of repeat units (increasing length of the chromophores). The *extension of the conjugation* is opposed in these cases by a decrease in the *intramolecular charge transfer* (ICT) which is involved in the electron excitation.

In this chapter, the emphasis in *materials science* is laid on *nonlinear optics* and *electroluminescence*; however, applications of OPVs and OPEs can be found in many more areas of materials science: field-effect transistors (FET), semiconductors (doped), photoconductors, solar cells, photovoltaic devices, optical brighteners, laser dyes, optical switching, imaging techniques, photoresists, liquid crystals, etc.

Push–pull systems with π linkers of the OPV or OPE type represent in materials science a new and highly promising class of conjugated oligomers. In addition to D–π–A chains, which are characterized by electroneutral resonance structures with benzene rings and zwitterionic resonance structures with *p*-quinoid substructures (Scheme 11.22), series like **61** should be studied, in which the zwitterionic resonance structure is compatible with aromatic subunits (Scheme 11.25) [1z, 216].

The interesting optical, electrical and optoelectronic properties of conjugated oligomers provoked a real boom in investigations in previous years. Nevertheless, many more systematic studies are necessary to reach the ultimate goal of tailored nanoparticles on the molecular basis of conjugated oligomers.

11.6
Experimental: Selected Procedures

11.6.1
General Procedure for the Extension of an OPV Aldehyde by Two Styryl Units (Scheme 11.7) [2]

Solutions of 3.6 g (32 mmol) of $KOC(CH_3)_3$ in 100 mL of dry DMF, 1.6 mmol of aldimine **8**, **11** or **13** in 100 mL of dry DMF and 0.80 g (1.6 mmol) of acetal **9** in 50 mL of dry DMF were prepared and degassed. The solution/suspension of the aldimine was added within 5 min to the base before the solution of **9** was added dropwise over 20 min into the reaction vessel. All these procedures were performed under argon. TLC control (SiO_2; DMF) revealed the complete consumption of **9** within 4–6 h. Raising the temperature from 25 to 80 °C reduced the reaction time to less than 1 h, but led to some side products. The mixture was poured on crushed ice, treated with 50 mL of HCl (6 M) and extracted 3 times with 50 mL of CH_2Cl_2. The combined organic phases were washed with water and $NaHCO_3$, dried with Na_2SO_4, filtered and the solvents evaporated. Purification requires either recrystallization from EtOH or column chromatography (SiO_2, CH_2Cl_2/C_6H_{12}, 2:1). The red solid products **10** (mp 120 °C), **12** (mp > 250 °C) and **14** (mp > 250 °C) were obtained with yields of 82, 92 and 52%, respectively [2, 17].

11.6.2
all-(E)-2,2′,5,5′-Tetrapropoxy-4,4′-bis(2,5-dipropoxy-4-{2,5-dipropoxy-4-[2,5-dipropoxy-4-(2,5-dipropoxy-4-{2,5-dipropoxy-4-[2,5-dipropoxy-4-(2,5-dipropoxystyryl)styryl]styryl}-styryl)styryl]styryl}styryl)stilbene (2j) [2]

Nitrogen was purged through a suspension of 0.17 g (2.64 mmol) of Zn in 14 mL of THF at 0 °C, before 0.15 mL (0.25 g, 1.32 mmol) of $TiCl_4$ was added with a syringe. The mixture was heated to reflux for 30 min, cooled and treated with 0.27 mL of pyridine. A solution of 0.4 g (0.23 mmol) of **14** in 15 mL of dry THF was added through a syringe. After stirring for 1.5 h, the mixture was poured onto crushed ice and acidified with HCl (6 M). The red precipitate was filtered, washed with water, NaHCO$_3$, water and a small portion of CH$_2$Cl$_2$. Yield: 0.15 g (19%) of a red solid, which decomposed above 290 °C.

11.6.3
General Procedure for the Cation–Anion Coupling Reaction (Scheme 11.10) [37]

NaH (2.0 mmol) was added to a solution of benzylcyanide **26** (1.0 mmol) in 3 mL of dry DMF at −60 °C in an argon atmosphere. After stirring for 1 h, the cation **22** (1.0 mmol) was added (as iodide) in 2 mL dry DMF. The resulting mixture was stirred for 8 h at −60 °C and was then allowed to reach room temperature overnight. Filtration of the reaction mixture over SiO$_2$ followed by removal of the solvent under vacuum yielded the donor-acceptor stilbene **27a** (96%). For further purification **27a** was reprecipitated from a THF solution into petrol ether (bp 40–60 °C).

11.6.4
Preparation of 1,4-Bis(2,5-dipropoxyphenylethynyl)-2,5-dipropoxybenzene (1b) (Scheme 11.16) [136]

2,5-Dipropoxy-1,4-bis(trimethylsilylethynyl)benzene (36). Diiodo compound **35** (2.0 g, 4.5 mmol) was dissolved in dry piperidine (20 mL) and the system was flushed with nitrogen. Bis(triphenylphosphine)palladium dichloride (0.24 g, 0.55 mmol), copper(I) iodide (0.12 g, 0.63 mmol), triphenylphosphine (0.24 g, 0.91 mmol), and trimethylsilylacetylene (1.0 g, 9.9 mmol) were added. The mixture was stirred for 18 h and purified as described above. Yield: 1.7 g (98%) of colorless crystals, mp 101 °C.

1,4-Diethynyl-2,5-dipropoxybenzene (37). Compound **36** (26.2 g, 0.068 mol) was dissolved in THF (100 mL), and a mixture of methanol (100 mL) and sodium hydroxide (31 mL, $c = 5$ mol L^{-1}) was added. After 5 min, the solvents were evaporated and the residue was washed several times with *n*-hexane. Yield: 16.4 g (98%) of beige crystals, mp 122 °C.

1,4-Bis(2,5-dipropoxyphenylethynyl)-2,5-dipropoxybenzene (1b). Compounds **34** (0.51 g, 1.14 mmol) and **37** (0.5 g, 2.3 mmol) were dissolved in dry piperidine and the system was flushed with nitrogen. To this yellow solution were added bis(triphenylphosphine)palladium dichloride (0.06 g, 0.085 mmol), copper(I) iodide (0.05 g, 0.26 mmol), and triphenylphosphine (0.06 g, 0.23 mmol). The reaction was complete within 5 d and the mixture was filtered through silica gel with CH$_2$Cl$_2$. The obtained brown oil was purified by column chromatography on silica gel (toluene). Yield: 0.65 g (91%) of yellow crystals, mp 98 °C.

11.6.5
General Procedure for the Preparation of the Series 52c ($n = 1$–4) by Wittig–Horner Reactions (Scheme 11.23 and Table 11.4) [211]

Equimolar amounts of the appropriate aldehyde 54, 52c ($n = 1, 2, \ldots$) and the phosphonate 55, dissolved in dry DMF or THF, were dropped into a solution of potassium t-butoxide (slight excess) in the same solvent. The mixture was stirred at room temperature for 1 h and then poured onto crushed ice. The aqueous phase was extracted three times with dichloromethane and the combined organic phases were dried over Na_2SO_4. After removal of the solvent, the residue was purified by silica gel column chromatography and/or by recrystallization.

4-(E)-2-{4-[Bis(2-hexyloctyl)amino]phenyl}ethenyl)benzaldehydes 52c ($n = 1$) [211]. The crude material was chromatographed on silica gel (50 × 3 cm) with petroleum ether (bp 40–70 °C)/ethyl acetate 2:1 to give 87% of an orange oil.

4-{(E)-2-[4-((E)-2-{4-[Bis(2-hexyloctyl)amino]phenyl}ethenyl)phenyl]ethenyl}benzaldehydes 52c ($n = 2$) [211]. Recrystallization from ethyl acetate gave 77% of orange crystals; mp 98 °C.

4-[(E)-2-(4-{(E)-2-[4-((E)-2-{4-[Bis(2-hexyloctyl)amino]phenyl}ethenyl)phenyl]ethenyl}-phenyl)ethenyl]benzaldehydes 52c ($n = 3$) [211]. The crude product was purified by column chromatography on silica gel with toluene/petroleum ether (bp 40–70 °C) 2:1, yielding orange crystals; mp 165 °C.

4-((E)-2-{4-[(E)-2-(4-{((E)-2-[4-((E)-2-{4-[Bis(2-hexyloctyl)amino]phenyl}ethenyl)phenyl]-ethenyl}phenyl)ethenyl]phenyl}ethenyl)benzaldehyde 52c ($n = 4$) [211]. The crude product was purified by column chromatography on silica gel with CH_2Cl_2 as eluant, yielding 62% orange crystals; mp 195 °C.

Abbreviations

β	First optical hyperpolarizability
$\chi^{(n)}$	Optical susceptibility of nth order
ϕ_F	Fluorescence quantum yield
γ	Second optical hyperpolarizability
γ_{res}	Second optical hyperpolarizability for wavelengths λ_L of the applied laser light which correspond to $3\lambda_{max}$
η_E	Electroluminescence quantum yield
μ_{ind}	Induced dipole moment
A	Absorbance
ADIMET	Acyclic diyne metathesis
ADMET	Acyclic diene metathesis
CCDC	Cambridge Crystallographic Data Centre
DAOPE	Oligo(phenyleneethynylene) with terminal donor–acceptor substitution

DAOPV	Oligo(phenylenevinylene) with terminal donor–acceptor substitution
DFT	Density functional theory
EL	Electroluminescence
FET	Field-effect transistor
GPC	Gel permeation chromatography
IC	Internal conversion
ICT	Intramolecular charge transfer
ISC	Intersystem crossing
ITO	Indium tin oxide
L	Length of conjugated chain
LED	Light emitting diode
M	Transition moment
MO	Molecular orbital theory
n	Number of repeat units in oligomers
n_{ECL}	Effective conjugation length
N-HOMO (HOMO−1)	Second highest occupied orbital
NLO	Nonlinear optics
N-LUMO (LUMO+1)	Second lowest unoccupied orbital
OE	Oligoene
OEY	Oligoenyne
OLED	Organic light emitting diode
OP	Oligo(phenylene)
OPE	Oligo(phenyleneethynylene)
OPV	Oligo(phenylenevinylene)
OPVE	Oligo(phenylenevinyleneethynylene)
OY	Oligoyne
P	Optical polarization
PL	Photoluminescence
PPE	Poly(phenyleneethynylene)
PPV	Poly(phenylenevinylene)
SHG	Second harmonic generation
THG	Third harmonic generation
TIPS	Triisopropylsilyl
TLC	Thin-layer chromatography
TTF	Tetrathiafulvalene
VB	Valence bond theory

References

1 Selected books and review articles: (a) H.-H. Hörhold, M. Helbig, D. Raabe, J. Opfermann, U. Scherf, R. Stockmann, D. Weiß, *Z. Chem.* **1987**, *27*, 126–137; (b) J. L. Brédas, R. Silbly, *Conjugated Polymers*, Kluwer,

Dordrecht, **1991**; (c) K. MÜLLEN, *Pure Appl. Chem.* **1993**, *65*, 89–96, (d) W. R. SALANECK, I. LUNDSTRÖM, B. R. RÅNBY, *Conjugated Polymers and Related Materials,* Oxford University Press, Oxford, **1993**; (e) J. M. TOUR, *Chem. Rev.* **1996**, *96*, 537–553; (f) R. GIESA, *J. Macromol. Sci., Rev. Macromol. Chem. Phys.* **1996**, *C36*, 631–670; (g) J. S. MOORE, *Acc. Chem. Res.* **1997**, *30*, 402–413; (h) J. RONCALI, *Chem. Rev.* **1997**, *97*, 173–205; (i) A. KRAFT, A. C. GRIMSDALE, A. B. HOLMES, *Angew. Chem.* **1998**, *110*, 416–443; *Angew. Chem. Int. Ed.* **1998**, *37*, 402–428; (j) *Electronic Materials: The Oligomer Approach*, K. MÜLLEN, G. WEGNER (eds.), Wiley-VCH, Weinheim, **1998**; (k) T. M. SWAGER, *Acc. Chem. Res.* **1998**, *31*, 201–207; (l) F. DIEDERICH, L. GOBBI, *Top. Curr. Chem.* **1999**, *201*, 43–79; (m) P. F. H. SCHWAB, M. D. LEVIN, J. MICHL, *Chem. Rev.* **1999**, *99*, 1863–1933; (n) U. SCHERF, *Top. Curr. Chem.* **1999**, *201*, 163–222; (o) R. E. MARTIN, F. DIEDERICH, *Angew. Chem.* **1999**, *111*, 1440–1469; *Angew. Chem. Int. Ed.* **1999**, *38*, 1350–1377; (p) U. H. F. BUNZ, *Top. Curr. Chem.* **1999**, *201*, 131–161; (q) T. YAMAMOTO, *Bull. Chem. Soc. Jpn.* **1999**, *72*, 621–638; (r) U. H. F. BUNZ, *Chem. Rev.* **2000**, *100*, 1605–1644; (s) J. L. SEGURA, N. MARTÍN, *J. Mater. Chem.* **2000**, *10*, 2403–2435; (t) G. HADZIIOANNOU, P. F. VAN HUTTEN, *Semiconductivity Polymers*, Wiley-VCH, Weinheim, **2000**; (u) J. RONCALI, *Acc. Chem. Res.* **2000**, *33*, 147–156; (v) J. M. TOUR, *Acc. Chem. Res.* **2000**, *33*, 791–804; (w) A. MISHRA, R. K. BEHERA, P. K. BEHERA, B. K. MISHRA, G. B. BEHERA, *Chem. Rev.* **2000**, *100*, 1973–2011; (x) U. H. F. BUNZ, *Acc. Chem. Res.* **2001**, *34*, 998–1010; (y) S. SZAFERT, J. A. GLADYSZ, *Chem. Rev.* **2003**, *103*, 4175–4205; (z) H. MEIER, *Angew. Chem.* **2005**, *117*, 2536–2561; *Angew. Chem. Int. Ed.* **2005**, *44*, 2482–2506.

2 H. MEIER, D. ICKENROTH, *Eur. J. Org. Chem.* **2002**, 1745–1749.

3 F. DIEDERICH, Y. RUBIN, C. B. KNOBLER, R. L. WHETTEN, K. E. SCHRIVER, K. N. HOUK, Y. LI, *Science* **1989**, *245*, 1088–1090.

4 See Ref. [1l] and references therein.

5 H. MEIER, *Synthesis* **2002**, 1213–1227 and references therein.

6 A. DE MEIJERE, F. E. MEYER, *Angew. Chem.* **1994**, *106*, 2473–2506; *Angew. Chem. Int. Ed. Engl.* **1994**, *33*, 2379–2411.

7 H. MEIER, N. HANOLD, H. KALBITZ, *Synthesis* **1997**, 276–278.

8 H. MEIER, S. KIM, *Eur. J. Org. Chem.* **2001**, 1163–1167.

9 H. MEIER, H. C. HOLST, *Adv. Synth. Catal.* **2003**, *345*, 1005–1011.

10 H. C. HOLST, T. PAKULA, H. MEIER, *Tetrahedron* **2004**, *60*, 6765–6775.

11 H. MEIER, M. LEHMANN, H. C. HOLST, D. SCHWÖPPE, *Tetrahedron* **2004**, *60*, 6881–6888.

12 H. MEIER, M. LEHMANN, *Encyclopedia of Nanoscience and Nantechnology*, H. S. NALWA (ed.), ASP, Stevenson Ranch, California, USA **2004**, vol. *10*, pp. 95–106 and references therein.

13 H. MEIER, M. LEHMANN, *Angew. Chem.* **1998**, *110*, 666–669; *Angew. Chem. Int. Ed.* **1998**, *37*, 643–645.

14 M. LEHMANN, B. SCHARTEL, M. HENECKE, H. MEIER, *Tetrahedron* **1999**, *55*, 13377–13394.

15 H. MEIER, M. LEHMANN, U. KOLB, *Chem. Eur. J.* **2000**, *6*, 2462–2469.

16 M. LEHMANN, I. FISCHBACH, H. W. SPIESS, H. MEIER, *J. Am. Chem. Soc.* **2004**, *126*, 772–784.

17 U. STALMACH, H. KOLSHORN, I. BREHM, H. MEIER, *Liebigs Ann.* **1996**, 1449–1456.

18 H. MEIER, U. STALMACH, H. KOLSHORN, *Acta Polym.* **1997**, *48*, 379–384.

19 V. GEBHARDT, A. BACHER, M. THELAKKAT, U. STALMACH, H. MEIER, H.-W. SCHMIDT, D. HAARER, *Synth. Met.* **1997**, *90*, 123–126.

20 U. STALMACH, H. DETERT, H. MEIER, V. GEBHARDT, D. HAARER, A. BACHER, H.-W. SCHMIDT, *Opt. Mater.* **1998**, *9*, 77–81.

21 U. STALMACH, D. SCHOLLMEYER, H. MEIER, *Chem. Mater.* **1999**, *11*, 2103–2106.

22 V. GEBHARDT, A. BACHER, M. THELAKKAT, U. STALMACH, H. MEIER,

H.-W. Schmidt, D. Haarer, *Adv. Mater.* **1999**, *11*, 119–123.

23 D. Oelkrug, J. Gierschner, H.-J. Egelhaaf, L. Lüer, A. Tompert, K. Müllen, U. Stalmach, H. Meier, *Synth. Met.* **2001**, *121*, 1693–1694.

24 An experimental procedure is described in Section 11.6.

25 T. Hrenar, R. Mitrić, Z. Meić, H. Meier, U. Stalmach, *J. Mol. Struct.* **2003**, *661–662*, 33–40.

26 H. Meier, H. Kretzschmann, M. Lang, W. Fraß, C. Albrecht, K. März, *J. Prakt. Chem.* **1994**, *336*, 297–302.

27 H. Meier, *Angew. Chem.* **1992**, *104*, 1425–1446; *Angew. Chem. Int. Ed. Engl.* **1992**, *31*, 1399–1420.

28 T. Maddux, W. Li, L. Yu, *J. Am. Chem. Soc.* **1997**, *119*, 844–845.

29 R. A. Wessling, *J. Polym. Sci., Polym. Symp.* **1985**, *72*, 55–66.

30 H. G. Gilch, W. L. Wheelwright, *J. Polym. Sci. A* **1966**, *4*, 1337–1349.

31 F. Louwet, D. Vanderzande, J. Gelan, J. Mullens, *Macromolecules* **1995**, *28*, 1330–1331.

32 See also D. Raabe, H.-H. Hörhold, U. Scherf, *Makromol. Chem. Rapid Commun.* **1986**, *7*, 613–619.

33 V. P. Conticello, D. L. Gin, R. H. Grubbs, *J. Am. Chem. Soc.* **1992**, *114*, 9708–9710.

34 Y.-J. Miao, G. C. Bazan, *Macromolecules* **1994**, *27*, 1063–1064.

35 E. Thorn-Csányi, P. Kraxner, *Macromol. Rapid Commun.* **1995**, *16*, 147–153.

36 R. Peetz, A. Strachota, E. Thorn-Csányi, *Macromol. Chem. Phys.* **2003**, *204*, 1439–1450.

37 G. Klärner, C. Former, X. Yan, R. Richert, K. Müllen, *Adv. Mater.* **1996**, *8*, 932–935.

38 T.-L. Chan, H.-F. Chow, S. Fong, M.-K. Leung, J. Tu, *J. Chem. Soc., Chem. Commun.* **1994**, 1919–1920.

39 (a) G. Drefahl, G. Plötner, *Chem. Ber.* **1958**, *91*, 1274–1280; (b) G. Drefahl, G. Plötner, *Chem. Ber.* **1961**, *94*, 907–914.

40 G. Drehfahl, R. Kühmstedt, H. Oswald, H.-H. Hörhold, *Makromol. Chem.* **1970**, *131*, 89–103.

41 J. Heinze, J. Mortensen, K. Müllen, R. Schenk, *J. Chem. Soc., Chem. Commun.* **1987**, 701–703.

42 See for example: (a) R. Schenk, H. Gregorius, K. Meerholz, J. Heinze, K. Müllen, *J. Am. Chem. Soc.* **1991**, *113*, 2634–2647; (b) K. Meerholz, H. Gregorius, K. Müllen, J. Heinze, *Adv. Mater.* **1994**, *6*, 671–674; (c) S. Barth, H. Bässler, T. Wehrmeister, K. Müllen, *J. Chem. Phys.* **1997**, *106*, 321–327; (d) B. Tian, G. Zerbi, R. Schenk, K. Müllen, *J. Chem. Phys.* **1991**, *95*, 3191–3197; (e) B. Tian, G. Zerbi, K. Müllen, *J. Chem. Phys.* **1991**, *95*, 3198–3207.

43 S. Wachsmann-Hogiu, L. A. Peteanu, L. A. Liu, D. J. Yaron, J. Wildeman, *J. Phys. Chem. B* **2003**, *107*, 5133–5143.

44 M. S. Wong, Z. H. Li, M. F. Shek, K. H. Chow, Y. Tao, M. D'Iorio, *J. Mater. Chem.* **2000**, *10*, 1805–1810.

45 M. S. Wong, Z. H. Li, M. F. Shek, M. Samoc, A. Samoc, B. Luther-Davies, *Chem. Mater.* **2002**, *14*, 2999–3004.

46 D. S. Seferos, D. A. Banach, N. A. Alcantar, J. N. Israelachvili, G. C. Bazan, *J. Org. Chem.* **2004**, *69*, 1110–1119.

47 S. Kubatkin, A. Danilov, M. Hjort, J. Cornil, J.-L. Brédas, N. Stuhr-Hansen, P. Hedegård, T. Bjørnholm, *Nature* **2003**, *425*, 698–701.

48 N. Stuhr-Hansen, J. B. Christensen, N. Harrit, T. Bjørnholm, *J. Org. Chem.* **2003**, *68*, 1275–1282.

49 J. Arient, *Collect. Czech. Chem. Commun.* **1981**, *46*, 101–106.

50 H. E. Katz, S. F. Shane, W. L. Wilson, M. L. Schilling, S. B. Ungashe, *Mater. Res. Soc. Symp. Proc.* **1994**, *328*, 361–369; *Chem. Abstr.* **1994**, *121*, 144337.

51 H. Detert, E. Sugiono, *Synth. Met.* **2002**, *127*, 237–239.

52 H. Detert, E. Sugiono, *Macromol. Symp.* **2002**, *181*, 39–45.

53 E. Sugiono, T. Metzroth, H. Detert, *Adv. Synth. Catal.* **2001**, *343*, 351–359.

54 F. Babudri, G. M. Farinola, L. C. Lopez, M. G. Martinelli, F. Naso, *J. Org. Chem.* **2001**, *66*, 3878–3885.

55 R. E. Gill, P. F. van Hutten, A. Meetsma, G. Hadziioannou, *Chem. Mater.* **1996**, *8*, 1341–1346.

56 N. N. Barashkov, D. J. Guerrero, H. J. Olivos, J. P. Ferraris, *Synth. Met.* **1995**, *75*, 153–160.

57 C. Melzer, V. V. Krasnikov, G. Hadziioannou, *J. Polym. Sci. B* **2003**, *41*, 2665–2673.

58 D. M. Johansson, M. Theander, T. Granlund, O. Inganaes, M. R. Andersson, *Macromolecules* **2001**, *34*, 1981–1986.

59 L. Ouoli, V. V. Krasnikov, U. Stalmach, G. Hadziioannou, *Adv. Mater.* **1999**, *11*, 1515–1518.

60 H. Detert, E. Sugiono, *Synth. Met.* **2003**, *138*, 181–183.

61 H. Detert, O. Sadovski, *Synth. Met.* **2003**, *138*, 185–188.

62 J. E. Wong, S. Schrader, H. Detert, S. Katholy, L. Brehmer, *Mater. Sci. Eng. C* **2002**, *22*, 413–417.

63 H. Detert, D. Schollmeyer, E. Sugiono, *Eur. J. Org. Chem.* **2001**, 2927–2938.

64 H. Detert, E. Sugiono, *J. Phys. Org. Chem.* **2000**, *13*, 587–590.

65 H. Detert, E. Sugiono, *Synth. Met.* **2000**, *115*, 89–92.

66 R. Peetz, O. Narwark, O. Herzog, S. Brocke, E. Thorn-Csányi, *Synth. Met.* **2001**, *119*, 539–540.

67 O. Narwark, A. Gerhard, S. C. J. Meskers, S. Brocke, E. Thorn-Csányi, H. Bässler, *Chem. Phys.* **2003**, *294*, 17–30.

68 L. Liao, Y. Pang, L. Ding, F. E. Karasz, *Polym. Prepr.* **2003**, *44*, 430–431.

69 O. Narwark, S. C. J. Meskers, R. Peetz, E. Thorn-Csányi, H. Bässler, *Chem. Phys.* **2003**, *294*, 1–15.

70 See also Ref. [35] and references therein.

71 M. J. Gómez-Escalonilla, F. Langa, J.-M. Rueft, L. Oswald, J.-F. Nierengarten, *Tetrahedron Lett.* **2002**, *43*, 7507–7511.

72 M. R. Wasielewski, B. A. Yoon, M. Fuller, G. P. Wiederrecht, M. P. Niemczyk, W. A. Svec, *Mater. Res. Soc. Symp. Proc.* **2000**, *597*, 211–219; *Chem. Abstr.* **2001**, *134*, 18178.

73 E. Peeters, R. A. J. Janssen, S. C. J. Meskers, E. W. Meijer, *Polym. Prepr.* **1999**, *40*, 519–520.

74 E. Peeters, A. M. Ramos, S. C. J. Meskers, R. A. J. Janssen, *J. Chem. Phys.* **2000**, *112*, 9445–9454.

75 A. P. H. J. Schenning, A. C. Tsipis, S. C. J. Meskers, D. Beljonne, E. W. Meijer, J. L. Brédas, *Chem. Mater.* **2002**, *14*, 1362–1368.

76 P. Leclère, M. Surin, P. Jonkheijm, O. Henze, A. P. H. J. Schenning, F. Biscarini, A. C. Grimsdale, W. J. Feast, E. W. Meijer, K. Müllen, J. L. Brédas, R. Lazzaroni, *Eur. Polym. J.* **2004**, *40*, 885–892.

77 A. Gesquière, P. Jonkheijm, A. P. H. J. Schenning, E. Mena-Osteritz, P. Bäuerle, S. De Feyter, F. C. De Schryver, E. W. Meijer, *J. Mater. Chem.* **2003**, *13*, 2164–2167.

78 H. Wang, H. H. Wang, V. S. Urban, K. C. Littrell, P. Thiyagarajan, L. Yu, *J. Am. Chem. Soc.* **2000**, *122*, 6855–6861.

79 H. Wang, M.-K. Ng, L. Yu, *Polym. Prepr.* **2001**, *42*, 439–440; *Chem. Abstr.* **2001**, *135*, 372110.

80 H. Wang, W. You, P. Jiang, L. Yu, H. H. Wang, *Chem. Eur. J.* **2004**, *10*, 986–993.

81 T. Goodson III, W. Li, A. Gharavi, L. Yu, *Adv. Mater.* **1997**, *9*, 639–643.

82 C. Xue, F.-T. Luo, *J. Org. Chem.* **2003**, *68*, 4417–4421.

83 D. A. Bussian, M. A. Summers, B. Liu, G. C. Bazan, S. K. Buratto, *Chem. Phys. Lett.* **2004**, *388*, 181–185.

84 A. Menon, M. Galvin, K. A. Walz, L. Rothberg, *Synth. Met.* **2004**, *141*, 197–202.

85 L. P. Candeias, J. Wildeman, G. Hadziioannou, J. Warman, *J. Phys. Chem. B* **2000**, *104*, 8366–8371.

86 W. Li, H. Wang, L. Yu, T. L. Morkved, H. M. Jaeger, *Macromolecules* **1999**, *32*, 3034–3044.

87 Y. Geng, A. C. A. Chen, J. J. Ou, S. H. Chen, K. Klubek, K. M. Vaeth, C. W. Tang, *Chem. Mater.* **2003**, *15*, 4352–4360.

88 R. E. Gill, P. F. van Hutten, A. Meetsma, G. Hadziioannou, *Chem. Mater.* **1996**, *8*, 1341–1346.

89 D. A. dos Santos, D. Beljonne, J. Cornil, J. L. Brédas, *Chem. Phys.* **1998**, *227*, 1–10.
90 U. Stalmach, H. Detert, *J. Prakt. Chem.* **2000**, *342*, 10–16.
91 H. Detert, E. Sugiono, *Synth. Met.* **2001**, *122*, 15–17.
92 H. Detert, E. Sugiono, *Synth. Met.* **2001**, *122*, 19–21.
93 H. Detert, E. Sugiono, G. Kruse, *J. Phys. Org. Chem.* **2002**, *15*, 638–641.
94 H. Detert, D. Schollmeier, *Synthesis* **1999**, 999–1004.
95 M. Pomerantz, J. Wang, S. Seong, K. P. Starkey, L. Nguyen, D. S. Marynick, *Macromolecules* **1994**, *27*, 7478–7485.
96 H. Wang, M.-K. Ng, L. Wang, L. Yu, B. Lin, M. Meron, Y. Xiao, *Chem. Eur. J.* **2002**, *8*, 3246–3253.
97 P. Jonkheijm, F. J. M. Hoeben, R. Kleppinger, J. van Herrikhuyzen, A. P. H. J. Schenning, E. W. Meijer, *J. Am. Chem. Soc.* **2003**, *125*, 15941–15949.
98 M. G. Kuzyk, *J. Chem. Phys.* **2003**, *119*, 8327–8334.
99 M. Rumi, J. E. Ehrlich, A. A. Heikal, J. W. Perry, S. Barlow, Z. Hu, D. McCord-Maughon, T. C. Parker, H. Roeckel, S. Thayumanavan, S. R. Marder, D. Beljonne, J.-L. Brédas, *J. Am. Chem. Soc.* **2000**, *122*, 9500–9510.
100 B. Strehmel, A. M. Sarker, H. Detert, *Chem. Phys. Chem.* **2003**, *4*, 249–259.
101 H. Detert, E. Sugiono, *Macromol. Symp.* **2002**, *181*, 39–45.
102 M. Gutíerrez-Nava, P. Masson, J.-F. Nierengarten, *Tetrahedron Lett.* **2003**, *44*, 4487–4490.
103 E. Sugiono, H. Detert, *Silicon Chemistry*, in press.
104 H. Detert, E. Sugiono, *Synth. Met.* **2002**, *127*, 233–235.
105 E. Zojer, M. Knupfer, Z. Shuai, J. Fink, J. L. Brédas, H.-H. Hörhold, J. Grimme, U. Scherf, T. Benincori, G. Leising, *Phys. Rev. B* **2000**, *61*, 16561–16569.
106 M. Helbig, H.-H. Hörhold, *Makromol. Chem.* **1993**, *194*, 1607–1618.
107 S. Wang, G. C. Bazan, S. Tretiak, S. Mukamel, *J. Am. Chem. Soc.* **2000**, *122*, 1289–1297.
108 J. W. Hong, B. S. Gaylord, G. C. Bazan, *J. Am. Chem. Soc.* **2002**, *124*, 11868–11869.
109 A. Hradsky, B. Bildstein, N. Schuler, H. Schottenberger, P. Jaitner, K.-H. Ongania, K. Wurst, J.-P. Launay, *Organometallics* **1997**, *16*, 392–402.
110 S. P. Dudek, H. D. Sikes, C. E. D. Chidsey, *J. Am. Chem. Soc.* **2001**, *123*, 8033–8038.
111 N. Ono, H. Tomita, K. Maruyama, *J. Chem. Soc., Perkin Trans. 1* **1992**, 2453–2456.
112 D. A. Bussian, M. A. Summers, B. Liu, G. C. Bazan, S. K. Buratto, *Chem. Phys. Lett.* **2004**, *388*, 181–185.
113 B. S. Gaylord, S. Wang, A. J. Heeger, G. C. Bazan, *J. Am. Chem. Soc.* **2001**, *123*, 6417–6418.
114 M. Stork, B. S. Gaylord, A. J. Heeger, G. C. Bazan, *Adv. Mater.* **2002**, *14*, 361–366.
115 B. Liu, B. S. Gaylord, S. Wang, G. C. Bazan, *J. Am. Chem. Soc.* **2003**, *125*, 6705–6714.
116 F. Giacalone, J. L. Segura, N. Martín, D. M. Guldi, *J. Am. Chem. Soc.* **2004**, *126*, 5340–5341.
117 A. Pogantsch, G. Heimel, E. Zojer, *J. Chem. Phys.* **2002**, *117*, 5921–5928.
118 D. Oelkrug, A. Tompert, H.-J. Egelhaaf, M. Hanack, E. Steinhuber, M. Hohloch, H. Meier, U. Stalmach, *Synth. Met.* **1996**, *83*, 231–237.
119 M. A. Summers, M. R. Robinson, G. C. Bazan, S. K. Buratto, *Synth. Met.* **2003**, *137*, 957–958.
120 P. F. Van Hutten, V. V. Krasnikov, G. Hadziioannou, *Acc. Chem. Res.* **1999**, *32*, 257–265.
121 C. Bubeck, in Ref. [1j], pp. 449–478.
122 A. Mathy, K. Ueberhofen, R. Schenk, H. Gregorius, R. Garay, K. Müllen, C. Bubeck, *Phys. Rev. B* **1996**, *53*, 4367–4376.
123 W. Werncke, M. Pfeiffer, T. Johr, A. Lau, H.-J. Jüpner, *Chem. Phys.* **1995**, *199*, 65–72.
124 G. P. Agrawal, C. Cojan, C.

Flytzanis, *Phys. Rev. B* **1978**, *17*, 776–789.
125 See for example Section 11.4.
126 C. Bubeck, H. Meier, in preparation.
127 H. Meier, U. Stalmach, M. Fetten, P. Seus, M. Lehmann, C. Schnorpfeil, *J. Inf. Recording* **1998**, *24*, 47–60.
128 M. D. Joswick, I. H. Campbell, N. N. Barashkov, J. P. Ferraris, *J. Appl. Phys.* **1996**, *80*, 2883–2890.
129 R. E. Gill, A. Hilberer, P. F. van Hutten, G. Berentschot, M. P. L. Werts, A. Meetsma, J.-C. Wittmann, G. Hadziioannou, *Synth. Met.* **1997**, *84*, 637–638.
130 H. J. Brouwer, V. V. Krasnikov, T. A. Pham, R. E. Gill, P. F. van Hutten, G. Hadziioannou, *Chem. Phys.* **1998**, *227*, 65–74.
131 P. Brendel, A. Grupp, M. Mehring, R. Schenk, K. Müllen, W. Huber, *Synth. Met.* **1991**, *45*, 49–57.
132 R. Schenk, H. Gregorius, K. Müllen, *Adv. Mater.* **1991**, *3*, 492–493.
133 M. Wohlgenannt, K. Tandon, S. Mazumdar, S. Ramasesha, Z. V. Vardeny, *Nature* **2001**, *409*, 494–497 and references therein.
134 J. S. Wilson, A. S. Dhoot, A. J. A. B. Seeley, M. S. Khan, A. Köhler, R. H. Friend, *Nature* **2001**, *413*, 828–831.
135 N. Dam, R. D. Scurlock, B. Wang, L. Ma, M. Sundahl, P. R. Ogilby, *Chem. Mater.* **1999**, *11*, 1302–1305.
136 H. Meier, D. Ickenroth, U. Stalmach, K. Koynov, A. Bahtiar, C. Bubeck, *Eur. J. Org. Chem.* **2001**, 4431–4443.
137 D. Ickenroth, S. Weissmann, N. Rumpf, H. Meier, *Eur. J. Org. Chem.* **2002**, 2808–2814.
138 H. Meier, unpublished results.
139 T. Kawase, N. Ueda, K. Tanaka, Y. Seirai, M. Oda, *Tetrahedron Lett.* **2001**, *42*, 5509–5511.
140 T. Kawase, H. R. Darabi, M. Oda, *Angew. Chem.* **1996**, *108*, 2803–2805; *Angew. Chem. Int. Ed.* **1996**, *35*, 2664–2666.
141 G. Drehfahl, G. Plötner, *Chem. Ber.* **1958**, *91*, 1280–1285.
142 S. Misumi, *Bull. Chem. Soc. Jpn.* **1961**, *34*, 1827–1832.
143 J. S. Schumm, D. L. Pearson, J. M. Tour, *Angew. Chem.* **1994**, *106*, 1445–1448; *Angew. Chem. Int. Ed. Engl.* **1994**, *33*, 1360–1363.
144 L. R. Jones II, J. S. Schumm, J. M. Tour, *J. Org. Chem.* **1997**, *62*, 1388–1410.
145 U. Ziener, A. Godt, *J. Org. Chem.* **1997**, *62*, 6137–6143.
146 H. Kukula, S. Veit, A. Godt, *Eur. J. Org. Chem.* **1999**, 277–286.
147 V. Hensel, A. Godt, R. Popovitz-Biro, H. Cohen, T. R. Jensen, K. Kjaer, I. Weissbuch, E. Lifshitz, M. Lahav, *Chem. Eur. J.* **2002**, *8*, 1413–1423.
148 K. Weiss, A. Michel, E.-M. Auth, U. H. F. Bunz, T. Mangel, K. Müllen, *Angew. Chem.* **1997**, *109*, 522–525; *Angew. Chem. Int. Ed. Engl.* **1997**, *36*, 506–509.
149 L. Kloppenburg, D. Song, U. H. F. Bunz, *J. Am. Chem. Soc.* **1998**, *120*, 7973–7974.
150 F. Koch, W. Heitz, *Macromol. Chem. Phys.* **1997**, *198*, 1531–1544.
151 H.-F. Chow, C.-W. Wan, K.-H. Low, Y.-Y. Yeung, *J. Org. Chem.* **2001**, *66*, 1910–1913.
152 S. Anderson, *Chem. Eur. J.* **2001**, *7*, 4706–4714.
153 J.-J. Hwang, J. M. Tour, *Tetrahedron* **2002**, *58*, 10387–10405.
154 J.-F. Nierengarten, T. Gu, T. Aernouts, W. Geens, J. Poortmans, G. Hadziioannou, D. Tsamouras, *Appl. Phys. A* **2004**, *79*, 47–49.
155 T. Gu, D. Tsamouras, C. Melzer, V. Krasnikov, J.-P. Gisselbrecht, M. Gross, G. Hadziioannou, J.-F. Nierengarten, *Chem. Phys. Chem.* **2002**, *3*, 124–127.
156 E. Arias, I. Moggio, D. Navarro, J. Romero, L. Larios, J. Le Moigne, D. Guillon, T. Maillou, V. Gonzalez, B. Geffroy, *Rev. Soc. Quim. Mexico* **2002**, *46*, 23–31.
157 E. Arias, T. Maillou, I. Moggio, D. Guillon, J. Le Moigne, B. Geffroy, *Synth. Met.* **2002**, *127*, 229–231.
158 I. Moggio, J. Le Moigne, E. Arias-Marin, D. Issautier, A. Thierry, D. Comoretto, G. Dellepiane, C. Cuniberti, *Macromolecules* **2001**, *34*, 7091–7099.

159 M. I. Sluch, A. Godt, U. H. F. Bunz, M. A. Berg, *J. Am. Chem. Soc.* **2001**, *123*, 6447–6448.
160 E. Arias-Marin, J. C. Arnault, D. Guillon, T. Maillou, J. Le Moigne, B. Geffroy, J. M. Nunzi, *Langmuir* **2000**, *16*, 4309–4318.
161 S. Huang, J. M. Tour, *J. Org. Chem.* **1999**, *64*, 8898–8906.
162 S. Huang, J. M. Tour, *J. Am. Chem. Soc.* **1999**, *121*, 4908–4909.
163 J. Polin, M. Buchmeiser, H. Nock, H. Schottenberger, *Mol. Cryst. Liq. Cryst. A* **1997**, *293*, 287–307.
164 J. Xu, C.-Z. Zhou, L. H. Yang, N. T. S. Chung, Z.-K. Chen, *Langmuir* **2004**, *20*, 950–956.
165 C.-Z. Zhou, T. Liu, J.-M. Xu, Z.-K. Chen, *Macromolecules* **2003**, *36*, 1457–1464.
166 See Ref. [153] and references therein.
167 A. Javier, C. S. Yun, J. Sorena, G. F. Strouse, *J. Phys. Chem. B* **2003**, *107*, 435–442.
168 F.-R. F. Fan, J. Yang, L. Cai, D. W. Price Jr., S. M. Dirk, D. V. Kosynkin, Y. Yao, A. M. Rawlett, J. M. Tour, A. J. Bard, *J. Am. Chem. Soc.* **2002**, *124*, 5550–5560.
169 J. M. Tour, A. M. Rawlett, M. Kozaki, Y. Yao, R. C. Jagessar, S. M. Dirk, D. W. Price, M. A. Reed, C.-W. Zhou, J. Chen, W. Wang, I. Campbell, *Chem. Eur. J.* **2001**, *7*, 5118–5134.
170 J. P. Novak, C. Nickerson, S. Franzen, D. L. Feldheim, *Anal. Chem.* **2001**, *73*, 5758–5761.
171 T. D. Dunbar, M. T. Cygan, L. A. Bumm, G. S. McCarty, T. P. Burgin, W. A. Reinerth, L. Jones II, J. J. Jackiw, J. M. Tour, P. S. Weiss, D. L. Allara, *J. Phys. Chem. B* **2000**, *104*, 4880–4893.
172 J. P. Novak, D. L. Feldheim, *J. Am. Chem. Soc.* **2000**, *122*, 3979–3980.
173 L. C. Brousseau III, J. P. Novak, S. M. Marinakos, D. L. Feldheim, *Adv. Mater.* **1999**, *11*, 447–449.
174 M. T. Cygan, T. D. Dunbar, J. J. Arnold, L. A. Bumm, N. F. Shedlock, T. P. Burgin, L. Jones II, D. L. Allara, J. M. Tour, P. S. Weiss, *J. Am. Chem. Soc.* **1998**, *120*, 2721–2732.
175 C. J. Yu, Y. Chong, J. F. Kayyem, M. Gozin, *J. Org. Chem.* **1999**, *64*, 2070–2079.
176 S. Creager, C. J. Yu, C. Bamdad, S. O'Connor, T. MacLean, E. Lam, Y. Chong, G. T. Olsen, J. Luo, M. Gozin, J. F. Kayyem, *J. Am. Chem. Soc.* **1999**, *121*, 1059–1064.
177 See Ref. [144] and references therein.
178 J. M. Tour, L. Jones II, D. L. Pearson, J. J. S. Lamba, T. P. Burgin, G. M. Whitesides, D. L. Allara, A. N. Parikh, S. Atre, *J. Am. Chem. Soc.* **1995**, *117*, 9529–9534.
179 R. P. Hsung, C. E. D. Chidsey, L. R. Sita, *Organometallics* **1995**, *14*, 4808–4815.
180 S. Anderson, P. N. Taylor, G. L. B. Verschoor, *Chem. Eur. J.* **2004**, *10*, 518–527.
181 T. Maillou, J. Le Moigne, V. Dumarcher, L. Rocha, B. Geffroy, J.-M. Nunzi, *Adv. Mater.* **2002**, *14*, 1297–1301.
182 M. Sirota, L. Fradkin, R. Buller, V. Henzel, M. Lahav, E. Lifshitz, *Chem. Phys. Chem.* **2002**, *3*, 343–349.
183 I. Aujard, J.-P. Baltaze, J.-B. Baudin, E. Cogné, F. Ferrage, L. Jullien, E. Perez, V. Prévost, L. M. Qian, O. Ruel, *J. Am. Chem. Soc.* **2001**, *123*, 8177–8188.
184 This includes free carboxylic acids and their salts.
185 J. G. Gonzalo, J. Esquivias, A. Lafuente, C. Díaz, *J. Org. Chem.* **2003**, *68*, 8120–8128.
186 J. G. Rodríguez, J. L. Tejedor, J. Esquivias, C. Díaz, *Tetrahedron Lett.* **2003**, *44*, 6375–6378.
187 J. G. Rodríguez, J. Esquivias, *Tetrahedron Lett.* **2003**, *44*, 4831–4834.
188 T. Maillou, J. Le Moigne, B. Geffroy, A. Lorin, A. Rosilio, V. Dumarcher, L. Rocha, C. Denis, C. Fiorini, J.-M. Nunzi, *Synth. Met.* **2001**, *124*, 87–89.
189 B. Kenda, F. Diederich, *Angew. Chem.* **1998**, *109*, 3357–3361; *Angew. Chem. Int. Ed.* **1998**, *37*, 3154–3158.
190 P. Wautelet, J. Le Moigne, V. Videva, P. Turek, *J. Org. Chem.* **2003**, *68*, 8025–8036.

191 P. Wautelet, L. Catala, A. Bieber, P. Turek, J.-J. André, *Polyhedron* **2001**, *20*, 1571–1576.
192 A. Khatyr, R. Ziessel, *J. Org. Chem.* **2000**, *65*, 3126–3134.
193 A. Khatyr, R. Ziessel, *Tetrahedron Lett.* **1999**, *40*, 5515–5518.
194 S. Höger, A.-D. Meckenstock, S. Müller, *Chem. Eur. J.* **1998**, *4*, 2423–2434.
195 J. Le Moigne, J. L. Gallani, P. Wautelet, M. Moroni, L. Oswald, C. Cruz, Y. Galerne, J. C. Arnault, R. Duran, M. Garrett, *Langmuir* **1998**, *14*, 7484–7492.
196 P. Wautelet, A. Bieber, P. Turek, J. Le Moigne, J.-J. Andre, *Mol. Cryst. Liq. Cryst. A* **1997**, *305*, 55–67.
197 L. Jones II, D. L. Pearson, J. S. Schumm, J. M. Tour, *Pure Appl. Chem.* **1996**, *68*, 145–148.
198 S. Rucareanu, O. Mongin, A. Schuwey, N. Hoyler, A. Gossauer, W. Amrein, H.-U. Hediger, *J. Org. Chem.* **2001**, *66*, 4973–4988.
199 Y. Liu, S. Jiang, K. S. Schanze, *Chem. Commun.* **2003**, 650–651.
200 C. Hortholary, C. Coudret, *J. Org. Chem.* **2003**, *68*, 2167–2174.
201 H.-Y. Chao, W. Lu, Y. Li, M. C. W. Chan, C.-M. Che, K.-K. Cheung, N. Zhu, *J. Am. Chem. Soc.* **2002**, *124*, 14696–14706.
202 A. Harriman, A. Khatyr, R. Ziessel, A. C. Benniston, *Angew. Chem.* **2000**, *112*, 4457–4460; *Angew. Chem. Int. Ed.* **2000**, *39*, 4287–4290.
203 P. Brodard, S. Matzinger, F. Vauthey, O. Mongin, C. Papamicaeel, A. Gossauer, *J. Phys. Chem. A* **1999**, *103*, 5858–5870.
204 O. Mongin, C. Papamicaël, N. Hoyler, A. Gossauer, *J. Org. Chem.* **1998**, *63*, 5568–5580.
205 O. Lavastre, L. Ollivier, P. H. Dixneuf, S. Sibandhit, *Tetrahedron* **1996**, *52*, 5495–5504.
206 P. Wautelet, M. Moroni, L. Oswald, J. Le Moigne, A. Pham, J.-Y. Bigot, S. Luzzati, *Macromolecules* **1996**, *29*, 446–455.
207 J. P. Novak, L. C. Brousseau III, F. W. Vance, R. C. Johnson, B. I. Lemon, J. T. Hupp, D. L. Feldheim, *J. Am. Chem. Soc.* **2000**, *122*, 12029–12030.
208 See also A. Montali, P. Smith, C. Weder, *Synth. Met.* **1998**, *97*, 123–126.
209 E. Arias-Marin, J. C. Arnault, T. Maillou, D. Guillon, J. Le Moigne, B. Geffroy, A. Lorin, J. M. Nunzi, A. Rosilio, *Mater. Res. Soc. Symp. Proc.* **2000**, BB 1.8/1–BB 1.8/7; *Chem. Abstr.* **2000**, *134*, 107665.
210 H. Meier, J. Gerold, H. Kolshorn, W. Baumann, M. Bletz, *Angew. Chem.* **2002**, *114*, 302–306; *Angew. Chem. Int. Ed.* **2002**, *41*, 292–295.
211 H. Meier, J. Gerold, H. Kolshorn, B. Mühling, *Chem. Eur. J.* **2004**, *10*, 360–370.
212 H. Meier, B. Mühling, H. Kolshorn, *Eur. J. Org. Chem.* **2004**, 1033–1042.
213 H. Meier, J. Gerold, D. Jacob, *Tetrahedron Lett.* **2003**, *44*, 1915–1918.
214 V. Francke, T. Mangel, K. Müllen, *Macromolecules* **1998**, *31*, 2447–2453.
215 K. Koynov, A. Bahtiar, C. Bubeck, B. Mühling, H. Meier, *J. Phys. Chem. B* **2005**, *109*, 10184–10188.
216 See also A. Abbotto, L. Beverina, S. Bradamante, A. Facchetti, C. Klein, G. A. Pagani, M. Redi-Abshiro, R. Wortmann, *Chem. Eur. J.* **2003**, *9*, 1991–2007.

12
Synthesis and Chemistry of Polycyclic Aromatic Hydrocarbons with Curved Surfaces: Buckybowls

Andrzej Sygula and Peter W. Rabideau

12.1
Introduction

The discovery of a novel form of elemental carbon – the fullerenes – in the mid-1980s initiated unprecedented research activity in the field of physics, chemistry, and material science related to these carbon cages [1]. While expectations for the immediate widespread applications of fullerenes in technology have yet to be fully realized, progress in the basic understanding of carbon-rich systems, insight into the nature of the aromaticity of nonplanar systems, and the development of novel synthetic protocols are undisputed results of this research.

The existence of fullerenes, consisting of sp^2-hybridized carbon networks of six- and five-membered rings, renewed an interest in the chemistry and physics of carbon-rich polycyclic aromatic hydrocarbons (PAHs) in general, and especially in the subclass of PAHs that have become known as fullerene fragments or "buckybowls." The latter curved-surface PAHs have carbon frameworks that can be identified on the buckminsterfullerene (C_{60}) surface. The simplest example is corannulene and it has been known for decades; all of the other known buckybowls have been prepared since the early 1990s.

The synthesis and study of buckybowls has been of interest for a number of reasons:

1. Buckybowls, in contrast to fullerenes, possess easily accessible convex and concave faces that allow for studies of *exo* vs. *endo* preferences of reactivity including metal complex formation.
2. Buckybowls of various sizes offer a spectrum of curved carbon surfaces with various degrees of pyramidalization, again offering a possibility for studies of curvature vs. reactivity relationships.
3. As fullerene fragments, buckybowls are potential substrates for the synthesis of fullerenes and nanotubes.
4. Buckybowls offer unique molecular architectures that may be used as scaffolds for nanometric size molecular systems including bowls for host/guest chemistry.

Carbon-Rich Compounds. Edited by Michael M. Haley and Rik R. Tykwinski
Copyright © 2006 WILEY-VCH Verlag GmbH & Co. KGaA, Weinheim
ISBN: 3-527-31224-2

5. If the expected industrial applications of fullerenes take place, the decomposition of "waste" fullerenes may lead to fragmentation, with buckybowls entering the environment. In such case, knowledge concerning their chemical and physiological properties becomes imperative.

Thus for these and other reasons, a significant amount of research began in the 1990s when buckybowls became more accessible, and a number of reviews have since appeared [2]. In this chapter we will concentrate mostly on recent progress not covered by previous review articles.

12.2 Synthesis of Buckybowls

12.2.1 Corannulene: A Synthetic Travel from "Wet" Chemistry ...

In 1966 Barth and Lawton reported the preparation of a novel [5]annulene (dibenzo[*ghi*,*mno*]fluoranthene) $C_{20}H_{10}$ (**1**) to which they assigned the trivial name "corannulene" [3]. This remarkable synthesis consisted of 17 steps starting from acenaphthene and building up the remaining rings one after another. Some of the crucial intermediates of the synthesis are shown in Scheme 12.1. Not surprisingly, the overall yield was quite low and only a very limited amount of **1** was available for further studies. Despite the scarcity, however, some important results were obtained: the ^{1}H NMR spectrum was recorded, exhibiting a single line signal indicat-

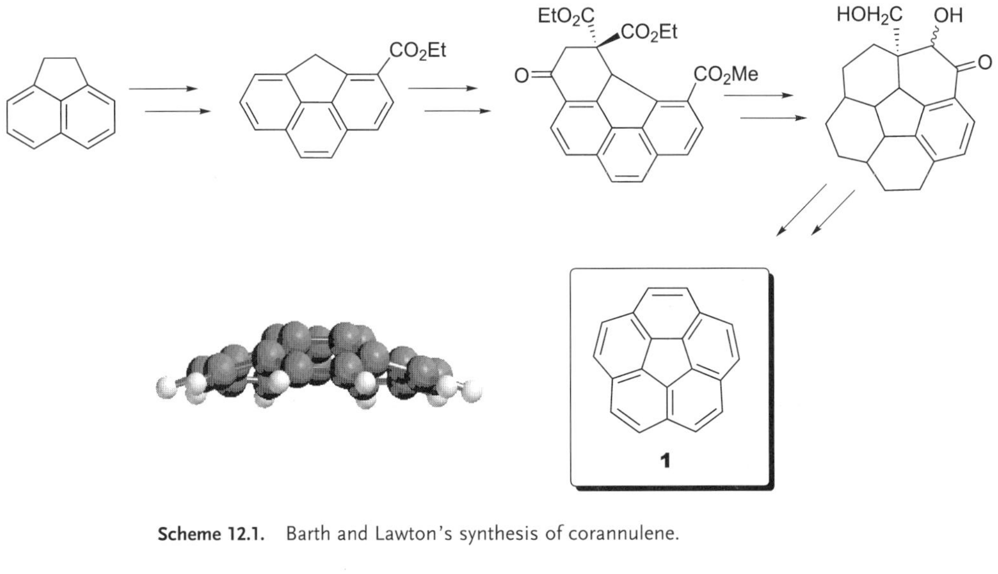

Scheme 12.1. Barth and Lawton's synthesis of corannulene.

ing the magnetic isochrony of all ten hydrogen atoms, and electrochemical studies showed a facile reversible reduction of **1** to its radical anion and to a dianion [4]. Most importantly, X-ray crystallographic structure determination was performed showing **1** to be a molecular bowl with a depth of ca. 0.8 Å as defined by an interplanar distance between the five "hub" carbon atoms and the ten "rim" CH carbon atoms [5]. The curvature of the system represents an obvious manifestation of the angle strain introduced by the presence of the central five-membered ring. The analogous [6]annulene (coronene) is planar while [7]annulene is saddle-shaped in the solid state [6].

The elegance and high symmetry of corannulene would likely have attracted a number of investigators were it not for the fact of limited accessibility due to the painfully laborious synthetic route. Several synthetic groups began work on alternative synthetic routes recognizing that fluoranthene derivatives with four of the corannulene rings already in place might serve as potential starting materials; however, these attempts were unsuccessful as no doubt were others that went unpublished [7].

The failure of classical ring closure methods – often effective for the formation of planar PAHs – may be attributed to the strain present in the bowl-shaped product (corannulene) as compared to the (almost) strainless starting materials (fluoranthenes). Hence the relatively high energy transition states leading to corannulene formation are avoided in favor of lower energy intermolecular processes including polymerizations.

12.2.2
... to Flash Vacuum Pyrolysis ...

The discovery of the fullerenes produced a resurgence of interest in the synthesis of corannulene. A major breakthrough occurred in 1991 when the Scott group reported the successful synthesis of corannulene wherein the double ring-forming step was achieved by flash vacuum pyrolysis (FVP) of diethynylfluoranthene **2**, which was synthesized in six steps from acenaphthenquinone (Scheme 12.2) [8]. The success of FVP was attributed to two factors: (i) the high temperature applied in pyrolysis (ca. 1000–1100 °C) assures that there is enough thermal energy delivered to the molecules to facilitate *intra*molecular ring closures involving high energy barriers, and (ii) the reaction takes place in the gas phase where competing *inter*molecular reactions leading to oligo- and/or polymerization are less likely to occur. The competing polymerization of diyne **2** before it sublimes into the hot zone was subsequently overcome by using the "masked acetylene" dichlorovinylfluoranthene **3** [9]. Compound **3**, synthesized in two steps from acenaphthenequinone, is more stable than **2** at elevated temperatures, and it easily sublimes into the hot zone where it is quickly converted into **2**, leading to **1** in yields for the pyrolytic step of ca. 35%.

A few other FVP-based procedures leading to corannulene from various precursors were discovered and published by several research groups (Scheme 12.3) [10] but the Scott procedure provides the highest yield in the final pyrolytic step.

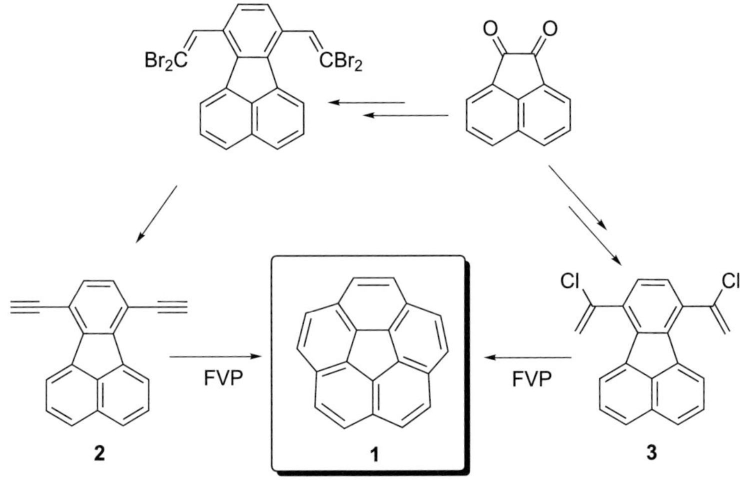

Scheme 12.2. Scott's FVP syntheses of corannulene.

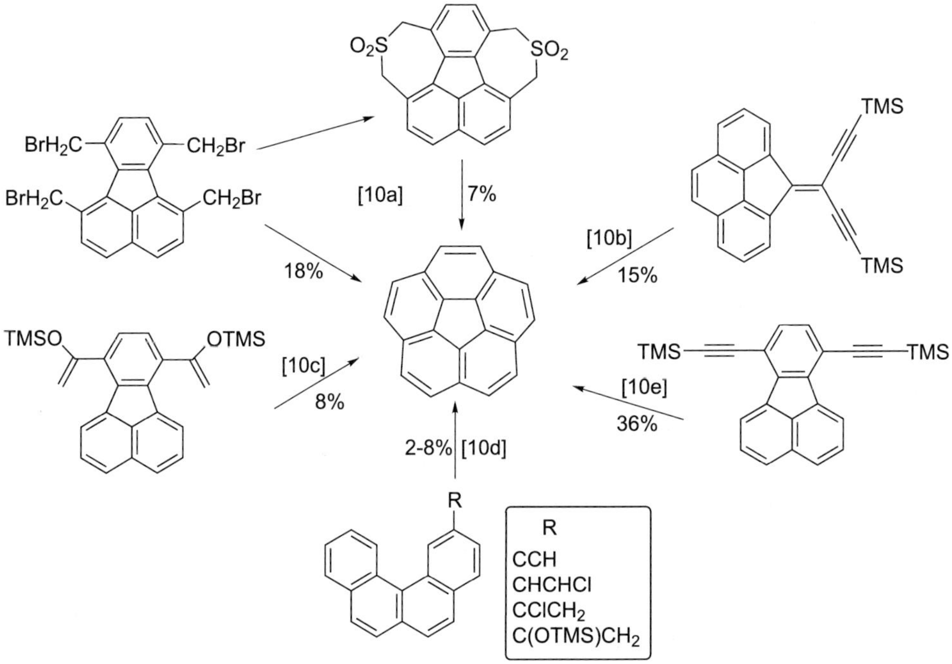

Scheme 12.3. Alternative FVP approaches to corannulene.

12.2.3
... and Back

Despite the indisputable success of the FVP procedure for the production of corannulene in good yields, it became clear that "wet" chemistry routes would still be required for the efficient synthesis of other buckybowls. The limitations of FVP methodology are as follows:

1. The yields for the synthesis of larger systems drop dramatically with size (*vide infra*); larger precursors are simply less volatile and the higher temperatures needed for sublimation cause significant decomposition of the precursor before it reaches the hot reaction zone.
2. The high temperatures applied for FVP preclude the presence of functional groups, limiting the possibilities for substituted products.
3. Technical difficulties arise with the scale-up to practical gram-scale reactions since FVP is a gas-phase process.
4. The synthesis of more strained systems by FVP requires temperatures in the hot zone well exceeding 1100 °C. Since the FVP apparatus is usually made of quartz, these high temperatures cause softening and degradation of this relatively expensive material.

For these reasons (*as well as the challenge to organic chemists*) the search for "classical" routes to corannulenes and related systems did not stop with the introduction of FVP methodology. The fluoranthene derivatives were still considered as attractive precursors for corannulenes due to the presence of three six-membered rings and a five-membered ring in the "right" orientation. In 1996 Siegel reported a successful solution synthesis of dimethylcorannulene 4 from tetrabromide 5 (Scheme 12.4) [11]. The double ring closure leading to tetrahydrodimethylcorannulene 6 was achieved by low-valent titanium coupling in good yield (55%). Notwithstanding the fact that subsequent dehydration of 6 to 4 by DDQ lowered the total yield of this two-step process to 18%, this route to buckybowls appeared to be an attractive alternative to FVP.

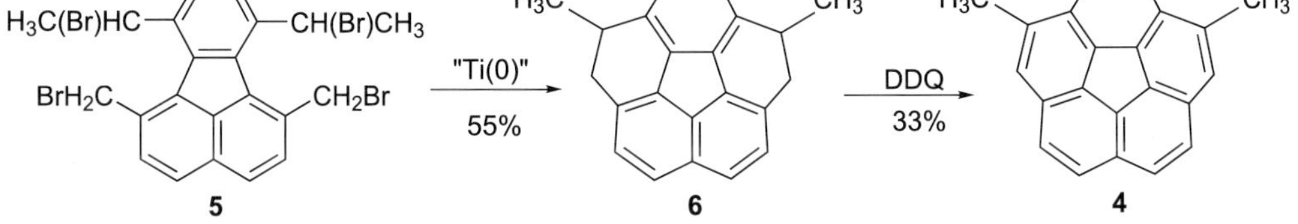

Scheme 12.4. Siegel's nonpyrolytic synthesis of dimethylcorannulene 4.

This reaction seems curious since previous attempts to produce the corannulene moiety based on fluoranthene chemistry had failed. This method is a variation of McMurry coupling originally devised for the coupling of two carbonyl groups to form a carbon–carbon double bond and extended by Olah to the coupling of bromomethyl groups [12]. This methodology was also found effective for intramolecular carbonyl coupling leading to the formation of cycloalkenes, including highly strained cyclopropenes and cyclobutenes, with good yields [12a]. Success in the formation of strained systems was attributed to high energy organotitanium intermediates and so the surprising formation of the strained corannulene framework by the intramolecular coupling of bromomethyl groups may also result from high energy organometallic intermediates.

We tried to apply Siegel's approach to the nonpyrolytic synthesis of semibuckminsterfullerene **7** by a four-fold ring closure through the same low-valent titanium intramolecular coupling of bromomethyl groups in the octabromo precursor **8** [13] (Scheme 12.5). However, despite numerous attempts employing various reductive systems in both THF and DME, no measurable amounts of the expected octahydro derivative of **7** were detected. Out of desperation, we explored the further radical bromination of **8** and produced dodecabromo derivative **9** with the four "outer" and the four "inner" methyl groups of the starting hydrocarbon converted to dibromomethyl and bromomethyl groups, respectively. X-ray structure determination showed the aromatic framework of **9** to be strongly twisted as an apparent result of steric overcrowding of the dangling groups. Presumably, as the result of a more strained starting material, low valent vanadium or titanium coupling was successful in this case leading to the formation of semibuckminsterfullerene **7**, albeit in modest yield (*ca.* 20%) [13].

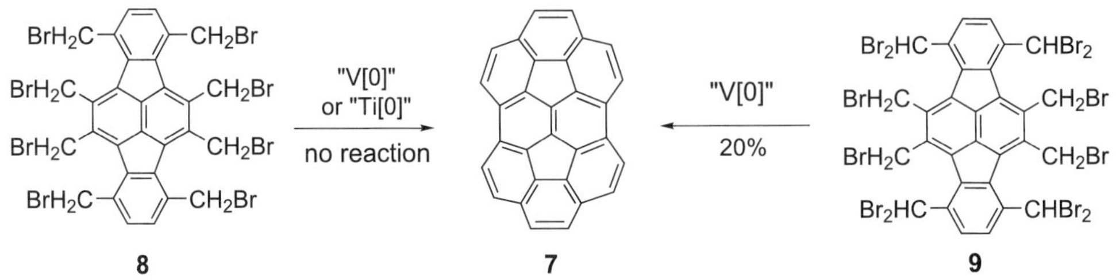

Scheme 12.5. Nonpyrolytic approaches to semibuckminsterfullerene **7**.

The success of this reaction reveals two important features: (i) it appears that dibromomethyl groups are more likely to couple with bromomethyl groups to form strained molecular frameworks as compared with the coupling of two bromomethyl units, and (ii) the additional Br atoms are lost after the coupling to produce aromatized products. Thus the buckybowl is produced in a single step rather than formed as a partially hydrogenated intermediate that can be difficult to aromatize [3, 11].

This result immediately raised the question as to whether the employment of dibromomethyl rather than bromomethyl coupling would improve the yields of a nonpyrolytic route to corannulene. To test this possibility we converted tetramethylfluoranthene **10** to its octabromo derivative **11** by a standard radical bromination procedure with 8.8 molar equivalents of NBS in refluxing benzene with a catalytic amount of dibenzoyl peroxide under sun lamp irradiation (Scheme 12.6). The low-valent vanadium coupling (slow syringe pump addition of a solution of **11** in DME to the refluxing suspension of $VCl_3/LiAlH_4$ in the same solvent) led to formation of corannulene in an impressive yield of 70–75% [14a]. The only identifiable sideproduct of the reaction was ca. 5% bromocorannulene that could be separated by column chromatography or converted to corannulene by a treatment of the mixture with butyllithium in THF at $-78\,°C$ followed by quenching with water [14]. A similar procedure employing low-valent titanium coupling was independently discovered and published by Siegel's research group [14b,c].

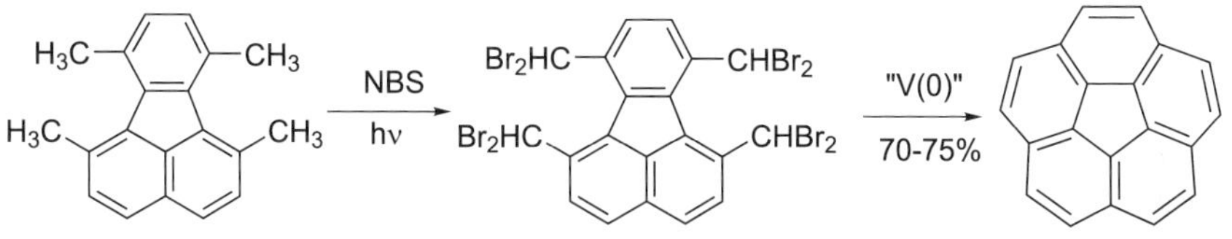

Scheme 12.6. Synthesis of corannulene through a low-valent vanadium coupling of dibromomethyl groups.

12.2.4
Practical Synthesis of Corannulene by Intramolecular Carbenoid Coupling of Dibromomethyl Groups

Since the above low-valent Ti/V coupling is a variation of McMurry coupling, we became interested in the possibility of corannulene core formation by the original McMurry coupling starting from tetraaldehyde **12**. An obvious approach for the synthesis of this unknown fluoranthene derivative was the hydrolysis of the dibromomethyl groups in **11** which we had already prepared. However, refluxing aqueous acetone solutions of **11** with Na_2CO_3 failed to produce any products exhibiting characteristic CHO ^1H NMR absorptions, but rather a high melting material consisting of a mixture of products, as evidenced by the complicated ^1H NMR pattern of aromatic protons within the range 7.1–8.1 ppm [15]. ^{13}C NMR and GC/MS analysis showed that instead of conversion of **11** to **12**, the attempted hydrolysis quite unexpectedly led directly to the formation of the corannulene core resulting in a mixture of brominated corannulenes $C_{20}H_{10-n}Br_n$ ($n = 1, 2, 3$ and 4) (Scheme 12.7)!

Debromination of the reaction mixture with butyllithium in THF at $-78\,°C$ followed by quenching with water produced corannulene in 50–55% yield for the two steps combined [15].

After some experimentation with the reaction conditions we found that replacement of Na_2CO_3 with NaOH accelerates the reaction significantly, but otherwise does not change the outcome or yield. On the other hand replacement of aqueous acetone by aqueous dioxane changes the reaction considerably since brief (15 min)

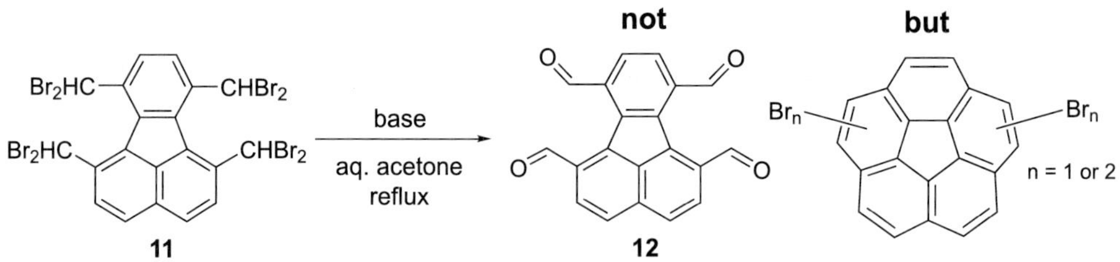

Scheme 12.7. Attempted hydrolysis of **11** leading to the formation of corannulene core.

reflux of octabromide **11** in aqueous dioxane with NaOH cleanly produced 1,2,5,6-tetrabromocorannulene **13** in an impressive 83% isolated yield [15, 16] (Scheme 12.8). Furthermore, **13** is rather insoluble in common organic solvents so it precipitates out from the aqueous dioxane and can be simply filtered off, washed and dried to obtain reasonably pure product.

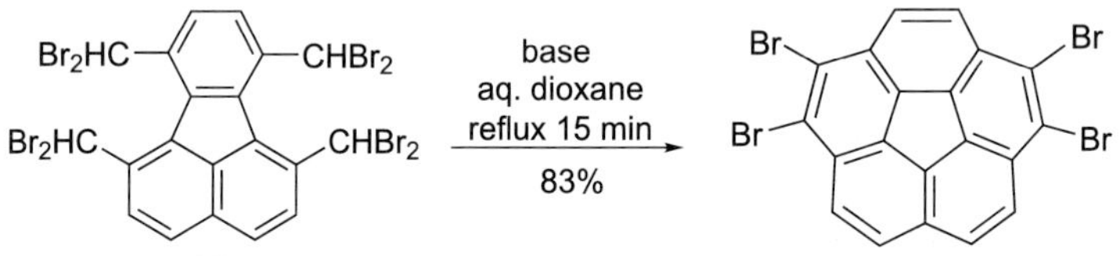

Scheme 12.8. A practical synthesis of tetrabromocorannulene **13**.

This is a simple, low cost procedure that, in contrast to the low-valent metal methods, does not require dry/degassed solvents, inert atmosphere, or time consuming, high dilution applications. The resulting tetrabromocorannulene can be easily converted to a number of derivatives by standard coupling procedures or it can be conveniently debrominated to the parent corannulene in 91% yield by simply refluxing with KI and Zn powder in EtOH [15, 16]. Thus, in our opinion, this is by far the most convenient large-scale synthesis of corannulenes available today – it has never failed in our laboratory even though it has been repeated countless times. Scale-up is not a problem and is only limited by the amount of octabromo precursor **11** available.

The mechanism of this novel transformation of octabromide **11** to tetrabromocorannulene **13** is open to speculation [16a]. The presence of base is required since prolonged reflux of **11** in aqueous dioxane without base, or in presence of dilute HCl, only results in unchanged starting material. Therefore the initial step almost certainly involves abstraction of the acidic benzylic proton by base with the formation of benzyl anion (see below). We proposed two possible mechanisms (Scheme 12.9) leading to the formation of a new carbon–carbon bond between the two proximal dibromomethyl groups: (i) α-elimination of HBr leading to carbene formation,

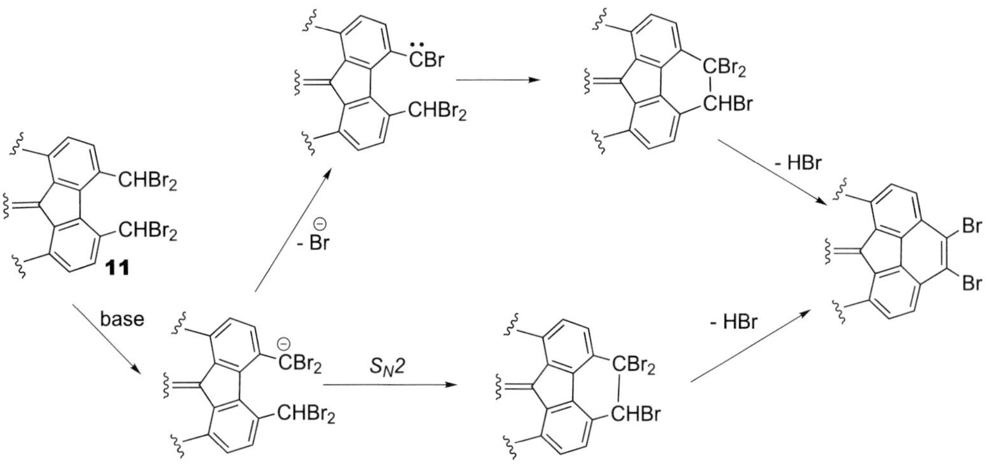

Scheme 12.9. Possible ring forming mechanisms leading to the formation of **13**.

or (ii) nucleophilic substitution. After the C–C single bond-forming step is completed, elimination of HBr will convert the 1,3-cyclohexadiene rings to the aromatized dibromobenzene product. A repetition of the process on the other side of the fluoranthene precursor **11** completes the conversion.

At first glance, it would appear remarkable that such a simple reaction could produce the corannulene framework. However, with a closer look we can understand that this process does indeed fit the pattern of the other successful routes to corannulene outlined above. Simple molecular mechanics calculations offer some insight into the problem. The steric energy calculated at this level for corannulene is ca. 10 kcal mol^{-1} while fluoranthene and its 7,10-disubstituted derivatives (model substrates for the unsuccessful attempts to a double Friedel-Crafts acylation) have MM2 steric energies in the range −8 to 4 kcal mol^{-1}. On the other hand, steric energies of 1,6,7,10-tetrasubstituted derivatives (precursors for the successful intramolecular coupling) are in the range 5.3 to 14.3 kcal mol^{-1} which indicates that they are good precursors to corannulenes due to the release of the steric overcrowding in the final step of the synthesis [16a].

Other methods can also be used for the formation of the corannulene framework from octabromide **11**. Following a suggestion from Professor Ben Plummer we heated **11** in DMF with NaI and obtained a mixture of brominated corannulenes which was converted to corannulene in ca. 40% unoptimized yield of the two steps [16a]. Also, heating **11** in the same solvent with Ni powder followed by debromination produced a similar result with ca. 75% yield of corannulene [16a]. The latter method appeared to be a milder alternative for our "original" NaOH/aq. dioxane coupling procedure. We utilized this method for the efficient synthesis of 1,2-dicarbmethoxycorannulene **14** from the respective hexabromo precursor and obtained a good yield of 60% [17] (Scheme 12.10). It is worth noting that this protocol applied to the coupling of dibromomethyl with bromomethyl groups leads to the

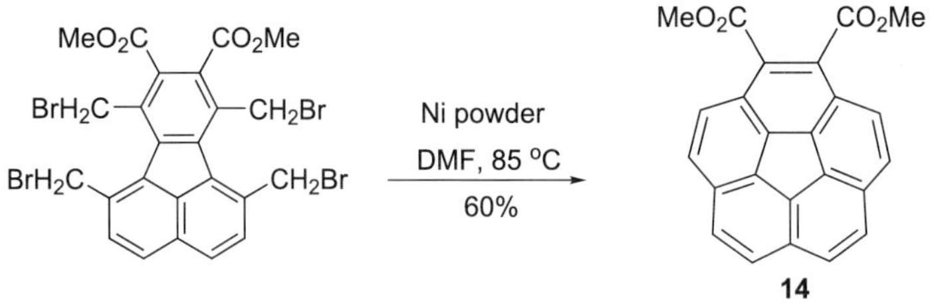

538 *12 Synthesis and Chemistry of Polycyclic Aromatic Hydrocarbons with Curved Surfaces: Buckybowls*

Scheme 12.10. Nickel powder induced formation of the corannulene framework.

formation of the corannulene framework without any remaining bromine atoms. The aq. dioxane/NaOH method gave less satisfactory results in this case for the obvious reason – hydrolysis of the carbomethoxy groups under basic conditions led to the formation of a mixture of dibromo dicarbomethoxy-**1** with diacid and acid ester sideproducts.

12.2.5
Beyond Corannulene

12.2.5.1 On the Surface of C_{60}

Sumanene (**15**), a $C_{21}H_{12}$ hydrocarbon with C_{3v} symmetry was recognized some time ago as a potential synthon for the synthesis of fullerene fragments [18]. Even though sumanene is not formally a member of the buckybowl family since its three carbon atoms are sp^3 hybridized, its significant curvature makes it an interesting system for further studies. Several synthetic approaches to **15** failed, including FVP-based procedures [18]. Only recently a Japanese research team reported a successful four-step nonpyrolytic synthesis of sumanene (Scheme 12.11) in

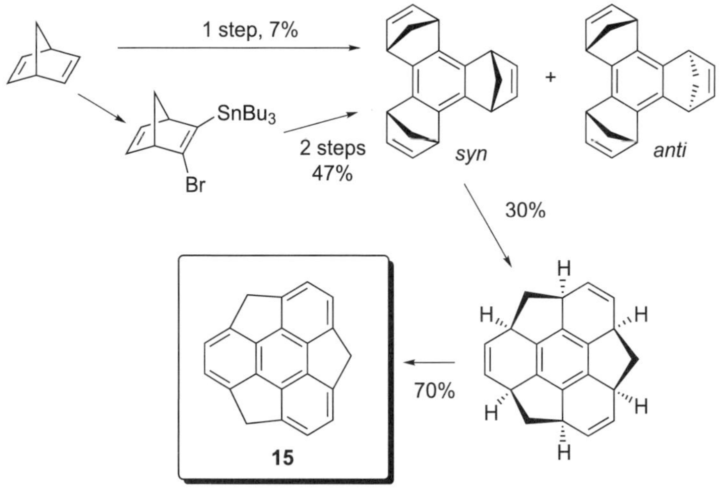

Scheme 12.11. Nonpyrolytic synthesis of sumanene.

which the crucial step was achieved by Ru-catalyzed tandem ring-opening and ring-closing metathesis reaction of *syn*-benzotris(norbornadiene) leading to hexahydrosumanene which was subsequently oxidized by DDQ to give **15** [19]. ^{1}H NMR evidence shows that **15** is indeed bowl-shaped with the bowl-to-bowl inversion barrier significantly higher than corannulene [18].

Much earlier we became interested in the synthesis of larger buckybowls containing the corannulene subunit, especially to investigate whether or not rigid buckybowls could be prepared since **1** undergoes rapid bowl-to-bowl inversion. Using a modification of the FVP synthesis of corannulene, our group prepared cyclopentacorannulene **16** in which the extra –CH=CH– fragment clips two adjacent benzene rings together (Scheme 12.12) [20]. Interestingly, the pyrolytic conditions caused a loss of the two benzylic hydrogen atoms in the precursor leading to the formation of unsaturated **16** rather than the expected dihydro derivative **17**. Corannulene was also formed as a minor product, an apparent result of loss of the ethane bridge. Both hetero- and homogeneous Ni-catalyzed hydrogenation quantitatively converted **16** to the dihydro derivative **17** which exhibited an AA'BB' pattern for the alicyclic protons in 1H NMR spectrum at both ambient and elevated temperatures up to 140 °C indicating slow bowl-to-bowl inversion. This was a remarkable finding considering that corannulene itself inverts more than 200 000 times per second at ambient temperatures. X-ray crystal structure determination of **16** showed that addition of the extra five-membered ring to the rim of corannulene significantly increases the curvature of the system. In fact, the pyramidalization of the carbon atom in **16** in the region of maximum curvature is quite comparable to the pyramidalization of carbon atoms in buckminsterfullerene. In addition, we found an interesting type of long range bowl π–π stacking of the molecules in a concave to convex (endo-exo) orientation (Fig. 12.1) [21], an arrangement which is absent in corannulene itself [5].

Later we also managed to synthesize **16** by a nonpyrolytic route [14a]. The synthesis started with known 1,3,6,8-tetramethylnaphthalene **18**, and subsequent low-valent vanadium coupling of the dibromomethyl groups of dodecabromide **19** gave

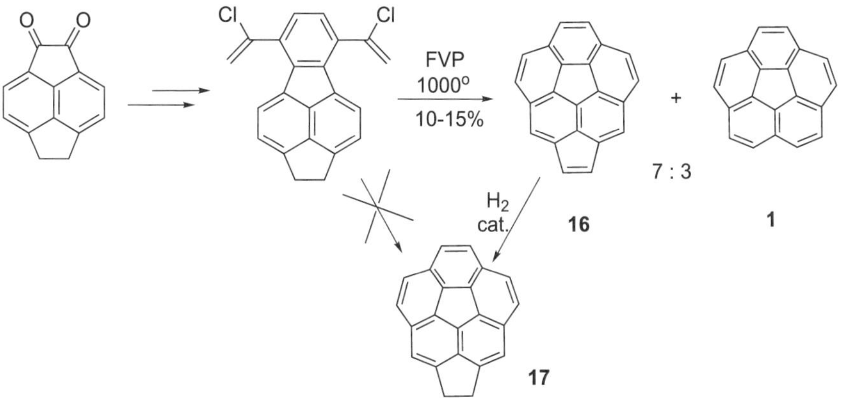

Scheme 12.12. Synthesis of cyclopentacorannulenes **16** and **17**.

540 | *12 Synthesis and Chemistry of Polycyclic Aromatic Hydrocarbons with Curved Surfaces: Buckybowls*

(a)

(b)

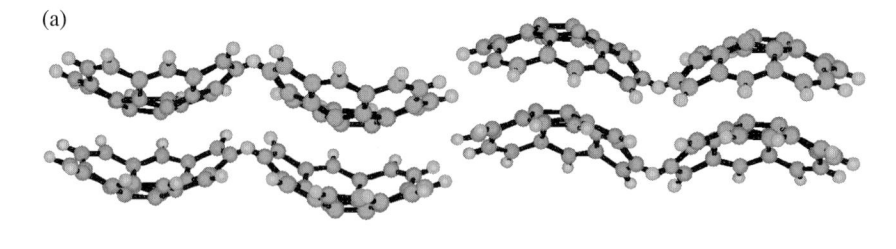

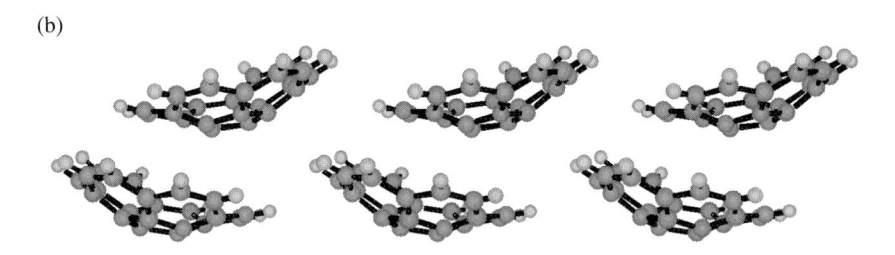

Figure 12.1. Crystal packing in **16** as viewed approximately along the crystallographic *a*-axis (a) and *c*-axis (b).

Scheme 12.13. Synthesis of cyclopentcorannulene **16** by a low valent vanadium coupling.

16 with a moderate yield in the last step of 20–30% (Scheme 12.13). While the yield is much lower than for corannulene, the nonpyrolytic route to **15** is more attractive than FVP because, in addition to the slightly higher yield of the former approach, FVP produces a mixture of **16** and corannulene and separation of the two proved to be difficult [20]. It is not surprising that formation of the more strained systems (i.e., **1** vs. **7** and **16**) leads to a decrease in yields of the ring closure step of the syntheses.

Even larger buckybowls were synthesized by the FVP methodology. Our group succeeded in the synthesis of the first "semibuckminsterfullerenes", i.e., buckybowls representing one half of the buckminsterfullerene C_{60} ($C_{30}H_{12}$; **7** and **20**) by FVP routes (Scheme 12.14) [22]. Both these compounds were later synthesized by others also using FVP for the final step [23]. As discussed earlier we also managed to obtain **7** by the nonpyrolytic low valent vanadium coupling of dodecabro-

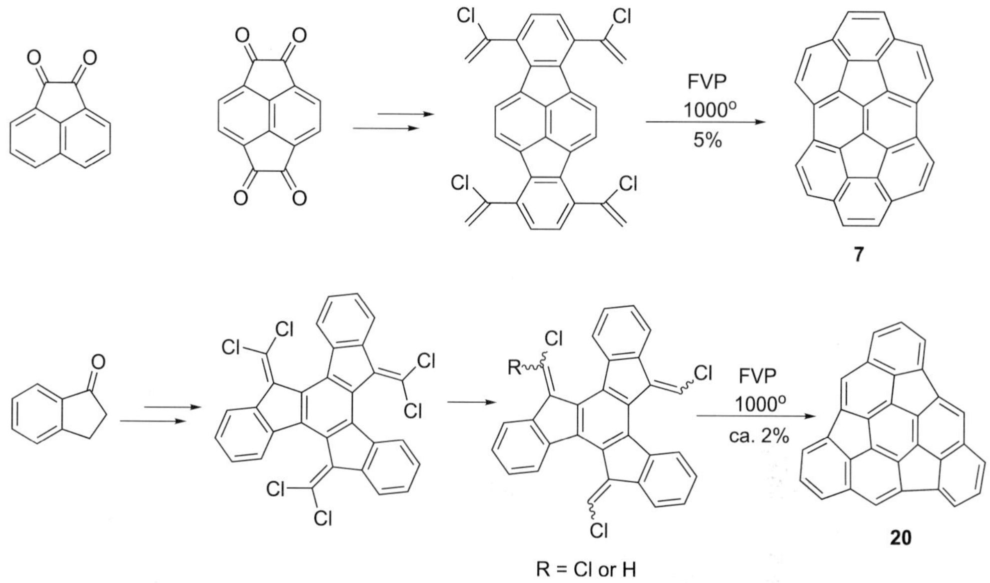

Scheme 12.14. FVP-based syntheses of semibuckminsterfullerenes **7** and **20**.

mide **9** in a modest yield of 20% [14a]. Recently we applied the carbenoid coupling methodology to **9** (Scheme 12.15) which led to formation of tetrabromo derivative **21** with a yield of 26% [24]; **21** was then converted to its tetramethyl derivative **22** which was characterized by X-ray crystal structure determination. This represented the first crystal structure study of a semibuckminsterfullerene since previous attempts to obtain X-ray quality crystals of parent **7** failed, probably due to multiple twinning [24]. The structure determination of **22** showed that the pyramidalization of the two central carbon atoms of the bowl is quite comparable to the pyramidalization of carbon atoms in C_{60}. The packing of **22** in the crystal represents yet another pattern since **22** co-crystallizes with CCl_4 (Fig. 12.2). The bowls of the hydro-

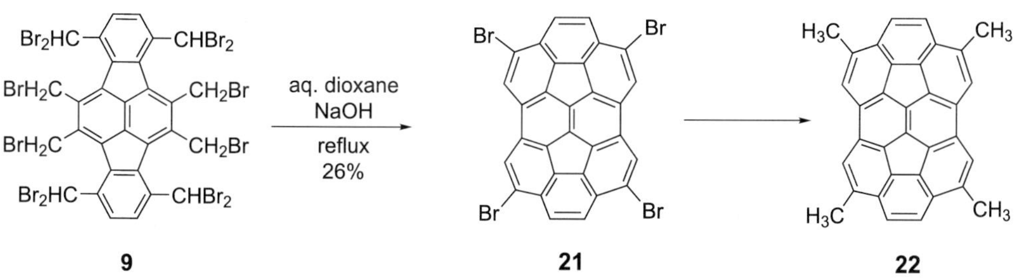

Scheme 12.15. Nonpyrolytic synthesis of tetramethylsemibuckminsterfullerene **22**.

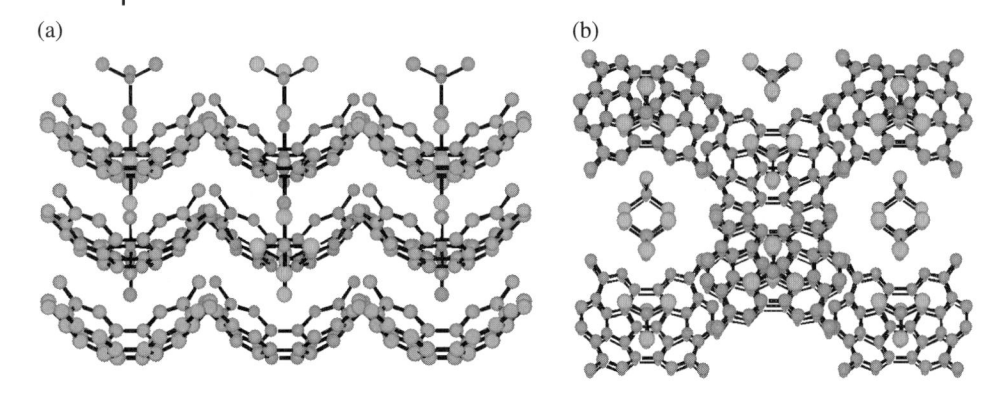

Figure 12.2. Crystal packing patterns in **22**∗2CCl$_4$ as viewed along the crystallographic b-axis (a) and c-axis (b). Hydrogen atoms omitted for clarity.

carbon are stacked in the concave–convex fashion, but they are separated by one of the two solvating CCl$_4$ molecules which prevents close contacts between the sp^2-carbon networks [24].

Very recently the crystal structure of the other semibuckminsterfullerene **20** was determined by X-ray study in the Petrukhina laboratory [25]. Unfortunately the random distribution of both enantiomers of the chiral hydrocarbon precluded the precise determination of the molecular geometry, but allowed observation of the packing patterns in two different polymorphs that revealed an *endo–exo* stacking of the bowls of **20** in one of them [25].

Rim-benzoannulated corannulenes have been of interest since they represent an obvious extension of the fullerene surface and could potentially serve as precursors for larger bowl syntheses. Several of these derivatives including benzocorannulene **23** [26], dibenzo[a,d]corannulene **24** [26a], dibenzo[a,g]corannulene **25** [16, 27] and tribenzo[a,d,j]corannulene **26** [26] were recently synthesized by FVP and/or non-pyrolytic methods. X-ray crystal structure determination of **25** was very recently achieved [28] on a crystal grown by sublimation and revealed an assembling of

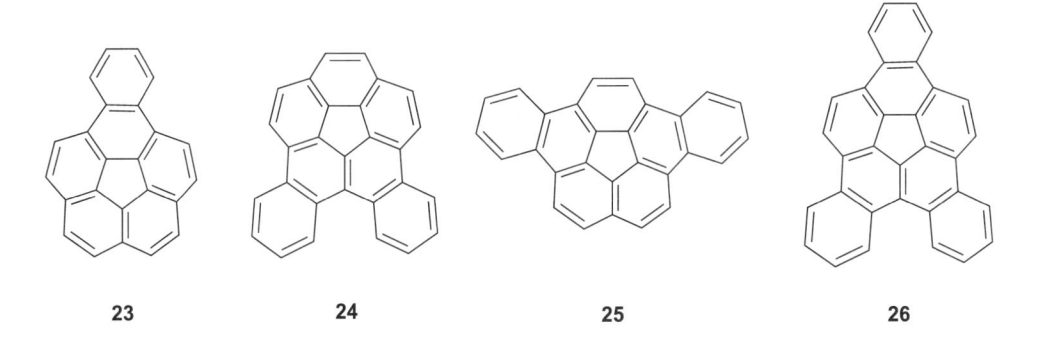

23   **24**   **25**   **26**

the bowls of **25** into columnar *exo-endo* stacks analogous to those found in **16** and one of the polymorphs of **20** (see above).

Clinton and Rabideau succeeded in the formation of $C_{32}H_{12}$ buckybowl **27** employing FVP for the final ring closure step, although with a very low yield (Scheme 12.16(a)) [29]. An even larger buckybowl (circumtrindene; $C_{36}H_{12}$, **28**) was prepared by the Scott group from the commercially available decacyclene using temperatures that exceeded the usual range of FVP, but only with 0.2–0.6% yield (Scheme 12.16(b)) [2a, 30a]. A subsequent approach to improve the yield by using the halogenated decacyclenes as FVP precursors dramatically increased the yield of the pyrolytic step to 25–27%, but in turn required a significant effort in the synthesis of the pyrolytic precursor [24b]. The crystal structure of **28** was later investigated [31] and exhibited the *endo–exo* stacking of the bowls, similar to the crystal packing in **16**, **20** and **25**.

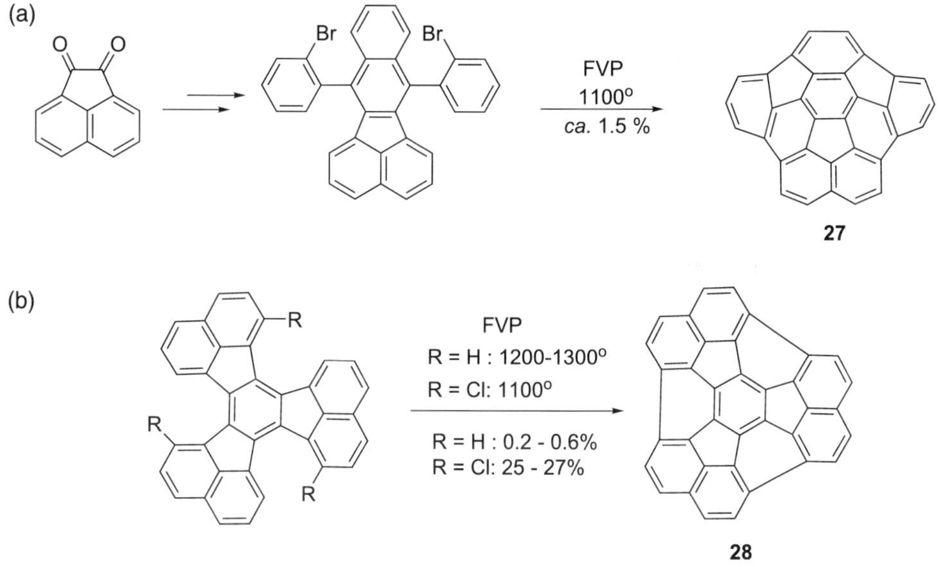

Scheme 12.16. FVP based syntheses of large buckybowls **27** and **28**.

Of course the *piece de resistance* was the FVP of $C_{60}H_{27}Cl_3$ to produce C_{60} in an estimated yield of 0.1–1.0%, as reported by de Meijere, Scott and others (Scheme 12.17) [32]. Even though this procedure seems to be of limited practical use because of the low yield and technical difficulties associated with the FVP step, it still represents a remarkable achievement, proving that fullerenes can be formed in a controlled way without tearing the carbon network into small pieces and reassembling them later into a carbon cage.

Scheme 12.17. Formation of buckminsterfullerene C_{60} by FVP.

An interesting example of the formation of large curved surface conjugated systems came from the recently published work of Nakamura et al. [33]. The authors formed the bowl-shaped benzenoid systems by selective detraction of [60]fullerene conjugation. The regio-controlled pentaaddition of methyl copper followed by addition of phenylcopper to buckminsterfullerene converted some carbon atoms to sp3-hybridized centers while leaving the others sp2 hybridized. By this approach the authors managed to produce interesting models of hoop- and bowl-shaped benzenoid systems embedded into a cage of modified fullerene [27]. One of the bowl-shaped systems produced this way (**29**) is structurally related to **7**.

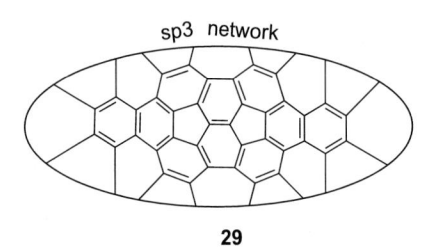

12.2.5.2 Molecular Architectures with Corannulene Scaffolds

The large scale synthesis of corannulene **1** and its tetrabromo derivative **13** provide opportunities for further elaboration to construct larger systems, including novel molecular architectures. We found a convenient method for the monoalkylation of corannulene in good to excellent yields by addition of organolithium reagent followed by dehydrogenation of the resulting 1-alkyl-1,2-dihydrocorannulenes (Scheme 12.18) [34]. Also, several derivatives of corannulene were obtained in our laboratory with reasonable yields from 1,2,5,6-tetrabromocorannulene **13** [15, 16]. Scheme 12.19 shows some of the products which can be formed from that easily obtainable synthon.

Since the chlorination behavior of **1** has been known for some time, a number

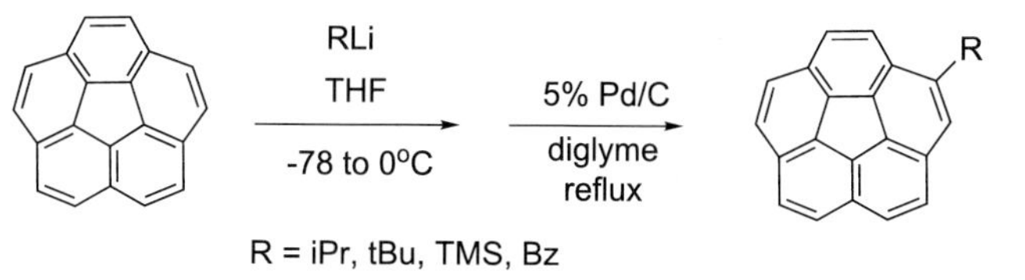

12.2 Synthesis of Buckybowls

Scheme 12.18. Alkylation of corannulene by addition of organolithium reagent.

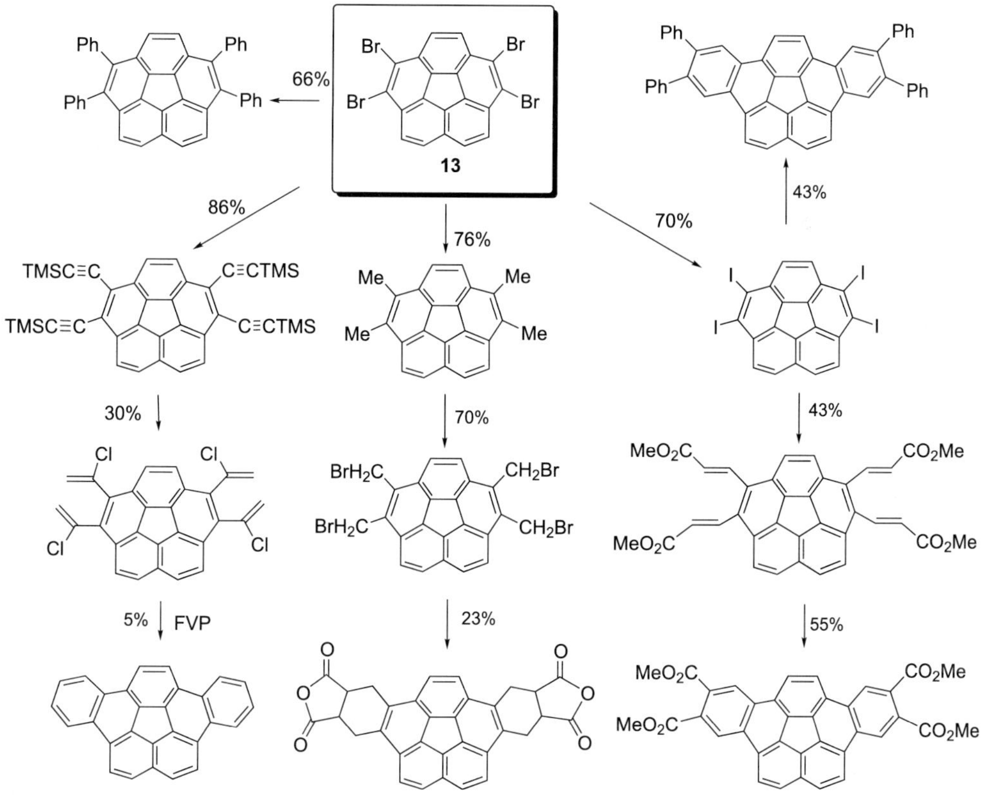

Scheme 12.19. Syntheses of corannulene derivatives utilizing **13** as a synthon.

of chlorinated corannulenes, like 1,3,5,7,9-pentachlorocorannulene **29**, became readily available [2d, 14c, 35]. This opens the field for the synthesis of high symmetry derivatives of **1**. Also, perchlorination of corannulene produces decachloro derivative **30**, yet another attractive substrate for further derivatization [2d, 28]. Siegel achieved conversions of both **29** and **30** to the corresponding penta- and decamethylated derivatives **31** and **32** with modest yields of ca. 30% [14b,c]. Scott and coworkers applied nucleophilic aromatic substitution with sodium thiophenolates to **29** and produced the pentakis derivatives **33** and **34** [28b]. Subsequent ^{1}H NMR titration studies demonstrated that both **29** and **30** formed 1:1 complexes with C_{60} with association constants of *ca.* 450 and 370 M^{-1}, respectively. The complex formation was facilitated by the preferable location of the rim substituents, induced by the curvature of the corannulene bowl. These electron-rich benzenoid substituents interact with the cage of C_{60}, forming strong complexes. Quite recently a synthesis of two more *sym* pentakis-(alkylthio)corannulenes **35** and **36** was reported by the same group [36]. The C_{60} complexing properties by these sulfides are under investigation.

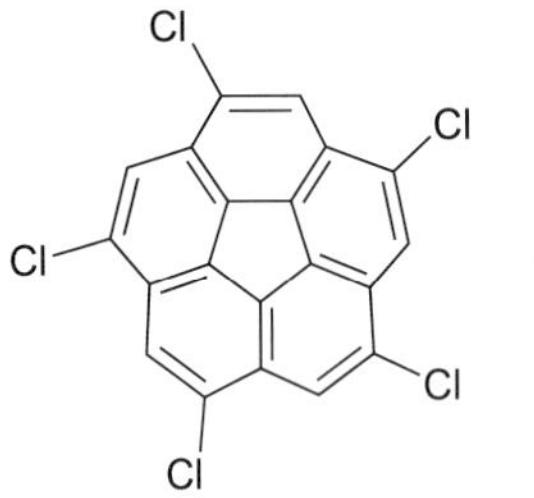

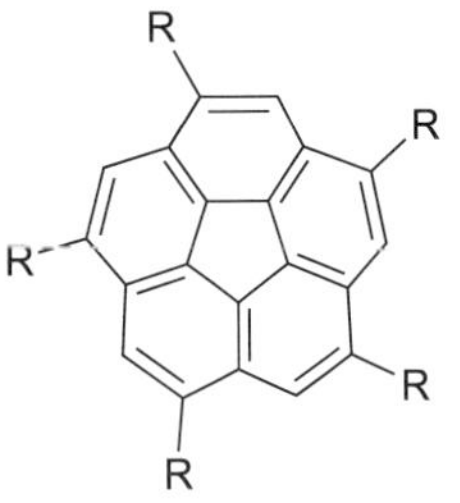

33: R = S—⟨C$_6$H$_4$⟩—OCH$_3$

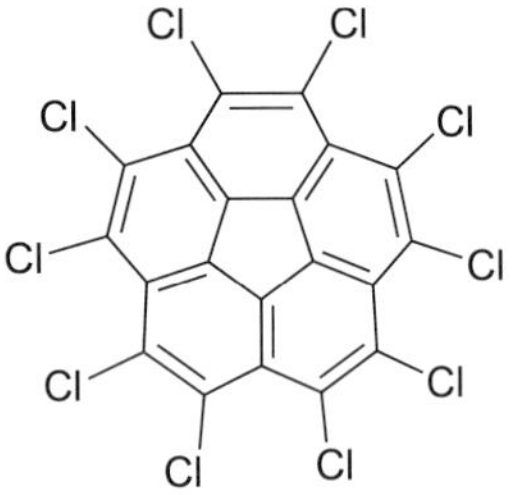

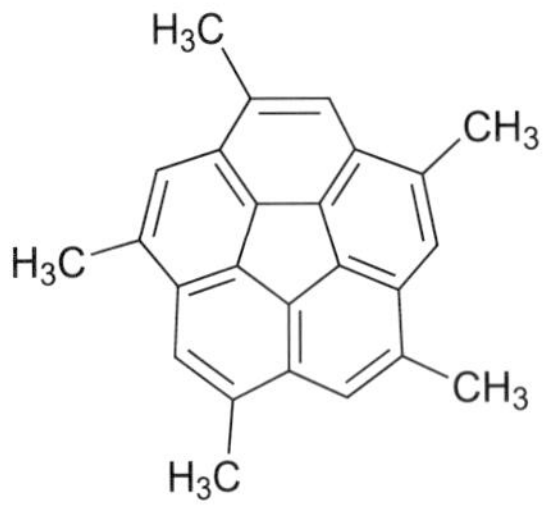

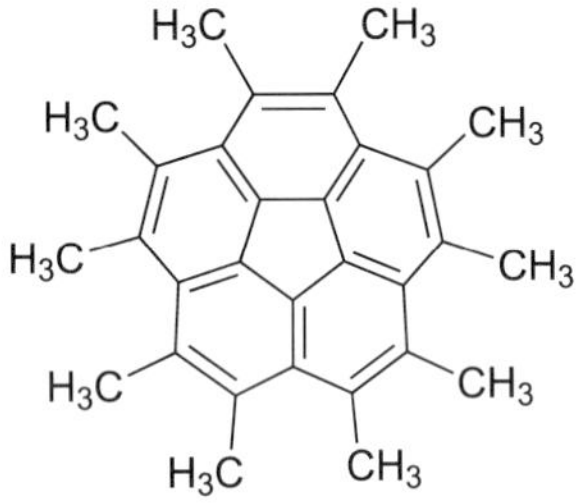

34: R = S—⟨Naphthyl⟩

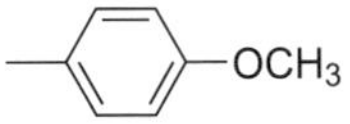

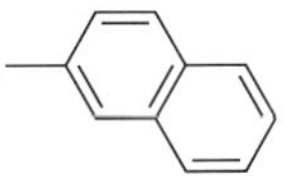

35: R = SCH$_2$CH$_2$CH$_3$

36: R = S(CH$_2$)$_{11}$CH$_3$

37: R = C$_2$H$_5$, C$_8$H$_{17}$

38: R = Ph, *p*-C$_6$H$_4$-*n*-octyl, C$_6$H$_4$-*o*-Me, a-Naphthyl, *etc.*

39: R = CC(TMS)

40: R = SPh, S(C$_6$H$_4$)-p-SPh, SC$_2$H$_5$

41: R = OC$_2$H$_4$OC$_2$H$_4$OCH$_3$

The synthesis of a plethora of *sym*-pentakis substituted corannulenes was recently published by Siegel et al. [37]. The authors introduced various substituents, including sp^3 (**37**), sp^2 (**38**) and sp hybridized carbon atoms (**39**) directly connected to **1** as well as some -SR (**40**) and -OR (**41**) derivatives [37]. The UV/vis absorption and emission properties of the derivatives were also studied, as well as the inversion barriers of some of the derivatives (*vide infra*). The Stokes' shifts were found to be in the range of 130 nm in all derivatives investigated [37].

12.2.5.3 Large Systems with Corannulene Fragments

An alternative approach to large systems with curved surfaces is to introduce the corannulene unit(s) to the rim of other systems. With the nonpyrolytic methods now available this approach became more feasible since the size of the targeted product is no longer of major importance. As an example we present here recent results from our laboratory leading to the synthesis of a large system with two corannulene units attached to the central barrelene fragment [38]. Our synthesis (Scheme 12.20) started with the successful double Diels-Alder addition of cyclopentadienone 42 to benzoquinone, which led to anthraquinone 43. Reduction with aluminum cyclohexoxide in cyclohexanone produced anthracene 44. Diels-Alder reaction with dimethyl acetylenedicarboxylate gave the corresponding barrelene 45 which was then brominated to dodecabromide 46. We were delighted to learn that our previously discovered protocol leading to the formation of the corannulene network upon heating the brominated precursor with Ni powder in DMF worked this time on both sides of 46 producing dicorannulenobarrelene 47 in the satisfactory yield of 40% [38]. Crystal structure determination showed that 47 exists in its di-exo conformation 47a. However, our hybrid density functional theorem (HDFT) calculations revealed that the two other conformers 47b and 47c are only slightly

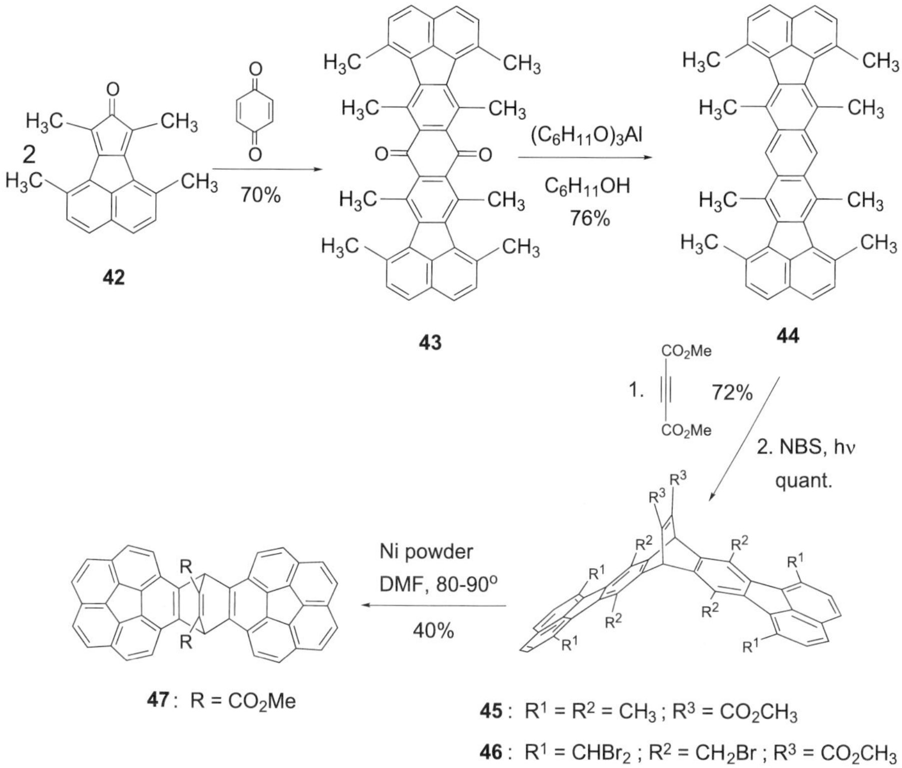

Scheme 12.20. Synthesis of dicorannulenobarrelene 47.

higher in energy than **47a**, by 1.1 and 0.5 kcal mol^{-1}, respectively. Therefore, we were not surprised that the ^{1}H NMR spectroscopy below the coalescence temperature for corannulene bowl-to-bowl inversion revealed all three isomers with approximate populations of 82:5:13 [38].

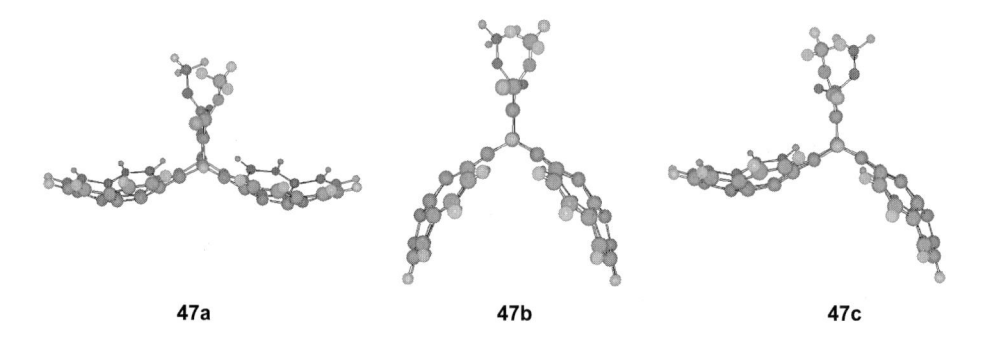

47a   **47b**   **47c**

Barrelene **47** also offered a surprise. Since corannulene can be exposed to air and light for lengthy periods without any noticeable change in its spectrum, we expected **47** to be rather stable as well. However, we noticed gradual changes in its spectrum when the NMR sample was left on the bench. We later learned that light was responsible for the spectral changes and brief (15 min) irradiation of the sample with a sun lamp converted **47** to a new species which was identified as dicorannulenobullvalene **48** [39]. This photochemical barrelene–semibullvalene conversion, known since 1966 [40], represents a sub-class of the di-π-methane rearrangement [41]. However, the conversion of **47** to **48** is, as far as we are aware, the first reported example induced by visible light since all previous conversions were achieved under UV irradiation. The unusual facility of the photoisomerization of **47** is likely due to better stabilization of the biradical intermediates postulated for this type of rearrangement by the large conjugated corannulene units [39].

47   visible light   **48**
R = CO$_2$CH$_3$

HDFT calculations for the three possible conformers **48a–c** disclosed a slightly different picture than in the case of the conformers of **47**. For this case the *endo-exo* conformer **48b** was predicted to have the lowest energy with di-*exo* **48a** and di-*endo* **48c** being less stable by 0.9 and 4.9 kcal mol^{-1}, respectively [39]. While we were not successful in growing X-ray quality crystals, we strongly believe that **48b** represents the preferred conformation in both solid state and solution considering the relatively large calculated energy differences between the conformers as well as the success of the theoretical model applied previously for **47**. A low temperature ^{1}H NMR study of **48** showed that below the coalescence temperature only one conformer is detectable, presumably **48b** [39]. This finding further supported the validity of the theoretical model used for the conformational preference studies of both **47** and **48**.

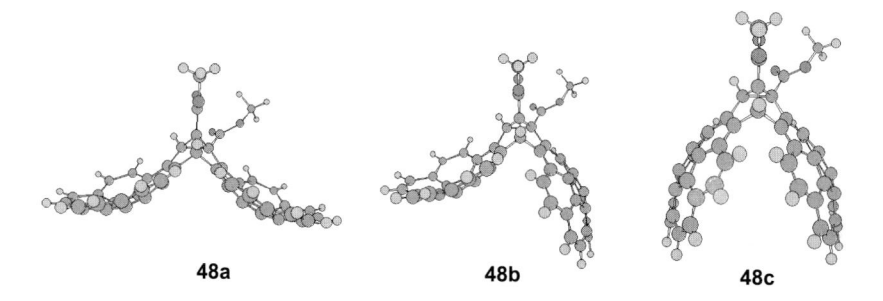

48a **48b** **48c**

The different conformational preferences of the isomeric molecules **47** and **48** clearly show that we can design the shape of the whole molecular assembly by changing the size and/or shape of the central subunit to which the corannulene moieties are attached. This is of premium importance for the potential synthesis of molecular clips, tweezers, bowls and other *nano*sized molecular assemblies with interesting supramolecular properties.

12.3
Bowl-to Bowl Inversion in Buckybowls

While it may be presumed that corannulene undergoes bowl to bowl inversions through a planar transition state, this process cannot be observed by NMR spectroscopy due to the high symmetry of the system. However, such observations can be achieved with properly substituted corannulenes. Thus the presence of the group with diastereotopic protons like –CH$_2$X or C(CH$_3$)$_2$X would give a simple ^{1}H NMR spectrum under fast inversion (on the NMR time scale) of the bowl due to exchange of the diastereotopic nuclei. On the other hand, if the inversion is slow, two separate signals or groups of signals would be observed. The barrier for inversion (ΔG^*) can be easily calculated from the coalescence temperature (T_c) of the two signals and their separation (Δv [Hz]) in the slow exchange region. An exemplary variable-temperature ^{1}H NMR spectra of the benzylic hydrogen atoms of benzylcorannulene is shown in Fig. 12.3.

The first determination of the barrier for inversion in corannulene was reported in 1992 by Scott's group. The coalescence temperature for the two diasteromeric

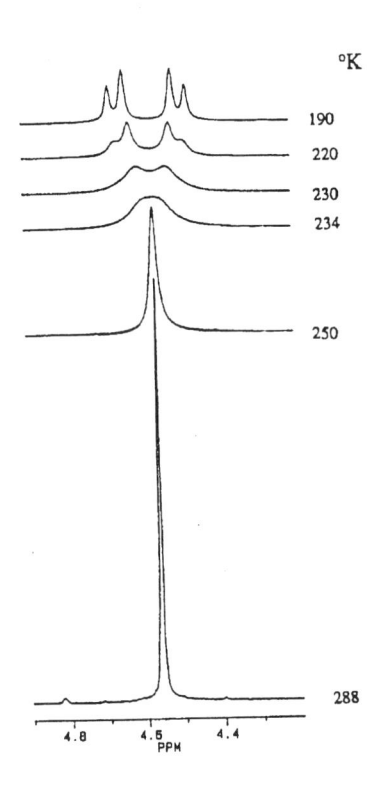

Figure 12.3. Variable-temperature ^{1}H NMR spectra of the benzylic region of benzylcorannulene in THF-d_8.

methyl groups in corannulenyldimethylcarbinol (**49**) was found to be 204 K (−64 °C), from which the barrier of 10.2 kcal mol^{-1} was calculated [42]. At the same time the inversion barriers for 1,2-dihydrocorannulene (**50**) and 1,2,5,6-tetrahydrocorannulene (**51**) were estimated by Siegel et al. as 8.5 and 6–7 kcal mol^{-1}, respectively [10a]. These numbers clearly show that despite its significant nonplanarity the bowl of corannulene is surprisingly flexible. Indeed, the measured barrier for **49** is very close to the chair-to-chair inversion barrier of cyclohexanes. Due to significant progress in synthetic approaches to corannulenes several derivatives of **1** have been investigated and their barrier for inversion measured. Table 12.1 summarizes the results of these studies. Most of the recent determinations were reported by Siegel's group [37, 43].

Table 12.1. Barriers for ring inversion in substituted corannulenes.

	Substituent(s)	Barrier (kcal mol^{-1})	Ref.
1	C(CH$_3$)$_2$OH	10.2	42
2	CH(CH$_3$)$_2$	11.3	20b
3	CH$_2$Ph	11.2	20b
4	CH$_2$Br	11.0	43
5	1,6-bis(CH$_2$Br)	10.5	43
6	2,5-bis(CH$_2$Br)	10.4	43
7	2,3-bis(CH$_2$Br)	9.1	43
8	2,3-diPh	9.4	43
9	2,7-bis(CH$_2$Br)-4,5-diCl	8.7	43
10	2,3-bis(2-(2-methoxyethoxy)ethoxy	9.9	43
11	1,3,5,7,9-pentaethyl	11.0	37
12	1,3,5,7,9-pentakis(1-thiapropyl)	10.5	37
13	1,3,5,7,9-pentakis(1,4,7-trioxaoctyl)	11.5	37
14	1,3,5,7,9-penta(manisyl)	12.0	37

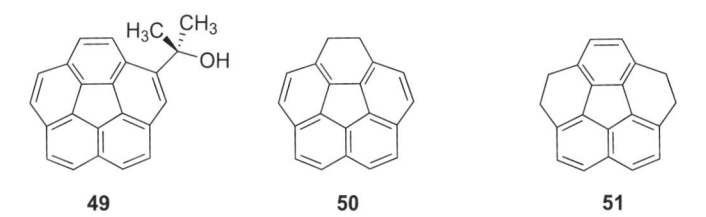

49   **50**   **51**

As is evident from Table 12.1, the inversion barriers for the mono- and polysubstituted corannulenes are all rather similar, falling in the range 8.7–12.0 kcal mol^{-1}, thus indicating fast bowl to bowl inversion at ambient temperatures. In the early stages of our own research on corannulene, we asked the question how much further one should go on the surface of fullerene to make a buckybowl with the barrier for inversion large enough to virtually stop the inversion, at least at ambient temperatures. This question prompted us to develop a route to cyclopentacorannulene **16** [20]. As we mentioned earlier, dihydrocyclopentacorannulene **17** exhibits an AA'BB' spin pattern in its ^{1}H NMR spectrum at room temperature as well as at temperatures up to 140 °C, indicating slow inversion of the bowl, in strong contrast to the parent **1**. From these data we were able to estimate a lower limit for the inversion barrier in **17** at ca. 19 kcal mol^{-1}. In addition, we applied the spin polarization transfer method which allows the NMR determination of the kinetics of processes too slow to be studied by the classical coalescence method [44]. This technique did not give us the exact answer either, since even at 126 °C the inversion process in **17** was too slow, but it raised the lower limit for the barrier to ca. 26 kcal mol^{-1} [20a]. We wondered if we could take advantage of the bowl shape of **16** and achieve some stereoselectivity of its deuterogenation to produce unequal amounts of exo- and endo-dideutero products. To our surprise we found out that deuterogenation of **16** is highly π-facial stereoselective leading exclusively to the *exo*-dideuterio derivative **52**. The formation of a single product represented the first example of the use of buckybowls' curvature to drive the sterochemical outcome of a reaction. Equilibration of the *endo*- and *exo*-isotopomers of **52** allowed determination of ΔG^* for ring inversion as 27.67–27.61 kcal mol^{-1} over the range 55–99 °C, with ΔH^* determined to be 27.3 kcal mol^{-1} with a small ΔS^* of −1.1 e.u. [20b]. Hence the addition of a single five-membered ring to the rim of corannulene more than doubles the barrier for inversion, and virtually stops inversion of the system at room temperature.

We also studied the kinetics of inversion of **53**, a larger dibenzoannulated analog of **17** [45]. In contrast to **17**, variable temperature ^{1}H NMR spectra of **53** showed changes in the AA'BB' spectral region over the temperature range 160–184 °C. While the coalescence temperature could not be reached, we were able to determine the inversion barrier employing line-shape analysis. The ΔG^* of ca. 23.5 kcal mol^{-1} is significantly lower than that for the simpler analog **17**, demonstrating that benzoannulation of corannulene indeed lowers the barriers for its inversion. Similar results came from the unpublished work of Scott's research group since the estimated barriers for bowl-to-bowl inversion of the dimethylcarbinols of benzocorannulene [26c] and dibenzo[a,g]corannulene [27a] were 9.0 and less than 7.0 kcal mol^{-1}, respectively, notably lower than the inversion barrier of the dimethylcarbinol derivative of corannulene **49** (10.2 kcal mol^{-1}, [42]).

A series of derivatives of corannulenes with an external three-membered tether clipping positions 2 and 3 was recently investigated by Siegel and coworkers (Scheme 12.21) [43]. Since the six-membered external ring is more flexible than the five-membered analog in **16**, **52** and **53**, it is not surprising that the inversion

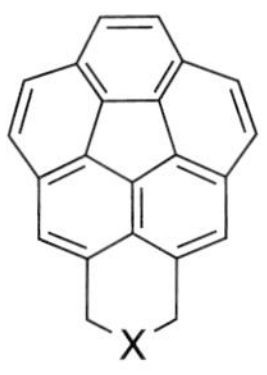

	X	Barrier (kcal mol⁻¹)
54:	C(CO$_2$Et)$_2$	15.5
55:	NPh	16.7
56:	O	17.3
57:	S	13.9
58:	Se	13.0

Scheme 12.21. Inversion barriers for the derivatives of corannulene with an external six-membered ring [35].

barriers of 54–58 fall in between the barriers of corannulenes (Table 12.1) and the cyclopentacorannulenes 52 and 53.

The same research group proposed a correlation of the inversion barrier in corannulenes with the degree of their curvature [43], expressed by the bowl-depth, i.e., the distance between the central five-membered ring and the average plane of the ten rim carbon atoms of the corannulene fragment. A very good structure–energy correlation was found for both experimental and computed data assuming a quartic relationship of the inversion energy with equilibrium bowl depth [43].

12.4
Organometallic Derivatives of Buckybowls

12.4.1
Alkali Metal Reduction

One of the very first observations reported for corannulene (**1**) was that it can easily be converted to its radical anion electrochemically or by exposure of the hydrocarbon to alkali metals in anhydrous solvents. The radical, which is easily identified by the appearance of a green color, was characterized by EPR and visible spectroscopy in 1967 [4]. In addition, further reduction of **1** was also observed, even though the overall process was not fully understood. More recent studies showed that **1** can be reduced by alkali metals up to the stage of a tetraanion. NMR and EPR spectroscopic studies allowed the identification and characterization of all four reduced states of **1**, i.e., radical anion, dianion, radical trianion and tetraanion [46]. The degree of reduction achieved depends strongly on the alkali metal. Thus while **1** is reduced with metallic K to the radical trianion stage only and prolonged exposure to the metal does not appear to produce further reduction, it is easily reduced to its tetraanion with Li metal. This rather surprising difference in reductive behavior (K has been found to be more effective than Li in the formation of highly charged organic polyanions [47]) may be attributed to the formation of a supramolecular dimer of the tetralithio derivative of **1** (see *infra*) that stabilizes the resulting overcharged system. Two of the four possible reduced corannulene intermediates, the mono- and trianion, are odd-electron species and were thus characterized by EPR and UV/VIS spectroscopy; the dianion and tetraanion are even-electron species [46]. However, since the corannulene framework exhibits C_{5v} symmetry, both its HOMO and LUMO orbitals are doubly degenerate with E_2 and E_1 symmetries, respectively. Therefore formation of the dianion by distribution of the two additional electrons without distortion of the geometry of the framework may lead to the formation of four new states: a triplet $^3A_2$ and singlet states $^1A_1$ and $^1E_2$ [48]. On the other hand Jahn–Teller distortion of the geometry of the dianion to C_s may lead to the formation of two new singlet states. (Lowering of the symmetry will also be forced by interactions of the dianion with two countercations as contact ion pairs with metallated **1**).

Our computational study suggests that the triplet $^3A_2$ is favored over singlet states for the isolated dianion with C_{5v} symmetry, while the singlet states become slightly more stable after distortion of the molecular framework to C_s symmetry [48]. These preliminary results are in accord with ^1H NMR studies showing a single broad signal for dilithio and dipotassiocorannulene at temperatures below 230 K, consistent with the existence of a low-lying, thermally accessible triplet state of the dianion [46].

It also appears that the addition of alkali metals may have an effect on the bowl-to-bowl inversion of **1**. An interesting study of the alkali metal reduction of **59**, in which two corannulene moieties are tethered by an eight carbon atom chain, allowed for the estimation of inversion barriers in the corannulene dianion [49]. The barrier for the doubly charged corannulene moieties (8.8 and 9.2 kcal mol^{-1} at 186 and 196 K for K and Cs salts, respectively) is significantly lower than for neutral corannulene (ca. 11 kcal mol^{-1}), and closely agrees with our theoretical prediction (7.9–9.2 kcal mol^{-1} [48]) published earlier.

59

In contrast with either Na or K, exposure of **1** to excess Li leads all the way to the diamagnetic "tetraanion." The combined effort of three research groups led to the discovery of an unusually stable dimer for the tetraanion of corannulene [50]. Li NMR studies showed that four of eight lithium countercations are sandwiched between the two organic negatively charged units while the remaining four are located outside the assembly. Variable temperature studies indicated considerable stability for the dimer which does not seem to dissociate even at ambient temperatures although "inside–outside" exchange of the Li cations is observed. Perhaps the fact that tetraanions are not observed with Na or K is due to the inability of these larger cations to form a similar dimeric structure.

The experimental data available for reduced corannulene suggest that only tetralithio-**1** exists as a dimer while the other observed species – from monoanion radicals to trianion radicals – exist as monomers in ether solutions. Interestingly, octalithio-**59** (two tetraanions tethered by an octamethylene chain) was found to form an "internal dimer" by intramolecular stacking of the two corannulene tetraanions resulting in the formation of Li ion-bound sandwiches of the type seen in tetralithiocorannulene [49].

The esthetically attractive "an anion inside a trianion" model was proposed for the tetraanion of corannulene based on the early models of electron distribution in corannulene [51]. This view, however, has been a matter of some controversy, since while some of the spectral characteristics of the tetraanion seem to be explained well by this model, the charge-^{13}C NMR relations as well as the results of early *ab initio* calculations suggest that more than three extra electrons go to the rim of corannulene upon the tetraanion formation [52].

Anions of other buckybowls were also characterized by NMR spectroscopy. For example the dianions and tetranions of dibenzocorannulene **25** and dibenzocyclopentacorannulene **60** were reported by Rabinowitz et al. [53]. In these cases, however, the tetraanion stage of reduction could be achieved only with K metal. No sandwich type dimerization similar to tetralithio-**1** was observed for these larger systems.

12.4.2
Transition Metal Complexes

The potential formation of transition metal complexes with buckybowls has been of interest since a variety of transition metal complex units were reported to coordinate to buckminsterfullerene C$_{60}$ [54]. Interestingly, in the cases of C$_{60}$ complexes the metal is η^2-coordinated to two carbon atoms shared between two 6-membered rings and no η^6-coordinated haptomers were ever reported.

In contrast, the successful formation of transition metal complexes of buckybowls turned out to be much more challenging. For some time there was only one report of an X-ray characterized buckybowl metal compound (**61**); the reaction of (Ph$_3$P)$_2$Pt(H$_2$C=CH$_2$) with semibuckminsterfullerene **20** [55]. In this case we found that the Pt(0) inserted into a C–C bond of a strained 5-membered ring and we suggested that the relief of steric strain going from **20** to **61** was the driving force for the reaction. Later the theoretical study was published fully supporting our assumption [56].

The first spectroscopic characterization of a corannulene complex, published by O'Connor, Siegel and coworkers, described the formation of $(\eta^5\text{-}C_5Me_5)Ru(\eta^6\text{-}C_{20}H_{10})^+$ (**62**) by the reaction of $(\eta^5\text{-}C_5Me_5)Ru(NCMe)_3^+$ with **1** [57]. The complex was not characterized by X-ray but its ^1H and ^{13}C NMR spectra clearly establish the $\eta^6\text{-}C_{20}H_{10}$ structure. Later, in cooperation with the Angelici group, we reported the formation of iridium complexes of both corannulene (**63**) and 1,2,5,6-tetramethyl-corannulene (**64**) by the reaction of the hydrocarbons with [Cp*Ir(OCMe$_2$)$_3$](BF$_4$)$_2$ [58]. As with the previous study, NMR provided evidence for η^6-coordination of the metal with the organic moiety. Furthermore, an interesting migration behavior of the Cp*Ir^{2+} fragment was observed in the case of the tetramethylcorannulene complex (Scheme 12.22). Thus, while initially all three haptomers of **64** (A–C)

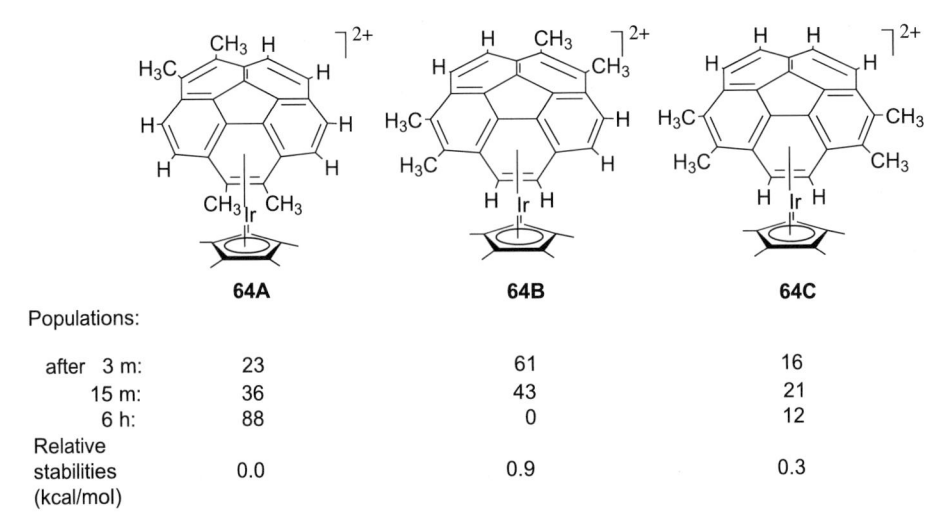

Populations:	64A	64B	64C
after 3 m:	23	61	16
15 m:	36	43	21
6 h:	88	0	12
Relative stabilities (kcal/mol)	0.0	0.9	0.3

Scheme 12.22. Three haptomers of [(Cp*Ir)$_2$]C$_{24}$H$_{18}$(BF$_4$)$_2$ (**64**) with their populations and calculated relative stabilities.

were observed in the ratio of ca. 2:3:2, the population changed with time and at equilibrium only **64A** and **64C** survived in the ratio of ca. 7:1. Our B3LYP calculations with 6-31G(d) basis sets for C and H and a pseudopotential double-zeta basis set for Ir on a model set of dications of **64A–C** correctly found that **64A** is the lowest energy haptomer (found as the major component at equilibrium), followed by **64C**, with **64B** as the least stable [59].

Unfortunately, X-ray quality crystals of **64** could not be obtained and *exo* vs. *endo* (convex vs. concave) preference for the location of the transition metals could not be answered simply based on NMR data.

More recently O'Connor, Siegel et al. reported the formation of isomers of a Cp*Ru$^+$ complex of dihydrocyclopentacorannulene **17**, Cp*Ru(η^6-C$_{22}$H$_{12}$)$^+$ (**65**), which were characterized in solution by their NMR spectra [60]. Also in this case the authors observed the migration of the Cp*Ru$^+$ unit leading to formation of the thermodynamically favored isomer (Scheme 12.23). Originally only **65B** was formed at −80 °C, but upon warm-up the reaction mixture changed and exhibited a 2:1 ratio of **2B** and **2C** haptomers; **65A** was never observed. Theoretical calculations provided strong support for the experimental findings showing that **65B** is indeed the most stable complex of Cp*Ru$^+$ with **17**, while **65A** is the least stable by ca. 3 kcal mol^{-1} explaining its absence in the reaction mixture [60]. The energy of the cation η^6 complexation is a function of the nonplanarity of the benzene rings. Crystal structure determination of cyclopentacorannulene **16** showed [21] that the degree of benzene ring puckering is highest in the ring adjacent to the external five-membered ring (Ru-coordinated in **65A**), and diminishes gradually for the second and third ring (metal coordinated in **65B** and **65C**, respectively).

Only very recently some crystal structure determinations of buckybowl metal complexes became available. Petrukhina and Scott reported successful preparation of molecular solids by gas-phase co-deposition of **1** with Rh$_2$(O$_2$CCF$_3$)$_4$ which were characterized by X-ray [61]. The resulting solids consisted of 1D and 2D networks of Rh$_2$(O$_2$CCF$_3$)$_4$ and corannulene units with the [Rh$_2$] fragments η^2-coordinated

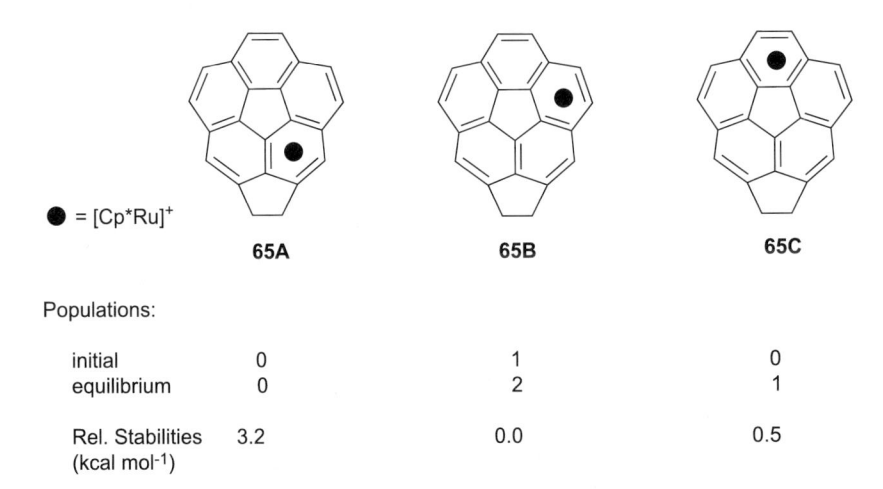

● = [Cp*Ru]$^+$

	65A	65B	65C
Populations:			
initial	0	1	0
equilibrium	0	2	1
Rel. Stabilities (kcal mol^{-1})	3.2	0.0	0.5

Scheme 12.23. Equilibration of the haptomers of [(Cp*Ru)C$_{22}$H$_{12}$]$^+$ (**65**).

to both convex and concave sides of **1** which exhibits virtually no structural changes as compared to pure corannulene. These structures, apparently very weakly bound, are not likely to represent the structures of the transition metals of **1** in solution; exposure of the complexes to organic solvents causes an immediate dissociation to the starting materials. Later the same group reported vacuum co-deposition formation and X-ray crystal structure determination of the complex of semibuckminsterfullerene **20** with $Rh_2(O_2CCF_3)_4$ in which the [Rh_2] fragments are also η^2-coordinated to the rim double bonds of the organic part on both convex and concave sides [25]. Again, the resulting complex is very weakly bound and dissociates into **20** and rhodium salt when dissolved in common organic solvents.

Similar types of molecular solids were very recently characterized by X-ray by Siegel and coworkers [62]. The co-crystallization of corannulene with various silver salts led to the formation of molecular networks in which the metal is η^2-coordinated to the rim of **1**. No statistically relevant deviations in the corannulene bond lengths of the complexes were found as compared to isolated **1** [62].

Another example of the molecular solids formed by vacuum co-deposition of $Rh_2(O_2CCF_3)_4$ with dibenzo[a,g]corannulene (**25**) was just reported along with the crystal structure determination of **25** alone [28]. The complex, $\{[Rh_2(O_2CCF_3)_4]_3(25)_2\}$, exhibited a one-dimensional extended structure with $\eta^2:\eta^2:\eta^2$ rim coordination of **25**. Also in this case the complex is too weakly bound to survive dissolving in organic solvents. The bond lengths of the ligand in the complex are not significantly changed from the isolated **25**. However, some bowl depth reduction of **25** in the complex as compared to the isolated dibenzo[a,g]-corannulene was observed [28].

Finally, our collaboration with the Angelici group produced X-ray quality crystals grown from solutions of ruthenium complexes of **1**. In all characterized crystals of mono and di-metallated corannulenes, we observed η^6 coordination of the metal with the organic moiety. The first characterized complex was that of $[(Cp^*Ru)_2(\mu_2\text{-}\eta^6,\eta^6\text{-}C_{20}H_{10})][PF_6]_2$ ($Cp^* = \eta^5\text{-}C_5Me_5$) (**66**). The structure determination showed that the two $\{Cp^*Ru\}^+$ fragments were η^6 bound to nonadjacent rings on opposite sides of the corannulene [63]. In contrast to the previously described Rh co-deposition complexes [25, 28, 61] and Ag complexes [62] **66** exhibits significant changes of the geometry of the corannulene subunit induced by complexation. The most pronounced change in the bond lengths was observed for the CH=CH rim bonds in the metal-coordinated rings; these bonds are elongated to ca. 1.45 Å as compared to 1.39 Å in corannulene itself, while the remaining three rim CH=CH bonds are in the range 1.37–1.38 Å. The Ru–C distances in the co-ordinated rings of corannulene were in the range 2.17–2.35 Å. The curvature of the corannulene ligand upon coordination of two $\{Cp^*Ru\}^+$ units was significantly reduced to approximately half of the curvature of corannulene.

Quite recently we also succeeded in X-ray structure determination of the mono-metallated complex of **1** with $(Cp^*Ru)^+$ [64]. As in the bis-ruthenium analog the metal was η^6 coordinated to one of the benzene rings of **1** on its exo (convex) face. This is the first experimental evidence demonstrating the *exo* preference of η^6 co-ordination of $(Cp^*Ru)^+$ to corannulene. Also, another bis-metallated complex, $[(Cp^*Ru)_2(\mu_2\text{-}\eta^6,\eta^6\text{-}C_{20}H_{10})][SbF_6]_2$ was prepared and characterized by X-ray [64].

In this case a subtle difference in composition as compared to **66** (SbF$_6$ vs. PF$_6$ counteranion) caused a significant change in the structure of the corannulene subunit which lost its bowl shape and became almost planar. The structural details of these complexes will be discussed in the forthcoming paper [64].

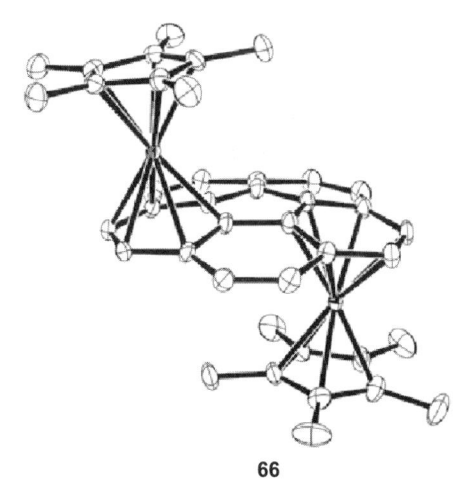

66

A quite different approach to the complexation of corannulene derivatives with transition metals was applied by Chin [65]. His group hydrogenated **1** to its octahydro derivative **67** which was subsequently deprotonated by BuLi to the respective fluorene-type anion **68**. This species was then used for the formation of two complexes with (CpZrCl$_2$)$^+$ and [Re(CO)$_3$]$^+$. In both cases X-ray structure determination showed η^5 coordination of the metal to the central ring on the *exo* (convex) side of the very shallow bowl of **68**. However, these complexes relate more to the fluorene anion chemistry than to buckybowls.

1   **66**   **67**

In principle, understanding the coordination of metal complexed to buckybowls involves many of the same considerations as those for C$_{60}$. Most theoretical calculations involving metal ion binding to buckybowls have involved the main group ions Li$^+$, Na$^+$, and K$^+$ [66]. In general, these calculations indicate that the metal cation favors η^6-coordination over η^5-coordination by 1–2 kcal mol^{-1}, and coordination to the convex (outside) surface of corannulene is preferred over the concave (inside) surface of corannulene (**1**) by 1–5 kcal mol^{-1}. In contrast, theory suggests that the concave location of the metal cations coordinated with semibuckminsterfullerene **20** is slightly preferred over the convex coordination [66f, 67]. In the case of transition metals the calculations by Dunbar for corannulene binding to

Ti$^+$, Cr$^+$, Ni$^+$, and Cu$^+$ ions indicated that Ti$^+$ and Ni$^+$ prefer η^6-coordination to the convex face, but Cr$^+$ binds about equally well to the 6-membered (η^6) and 5-membered (η^5) rings, and Cu$^+$ prefers η^2-coordination to edge sites [66c]. In all cases, binding on the convex (outside) surface was favored over the concave (inside) surface by 5–11 kcal mol^{-1}. Calculations on metal complexes of buckybowls with ligands including (η^5-C$_5$H$_5$)Ru(η^6-corannulene)$^+$ [57] and haptomers of (η^5-C$_5$H$_5$)Ru(η^6-CPC)$^+$ [60] as well as our results for (η^5-C$_5$Me$_5$)Ru(η^6-corannulene)$^+$ [64], (η^5-C$_5$Me$_5$)Ru(η^6-tetramethylcorannulene)$^+$ [59] and [(η^5-C$_5$Me$_5$)Ru]$_2$(η^6-corannulene)$^{2+}$ [64] showed that coordination on the convex (*exo*) surface is preferred over the concave (*endo*) surface by ca. 6 kcal mol^{-1}.

12.5
Conclusions

Curved surface polynuclear aromatics represent an exciting new class of compounds that have great potential for the production of unique molecular architectures. However, these opportunities will only be realized through the development of simple, large-scale synthetic routes to small and medium-sized bowls that may then be employed to lead to larger and more complex systems, including molecular assemblies of nanometric sizes ("From Molecules to Materials"). The fact that corannulene, once rare and only produced by lengthy methods or in the gas phase, can now be easily prepared in solution on a large scale suggests that these synthetic goals can and will be reached. We look forward to the exciting outcome.

12.6
Experimental: Selected Procedures

12.6.1
Preparation of Octabromide 11 from 10

2.6 g of **10** (10 mmol), 15.0 g of NBS (84 mmol) and 150 mg of dibenzoyl peroxide in 200 mL of benzene was refluxed and irradiated with a sun lamp for 24 h. The solvent was removed under reduced pressure, the resulting dark solid was washed well with H$_2$O and EtOH. Crystallization from EtOH gave 6.6 g (74%) octabromide **11** as a yellow solid.

12.6.2
Preparation of Corannulene from 11 by Low-valent Vanadium Coupling

A solution of 2.29 g of **11** in 100 mL of dry/degassed DME was added through a syringe pump to the refluxing suspension of 8.4 g of VCl$_3$ and 0.76 g of LiAlH$_4$ in DME at a rate of ca. 1 mL h^{-1}. After the addition was complete, the mixture was additionally refluxed for 8 h. The reaction mixture was then diluted with benzene and filtered through a pad of florosil. The solvents were removed under reduced pressure and the crude product chromatographed on silica gel with cyclohexane. The first fraction collected gave 44 mg (5%) of bromocorannulene, and the second fraction gave 480 mg of corannulene **1** (74%).

12.6.3
Preparation of Tetrabromocorannulene from 11 by Carbenoid Coupling

3 g of NaOH pellets was added to a stirred suspension of 6.5 g of octabromide **11** in a mixture of 250 ml of dioxane and 100 ml H_2O. The reaction mixture was refluxed for 15 min (the originally formed red color faded during that time), then cooled, poured into water and acidified with HCl. The yellow precipitate was filtered off and dried. Crystallization of the resulting solid from xylenes with activated charcoal gave 3.42 g of colorless **13** (83%).

12.6.4
Preparation of 43 by Double Diels–Alder Reaction

520 mg of cyclopentanone **42** (2 mmol) and 216 mg of 1,4-benzoquinone (2 mmol) were suspended in 15 mL of DMF and stirred at rt. After 30 min 0.5 mL of Et_3N was added and the stirring continued for 2 h. Another portion of **42** (520 mg, 2 mmol) was added to the reaction mixture in portions over 1 h, and the stirring was continued overnight. 10 drops of DBU was then added and the reaction mixture was stirred for another 8 h, then poured into water, acidified with dilute HCl and filtered. The separated solid was washed with EtOH, dried and subsequently refluxed with ca. 20 mL of toluene. The cooled suspension was filtered yielding 790 mg (70%) of **43**, sufficiently pure for the next step.

12.6.5
Preparation of 44 by Aluminum Cyclohexoxide Reduction of 43

2 g of aluminum foil was refluxed for 3 h in 70 mL of cyclohexanol to which ca. 20 mg of $HgCl_2$ and 2 mL of CCl_4 were added. 1.00 g of the anthraquinone **43** (1.8 mmol) was added to the dark solution and the reaction mixture was refluxed for 4 d. Most of the cyclohexanol was then removed under reduced pressure, H_2O was added and the mixture was acidified with HCl. The red solid was filtered, washed with H_2O and EtOH and dried to give 720 mg (76%) of the crude air-sensitive product (**44**) exhibiting a strong green fluorescence in solutions.

12.6.6
Preparation of 45

540 mg of **44** (1 mmol) was suspended in 15 mL of 1,1,2,2-tetrachloroethane with 570 mg of DMAD (4 mmol) and the reaction mixture was refluxed for 2 h. The sol-

vent was removed under reduced pressure and the product was chromatographed on silica with DCM, yielding 490 mg of **45** (72%).

12.6.7
Preparation of Dodecabromide 46

340 mg (0.5 mmol) of **45** was refluxed and irradiated for 4 h with a sun lamp with 1.42 g (8 mmol) of NBS and ca. 5 mg of dibenzoyl peroxide in 70 mL of CCl_4. The solvent was then removed, the resulting solid was washed thoroughly with H_2O and aq. EtOH and dried under vacuum, yielding 810 mg (quantitative) of dodecabromide **46**, sufficiently pure for the next step.

12.6.8
Preparation of Dicorannulenobarrelene 47

814 mg of **46** (0.5 mmol) was dissolved in 20 mL of dry DMF with 1.2 g of Ni powder. The mixture was purged with N_2 for 30 min and heated on an oil bath under N_2 for 16 h at 80–90 °C. DMF was then removed under reduced pressure, the dark residue was treated with dilute HCl, then extracted with DCM. The crude product was chromatographed on silica with DCM to provide 135 mg (40%) of pure **47**.

Acknowledgments

We would like to acknowledge the financial support from U.S. Department of Energy, Office of Science, Office of Basic Energy Services, Chemical Science Division and from Mississippi State University through the start-up grant for AS.

Abbreviations

B3LYP	Becke's three-parameter hybrid exchange functional with the Lee-Yang-Parr correlation functional
Bu	butyl
Bz	benzyl
Cp	η^5-cyclopentadienyl
Cp*	η^5-pentamethylcyclopentadienyl
DBU	1,8-diazabicyclo[5.4.0]undec-7-ene
DCM	dichloromethane
DDQ	2,3-dichloro-5,6-dicyano-1,4-benzoquinone
DMAD	dimethyl acetylenedicarboxylate
DME	1,2-dimethoxyethane
DMF	N,N'-dimethylformamide
Et	ethyl
FVP	flash vacuum pyrolysis
HDFT	hybrid density functional theorem

HOMO   highest occupied molecular orbital
LUMO   lowest unoccupied molecular orbital
Me   methyl
MM   molecular mechanics
NBS   *N*-bromosuccinimide
PAH   polycyclic aromatic hydrocarbons
Ph   phenyl
Pr   propyl
THF   tetrahydrofuran
TMS   trimethylsilyl

References

1 (a) H. W. Kroto, J. R. Heath, S. C. O'Brien, R. F. Curl, R. E. Smalley, *Nature* **1985**, *318*, 162–163; (b) *Fullerenes: Chemistry, Physics and Technology*, K. M. Kadish, R. S. Ruoff (eds.), Wiley, New York, **2000**.

2 (a) L. T. Scott, *Angew. Chem.* **2004**, *116*, 5102–5116, *Angew. Chem. Int. Ed.* **2004**, *43*, 4994–5007; (b) L. T. Scott, H. E. Bronstein, D. V. Preda, R. B. M. Ansems, M. S. Bratcher, S. Hagen, *Pure Appl. Chem.* **1999**, *71*, 209–219; (c) G. Mehta, H. S. P. Rao, *Tetrahedron* **1998**, *54*, 13325–13370; (d) L. T. Scott, *Pure Appl. Chem.* **1996**, *68*, 291–300; (e) P. W. Rabideau, A. Sygula, *Acc. Chem. Res.* **1996**, *29*, 235–242; (f) P. W. Rabideau, A. Sygula, in *Advances in Theoretically Interesting Molecules*, Vol. 3, R. P. Thummel (ed.), JAI Press Inc., Greenwich, **1995**, pp. 1–36.

3 (a) W. E. Barth, R. G. Lawton, *J. Am. Chem. Soc.* **1966**, *88*, 380–381; (b) W. E. Barth, R. G. Lawton, *J. Am. Chem. Soc.* **1971**, *93*, 1730–1745.

4 J. Janata, J. Gendell, C.-H. Ling, W. E. Barth, L. Backes, J. B. Mark Jr., R. G. Lawton, *J. Am. Chem. Soc.* **1967**, *89*, 3056–3058.

5 J. C. Hanson, C. E. Nordman, *Acta Crystallogr. Sect. B.* **1976**, *32*, 1147–1153.

6 K. Yamamoto, T. Harada, Y. Okamoto, H. Chikamatsu, M. Nakazaki, Y. Kai, T. Nakao, M. Tanaka, S. Harada, N. Kasai, *J. Am. Chem. Soc.* **1988**, *110*, 3578–3584.

7 (a) J. T. Craig, M. D. Robins, W. *Aust. J. Chem.* **1968**, *21*, 2237–2245; (b) R. H. Jacobson, PhD Dissertation, University of California, Los Angeles, 1986; (c) J. R. Davy, M. N. Iskander, J. A. Reiss, *Tetrahedron Lett.* **1978**, 4085–4088.

8 L. T. Scott, M. M. Hashemi, D. T. Meyer, H. B. Warren, *J. Am. Chem. Soc.* **1991**, *113*, 7082–7084.

9 L. T. Scott, P.-C. Cheng, M. M. Hashemi, M. S. Bratcher, D. T. Meyer, H. B. Warren, *J. Am. Chem. Soc.* **1997**, *119*, 10963–10968.

10 (a) A. Borchardt, A. Fuchicello, K. V. Kilway, K. K. Baldridge, J. S. Siegel, *J. Am. Chem. Soc.* **1992**, *114*, 1921–1923; (b) G. Zimmermann, U. Nuechter, S. Hagen, M. Nuechter, *Tetrahedron Lett.* **1994**, *35*, 4747–4750; (c) C. Z. Liu, P. W. Rabideau, *Tetrahedron Lett.* **1996**, *37*, 3437–3440; (d) G. Mehta, G. Panda, *Tetrahedron Lett.* **1997**, *38*, 2145–2148; (e) H.-J. Knoelker, A. Braier, D. J. Broecher, P. G. Jones, H. Piotrowski, *Tetrahedron Lett.* **1999**, *40*, 8075–8078.

11 T. J. Seiders, K. K. Baldridge, J. S. Siegel, *J. Am. Chem. Soc.* **1996**, *118*, 2754–2755.

12 (a) J. E. McMurry, *Chem. Rev.* **1989**, *89*, 1513–1524 and references therein; (b) G. A. Olah, G. K. S. Prakash, *Synthesis* **1976**, 607–609.

13 A. Sygula, P. W. Rabideau, *J. Am. Chem. Soc.* **1998**, *120*, 12666–12667.

14 (a) A. Sygula, P. W. Rabideau, *J. Am. Chem. Soc.* **1999**, *121*, 7800–7803; (b)

T. J. Seiders, K. K. Baldridge, E. L. Elliott, G. H. Grube, J. S. Siegel, *J. Am. Chem. Soc.* **1999**, *121*, 7439–7440; (c) T. J. Seiders, E. L. Elliott, G. H. Grube, J. S. Siegel, *J. Am. Chem. Soc.* **1999**, *121*, 7804–7813.

15 A. Sygula, P. W. Rabideau, *J. Am. Chem. Soc.* **2000**, *122*, 6323–6324.

16 (a) A. Sygula, G. Xu, Z. Marcinow, P. W. Rabideau, *Tetrahedron* **2001**, *57*, 3637–3644; (b) G. Xu, A. Sygula, Z. Marcinow, P. W. Rabideau, *Tetrahedron Lett.* **2000**, *41*, 9931–9934.

17 A. Sygula, S. D. Karlen, R. Sygula, P. W. Rabideau, *Org. Lett.* **2002**, *4*, 3135–3137.

18 (a) G. Mehta, S. R. Shah, K. Ravikumar, *J. Chem. Soc., Chem. Commun.* **1993**, 1006–1008; (b) S. Cossu, O. De Lucchi, *Phosphorus, Sulfur Silicon Relat. Elem.* **1999**, *41*, 153–154.

19 H. Sakurai, T. Daiko, T. Hirao, *Science* **2003**, *301*, 1878.

20 (a) A. H. Abdourazak, A. Sygula, P. W. Rabideau, *J. Am. Chem. Soc.* **1993**, *115*, 3010–3011; (b) A. Sygula, A. H. Abdourazak, P. W. Rabideau, *J. Am. Chem. Soc.* **1996**, *118*, 339–343.

21 A. Sygula, H. E. Folsom, R. Sygula, A. H. Abdourazak, Z. Marcinow, F. R. Fronczek, P. W. Rabideau, *J. Chem. Soc.,Chem. Commun.* **1994**, 2571–2572.

22 (a) P. W. Rabideau, A. H. Abdourazak, H. E. Folsom, Z. Marcinow, A. Sygula, R. Sygula, *J. Am. Chem. Soc.* **1994**, *116*, 7891–7892; (b) A. H. Abdourazak, A. Z. Marcinow, A. Sygula, R. Sygula, P. W. Rabideau, *J. Am. Chem. Soc.* **1995**, *117*, 6410–6411.

23 (a) G. Mehta, G. Panda, *J. Chem. Soc., Chem. Commun.* **1997**, 2081–2082; (b) G. Mehta, G. Panda, P. V. V. S. Sarma, *Tetrahedron Lett.* **1998**, *39*, 5835–5836; (c) S. Hagen, M. S. Bratcher, M. S. Erickson, G. Zimmermann, L. T. Scott, *Angew. Chem.* **1997**, *109*, 407–409, *Angew. Chem. Int. Ed.* **1997**, *36*, 406–408.

24 A. Sygula, Z. Marcinow, F. R. Fronczek, I. Guzei, P. W. Rabideau, *Chem. Commun.* **2000**, 2439–2440.

25 M. A. Petrukhina, K. W. Andreini, L. Peng, L. T. Scott, *Angew. Chem.* **2004**, *116*, 5593–5597, *Angew. Chem. Int. Ed.* **2004**, *43*, 5477–5481.

26 (a) G. Mehta, P. V. V. S. Sarma, *Chem. Commun.* **2000**, 19–20; (b) B. McMahon, B.S. Thesis, Boston College, 1997; (c) C. C. McComas, B.S. Thesis, Boston College, 1996.

27 (a) M. S. Bratcher, Ph.D. Thesis, Boston College, **1997**; (b) H. A. Reisch, M. S. Bratcher, L. T. Scott, *Org. Lett.* **2000**, *2*, 1427–1430.

28 M. A. Petrukhina, K. W. Andreini, V. M. Tsefrikas and L. T. Scott, *Organometallics* **2005**, *24*, 1394–1397.

29 M. D. Clayton, P. W. Rabideau, *Tetrahedron Lett.* **1997**, *38*, 741–744.

30 (a) L. T. Scott, M. S. Bratcher, S. Hagen, *J. Am. Chem. Soc.* **1996**, *118*, 8743–8744; (b) R. B. M. Ansems, L. T. Scott, *J. Am. Chem. Soc.* **2000**, *122*, 2719–2724.

31 D. M. Forkey, S. Attar, B. C. Noll, R. Koerner, M. M. Olmstead, A. L. Balch, *J. Am. Chem. Soc.* **1997**, *119*, 5766–5767.

32 L. T. Scott, M. M. Boorum, B. J. McMahon, S. Hagen, J. Mack, J. Blank, H. Wegner, A. de Meijere, *Science* **2002**, *295*, 1500–1503.

33 (a) Y. Matsuo, K. Tahara, M. Sawamura, E. Nakamura, *J. Am. Chem. Soc.* **2004**, *126*, 8725–8734; (b) E. Nakamura, K. Tahara, Y. Matsuo, M. Sawamura, *J. Am. Chem. Soc.* **2003**, *125*, 2834–2835.

34 A. Sygula, R. Sygula, F. F. Fronczek, P. W. Rabideau, *J. Org. Chem.* **2002**, *67*, 6487–6492.

35 (a) P.-C. Chang, Ph.D. Dissertation, Boston College, Boston, MA, **1996**; (b) S. Mizyed, P. E. Georghiou, M. Bancu, B. Cuadra, A. K. Rai, P. Cheng, L. T. Scott, *J. Am. Chem. Soc.* **2001**, *123*, 12770–12774.

36 M. Bancu, A. K. Rai, P. Cheng, R. D. Gilardi, L. T. Scott, *Synlett* **2004**, 173–176.

37 G. H. Grube, E. E. Elliot, R. J. Steffens, C. S. Jones, K. K. Baldridge, J. S. Siegel, *Org. Lett.* **2003**, *5*, 713–716.

38 A. Sygula, R. Sygula, A. Ellern, P. W. Rabideau, *Org. Lett.* **2003**, *5*, 2595–2597.

39 A. Sygula, R. Sygula, P. W. Rabideau, *Tetrahedron Lett.* **2005**, 1189–1192.

40 (a) H. E. Zimmerman, G. L. Grunewald, *J. Am. Chem. Soc.* **1966**, *88*, 183–184; (b) E. Ciganek, *J. Am. Chem. Soc.* **1966**, *88*, 2882–2883; (c) J. Chen, J. R. Scheffer, J. Trotter, *Tetrahedron* **1992**, *48*, 3251–3274.

41 H. E. Zimmerman, D. Armesto, *Chem. Rev.* **1996**, *96*, 3065–3112 and references therein.

42 L. T. Scott, M. M. Hashemi, M. S. Bratcher, *J. Am. Chem. Soc.* **1992**, *114*, 1920–1921.

43 T. J. Seiders, K. K. Baldridge, G. H. Grube, J. S. Siegel, *J. Am. Chem. Soc.* **2001**, *123*, 517–525.

44 F. W. Dahlquist, K. J. Longmur, R. B. Du Vernet, *J. Magn. Reson.* **1975**, *17*, 406–410.

45 Z. Marcinow, A. Sygula, A. Ellern, P. W. Rabideau, *Org. Lett.* **2001**, *3*, 3527–3529.

46 M. Baumgartner, L. Gherghel, M. Wagner, A. Weitz, M. Rabinovitz, P.-C. Cheng, L. T. Scott, *J. Am. Chem. Soc.* **1995**, *117*, 6254–6257.

47 F. Gerson, W. Huber, *Acc. Chem. Res.* **1987**, *20*, 85–90.

48 A. Sygula, P. W. Rabideau, *Theochem* **1995**, *333*, 215–226.

49 E. Shabtai, R. E. Hoffman, P.-C. Cheng, E. Bayrd, D. V. Preda, L. T. Scott, M. Rabinovitz, *Perkin Trans. 2* **2000**, 129–133.

50 A. Ayalon, A. Sygula, P.-C. Cheng, M. Rabinovitz, P. W. Rabideau, L. T. Scott, *Science* **1994**, *265*, 1065–1067.

51 A. Ayalon, M. Rabinovitz, P.-C. Cheng, L. T. Scott, *Angew. Chem.* **1992**, *104*, 1691–1692, *Angew. Chem. Int. Ed.* **1992**, *31*, 36–37.

52 (a) P. W. Rabideau, Z. Marcinow, R. Sygula, A. Sygula, *Tetrahedron Lett.* **1993**, 6351–6354; (b) Z. Zhou, *J. Phys. Org. Chem.* **1995**, *8*, 103–107.

53 A. Weitz, E. Shabtai, M. Rabinovitz, M. S. Bratcher, C. C. McComas, M. D. Best, L. T. Scott, *Chem. Eur. J.* **1998**, *4*, 234–239.

54 A. L. Balch, M. M. Olmstead, *Chem. Rev.* **1998**, *98*, 2123–2165.

55 R. M. Shaltout, R. Sygula, A. Sygula, F. R. Fronczek, G. G. Stanley, P. W. Rabideau, *J. Am. Chem. Soc.* **1998**, *120*, 835–836.

56 F. Nunzi, A. Sgamellotti, *Organometallics* **2002**, *21*, 2219–2225.

57 T. J. Seiders, K. K. Baldridge, J. M. O'Connor, J. S. Siegel, *J. Am. Chem. Soc.* **1997**, *119*, 4781–4782.

58 C. M. Alvarez, R. J. Angelici, A. Sygula, R. Sygula, P. W. Rabideau, *Organometallics* **2003**, *22*, 624–626.

59 A. Sygula, unpublished results.

60 T. J. Seiders, K. K. Baldridge, J. M. O'Connor, J. S. Siegel, *Chem. Commun.* **2004**, 950–951.

61 M. A. Petrukhina, K. W. Andreini, J. Mack, L. T. Scott, *Angew. Chem.* **2003**, *115*, 3497–3501, *Angew. Chem. Int. Ed.* **2003**, *42*, 3375–3379.

62 E. L. Elliot, G. A. Hernandez, A. Linden, J. S. Siegel, *Org. Biomol. Chem.* **2005**, *3*, 407–413.

63 P. A. Vecchi, C. M. Alvarez, A. Ellern, R. J. Angelici, A. Sygula, R. Sygula, P. W. Rabideau, *Angew. Chem.* **2004**, *116*, 4597–4600, *Angew. Chem. Int. Ed.* **2004**, *43*, 4497–4500.

64 P. A. Vecchi, C. M. Alvarez, A. Ellern, R. J. Angelici, A. Sygula, R. Sygula, P. W. Rabideau, *Organometallics* **2005**, *24*, 4543–4552.

65 (a) R. M. Chin, M. S. Jarosh, J. D. Russel, R. J. Lachicotte, *Organometallics*, **2002**, *21*, 2027–2029; (b) R. M. Chin, B. Baird, M. Jarosh, S. Rassman, B. Barry, *Organometallics*, **2003**, *21*, 4429–4432.

66 (a) F.-G. Klaerner, J. Panitzky, D. Preda, L. T. Scott, *J. Mol. Model.* **2000**, *6*, 318–327; (b) M. V. Frash, A. C. Hopkinson, D. K. Bohme, *J. Am. Chem. Soc.* **2001**, *123*, 6687–6695; (c) R. C. Dunbar, *J. Phys. Chem. A* **2002**, *106*, 9809–9819; (d) R. Faust, K. P. C. Vollhardt, *J. Chem. Soc., Chem. Commun.* **1993**, 1471–1473; (e) J. Plater, H. S. Rzepa, F. Stoppa, S. Stossel, *J. Chem. Soc., Perkin Trans. 2* **1994**, 399–400; (f) A. Sygula, P. W. Rabideau, *J. Chem. Soc., Chem. Commun.* **1994**, 2271–2272.

67 R. B. M. Ansems, L. T. Scott, *J. Phys. Org. Chem.* **2004**, *17*, 819–823.

13
Reduction of Carbon-rich Compounds

Tamar Sternfeld and Mordecai Rabinovitz

13.1
Introduction

Studies of metal reduction of polycyclic aromatic hydrocarbons (PAHs) [1] were reported as early as 1867 by Berthelot [2], where he described the fusion of metallic potassium with naphthalene 1. Studies in solution of electron transfer from Na and K to PAHs started as early as 1913 with Schlenk et al. [3] who reported that when anthracene reacted with the alkali metal in ether, 1:1 and 2:1 metal: anthracene adducts were obtained. Not understanding what the 1:1 adduct really was, it was termed "radical" and later on it was suggested by Hückel that the "radical" is a singly charged species [4].

In this chapter we concentrate on reduction processes of carbon-rich systems. The formation of anions and radical anions of π-conjugated monocyclic systems, cyclophanes, bowls and fullerenes is described. Carbon-rich compounds can be reduced directly by contact with alkali metals: Li, Na, K, Rb and Cs, which have a low reduction potential. Proton, carbon and lithium NMR and EPR spectroscopies are the main methods used to gain a better understanding of the mono- and polycyclic systems in solution. Special attention will be given to modes of electron delocalization, aromaticity, anti-aromaticity, as well as aggregation, bond formation and bond cleavage processes of diamagnetic electron transfer products. Electrochemical reductions will be briefly discussed.

Upon reaction with a metal, the NMR spectrum of a neutral hydrocarbon first disappears, due to one-electron transfer from the metal. This step can be studied by EPR spectroscopy [1a, 5]. The reduction of carbon-rich compounds usually occurs stepwise by one-electron transfers as depicted in Scheme 13.1 [1b]. The same set of events occurs in electrochemical processes and in alkali metal reductions. This reaction has become the main synthetic route to π-conjugated anions.

The electron transfer method and the reducing metal are very significant, since they can influence the reduction products. This affects the reduction state of the anions formed, the electronic and magnetic spectra, and the dynamic and structural changes that the negatively charged molecules undergo [1b–d].

Carbon-Rich Compounds. Edited by Michael M. Haley and Rik R. Tykwinski
Copyright © 2006 WILEY-VCH Verlag GmbH & Co. KGaA, Weinheim
ISBN: 3-527-31224-2

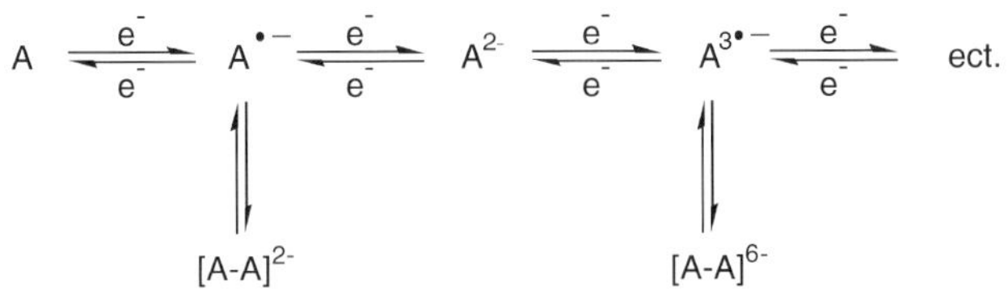

$$A \underset{e}{\overset{e^-}{\rightleftharpoons}} A^{\bullet -} \underset{e}{\overset{e^-}{\rightleftharpoons}} A^{2-} \underset{e}{\overset{e^-}{\rightleftharpoons}} A^{3\bullet -} \underset{e}{\overset{e^-}{\rightleftharpoons}} \quad \text{ect.}$$

$$\quad\quad\quad\quad \updownarrow \quad\quad\quad\quad\quad\quad\quad\quad \updownarrow$$

$$\quad\quad\quad [A\text{-}A]^{2-} \quad\quad\quad\quad\quad\quad [A\text{-}A]^{6-}$$

Scheme 13.1. Electron transfer process upon reduction.

13.2
Electron Transfer Methods and Detection of the Products

13.2.1
Reduction via Alkali Metals

The reduction using alkali metals is carried out in dry aprotic solvents, usually THF, and in an inert atmosphere (vacuum). The metals used are Li (wire), Na, K, Rb and Cs (mirror). The PAHs can accept several charges (up to 6 electrons). Consecutive paramagnetic and diamagnetic states can be observed by EPR and NMR spectroscopies, respectively. Many examples of reduction with alkali metals are discussed in depth here. Corannulene (**2**) offers an excellent example of the application of this reduction method. **2** was reduced by Li [6, 7], Na, K, Rb and Cs [8], and four reduction steps were observed, each having a different color (with the exception of sodium, which gave only two reduction steps). The four reduction steps yielded mono-, di-, tri- and tetra-anions, which were characterized by EPR or NMR [6–9].

Reduction of PAHs in a solution of alkali metal in liquid ammonia, known as the Birch reaction [10], is a conventional synthetic way to reduce polyaromatic rings to their dihydro derivatives [11]. Highly basic dianions are produced as intermediates, and are protonated by the ammonia to afford monoanions. In most cases, the monoanions persist in ammonia, and are quenched when a stronger proton source is added [12]. The choice of alkali metal affects the kinetics of the protonation of the monoanion intermediate.

13.2.2
The Application of Electron Shuttles

One obstacle in the reduction of carbon-rich molecules using alkali metals is their low solubility, especially when the compounds are large; however, the anions of such compounds are soluble in THF. One way to overcome this difficulty is by using another carbanion as an intermediate reagent to aid the electron transfer. In this manner, the soluble carbanion can effectively transfer electrons to the compounds of low solubility and thus reduce them. This method was used already in 1960 by both Horner and Normant, who used the radical anion of **1** for a variety of reduction reactions, such as the Wurtz synthesis [13]. The sodium salt of **1** is more effective than sodium itself in yielding new carbanions.

The new carbon allotropes, fullerenes [14] and nanotubes [15], can accept a large number of electrons to their π-system; however, the neutral compounds are hardly soluble, especially in THF. Because of the low solubility of fullerenes, their complete reduction to hexaanions with Li is possible only under extreme conditions, such as the use of an ultrasound bath [16]. When the reduction of fullerenes is performed in the presence of a small amount of **2**, moderate conditions are required, e.g., low temperature and without sonication [17]. Corannulene, which is a fragment of fullerenes, is highly soluble in THF and is easily reduced. The resulting corannulene anion serves as an efficient electron shuttle, which assists in complete reduction of the practically insoluble fullerenes. In a similar way, single walled nanotubes were reduced by charge transfer by the use of small aromatic compounds, such as naphthalene, fluorenone and anthraquinone [18].

1

2

13.2.3
Potassium Graphite Intercalates as Reducing Agents

Graphite reacts with alkali metals – potassium, cesium and rubidium – to form lamellar compounds with different stoichiometries. The most widely known intercalate is the potassium-graphite which has the stoichiometry of C_8K. In this intercalate the space between the graphite layers is occupied by K atoms. C_8K functions as a reducing agent in various reactions such as: reduction of double bonds in α,β-unsaturated ketones [19], carboxylic acids and Schiff bases; alkylation of nitriles [20], esters and imines [21]; reductive cleavage of carbon–sulfur bonds in vinylic and allylic sulfones [22]. The detailed reaction mechanism of C_8K is not known, and the special properties which are ascribed to the intercalate come either from the equilibrium between K^+/K^0 [23], or topochemical observations (the layer structure) [24].

Using C_8K as a reducing agent has a number of advantages such as: (i) selectivity; (ii) the ability to react at room temperature; (iii) high yields; (iv) short reaction times; and (v) even unique reactivity. The heterogeneous graphite can be easily separated, thus simplifying the work-up procedure.

Glocking and Kingston [25], and later Rabinovitz [26], showed that the reduction of phenyl halides (to obtain biphenyl) with C_8K afforded 85–99% yield, in contrast

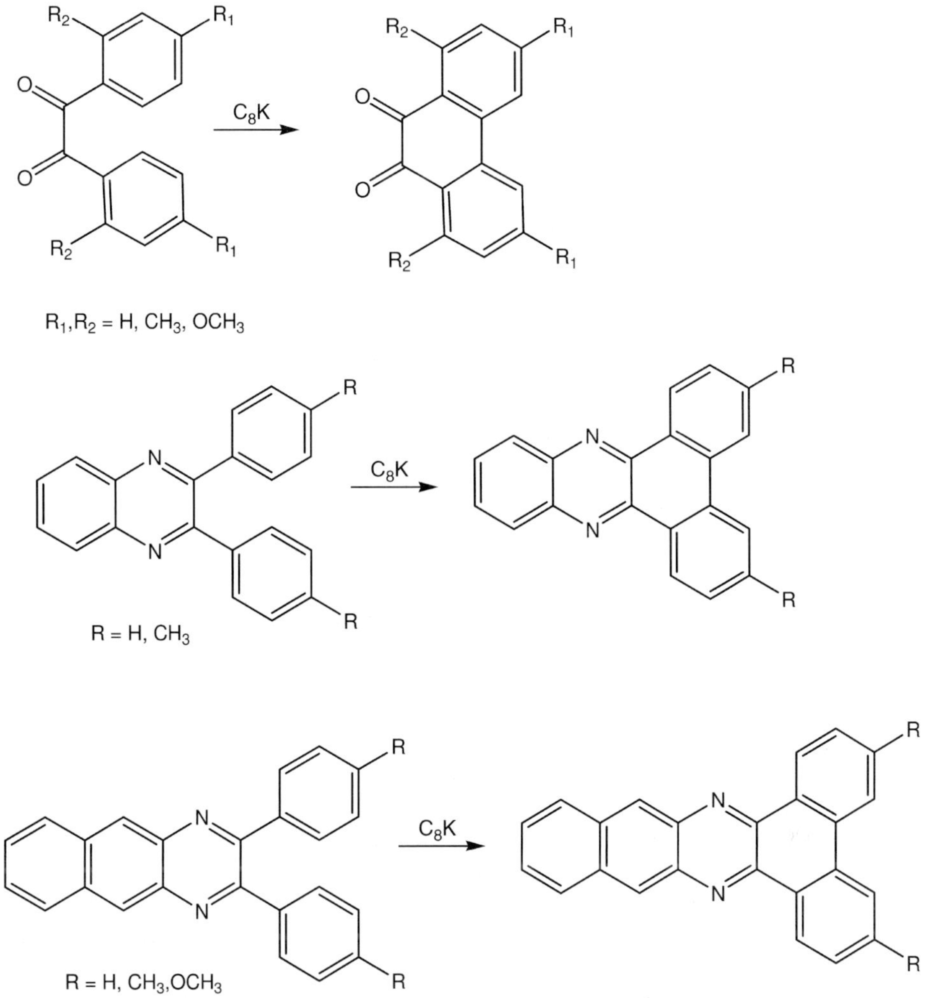

Scheme 13.2. Ring closure of α-diketones and nitrogen-containing heterocyclic compounds.

to the typically low yields in other coupling reactions. A unique reactivity is observed when C_8K is applied to reduce carbonyl compounds. For example, benzophenones undergo reductive coupling reactions to yield the corresponding pinacols [27]. In the reduction reactions of α-diketones and nitrogen-containing heterocyclic compounds, the reaction affords polycyclic systems via a ring closure process [28] (Scheme 13.2).

The reduction of substituted naphthalene with C_8K demonstrates the high selectivity of the reduction, as only the 1,4-dihydronaphthalene is obtained [29]. This method can be used as an alternative route to the Birch reduction and its modifications, which produce mixtures of 1,4-dihydronaphthalene, 1,2-dihydronaphthalene and tetrahydronaphthalene [30]. Additional advantages of using C_8K are the moderate conditions required in the reduction. The product composition does not change when a large excess of C_8K is used when the reaction temperature is elevated, or when a more convenient solvent is used (THF instead of liquid ammonia).

C_8K was also used in the reduction of PAHs, and the resulting anions were identical to those obtained by a direct reduction with a K mirror. The lower reactivity of C_8K, however, makes it a more selective reagent [31].

13.2.4
Electrochemical Studies

Cyclic voltammetry (CV) can provide information about the thermodynamics of the redox process, kinetics of heterogeneous electron transfer reactions and coupled chemical reactions [32]. The reversible electron transfer steps inform us about the compound's ability to accept electrons; however, experimental conditions, such as solvent and temperature also influence the voltammogram. The structure of the lowest unoccupied molecular orbital (LUMO) levels of the compound can be determined from the number of CV waves and reduction potentials ($E_{1/2}$). Moreover, the CV can serve as a "spectroscopy" as demonstrated by Heinze [32], since the characteristic shapes of the waves and their unequivocal positions on the potential scale are effectively a fingerprint of the individual electrochemical properties of the redox system.

Recently, the electrochemistry of fullerenes and their derivatives has gained much attention [33]. C_{60}, C_{70} and higher fullerenes were reduced electrochemically, and six reduction waves were observed for both C_{60} and C_{70} [34], as well as for most of the higher fullerenes [35]. The energy levels that were obtained from these experiments were mostly in line with MO calculations. The electrochemistry of numerous fullerene derivatives was studied to compare their electron affinities and energy levels with their parent fullerenes. Electrochemically induced isomerizations can be observed in CV, as is the case in the rearrangement of fulleroids to methanofullerenes [36].

An electrochemical study of the redox properties of 23 PAHs was recently reported by Jenneskens [37]. The PAHs show a good linear correlation between their first reduction potential, which was measured by CV ($E_{1/2}$), and their standard Hückel LUMO energy ($-\varepsilon_{LUMO}/\beta$). Moreover, the one-electron reduction study allowed the determination of the effect of 5-membered rings (5MR) on PAH molecules. It was shown that the annelation of a peripheral 5MR to the molecular perimeter enhances its electron affinity, since the ring acts as electron-withdrawing *peri*-substituent.

Electrochemistry can also be used as a synthetic tool, as it can provide electrons to the reaction and simultaneously measure the reaction progress *in situ*. A few methanofullerenes were synthesized by electrosynthesis, generating C_{60}^{2-}, which is a strong base [38].

13.2.5
Ion Solvation Equilibria in π-Conjugated Reduced Systems

The reduction process of polycycles by alkali metals converts the neutral atoms to anions. The electron transfer is best achieved in ethereal solvents. This enables the stabilization of the metal cation by coordination to the oxygen atoms of the solvent. The hydrocarbon anion and the cation are linked together by electrostatic forces in which the solvent molecules are also involved; therefore, the ion-solvation equilibrium should be considered [39]. The limiting cases in this equilibrium are free ions and contact ion-pairs (CIP), and in between there are several forms of solvent separated ion-pairs (SSIP) [40]. In reality, anionic species of aromatic hydrocarbons in ethereal solvents exist between CIP and SSIP states. Four major factors influence the ion solvation equilibrium of alkali metal reduced π-conjugated hydrocarbons, as observed by ^{1}H and ^{7}Li NMR spectroscopies [39, 41]. (i) Charge delocalization of the anion: localized charge density encourages the formation of contact ion-pairs. In the case of π-conjugated polycyclic systems solvent separated ion-pairs are preferred. (ii) Size of the cation: hard lithium cations prefer to form SSIP unless the lithium cation resides inside a molecular cage or in between layers of anions. (iii) Solvent: ethereal solvents having a high dielectric constant prefer the formation of SSIP. (iv) Temperature: generally speaking CIP are preferred from entropy arguments (less ordered than SSIP). Higher temperatures encourage the formation of CIP [42].

The following observation emphasizes the influence of the temperature on the ion solvation equilibrium. The reduction product of naphthalene **1** with Li metal in 2-methyltetrahydrofuran is temperature dependent [43]. At −120 °C only the radical anion (**1**$^{\cdot-}$) could be observed by EPR, while at higher temperatures the paramagnetism disappears and the dianion (**1**$^{2-}$) is detected. This reaction must be endothermic, it therefore seems that disproportionation is driven by entropy and not by energy, due to the ion pair solvation equilibrium. It is noteworthy that the dianion **1**$^{2-}$ cannot be observed by NMR spectroscopy due to its special electronic structure [44].

13.2.6
Magnetic Resonance Spectroscopy – Tool of Choice for the Study of Reduced Carbon-rich Compounds

13.2.6.1 NMR
The stability of anionic systems is governed by several factors: (i) carbon hybridization, (ii) effective overall π-conjugation, (iii) inductive effects, (iv) aromatic stabilization and (v) environmental factors, e.g., ion-solvation equilibria.

The ^{1}H, ^{13}C and ^{7}Li NMR study of polycyclic anions includes 1D and 2D methods that enable an unequivocal structure elucidation of the molecules. Through-bond and through-space interactions as well as correlation spectroscopy allow the study of various aspects of Li reduced polycyclic hydrocarbons described in the following sections. The carbon and proton chemical shifts allow elucidation of the charge delocalization [39, 45], structure of the system under study, the mode of conjugation [46], and the magnetic anisotropy of the system.

It is possible to use the carbon chemical shifts to get information about the charge density at each carbon atom. The link between chemical shift and charge density has been studied both experimentally and theoretically [39, 47, 48]. A linear dependence between charge density and NMR chemical shifts was suggested by Fraenkel [48]. An empirical relationship has been formulated as follows:

$$\Delta\delta = K\Delta q_\pi \tag{1}$$

where $\Delta\delta$ is the chemical shift difference relative to the neutral polycycle; Δq_π is the charge density difference; K is a constant ($K_H = 10.7$ ppm e^{-1}; $K_C = 160$ ppm e^{-1}) [49]. In reality, K (70 < K < 200) depends on the class of the systems under study [50]. A generalized equation was suggested by Karplus [51] that takes into account the paramagnetic term (σ_{para}) of the basic Ramsey equation [52] that links the three components of the shielding of a nucleus:

$$\sigma_{total} = \sigma_{dia} + \sigma_{para} + \sigma' \tag{2}$$

The effect of environmental factors is included in σ', and σ_{dia} represents the diamagnetic effects. The Karplus equation relates the paramagnetic term to the carbon type, bond polarity and average excitation energy (ΔE) of the anion. When ΔE is small then the paramagnetic term becomes significant [53]. An improved relationship between the total carbon shift difference and the number of negative charges on the anion was reported by Müllen giving greater weight to anion anisotropy [50]. This equation better evaluates the degree of charging and the mode of charge delocalization.

13.2.6.2 EPR

Electron paramagnetic resonance (EPR) spectroscopy has played a major role in the study of charged organic compounds, as it can be used in the identification of the radical anions formed [54]. The equation describing the absorption (or emission) of microwave energy between two spin states is:

$$\Delta E = h\nu = g\beta B \tag{3}$$

where ΔE is the energy difference between the two spin states, h is the Planck constant, ν is the microwave frequency, g is the Zeeman splitting factor, β is the Bohr magneton, and B is the applied magnetic field. Four main values can be deduced from a typical EPR spectrum: (i) the resonance field position (g factor); (ii) the spectrum structure (hyperfine network); (iii) the line width (related to the spin relaxation process); (iv) the intensity (the integral of the absorption signal). The spin densities of the π-system can be calculated from the EPR hyperfine splitting [55], using the well-known McConnell relationship:

$$a_{iH} = Q_{CH}{}^H \rho_i{}^\pi \tag{4}$$

where a_{iH} is the hyperfine splitting due to the proton residing on carbon atom i, ρ_i^π is the π-electron spin density at carbon atom i, and Q_{CH}^H is a constant.

13.3
Monocyclic π-Conjugated Anions

13.3.1
Annulenes: General

Annulenes [56] are defined as monocyclic conjugated π-systems of the general formula $(CH)_{2i}$ [57]. The neutral compounds will possess [4n+2] π-electrons or [4n] π-electrons, when i is an odd or even number, respectively. The [4n+2]annulenes and [4n]annulenes are predicted to have different spectroscopic properties, as expressed in terms of aromaticity (Hückel rule [58]), anti-aromaticity [59] and nonaromaticity. [4n+2]annulenes have degenerate or nearly degenerate LUMO levels that are separated by large energy gaps from the degenerate (or nearly degenerate) highest occupied molecular orbital (HOMO) levels [60]. These levels are populated by electrons in a closed shell ground state. Such electronic structures are characteristic of aromatic systems. When the orbitals are filled by [4n] π-electrons, biradicals are expected to form; however, in such systems a degenerate ground state is not retained. A pseudo Jahn–Teller distortion leads to closed shell [4n]annulenes, which are associated with bond-length alternation (BLA) [61].

Annulenes have been utilized in the investigation of the effect of adding (or removing) electrons to π-conjugated systems [1b]. Annulenes can be used in demonstrating the validity of the Hückel theory and the experimental criteria of aromaticity [57], because the molecular frame can hold different numbers of electrons.

Molecular orbital (MO) diagrams show that when a [4n]annulene is reduced to a [4n+2]dianion, the LUMO level becomes fully populated, resulting in a stable ground state and an aromatic system. Conversely, the reduction of a [4n+2]annulene affords a [4n]dianion, that possesses a low-lying LUMO, which can undergo a pseudo-Jahn–Teller distortion. These [4n]annulenes are predicted to be anti-aromatic [1b, 56–61].

In this section we describe the reduction of some "classical annulenes", [8]-annulene **3**, [12]annulene **4**, [16]annulene **5** and [18]annulene **6**. In all of these systems the reduction is accompanied by changes in the molecular frame. We shall concentrate on the NMR spectra of the diamagnetic anions and the EPR spectra of the [4n+1]-radical anions, both of which serve as probes for the delocalization and planarity of these π-systems. The Hückel theory [58], which only considers the π-energy, neglecting the σ-frame, can only describe planar, strain-free π-perimeters; therefore, it is important to analyze the structure of the annulenes and their anions.

Most annulenes are not completely planar, and except for the unique case of benzene, annulenes do not conform to the full D_{nh} symmetry [56, 57]. The adoption of nonplanar conformations is not only typical of the [4n]annulenes, but also occurs in [4n+2]annulenes. In annulenes larger than benzene, the planar conformation is unfavorable, and even rings with all-cis peripheral bonds suffer from increasing angle strain.

To exclusively measure the "pure electronic effects" of reduction, it is necessary to study rigid and planar compounds. The reduction of a few types of "rigid annulenes" has been reported in the literature. Bridged annulenes, e.g., bridged [10]-annulene **7** and bridged [14]annulene **8**, are rigid annulenes, where reduction and oxidation have been actively studied. In these systems the bridge provides an effective means for holding the annulene ring in a specific geometry [62]. When the bridge is aliphatic and its protons are located above the center of the ring, they show high-field NMR shifts for aromatic [4n+2]annulene components, and low-field shifts when anti-aromatic [4n]annulene components are present. Other rigid annulene derivatives were also reduced to anions, such as dihydropyrenes **9**, which are annulenic in the perimeter [63]. A detailed description is given below concerning the reduction of dehydroannulenes and cyclophanes. In these two families, a series of homologous compounds was investigated, and the electronic effects were demonstrated.

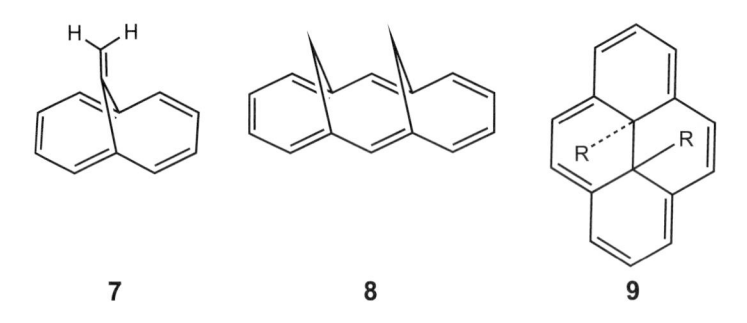

7   **8**   **9**

13.3.2
"Classical Annulenes"

13.3.2.1 Cyclooctatetraene–[8]Annulene

Cyclooctatetraene, i.e., [8]annulene **3**, was the first [4n] π-electron hydrocarbon to be studied [57, 64]. It was found that **3** adopts a "tub" conformation, that minimizes the electronic interaction between the double bonds, and makes the system less conjugated and thus "nonaromatic" [65]. Compound **3** undergoes a dynamic process of ring inversion, which involves D_{4h} transition states that have been intensively studied [66].

Compound **3** was reduced to its dianion by Katz in his pioneering work concerning the reduction of annulenes [67]. Compound **3** was reacted with two moles of either Li or K, and in both cases the NMR spectrum showed a single sharp peak at low-field, which is characteristic of a planar aromatic ring (3^{2-}). It was therefore concluded that reduction formed an aromatic anion, which consisted of an eight-membered ring with 10 π-electrons, thus fulfilling the [4n+2] π-electron requirement.

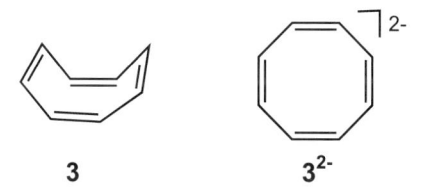

3   **3^{2-}**

The radical anion of **3** [67c,d] also attracted much attention and the main issue in its study was whether or not it was planar. The nine lines in the EPR spectrum of the **3** radical anion imply that the unpaired electron is equally distributed over the eight carbon atoms; however, the splitting constant suggests the presence of a planar octagon with bent bonds. Borden and Hammons [68] showed theoretically that the ground state of the **3** radical anion can be represented by a degenerate pair of Jahn–Teller distorted D_{4h} BLA structures. They also showed that the transition state can be represented by a pair of D_{4h} structures with alternating bond angles (ABA).

The kinetic studies of the electron exchange between the radical anion and the dianion of **3**, compared with the reaction between the radical anion and the neutral molecule [67c], provide insight into the structure of the radical anion. The reaction rate was found to be dependent on the counter-cation, in the order Li > Na > K. The electron exchange reaction occurs rapidly between the radical anion and the dianion, and much more slowly between the radical anion and the neutral molecule. This behavior is different from that observed in most aromatic hydrocarbons. Naphthalene shows extremely fast exchange between the radical anion and the neutral molecule, and a slower one with the dianion [69]. The exchange rates of the radical anion of **3** suggest that its carbon skeleton is similar to that of the dianion, and not that of the neutral molecule [67c].

Many studies in this area focus on the reduction of substituted **3**, for example, the reduction of the four isomers of dimethylcyclooctatetraene [70]. These were reduced by K, and all the formed dianions were characterized as planar or essentially planar. The steric interactions and planarity of the radical anions of *tert*-butoxy-**3** were also studied [71]. Another interesting derivative of **3** is 1,5,22,26-tetraoxa-[5,5]-(2,8)-dibenzo[*a,e*]cyclooctatetraenophane **10** [72], which is the first macrocycle incorporating two **3** units. A pair of diasteriomeric tetraanions were observed in the NMR when **10** was reduced with alkali metals. The NMR spectra and cyclovoltammetric studies suggest that the two **3** subunits in **10**$^{4-}$ are flat dianions.

13.3.2.2 [12]Annulene

The [12]annulene (**4**) system is thermodynamically unstable due to bond angle strain, steric interactions between the inner protons and the number of π-electrons [56, 57]; therefore, this molecule has not been studied intensively. A possible configuration for **4** is one in which the double bonds are alternatively arranged in a cis and trans (tri-trans) fashion; however, this conformation deviates strongly from planarity [73].

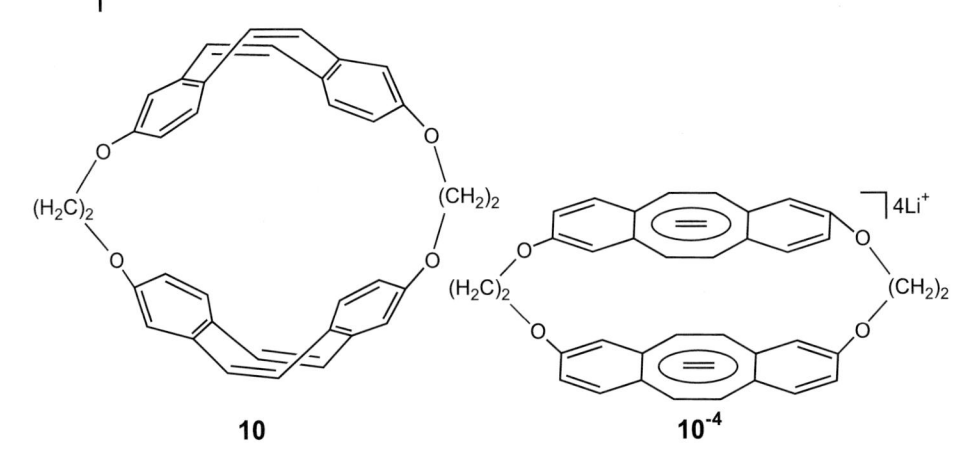

Compound **4** was reduced electrochemically by Oth [74], and two reversible reduction steps were observed. Reduction with Li metal yields two anions: a radical anion that was observed by EPR, and a dianion that was observed by NMR. The dianion is much more stable than the neutral molecule, as expected from a 14 π-electron system, i.e., [4n+2] π-electrons. The ^{1}H NMR chemical shifts of the inner protons appear at high field and those of the outer protons are at low field, as expected from the ring current model. The NMR spectrum was unaffected by temperature changes, indicating that the dianion is highly stable, and that the free energy for the dynamic process ($\Delta G^\ddagger$) is higher than 18 kcal mol^{-1} ($\Delta G^\ddagger_{273}$, neutral = 6.6 kcal mol^{-1}) [73]. Both NMR data and Hückel molecular orbital (HMO) calculations indicate the presence of high resonance energy in **4**$^{2-}$.

Stevenson made a comprehensive study of the radical anions of **4** [75]. The EPR spectrum of the anion of the tri-trans isomer (**4a**) shows that its structure is similar to the dianion, but the narrow spectral line width (9.8 G) indicates a severe distortion from planarity.

The reduction of the di-trans isomer (**4b**) with K was recently reported [76]. Its radical anion has a different EPR spectrum (13.5 G) from that of the tri-trans isomer. The well resolved spectrum reveals that most of the spin density resides on the planar side of the ring, rather than on the twisted side (Fig. 13.1(a)). Warming the solution results in the formation of the radical anion of heptalene, which has a

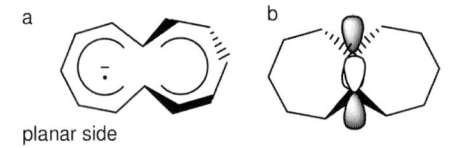

Figure 13.1. Di-*trans* [12]annulene radical anion **4b**, (a) spin density of the radical anion; (b) the anion radical of di-*trans*-[12]annulene revealing the overlap of the p$_z$ orbitals on carbons 1 and 7.

"normal" EPR spectrum, and normal total spectral width (26.5 G), as expected for a planar hydrocarbon radical anion. The observed rearrangement is due to the overlap of the p_z orbitals of carbons 1 and 7, which evolve into a new σ-bond, with the loss of a hydrogen molecule (Fig. 13.1(b)).

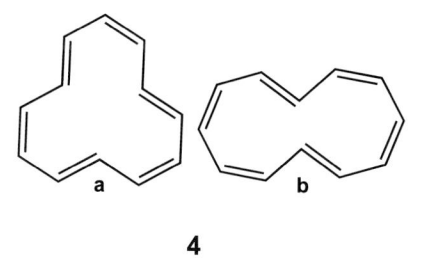

13.3.2.3 [16]Annulene

The [4n] π-electron [16]annulene (**5**) has been extensively studied, and a special focus has been placed on the dynamic equilibrium found between the two different configurations, which differ in the number and sequence of cis and trans double bonds [77]. Each of these configurations exhibits single and double bond alternations, and undergoes conformational changes and fast bond shifts. The planarity of this annulene (as well as **6**) is affected by the steric interactions of the inner protons. Recently, some new conformers of **5** and **6** were observed and analyzed [78].

Adding one or two electrons to **5** influences the energy, geometry and dynamic behavior of the system [79]. The dianion of **5** contains [4n+2] π-electrons, and displays aromatic behavior, as deduced from its NMR spectrum. The diamagnetic character associated with the delocalized 18 π-electrons in the reduced annulene manifests itself in the extremely high-field shift of the inner protons (−8.07 ppm) and the low-field shift of the outer protons (8.77, 7.40 ppm) [79]. The symmetry of the dianion is most likely D_{4h}, although D_4 and D_{2d} cannot be ruled out.

The dianion shows great stability at elevated temperatures. No dynamic process was observed over a wide range of temperatures (−100 °C < T < 140 °C), which reflects the high resonance energy associated with the system [79]. The deuterated [16]annulene-d_{15} was synthesized and reduced, and it was found that the antiaromatic and aromatic characters of the neutral and dianionic systems are augmented upon deuteration [80]. It was shown that the resonance of the inner proton of the dianion is shifted up-field by $\delta = 0.09$ ppm. The deuteration allows flattening of the ring, since the C–D bond is shorter than the C–H bond, which decreases the interaction between the inner hydrogens.

The radical monoanion was monitored by EPR. The EPR data, based on symmetry and HMO considerations, led to important conclusions. The radical-anion does not exhibit bond alternation, but rather it exhibits delocalization of the 17 π-electrons and has a planar geometry; therefore, a high resonance energy was associated with the radical-anion **5**$^{\cdot -}$.

Vincow and Concepcion reported a convenient approach to the synthesis of the radical anion of **5**, by the reduction of the syn [2+2] dimer of cyclooctatetraene (**11**) [81]. It was found that when **11** is reduced by an alkali metal at room temperature, the radical anion 5·⁻ is formed [82]. The electron rich **3** radical anion reacts with the electron deficient **3** to yield 5·⁻. The mechanism of this process was explored by adding D_2O to the solution and studying the species formed and the isotopic exchange by both EPR and NMR [83]. The reduced **3** solution contained a mixture of deuterated **3** radical anion, **3** dianion, 5·⁻, and 5²⁻. The kinetic studies allowed the proposal of a complete mechanism.

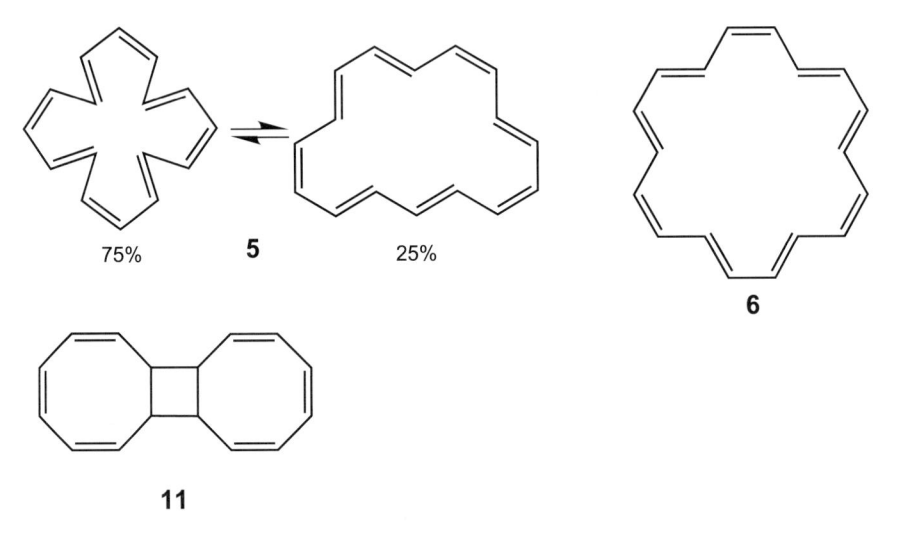

13.3.2.4 [18]Annulene

The reduction of [18]annulene (**6**) demonstrates the transformation of a [4n+2] π-system to a [4n] π-system. The large size of **6** permits only weak interactions between the six inner hydrogens, leading to an essentially planar structure [78]. The annulene possesses conformational mobility associated with the exchange of the position of the inner and outer hydrogens, which decreases the aromatic stabilization [84]. In a recent paper, Schleyer established that **6** shows C–C bond alternation and can adopt lower symmetries than D_{6h} [85]; nonetheless, **6** has aromatic properties, as depicted from its ¹H NMR spectrum and the aromatic stabilization energy [86].

The radical anion of **6** was first observed by Sondheimer [87] and recently by Stevenson [88]. The main conclusion from these studies is that the structure is nonplanar, because of a Jahn–Teller distortion. The large total spectral width in the radical anion indicates that the carbons with negative spin densities contribute extensively to the coupling. The complex EPR spectrum eliminates the possibility of pseudorotation on the EPR time scale, which could lead to uniform spin densities. The rotation is prevented by the nonplanarity of the radical anion.

The dianion **6**²⁻ was prepared and analyzed by Sondheimer, and was characterized as a nonaromatic species [87]. The temperature dependence of the ¹H NMR spectrum clearly shows an equilibrium between the two conformers of **6**²⁻. The conformers have lower symmetry than the neutral molecule, and are most likely nonplanar and exhibit π-bond alternation. The ^{1}H NMR chemical shifts could be rationalized in terms of a paramagnetic ring current induced in a cyclic [4n] π-electron system, which is associated with π-bond delocalization, i.e., anti-aromatic character. The nonplanarity and the bond alternation reduce the paramagnetism.

13.3.3
Dehydroannulenes

Dehydroannulenes are annulenes which contain one or more triple bonds. The triple bonds enforce rigidity on the annulene skeleton and can lead to planar systems. A problem that could arise in such compounds is that the triple bond (orthogonal to the π-system) can disturb the electronic configuration.

Sondheimer obtained the first dianion of a dehydroannulene by reducing 1,3,7,9,13,15,19,21-octadehydro[24]annulene (**12**) with K [89]. Earlier, only the reductions to radical anions had been reported, as is the case in the reduction of 1,8-bisdehydro[14]annulene (**13**) by Wratten [90], and the reduction of 1:2,5:6, 9:10-tribenzo-3,4,7,8,11,12-tridehydro[12]annulene (**14**) by Staab and Graf [91]. Neutral **12** is nonplanar and nonaromatic, probably due to the presence of four cis double bonds and the rigidity imposed by the four 1,3-diyne units. The EPR pattern of the radical anion shows that it has a planar structure and that the π-electrons are delocalized. The ^{1}H NMR spectrum of the dianion, which contains 26 π-electrons, shows low-field signals, indicating that the π-system is delocalized over the planar system, i.e., an aromatic system.

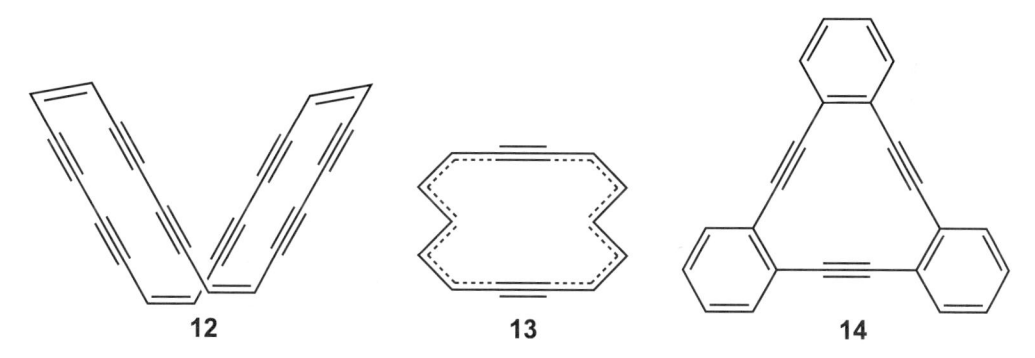

12   **13**   **14**

Müllen performed a systematic study of the changes in the ring current of neutral, dianionic and tetraanionic annulenes having different ring sizes but identical architectures [92]. This was performed by the reduction of the following series – tetra-*t*-butyl-didehydro[14]annulene (**15**), tetra-*t*-butyl-didehydro[18]annulene (**16**), tetra-*t*-butyl-didehydro[22]annulene (**17**), and tetra-*t*-butyl-didehydro[26]annulene (**18**) [93].

Annulenes **15–18** have well-defined ring configurations and conformations, and are essentially planar. These observations make this family a good model for studying the nature of π-bond delocalization and the influence of ring size on the reduction process. The homologous [4n+2]annulene series **15–18** was reduced by Li,

Na, K, Rb and Cs [92]. The [4n]annulene dianions 15^{2-}–18^{2-} were first to be observed, followed by the [4n+2]annulene tetraaanions 16^{4-}–18^{4-}. As expected, the NMR spectra of all the above species show diamagnetic ring currents for the [4n+2]annulenes, i.e., aromaticity, and paramagnetic shifts for the [4n]annulenes, i.e., anti-aromaticity. This means that in the neutral and tetraanionic compounds, the outer protons are deshielded and the inner ones are shielded. The opposite effect is observed for the dianions. For example, the ^{1}H NMR chemical shifts of the outer protons are 7.93, 1.20 and 9.84 ppm in **18**, **18^{2-}** and **18^{4-}**, respectively, whereas the inner protons resonate at 1.95, 23.39 and 1.17 ppm, respectively. The chemical shift difference between the outer and the inner protons is a function of reduction state and the ring size. Obviously, the diamagnetic ring current effect in tetraanions is always larger than in the corresponding neutral annulenes, which means that the π-delocalization energy is higher in the tetraanions. When the effective ring current effect was analyzed relative to the ring size, it was found that it decreases with increasing ring size [92]. This trend was found in both diamagnetic and paramagnetic ring currents; moreover, it was found that the decrease in the ring current was most rapid in the dianions.

The systems 1,5-bisdehydro[12]annulene (**19**) [94] and 1,5,9-tridehydro[12]-annulene (**20**) [95] are paratropic, nearly planar and contain 12 out-of-plane π-electrons. These two systems were reduced with K, and in both cases a radical anion was observed by EPR and a dianion by NMR [96]. The EPR spectra appeared immediately after reduction at low temperature and it was shown that the radical anion is in equilibrium with the neutral molecule. This is in contrast with what was observed for **3** and its radical anion, and reflects the fact that both **19** and **20** do not have to overcome a barrier of ring flattening when reduced. The NMR spectra of both dianions, which contain 14 π-electrons, clearly show their aromaticity. The inner proton of 19^{2-} appears at high field (-6.88 ppm), whereas in the neutral molecule it appears at $\delta = 10.9$ ppm [94]. The chemical shift difference of the outer protons is less marked. In **20** there is only one type of outer proton, which appears at $\delta = 4.45$ ppm [95a], and is shifted to low field in the dianion ($\delta = 6.74$ ppm) [96]. The dynamic process of exchanges between the intenal and outer protons of neutral **19** was not observed in the diaanion, just as for **5** [96].

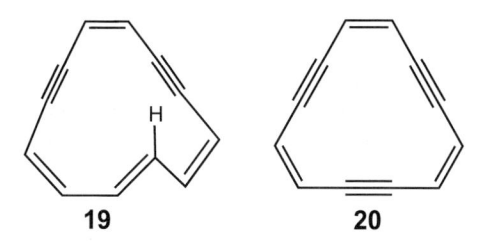

13.3.4
Cyclophanes

Cyclophanes [97], in which the aromatic units are linked at the 1,4 positions by unsaturated bridges, can be viewed as derivatized annulenes. They therefore provide a playground for the study of the aromaticity of annulenes. This family of carbon-rich systems behaves like rigid annulenes with well-defined ring configurations; therefore their reduction is the choice for systematic study of the effect of the number of the π-electrons on aromaticity.

The neutral systems exhibit typical ^{1}H NMR shifts for aromatic and olefinic protons and there is no indication of the existence of peripheral ring-current effects. The benzene rings rotate rapidly around the neighboring single bonds on the NMR time scale, exchanging the internal and outer protons. The X-ray structure of [2.2.2.2]paracyclophanetetraene (**21**) shows that the molecule has an essentially planar arrangement, in which the individual benzene rings are tilted toward the main molecular plane by 27–40° [98].

The first alkali metal reduction study of cyclophanes was conducted by Müllen, who reduced **21** with Li [99]. Two anions were observed in the NMR spectra: a dianion and a tetraanion. The chemical shifts of both these anions showed that a peripheral conjugation is present. The reduction drastically slowed down the rotation of the benzene fragments, because of an increase in the π-bond order of the single bonds, thus allowing spectroscopic discrimination between the inner and outer protons.

The ^{1}H NMR of the dianion showed a signal at low field ($\delta = 9.26$) that was assigned to the outer protons and one at very high field ($\delta = -7.07$) that was assigned to the inner protons. In the tetraanion, the signals of the inner and outer protons exchanged locations (high field for the outer protons and low field for the inner protons). The NMR data are in agreement with the number of π-electrons in the periphery: 26 π-electrons, a Hückel number for the dianion that shows diamagnetic ring currents, and 28 π-electrons ([4n]) for the tetraanion, that shows paramagnetic ring currents.

The reduction of a series of cyclophanes (**21–29**), which are closely related but structurally different, helped understanding of the effects of the ring size, the nature of the aromatic unit, planarity and heteroatoms on the ring current [100]. A prerequisite in all these model cyclophanes was the existence of [4n] π-electrons in the perimeter; 24 π-electrons in **21**, **24**, **25** and **26**, 28 π-electrons in **22** and **27**, 32 π-electrons in **23** 36 π-electrons in **28** and 20 π-electrons in **29**. All of the neutral cyclophanes **21–29** show chemical shifts that are characteristic of olefinic and aromatic protons and only weak indications of anisotropy effects can be observed.

The picture is completely different in all the corresponding dianions, **21^{2-}–29^{2-}**. They exhibit ^{1}H NMR shifts at very low field for the outer protons ($\delta = 9$–10 ppm), and very high field for the inner protons ($\delta = -6$ to -10 ppm). This demonstrates the presence of high diamagnetic ring currents in the [4n+2] π-perimeters. Two tetraanions were observed in the reduction of **21** and **26**. The position of their ^{1}H NMR signals are reversed relative to the dianions – high-field shifts for the outer protons, and low-field shifts for the inner ones, as a result of their antiaromatic character.

The chemical shift differences between the outer and inner protons can serve as an index for comparing the ring currents effect in the systems, and could also reveal structural effects. The inner–outer shift difference increases upon going from **21^{2-}** to **22^{2-}** to **23^{2-}**, all having the same type and number of aromatic moieties, differing only in the ring size. The relative ring current effect, which is defined as $\Delta\delta/S$ (where $\Delta\delta$ is the inner–outer difference and S is the ring area), decreases with the increase in ring size (1.00, 0.81 and 0.73 for **21^{2-}**, **22^{2-}** and **23^{2-}**, respectively), similar to the behavior of the dehydroannulenes. On the other hand, comparing the inner–outer shift difference of **21^{2-}** with the less planar cyclophane **24^{2-}**, the benzo-condensed derivative **26^{2-}**, and the thiophene derivative **25^{2-}**, only results in slight changes. This leads to the important conclusion that the ring current effect in dianions, which are formed from cyclophanes with different numbers and types of aromatic subunits, is similar.

To study the electronic effect as a function of ring size, the following homologous series, **21**, [2.2.0.2.2.0]paracyclophane-1,9,23,31-tetraene (**30**), [2.2.0.0.2.2.0.0]-paracyclophane-1,9,29,37-tetraene (**31**), [2.0.2.0.2.0.2.0]paracyclophane-1,15,29,43-tetraene (**32**), was reduced, with both Li and K [101]. All of these cyclophanes exhibit aromatic or anti-aromatic character, as a function of the number of added

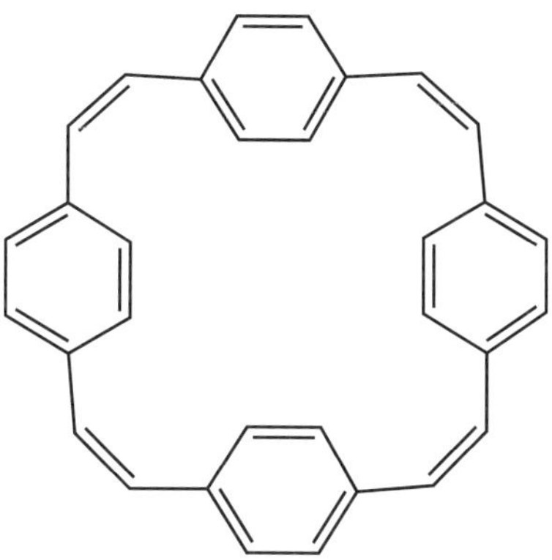

13.3 *Monocyclic π-Conjugated Anions* | 583

21

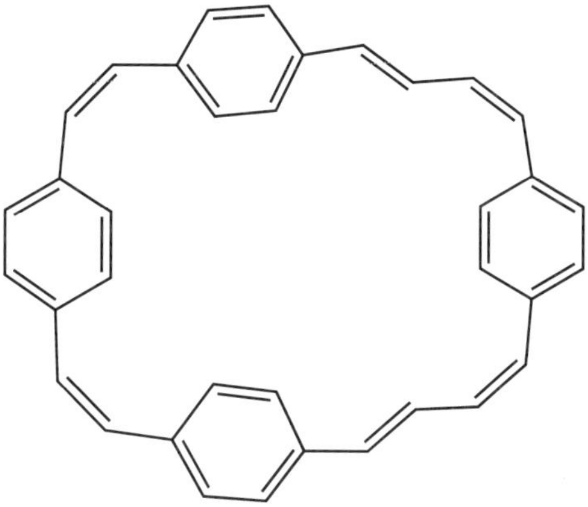

22

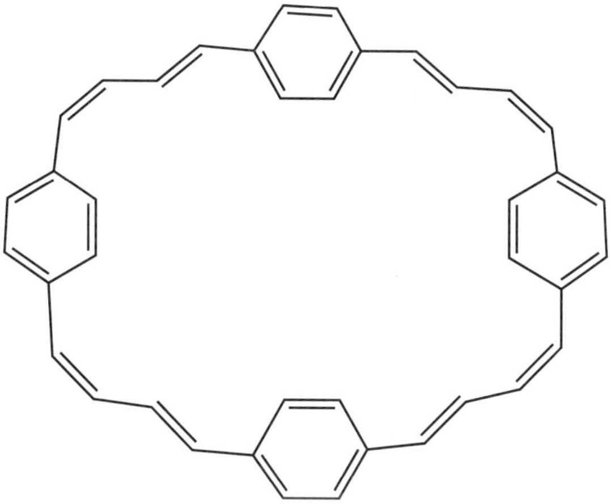

23

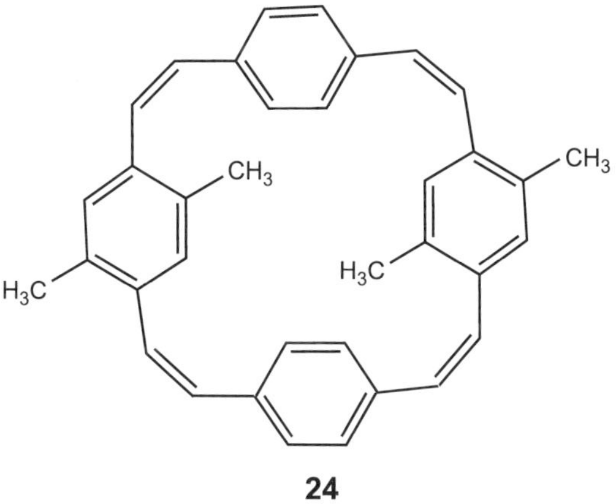

24

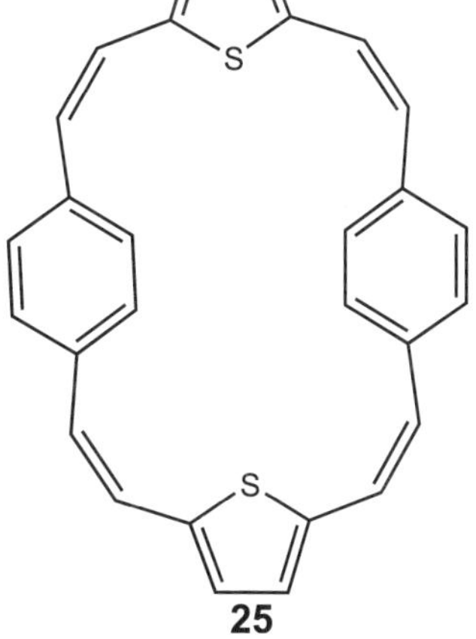

25

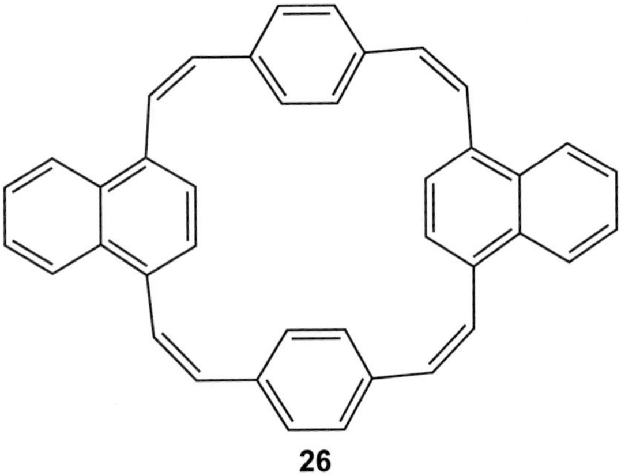

26

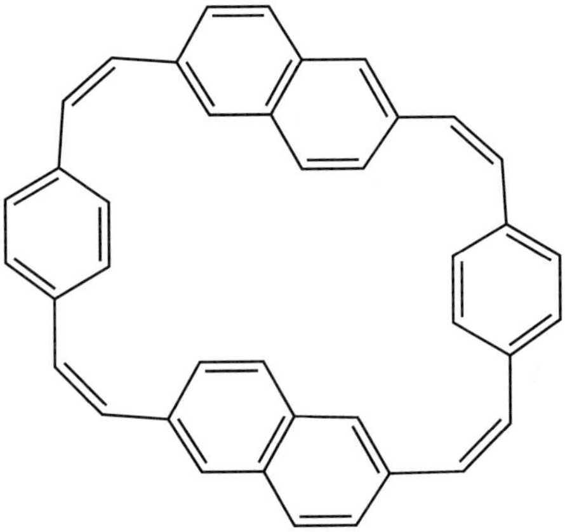

27

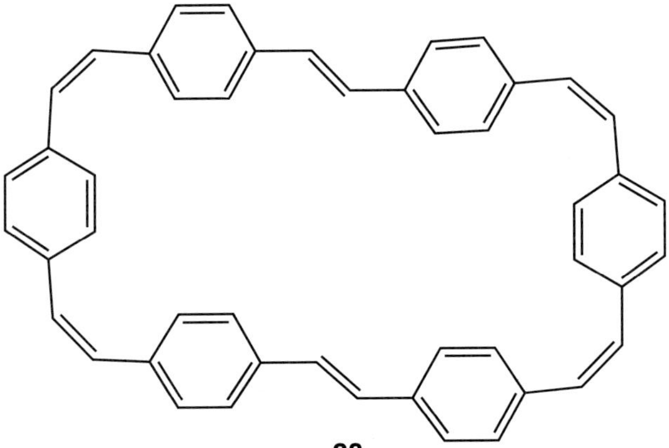

28

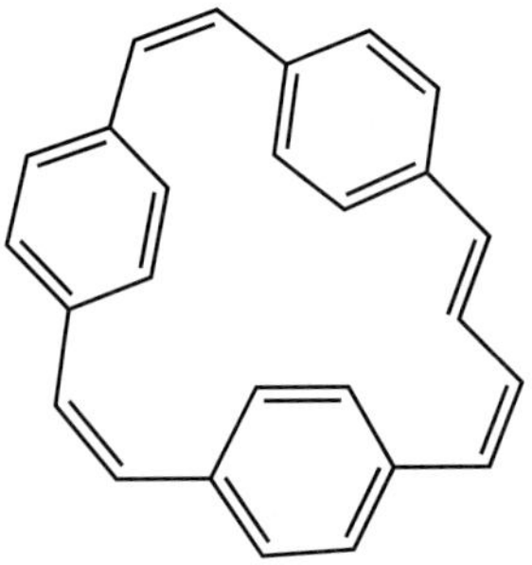

29

electrons. The ^{1}H NMR spectra of the dianions, which all have [4n+2] π-electrons, show low-field shifts for the outer protons (δ = 9.64, 9.90, 10.69 and 10.65 for **21**$^{2-}$, **30**$^{2-}$, **31**$^{2-}$ and **32**$^{2-}$, respectively) and high-field shifts for the inner protons. By

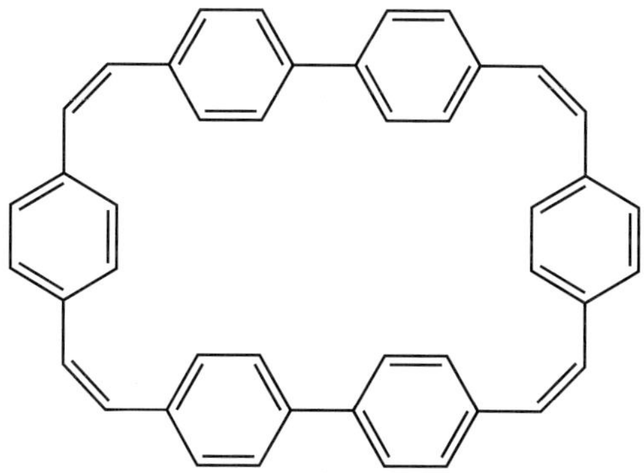

13.3 Monocyclic π-Conjugated Anions

30

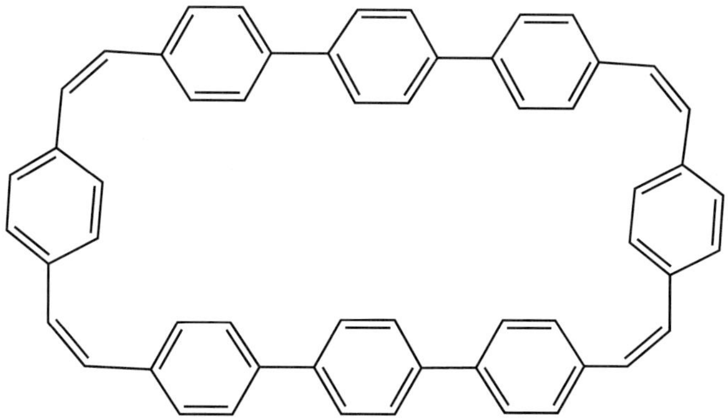

31

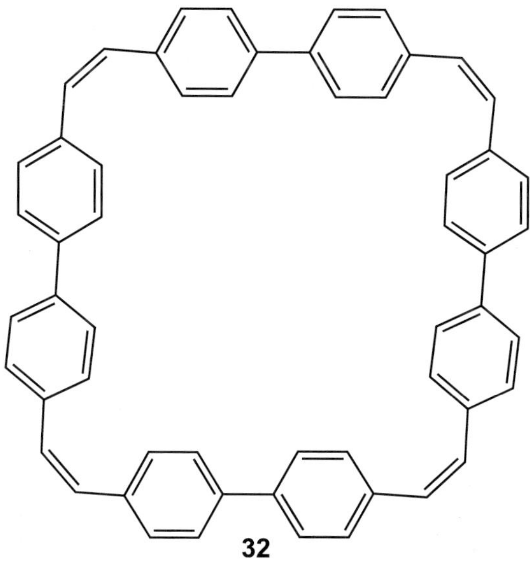

32

contrast, the tetraanions that contain [4n] π-electrons have outer protons that appear at high field (δ = 0.92, 0.02, 0.55 and −0.54 for 21^{4-}, 30^{4-}, 31^{4-} and 32^{4-}, respectively) and inner protons that appear at extremely low field (δ = 16.44, 18.7, 16.56 and 20.21 for 21^{4-}, 30^{4-}, 31^{4-} and 32^{4-}, respectively). There is a dependence between the proton chemical shift and the size of the cyclophane in the dianions and tetraanions. The only hexaanion that was observed in this series was the hexaanion of **32**. It shows a diamagnetic ring current, which is expressed by the high-field shift of the inner protons, and low-field shift of the outer protons, as expected from a system with 46 π-electrons.

The size and the geometry of the rings influence the degree of reduction, which means that a correlation exists between the ring size and the highest reduction state ($21^{4-}/4K^+$, $30^{5-}/5K^+$, $31^{6-}/6K^+$, and $32^{5-}/5K^+$). Some differences were observed between the Li and K reduction processes. Small rings yield stable anions with Li^+, while the larger systems prefer the heavier metal. When **21** was reduced with Li a stable dianion and tetraanion were formed; however, the reduction with K led to the formation of the tetraanion only after extended reaction time. The opposite behavior was found in the reduction of **30**. While only a dianion was formed with Li, K gave a tetraanion and even a pentaanion. The explanation for this effect is mainly based on the differences in the ion pairing equlibria and the capability of Li^+ to aggregate and stabilize anions, but not on the difference in the reduction potential.

13.4
Reduction of Nonplanar Polycyclic Aromatic Hydrocarbons

13.4.1
Nonplanar PAHs

The study of the aromaticity of nonplanar systems attracts considerable attention, as they depart from the original Hückel definition for aromaticity, where planarity of the ring is a prerequisite. Hopf [102] divided nonplanar aromatic compounds into three principal categories: (i) angularly annelated benzene rings (e.g. helicenes), (ii) systems in which non-six-membered rings are incorporated into the aromatic skeleton (e.g. buckybowls), and (iii) bridged systems (e.g. cyclophanes). The discovery of fullerenes [14] and the subsequent isolation of C_{60} (**33**) [103] have rekindled interest in the study of curved PAHs, especially those that can be mapped onto the surface of C_{60} or other fullerenes, as well as their aromatic behavior. The progression from smaller to larger charged fragments of the fullerene surface provides a better understanding of fullerene-like anions [104].

The reduction of two types of nonplanar PAHs is discussed here. We first describe the reduction of curved PAHs that contain five-membered rings (5MR), which cause deviation from planarity. This group is represented by corannulene (**2**) and its derivatives, as **2** is the smallest curved subunit of buckminsterfullerene.

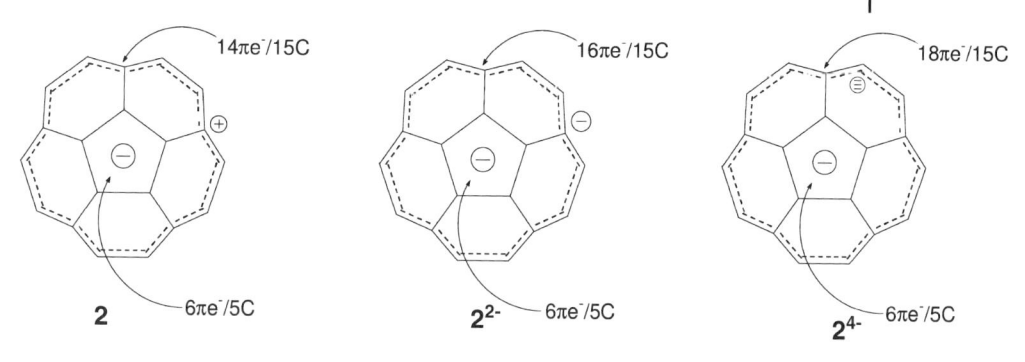

Figure 13.2. "Annulene within an annulene" π-topology of corannulene **2**.

The second group is the cyclophanes of pyrene, where a planar system is forced to bend by linking both ends of the pyrene with an alkane chain. This highly strained curved group represents a compromise between strain and conjugation, and the nature of these systems' aromaticity is of great interest.

13.4.2
PAHs with Five-membered Rings

13.4.2.1 Corannulene

Corannulene (**2**) was first synthesized by Barth and Lawton in 1966 [105] in a 16-step process. Later on, other convenient synthetic methods for **2** were reported by Scott [106], Siegel [107], Zimmermann [108] and Rabideau [109]. The new methods made **2** more accessible for experimental studies on the bowl-shaped hydrocarbon. The ^{1}H NMR spectrum of **2** consists of one line ($\delta = 7.93$), indicating a diamagnetic ring current about the perimeter. The model "annulene-within-an-annulene", suggested by Barth and Lawton [105, 110], describes **2** as a system that is composed of an inner ring with 6 π-electrons, and an outer ring with 14-π electrons (Fig. 13.2). The lowest unoccupied molecular orbital (LUMO) of **2** lies low in energy and is doubly degenerate [104]. This accounts for the ease with which it can accept up to four electrons.

13.4.2.2 Reduction of Corannulene with Lithium

The reduction of **2** with Li is accompanied by a series of three color changes, [6, 7, 9], representing four anions, which can be observed by EPR for the paramagnetic mono- and trianions, and by NMR for the diamagnetic di- and tetraanions. The detection and characterization of all four reduction stages with Li metal allows the delineation of a full reduction path for **2**.

The corannulene monoanion **2**$^{•-}$ which is green in THF solution, has an EPR spectrum that contains 11-line hyperfine patterns [111] ($a_H = 0.157$ mT [111]/ 0.156 mT [112], $g = 2.0027$ [111]/2.0029 [112]) for the ten equivalent protons. At low temperatures the spectrum becomes more complicated. The radical anion **2**$^{•-}$

did not undergo microwave power saturation, even at 200 K, preventing ENDOR studies. This behavior was attributed to the high symmetry of $2^{\cdot -}$ [9].

Upon further reduction, the EPR intensity decreased, and the sample became purple and diamagnetic. The corannulene dianion 2^{2-} was first thought to be paramagnetic due to the doubly-degenerate, low-lying LUMO [1d]; however, ^{1}H and ^{13}C NMR spectra were detected due to a Jahn–Teller distortion [9]. The extreme high-field shifts of the protons ($\delta = -5.6$) are typical of anti-aromatic systems [62], and can be explained by the "annulene-within-an-annulene" π-topology model, where a cyclopentadienyl anion (6e/5C) is placed in the center of an antiaromatic (16e/15C) annulene perimeter. The temperature-dependent behavior of the proton spectrum, which cannot be seen at high temperatures ($T \geq 230$ K), was attributed to the mixing of a low-lying triplet state with the electronic structure of 2^{2-} [9]. Despite the two extra electrons in its backbone, the ^{13}C NMR spectrum of 2^{2-} showed three signals ($\delta = 120$, 154, and 204 ppm) at low field relative to those of the neutral compound. The extremely low field signal of the quaternary hub carbon ($\delta = 204$ ppm) was explained by the strong deshielding effect of the outer antiaromatic ring current. Quenching of the dianion with water afforded dihydrocorannulene as the major product, where the two protons reside on the perimeter [8].

In the third stage of the reduction, where the solution was brown, corannulene trianion ($2^{3\cdot -}$) was detected by EPR. Counterion couplings were observed in the trianion radical Li species [113]. The trianion radical of **2** with Li ($2^{3\cdot -}$) shows a highly resolved EPR spectrum where additional ^{7}Li (and ^{6}Li [113]) couplings are evident [$g = 2.0025$, $a_H = 1.62$ mT, ^{7}Li ($I\ 5\ 3/2$) $= 0.04$–0.10 mT]. Varying the temperature from 200–290 K caused reversible changes in the appearance of the EPR spectrum, presumably as a result of tightening and loosening of the associated Li cations [9].

The most fascinating stage in the reduction of **2** is the fourth stage (**34**), in which an octaanionic dimer is formed. The ^{1}H NMR spectrum of **34** consisted of a single line at $\delta = 6.95$ ppm, which is in line with an aromatic system. The ^{13}C NMR spectrum which corresponds to this stage shows three signals ($\delta = 86.8$, 95.1, and 112.4 ppm) at a higher field than those of the neutral hydrocarbon ($\delta = 127.9$, 132.3 and 136.9 ppm). The π-topology model of "annulene-within-an-annulene" accurately describes the tetraanion as an aromatic cyclopentadienyl anion (6e/5C) surrounded by an aromatic (18e/15C) annulenyl trianion, which accounts for its stability. Quench experiments with water afforded tetrahydrocorannulene as a major product, thus confirming the reduced state of the system. Analysis of the oxidation product by NMR spectroscopy showed clearly that one proton is attached to the hub while the other three are attached to the perimeter, hence supporting the annulene model [8].

The tetraanion **34** has been characterized as a sandwich compound, in which four Li cations "glue" together two tetraanion decks of **2**, and four external Li cations that are attached the exterior. The ^{7}Li NMR spectra show the formation of a dimer. The ^{7}Li NMR spectrum of **34** recorded at a low temperature (210 K) [6] features two signals of equal intensity (δ_{Li} −4.5 and −11.7 ppm). These chemical shifts represent two different types of Li cations, one sandwiched between the two tetraanion units (CIP) and one on the outside. At higher temperatures (e.g. 260 K), the Li cations all undergo exchange, resulting in a single line at $\delta = -8.1$ ppm. The addition of LiBr to the solution causes an averaging of its signals with that of the broad (SSIP) ^{7}Li signal, but the sharp ^{7}Li signal (CIP) remains unchanged, indicating that it is tightly bound [6].

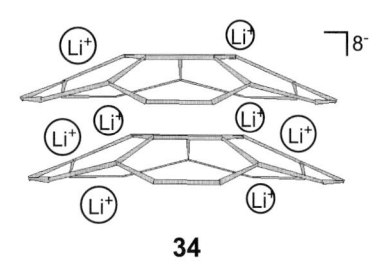

34

Diffusion measurements [114] on corannulene show a significant decrease in the self-diffusion coefficient of the tetraanion, **34**, further confirming self-aggregation. The formation of the dimer takes place only at the tetraanionic stage [9]. No change in the characteristic hyperfine pattern could be observed in the EPR spectra of the trianion radical over the entire reduction process, which eliminated a dimerization process for this anion. As far as the dianion is concerned, based on its magnetic behavior, it was concluded that 2^{2-} exists as a monomer with two associated lithium cations. The ^{7}Li NMR spectrum featured only one broad absorption around $\delta = 3.3$–3.7 ($T = 173$–220 K), showing only a small temperature effect, typical for loose ion-pair structures (SSIP), and singlet triplet mixing [115].

13.4.2.3 Reduction of Corannulene with Different Alkali Metals

As a result of the above mentioned studies on the anions of **2**, it was first thought that Li is the only alkali metal that can produce the highly charged tetraanion due to its tendency to aggregate. Later, ^{1}H and ^{13}C NMR spectra were also detected for the diamagnetic anions of **2** reduced with Na, K, Rb and Cs [8]. Di- and tetraanions were observed for all alkali metals other than sodium (sodium was the only metal where the reduction stopped at the dianion stage). The UV–vis spectra of the dianionic solutions with different counterions, showed only small differences (up to 4 nm). This indicates that the cations are loosely bonded. In contrast to the highly stable tetraanion with Li, the tetraanions with K, Rb, and Cs could only be detected after a long period of reduction. The spectra of these tetraanions yielded very broad signals, and could be obtained only at low temperatures, indicating the presence of equilibria with a paramagnetic species (trianion radical) or an energetically close triplet state. The NMR chemical shifts of the reduced species with K, Rb and Cs are similar, yet are different from the Li reduction product, suggesting the existence of different types of solvation states. The paramagnetic species, mono- and tri-anions formed by the reduction with K, were observed by EPR, and the differences that were seen with K$^+$ and Li$^+$ as counterions were negligibly small [9].

To clarify the temperature dependence of the spectrum of $2^{·-}$/Li$^+$, Sato measured the EPR spectrum of $2^{·-}$/Na$^+$ [112]. Two interpretations for the temperature-dependent complicated spectra were considered. The first being the possibility of the formation of ion pairs consisting of the alkali metal and corannulene anion, similar to $2^{3 \cdot -}$ [9]. The second being that the Jahn–Teller (JT) effect will lower the molecular structure symmetry, and thus the hydrogen atoms may become nonequivalent. Since the spin–spin interaction between the odd electron of the corannulene anion and the metal nucleus sodium ($I = 3/2$) gave rise to the same hyperfine splitting as Li, it was concluded that the temperature dependence is due to the occurrence of the JT effect. The resulting spectra can be interpreted in terms of pseudorotation about the JT crossing and fifth-order anharmonic term.

13.4.2.4 Substituted Corannulenes

The monosubstituted corannulene derivatives tert-butylcorannulene (**35**) and isopropylcorannulene (**36**) have lower symmetry than corannulene; therefore their reduction provided evidence for the formation of the dimers of corannulene tetraanion. The dimers of monosubstituted corannulene tetraanions are expected to exhibit supramolecular stereochemistry, existing in meso and d,l dimeric forms.

Reduction of **35** with Li led to two sets of alkyl groups in almost equal abundance, thus indicating the presence of tightly bound dimers. In the reduction of **36** the two methyl groups in each isopropyl substituent becomes diasteriotopic, so each dimer gives rise to two methyl hydrogen doublets in the ^{1}H NMR spectrum, even at room temperature [6].

Compelling evidence for dimerization came from the successful detection of a "mixed dimer" between $2^{4-}/4Li^+$ and $35^{4-}/4Li^+$ [6]. In this experiment, a 5:1 mixture of **2** and **35** was reduced and detected by NMR. The NMR spectra contained the dominant singlet of self-dimer of 2^{4-} (**34**), which was stoichiometrically favored. On the other hand, the signals of the self-dimers of 35^{4-}, the stoichiometrically unfavorable species, hardly appeared. In addition, a new set of signals also appeared in the spectra that could not be attributed to the self-dimer of either of the tetraanions. The new peaks were assigned as the signals of the mixed dimer, $2^{4-}/35^{4-}/8Li^+$.

In the reduction of **35** to its radical anion $35^{\cdot-}$ the chromophoric properties were not changed significantly ($g = 2.0027$), and the ENDOR measurements allowed resolution of at least eight different proton hyperfine couplings [9].

The chiral penta-substituted corannulene, 1,3,5,7,9-penta-tert-butylcorannulene (**37**), which shows anomalous dynamic behavior [116], was reduced with Li and four reduction stages were observed [117]. The behavior of the anti-aromatic dianion and the aromatic tetraanion resemble that of the parent corannulene. In the final stage of the reduction, three distinct tetraanionic species could be detected. Two of the species are sandwich-type dimers, as in the case of corannulene, and are diastereomers, due to the chirality of **37** (RR/SS and meso – RS/SR). The third species was assigned as a tetraanionic monomer, which slowly disappears.

Following the dimerization of corannulene, a system of two tethered corannulene units, 1,8-dicorannulenyloctane (**38**) [118], was reduced with alkali metals to study the formation of an intramolecular sandwich [8]. The reduction was done with Li, Na, K, Rb and Cs, and the two diamagnetic anions tetraanion and octa-

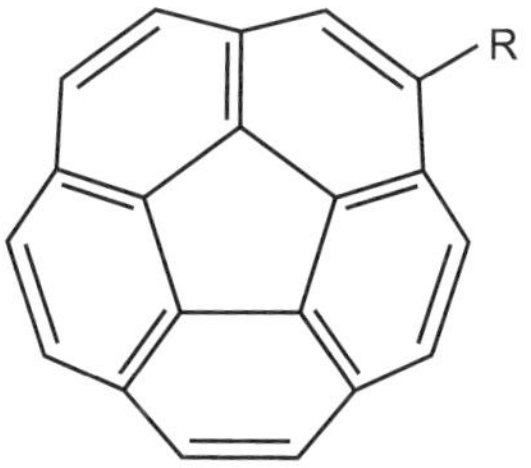

35 R = *t*-Bu
36 R = *i*-Pr

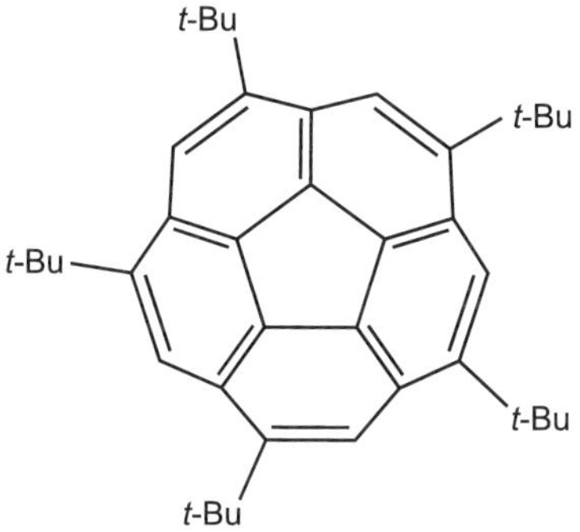

37

anion (total charge) were characterized by their NMR spectra (except for sodium, which stopped at the tetraanion stage). Except for slight differences in the chemical shifts, the reduction of **38** with these alkali metals gave similar spectra.

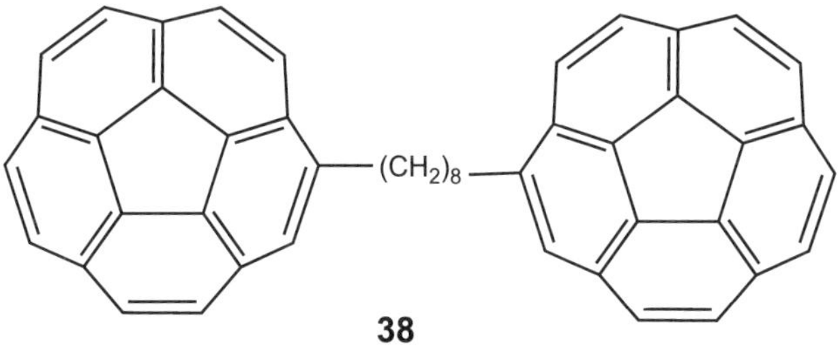

38

All the reduction steps are similar to those observed in the reduction of **2**. The tetraanion of **38** parallels the dianion of corannulene and represents two antiaromatic doubly charged corannulene moieties linked by an alkane chain. The protons of the corannulene moiety in **38**$^{4-}$ appeared in the ^{1}H NMR spectrum at high field ($\delta_H = -3.5$ to -4.7 ppm), due to the anti-aromatic character of the system. The protons of the methylene units also appeared at relatively high field for alkane protons ($\delta_H = -1.94$ to 0.28), due to the influence of the induced paramagnetic ring current. The bowl-to-bowl inversion barrier of the tethered corannulene dianions in **38**$^{4-}$ was determined by variable temperature ^{1}H NMR spectroscopy [8]. This was the first time that the barrier for a bowl-shaped charged species was determined. The barrier was found to be 8.8 ± 0.3 kcal mol^{-1} with potassium counter ions and 9.2 ± 0.3 kcal mol^{-1} with cesium counter ions. As expected, the reduction decreases the energy barrier for bowl-to-bowl inversion relative to that of the neutral **38** (10.9 ± 0.3 kcal mol^{-1}).

The second diamagnetic anion is the octaanion **38**$^{8-}$, which consists of two sets of ^{1}H NMR peaks. One set is between $\delta_H = 6.7$ and 7.2 ppm, in the same region as the signal of the **34** itself, therefore they were assigned to the protons of the corannulene fragments. The second set contained four broad peaks in the range $\delta_H = 2.2$ to 3.2 ppm assigned to the protons of the tether. The ^{7}Li NMR spectrum of **38**$^{8-}$ is very similar to that of **34** (dimer of **2**$^{4-}$). The chemical shifts of the two

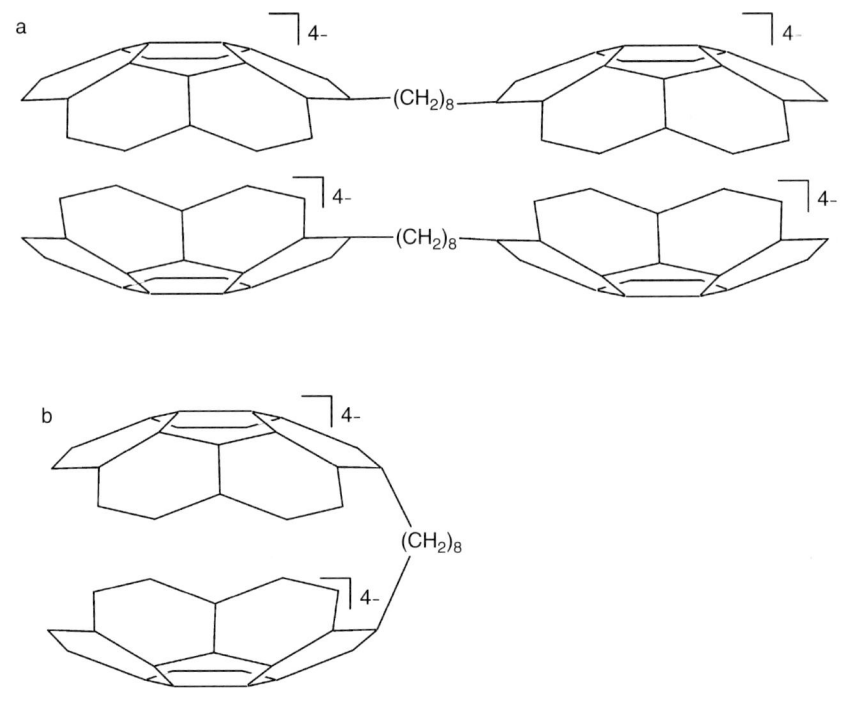

Figure 13.3. (a) Intermolecular dimer and (b) intramolecular dimer of 38^{8-}.

signals are very similar (δ_{Li} (38^{8-}) = −4.3 and −11.5 ppm), as are the barriers for the Li$^+$ exchange.

The dimer formed in the reduction of **38** to 38^{8-} was described as an intramolecular dimer, and not as an intermolecular one (Fig. 13.3). This characterization was based on the fact that no mixed dimer was observed for 38^{8-} and 2^{4-}. Diffusion coefficients [42, 114] were measured for both **34** and 38^{8-} dimer. Similar diffusion constants were measured for both systems, indicating that both have similar sizes, which is consistent with intramolecular sandwich formation in 38^{8-}.

13.4.2.5 Extended Corannulenes

The PAHs indenocorannulene (**39**), dibenzo[a,g]corannulene (**40**), and dibenzo[a,g]cyclopenta[h,l]corannulene (**41**) are three examples of extended corannulenes, which contain a central corannulene system fused to five- and six-membered rings. The reduction of these systems [119, 120] focused on the following issues. First, what is the aromaticity of these curved PAHs anions, will they behave like large polycyclic systems, or have annulenic character? Secondly, what is the possibility of aggregation and dimerization in these systems? The third subject of interest was the effect of different alkali metals on the reduction process.

The reduction of **39** with K yields a unique four-step reduction process in which alternate dimerization and bond cleavage occurs [119]. Four different diamagnetic

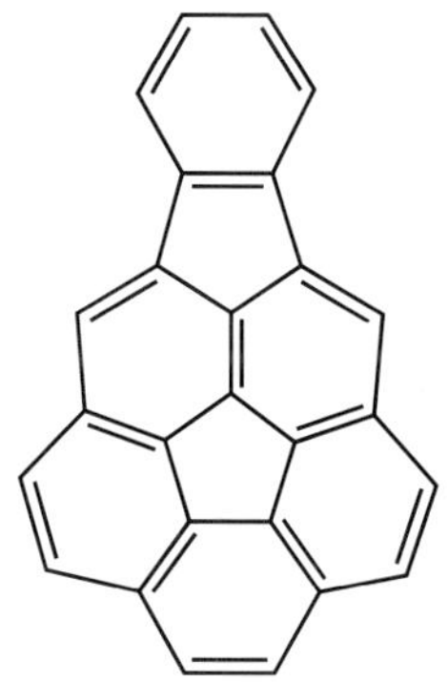

39

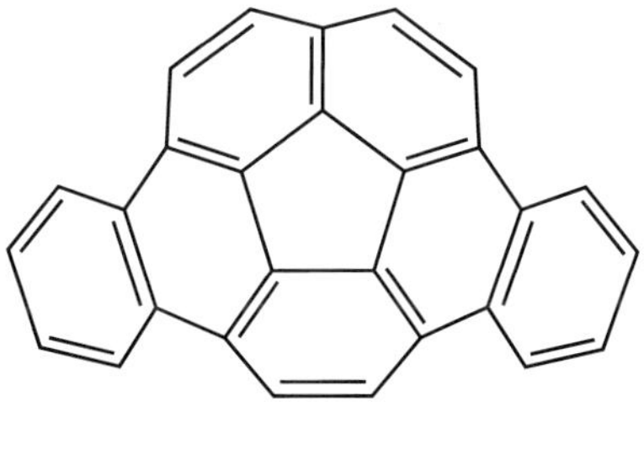

40

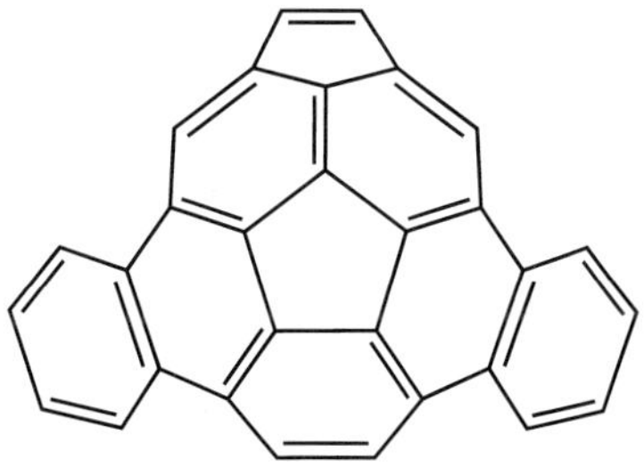

41

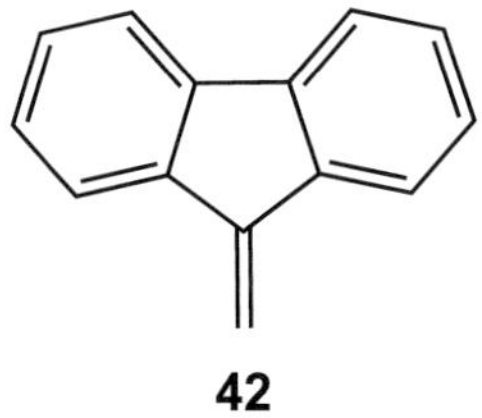

42

anions were observed when **39** was reduced with K (Scheme 13.3). NMR studies of all these anions suggest that the anions formed in the first and third stages are dimers of **39**, in which a new σ-bond connects two identical C3' carbon atoms. The second and fourth stages of reduction are the dianion and tetraanion monomers of **39** (**39b**, **39d**) that are formed by a reductive bond cleavage. The principle evidence for this process comes from the following NMR data: the symmetry of the system alternates from C_s to C_1 as the reduction proceeds. This is observed in both ^1H and ^{13}C NMR spectra. The up-field shift and $^1J_{C,H}$ coupling constant of carbon C3' in **39a** and **39c** indicate the sp^3 hybridization of this atom. An extra long-range ^1H–^{13}C NMR interaction in **39a** and **39c** indicates that a new bond between two C3' carbon atoms is formed. Long-range coupling between H3' and its symmetric equivalent ($^3J_{H3', \cdot H3'}$) was observed from the ^{13}C NMR satellite signals in the short range ^{13}C–^1H correlation experiments (HSQC). The dimerization is explained by the presence of a dibenzofulvene (**42**) subunit in **39**. It has been shown that such a unit can undergo dimerization when reduced [121].

Since the bowl shape of **39** does not undergo a dynamic process on the NMR timescale, each half of the dimer imparts chirality. This property leads to the interesting possibility of multiple diasteromeric reduced species, depending on the position of the connection as shown in Fig. 13.4. Although it was not possible to fully assign the stereochemistry of the dimers, it is apparent from the $^3J_{H3, \cdot H3'}$ coupling constant (10.0 and 10.5 respectively, for **39a** and **39c**) that they adopt *anti* conformation.

The reductive dimerization/bond-cleavage of **39** represents the first case in which a large nonplanar PAH undergoes such a process. In addition, this is the

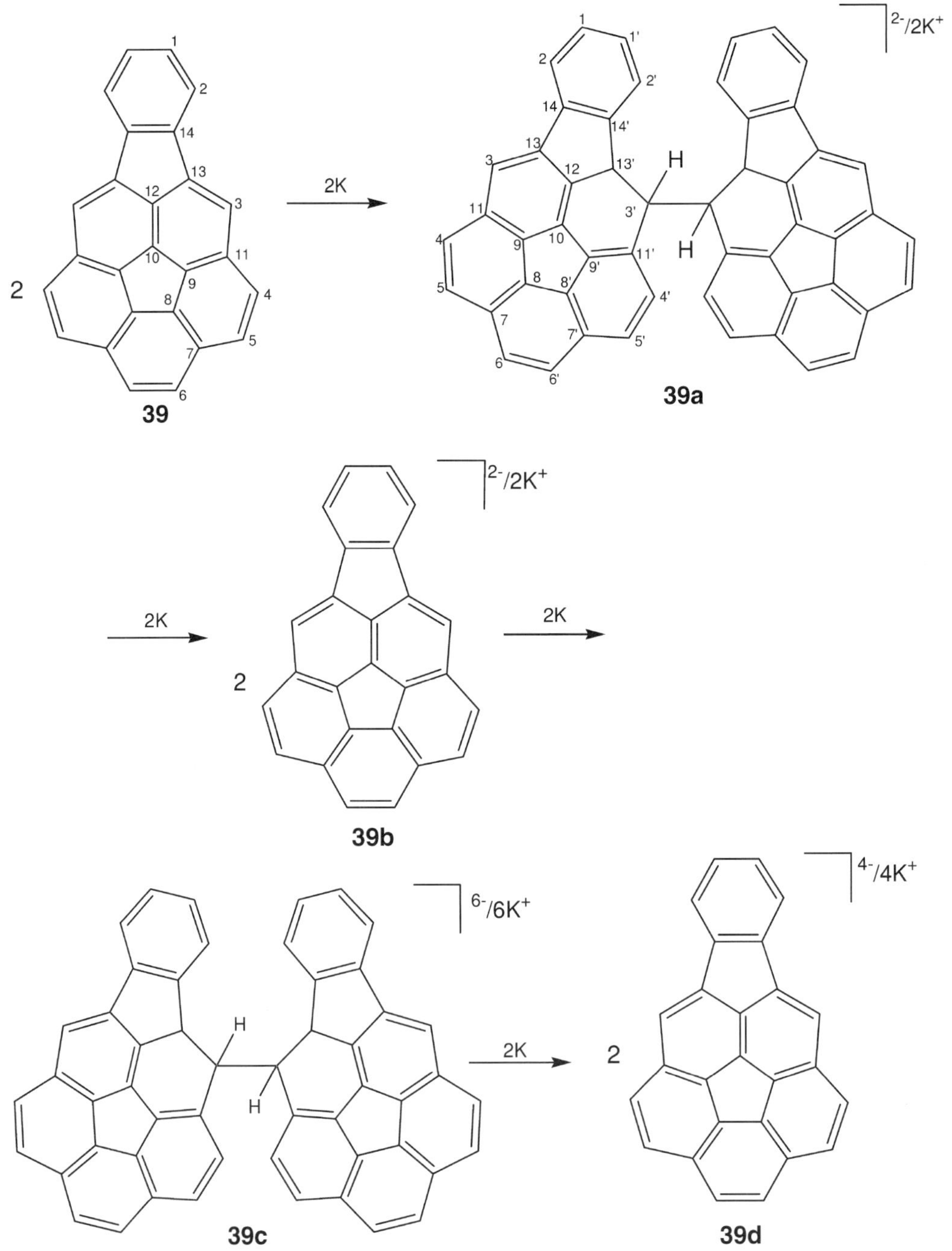

Scheme 13.3. Reduction of indenocorannulene (**39**).

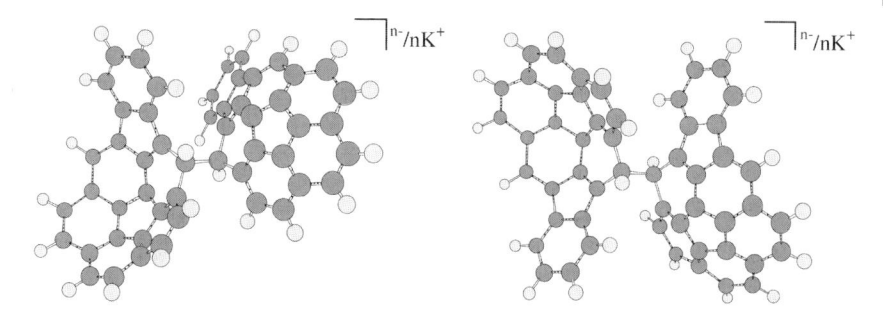

Figure 13.4. Possible structure of the dimers of indenocorannulene (**39a**, **39c**).

first time that such a process has been extended beyond a single dimerization/bond-cleavage stage [122].

Compounds **40** and **41** do not contain the dibenzofulvene subunit, which is responsible for the dimerization process in **39**; therefore, their reduction produces only monomeric anions. The LUMO level in **40** and **41** is not doubly degenerate like that of corannulene, but there is rather a small energy gap between the LUMO and the NLUMO, which indicates that the formation of tetraanions of **40** and **41** should be possible [120].

In the reduction of both **40** and **41** with Li, only three reduction steps were observed, in contrast to **2** where a fourth reduction step was observed; therefore, it seems that the reduction of **40** and **41** with Li does not lead to the formation of a dimer. The reduction of **40** and **41** with K takes place in four steps that include two diamagnetic species, dianion and tetraanion, which were observed and analyzed by NMR. The ^{1}H and ^{13}C NMR of $40^{2-}/2K^+$ are identical to those observed in the reduction with Li ($40^{2-}/2Li^+$); however, there are some differences in the ^{1}H and ^{13}C NMR spectra of the reduction product of **41** with K ($41^{2-}/2K^+$) compared to $41^{2-}/2Li^+$. These changes were attributed to the smaller cation size of lithium and its higher charge density. The most significant change was that in $41^{2-}/2K^+$ where a less negative charge was observed at the external 5MR.

As far as the aromatic character of the anions formed is concerned, the ^{1}H NMR of 40^{2-} is shifted to high field, demonstrating the anti-aromaticity of the dianion, though the effect is less pronounced than in 2^{2-}. On the other hand, the dianion 41^{2-} was found to be aromatic. The two tetraanions 40^{4-} and 41^{4-} showed the opposite behavior; the center of gravity (CG) of 40^{4-} was shifted to a lower field than that of 40^{2-} ($\delta = 3.51$ ppm) indicating the aromatic character of the tetraanion. In 41^{4-} the CG of the ^{1}H NMR spectrum appears at a higher field than that of 41^{2-} ($\delta = 5.58$ and 6.93 ppm, respectively), indicating the presence of a weak anti-aromatic effect.

The π-topology model of "an annulene within an annulene" was tested in **40**, **41** and their anions. **40** can be viewed as having six π-electrons in the central 5MR and 22 π-electrons delocalized around the rim, thus explaining its aromatic character. On the other hand, **41** is predicted to be an anti-aromatic system because 24 π-

electrons can be delocalized around its rim; however, this is inconsistent with the aromatic character of **41**, as deduced from the low field ^{1}H NMR shifts. Two explanations were suggested to solve this contradiction. Either the two π-electrons of the etheno bridge do not participate in the overall delocalization on the rim of the outer ring, or the annulene model does not hold in the case of **41**.

The total change in the carbon chemical shifts upon reduction ($\Sigma\Delta\delta$, Eq. (1)) shows smaller values for both **40** and **41** compared to corannulene, although negative charge was added ($\Sigma\Delta\delta = 336$, 50, -178, for $\mathbf{2}^{2-}/2\text{Li}^+$, $\mathbf{40}^{2-}/2\text{Li}^+$, $\mathbf{41}^{2-}/2\text{Li}^+$, respectively). This comparison shows that the annulenic character of dianions of **40** and **41** is less pronounced than that in **2**. It also suggests that the annulenic character in $\mathbf{40}^{2-}$, which contains two benzannelated rings, is more significant than in $\mathbf{41}^{2-}$, which contains an additional 5MR.

13.4.3
Strain in PAHs: The Story of Curved Pyrene

The effect of curvature on the aromatic character of PAHs was studied by comparing planar pyrene (**43**) and its curved analogs. The homologous series of [*n*](2,7)pyrenophanes (**44–47**, *n* = 7–10) and [2]metacyclo[2](2,7)pyrenophane (**48**) was synthesized by Bodwell et al. [123]. In these molecules, the pyrene unit is strongly distorted from planarity, as a consequence of tethering the 2 and 7 positions by alkane chains. The degree of distortion and the resultant strain is controlled by the length and type of the tether; the shorter the tether, the greater the strain. The bending angle of the pyrene component in such systems can be larger than that of the pyrene unit in C$_{70}$ (**49**) [123b], depending on the type of the tether.

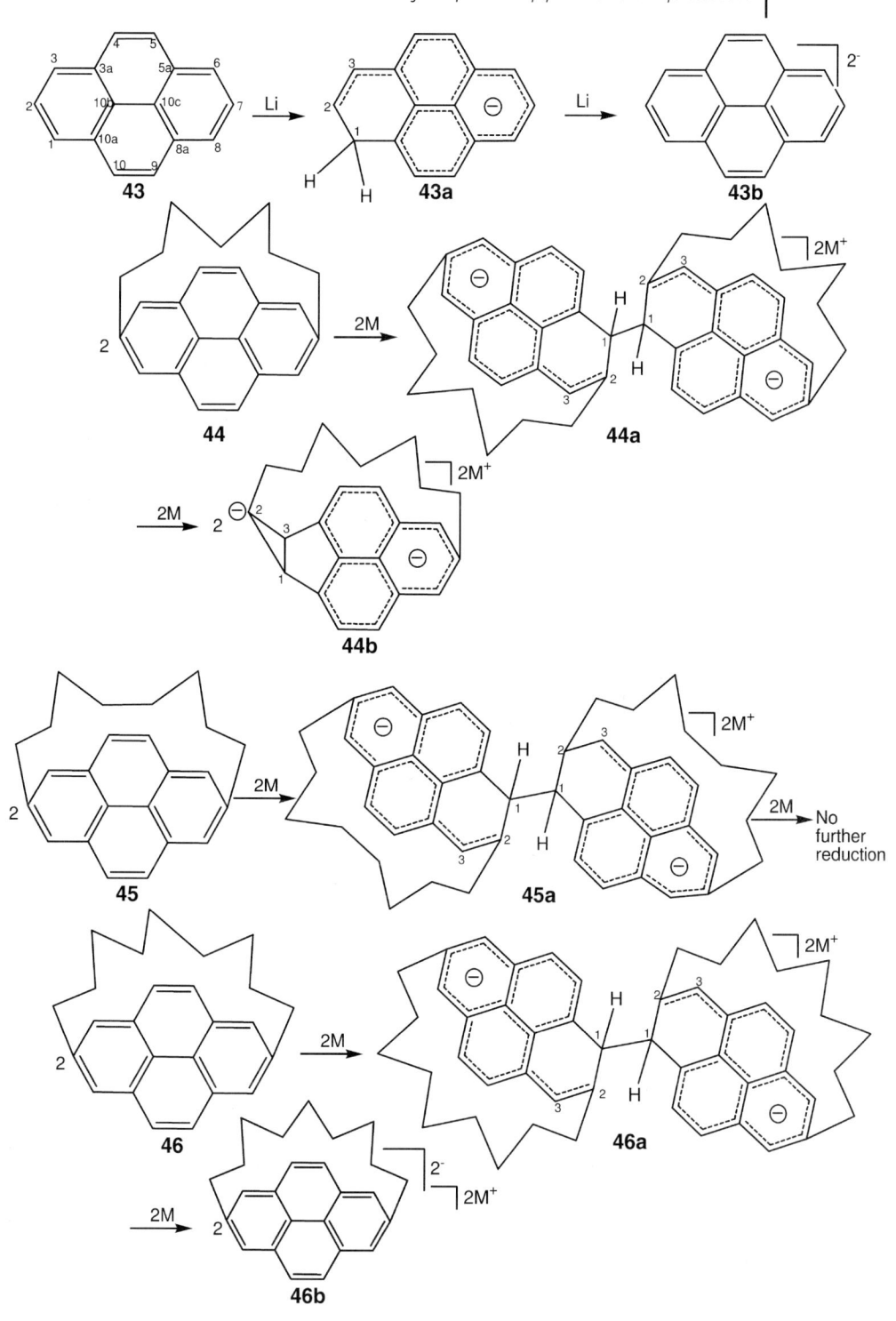

13.4 Reduction of Nonplanar Polycyclic Aromatic Hydrocarbons | 597

Scheme 13.4. Reduction products of tethered pyrenes – the influence of length of tether.

The study of the neutral systems showed that enforcing nonplanarity on the pyrene nucleus does little to diminish the anisotropy effect of the bent systems. The pyrene unit still maintains most of its aromatic character.

All the systems **44** to **48** were reduced with Li and K (Scheme 13.4). Since both metals yielded similar results, it was suggested that the processes are not alkali metal-dependent [124].

13.4.3.1 First Reduction Process – Dimerization

The first reduction process is reductive dimerization [125], similar to the process that was observed in the reduction of **39** [119]. This step is common to all of the studied [n](2,7)pyrenophanes, which led to the conclusion that in the one electron reduction the degree of strain in the system does not play a crucial role (Scheme 13.4). Only protonation of the radical monoanion was reported for the analogous planar systems, **43** [126] and dimethylpyrene (**49**) [125]. The characterization of the dimers was achieved using the symmetry of the spectral lines and the NMR data. The high-field absorption of carbon atom C1 in species **44a–48a**, and the magnitude of its $^1J_{C,H}$ coupling constant were consistent with an sp^3-hybridized carbon atom. The long-range $^1H-^{13}C$ NMR correlation experiments and the $^3J_{H1,\cdot H1}$ coupling constant indicated that a reaction had taken place between two identical C1 carbon atoms of separate molecules, forming sp^3-hybridized carbons and a new σ-bond, which in turn lowers the symmetry of the system. The formation of this new bond was rationalized in terms of high spin density for the odd electron in the monoanion radical in position C1 (and its symmetrical counterparts) [125].

13.4.3.2 The Second Reduction Stage

In contrast to the first reduction process, where the length of the tether does not influence the nature of the reduction product, the outcome of the second reduction step depends on the strain in the system. Three different cases were reported.

1. Pyrene tethered by seven carbon atoms: A surprising structure was found for the dianions of **44** and **48**, that have the shorter bridge (seven carbon atoms). A new intramolecular σ-bond is formed, that transforms one of the benzene rings into a "cyclopropano-cyclopentane" (bicyclo[3.1.0]) ring system [127]. The ^1H and ^{13}C NMR spectra of these dianions show that their symmetry is lower than that of neutral **44** and **48** but higher than that of the dimers **43a–48a**. These unexpected products could explain all of the unexpected NMR data. The high field shift of carbon atom C1 (δ_C = 33.9 and 33.7 ppm for **44b** and **48b**, respectively) and its one-

bond CH-coupling ($^1J_{C1,H1}$ = 163.2 and 162.3 Hz for **44b** and **48b**, respectively) are consistent with a strained sp^3-hybridized carbon, as is the case in a cyclopropane ring [45]. The $^3J_{H1,·H1}$ coupling constant (9.1 and 8.7 Hz for **44b** and **48b**, respectively) and the correlation seen in the long-range correlation experiments between hydrogen atom H1 and carbon atom C2 resulted from the formation of this new σ-bond. The high-field shift of carbon atom C2 (δ_C = 29.2 and 30.1 for **44b** and **48b**, respectively) results from charge localization on C2.

The cyclization also yields an aromatic phenalenyl anion moiety. This was demonstrated by the high-field shift of carbon atoms C8, C9, and C10a (and their symmetrically equivalent atoms) and hydrogen atoms H8 and H9, which are part of the periphery of phenalene. This conclusion was supported by the similarity between the carbon chemical shifts of the phenalene moiety of **44b** and **48b** and those of the 2-methylphenalenyl anion **50** [128].

It was suggested that the formation of the new σ-bond allows the system to avoid anti-aromaticity by separating the two charges, as well as releasing some of the strain in the pyrene moiety by producing a three membered ring and leaving behind a relatively flat aromatic "phenalenyl" anion.

2. Pyrene tethered by nine and ten carbon atoms: Compounds **46** and **47**, which have relatively long tethers, behave like **43**; therefore, their two-electron reduction yielded anti-aromatic species **46b** and **47b**. The spectra of the dianions were broad and could be obtained only at low temperature. This is typical for systems that have a singlet–triplet equilibrium, with a higher contribution of the triplet state [124].

The high field shift of the pyrene units protons (δ = 0.64–2.61 ppm, for both **46b** and **47b**) is direct evidence for the anti-aromaticity of the species. Another strong indication of the anti-aromaticity of these dianions comes from the chemical shift of the tether hydrogens, which function as probes for the anisotropy effect in the system. The best example comes from the geminal hydrogens H14 and H14′ in **46b**/2K$^+$, which appear at δ = 9.93 and 3.22 ppm, respectively.

3. Pyrene tethered by eight carbon atoms: The threshold for the cyclization is a tether of eight carbon atoms **45**, where no two-electron reduction of the pyrene occurs even, after prolonged contact with either Li or K metals. It was suggested that the strain in **45** is large enough not to permit either an anti-aromatic state, or the formation of an intramolecular σ-bond [124].

13.5
Reduced Fullerenes

13.5.1
General

Fullerenes were discovered in 1985 by Kroto and Smalley [14]. They are the most carbon rich compounds, and their chemical properties have attracted much attention [129]. The most familiar and best investigated fullerene is C$_{60}$ (**33**), and it is also the most symmetrical one (I_h). Many studies have also been conducted on the next fullerene of the series, C$_{70}$ **51**. Beside these two well-known fullerenes,

13 Reduction of Carbon-rich Compounds

a LUMO + 1 — — —
     LUMO   — — —

b LUMO   — — —
                — —
       HOMO   _ — _

HOMO — — — — —

Figure 13.5. Partial diagram of the energy levels of (a) C_{60} (a) and (b) C_{70}.

higher fullerenes (C_n, $n > 70$) have also been prepared, separated and characterized [130]; however, low yields have prevented extensive research. Fullerenes are found to be highly reactive, and they undergo many addition reactions, producing a variety of fullerene derivatives [129].

One important characteristic of fullerenes is their ability to accept a large number of electrons. This ability was predicted in the 1980s when the first calculations were performed on these systems. It was found that fullerenes have low HOMO–LUMO gaps and multiple degenerate energy levels, as is the case in the triply degenerate LUMO of C_{60} [131] (Fig. 13.5). The reduction of fullerenes thus became one of their most studied facets. These studies focused on the ability of fullerenes to accept a large number of electrons (six), the unique aromatic properties of the anions formed, and the use of reduction as a synthetic route for fullerene derivatives.

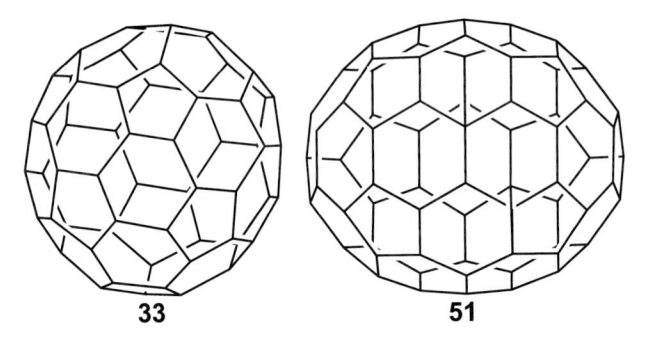

33   51

We shall concentrate on the formation of fullerene anions and their characterization, especially by the study of their magnetic properties. We focus on the study of the aromaticity of the fullerene anions of different sizes and shapes, and compare their aromaticity with that of the neutral fullerenes. A few examples of fullerene derivatives and metallofullerenes are given, and the influence of the added functional group or transition metal on the reduction process is discussed. In addition, some recent papers concerning the reduction of carbon nanotubes are briefly described.

13.5.2
Reduction Stages of Fullerenes

Electron affinity measurement [132] and reduction experiments [133] showed that fullerenes can accept a large number of electrons. The anions formed were charac-

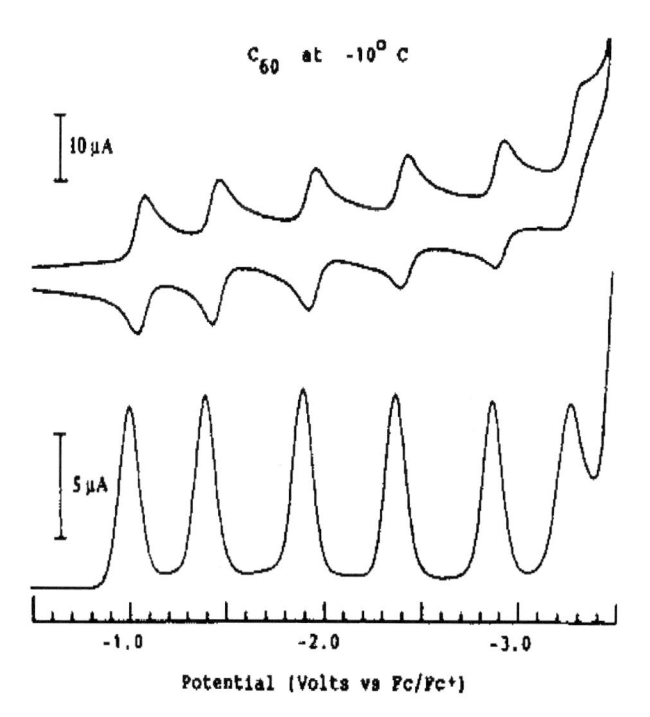

Figure 13.6. Cyclic and differential pulse voltammograms of C_{60} in CH_3CN/toluene at -10 °C showing six successive reversible reductions to C_{60}^{6-}. Reproduced with permission from *J. Am. Chem. Soc.* **1992**, *114*, 3978–3980. Copyright 1992 Am. Chem. Soc.

terized by absorption spectroscopies, e.g., NIR (the near-infrared region has proven to be the most useful and informative for fullerene ions), Raman and IR spectroscopies and by EPR and NMR methods. In his review in 2000, Reed broadly described the properties of fullerene anions, and summarized the observations and properties of each reduction step of C_{60} [133].

Electrochemistry holds a central place in the study of fullerene anions. After several experiments, the optimized electrochemical conditions made it possible to observe the fifth and sixth reduction waves on the cyclic voltammetry (CV) time scale (Fig. 13.6) [34]. It was found that the medium highly influences the redox potential [133, 134]. The counter cations influence the kinetics, whereas the solvent affects the redox potentials substantially. For a number of fullerenes, a good correlation was found between the calculated HOMO–LUMO gap and the difference between the first reduction potential and the first oxidation potential in the CV [135].

Fullerenes have been reduced in the solid state, using the alkali metals Li, Na, K, Rb and Cs, which have the ability to intercalate in the vapor phase into solid fullerenes [133]. The discovery of the superconductivity of metal intercalated C_{60} [136] attracted much attention to such materials; therefore, the metal intercalation of fullerenes has been the subject of several reviews, which discuss the effect of the anions on their physical properties [137].

In solution, the reduction of fullerenes is typically performed in etheral solvents (e.g., tetrahydrofuran, dimethoxyethane) [138] or liquid ammonia [139]. Using Li as a reducing agent it is possible to reach the highest reduction step, the hexaanion. With the other alkali metals this was observed only when naphthalide salt was added [140]. The reduction of C_{60} and all the higher fullerenes to their hexaanions was first made possible by sonication with excess Li [16] and later by adding a small amount of **2** as an electron shuttle (*vide supra*).

13.5.2.1 Anions of C_{60}

Six reduction waves were observed in the CV of C_{60} (Fig. 13.6) [34]. The first reduction potential is very low (−0.98 V vs. ferrocene/ferrocenium) [34a], and reflects the ease of reduction of C_{60}. The next waves have relatively even spacing of around 0.45 V, as predicted by the triply degenerate LUMO levels (Fig. 13.5(a)) ($E_{1/2} = -0.98, -1.37, -1.87, -2.35, -2.85$, and -3.26 V) [34a]. Each of the C_{60}^{n-} anions ($n = 1-6$) has a distinctive NIR spectrum, which shows some solvatochromic effect, as reported by Lawson [141], Heath [139a], and Baumgarten [138c]. In contrast, only the low charged anions were characterized by vibrational spectroscopies (IR and Raman) [133]. NMR and EPR measurements of the anions of C_{60} led to the conclusion that the only diamagnetic anion is C_{60}^{6-} and that the others, C_{60}^{n-} ($n = 1-5$), are paramagnetic, or are in a singlet–triplet equilibrium. The review by Eaton from 1996 summarizes the EPR spectra for the first three anions of C_{60} [142].

The monoanion $C_{60}^{\cdot-}$ has a low g value, close to 1.999 [142], which does not show any significant change with temperature [133]. This was attributed to spin–orbit coupling in the degenerate ground states [143, 144]. The characteristics of the EPR signals of $C_{60}^{\cdot-}$ and C_{60}^{3-} are consistent with a Jahn–Teller distortion of the C_{60} anions [142]. In addition to these broad signals, sharp signals also appeared in some of the spectra of these anions. These were attributed to species with lower symmetry; namely dimers and substituted radicals, in which the degeneracy of the LUMO level has been broken. The EPR spectrum of C_{60}^{2-} contains a sharp signal [133, 142, 145], which is consistent with a singlet ground state (a close-lying triplet excited state, which makes it EPR detectable).

C_{60}^{4-} is a "hole equivalent" of the C_{60}^{2-} ion, and like C_{60}^{2-} sharp signals were observed for it; its ground state was assigned as a singlet [146]. The appearance of the EPR signals of C_{60}^{2-} and C_{60}^{4-} suggests the presence of a low-lying $S = 1$ excited state which is EPR detectable.

Reed was able to obtain ^{13}C NMR spectra for C_{60}^{n-}, $n = 1-4$, by using different solvents as well as different alkali metals and increasing the solubility by adding crown ethers or cryptands [145a, 146, 147]. The NMR signals were described as broad signals, and the chemical shifts of all of them appeared at low field relative to neutral C_{60} ($\delta = 143$-ppm). The down-field shifts were ascribed to the effects of paramagnetic electronic states.

Olah reduced C_{60} and C_{70} with Li metal [16], and obtained diamagnetic anions,

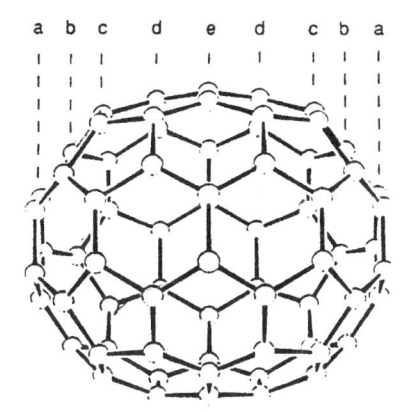

Figure 13.7. Schematic diagram of C_{70} depicting the different carbon atoms a–e. *J. Chem. Soc. Chem. Commun.* **1990**, 1424 – Reproduced with permission of The Royal Society of Chemistry.

which were assigned as the hexaanions relying on the CV results and the structure of the LUMO levels. A sharp singlet appeared in the ^{13}C NMR spectrum of C_{60}^{6-} at $\delta = 156.7$ ppm; it is deshielded by 14 ppm relative to neutral C_{60}, which was an unexpected result considering the high negative charge that was added. Solid-state ^{13}C NMR chemical shift data on metal-intercalated M_xC_{60} have also been published. Solution and solid-state results show similar NMR shifts [133].

13.5.2.2 Anions of C_{70}

The energy level structure of C_{70} (Fig. 13.5(b)) contains one LUMO level, followed closely by two degenerate LUMO+1 levels [131]; therefore, C_{70}, like C_{60}, was predicted to be reduced to a hexaanion. The ability of C_{70} to accept six electrons was demonstrated by the six reduction waves in CV experiments. The reduction waves had redox potentials similar to those of C_{60} [34, 35a, 135]; however, only the absorption spectra (NIR) of the anions C_{70}^{n-}, $n = 1$–4, were reported [133, 148]. The EPR spectra of $C_{70}^{\cdot -}$ and C_{70}^{2-} were reported by Baumgarten [138d], who showed that the dianion had a slightly larger g value than the mono-anion (2.0026, compared to 2.0022).

The ^{13}C NMR of the diamagnetic hexaanion of C_{70} (C_{70}^{6-}) was reported for the first time by Olah [16]. Five signals were observed, in a ratio of 1:2:1:2:1, as expected from the symmetry of C_{70} (Fig. 13.7). To conclusively determine the chemical shift of each type of carbon in C_{70}^{6-}, a 2D INADEQUATE NMR experiment (that correlates the ^{13}C NMR absorption of a carbon to that of its bonded neighbor) was performed, which led to the following assignment (Fig. 13.8): $\delta_a = 133.6$; $\delta_b = 158.3$; $\delta_c = 152.3$; $\delta_d = 138.1$; $\delta_e = 149.8$ [149].

Comparison between the ^{13}C NMR chemical shifts of neutral C_{70} [150] and its hexaanion [16, 149] showed a large shift for carbons **a** and **e**, that are located at the "poles" and the "equator" of the fullerene, respectively (Fig. 13.8). Carbons **a** showed a strong up-field shift ($\Delta\delta = -17$ ppm), due to the concentration of nega-

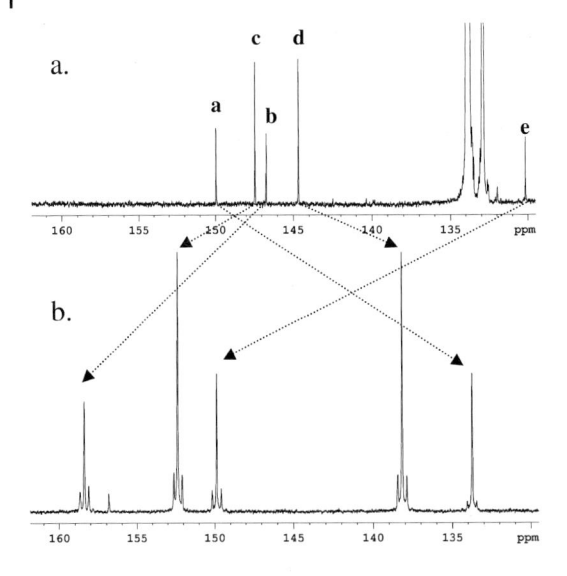

Figure 13.8. The ^{13}C NMR of C_{70} (in 1-methylnaphthalene) (a) and C_{70}^{6-} (b).

tive charge on the polar 5MRs. On the other hand, the **e** type carbons had a strong low-field shift ($\Delta\delta = 19$ ppm), which originates from a decrease in the charge distribution on the carbons at the equator of the C_{70}^{6-} surface. The most significant change in the magnitude of the $^1J_{C,C}$ coupling constants, due to reduction, was observed for $^1J_{A,B}$ which was reduced by 7 Hz. This confirms that the charge is located at the poles, leading to increased bond length between carbons **a** and **b** [149].

13.5.2.3 Reduced Higher Fullerenes

There are fewer reports concerning the reduction of higher fullerenes, since they are less widely available in preparative quantities. Although they have lower symmetries, calculations predict high degenerate energy levels for most of them [151]. CV experiments performed on C_{76}, C_{78}, C_{82}, C_{84} and C_{86} showed that they all exhibit multiple reduction waves (four or six) [35, 135]. Some of the higher fullerenes are easier to reduce to the monoanionic state, but there is no strict correlation between the reduction potential and the fullerene size.

The anions of C_{76}-D_2 (**52**), C_{78} (C_{2v} (**53**) and D_3 (**54**)) and C_{84} (D_2 (**55**) and D_{2d} (**56**)) were measured by EPR and NMR spectroscopies. C_{78} and C_{84} were reduced as a mixture of isomers, and in some of the measurements the different isomers could be distinguished. The EPR measurements by Pénicaud [152] showed that the g values of all of the monoanions were higher than that of C_{60}^- (2.0023, 2.0018 and 2.0008 for C_{76}^-, C_{78}^-, C_{84}^-, respectively, in benzonitrile). In tetrachloroethane the two isomers of C_{78}^- could be distinguished by their different EPR characteristics [152].

Kadish, Ruoff and Jones succeeded in observing the three reduction stages (1–3) of the two isomers of C_{84} by EPR spectroscopy [35c,d]. The EPR spectra were measured in frozen pyridine, and five paramagnetic anions appeared consecutively. The isomer **56** was the easiest to reduce. The assignment of the isomer type was based upon symmetry considerations, which predict a doubly degenerate LUMO for the isomer D_{2d} symmetry and a nondegenerate LUMO for the D_2 isomer. The dianion $[C_{84}\text{-}D_{2d}]^{2-}$ was predicted to behave like $C_{60}{}^{2-}$ with a singlet ground state and a thermally accessible triplet excited state (the singlet–triplet energy separation was found to be 0.022 eV), however, the dianion **56**$^{2-}$ was not EPR active. Moreover, Pénicaud characterized by EPR ten different isomers of the monoanion $C_{84}{}^-$ (four of them were never characterized before) and all of them exhibited different narrow lines [151e].

The ^{13}C NMR spectra of the diamagnetic anions of the aforementioned fullerenes all showed narrow signals compatible with their symmetries [153]. An indication about their reduction states was obtained by comparing their ^{13}C NMR spectra with those calculated by density functional theory (DFT) and MO energy considerations. Accordingly, all of the higher fullerenes were assigned as hexaanions. Similar to C_{60} and C_{70} [16], the higher fullerenes can also accept six electrons to their π-systems and form stable diamagnetic hexaanions.

An interesting phenomenon was observed in the reduction of the two isomers of C_{78}. ^{13}C NMR measurements at different temperatures distinguished between the two isomers. The room-temperature spectrum contains only 13 major absorptions, which were assigned to the D_3 isomer **54**. Reduction of the temperature resulted in the appearance of another set of lines, and at 170 K it was possible to see an extra 21 signals, which were assigned to the C_{2v} isomer **53**. It was suggested that this behavior is due to some triplet character and the smaller HOMO–LUMO gap in the multiply charged anion of **53** (calculated to be 1.05 and 2.02 eV, for the hexanaions of **53** and **54**, respectively) [153].

13.5.3
The Aromatic Character of Reduced Fullerenes

The discovery of fullerenes offered a new challenge in the field of the study of the aromaticity of π-conjugated systems. These carbon allotropes embody completely conjugated spheroidal π-systems, so the carbon skeletons are boundaryless and thus large numbers of Kekulé structures can be drawn.

The aromaticity of fullerenes has been investigated theoretically and substantiated experimentally by NMR [154]. The best experimental method for studying the aromaticity of fullerenes was devised by Saunders and Cross, who encapsulated helium atoms inside fullerenes and measured their NMR spectra [155]. ^{3}He NMR measurements show that C_{60} exhibits the least aromatic character whereas C_{70} has the highest, and that the aromaticity of higher fullerenes lies between these two extremes. This is in contradiction to the initial suggestion in 1985 that C_{60} is covered by a sea of π electrons that make it aromatic [14a, 156], and to the assumption of high aromaticity due to the large number of possible Kekulè structures [157]. The reduction of fullerenes is not expected to modify their shape or symmetry; therefore, the effect of added electrons is expected to manifest itself chiefly by their aromaticity.

The aromaticity changes upon reduction of fullerenes of different sizes and shapes depend on several parameters. Their size, symmetries, as manifested by ^{13}C NMR and moreover, the ^{3}He NMR of endohedral ^{3}He atoms afford an insight into their aromaticity. The reduction of bridged C_{60} and C_{70} is also discussed. The bridged fullerenes contain methylene bridges, whose hydrogens can serve as "built in probes of aromaticity" of the ring currents of individual rings.

13.5.3.1 Endohedral Helium in Fullerenes – an Internal Probe

He@C_{60}^{6-} and He@C_{70}^{6-}: The ^{3}He NMR spectra of endohedral helium atoms were used to monitor the aromaticity of the fullerenes as a function of electron addition upon reduction. The reduction of ^{3}He@C_{60} and ^{3}He@C_{70} to their hexaanions was followed by a study of their ^{3}He shifts [17]. The reduction was carried out with Li metal (in the presence of 2), and the ^{13}C NMR spectra of the anions were found to be identical to those of "empty" C_{60}^{6-} and C_{70}^{6-}. The ^{3}He atom inside the C_{60}^{6-} was found to be strongly shielded compared to that of the neutral ^{3}He@C_{60} (−48.7 and −6.3 ppm [155a], respectively). The chemical shift of the ^{3}He encapsulated in C_{60}^{6-} was the highest field shift ever reported for ^{3}He in fullerene. The ^{3}He NMR of ^{3}He@C_{70}^{6-} was shifted to the opposite direction. The ^{3}He signal appeared at low field (+8.3 ppm), deshielded by 37.1 ppm, from the neutral compound. The ^{3}He NMR chemical shifts are the most compelling evidence for the high aromaticity of C_{60}^{6-} and the ability of electrons to move freely about the surface of a spheroidal π-system (Fig. 13.9) [17].

These NMR experiments confirmed the conclusions of several calculations regarding fullerene anions. Intensive computational research was done into the spherical aromaticity of fullerene and fullerene anions and is summarized in the 2001 review by Bühl and Hirsch [154]. The London calculations on C_{60}^{6-} predicted only diatropic ring currents, a large diamagnetic susceptibility and a huge endohe-

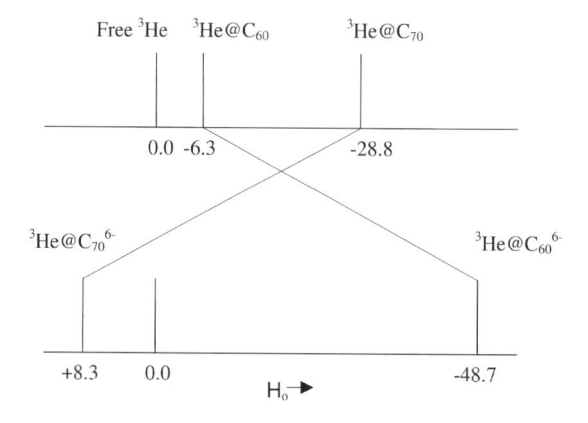

Figure 13.9. ^{3}He NMR chemical shifts of ^{3}He@C_{60} and ^{3}He@C_{70} and their respective hexaanions.

dral shielding, indicative of a highly aromatic compound [158]. On the other hand, for C_{70} it was predicted that a large part of its diatropicity would be lost upon the formation of the hexaanion. The same trends were apparent from *ab initio* Hartree-Fock calculations of endohedral shieldings. Values between −58 and −64 ppm for C_{60}^{6-} and −11 ppm for C_{70}^{6-} were calculated [159], in good agreement with experiment.

Endohedral helium of higher fullerenes: The reduction of ^{3}He@C_{60} and ^{3}He@C_{70} to their hexaanions was soon followed by the reduction of ^{3}He@C_{76}, C_{78}(D_3, C_{2v}', C_{2v}), C_{84}(D_2, D_{2d}) symmetries to their hexaanions [160]. The ^{3}He NMR signals of all the anions of the higher fullerenes yielded chemical shifts between two extremes: the high-field shift of ^{3}He@C_{60}^{6-}, and the low-field shift of He@C_{70}^{6-}; therefore an "aromaticity scale" for the fullerene hexaanions was suggested as follows: $C_{60}^{6-} > C_{78}^{6-}(D_3) > C_{84}^{6-}$ (two isomers) $> C_{76}^{6-} > C_{78}^{6-}(C_{2v}') > C_{78}^{6-}(C_{2v}) > C_{70}^{6-}$ [161].

Comparison of the ^{3}He NMR chemical shifts of the neutral fullerenes and their anions shows that the changes in the chemical shifts are different from one fullerene to another. While the ^{3}He bands of C_{78}-D_3, C_{84} and C_{76} are shifted to a higher field as a result of their reduction (similar to C_{60}), those of the two isomers of C_{78} (C_{2v} and C_{2v}') are shifted to a lower field (similar to C_{70}). These changes point to an increase or decrease, respectively, in the aromaticity of the fullerenes. Accordingly, it was concluded that the aromaticity of fullerenes and their anions are not simply related to the number of carbons or the number of electrons in the π-system. This was nicely demonstrated by the three C_{78} isomers which behave differently. The aromatic character of C_{78}-D_3 increased, and that of the two C_{2v} isomers decreased. The differences in the aromaticity of the isomers of C_{78}^{6-} are not only visible from the ^{3}He chemical shifts, but were also demonstrated by their kinetic behavior. The two isomers of C_{78}^{6-}-D_3 and C_{2v}', which are found to be more aromatic than C_{78}^{6-}-C_{2v}, also have a faster reduction rate.

Two helium atoms inside fullerenes: Another signal that was measured in the ^{3}He NMR spectra is that of two helium atoms inside a fullerene, e.g. He$_2$@C_n^{6-}. The di-helium peaks are measurably smaller than the mono-helium shifts, and appear at a slightly higher or lower field than that of the mono-helium. A general trend was observed in the ^{3}He NMR shifts. The "highly aromatic" fullerene anions, which show a high-field ^{3}He shift (C_{60}^{6-}, C_{84}^{6-} and C_{76}^{6-}), also show helium signals of ^{3}He$_2$@C_n^{6-} that are shifted to a lower field. On the other hand, when the mono-helium signal is shifted to low-field (C_{70}^{6-}, C_{78}^{6-}-C_{2v}), the di-helium signals have a higher shift [160, 161].

13.5.3.2 ^{13}C NMR of Reduced Fullerenes

The ^{13}C NMR spectra of all of the diamagnetic fullerene anions are shifted to low field, in comparison to the spectra of the neutral species, despite the negative charges being added. The shifts of the centers of gravity, compared with the neutral spectra, are 14.0, 0.9, 6.8, 8.5, 5.5 and 8.2 ppm for C_{60}, C_{70}, C_{76}-D_2, C_{78}-D_3, C_{78}-C_{2v}, C_{84} (D_2+D_{2d}), respectively [16, 153].

The degree of deshielding is different from one fullerene to another, where the

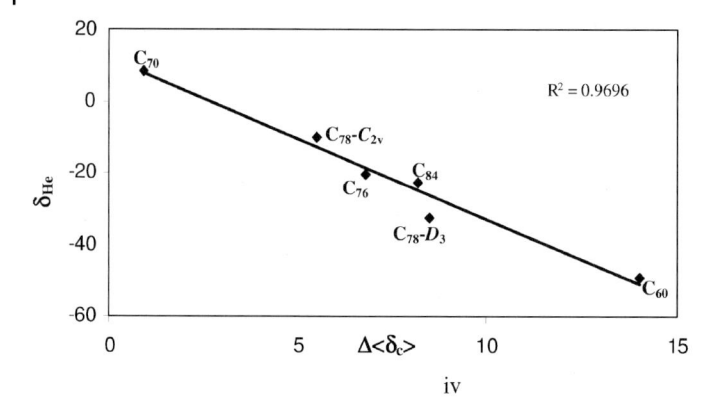

Figure 13.10. Correlation of the ^{3}He NMR shift (δ_{He}) and the degree of ^{13}C NMR deshielding ($\Delta\langle\delta_c\rangle$) of fullerenes.

highest and the lowest degrees of deshielding are for $C_{60}{}^{6-}$ and $C_{70}{}^{6-}$, respectively. A correlation between the ^{3}He NMR [160] shifts and the degree of ^{13}C NMR deshielding [16, 153] was found (Fig. 13.10) [162]. This suggests that the aromatic character could be deduced from the carbon shifts. Accordingly, the difference between the ^{13}C NMR centers of gravity of the neutral and anionic fullerene ($\Delta\langle\delta_C\rangle$) demonstrates the relative aromaticity of the fullerene anion.

13.5.3.3 Methylene Bridged Fullerenes – an External Probe

More detailed characterization of the magnetic properties of each fullerene is possible by probing the separate contributions of the individual rings. Ring current calculations have shown that the six-membered rings (6MRs) exhibit diamagnetic ring currents while the five-membered rings (5MRs) exhibit paramagnetic ring currents [158b,c, 163]. The low aromaticity of C_{60} has been attributed to a near cancellation of these two opposing contributions. The higher aromaticity of C_{70} results from the increase in the number of 6MRs, while the number of 5MRs remains the same. London and NICS calculations of $C_{60}{}^{6-}$, however, show diatropic ring currents in both 5MRs and 6MRs, which explains its high aromaticity [158, 159, 163].

Monitoring the local ring currents of the 5MRs and 6MRs in reduced fullerenes was done by measuring the ^{1}H NMR spectra of the bridged fullerenes, thus enabling observation of the aromaticity of the particular rings from above. Five isomers of $C_{61}H_2$ and $C_{71}H_2$ were reduced with Li and analyzed by NMR methods: two isomers of $C_{61}H_2$ (**57** and **58**) [164] and three isomers of $C_{71}H_2$ (**59**, **60** and **61**) [165]. In all these bridged fullerenes, the protons are located above the centers of the rings, either one above a 5MR and the other above a 6MR (noted as [5,6] e.g. **57**, **59** and **60**) or above two 6MRs ([6,6] e.g. **58**, **61**).

Bridged $C_{60}{}^{6-}$: The ^{1}H NMR of **57**$^{6-}$ showed two doublets at 2.74 and 1.34 ppm, as expected from its two different protons [166]. The spectrum of **58**$^{6-}$ contained one singlet at $\delta = 2.33$ [167], since both protons are identical. The **58**$^{6-}$ signal is shielded by 1.6 ppm compared with the signal of neutral **58**. Based on the ^{1}H NMR spectrum of **58**$^{6-}$, in which both protons sit above 6MRs, the two doublets of **57**$^{6-}$ at 2.74 and 1.34 ppm were assigned as Ha (above 6MR) and Hb (above 5MR), respectively. While there was almost no change in the chemical shift of Ha ($\Delta\delta = -0.13$ ppm), the signal of Hb, showed a dramatic up-field shift ($\Delta\delta = -5.01$ ppm), compared with the neutral state. The additional electrons are therefore located mainly in the 5MRs, thus converting them from paratropic to diatropic rings, while the ring currents of the 6MRs experience little change. The NMR results are in line with calculations, and explain the increase in the aromaticity of C_{60} upon reduction [158, 159b, 163].

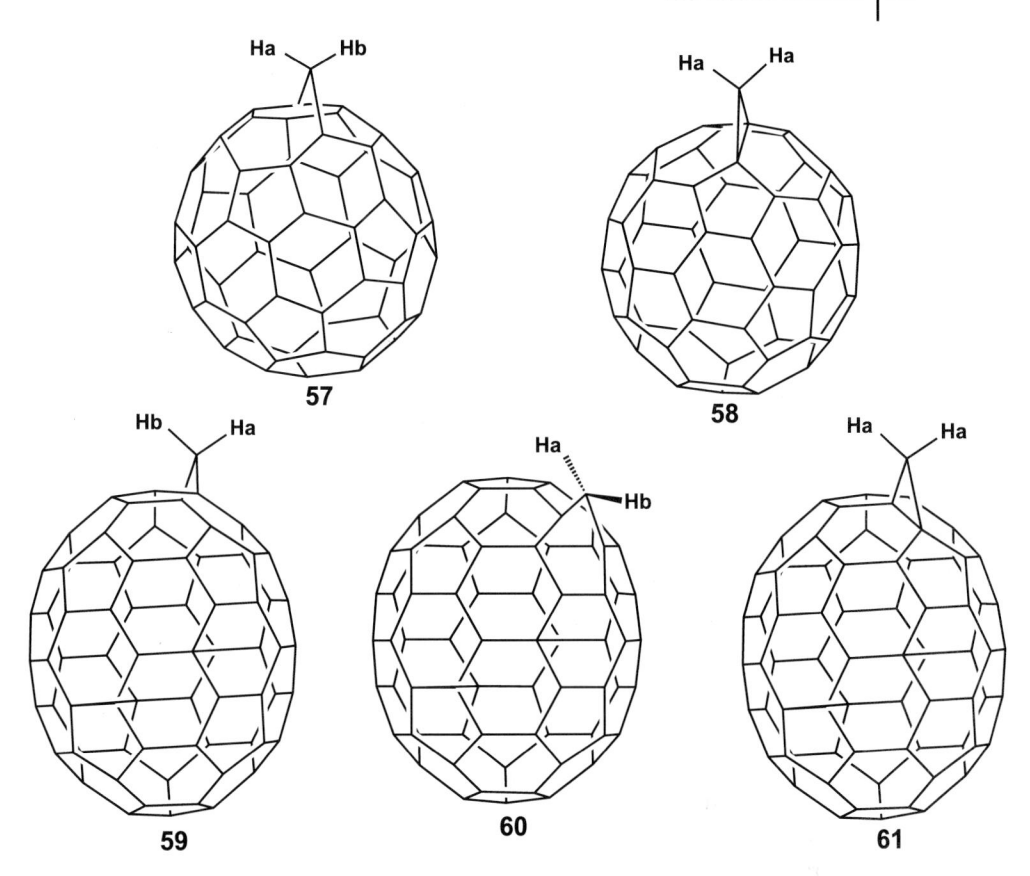

Bridged $C_{70}{}^{6-}$: The isomers **59** and **60** were reduced as a mixture and were distinguished according to their peak area ratio [168]. The ^{1}H NMR spectrum of **59**$^{6-}$ contains two doublets at 2.27 ppm (Ha) and −0.25 ppm (Hb). The ^{1}H NMR spectrum of **60**$^{6-}$ contains two doublets, one at 2.34 ppm (Ha), and the other, which lies under the solvent peak, at 3.6 ppm (Hb). The full assignment of the ^{1}H NMR chemical shifts of the two isomers **59**$^{6-}$ and **60**$^{6-}$ was possible because the proton Ha of both isomers is located above the same 6-MR of the C_{70} framework. The substantial upfield shift of both Hb peaks, relative to the neutral fullerene, indicates an increased diamagnetism of the 5MRs in C_{70}^{6-}. While the chemical shifts of protons Ha are very similar in both **59**$^{6-}$ and **60**$^{6-}$, those of the Hb are quite different. In **59**$^{6-}$, where the bridge is positioned at the C_{70} pole, the chemical shift of Hb is shifted to higher field than in **60**$^{6-}$. This means that the added electrons are located on 5MRs at the pole.

In isomer **61** the two protons are located over equivalent 6-MRs; therefore, the ^{1}H NMR of **61**$^{6-}$ contains one singlet. The signal appeared at 2.77 ppm, shielded by 0.11 ppm compared to neutral **61** [167].

13.5.4
Fullerene Derivatives

Reduction of fullerenes is a method of choice for preparing fullerene adducts, since the fullerene anions exhibit nucleophilic behavior. Fullerene derivatives, such as alkyl fullerenes, were prepared by chemical reduction of the fullerene [169] or via organic electrosynthesis [38], mostly via C_{60}^{2-} [147c].

Negatively charged fullerenes can undergo dimerization. It is also possible to prepare fullerene dimers with several spacers separating the two units [170]. Reduction of fullerene dimers is interesting due to the possibility of electronic communication between the two units. Electronic communication was reported by Balch in the dimer $C_{120}O$ (**62**) [171]. The EPR spectrum of **62**$^{2-}$ is different from that of C_{60}^{-} and shows characteristics that suggest electronic interaction between the two fullerene moieties. The EPR spectrum of **62**$^{2-}$ consists of a doublet-like feature ($g = 2.0016$) that is surrounded by features characteristic of a triplet state ($g = 2.0030$). The appearance of triplet features indicates the existence of a state with significant interaction between the two electrons that are located on the two separate fullerene cages. This means that communication exists between the π–π overlap in the region where the two cages are held closest.

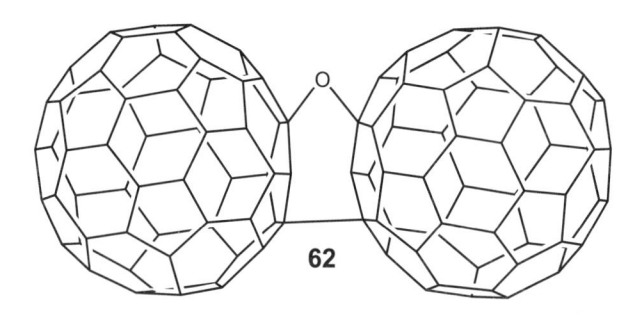

62

Promising research was carried out on the reduction of fullerene derivatives, which is expected to tune the sphere's electronic properties and investigate how the nature, geometry, structure, and number of adducts influence the electrochemical behavior of this fullerene [33]. The nature and the number of the adduct groups could both increase and decrease the electron affinity, as observed in CV experiments. Hydrofullerene [172] and epoxyfullerene [173] are two examples in which a decrease in the electron affinity was measured. Both of these decompose to the parent C_{60} after one to three reduction waves.

Increasing the electron affinity of C_{60} is possible by derivatization with electron-withdrawing groups. The effect of these groups was attributed to an orbital through-space interaction phenomenon referred to "periconjugation" [174]. Fullerenes functionalized with cyano [174c] or imino [175] groups are highly electronegative and the potentials of their first reduction peaks are less negative than that of C_{60}.

13.5.4.1 Metallofullerene Complexes

The reaction of transition metal complexes with fullerenes results in the formation of an interesting array of new compounds. Park and Shapley had independently prepared various C_{60}–metal cluster complexes and demonstrated that a variety of cluster frameworks such as Ru_3 [176], Os_3 [177], Rh_6 [178] can bind to C_{60}. Several reviews summarize these types of reactions and their properties [179].

An important characteristic of these metallofullerene complexes is their unique electrochemical properties. The comparison between the reduction potential of these complexes with those of free C_{60} gives an indication of the localization and delocalization of the added electrons. In some mono- [180] and trimetallic C_{60} complexes the first reduction wave is slightly more negative than that of free C_{60} (for example, for $[Os_3(CO)_{11}(\eta^2\text{-}C_{60})]$ $E_{1/2} = -1.08$ V, compared to $E_{1/2} = -1.05$ V for free C_{60}) [181]. This is consistent with metal-to-C_{60}-back-donation, and provides a direct indication of charge transfer upon complex formation.

Metallobisfullerene complexes are predicted to have intercage electronic communication between C_{60} cages. The CV of $[Rh_6(CO)_5(dppm)_2(CNCH_2C_6H_5)(\mu_3\text{-}\eta^2,\eta^2,\eta^2\text{-}C_{60})_2]$ (**63**) exhibits six well-separated reversible, one-electron redox waves localized at the C_{60} cages [178]. Each redox wave corresponds to sequential, pairwise addition of six electrons into the two C_{60} centers to form eventually $C_{60}{}^{3-}$–Rh_6–$C_{60}{}^{3-}$. The large peak separations $[(E_{1/2}{}^1, E_{1/2}{}^2) = 0.19$ V, $(E_{1/2}{}^3, E_{1/2}{}^4) = 0.24$ V and $(E_{1/2}{}^5, E_{1/2}{}^6) = 0.29$ V] in the three redox pairs of the two C_{60} ligands reflect an unusually strong electronic communication between the two C_{60} centers via the Rh_6 spacer.

13.5.5
Nanotubes

The unique electronic and mechanical properties of carbon nanotubes have been intensively investigated, with the intention of using these materials in nanodevice engineering [182]. The chemical properties of carbon nanotubes lie between those of graphite, which is almost chemically inert, and fullerenes, that are functionalized rather easily.

A few studies concerning the reduction of carbon nanotubes, both single-walled nanotubes (SWNTs) and multi-walled nanotubes (MWNTs), have been reported. Charging SWNTs makes it possible to modify their conductive nature by tuning their Fermi level, as shown by Petit [18]. The reduction of the SWNTs, e.g. n-

63

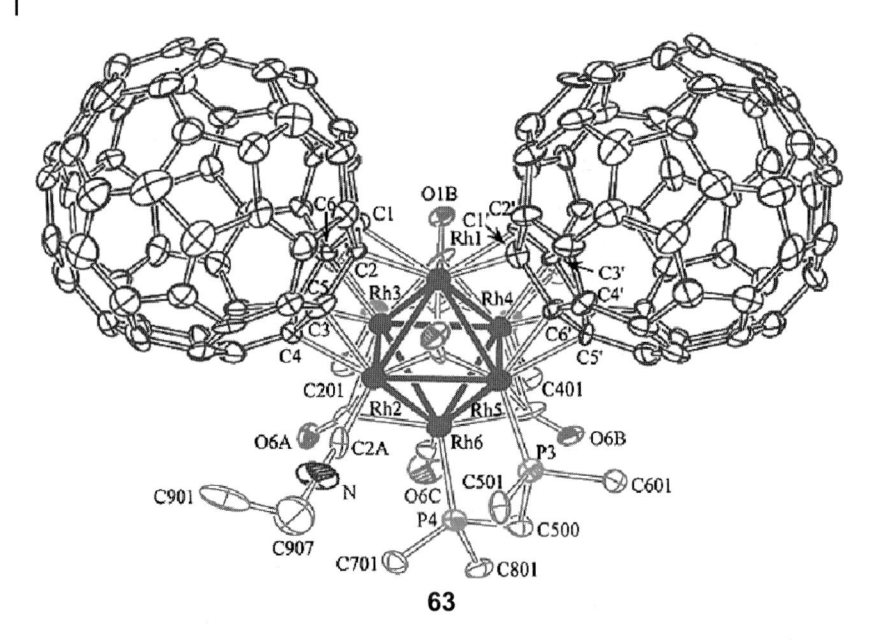

doping, was done in THF with Li, where the charge transfer to the nanotubes was performed by the radical anions of small aromatic compounds (naphthalene, fluorenone and anthraquinone).

Reducing nanotubes is used as a route to functionalize them, to modify their surfaces and to increase their solubility. Hydrogenation of MWNT and SWNT was performed via a modified Birch reduction, with Li and methanol in liquid ammonia [183]. The hydrogen content in the nanotubes, estimated from GC/MS measurements, corresponds to $C_{11}H$. SWNTs also undergo reductive alkylation in liquid ammonia in the presence of Li and alkyl halides, where the functionalized nanotubes are soluble in organic solvents [184]. Recently it was shown that upon reduction with alkali metals, SWNTs form polyelectrolyte salts that are soluble in organic solvents without any sonication or any functionalization [185]. The chemical formula of the nanotube salts was found to be $M(THF)C_{10}$ (M = Li, Na), hence one negative charge is added per ten carbon atoms. This ratio is very close to the formula that was obtained in the hydrogenation of the nanotubes ($C_{11}H$) [183].

13.6
Conclusions

The stepwise transfer of electrons from alkali metals to π-conjugated systems enables the study of the relationship between the number of π-electrons and the aromatic/antiaromatic properties of the systems *vis á vis* the Hückel rule. The magnetic criterion for aromaticity serves as a probe for the aromatic nature of the systems under study.

The notion of aromaticity and anti-aromaticity is extended to polycyclic aromatic hydrocarbons PAHs, especially to nonplanar π-systems, such as bowl-shaped and curved PAHs. These systems represent a unique class of organic molecules because they combine the fascination of fullerene building blocks with the outstanding material properties of graphite and conducting polymers. The planarity of PAHs is often presumed to be their most significant geometric characteristic, however, molecules that are at variance with this structural truism have been synthesized and their reduction is described in this chapter.

A much-studied aspect of such curved systems, neutral and anions, is their aromatic behavior, for such systems can help clarify the nature of aromaticity as they manifest a compromise between strain and conjugation. Reducing such PAHs yields negatively charged π-conjugated molecules and their properties rely on the number of added electrons. They not only have different aromatic character but they can undergo structural changes. At the extreme are ball-shaped molecules that have 60+ π-electrons that can be aromatic or anti-aromatic depending on their reduction state.

Alkali metal reduction of PAHs allows, in some cases, the formation of new inter- and intra-molecular σ-bonds as well as the formation of supramolecular dimers and aggregates. The outcome of the reduction depends in many cases on the alkali metal and the ion-solvation equilibria.

13.7
Experimental: Selected Procedures

13.7.1
Reduction with Li and K

The samples were reduced in 5 mm-diameter NMR glass tubes equipped with an upper reduction chamber. The compounds (3–5 mg) were introduced into the lower chamber of the tube under argon atmosphere. The alkali metal (kept in paraffin oil, cleansed from the oxidized layer and rinsed in petroleum ether 40–60 °C) was introduced under argon to the reduction chamber as a lithium wire or a piece of potassium. The tube was then placed under high vacuum and dried by flame. In the case of potassium, the metal was sublimed several times, creating a potassium mirror on the reduction chamber. Approximately 1 mL of anhydrous THF-d_8 (dried over a sodium/potassium alloy under high vacuum) was vacuum transferred to the NMR tube and was degassed several times. Finally the tube was flame-sealed under high vacuum (3×10^{-4} mbar).

The reduction takes place when the THF-d_8 solution is brought into contact with the metal by inverting the sample in solid dry ice. Reduction is stopped by returning the sample to the upright position thus separating the metal from the solution.

A full assignment of the ^{1}H and ^{13}C NMR spectra is performed for each diamagnetic species (including the neutral compound). The following 2D NMR methods are used: COSY, NOESY, HSQCSI, HMBC and INADEQUATE. The NMR studies were carried out at low temperatures.

For the reduction of fullerenes a small amount (0.5–1 mg) of corannulene is added to the sample.

13.7.2
Quench Reaction with Water

The anionic systems are quenched with H_2O by breaking and removing the reduction chamber in a glovebox, and pouring the solution into a vial containing a small amount of water. The product is extracted with dichloromethane and dried over $MgSO_4$, and then the solvent is evaporated. The dry quench product is analyzed using NMR spectroscopy and MS.

13.7.3
Reduction with C_8K

The Potassium-Graphite Intercalate (C_8K) is prepared as follows: In a three-necked round bottomed flask flushed with argon and equipped with a magnetic stirrer, graphite (1–2 g) is flame dried and heated to 200 °C. After 15 min, K (0.5 g) is added in small pieces and the stirring at 200 °C is continued until the bronze-colored C_8K is formed. The reagent is cooled to room temperature and kept under argon.

The reduction of PAHs with C_8K (reduction of substituted naphthalenes) is carried out as follows: The compounds (1 mmol) in dry THF (50 mL) are added to freshly prepared C_8K (12.5 mmol). The mixture is stirred at 0 °C for a few hours under argon, and quenched with water (10 mL). The reaction mixture is then filtered and washed with water and dichloromethane, and dried over $MgSO_4$.

Acknowledgments

We thank Dr. I. Aprahamian for his help during the preparation of the manuscript.

Abbreviations

ABA	alternating bond angle
BLA	bond-length alternation
CG	center of gravity
CIP	contact ion pair
COSY	correlation spectroscopy
CV	cyclic voltammetry
DFT	density functional theory
ENDOR	electron nuclear double resonance
GC	gas chromatography
HMBC	heteronuclear multiple-bond coherence

HOMO	highest occupied molecular orbital
HSQC	heteronuclear single-quantum correlation
INADEQUATE	incredible natural abundance double quantum transfer experiment
LUMO	lowest unoccupied molecular orbital
MO	molecular orbital
MS	mass spectrometry
MWNT	multi-walled nanotubes
NLUMO	second lowest unoccupied molecular orbital
NOESY	nuclear Overhauser effect spectroscopy
PAH	polycyclic aromatic hydrocarbon
SSIP	solvent separated ion pair
SWNT	single-walled nanotubes
THF	tetrahydrofuran

References

1 (a) M. Szwarc, *Ions and Ion Pairs in Organic Reactions*, Wiley-Interscience, New York, **1974**; (b) K. Müllen, *Chem. Rev.* **1984**, *84*, 603–646; (c) M. Rabinovitz, *Top. Curr. Chem.* **1988**, *146*, 99–169; (d) R. Benshafrut, E. Shabtai, M. Rabinovitz, L. T. Scott, *Eur. J. Org. Chem.* **2000**, 1091–1106.

2 M. Berthelot, *Ann. Chim.* **1867**, *12*, 155–159.

3 (a) W. Schlenk, J. Appenrodt, A. Michael, A. Thal, *Chem. Ber.* **1914**, *47*, 473–490; (b) W. Schlenk, E. Bergmann, *Liebigs Ann. Chem.* **1928**, *464*, 1–21.

4 W. Hückel, H. Bretschneider, *Liebigs Ann. Chem.* **1939**, *540*, 157–189.

5 (a) D. Lipkin, D. E. Paul, J. Townsend, S. I. Weissman, *Science* **1953**, *117*, 534–535; (b) S. I. Weissman, J. Townsend, D. E. Paul, G. E. Pake, *J. Chem. Phys.* **1953**, *21*, 2227–2228; (c) G. J. Hoijtink, J. van Schooten, E. de Boer, W. IJ. Aalbersberg, *Rec. Trav. Chim.* **1954**, *73*, 355–375; (d) D. E. Paul, D. Lipkin, S. I. Weissman, *J. Am. Chem. Soc.* **1956**, *78*, 116–120.

6 A. Ayalon, A. Sygula, P.-C. Cheng, M. Rabinovitz, P. W. Rabideau, L. T. Scott, *Science* **1994**, *265*, 1065–1067.

7 A. Ayalon, M. Rabinovitz, P.-C. Cheng, L. T. Scott, *Angew. Chem.* **1992**, *104*, 1691–1692; *Angew. Chem. Int. Ed. Engl.* **1992**, *31*, 1636–1637.

8 E. Shabtai, R. E. Hoffman, P.-C. Cheng, E. Bayrd, D. V. Preda, L. T. Scott, M. Rabinovitz, *J. Chem. Soc. Perkin Trans. 2* **2000**, 129–133.

9 M. Baumgarten, L. Gherghel, M. Wagner, A. Weitz, M. Rabinovitz, P.-C. Cheng, L. T. Scott, *J. Am. Chem. Soc.* **1995**, *117*, 6254–6257.

10 (a) A. J. Birch, *J. Chem. Soc.* **1944**, 430–436; (b) A. J. Birch, *Q. Rev. Chem. Soc.* **1950**, *4*, 69–93; (c) A. J. Birch, D. Nasipuri, *Tetrahedron* **1959**, *6*, 148–153.

11 (a) R. G. Harvey, *Synthesis* **1970**, 161–172; (b) E. M. Kaiser, *Synthesis* **1972**, 391–415.

12 P. W. Rabideau, *Tetrahedron* **1989**, *45*, 1579–1603.

13 (a) H. Normant, B. Angelo, *Bull. Soc. Chim. Fr.* **1960**, 354–359; (b) L. Horner, H. Güsten, *Liebigs Ann. Chem.* **1962**, *652*, 99–107; (c) H. Güsten, L. Horner, *Angew. Chem.* **1962**, *74*, 586; *Angew. Chem. Int. Ed. Engl.* **1962**, *1*, 455.

14 (a) H. W. Kroto, J. R. Heath, S. C. O'Brien, R. F. Curl, R. E. Smalley, *Nature* **1985**, *318*, 162–163;

(b) H. W. Kroto, A. W. Allaf, S. P. Balm, *Chem. Rev.* **1991**, *91*, 1213–1235.

15 S. Iijima, *Nature* **1991**, *354*, 56–58.

16 J. W. Bausch, G. K. S. Prakash, G. A. Olah, D. S. Tse, D. C. Lorents, Y. K. Bae, R. Malhorta, *J. Am. Chem. Soc.* **1991**, *113*, 3205–3206.

17 E. Shabtai, A. Weitz, R. C. Haddon, R. E. Hoffman, M. Rabinovitz, A. Khong, R. J. Cross, M. Saunders, P.-C. Cheng, L. T. Scott, *J. Am. Chem. Soc.* **1998**, *120*, 6389–6393.

18 P. Petit, C. Mathis, C. Journet, P. Bernier, *Chem. Phys. Lett.* **1999**, *305*, 370–374.

19 M. Contento, D. Savoia, C. Trombini, A. Umani-Ronchi, *Synthesis* **1979**, 30–32.

20 D. Savoia, C. Trombini, A. Umani-Ronchi, *Tetrahedron Lett.* **1977**, *18*, 653–656.

21 D. Savoia, C. Trombini, A. Umani-Ronchi, *J. Org. Chem.* **1978**, *43*, 2907–2910.

22 (a) D. Savoia, C. Trombini, A. Umani-Ronchi, *J. Chem. Soc., Perkin Trans. 1* **1977**, 123–125; (b) P. O. Ellingsen, K. Undeheim, *Acta. Chem. Scand. B* **1979**, *33*, 528–530.

23 (a) L. B. Ebert, *J. Mol. Cat.* **1982**, *15*, 275–296; (b) L. B. Ebert, L. Matty, *Mater. Res. Bull.* **1980**, *15*, 251–260; (c) L. B. Ebert, *Carbon* **1985**, *23*, 585–587.

24 R. Schlögel, H. P. Boehm, *Carbon* **1984**, *22*, 351–358.

25 F. Glocking, D. Kingston, *Chem. Ind.* **1961**, 1037–1037.

26 M. Rabinovitz, D. Tamarkin, *Synth. Commun.* **1984**, *14*, 377–379.

27 (a) D. Tamarkin, M. Rabinovitz, *Synth. Met.* **1984**, *9*, 125–128; (b) D. Tamarkin, M. Rabinovitz, *J. Org. Chem.* **1987**, *52*, 3472–3474.

28 (a) M. Rabinovitz, D. Tamarkin, *Synth. Met.* **1988**, *23*, 487–491; (b) R. Setton, F. Beguin, S. Piroelle, *Synth. Met.* **1982**, *4*, 299–318.

29 I. S. Weitz, M. Rabinovitz, *J. Chem. Soc., Perkin Trans. 1* **1993**, 117–120.

30 P. W. Rabideau, D. L. Huser, *J. Org. Chem.* **1983**, *43*, 4266–4271.

31 A. Weitz, M. Rabinovitz, *Synth. Met.* **1995**, *74*, 201–205.

32 J. Heinze, *Angew. Chem.* **1984**, *96*, 823–840; *Angew. Chem., Int. Ed. Engl.* **1984**, *23*, 831–847.

33 L. Echegoyen, L. E. Echegoyen, *Acc. Chem. Res.* **1998**, *31*, 593–601.

34 (a) Q. Xie, E. Perez-Cordero, L. Echegoyen, *J. Am. Chem. Soc.* **1992**, *114*, 3978–3980; (b) Y. Ohsawa, T. Saji, *J. Chem. Soc., Chem. Commun.* **1992**, 781–782; (c) D. Dubois, K. M. Kadish, S. Flanagan, L. J. Wilson, *J. Am. Chem. Soc.* **1991**, *113*, 7773–7774.

35 (a) C. Boudon, J.-P. Gisselbrecht, M. Gross, A. Herrmann, M. Rüttimann, J. Crassous, F. Cardullo, L. Echegoyen, F. Diederich, *J. Am. Chem. Soc.* **1998**, *120*, 7860–7868; (b) P. Boulas, M. T. Jones, K. M. Kadish, R. S. Ruoff, D. C. Lorents, D. T. Tse, *J. Am. Chem. Soc.* **1994**, *116*, 9393–9394; (c) P. L. Boulas, M. T. Jones, R. S. Ruoff, D. C. Lorents, R. Malhotra, D. S. Tse, K. M. Kadish, *J. Phys. Chem.* **1996**, *100*, 7573–7579; (d) J. Crassous, J. Rivera, N. S. Fender, L. Shu, L. Echegoyen, C. Thilgen, A. Herrmann, F. Diederich, *Angew. Chem.* **1999**, *111*, 1716–1721; *Angew. Chem., Int. Ed.* **1999**, *38*, 1613–1617; (e) M. R. Anderson, H. C. Dorn, S. A. Stevenson, S. M. Dana, *J. Electroanal. Chem.* **1998**, *444*, 151–154; (f) P. B. Burbank, J. R. Gibson, H. C. Dorn, M. R. Anderson, *J. Electroanal. Chem.* **1996**, *417*, 1–4.

36 (a) F. Arias, L. Echegoyen, S. R. Wilson, Q. Lu, *J. Am. Chem. Soc.* **1995**, *117*, 1422–1427; (b) F. Arias, Q. Xie, Y. Wu, Q. Lu, S. R. Wilson, L. Echegoyen, *J. Am. Chem. Soc.* **1994**, *116*, 6388–6394; (c) M. Eiermann, F. Wudl, M. Prato, M. Maggini, *J. Am. Chem. Soc.* **1994**, *116*, 8364–8365; (d) F. Paolucci, M. Marcaccio, S. Roffia, G. Orlandi, F. Zerbetto, M. Prato, M. Maggini, G. Scorrano, *J. Am. Chem. Soc.* **1995**, *117*, 6572–6580.

37 C. Koper, M. Sarobe, L. W. Jenneskens, *Phys. Chem. Chem. Phys.* **2004**, *6*, 319–327.

38 (a) C. Caron, R. Subramanian, F. D'Souza, J. Kim, W. Kutner, M. T. Jones, K. M. Kadish, *J. Am. Chem.*

Soc. **1993**, *115*, 8505–8506; (b) K.-M. Mangold, W. Kutner, L. Dunsch, J. Frohner, *Synth. Met.* **1996**, *77*, 73–76.

39 *Comprehensive Carbanion Chemistry*, Part A, E. Buncel and T. Durst (eds.), Elsevier, Amsterdam, **1980**.

40 C. Reichardt, *Solvents and Solvent Effects in Organic Chemistry*, VCH, Weinheim, **1988**.

41 (a) T. E. Hogen-Esch, J. Smid, *J. Am. Chem. Soc.* **1966**, *88*, 307–318; (b) T. E. Hogen-Esch, J. Smid, *J. Am. Chem. Soc.* **1969**, *91*, 4580–4581.

42 R. E. Hoffman, E. Shabtai, M. Rabinovitz, V. S. Iyer, K. Müllen, A. K. Rai, E. Bayrd, L. T. Scott, *J. Chem. Soc., Perkin Trans. 2* **1998**, 1659–1664.

43 (a) K. H. J. Buschow, G. J. Hoijtink, *J. Chem. Phys.* **1964**, *40*, 2501–2504; (b) M. Szwarc, *Carbanions, Living Polymers and Electron Transfer Processes*, Wiley-Interscience, New York, **1968**.

44 (a) A. Minsky, A. Y. Meyer, M. Rabinovitz, *Tetrahedron Lett.* **1982**, *23*, 5351–5354; (b) A. Minsky, A. Y. Meyer, R. Poupko, M. Rabinovitz, *J. Am. Chem. Soc.* **1983**, *105*, 2164–2172; (c) A. Minsky, A. Y. Meyer, M. Rabinovitz, *Angew. Chem.* **1983**, *95*, 45–46; *Angew. Chem., Int. Ed. Engl.* **1983**, *22*, 45–46; (d) A. Minsky, A. Y. Meyer, M. Rabinovitz, *Tetrahedron* **1985**, *41*, 785–791.

45 H. Günther, *NMR Spectroscopy – An Introduction*, Wiley, Chichester, U.K., **1980**.

46 F. Gerson, K. Müllen, E. Vogel, *Angew. Chem.* **1971**, *83*, 1014–1015; *Angew. Chem., Int. Ed. Engl.* **1971**, *10*, 920–921.

47 (a) G. J. Martin, M. L. Martin, S. Odiot, *Org. Magn. Reson.* **1975**, *7*, 2–17; (b) H. Baumann, H. Olsen, *Helv. Chim. Acta* **1980**, *61*, 2202–2211. (d) S. Fliszár, G. Cardinal, M.-T. Béraldin, *J. Am. Chem. Soc.* **1982**, *104*, 5287–5292.

48 G. Fraenkel, R. E. Carter, A. McLachlan, J. H. Richard, *J. Am. Chem. Soc.* **1960**, *82*, 5846–5850.

49 (a) P. C. Lauterbur, *J. Am. Chem. Soc.* **1961**, *83*, 1838–1846; (b) P. C. Lauterbur, *Tetrahedron Lett.* **1961**, *2*, 274–279; (b) H. Spiesecke, W. G. Schneider, *Tetrahedron Lett.* **1961**, *2*, 468–472; (c) T. Schaefer, W. G. Schneider, *Can. J. Chem.* **1963**, *41*, 966–982.

50 B. Eliasson, U. Edlund, K. Müllen, *J. Chem. Soc., Perkin Trans. 2* **1986**, 937–940.

51 (a) M. Karplus, T. P. Das, *J. Chem. Phys.* **1961**, *34*, 1683–1692; (b) M. Karplus, J. A. Pople, *J. Chem. Phys.* **1963**, *38*, 2803–2807; (c) M. Rabinovitz, I. Willner, A. Minsky, *Acc. Chem. Res.* **1983**, *16*, 298–304.

52 (a) N. F. Ramsey, *Phys. Rev.* **1950**, *78*, 699–703; (b) A. Saika, C. P. Slichter, *J. Chem. Phys.* **1954**, *22*, 26–28; (c) J. Mason, *J. Chem. Soc. A* **1971**, 1038–1047.

53 C. W. Haigh, R. B. Mallion, *Prog. Nucl. Magn. Reson. Spectrosc.* **1979**, *13*, 303–344.

54 F. Gerson, *High Resolution ESR Spectroscopy*, Wiley, New York, **1970**.

55 (a) H. M. McConnel, *J. Am. Chem. Soc.* **1956**, *24*, 764–766; (b) J. R. Bolton, *J. Chem. Phys.* **1965**, *43*, 309–310.

56 R. D. Kennedy, D. Lloyd, H. McNab, *J. Chem. Soc., Perkin Trans 1* **2002**, 1601–1621.

57 V. I. Minkin, N. M. Glukhovtsev, B. Ya. Simkin, *Aromaticity and Antiaromatcity: Electronic and Structural Aspects*, Wiley, New York, **1994**.

58 (a) E. Hückel, *Z. Phys.* **1931**, *70*, 204–286; (b) E. Hückel, *Z. Phys.* **1931**, *72*, 310–337; (c) E. Hückel, *Z. Physik* **1932**, *76*, 628–48; (d) E. Hückel, *Z. Phys.* **1933**, *83*, 632–668.

59 K. B. Wiberg, *Chem. Rev.* **2001**, *101*, 1317–1331.

60 (a) A. Streitweiser Jr., *Molecular Orbital Theory for Organic Chemists*, Wiley, New York, **1961**; (b) M. J. S. Dewar, *The Molecular Orbital Theory of Organic Chemistry*, McGraw-Hill, New York, **1969**.

61 (a) G. Binsch, I. Tamir, R. D. Hill, *J. Am. Chem. Soc.* **1969**, *91*, 2446–2450; (b) G. Binsch, I. Tamir, *J. Am. Chem. Soc.* **1969**, *91*, 2450–2455.

62 (a) K. Müllen, T. Meul, E. Vogel, U. Kürschner, H. Schmickler, O.

WENNERSTRÖM, *Tetrahedron Lett.* **1985**, *26*, 3091–3094; (b) K. MÜLLEN, T. MEUL, P. SCHADE, H. SCHMICKLER, E. VOGEL, *J. Am. Chem. Soc.* **1987**, *109*, 4992–5003.

63 R. H. MITCHELL, C. E. KLOPFENSTEIN, V. BOEKELHEIDE, *J. Am. Chem. Soc.* **1969**, *91*, 4931–4932.

64 (a) R. WILLSTÄTTER, E. WASER, *Chem. Ber.* **1911**, *44*, 3423–3445; (b) R. WILLSTÄTTER, M. HEIDELBERGER, *Chem. Ber.* **1913**, *46*, 517–527.

65 (a) M. TRAETTEBERG, *Acta Chem. Scand.* **1966**, *20*, 1724–1726; (b) O. BASTIANSEN, L. HEDBERG, K. HEDBERG, *J. Chem. Phys.* **1957**, *27*, 1311–1317.

66 (a) P. G. WENTHOLD, D. A. HROVAT, W. T. BORDEN, W. C. LINEBERGER, *Science* **1996**, *272*, 1456–1459; (b) S. KATO, H. S. LEE, R. GAREYEV, P. G. WENTHOLD, W. C. LINEBERGER, C. H. DEPUY, V. M. BIERBAUM, *J. Am. Chem. Soc.* **1997**, *119*, 7863–7864.

67 (a) T. J. KATZ, *J. Am. Chem. Soc.* **1960**, *82*, 3784–3785; (b) T. J. KATZ, W. H. REINMUTH, D. E. SMITH, *J. Am. Chem. Soc.* **1962**, *84*, 802–808; (c) H. L. STRAUSS, T. J. KATZ, G. K. FRAENKEL, *J. Am. Chem. Soc.* **1963**, *85*, 2360–2364; (d) T. J. KATZ, H. L. STRAUSS, *J. Chem. Phys.* **1960**, *32*, 1873–1875.

68 J. H. HAMMONS, D. A. HROVAT, W. T. BORDEN, *J. Am. Chem. Soc.* **1991**, *113*, 4500–4505.

69 R. L. WARD and S. I. WEISSMAN, *J. Am. Chem. Soc.* **1957**, *79*, 2086–2090.

70 L. A. PAQUETTE, S. V. LEY, R. H. MEISINGER, R. K. RUSSELL, M. OKU, *J. Am. Chem. Soc.* **1974**, *96*, 5806–5815.

71 C. D. STEVENSON, L. J. HEINLE, J. P. DAVIS, R. C. REITER, *J. Am. Chem. Soc.* **2002**, *124*, 2704–2708.

72 W. HEINZ, H.-J. RÄDER, K. MÜLLEN, *Tetrahedron Lett.* **1989**, *30*, 159–162.

73 J. F. M. OTH, J.-M. GILLES, G. SCHRÖDER, *Tetrahedron Lett.* **1970**, *11*, 67–72.

74 J. F. M. OTH, G. SCHRÖDER, *J. Chem. Soc. B* **1971**, 904–907.

75 G. R. STEVENSON, R. CONCEPCION, R. C. REITER, *J. Org. Chem.* **1983**, *48*, 2777–2778.

76 M. N. GARD, R. C. REITER, C. D. STEVENSON, *Org. Lett.* **2004**, *6*, 393–396.

77 J. F. M. OTH, J.-M. GILLES, *Tetrahedron Lett.* **1968**, *9*, 6259–6264.

78 C. D. STEVENSON T. L. KURTH, *J. Am. Chem. Soc.* **2000**, *122*, 722–723.

79 J. F. M. OTH, H. BAUMANN, J. M. GILES, G. SCHRÖEDER, *J. Am. Chem. Soc.* **1972**, *94*, 3498–3512.

80 C. D. STEVENSON, T. L. KURTH, *J. Am. Chem. Soc.* **1999**, *121*, 1623–1624.

81 J. G. CONCEPCION, G. VINCOW, *J. Phys. Chem.* **1975**, *79*, 2037–2041.

82 G. R. STEVENSON, R. C. REITER, J. B. SEDGWICK, *J. Am. Chem. Soc.* **1983**, *105*, 6521–6522.

83 C. D. STEVENSON, R. D. BURTON, S. J. PETERS, R. C. REITER, *J. Org. Chem.* **1993**, *58*, 5838–5842.

84 R. H. MITCHELL, *Chem. Rev.* **2001**, *101*, 1301–1316.

85 C. S. WANNERE, K. W. SATTELMEYER, H. F. SCHAEFER III, P. VON R. SCHLEYER, *Angew. Chem.* **2004**, *116*, 4296–4302; *Angew. Chem., Int. Ed.* **2004**, *43*, 4200–4206.

86 (a) C. S. WANNERE, P. VON R. SCHLEYER, *Org. Lett.* **2003**, *5*, 865–868; (b) C. H. CHOI, M. KERTESZ, *J. Chem. Phys.* **1998**, *108*, 6681–6688.

87 J. F. M. OTH, E. P. WOO, F. SONDHEIMER, *J. Am. Chem. Soc.* **1973**, *95*, 7337–7345.

88 T. L. KURTH, E. C. BROWN, C. M. HATTAN, R. C. REITER, C. D. STEVENSON, *J. Phys. Chem. A* **2002**, *106*, 478–481.

89 R. M. MCQUILKIN, P. J. GARRATT, F. SONDHEIMER, *J. Am. Chem. Soc.* **1970**, *92*, 6682–6683.

90 N. M. ATHERTON, R. MASON, R. J. WRATTEN, *Mol. Phys.* **1966**, *11*, 525–529.

91 H. A. STAAB F. GRAF, *Tetrahedron Lett.* **1966**, *7*, 751–757.

92 K. MÜLLEN, W. HUBER, T. MEUL, M. NAKAGAWA, M. IYODA, *J. Am. Chem. Soc.* **1982**, *104*, 5403–5411.

93 M. NAKAGAWA, *Pure Appl. Chem.* **1975**, *44*, 885–924.

94 R. WOLOVSKY, F. SONDHEIMER, *J. Am. Chem. Soc.* **1965**, *87*, 5720–5727.

95 (a) K. G. UNTCH, D. C. WYSOCKI, *J. Am. Chem. Soc.* **1966**, *88*, 2608–

2610; (b) F. Sondheimer, R. Wolovsky, P. J. Garratt, I. C. Calder, *J. Am. Chem. Soc.* **1966**, *88*, 2610–2610.

96 P. J. Garratt, N. E. Rowland, F. Sondheimer, *Tetrahedron* **1971**, *27*, 3157–3162.

97 (a) *Cyclophanes*, P. M. Keehn, S. M. Rosenfeld (eds.), Academic Press, New York, **1983**; (b) F. Diederich, *Cyclophanes*, The Royal Society of Chemistry, Cambridge, **1991**; (c) *Modern Cyclophanes Chemistry*, R. Gleiter, H. Hopf (eds.), Wiley-VCH, Weinheim, **2004**.

98 E. Ljungström, O. Lindqvsit, O. Wennerström, *Acta. Crystallogr. Sect. B* **1978**, *34*, 1889–1893.

99 W. Huber, K. Müllen, O. Wennerström, *Angew. Chem.* **1980**, *19*, 624–625; *Angew. Chem., Int. Ed. Engl.* **1980**, *92*, 636–637.

100 K. Müllen, H. Unterberg, W. Huber, O. Wennerström, U. Norinder, D. Tanner, B. Thulin, *J. Am. Chem. Soc.* **1984**, *106*, 7514–7522.

101 E. Shabtai, M. Rabinovitz, B. König, B. Knieriem, A. de Meijere, *J. Chem. Soc., Perkin Trans. 2* **1996**, 2589–2595.

102 H. Hopf, *Classics in Hydrocarbon Chemistry*, Wiley-VCH, Weinheim, **2000**.

103 W. Krätschmer, L. D. Lamb, K. Fostiropoulos, D. R. Huffman, *Nature* **1990**, *347*, 354–358.

104 P. W. Rabideau, A. Sygula, *Acc. Chem. Res.* **1996**, *29*, 235–242.

105 (a) W. E. Barth, R. G. Lawton, *J. Am. Chem. Soc.* **1966**, *88*, 380–381; (b) W. E. Barth, R. G. Lawton, *J. Am. Chem. Soc.* **1971**, *93*, 1730–1745.

106 (a) L. T. Scott, M. M. Hashemi, D. T. Meyer, H. B. Warren, *J. Am. Chem. Soc.* **1991**, *113*, 7082–7084; (b) L. T. Scott, P.-C. Cheng, M. M. Hashemi, M. S. Bratcher, D. T. Meyer, H. B. Warren, *J. Am. Chem. Soc.* **1997**, *119*, 10963–10968.

107 A. Borchardt, A. Fuchicello, K. V. Kilway, K. K. Baldridge, J. S. Siegel, *J. Am. Chem. Soc.* **1992**, *114*, 1921–1923.

108 G. Zimmermann, U. Nuechter, S. Hagen, M. Nuechter, *Tetrahedron Lett.* **1994**, *35*, 4747–4750.

109 (a) A. Sygula, P. W. Rabideau, *J. Am. Chem. Soc.* **2000**, *122*, 6323–6324; (b) A. Sygula, P. W. Rabideau, *J. Am. Chem. Soc.* **1999**, *121*, 7800–7803.

110 (a) P. W. Rabideau, Z. Marcinow, R. Sygula, A. Sygula, *Tetrahedron Lett.* **1993**, *34*, 6351–6354; (b) Z. Zhou, *J. Phys. Org. Chem.* **1995**, *8*, 103–107.

111 J. Janata, J. Gendell, C.-Y., Ling, W. Barth, L. Backes, H. B. Mark, Jr., R. G. Lawton, *J. Am. Chem. Soc.* **1967**, *89*, 3056–3058.

112 T. Sato, A. Yamamoto, H. Tanaka, *Chem. Phys. Lett.* **2000**, *326*, 573–579.

113 G. Zilber, V. Rozenshtein, P.-C. Cheng, L. T. Scott, M. Rabinovitz, H. Levanon, *J. Am. Chem. Soc.* **1995**, *117*, 10720–17025.

114 Y. Cohen, A. Ayalon, *Angew. Chem.* **1995**, *107*, 888–890; *Angew. Chem., Int. Ed. Engl.* **1995**, *34*, 816–818.

115 (a) R. H. Cox, H. W. Terry Jr., L. W. Harrison, *J. Am. Chem. Soc.* **1971**, *93*, 3297–3298; (b) H. Günther, D. Moskau, P. Best, D. Schmalz, *Angew. Chem.* **1987**, *99*, 1242–1250; *Angew. Chem., Int. Ed. Engl.* **1987**, *25*, 1212–1220.

116 G. H. Grube, E. L. Elliott, R. J. Steffens, C. S. Jones, K. K. Baldridge, J. S. Siegel, *Org. Lett.* **2003**, *5*, 713–716.

117 A. Weitz, M. Rabinovitz, P.-C. Cheng, L. T. Scott, *Synth. Met.* **1997**, *86*, 2159–2160.

118 P.-C. Cheng, Dissertation, Boston College, **1996**.

119 I. Aprahamian, R. E. Hoffman, T. Sheradsky, D. V. Preda, M. Bancu, L. T. Scott, M. Rabinovitz, *Angew. Chem.* **2002**, *114*, 1788–1791; *Angew. Chem., Int. Ed.* **2002**, *41*, 1712–1715.

120 A. Weitz, E. Shabtai, M. Rabinovitz, M. S. Bratcher, C. C. McComas, M. D. Best, L. T. Scott, *Chem. Eur. J.* **1998**, *4*, 234–239.

121 (a) E. D. Bergmann, *Chem. Rev.* **1968**, *68*, 41–84; (b) M. Baumgarten, K. Müllen, *Top. Curr. Chem.* **1994**, *169*, 1–103.

122 U. Edlund, B. Eliasson, *J. Chem. Soc., Chem. Commun.* **1982**, 950–952.

123 (a) G. J. Bodwell, J. N. Bridson, T. J. Houghton, J. W. J. Kennedy, M. R. Mannion, *Angew. Chem.* **1996**, *108*, 1418–1420; *Angew. Chem., Int. Ed. Engl.* **1996**, *35*, 1320–1321; (b) G. J. Bodwell, J. N. Bridson, T. J. Houghton, J. W. J. Kennedy, M. R. Mannion, *Chem. Eur. J.* **1999**, *5*, 1823–1827; (c) G. J. Bodwell, J. J. Fleming, M. R. Mannion, D. O. Miller, *J. Org. Chem.* **2000**, *65*, 5360–5370; (d) G. J. Bodwell, J. N. Bridson, M. K. Cyrański, J. W. J. Kennedy, T. M. Krygowski, M. R. Mannion, D. O. Miller, *J. Org. Chem.* **2003**, *68*, 2089–2098.

124 I. Aprahamian, G. J. Bodwell, J. J. Fleming, G. P. Manning, M. R. Mannion, T. Sheradsky, R. J. Vermeij, M. Rabinovitz, *J. Am. Chem. Soc.* **2004**, *126*, 6765–6775.

125 I. Aprahamian, G. J. Bodwell, J. J. Fleming, G. P. Manning, M. R. Mannion, B. L. Merner, T. Sheradsky, R. J. Vermeij, M. Rabinovitz, *Angew. Chem.* **2003**, *115*, 2651–2654; *Angew. Chem., Int. Ed.* **2003**, *42*, 2547–2550.

126 C. Schnieders, K. Müllen, W. Huber, *Tetrahedron* **1984**, *40*, 1701–1711.

127 I. Aprahamian, G. J. Bodwell, J. J. Fleming, G. P. Manning, M. R. Mannion, T. Sheradsky, R. J. Vermeij, M. Rabinovitz, *J. Am. Chem. Soc.* **2003**, *125*, 1720–1721.

128 M. A. Hempenius, W. Heinen, P. P. J. Mulder, C. Erkelens, H. Zuilhof, J. Lugtenburg, J. Cornelisse, *J. Phys. Org. Chem.* **1994**, *7*, 296–302.

129 (a) W. E. Billups, R. S. Ruoff, *Buckministerfullerenes*, Wiley, New York, **1993**; (b) K. M. Kadish, R. S. Ruoff, *Recent Advances in the Chemistry and Physics of Fullerenes and Related Materials*, The Electrochemical Society, New Jersey, **1994**; (c) *Fullerene and Related Structures*, A. Hirsch (ed.), Springer-Verlag, Berlin, **1999**.

130 F. Diederich, R. L. Whetten, *Acc. Chem. Res.* **1992**, *25*, 119–126.

131 (a) R. C. Haddon, L. E. Brus, K. Raghavachari, *Chem. Phys. Lett.* **1986**, *125*, 459–464; (b) R. L. Disch, J. M. Schulman, *Chem. Phys. Lett.* **1986**, *125*, 465–466; (c) P. W. Fowler, J. Woolrich, *Chem. Phys. Lett.* **1986**, *127*, 78–83.

132 (a) S. H. Yang, C. L. Pettiette, J. Coneicao, O. Cheshnovsky, R. E. Smalley, *Chem. Phys. Lett.* **1987**, *139*, 233–238 (b) L.-S. Wang, J. Coneicao, C. Jin, R. E. Smalley, *Chem. Phys. Lett.* **1991**, *182*, 5–11; (c) X.-B. Wang, C.-F. Ding, L.-S. Wang, *J. Chem. Phys.* **1999**, *110*, 8217–8220; (d) O. V. Boltalina, L. N. Sidorov, A. Ya. Borshchevsky, E. V. Sukhanova, E. V. Skokan, *Rapid Commun. Mass. Spectrom.* **1993**, *7*, 1009–1011.

133 C. A. Reed, R. D. Bolskar, *Chem. Rev.* **2000**, *100*, 1075–1120 and references therein.

134 (a) W. R. Fawcett, M. Opallo, M. Fedurco, J. W. Lee, *J. Am. Chem. Soc.* **1993**, *115*, 196–200; (b) D. Dubois, G. Moninot, W. Kutner, M. T. Jones, K. M. Kadish, *J. Phys. Chem.* **1992**, *96*, 7137–7145; (c) M. V. Mirkin, L. O. S. Bulhoes, A. J. Bard, *J. Am. Chem. Soc.* **1993**, *115*, 201–204; (d) R. G. Compton, R. A. Spackman, D. J. Riley, R. G. Wellington, J. C. Eklund, A. C. Fisher, M. L. H. Green, R. E. Douthwaite, A. H. H. Stephens, J. Turner, *J. Electroanal. Chem.* **1993**, *344*, 235–247.

135 Y. Yang, F. Arias, L. Echegoyen, L. P. F. Chibante, S. Flanagan, A. Robertson, L. J. Wilson, *J. Am. Chem. Soc.* **1995**, *117*, 7801–7804.

136 R. C. Haddon, A. F. Hebard, M. J. Rosseinsky, D. W. Murphy, S. J. Duclos, K. B. Lyons, B. Miller, J. M. Rosamilia, R. M. Fleming, A. R. Kortan, S. H. Glarum, A. V. Makhija, A. J. Muller, R. H. Eick, S. M. Zahurak, R. Tycko, G. Dabbagh, F. A. Thiel, *Nature* **1991**, *350*, 320–322.

137 M. J. Rosseinsky, *Chem. Mater.* **1998**, *10*, 2665–2685.

138 (a) S. G. Kukolich, D. R. Huffman, *Chem. Phys. Lett.* **1991**, *182*, 263–265; (b) A. J. Schell-Sorokin, F. Mehran, G. R. Eaton, S. S. Eaton, A. Viehbeck, T. R. O'Toole, C. A. Brown, *Chem. Phys. Lett.* **1992**, *195*,

225–231; (c) M. Baumgarten, A. Gügel, L. Gherghel, *Adv. Mater.* **1993**, *5*, 458–461; (d) M. Baumgarten, L. Gherghel, *Appl. Magn. Reson.* **1996**, *11*, 171–182; (e) K. Hiraoka, I. Kudaka, S. Fujimaki, H. Shinohara, *Rapid Commun. Mass Spectrom.* **1992**, *6*, 254–256.

139 (a) W. K. Fullagar, I. R. Gentle, G. A. Heath, J. W. White, *J. Chem. Soc., Chem. Commun.* **1993**, 525–527; (b) D. R. Buffinger, R. P. Ziebarth, V. A. Stenger, C. Recchia, C. H. Pennington, *J. Am. Chem. Soc.* **1993**, *115*, 9267–9270.

140 Y. Ederle, C. Mathis, *Macromolecules* **1997**, *30*, 4262–4267.

141 D. R. Lawson, D. L. Feldheim, C. A. Foss, P. K. Dorhout, C. M. Elliott, C. R. Martin, B. Parkinson, *J. Electrochem. Soc.* **1992**, *139*, L68–L71.

142 S. S. Eaton, G. R. Eaton, *Appl. Magn. Reson.* **1996**, *11*, 155–170.

143 F. J. Adrian, *Chem. Phys.* **1996**, *211*, 73–80.

144 T. Kato, T. Kodama, M. Oyama, S. Okazaki, T. Shida, T. Nakagawa, Y. Matsui, S. Suzuki, H. Shiromaru, K. Yamauchi, Y. Achiba, *Chem. Phys. Lett.* **1991**, *186*, 35–39.

145 (a) P. D. W. Boyd, P. Bhyrappa, P. Paul, J. Stinchcombe, R. O. Bolskar, Y. Sun, C. A. Reed, *J. Am. Chem. Soc.* **1995**, *117*, 2907–2914; (b) D. Dubois, M. T. Jones, K. M. Kadish, *J. Am. Chem. Soc.* **1992**, *114*, 6446–6451; (c) C. Bossard, S. Rigaut, D. Astruc, M. H. Deliville, G. Felix, A. Fevrierbouvier, J. Amiell, S. Flandrois, P. Delhaes, *J. Chem. Soc., Chem. Commun.* **1993**, 333–334.

146 Y. Sun, C. A. Reed, *J. Chem. Soc., Chem. Commun.* **1997**, 747–748.

147 (a) W. K. Fullagar, I. R. Gentle, G. A. Heath, J. W. White, *J. Chem. Soc., Chem. Commun.* **1993**, 525–527. (b) J. Stinchcombe, A. Pénicaud, P. Bhyrappa, P. D. W. Boyd, C. A. Reed, *J. Am. Chem. Soc.* **1993**, *115*, 5212–5217.

148 (a) D. R. Lawson, D. L. Feldheim, C. A. Foss, P. K. Dorhout, C. M. Elliott, C. R. Martin, B. Parkinson, *J. Phys. Chem.* **1992**, *96*, 7175–7177; (b) D. V. Konarev, N. V. Drichko, A. Graja, *J. Chim. Phys.* **1998**, *95*, 2143–2156.

149 T. Sternfeld, R. E. Hoffman, I. Aprahamian, M. Rabinovitz, *Angew. Chem.* **2001**, *113*, 469–471; *Angew. Chem., Int. Ed.* **2001**, *40*, 455–457.

150 (a) R. D. Johnson, G. Meijer, J. R. Salem, D. S. Bethune, *J. Am. Chem. Soc.* **1991**, *113*, 3619–3621; (b) R. D. Johnson, D. S. Bethune, C. S. Yannoni, *Acc. Chem. Res.* **1992**, *25*, 169–175.

151 (a) J. C. Niles, X.-Q. Wang, *J. Chem. Phys.* **1995**, *103*, 7040–7047; (b) X.-Q. Wang, C. Z. Wang, B. L. Zhang, K. M. Ho, *Chem. Phys. Lett.* **1992**, *200*, 35–38; (c) J. R. Colt, G. E. Scuseria, *Chem. Phys. Lett.* **1992**, *199*, 505–512; (d) S. Saito, S. Okada, S. Sawada, N. Hamada, *Phys. Rev. Lett.* **1995**, *75*, 685–688; (e) J. A. Azamar-Barrios, T. J. S. Dennis, S. Sadhukan, H. Shinohara, G. E. Scuseria, A. Pénicaud, *J. Phys. Chem. A* **2001**, *105*, 4627–4632.

152 J. A. Azamar-Barrios, P. E. Muñoz, A. Pénicaud, *J. Chem. Soc., Faraday Trans.* **1997**, *93*, 3119–3123.

153 T. Sternfeld, C. Thilgen, Z. Chen, S. Seifken, P. v. R. Schleyer, W. Thiel, F. Diederich, M. Rabinovitz, *J. Org. Chem.* **2003**, *68*, 4850–4854.

154 M. Bühl, A. Hirsch, *Chem. Rev.* **2001**, *101*, 1153–1184.

155 (a) M. Saunders, H. A. Jimenez-Vazquez, R. J. Cross, S. Mroczkowski, D. I. Freedberg, F. A. L. Anet, *Nature* **1994**, *367*, 256–258; (b) M. Saunders, H. A. Jiménez-Vázquez, R. J. Cross, R. J. Poreda, *Science* **1993**, *259*, 1428–1430; (c) R. Shimshi, R. J. Cross, M. Saunders, *J. Am. Chem. Soc.* **1997**, *119*, 1163–1164; (d) M. Saunders, R. J. Cross, H. A. Jiménez-Vázquez, R. Shimshi, A. Khong, *Science* **1996**, *271*, 1693–1697; (e) M. Saunders, H. A. Jiménez-Vázquez, R. J. Cross, S. Mroczkowski, M. L. Gross, D. E. Giblin, R. J. Poreda, *J. Am. Chem. Soc.* **1994**, *116*, 2193–2194; (f) R. Shimshi, A. Khong, H. A. Jiménez-

Vázquez, R. J. Cross, M. Saunders, *Tetrahedron* **1996**, *52*, 5143–5148; (g) M. Saunders, H. A. Jiménez-Vázquez, R. J. Cross, W. E. Billups, C. Gesenberg, A. Gonzalez, W. Luo, R. C. Haddon, F. Diederich, A. Herrmann, *J. Am. Chem. Soc.* **1995**, *117*, 9305–9308.

156 H. W. Kroto, *Pure Appl. Chem.* **1990**, *62*, 407–415.

157 D. J. Klein, T. G. Schmaltz, G. E. Hite, W. A. Seitz, *J. Am. Chem. Soc.* **1986**, *108*, 1301–1302.

158 (a) V. Elser, R. C. Haddon, *Nature* **1987**, *325*, 792–794; (b) A. Pasquarello, M. Schluter, R. C. Haddon, *Science* **1992**, *257*, 1660–1661; (c) A. Pasquarello, M. Schluter, R. C. Haddon, *Phys. Rev. A* **1993**, *47*, 1783–1789.

159 (a) M. Bühl, W. Thiel, H. Jiao, P. v. R. Schleyer, M. Saunders, F. A. L. Anet, *J. Am. Chem. Soc.* **1994**, *116*, 6005–6006; (b) M. Bühl, *Chem. Eur. J.* **1998**, *4*, 734–739.

160 T. Sternfeld, M. Saunders, R. J. Cross, M. Rabinovitz, *Angew. Chem.* **2003**, *115*, 3244–3247; *Angew. Chem., Int. Ed.* **2003**, *42*, 3136–3139.

161 T. Sternfeld, R. E. Hoffman, M. Saunders, R. J. Cross, M. S. Symala, M. Rabinovitz, *J. Am. Chem. Soc.* **2002**, *124*, 8786–8787.

162 T. Sternfeld, Dissertation, The Hebrew University of Jerusalem, **2003**.

163 R. C. Haddon, *Science* **1993**, *261*, 1545–1550.

164 (a) T. Suzuki, Q. C. Li, K. C. Khemani, F. Wudl, *J. Am. Chem. Soc.* **1992**, *114*, 7301–7302; (b) A. B. Smith, R. M. Strongin, L. Brard, G. T. Furst, W. J. Romanow, K. G. Owens, R. C. King, *J. Am. Chem. Soc.* **1993**, *115*, 5829–5830.

165 (a) A. B. Smith, R. M. Strongin, L. Brard, G. T. Furst, W. J. Romanow, K. G. Owens, R. J. Goldschmidt, R. C. King, *J. Am. Chem. Soc.* **1995**, *117*, 5492–5502; (b) A. B. Smith, R. M. Strongin, L. Brard, G. T. Furst, W. J. Romanow, K. G. Owens, R. J. Goldschmidt, *J. Chem. Soc., Chem. Commun.* **1994**, 2187–2188.

166 T. Sternfeld, F. Wudl, K. Hummelen, A. Weitz, R. C. Haddon, M. Rabinovitz, *Chem. Commun.* **1999**, 2411–2412.

167 T. Sternfeld, C. Thilgen, R. E. Hoffman. M. del R. Colorado Heras, F. Diederich, F. Wudl, L. T. Scott, J. Mack, M. Rabinovitz, *J. Am. Chem. Soc.* **2002**, *124*, 5734–5738.

168 T. Sternfeld, R. E. Hoffman, C. Thilgen, F. Diederich, M. Rabinovitz, *J. Am. Chem. Soc.* **2000**, *122*, 9038–9309.

169 (a) S. Fukuzumi, T. Suenobu, T. Hirasaka, R. Arakawa, K. M. Kadish, *J. Am. Chem. Soc.* **1998**, *120*, 9220–9227; (b) R. Subramanian, K. M. Kadish, M. N. Vijayashree, X. Gao, M. T. Jones, M. D. Miller, K. L. Krause, T. Suenobu and S. Fukuzumi, *J. Phys. Chem.* **1996**, *100*, 16327–16335; (c) K. M. Kadish, X. Gao, E. Van Caemelbecke, T. Hirasaka, T. Suenobu, S. Fukuzumi, *J. Phys. Chem. A* **1998**, *102*, 3898–3906; (d) F. Cheng, Y. Murata, K. Komatsu, *Org. Lett.* **2002**, *4*, 2541–2544; (e) E. Allard, J. Delaunay, F. Cheng, J. Ordúna, J. Garín, *Org. Lett.* **2001**, *3*, 3503–3506.

170 (a) T. Susuki, Q. Li, K. Khemani, F. Wudl, Ö. Almarson, *J. Am. Chem. Soc.* **1992**, *114*, 7300–7301; (b) C. Boudon, J.-P. Gisselbrecht, M. Gross, L. Isaacs, H. L. Anderson, R. Faust, F. Diederich, *Helv. Chim. Acta* **1995**, *78*, 1334–1344; (c) D. V. Konarev, S. S. Khasanov, I. I. Vorontsov, G. Saito, A. Otsuka, *Synth. Met.* **2003**, *135*, 781–782.

171 A. L. Balch, D. A. Costa, W. R. Fawcett, K. Winkler, *J. Phys. Chem.* **1996**, *100*, 4823–4827.

172 (a) T. F. Guarr, M. S. Meier, V. K. Vance, M. Clayton, *J. Am. Chem. Soc.* **1993**, *115*, 9862–9863; (b) P. Boulas, F. D'Souza, C. C. Henderson, P. A. Cahill, M. T. Jones, K. M. Kadish, *J. Phys. Chem.* **1993**, *97*, 13435–13437.

173 (a) T. Suzuki, Y. Maruyama, T. Akasaka, W. Ando, K. Kobayashi, S. Nagase, *J. Am. Chem. Soc.* **1994**, *116*, 1359–1363; (b) K. Winkler, D. A. Costa, A. L. Balch, W. R. Fawcett, *J.*

Phys. Chem. **1995**, *99*, 17431–17436; (c) M. FEDURCO, D. A. COSTA, A. L. BALCH, W. R. FAWCETT, *Angew. Chem.* **1995**, *107*, 220–222; *Angew. Chem., Int. Ed. Engl.* **1995**, *34*, 194–196.

174 (a) M. EIERMANN, R. C. HADDON, B. KNIGHT, Q. C. LI, M. MAGGINI, N. MARTIN, T. OHNO, M. PRATO, T. SUZUKI, F. WUDL, *Angew. Chem.* **1995**, *107*, 1733–1735; *Angew. Chem., Int. Ed. Engl.* **1995**, *34*, 1591–1594; (b) T. OHNO, N. MARTIN, B. KNIGHT, F. WUDL, T. SUZUKI, H. YU, *J. Org. Chem.* **1996**, *61*, 1306–1309; (c) M. KESHAVARZ, B. KNIGHT, R. C. HADDON, F. WUDL, *Tetrahedron* **1996**, *52*, 5149–5159.

175 J. ZHOU, A. RIEKER, T. GRÖSSER, A. SKIEBE, A. HIRSCH, *J. Chem. Soc., Perkin Trans. 2* **1997**, 1–5.

176 (a) H.-F. HSU, J. R. SHAPLEY, *J. Am. Chem. Soc.* **1996**, *118*, 9192–9193; (b) H.-F. HSU, J. R. SHAPLEY, *J. Organomet. Chem.* **2000**, *599*, 97–105.

177 (a) J. T. PARK, H. SONG, J.-J. CHO, M.-K. CHUNG, J.-H. LEE, I.-H. SUH, *Organometallics* **1998**, *17*, 227–236; (b) H. SONG, K. LEE, J. T. PARK, M.-G. CHOI, *Organometallics* **1998**, *17*, 4477–4483; (c) H. SONG, K. LEE, J. T. PARK, H. Y. CHANG, M.-G. CHOI, *J. Organomet. Chem.* **2000**, *599*, 49–56.

178 K. LEE, H. SONG, B. KIM, J. T. PARK, S. PARK, M.-G. CHOI, *J. Am. Chem. Soc.* **2002**, *124*, 2872–2873.

179 (a) A. L. BALCH, M. M. OLMSTEAD, *Chem. Rev.* **1998**, *98*, 2123–2166; (b) A. STEPHENS, M. L. H. GREEN, *Adv. Inorg. Chem.* **1997**, *44*, 1–43; (c) P. MATHUR, I. J. MAVUNKAL, S. B. UMBARKAR, *J. Cluster Sci.* **1998**, *9*, 393–415. (d) K. LEE, H. SONG, J. T. PARK, *Acc. Chem. Res.* **2003**, *36*, 78–86.

180 (a) R. S. KOEFOD, M. F. HUDGENS, J. R. SHAPLEY, *J. Am. Chem. Soc.* **1991**, *113*, 8957–8958; (b) R. S. KOEFOD, C. XU, W. LU, J. R. SHAPLEY, M. G. HILL, K. R. MANN, *J. Phys. Chem.* **1992**, *96*, 2928–2930; (c) S. A. LERKE, B. A. PARKINSON, D. H. EVANS, P. J. FAGAN, *J. Am. Chem. Soc.* **1992**, *114*, 7807–7813.

181 J. T. PARK, J.-J. CHO, H. SONG, C.-S. JUN, Y. SON, J. KWAK, *Inorg. Chem.* **1997**, *36*, 2698–2699.

182 (a) S. J. TRANS, A. R. M. VERSCHUEREN, C. DEKKER, *Nature* **1998**, *393*, 49–52; (b) H. DAI, E. W. WONG, C. M. LIEBER, *Science* **1996**, *272*, 523–526; (c) E. W. WONG, P. E. SHEEHAN, C. M. LIEBER, *Science* **1997**, *277*, 1971–1975; (d) P. PONCHARAL, Z. L. WANG, D. UGARTE, W. A. DE HEER, *Science* **1999**, *283*, 1513–1516; (e) R. H. BAUGHMAN, A. A. ZAKHIDOV, W. A. DE HEER, *Science* **2002**, *297*, 787–792.

183 S. PEKKER, J.-P. SALVETAT, E. JAKAB, J.-M. BONARD, L. FORRÓ, *J. Phys. Chem. B* **2001**, *105*, 7938–7943.

184 F. LIANG, A. K. SADANA, A. PEERA, J. CHATTOPADHYAY, Z. GU, R. H. HAUGE, W. E. BILLUPS, *Nano Letters* **2004**, *4*, 1257–1260.

185 A. PENICAUD, P. POULIN, A. DERRE, E. ANGLARET, P. PETIT, *J. Am. Chem. Soc.* **2005**, *127*, 8–9.

Index

a
ab initio methods 339
acenaphthene 5
acenaphthenquinone 531
acenes 90
acepleiadiene 18
activationless rate constant 62
acyclic diyne metathesis 503
adamantane 334
aggregation effects, determination of 490
algebraic structure count 145
alizarin 3
alkyne, electron-donating effect 442
alkynyl complexes, rhenium(I)
– diimine alkynyl complexes 428
– EPR spectra 427
– HOMO-LUMO energy gap 427
– luminescence 430
– ^{3}MLCT emission bands 430
– molecular orbital calculations 430
– photophysical properties 428
– structure properties studies 430
– synthesis 427
– tricarbonyl diimine system 430
– wire-like 428
alkynyl complexes, copper(I)
– bonding modes 448
– comparison of Cu(I) and Ag(I) clusters 449
– DFT calculations 453
– electrochemical properties 453
– electronic absorption 449
– hexanuclear complex 452
– luminescence 448
– tetranuclear complex 450
– trinuclear complex 449
– X-ray crystal structures 448
alkynyl complexes, gold(I)
– catenane 455
– chemosensing 455
– depolymerization 453
– geometry 453
– intermolecular Au···Au contacts 454
– intramolecular Au···Au contacts 454
– linear geometry 454
– luminescence 453
– metalloreceptors 455
– NLO properties 454
– photophysical properties 454
– quantum yields 456
– spectroscopic properties 455
– supramolecular assemblies 457
– synthesis 453
– tetranuclear 456
– vibrational structure 455
– wire-like 455
– X-ray crystallography 453, 456
alkynyl complexes, silver(I)
– bonding modes 448
– comparison of Cu(I) and Ag(I) clusters 449
– DFT calculations 453
– electrochemical properties 453
– hexanuclear complex 452
– luminescence 449
allene 334
allene and propyne, computational comparison 350
allene and propyne bond lengths 352
allenophane 266
– acetylenic 266
allotrope of carbon 198, 295
α,ω-polyacetylenes 295
amplified helical chirality 255
angular phenylenes
– cobalt-catalyzed cycloisomerization 149
– synthesis of 146, 149
[8]annulene
– EPR spectrum 575
– radical anion 575
– reduction 574ff
– substituted 575

[12]annulene
– conformation 576
– EPR spectrum of the anion 576
– NMR chemical shifts 576
– radical anion 576
– reduction 575
[16]annulene 13
– C–C bond alternation 578
– configurations 577
– conformational mobility 578
– dianion 577–578
– EPR data 577
– geometry 577
– NMR spectrum 577
– radical anion 578
– radical monoanion 577
– reduction 577f
– single and double bond alternations 577
[18]annulene 14
[22]annulene 234
annulene 12
– [4n+2]annulenes 573
– [4n]annulenes 573
– bridged 574
– computational studies 364
– double helical 251
– Hückel theory 573
– reduction 573
– synthesis 305ff
annulene, anions
– anti-aromaticity 580
– aromaticity 580
– diamagnetic ring currents 580
– paramagnetic shifts 580
– π-delocalization energy 580
annulene-within-an-annulene 588, 595
anthanthrene 7
anthracene 1, 3, 90, 334
– reaction with alkali metal 566
anthraceneotetraphene 9
anthrazine 8
anti-Clar's rule 145
Archimedean solid 159
archimedene 159, 180
aromaticity
– characteristics 140
– continuum 140
– definition 140
– experimetal criteria 573
arylallenes 5
atropisomers 239
attenuation coefficient 34
Austin Model 1, 340
average bond length alternation 175

axial chirality 266
azafullerene
– dimer 393
– dyad with corannulene 396
– dyad with pyrene 396
– monomeric cation 394
– synthesis 393
azafullerene cation
– arylation 395
– ^{13}C NMR spectrum 394
– electrophilic aromatic substitution 395
– formation 394
– nucleophilic trapping 395
– reactions 395
– reaction with enols 396
– trapping with alcohol 396
– X-ray analysis 394

b

Baeyer, A. 2, 8
barrelene 547
basis sets
– choice of 344
– correlation-consistent 345
– Pople-type 345
Becke exchange 352
benzene 334
– resonance energy 140
– σ–π balance 142
benzocorannulene 542
benzocyclobutadiene 16
1.2-benzopentacene 8
benzopentalene 171
benzopentaphene 9
benzo[*ghi*]perylene, synthesis of 95
benzopyrene 9
benzylcorannulene 549
benzyne 16
Bergman cyclization 358
Bergman reaction, computational study 359ff
Berthelot, M. 3
bianthracenyl 92
bicyclophanes 244
binaphthyl 230
1,1'-binaphthyl dimer 254
binuclear complexes, thiophene-bridge 71
biolabel 47
biophotonics 28
biphenyl 1, 4, 16
biphenylene 10, 16
– aromatic character 141
– bond alternation 142
– crystal structure 142

– electronic spectrum   142
– NMR spectrum   142
– reactivity   142
– resonance forms   141
– synthesis   141, 146
bipyrenyl   92
2,2′-bipyridine   38
bipyrimidine   64
biradicals
– DNA-cleaving   358
– spin-separated   360
Birch reaction   567
bisanthene   10, 19
Boekelheide   V, 17
Bohr magneton   572
bond alternation   142
branched phenylenes
– branched/angular   155
– branched/linear   155
– C_3-symmetric   158
– synthesis of   147, 155
Breslow homocoupling protocol   302
bridged fullerenes
– hexaanion   608
– reduction of   608
bromofluorobenzene   16
Brückner orbitals   362
Brückner-double coupled cluster approach   361
Brückner-type orbitals   347
Buckminsterfullerene, chemical synthesis   93
buckybowl metal complex   557
buckybowls   529, 586
– convex vs concave coordination of metals   559
– corannulene   530
– dianions   555
– metal ion binding   559
– tetraanions   555
– transition metal anions   555
bucky-onions   231
1,3-butadiyne   298
1,4-donor-substituted   299
– ^{13}C NMR data   324
– chemical shift of the sp carbons   324
– dithiadiynes   300
– rotational barrier   320
– synthesis   296ff
– via cumulenes   299
– via reductive elimination   298
1,3-butadiyne, cyclic   301
1,4-donor-substituted   301
– acid-induced transannular reaction   311
– as rigid framework   301

– axial substituents   318
– bicyclic ring system   302
– boat-chair equilibrium   326
– ^{13}C NMR data   324
– cage system   302
– cavities   316
– chair conformation   316–317
– columnar structure   321
– cyclophanes   301
– diaza   300
– dithia   300
– dynamic NMR studies   325
– electronic absorption spectrum   326
– elimination to annulene   308
– guest molecules   321
– highly strained cyclotetraynes   316
– hydrochlorination reaction   314
– hydrogenation   309
– hydrophobic cage system   302
– intracavity complexes   301
– long wavelength band   327
– monocyclic receptor   301
– nitrogen-containing   317
– prototropic rearrangements   304
– quantum chemical calculations   320
– rearrangement to annulene   305
– silyl   300
– solid-state behavior based on guest   323
– solvent accessible volume   324
– spectroscopic properties   316
– stereoelectronic effects   318
– structural investigation   316
– synthesis   296ff
– transannular reaction   305
– transannular reductive cyclization   309
– vibrational bands   327
– X-ray structural data   312, 316
1,3-butadiyne   295
butatrienyl cation   313
tert-butylcorannulene
– radical anion   590
– reduced dimer   590
– reduction of   590

c

C_{20} isomers, computational comparison   356
C_{60} fullerene   529
– 1,2- vs. 1,4-adduct   390
– 1,4-adduct   390
– addition of electrophiles   389
– aromatic character   605
– calculated HOMO-LUMO   601
– carbanions of   389
– cyclic voltammetry   601

C_{60} fullerene (cont.)
- derivatization 544
- diamagnetic anion 602
- dianion 602
- electron affinity measurement 600
- energy levels 600
- first reduction potential 602
- Friedel-Crafts type reactions 389
- helium atom, insertion of 408
- hexaanion 568, 601, 603
- HOMO-LUMO gap 600
- metal cluster complex 611
- monoanion 602
- monometalic 611
- NICS calculations of 608
- oxidation potential 384
- photooxygenation 402
- precursors 243
- protonation 387
- radical cation 384
- reduction experiments 600
- solid state reduction 601
- superconductivity 601
- synthesis of 543
- tetraanion 602
- trimetalic 611
C_{60} fullerene cation
- absorption maxima 385
- from fullerene dimer 386
- from hydrofullerene 386
- from RC_{60}-Cl 390
- ^{3}He-encapsulation 388
- ^{13}C NMR spectra 385
- NMR spectroscopy 392
- oxidation 385
- pentaarylated 387
- protonated 387
- reaction of 388
- reduction of 599
- stabilities 392
- UV-vis-NIR spectroscopy 392
C_{70} fullerene
- addition patterns 390
- aromatic character 605
- CV experiments 603
- diamagnetic hexaanion 603
- energy levels 600
- equator 603
- Friedel-Crafts type reactions 389
- hexaanion 603
- poles 603
- pyramidalized carbons 390
- redox potentials 603
- reduction of 600

C_{70} fullerene cation
- calculated energies 387
- formation 387
- from RC_{70}-Cl 390
- ^{13}C NMR data 387
- NMR spectroscopy 392
- pyramidalization 387
- reaction of 388
- stabilities 392
- UV-vis-NIR spectroscopy 392
C_{76} fullerene 604
- oxidation potential 383
- radical cation 383
C_{78} fullerene 604
C_{82} fullerene 604
C_{84} fullerene 604
Cadiot-Chodkiewicz coupling 4, 239, 296
cage structures 243
calixarenes 230
carbomers *see also* expanded molecules and supermolecules
- carbo-aromatics 364
- carbo-benzene 364
- carbo-cubane 364
- carbo-cycloalkanes 364
- carbo-cyclohexatriene 364
- carbo-methane 364
- definition 364
- polyhedrane 366
carbon allotrope 162
carbon nanotubes 27
- electrical resistance 71
- electronic properties 611
- Friedel-Crafts type addition 398
- hydrogenation 612
- mechanical properties 611
- modified Birch reduction 612
- reduction 568, 611f
- sidewall functionalization 398
carbon sheets 334
carbon-rich, definition 1
carbon-rich-macrocyles, chiral 230
carborane anion 384
carbyne 198
- definition 421
- HOMO-LUMO gap 423
carcinogenic activity 8
carotene 30
catenate 304
cavitand 304
CD exciton chirality 273
chalcogen-chalcogen contacts 321ff
chaoite *see* carbyne
charge-separated state 30

chemosensors 455
chiral building blocks 255
chiral cyclophanes see cyclophanes, chiral
chiral helicates 271
chiral induction 270
chiral macrocycle
– allenes 266
– azo-linked 264
– barrier to enantiomerization 231
– chiral diamines 267
– chiral side chain 268
– conformationally rigid 237
– coordination of metal ions 262
– diastereomeric conformers 241
– enantioselective sensor 262
– helical conformation 231
– helicene based 263
– intra- and intermolecular aggregation 264
– Ni(II) complex 261
– NMR spectroscopy 241, 264
– oligomeric 264
– saddle-shaped conformation 234
– self-assembly 268
– properties 268
– spiroindane 265
– synthetic strategy 229
– vapor pressure osmometry 264
– X-ray crystallographic analysis 231, 234
chiral macrocycle, metal containing
– Ag-complexes 270
– binding of small organic guests 270
– catenated 271
– CD exciton chirality 273
– CD spectroscopy 270
– chiral diamine ligand 271
– chiral induction 270
– (S,S)-chiraphos 284
– diaza ligand 270
– dimeric complexes 271
– dinuclear metal complex 275
– DIOP ligand 270
– enantioselective binding 275
– enantioselective catalysis 281
– in multilayer LED 281
– iodonium-transition metal 270
– ligand exchange 284
– luminescence quenching 275
– mesoscopic 284
– metal-acetylide 278, 280
– nanoscopic 284
– Pd-complex 271
– Pt(II)-diimine complex 281
– (R)-BINAP 268
– self-assembly 268

– X-ray crystallographic analysis 271, 278
– zirconocene-containing 278
chiral tube 244
chirality transfer 255
chiroptical molecular switch 71
chlorinated fullerenes see fullerenes, chlorinated
chrysene 2
circular phenylenes see phenylenes, circular
[5]circulene see corannulene
[7]circulene 22
circulenes 91
circumanthracene, synthesis 94
circumanthrene 19
circumtrindene 543
Clar, E. 7
coal-tar 1, 8
columnar superstructures 120
computational chemistry 334
computational chemistry
– allene and propyne 350
– basic aspects 338
– basis sets 344
– C_{20} isomers 356
– cumulenes 351
– cyclization reaction 358
– cyclo[n]carbons 354
– density functional theory 348
– IR, Raman, and NMR chemical shifts 356
– optimized geometries, importance of 357
– polyynes 351
– self-interaction errors 353
– symmetry 344
– validation 348
computational method, validation 349
computational methodology 334
configuration state functions 346
conformational change, redox-induced 252
conjugation effect 513
contact ion-pairs 571
Cook, J. W. 7
coordinative bonding 52
Cope reaction
– mechanism 363
– non-concerted 363
Cope rearrangement, computational study 359ff
Cope-like reaction 363
copper-catalyzed dehydrohalogenation 426
corannulene 22, 91, 586
– addition of alkali metals 554
– addition of organolithium reagent 544
– barrier for inversion 549
– benzoannulation 552

corannulene (cont.)
- bis-metallated complex 558
- bowl-to-bowl inversion 539, 549
- chlorinated 545
- co-crystallization with silver salts 558
- co-deposition with $Rh_2(O_2CCF_3)_4$ 557f
- convex vs. concave coordination of metals 559
- crystal packing 540
- cyclopentacorannulene 539
- dianion 553, 588
- dimethylcorannulene 533
- electrochemical studies 531
- endo- and exo-isotopomers 552
- fluorescence 546
- hydrogenation 559
- inversion barrier for dianion 554
- mechanism of formation 536
- methylated 545
- monoanion 587
- NMR spectrum 530, 587
- nonpyrolytic synthesis 533
- octanionic dimer 588
- pentakis (alkylthio) 546
- π-facial stereoselective 552
- π-π stacking 539
- radical 553
- reduction of 568, 587, 589
- rim-benzoannulated 542
- ruthenium complex 558
- structure-energy correlation 553
- sym-pentakis 546
- synthesis of 530, 535
- tetraanion 553f, 588
- tetralithio derivative 553
- trianion 588
- UV-vis absorption 546
- via flash vacuum pyrolysis 531
- X-ray crystallographic structure 531
corannulene derivatives, synthesis of 545
corannulene, extended 592
- anions 593
- bowl shape 593
- diastereomeric reduced species 593
- reduction of 592
- reductive bond cleavage 593
corannulene, substituted, reduction of 590
coronene 10, 19, 91, 531
- synthesis of 117
coupling element 41
cross conjugation 142
cross-coupling, Ni-catalyzed 251
crystal-to-crystal reaction 200

cubane 301, 334, 344, 366
cubane, expanded 364
- computational studies 366
- strain energy 366
cumulene, cyclic 357
cyanine limit 509
cyclic diynes 295
cyclic voltammetry 570
cyclobutadiene
- crystallographic characterization 140
- isolation of 140
- persilylated 140
- resonance energy 140
cyclodextrins 230
cyclohexatriene 163
cyclohexatrienoid bond alternation 175
cyclooctatetraene see [8]annulene
cyclooctatetrathiophene 251, 253
cyclopentacorannulene 539
- crystal structure determination of 557
cyclopentadienones 107
cyclopentenophenanthrene 9
cyclophane 301, 574, 586
see also macrocycle
- alkali metal reduction 581
- definition 581
- double helical alkynyl 259
- doubly folded 259
- helical 240
- helical racemic 236
- meta 241
- NMR shifts 581
- ortho 304
- para 242, 327
- reduction of 581
- perfluorophenyl-phenyl 250
- thiophene 250
cyclophane, chiral
- CD spectroscopy experiments 261
- chirality transfer 261
- synthetic strategy 229
- X-ray crystallography 235, 261
cyclophane, helical
- 2,2'-bipyridine based 246
- coordination polymers 248
- energy-minimized structure 248
- exotopic binding sites 248
- fluorescence emission quenching 248
- ion sensor 246
- metal binding 247
- phenanthroline subunit 247
- sensor for metal ions 248
- UV-vis and fluorescence measurement 247
- variable temperature ^{13}C NMR 247

cyclophane receptor
– binding affinity   256
– diastereoselectivity   256
– enantioselectivity   256
cyclophane receptor, chiral
– carbohydrate recognition   255
– complexation of disaccharide   257
– host-guest interaction   258
– NMR spectroscopy   257
– NMR titration studies   255
– saccharide guests   258
– symmetry   255
– tetraanionic analogs   257
cyclophane, reduced
– anisotropy effects   582
– anti-aromatic character   582
– aromatic character   582
– dianion   582
– electronic effect vs ring size   582
– NMR shifts   582
– ring currents   582
– tetraanion   582
cyclotrimerization of acetylene   3

d

Davidov splitting   491
decacyclene   5, 16, 543
1,5,9-decatriyne   305
dehydro[4n+2]annulenes   295
dehydroannulene   13, 574
– definition   579
– dianion   579
– π-bond delocalization   579
– radical anions   579
– reduction   579
– ring current   579
dehydrobenzannulene   232
dehydrobenzo[12]annulene   14, 161
dehydrobenzo[18]annulene   14
dehydrobenzoannulene   13
dendrimer   437
– ethynylene   425
dendrons   128
density functional theory   340, 342, 348
– B3LYP   348
– carbon-rich structures   348
– description   348
– energy single points, RMS error   361
– functionals   348
– open-shell character   348
– self-interaction errors   353
density of states   41
Dexter-type electron exchange   38

diacetylene   15
– aromatic substituted   208
– catechol   208
– crystal structure   202
– diacids   206
– dipyridyl   208
– gamma irradiation   208
– hydrogen bonded host-guest complex   202
– mono substituted   208
– 1,4-polymerization   199f
– polymerization   202
– supramolecular structural features   203
– thermal annealing   206
– triple helix   210
– ultraviolet irradiation   208
diacetylene polymerization
– host-guest strategy for   213
– supramolecular strategies for   201
diadamantane   370
diamondoid radical cations   370
diamantane   367
diamantyl   370
diamine ligand, chiral   271
diamond, thermodynamic stabilty   337
diamondoids   334, 349, 367
– computational studies   368
– from nature   368
diamondoidyl cations   368
diamondoidyl radicals   368
dianthracene   3
diazacyclotetraynes   300
dibenz[a,h]anthracene   94
dibenzanthracene   9
7.8-dibenzobiphenylene   16
dibenzo[a,g]corannulene   592
dibenzocorannulene   555
dibenzo[a,g]cyclopenta[h,l]corannulene   592
dibenzo[b,k]chrysene, synthesis of   95
dibenzoctacene   19
dibenzocyclopentacorannulene   555
dibenzofulvene   593
dibenzoheptacene   19
dibenzopentacene   8
dibenzopentalene   18
dibenzoperylene   10
1.2,6.7-dibenzopyrene   16
2-dicarbomethoxycorannulene   537
dicorannulenobarrelene
– bowl-to-bowl inversion   548
– conformers   547
– dicorannulenobullvalene, conversion to   548
– synthesis   547
dicorannulenobullvalene   548

1,8-dicorannulenyloctane
- intramolecular sandwich 590
- reduction of 590
diene metathesis 484
diene polymerization, structural parameters for 220
diethynylbinaphthalene 271
dihydrocorannulene 588
dihydrocyclopentacorannulene
- bowl-to-bowl inversion 551
- CpRu complex of 557
4-dihydronaphthalene 569
dihydropentacene 8
dihydropentalene 8
2,2'-dilithiobiphenyl 16
dimethylpyrene 598
dinuclear complex 35
1,4-diphenylbutadiyne 4
diphenylpolyacetylene 15
diphenyltetrayne 15
diphenyltriyne 15
dipole-dipole mechanism 45
dithienylethenes 72
donor-acceptor, electronic coupling 44
donor-acceptor dyads, hydrogen-bonded 53

e

effective conjugation length 507
- definition 490
Eglinton, G. 12, 296
Eglinton and Galbraith coupling 4
Eglinton coupling 235, 296, 421
electrochemistry 570
electromechanical actuator 252
electron acceptor
- anthraquinone 51
- C_{60} 34
- 1,1'-dicyanovinyl 32
- naphthalene diimide 54
- Os-based 37
- Os(II) 49
- Os-terpy 70
- quinones 32
electron correlation, omission of 342
electron correlation energy, definition 342
electron correlation methods
- configuration interaction 346
- coupled cluster 347
- for carbon-rich compounds 346
- improvements in quantum chemical accuracy 347
- Moller-Plesset perturbation theory 346
- need for 345
- vs Hartree-Fock theory 346

electron donor
- 2,6-diacylaminopyridylporphyrin 54
- 1,4-dimethoxynaphthalene 32
- porphyrins 32
- Ru-based 37
- Ru(II) 48
- Ru-terpy 70
- tetrathiafulvalene 34
electron paramagnetic resonance spectroscopy
- intensity 572
- line width 572
- resonance field position 572
electron shuttle 602
electron transfer in DNA 55
electron-exchange mechanism 45
electron-hole capture 495
electronic coupling 31
electronic coupling matrix element 41
electronic switch 27
electron-transfer reaction, reversible 53
electrostatic effects 52
electrosynthesis 570
emission lifetime 40
emission quantum yield 40
endohedral fullerene see fullerene, endohedral
endoperoxide 167
enediyne 358
- computational study 359ff
energy transfer, hopping mechanism 69
enyne-allene 358
- computational study 359ff
epoxy fullerene 610
essential state model 494
excited state, long lived 55
exciton 496
exciton-photon coupling 495
expanded molecules see also carbomers
- formation of 363

f

Faraday, M. 3
Fermi Golden rule 41
Fermi levels 26
field effect transistor 96, 492
- schematic 124
flash vacuum pyrolysis, limitations of 533
fluoranthene 1, 16
fluorene 1
fluorescence, time-resolved 32
fluorocyclene 5
Förster mechanism 35
Förster-type coulombic interactions 35
Förster-type triplet energy transfer 40
foldamer 424

Franck-Condon 41
Friedel-Crafts dealkylation 99
fullerene
– aromaticity of 605
– cyclic voltammetry 570
– electrochemistry 570
– electron affinities 570
– energy level 570
– methylene bridged 608
– reduced 599
– reduction of 568
fullerene anion
– magnetic properties 599
– nucleophilic behavior 610
fullerene cation
– from fullerenols 384
– monoalkylated 384
– monoprotonated 384
– pentaarylated 384
fullerene, chlorinated
– free energy of activation for solvolysis 393
– Friedel-Crafts reactions of 391
– S_N1 solvolysis 391
– S_N2-type mechanism 391
fullerene derivatives
– electrosynthesis 610
– reduction of 610
– via reduction 610
fullerene, endohedral
– argon 409
– closure of the orifice 412
– definition 398
– dimerization of 413
– formation of 399
– $^3He@C_{60}$ 408
– helium atom 408
– hexaanions 606
– hydrogen molecule 408
– medical science 399
– molecular electronics 399
– molecular surgery 412
– neon 409
– NMR studies of anions 606
– reduction of 606
– single crystal synchrotron X-ray analysis 410
– solid-state 1H NMR study 410
– synthesis of 412
– water molecule 411
fullerene-heterofullerene dyad 397
fullerene, higher
– CV experiments 604
– endohedral helium 607
– hexaanions 605, 607
– reduction of 604
fullerene, reduced
– ^{13}C NMR of 607
– local ring currents 608
fullerenols 391
fulvalene 12
fused carbon rings 18

g

γ-motif 99
Gaussian-type-orbitals 344
GIAO-MNDO 341
Glaser, C. 1, 296
Glaser coupling 4, 296
Graebe, C. 1f
graphene 117, 334, 345, 349
graphite 349
– potassium-graphite 568
– propeller 111
– reaction with alkali metals 568
– ribbon 110ff
– segment 90
– sheet 112
– thermodynamic stability 337
graphite ribbon, electronic and vibrational properties 111
graphitic disc
– charge transport properties 114
– giant 109
– optical absorption properties 114
– Raman spectroscopy 108
– single-crystal analysis 108
– solid state UV-vis absorption 108
graphitic molecule
– acene-like 114
– armchair peripheries 114
– chemical reactivity 115
– cove peripheries 114
– cove-type edge 115
– electronic properties 115
– quantum chemical calculations 116
– quinoidal 114
– self-assembly 115
– synthesis of 116
– UV-vis and fluorescence spectra 115
– X-ray structure analysis 117
– zig-zag periphery 115
graphyne 162

h

H aggregates 491
$H_2@C_{60}$ 413
halogenation/dehydrohalogenation 5
Hammond postulate 392

Hartree-Fock theory 340
– basic description 342
– electron correlation energy 342
– electron density 342
– isodesmic equations 342
– limitations 342
– oligoacenes 343
– open-shell species 343
– polyacenes 343
– spin contamination 343
– spin polarization 343
– unrestricted 343
Hay, A.S. 296
Hay coupling 4, 296, 421, 503
^{3}He@C$_{60}$ 408
^{3}He@C$_{60}$, reduction of 606
^{3}He@C$_{70}$, reduction of 606
He$_2$@C$_n$, hexaanions 607
Heck reaction 480
helical chirality 230
helical phenylenes *see* heliphenes
helical structures 423
[4]helicene 9, 16
[5]helicene 9, 263
[6]helicene 9f, 20
[7]helicene 10, 20
[8]helicene 20
[9]helicene 20
helicene 9, 19, 90, 252
helicene, fused 241
– macrocycle 241, 263
– X-ray crystallographic analysis 253
heliphenes 149
– barrier to enantiomerization 180
– chiroptical properties 180
– enantiomer separation 180
– helical strain 180, 185
– helix parameters 182
– synthesis of 150
– X-ray structures 180
heptacene 90
heptafulvalene 20
heptapentafulvalene 22
Hessian matrix 344
heterosuperbenzene 119
hexabenzocoronene 19
hexacene 90, 343
hexachalcogenacyclohexaynes 319
1,5-hexadyne 14
hexahydrosumanene 539
hexamantane 370
hexa-peri-hexabenzocoronene 91
– arrangement on HOPG 129
– dimers and trimers 105

– halogenation 117
– hydrogenation 117
– nitration 117
– one-dimensional charge carrier mobility 99
– phosphorescence 97
– scanning tunneling microscopy 129
– scanning tunneling spectroscopy 129
– self-assembly with electron acceptor 130
– single crystal structure 99
– sublimation 99
– sulfonation 117
– synthesis of 96, 98
– UV-vis absorption spectra 97
– vacuum deposition 129
hexa-peri-hexabenzocoronene, substituted
– aggregation 128
– alignment of 126
– alkyl and alkylphenyl 121
– alkylchloride 99
– alkylester 99
– alkylphenyl 99
– amphiphilic 128
– atomic force microscopy 122
– bay position 100
– bromine functionality 100
– Buchwald coupling 104
– charge carrier mobility 120
– coaxial columnar stacking 103
– columnar liquid crystalline phase 99, 103
– columnar plastic crystalline phase 122
– columnar superstructures 121
– complexes 101f
– control of alignment 123
– cyclophane 118
– double-concave conformation 101
– electronic paramagnetic resonance 104
– energy/electron transport 100
– FET 124, 127
– field-effect mobility 127
– film formation capability 121
– heteroatoms 119
– hole-transport mobility 125
– homeotropically arranged HBC films 125
– hydrogenation 117
– hydrophobic 125
– intermolecular association 105
– iodine 103
– metal coordination 119
– on HOPG 130
– one-dimensional charge carrier mobility 99
– peralkoxylated 100
– phase transition temperatures 121
– photodiodes 126
– photovoltaic device, schematic 124

– self-assembly   121, 128
– single crystal structure   99
– solution processing   126
– Sonogashira coupling   103
– steric congestion   100
– synthetic limitations   102
– transition metal-catalyzed coupling reactions   100
– unsymmetric diode-like behavior   119
– UV-vis-NIR spectroscopy   104
– visualization by STM   118
hexaphene   90
hexaphenylene   16
hexaphenylpentafulvalene   21
hexaphenylpentalene   18
high resolution transmission electron microscopy   113
higher fullerenes   see fullerenes, higher
higher phenylenes, synthesis of   145ff
highly oriented pyrolytic graphite   122
homoaromaticity   364
homochiral macrocycles   230, 255
homoconjugation   364
Horner reaction   480
Huang-Rhys factor   41
Hückel theory   573
Hückel valence theory   339
Hückels rule   140
Hush theory   31
hydrocarbon stability   337
hydrocarbons, classification via hybridization   336
hydrofullerene   610
hydrogen bond, carboxylic acid-pyridine   214
hydrogen bonding, multi-point   52
hyperconjugation   32

i

imaginary frequencies   344
INADEQUATE   405
indenocorannulene   592
indium tin oxide   496
intermolecular oxidative cyclodimerization   248
intersystem crossing   39, 496
intervalence charge-transfer   31
intramolecular charge transfer   34, 509
intramolecular cyclodehydrogenation, Scholl-type   95
intramolecular triplet energy transfer, effect of distance   40
4-iodobiphenyl   6
4-iodoterphenyl   6
ion-solvation equilibrium   571
isomerization, photo-induced   72

j

J aggregates   491
Jahn-Teller distortion   573, 578, 602

k

Karplus equation   572
kedarcidin   309
Kekule structure   145, 605
kekulene   91
Klinger, H.   3
Knoevenagel condensation   115, 480

l

λ-conformation   286
Langmuir-Blodgett technique   128
Langmuir technique   128
laser dyes   492
ligand
– bipyridine   38
– bridging   35
– chiral   230
– exchange   284
– multitopic   52
– polytopic   35
light harvesting   424
– dendrimer   55
light-active donor group   32
light-emitting diode   28, 96
light-induced electron transfer   28, 64
light-induced energy transfer   28
linear phenylenes, preparation of   146
linear polarizability   493
liquid crystal displays   96
liquid crystalline phase, columnar   120
liquid crystals   492
Löb, W. C.   5
logic gate   28
London dispersion forces   341
long-distance vectorial energy transfer   77
long-range electron transfer   32
long-range electronic coupling   32
long-range energy transfer in DNA   55
Lothrop, W. C.   10
low-valent titanium intramolecular coupling   534

m

macrobicycles   305
macrocycle   see also cyclophane
– axial substituents   318
– bifunctional   303
– binaphthyl-derived   261f
– boat-chair equilibrium   326
– cage system   302
– cavities   316

macrocycle (cont.)
- chair conformation   316
- columnar structure   321
- dehydrogenation   244
- double helical   252
- highly strained cyclotetraynes   316
- multipolar   261
- oligophenylene   237
- photoelectron spectra   326ff
- saddle-shaped   240
- salophen-containing   261
- solvatochromic and halochromic effects   303
- spectroscopic properties   316
- stereoelectronic effects   318
- structural investigation   316
- supramolecular organization   319
- transannular reaction   310
- vibrational bands   327
- X-ray structures   316
macrocycle, helical, conformational rigidity   242
macrocycle, heterocyclic
- binding studies   250
- selective ion binding   250
- sensory applications   250
- templated synthesis   250
macrocycle, non-racemic
- barrier to racemization   254
- NMR spectroscopic experiments   254
- X-ray crystallographic analysis   254
macrocyclic squares, chiral   268
macroscopic susceptibility   508
many-body perturbation theory   346
Marcus inverted region   38, 41
Marcus-type rate   37
McConnell relationship   572
McMurry coupling   241, 259, 480
medium-frequency vibrational mode   41
memory-device, random-access   27
Merrifield resin   503
mesophase, columnar   96
metal alkynyl complexes
- alkynylating agents   426
- copper   448
- copper-catalyzed dehydrohalogenation   426
- gold   426, 453
- intraligand state   425
- ligand-field state   425
- ligand-to-ligand charge transfer state   425
- mercury   426
- metal-metal-to-ligand charge transfer state   425
- metal-to-ligand charge transfer state   425
- mixed-metal alkynyl complex   457
- palladium   426
- platinum   426, 432
- rhenium   427
- silver   448
- synthetic routes to   425
metallodendrimers   437
metallohelicates   52
metalloporphyrins   28
metal-to-ligand charge-transfer   39, 58
methylene, singlet states of   344
Meyer, G. M.   4
Misumi, S.   12, 15
mixed-metal alkynyl complexes
- branched complexes   463
- DFT calculations   463
- electro-switchable   460
- metalloligand approach   457
- osmium-ruthenium   457
- photophysical properties   458
- rhenium-copper   458
- rhenium-iron   458
- rhenium-iron(II)   460
- rhenium-palladium   463
- rhenium-silver   458
- σ- and π-donating ability   463
- synthesis   458ff
- trinuclear   463
mixed-valence metal alkynyl complexes   457
- iron   460
- rhenium   458
- rhodium   458
modified neglect of differential diatomic overlap (MNDO)   340
Möbius aromatics   341
molecular AND gate   30
molecular bridge   see also molecular wire
- acetylenic   42
- anthracene   48
- biphenyl   70
- cumulene-type   42
- design   26, 32, 40
- ethynylated phenylene units   69
- ethynylated thiophene   70
- naphthalene   28
- phenylene bridge   48
- poly(acetylene)   68
- poly(phenylene)   68
- poly(thiophene)   70
- resistivity   41
- semi-flexible   50
- short covalent   42
- sigma-bridged   67
- spiro linkage   64

– supramolecular 52
– triplet lifetime 39
molecular channel 304
molecular conductivity 31
molecular dyads, examples 49
molecular electronics 26ff
– organic-based 27
molecular grid 53
molecular ladder 53
molecular mechanics force field
– basic description 339
– deformation of carbon nanotubes 339
– harmonic vibrational frequencies 339
– heats of formation 339
molecular photonics 28
molecular polygons, chiral 283
molecular shuttle, light-activated 55
molecular square 53
molecular surgery 400
molecular switch 70
– calixarene 73
– calixquinone-based 74
– catenanes 71
– chiropticene molecules 71
– photocyclization quantum yield 72
– photostationary state consisting 72
– poly(ethylene glycol)-based 74
molecular-scale transistor 28
molecular wire see molecular bridge
molecular XOR gate 30
Moller-Plesset perturbation theory 346
molybdenum-alkylidene complex 484
molybdenum carbyne complex 503
monosubstituted diacetylene see terminal diacetylene
Monte Carlo computations 357
multiphoton emission 496
multireference second-order perturbation theory 358
Myers-Saito cyclization 358
Myers-Saito reaction, computational study 359ff

n

Nakagawa, M. 12, 14
nanodiamonds 337, 349
nanoparticle 27
nanowire 28
naphthalene 1, 334
– reduction with K 566
Negishi coupling 237
neocarzinostatin 309
NMR spectroscopy
– lithium 571
– of reduced PAHs 571
nonlinear optics, explanation 493
nucleus-independent chemical shift calculations 182, 341

o

octaphenylene 16
oligomers, defined length see oligomers, monodisperse
oligomers, monodisperse
– conformers 477
– cyclic 478
– dendritic 478
– OPE 499
– platinum alkynyl 437
– star-shaped 478
oligo(phenylene) 92, 478
– dendrimer 110
oligo(phenyleneethynylene) 423, 476
– AA + BB polycondensation 503
– AA + BB process 499
– absorption 504
– aggregation 507
– by bromination/dehydrobromination 500
– chromophores 504
– combinatorial protocol 503
– diyne metathesis 503
– donor-acceptor 509f
– effective conjugation length 507
– electroluminescence 508
– electrophores 504
– excimers 507
– fluorescence 504
– γ_{res} vs. length 508
– HOMO-LUMO gap 508
– intramolecular charge transfer 509
– iterative divergent/convergent strategies 502
– length per repeat unit 477
– nonlinear optics 508
– prediction of lambda values 504
– push-pull substituted 509
– rigid rod nanoparticles 502
– self-assembly 504
– Sonogashira reaction 499
– structure-function for D/A systems 511
– substitution patterns 504
– synthesis 499
– third harmonic generation 508
oligo(phenyleneethynylene), donor-acceptor
– push-pull effect 511
– structure-function studies 511
– synthesis 510
– UV-vis absorption 51

oligo(phenylenevinylene) 476
- aggregation effects 490
- applications in material science 492
- band scheme 497
- bandgap 490
- bidirectional growth strategy 481
- bipolarons 495
- calculations 491
- cation-anion coupling 484
- coupling technique 481
- Davidov splitting 491
- diene metathesis 484
- donor-acceptor 509f
- doping 495
- effective conjugation length 490
- essential state model 494
- fluorescence 488, 491
- Gilch route 484
- H aggregates 491
- intramolecular charge transfer 509
- isomerization 498
- J aggregates 491
- length per repeat unit 477
- nonlinear optics 492
- OLED 496
- photocrosslinking 498
- photopolymerization 498
- polarons 495
- prediction of values 489
- protection/deprotection technique 481
- push-pull substituted 509
- quinoid substructures 495
- radiative decay 491
- saturation domain, UV-vis 489
- scaling law 493
- *trans* selectivity 482
- solubility 479
- spin coating 498
- Strickler-Berg relationship 491
- structure-function for D/A systems 511
- structure property studies 488
- substitution patterns 487
- synthesis of 480
- thermal stability 498
- third harmonic generation 494
- UV-vis spectra 488
- Vanderzande route 484
- water/alkali soluble 487
- Wessling-Zimmermann route 484
oligo(phenylenevinylene), donor-acceptor
- AM1/INDO-S calculations 515
- bathochromic and hypsochromic contributions to transition energies 514
- HOMOs 515
- λ_{max} vs length 512
- LUMOs 515
- protonation 512
- push-pull effect 511
- structure-function studies 511
- synthesis of 510
- UV-vis absorption 511
oligoynes *see* polyynes
open cage fullerene
- 8-membered-ring orifice 401
- 12-membered-ring orifice 401
- 13-membered-ring orifice 404
- 14-membered-ring orifice 401
- 15-membered-ring orifice 406
- 16-membered-ring orifice 404, 406
- 19-membered-ring orifice 408
- 20-membered-ring orifice 406
- barrier to escape 409
- barrier to insertion 409
- examples 400
- helium atom, insertion of 408
- hydrogen, insertion of 409
- multiple cage scissions 407
- NMR measurements 405
- reaction with aromatic hydrazines 404
- reduction potential 402
- structural assignments 406
- sulfur atom insertion 405
- synthesis of 400ff
- water molecule insertion 411
- X-ray structure 403, 405
optical brighteners 492
organic light emitting diode 492, 495
- quantum yield 495
- schematic set-up 496
- single-layer 495
- turn-on current 496
organic photovoltaic device, design 125
organic semiconductors 96
ovalene 91
oxalamide 204
oxidative cyclodehydrogenation 93

p
π acceptor 35
Pauling point 350
π donor 35
pentacene 8, 19, 90, 343
pentafulvalene 20
pentahelicene *see* [5]helicene
pentamantane 337
pentaphene 90
periconjugation 611

pericyclynes 295
– computational studies 364
– homoaromaticity 364
– homoconjugation 364
peropyrene 19
perylene 5, 7, 19
petroleum products 1
phenanthrene 1, 90
phenanthryne 16
phenes 90
phenyl halides, reduction of 568
phenylenes
– algebraic structure count 145
– angular 149
– anti-doublebent 157
– barrier to planarization 180
– carbon-13 NMR spectroscopy 184
– C-C activation 172
– circular 145, 149
– comparative reactivity 162ff
– complexation 172, 174
– conjugated-circuit model 185
– correlation to polycyclic aromatic hydrocarbons 143ff
– CpCo complexes 172
– [4+2]cycloaddition 167
– cyclopropanation 164ff
– definitions 143
– deplanarization 180
– DFT calculations 187
– diatropism 183
– electronic spectra 185
– evaluation of their aromaticity 184
– experimental and calculated structural parameters 175
– flash vacuum pyrolysis 171
– ground-state energy 184
– heliphenes 149
– HOMO-LUMO gap 143, 186
– hydrogenation 162
– kinetic stability 184
– λ_{max} values 186
– magnetic properties 182ff
– metal-promoted ring openings 172
– mixed topologies 153
– NMR and structural data 175
– oxacyclopropanation 164ff
– paratropism 183
– photophysics 186ff
– physical properties 175ff
– reaction with dimethyl butynedioate 169
– reaction with electrophilic species 165
– reaction with metal fragments 172

– reaction with tetracyanoethane 168
– relative energies 185
– Shpol'skii spectroscopy 187
– singlet oxygen cycloaddition 168
– symmetry 157
– syn-doublebent 157
– topologies 143
– vibrational characteristics 187
– VT-NMR experiments 180
– X-ray data 175
– zigzag 149
phenylenes, circular
– antikekulene 160
– attempted syntheses 161
– superdelocalization 160
phloroglucin 304
phosphorescence lifetime 56
photoactive dyads 56
photochromic materials 72
photochromic reagent 72
photoconductors 492
photonic devices, molecular-scale 55
photonic wire 28
photosensitized oxidation 50
photostationary state 72
photosynthesis 28
photosynthetic reaction center 30, 52
photovoltaic device 492
– schematic 124
phthalocyanines 250
picene 1f
platinum(II) alkynyl complexes, phosphine systems
– branched molecules 437
– chain-length effects 432
– chelating phosphines 433
– chiral 433
– chloroplatinum complexes 437
– dendrimers 437
– electronic absorbtion spectra 432ff
– energy harvesting system 436
– energy transfer 440
– geometry 432
– ^{3}IL emission 436
– luminescence 434ff
– ^{3}MLCT emission 436
– molecular triangles 433
– molecular wire 432
– monodisperse oligomer 436
– nonlinear optical properties 433
– physical properties 433
– polymer 432
– structure-property studies 434
– triplet exciton 437

platinum(II) alkynyl complexes, polypyridine systems
– aggregates 445
– alkyne, sigma-donating ability 441
– biological properties 441
– diimine complexes 441
– dopants in OLED 446
– liquid crystal properties 441
– luminescence 440, 446
– nonlinear optical properties 441
– pendant crown ether 442
– phosphorescence 442
– photophysical properties 440–441
– polymorphic behavior 440
– self-assembly 445
– Stokes shifts 442
– stretching modes 445
– terpyridyl complexes 441
– vibronic-structured emission 445
– wire-like 442
– X-ray crystallography 445
polyacene 349
polyacetylene 198
polyadamantyl 367, 370
polyadmantane 349
polyarylene chemistry 90
polyaza macrocyclic 250
polycyclic aromatic hydrocarbons 529
– all-benzenoid 95
– circumtrindene 543
– corannulene 530
– curved pyrene 596
– dicorannulenobarrelene 547
– electronic properties 95
– nonplanar 586
– pyrenophanes 598
– semibuckminsterfullerene 534
– stability 95
– strain 596
polycyclic aromatic hydrocarbons, nonplanar, reduction of 586
polycyclic aromatic hydrocarbons, reduction of
– aggregation 566
– annulenes 573
– anthraquinone 568
– anti-aromaticity 566
– aromaticity 566
– charge density via NMR 572
– corannulene 568, 587
– cyclic voltammetry 570
– detection of products 567
– electron transfer 566f
– EPR spectroscopy 566, 572
– fluorenone 568
– in liquid ammonia 567
– ion solvation equilibrium 571
– K mirror 570
– kinetics 570
– naphthalene 568
– reduction with alkali metals 567
– ring closure 569
– solubility 567
– spin density via EPR 572
– stability of anionic system 571
– study by NMR spectroscopy 571
– thermodynamics 570
– via electron shuttles 567
– with potassium graphite 568
polycyclic hydrocarbons 7
– largest 110
polydiacetylene 15, 198
– combinatorial approach 214
– crystals 205
– formation of 205
– host-guest strategy 214
– optical reflectance spectra 205
– structural parameters for polymerization 200
– synthetic design 199
polyenic wire 32
polyhedra, chiral three-dimensional 270
polymers
– helical folding 423
– linear conjugated 198
polyphenylene 253
polytetrafluoroethene 126
polytopic ligand 63
polytriacetylene 198
– crystal structure 219
– isolation 219
– Raman spectroscopy 219
– strutural parameters for polymerization 215
– UV-vis spectrum 219
polyyne
– cyclic 357, 477
– electronic absorption spectra 423
– electronic structure 422
– endgroups 422
– strain energy 477
Pople-type basis sets 361
π orbitals 35
porphyrin 230
– oligomeric 28
porphyrin dimers 67
– free-base 67
potassium-graphite 568
π-π interactions 120

propellicene 263
propylcorannulene, reduction of 590
propyne 334
Pschorr, R. 2
pseudorubrene 6
π-stacked lamellae 128
π-stacking 52
– edge-to-phase 253
Pt(II) bis-acetylide
– energy transfer in 44
– osmium(II) tris(2,2′-bipyridine) 47
– photophysical properties 45
– porphyrin-based 45
– pyrene 47
– ruthenium(II) tris(2,2′-bipyridine) 47
pyrene 1, 5, 7, 91
– curvature 596
– cyclophanes of 587
– strain 596
pyrenophane
– dianions 598
– monoanion radical 598
– one electron reduction 598
– reduction of 596
– strained 598
– two electron reduction 599
pyridines, π-π stacking of 221

q

quadrupole interactions 250
quarterrylenes 93
quaterphenyl 16
quenching 496
quinones, polycycle 7

r

[3]radialene 20
[4]radialene 20
radialene 12, 20
– computational studies 364
Radziszewski, B. 2
Ramberg-Bäcklund reaction 485
Ramsey equation 572
rate of triplet energy transfer 35
reduced fullerene see fullerene, reduced
reduction
– detection of the products 567
– electron transfer methods 567
reductive cyclodehydrogenation 93
reorganization energy 41
resistivity, molecular bridge 41
revolveneynes 242
Riese, F. 3
rigid macrocycles 301
rim-benzoannulatedcorannulene 542

rotaxane 55
rubrene 5
ruffled benzimidazole-based ligands 250
Ru(II) poly(pyridine) 58ff
– complexes, binuclear 62
Ru(II)/Os(II) molecular dyad 52
– energy level diagram 50
– hydrogen-bonded 54
Ru(II)/Os(II) poly(pyridine)complexes, mixed-metal 67
Ru(II)/Ru(III) complex, mixed-valence 31, 53

s

σ-bonds, insulating nature 67
S_2 state, carotenoids 45
– energy diagram 47
– light-induced electron transfer 45
– light-induced energy transfer 45
– of metalloporphyrin 45
sandwich-herringbone stacking 99
saturation domain, UV-vis 489
scaling law 493
Schmittel cyclization 358
Schmittel reaction, computational study 359ff
Scholl, R. 5, 7
Schrödinger equation 342
second harmonic generation 493
second hyperpolarizability 478, 508
self-assembled monolayer 71
self-assembly 96
– barbiturate moiety 55
– metal-mediated 270
– molecular dyads 55
– of donor-acceptor dyads 53
– of photo-active arrays 53
– π-bonding interactions 55
– Ru(II) and Os(II) poly(pyridine) complex 54
– via complementary base pairing 54
self-consistent field see wavefunction theory
self-healing 96
semibuckminsterfullere 534, 540ff, 555
– complex of 558
– convex vs concave coordination of metals 559
– crystaly structure 541
– endo-exo stacking 542
– synthesis 541
semiconductors 492
semiempirical methods
– all-carbon structures 341
– basic description 340
– computational effort 341
– for heats of formation 341
– limitations 341
– nucleus-independant chemical shifts 341

sexiphenyl 7, 16
shape-persistent macrocycles 295
shear moduli 339
Siegrist reaction 480
singlet exciton 437
Slater-type-orbitals 344
small angle neutron scattering 437
smart materials 28
solar cells 96, 492
solvent separated ion-pair 571
solvophobic interactions 423
Sondheimer, F. 10, 295
Sonogashira coupling 150, 239, 421
– catalytic cycle 500
spherical aromaticity 606
spin contamination 343
Staab, H. 14
stabilomeric rearrangement 367
stationary point 344
Stephens-Castro coupling 161, 231
Stille coupling 421, 480
Stokes shift 442, 457, 546
Stone-Wales defects 363
Strickler-Berg relationship 491
sumanene 538
superbenzene 105
– nitrogen-containing 119
superdelocalization 160
super-exchange phenomenon 33
super-exchange theory 41
superfluorene 105
supermolecules *see* also carbomers
– formation of 363
– polyadamantyl 270
supernaphthalene 110
superphenylene 105, 107
supertriphenylene 110
supramolecular organization 319
supramolecular structure, triplet lifetime 57
supramolecular synthesis 199
– definition 214
– rational strategies for 201
Suzuki coupling 480
Suzuki-Miyaura reaction 487
synaxial 1,3-interactions 319
synthesis at the complex approach 77ff

t

terminal diacetylene
– helical network 211
– 1,4-polymerization of 208
– polymerization mechanism 212
– topochemical polymerization 211
– X-ray structure 208

ternaphthyl 2
terphenyl 2
terrylene 2, 19
tetrabenzobisanthene 19
tetrabenzocyclooctatetraene 142
tetrabenzoheptacene 19
tetrabenzopentacene 19
tetrabenzoperopyrene 96
tetrabromocorannulene 536, 544
tetracene 8, 19, 90
tetrachalcogenacyclotetraynes 319
tetrahydrotetracenequinone 8
tetramantane 337f
tetramethylcorannulene 556
tetramethylsemibuckminsterfullerene 541
tetraphene 7, 90
tetraphenyl 2
tetraphenylallene 5
tetraphenylcyclopentadienone 108
tetraphenyltetracene *see* rubrene
tetraphenylene 16, 254
tetrathiafulvalene 34
third harmonic generation 493
third order susceptibility 508
through-bond charge transfer 70
through-bond electron exchange 36
topochemical polymerization 15, 295
– definition 198
– diacetylene polymerization 15, 199
– diene polymerization 220
– terminal diacetylene 208
– triacetylene polymerization 215f
– triene polymerization 220
transannular reaction
– acid-induced 311
– HCl-induced 314
– mechanism 311, 314
– valence isomerization 316
transient absorption spectroscopy 32
transition moments 491
triacetylene
– via ^{60}Co gamma irradiation 219
– complex with vinylogous amide 216
– nicotinate and isonicotinate derivatives of 217
– polymerization of 215, 219
– supramolecular organization of 216
– supramolecular structure 217
– topochemical polymerization 216
– ultraviolet irradiation 219
– X-ray crystal structure 216, 219
triacetylene carboxylic acid 217
triad, donor-connector-acceptor 52
triamantane 368

tribenzocyclyne 161
triene polymerization
– host-guest approach 220
– molecular modeling 220
– structural parameters for polymerization 220
– supramolecular structure 220
trimethylenecyclopropane *see* [3]radialene
triphenylene 4, 120
triple bonds, repulsive interaction between 317
triple excitations, effect of 347
triplet energy transfer
– Dexter mechanism 66
– Förster mechanism 64
– intramolecular 56
– irreversible 57
– long-range 64ff
– long-range vs. stepwise 48
– maximum distance 69
– super-exchange interactions 64
– temperature dependence 64
– viability 66
triplet excition 437
triplet lifetime 55
– as function of temperature 61
– effect of bridge 60
– extending with acetylenic groups 63
– geometry of the bridge 64
– molecular bridge 39
– strategies for prolonging 59
triplet state, higher energy metal-centered 61
– ligand-localized 72
– spin-forbidden 455
– stabilization 56, 60
– thermally relaxed 44
triptycene 12, 16

truxene 92
twistophane 246
two-dimensional wide-angle X-ray scattering 121

u
Ullmann, F. 4
urea 202
ureylene dicarboxylic acid 202

v
valence bond theory 509
valence electron models 340
valence isomerization 316
validation 348
variational theorem 342
vibronic fine state 356
vinyl cation 313
vinylidene carbene 171
vinylogous amide 216
violanthrene 8

w
wavefunction theory, basic description 339
Wegner, G. 15, 295
Wiener index 145
Wittig, G. 10
Wittig reaction 480
Wurtz synthesis 568
Wurtz-Fittig coupling 370

z
Zeeman splitting factor 572
zethrene 92
zigzag phenylenes, synthesis of 151
zinc-dust distillation 2
zirconocene coupling, Negishi-type 278
zone-casting process 127